Using the M68HC11 Microcontroller

Using the M68HC11 Microcontroller

A GUIDE TO INTERFACING AND PROGRAMMING THE M68HC11 MICROCONTROLLER

John C. Skroder
Texas A&M University System
Texas Engineering Extension Service

Prentice Hall
Upper Saddle River, New Jersey
Columbus, Ohio

Library of Congress Cataloging-in-Publication Data

Skroder, John C.
   Using the M68HC11 microcontroller : a guide to interfacing and programming the M68HC11 microcontroller / John C. Skroder.
      p. cm.
   Includes index.
   ISBN 0-13-120676-1
   1. Programmable controllers. 2. Microprocessors. I. Title.
TJ223.P76S59 1997
629.8'9--dc20   96-10489
                                                                                        CIP

Editor: Charles E. Stewart, Jr.
Production Supervision: Custom Editorial Productions, Inc.
Design Coordinator: Julia Zonneveld Van Hook
Cover photo: ©1995 Jay Mallin
Cover Designer: Raymond Hummons
Production Manager: Deidra M. Schwartz
Marketing Manager: Debbie Yarnell

This book was set in Times Roman by Custom Editorial Productions, Inc., and was printed and bound by Quebecor Printing/Semline. The cover was printed by Phoenix Color Corp.

© 1997 by Prentice-Hall, Inc.
Simon & Schuster/ A Viacom Company
Upper Saddle River, New Jersey 07458

All rights reserved. No part of this book may be reproduced, in any form or by any means, without permission in writing from the publisher.

Printed in the United States of America

10 9 8 7 6 5 4 3 2 1

ISBN 0-13-120676-1

Prentice-Hall International (UK) Limited, *London*
Prentice-Hall of Australia Pty. Limited, *Sydney*
Prentice-Hall Canada Inc., *Toronto*
Prentice-Hall Hispanoamericana, S. A., *Mexico*
Prentice-Hall of India Private Limited, *New Delhi*
Prentice-Hall of Japan, Inc., *Tokyo*
Simon & Schuster Asia Pte. Ltd., *Singapore*
Editora Prentice-Hall do Brasil, Ltda., *Rio de Janeiro*

*I dedicate this book
to my dear mother,
Evelyn Skroder,
and to my late father,
Carl E. Skroder.*

Preface

As the title implies, this book provides comprehensive coverage of the M68HC11 microcontroller. A microcontroller is a computer on a chip, incorporating a central processing unit (CPU), memory, input/output devices, and timing devices. The M68HC11 is arguably the most widely used of all microcontrollers. Virtually all American-made automobiles and light trucks use the M68HC11. Appliances, instruments, industrial controllers, cameras, and communications equipment also incorporate this device. These widespread applications make M68HC11 study an eminently worthwhile activity.

Educational theorists provide a learning taxonomy that categorizes learning activities on a scale from primitive to sophisticated. Memorizing facts occupies the lowest level of the taxonomy. Applying learned facts to practical situations occupies the taxonomy's middle levels. Analyzing and evaluating facts and their applications constitute the taxonomy's highest levels. Of course, application, analysis, and evaluation provide the real joy and excitement in learning.

Most electronics textbooks concentrate on the learning taxonomy's lower levels—memorizing facts and concepts. This is unavoidable. In any technological pursuit, these facts and concepts form the tools of the trade. But *Using the M68HC11 Microcontroller* gives equal emphasis to application and analysis. This text includes sixty exercises. Thirty-three of these exercises are hands-on, operating at the application and analysis levels. The remaining twenty-seven cognitive exercises emphasize application where possible. Hands-on exercises use Motorola's popular EVB and EVBU boards, along with inexpensive and widely available external devices. This emphasis on application and analysis will significantly enhance the student's mastery of the M68HC11 and will increase his or her practical knowledge of microprocessors and microcontrollers in general.

Using the M68HC11 Microcontroller repeats a discernable pattern in its coverage of M68HC11 features. Each chapter introduces an M68HC11 feature by explaining its function and significance. Next, the chapter provides a comprehensive examination of M68HC11 on-chip resources relating to the discussed feature. These resources often include pin logic, control and status registers, and timing devices. Typically, a cognitive exercise reinforces the reader's understanding of these on-chip resources. The chapter then presents an application of the discussed feature, along with a hands-on exercise. The hands-on exercise gives the reader an opportunity to construct, program, analyze, and evaluate the application. Finally, each chapter summarizes its contents and provides a large number of review questions. Thus, *Using the M68HC11 Microcontroller* constitutes a complete learning package suitable for either classroom use or independent study.

A reader will enhance his or her understanding of M68HC11 features by having these skills:

- Doing basic mathematical operations on binary and hexadecimal numbers
- Writing truth or state tables for basic logic gates and flip-flops
- Translating between simple Boolean equations and implementing logic gates

Digital texts usually cover these concepts in early chapters. The reader can attain or resurrect these skills by consulting a good digital text at the library.

Acknowledgments

I could not have completed this project without considerable help and support. The following people have my heartfelt gratitude. Thanks to my dear wife, Toby, for proofreading, supply runs, and consistent understanding. I am grateful to Toby's wonderful mother, Dorothy Zellers, for the computer, confidence, and enthusiastic encouragement. Dick Helmer of Trinity University is my mentor. Thanks, Dick, for your ideas, prototypes, and many hours spent reviewing and critiquing the manuscript. Prentice Hall Senior Editor Charles Stewart and his staff kept me on track, furnished reviews, and produced the attractive product that you read here. Thanks to Holly Hodder, who originally conceived of this project and provided much early support. I owe much gratitude to the reviewers who worked with Prentice Hall. Their constructive criticism greatly improved this book. Thanks to Gordy Davies and his staff at Motorola for efficiently responding to my numerous permission requests. Charles Mitchell, my friend and colleague at Texas A&M, made test runs in his microprocessor classes on a number of this book's exercises. Thanks to Fernando Treviño for constructing a motor controller and discovering a program glitch.

I owe a large debt of gratitude to Tim Flem, Tamelee Straub-Haskins, and Melissa Ryan at Custom Editorial Productions, Inc. These folks turned a manuscript into the attractive and readable text that you see before you. Special thanks go to Julie Higgins for many hours of editing and insightful suggestions, all of which greatly improved the text.

Finally, thanks to all my close friends and relatives, whose interest and support made this project much easier.

I wish to make a final note. My late father, Carl E. Skroder, was professor of electrical engineering at the University of Illinois. He and a colleague completed a book for Prentice Hall almost fifty years ago to the day.

John Skroder

Brief Contents

CHAPTER 1	Introduction to the M68HC11	1
CHAPTER 2	M68HC11 Resets and Interrupts	85
CHAPTER 3	M68HC11 Parallel I/O	141
CHAPTER 4	Parallel I/O Using the Simple-Strobed and Full-Handshake Modes	175
CHAPTER 5	The M68HC11 Serial Communications Interface (SCI)	221
CHAPTER 6	The M68HC11 Serial Peripheral Interface (SPI)	263
CHAPTER 7	M68HC11 Free-Running Counter and Input Captures	301
CHAPTER 8	M68HC11 Output Compare Functions	343
CHAPTER 9	M68HC11 Forced Output Compares, Real-Time Interrupts, and Pulse Accumulator	407
CHAPTER 10	M68HC11 Analog-to-Digital Conversions and Fuzzy Inference	473
CHAPTER 11	M68HC11 Expanded Multiplexed Mode	551
APPENDIX A	The M68HC11EVB Board	571
APPENDIX B	The M68HC11EVBU Board	575
APPENDIX C	Connecting the EVB or EVBU to External Circuits	577
APPENDIX D	AS11 Assembler	581
APPENDIX E	M68HC11 Instruction Set	593
APPENDIX F	Parts/Equipment Listing	599
	Index	601

Contents

CHAPTER 1   **Introduction to the M68HC11**   1

     1.1   Goals and Objectives. 1
     1.2   M68HC11 On-Chip Devices 3
     1.3   The M68HC11's Central Processing Unit (CPU). 4
     1.4   BUFFALO ROM System Commands 11
            Exercise 1.1, Use BUFFALO System Commands
            to Load, Run, and Analyze Programs 18
     1.5   M68HC11 Addressing Modes 29
     1.6   M68HC11 Instructions 32
            Exercise 1.2, Load, Store, Push, and Pull Instructions 35
            Exercise 1.3, Bit Clear, Bit Shift, and Logic Instructions. 44
            Exercise 1.4, Arithmetic Instructions. 54
            Exercise 1.5, Compare and Program Control Instructions. 64
     1.7   Chapter Summary 73
            Exercise 1.6, Chapter Review. 77

CHAPTER 2   **M68HC11 Resets and Interrupts**   85

     2.1   Goal and Objectives 85
     2.2   M68HC11 Resets 86
            Exercise 2.1, M68HC11 Resets 91
     2.3   M68HC11 Interrupts. 97
            Exercise 2.2, M68HC11 Interrupts 120
     2.4   Chapter Summary 128
            Exercise 2.3, Chapter Review. 132

CHAPTER 3   **M68HC11 Parallel I/O**   141

     3.1   Goal and Objectives 141
     3.2   On-Chip Resources Used by the M68HC11 for Parallel I/O 142
     3.3   Parallel I/O Program Flow 142
     3.4   PORTB Pin Logic and Timing 145
     3.5   PORTC Pin Logic and Timing 147

		Exercise 3.1, M68HC11 PORTB and PORTC Pin Logic...................149
	3.6	The DL-1414 Alphanumeric Display 155
		Exercise 3.2, Interfacing the DL-1414 Alphanumeric Display 157
	3.7	An Example of Parallel Output Interfacing.........................158
		Exercise 3.3, Parallel Output Interfacing Laboratory................159
		Exercise 3.4, Parallel Output Interfacing II......................165
		Exercise 3.5, Parallel Input/Output Interfacing...................168
	3.8	Chapter Summary..170
		Exercise 3.6, Chapter Review...............................171

CHAPTER 4 Parallel I/O Using the Simple-Strobed and Full-Handshake Modes 175

	4.1	Goals and Objectives 175
	4.2	Simple-Strobed I/O..176
		Exercise 4.1, PIOC Control Words for Simple-Strobed Mode 180
		Exercise 4.2, Simple-Strobed I/O 181
	4.3	Full-Handshake I/O..191
		Exercise 4.3, PIOC Control Words for Full-Handshake Mode 201
		Exercise 4.4, Full-Handshake I/O Laboratory 202
		Exercise 4.5, Full-Handshake/Three-State Variation 208
	4.4	Chapter Summary..213
		Exercise 4.6, Chapter Review...............................214

CHAPTER 5 The M68HC11 Serial Communications Interface (SCI) 221

	5.1	Goal and Objectives 221
	5.2	Introduction to the M68HC11 SCI..............................222
	5.3	M68HC11 SCI Data Format..................................223
		Exercise 5.1, M68HC11 SCI Data Format 225
	5.4	On-Chip Resources Used by the M68HC11 SCI 226
		Exercise 5.2, M68HC11 SCI Baud Rate Generator 228
	5.5	M68HC11 SCI Initialization 236
		Exercise 5.3, SCI Initialization, and BSET/BCLR Source Code 237
	5.6	Transmit a Message—Typical Program Flow......................238
		Exercise 5.4, Transmitting a Message 240
	5.7	Receive a Message—Typical Program Flow.......................241
		Exercise 5.5, Receiving a Message 244
		Exercise 5.6, Serial Communications Interface I (EVBU Only)..........245
		Exercise 5.7, Serial Communications Interface II (EVBU Only) 249
	5.8	Chapter Summary..254
		Exercise 5.8, Chapter Review...............................256

CHAPTER 6 The M68HC11 Serial Peripheral Interface (SPI) 263

	6.1	Goal and Objectives 263
	6.2	M68HC11 On-Chip Resources Used by the SPI 264
		Exercise 6.1, SPCR Control Words 268
		Exercise 6.2, SPI Register Initialization 280
	6.3	Interfacing an M68HC11 Master with a Slave Shift Register 281
		Exercise 6.3, Interfacing an M68HC11 Master with an SN74LS91N Shift Register...............................283
	6.4	Interface an SPI Master with an SPI Slave 286
		Exercise 6.4, Interface an SPI Master with an SPI Slave 290
	6.5	Chapter Summary..293
		Exercise 6.5, Chapter Review...............................296

CHAPTER 7

M68HC11 Free-Running Counter and Input Captures   301

7.1	Goal and Objectives	301
7.2	The M68HC11's Free-Running Counter	301
	Exercise 7.1, Fifteen-Second Time-Out and Display	308
7.3	Introduction to Input Captures	312
7.4	On-Chip Resources Used by the Input Capture Function	312
	Exercise 7.2, Initialize the M68HC11E9 for Input Captures	316
7.5	Using Input Captures to Measure Time Between Events	318
	Exercise 7.3, Use Input Captures to Measure Time Between Events	321
7.6	Use Input Capture Pins for General-Purpose Interrupts	331
	Exercise 7.4, Use Input Capture Pins for General-Purpose Interrupts	333
7.7	Chapter Summary	338
	Exercise 7.5, Chapter Review	340

CHAPTER 8

M68HC11 Output Compare Functions   343

8.1	Goal and Objectives	343
8.2	On-Chip Resources Used by M68HC11E9 OC2–OC5 Output Compare Functions	343
	Exercise 8.1, Initialize the M68HC11E9 for OC2–OC5 Output Compares	350
8.3	Use Output Compares to Generate a Square Wave	351
	Exercise 8.2, Generate Square Waves with OC5	353
	Exercise 8.3, Generate Four Simultaneous Square Waves with OC2–OC5	357
8.4	Integrate Input Captures with Output Compares	360
	Exercise 8.4, Integrate Input Captures with Output Compares	372
8.5	On-Chip Resources Used by the OC1 Output Compare Function	379
8.6	Controlling All OC Pins with OC1	382
	Exercise 8.5, Modulus Thirty-Two Counter	384
8.7	Joint OC1/OCx Control of an Output Compare Pin	386
	Exercise 8.6, DC Motor Speed Control	391
8.8	Chapter Summary	397
	Exercise 8.7, Chapter Review	400

CHAPTER 9

M68HC11 Forced Output Compares, Real-Time Interrupts, and Pulse Accumulator   407

9.1	Goal and Objectives	407
9.2	M68HC11 Forced Output Compares	407
	Exercise 9.1, Two-Speed DC Motor Control	412
9.3	M68HC11 Real-Time Interrupts	416
	Exercise 9.2, Periodic Time Reference Pulses	420
9.4	Introduction to the M68HC11's Pulse Accumulator	422
9.5	M68HC11 Pulse Accumulator—Event-Counting Mode	423
	Exercise 9.3, Initialize the M68HC11 for Event-Counting Mode	426
	Exercise 9.4, DC Motor Speed Control and Tachometer	432
	Exercise 9.5, DC Motor Speed Regulation Using Tachometer Control	448
9.6	M68HC11 Pulse Accumulator—Gated Time Accumulation Mode	453
	Exercise 9.6, Initialize the M68HC11 for Gated Time Accumulation Mode	455
	Exercise 9.7, Measure Percent Duty Cycle	461
9.7	Chapter Summary	464
	Exercise 9.8, Chapter Review	467

CHAPTER 10　M68HC11 Analog-to-Digital Conversions and Fuzzy Inference　473

- 10.1　Goals and Objectives 473
- 10.2　The M68HC11 A/D Converter 474
  - Exercise 10.1, Analog-to-Digital Conversions 480
- 10.3　Using the M68HC11 for Fuzzy Inference 494
- 10.4　Motor Control Using the M68HC11 and Fuzzy Inference 512
  - Exercise 10.2, Motor Control Using the M68HC11 and Fuzzy Inference 533
- 10.5　Chapter Summary 539
  - Exercise 10.3, Chapter Review 543

CHAPTER 11　M68HC11 Expanded Multiplexed Mode　551

- 11.1　Goal and Objectives 551
- 11.2　M68HC11 Resources Used for Expanded Multiplexed Mode 552
- 11.3　Expanded Multiplexed Mode Timing 557
- 11.4　74HC373/74LS373 Octal Latch 558
- 11.5　MCM6264C Static RAM 559
  - Exercise 11.1, MCM6264C Cell Select 560
- 11.6　A Fully Decoded 8K-Byte Memory Expansion 562
  - Exercise 11.2, 8K-Byte Memory Expansion 563
- 11.7　Chapter Summary 565
  - Exercise 11.3, Chapter Review 567

APPENDIX A　The M68HC11EVB Board　571

APPENDIX B　The M68HC11EVBU Board　575

APPENDIX C　Connecting the EVB or EVBU to External Circuits　577

APPENDIX D　AS11 Assembler　581

- D.1　Introduction 581
- D.2　AS11 Functions 582
- D.3　Creating a Source Code File 583
- D.4　The List File and S-Records 588
- D.5　Use KERMIT to Download S-Records to the M68HC11 590
- D.6　List File for a More Complex Program 591

APPENDIX E　M68HC11 Instruction Set　593

APPENDIX F　Parts/Equipment Listing　599

Index　601

1 Introduction to the M68HC11

1.1 GOALS AND OBJECTIVES

This chapter introduces the M68HC11 microcontroller. What is a microcontroller, and how does it differ from a microprocessor? A microcontroller is actually a microcomputer on a single, fingernail-sized piece of silicon. A microcontroller incorporates a microprocessor as a key element. Since a microcontroller is actually a microcomputer, we should review the main elements and concepts of microcomputer operation.

A microcomputer consists of three main elements: memory, input/output (I/O), and microprocessor. The following paragraphs discuss these elements.

Microcomputer memory can consist of read-only memory (ROM), random-access memory (RAM), or data storage media such as tape, compact disk, or magnetic disk. ROM is nonvolatile, i.e., it retains information when powered down. A user or manufacturer stores data to a conventional ROM by blowing fusible links within its storage cells. Users can easily store data to electrically programmable ROM (EPROM), and can store or erase data in electrically erasable-programmable ROM (EEPROM). RAM is volatile memory, i.e., loses its data when powered down. Dynamic RAM (DRAM) uses capacitances to store binary-one bits as charges. These charges require restoration (refresh) every few milliseconds to avoid data loss. Static RAM (SRAM) incorporates bistable devices as storage cells. Thus, SRAM requires no refresh.

Memory generally organizes its bit storage cells into groups of eight. Thus, each group of eight cells can store an information byte. Each eight-bit cell group (byte) has a unique numerical address/identifier, consisting of four or more hex digits (sixteen or more binary bits).

Memory stores two kinds of information: program instruction opcodes, and data. An M68HC11 user compiles a program from hundreds of unique instruction opcodes. The programmer stores these opcodes to memory, at sequential addresses (in the order the computer should execute them). Storage data consists of operands, results, or lookup table (reference) information.

Thus, memory can include RAM, ROM, or storage media, and holds program, data, or look-up table information. Memory constitutes one of three main microcomputer elements.

A microcomputer's second major element consists of input/output (I/O). A microcomputer uses I/O devices to communicate with and manipulate its environment. I/O includes analog-digital converters, parallel communication devices, serial communication devices, and various timing mechanisms.

A microcomputer's third major element is its microprocessor. The microprocessor executes programs and controls overall computer operations. These activities consist of four major tasks:

First, the microprocessor synchronizes the movement of information throughout the computer.

Second, the microprocessor fetches (reads) instruction opcodes from memory.

Third, the microprocessor decodes instruction opcodes, converting them into triggers and enables.

Fourth, the microprocessor executes each program instruction. To execute instructions, the microprocessor accomplishes these tasks:

- Control data originations and destinations
- Fetch data from memory or I/O
- Execute arithmetic and logical operations on data
- Make decisions based upon current conditions
- Capture operands and results

Thus, most microcomputers consist of a discrete microprocessor chip, memory chip(s), and I/O devices—all intricately interconnected.

A microprocessor, interconnected with its I/O devices and memory, makes a mechanism such as a camera or engine controller bulky and fragile. Why not embed processor, I/O and memory on a single, fingernail-sized piece of silicon—make a microcomputer on a chip? A microcontroller is such a miniature, monolithic computer. A microcontroller can connect to off-chip (external) peripherals or expanded memory. But its big virtues derive from its monolithic construction. These virtues are small size, low cost, reliability, and durability. Designers apply microcontrollers to the gamut of control technology—from cameras and automated lathes, to washing machines and automobiles. Now, microcontrollers apply fuzzy inference to control problems which hitherto required mainframe or even supercomputer power.

This book describes the M68HC11 microcontroller. It comes in an impressive number of permutations and is designed to meet various requirements. Does the user need RAM or ROM or A-to-D? Does the user require lowest cost? Is the application simple? Is it sophisticated? There is an M68HC11 version to satisfy each user's requirements.

Motorola uses the following nomenclature to identify its M68HC11 family microcontrollers:[1]

```
MC 68HC x 11 xx
```

The nomenclature's "MC" prefix indicates that the microcontroller is a fully tested and qualified part. The "68HC" means that this microcontroller uses high speed, high density CMOS technology. The "x" preceding the "11" in the nomenclature is a single-digit number—either "7" or "8." A 7 indicates that on-board program memory is one-time, user-programmable EPROM. An 8 means that this memory is user-reprogrammable EEPROM. Absent either number, Motorola custom programs a ROM for the user or supplies a general-purpose operating system called BUFFALO.[2] The "11" in the nomenclature simply indicates that the device is a part of the M68HC11 family. Finally, the last two "x's" signify an alphanumeric that indicates the specific M68HC11 version. Each version has unique features. This book deals with the "A1" and the "E9" parts. The "A1" has 256 bytes of on-chip RAM, 512 bytes of on-chip EEPROM, and no ROM or EPROM. The "E9" has 512 bytes of on-chip RAM, 512 bytes of on-chip EEPROM, and 12K bytes of on-chip ROM. Each M68HC11 version incorporates M6801 processor architecture in its central processing unit (CPU).

This book's hands-on exercises assume that the reader uses a Motorola M68HC11 Universal Evaluation Board (M68HC11EVBU) or an M68HC11 Evaluation Board (M68HC11EVB).[3] The author further assumes that the "E9" and "A1" microcontrollers on these boards come with the BUFFALO ROM operating system. BUFFALO ROM includes

---

[1] Motorola Inc., *M68HC11 Reference Manual* (1990), p. 1-4.
[2] "BUFFALO" stands for Bit User Fast Friendly Aid to Logical Operation.
[3] See Appendices A, B, and C for detailed information on the EVB and EVBU boards.

nineteen system commands, a one-line assembler/disassembler, and a variety of utility subroutines.

Chapter 1 has several goals:

Introduce the reader to the M68HC11's CPU and internal devices.
Help the reader understand the BUFFALO ROM operating system, its system commands, and selected utility subroutines.
Help the reader understand and apply M68HC11 instructions.

By carefully studying this chapter and completing its exercises, the reader shall achieve these objectives:

Connect a terminal to the EVBU or EVB board and get a BUFFALO system prompt.
Use BUFFALO system commands to load, run, and analyze programs.
Develop, debug and demonstrate a program which uses load, store, push, and pull instructions.
Develop, debug and demonstrate a program which uses bit clear, bit shift, and logic instructions.
Develop, debug, and demonstrate programs which do multi-precision addition, subtraction, multiplication, and division.
Develop, debug, and demonstrate a program which uses compare instructions, program control instructions, and nested subroutines.

1.2 M68HC11 ON-CHIP DEVICES

Motorola's M68HC11 Universal Evaluation Board (M68HC11EVBU) uses an M68HC11E9. The "E9" incorporates a wider variety of internal memory and I/O devices than any of its fellow M68HC11 family members. An "E9" includes the following on-chip devices:

1. 512 bytes (two, 256-byte pages) of RAM.
2. 12K bytes (forty-eight, 256-byte pages) of ROM. Motorola stores the BUFFALO ROM operating system in this ROM.
3. 512 bytes of EEPROM (Electrically Erasable Programmable Read-Only Memory). A BUFFALO system command allows the user to copy data or program material to EEPROM. Absent BUFFALO, the user can easily connect and program the M68HC11 for data or program storage to this on-chip EEPROM.
4. Two bidirectional parallel ports (PORTC and PORTD). PORTC has an 8-bit capacity, and can operate with or without handshakes. PORTD includes six lines configurable for general-purpose I/O.
5. One input-only port (PORTE). The user can configure PORTE's eight lines as general-purpose inputs.
6. One output-only port (PORTB). This eight-bit port can operate with or without handshakes.
7. One port (PORTA) includes individual input-only, output-only, and I/O lines. This eight-bit port incorporates two bi-directional lines, three input-only lines, and three output-only lines.
8. A full-duplex UART.[4] This built-in UART, called the Serial Communications Interface (SCI), sends and receives serial message frames using an NRZ (Nonreturn to Zero) format. The SCI configures two PORTD pins as full-duplex input and output lines.
9. A synchronous serial interface. The Serial Peripheral Interface (SPI) provides synchronous, eight-bit serial data exchanges over PORTD lines.
10. An eight-bit analog-to-digital converter (ADC). This A-to-D converter can scale a single input or four inputs. It scales these inputs in groups of four, either one time or continuously. This ADC configures PORTE pins as its analog inputs.

---

[4]"UART" stands for Universal Asynchronous Receiver Transmitter. A UART converts between parallel and serial data, and adds or checks serial framing bits. See Chapter 5.

11. A pulse accumulator. The pulse accumulator counts voltage transitions (edges) applied by an external device to a PORTA pin. The pulse accumulator configures this PORTA pin as its PAI (Pulse Accumulator Input) pin. The pulse accumulator maintains its edge count in a special 8-bit register.
12. A sophisticated 16-bit timer system. The M68HC11's timer section bases its functions on a programmable, free-running, 16-bit counter. The user can program this counter to any of four count rates. Timer subfunctions include Input Captures and Output Compares. A user can configure up to four PORTA pins for input captures. An input capture mechanism records the free-running count, at the time an external device drives an input capture pin with a specified edge. The user can configure up to five PORTA pins for output compares. An output compare mechanism drives its assigned output compare pin with a specified edge, each time the free-running count matches a preestablished value.

The above list makes it clear that an M68HC11 incorporates useful, built-in features, and surprisingly large amounts of on-chip memory. A user who finds on-chip memory inadequate can easily connect up to 16K bytes externally. The M68HC11's wide variety of built-in serial and parallel I/O devices makes it easy to connect external peripherals.

1.3 THE M68HC11'S CENTRAL PROCESSING UNIT (CPU)

The central processing unit (CPU) provides the M68HC11's computing power. This CPU is, in essence, a 6801 microprocessor. Figure 1–1 provides a CPU functional diagram. This figure shows that the M68HC11's CPU divides into three functional areas, each surrounded by a dotted line: address handling, data handling, and instruction handling. Following sections describe these functions.

M68HC11 CPU Address Handling

Refer to Figure 1–1. The CPU's address handling function consists of five, 16-bit, parallel-in/parallel-out registers: four address pointers and one address latch. The address latch (A) drives the M68HC11's internal (on-chip) address bus. The address latch's contents may access a memory location, I/O device, or external device.[5] The CPU loads its address latch—directly or indirectly—from one of the four address pointers or from the internal data bus, depending upon a program instruction's addressing mode.

FIGURE 1–1

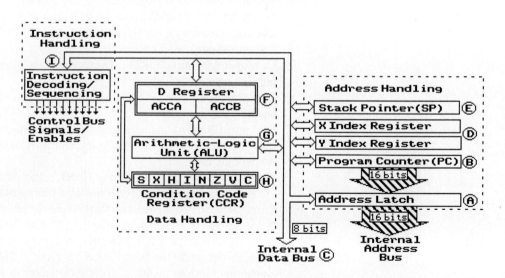

---

[5]When the M68HC11 operates in a mode called "expanded-multiplexed," it operates very much as a microprocessor does. In expanded-multiplexed mode, the M68HC11 drives multiplexed PORTB and PORTC pins with address information from the internal address bus. Thus, the M68HC11 accesses external devices and memory directly.

To fetch a program byte from memory, the CPU loads its address latch from the program counter (B). A user starts software execution by loading the program counter (B) with the program's starting address. The CPU fetches and executes program bytes in sequential order. As the CPU fetches each program byte, the program counter increments, always pointing at the next program byte in memory. Thus, the program counter (B) transfers its contents to the address latch each time the CPU fetches a program byte.

The CPU can access a data location (either memory or I/O) using direct or extended addressing mode. With direct or extended mode, the data location's address is part of the program. The CPU fetches this address information from program storage to the address latch—via the data bus (C). If the CPU accesses a data location using indexed addressing mode, the address latch gets its address information indirectly from one of the index registers (D). When it executes an indexed mode instruction, the CPU sums an index register's contents with a program byte, then transfers this sum to the address latch. If the CPU pushes a byte onto its stack or pulls a byte from the stack, the address latch drives the address bus with the contents of the stack pointer (E).

In summary, the CPU's address handling consists of an address latch and four address pointer registers. Each of these registers has a 16-bit capacity. The address latch drives the M68HC11's address bus with the address of a location to be accessed (effective address). The address latch gets this address information from the program counter—to access program bytes, from the data bus, or from an address pointer. Address information may come directly from program bytes (direct and extended addressing modes). Or it may be the sum of an index register's contents and a program byte. Address information may come from the stack pointer—if the CPU is pushing a byte onto, or pulling a byte off, the stack.

M68HC11 CPU Data Handling

Refer again to Figure 1–1. This figure's middle dotted-line box surrounds the CPU's data handling function. Data handling consists of the accumulators/D register (F), the ALU (G), and the condition code register (H).

The Arithmetic Logic Unit (ALU, G) executes mathematical and logical operations. The ALU performs these operations on data:

    Add/increment
    Subtract/decrement
    Add with carry
    Subtract with carry
    Multiply
    Integer divide
    Fractional divide
    Logical AND, OR, EOR
    One's and two's complement
    Arithmetic bit shifts
    Logic bit shifts
    Bit rotates

The ALU also performs arithmetic and logic operations on address information. When the CPU executes an indexed mode instruction, the ALU adds a program byte to the contents of an index register (D), and returns the sum to the address latch. When the CPU executes a relative mode instruction, the ALU adds a sign-extended program byte to the contents of the program counter (B) and returns the sum to the program counter. The ALU can add the contents of ACCB (F) to either of the index registers (D).

Finally, the ALU performs logical tests on condition code register (H) bits N, Z, V, and C. The ALU makes such a test to determine if the CPU should execute a conditional program branch. If this test produces a "yes" result, the ALU adds a sign-extended relative address (program byte) to the contents of the program counter (B), and returns the sum to the program counter. This sum represents the address to which program flow will branch.

> ### Numerical Prefixes "%" and "$"
>
> This book uses a "%" prefix to indicate a binary number. For example:
>
> %01001010 = 74
>
> A "$" prefix indicates a hexadecimal number:
>
> $48 = %01001010 = 74
>
> Where no prefix appears, assume a decimal number:
>
> 74 = $48 = %01001010

Accumulators/D register (F) serve four purposes:

1. These registers hold data to be manipulated by the ALU.
2. They hold results generated by the ALU.
3. A CPU load transfers data from memory or I/O to an accumulator.
4. A CPU store transfers data from one of these accumulators to memory or I/O.

Accumulator A (ACCA) and Accumulator B (ACCB) are separate, 8-bit registers. A number of M68HC11 instructions treat these accumulators as a single 16-bit register called the D register. ACCA forms the D register's most significant (MS) byte, and ACCB forms the D register's least significant (LS) byte.

Because they form termini for data reads, writes, and most ALU operations, the accumulators are the focus of CPU operations. Many M68HC11 instructions specify ACCA, ACCB, or D register as a terminus. For example:

`LDAA 150`

is BUFFALO one-line assembler source code. This line means "load ACCA with a byte of data from address $0150."

`STAB 3A`

means "store the contents of ACCB to memory location $003A."

`LDD 0`

means "load the D register with the contents of memory locations $0000 and $0001." This read copies the contents of location $0000 to ACCA and the contents of $0001 to ACCB.

The Condition Code Register (CCR) forms the third data handling component. This register holds three kinds of bits. First, the X and I bits mask (enable/disable) some sixteen different types of interrupts. Second, the S bit controls the M68HC11's response to a single instruction—the STOP instruction. Chapter 2 covers the X, I, and S bits in detail. Third, the H, N, Z, V, and C bits flag conditions resulting from CPU operations. These "flag" bits merit further discussion.

The C bit is a carry/borrow flag. The C flag sets under the following conditions:

An ALU arithmetic operation generates a result too large to be held by the result register (an accumulator or the D register). For example, the ALU adds addend $8F (from ACCA) to augend $A1 (from memory), and tries to return the sum—$130—to ACCA (the result

> ### Result Register Overflow Sets the C Flag
>
> ```
>   Addend (ACCA)      $8F    =   %10001111 = 143
> + Augend (memory)    $A1    =   %10100001 = 161
>   Sum              $(1)30   = %(1)00110000 = 304
>
>   C flag = 1, ACCA = $30
> ```

```
                    C Flag Indicates a Borrow
Minuend (ACCB/C flag)       $(1)3C = %(1)00111100 = 316
-Subtrahend (memory)          $6A =     %01101010 = 106
                              $D2 =     %11010010 = 210

C flag = 1, ACCB = $D2
```

register). This sum, $130, is too large for ACCA. ACCA holds 8 bits, i.e., results between $00 and $FF. Therefore, the ALU returns $30 to ACCA, and the C flag sets to indicate overflow from ACCA.

The ALU subtracts a higher number from a lower number.[6] For example, the ALU subtracts subtrahend $6A (memory) from ACCB's contents : $3C. To generate a difference, the ALU effectively borrows a one—to make the minuend equal $13C. The ALU completes the subtraction ($13C less $6A) and returns the difference, $D2, to ACCB. The C flag sets to indicate that the ALU borrowed a one to complete the subtraction.

The C flag operates as an extension bit in logical/arithmetic shift and rotate operations. For example, assume that a program calls for an ASLB operation. This instruction mnemonic stands for "arithmetic shift left ACCB." Figure 1–2, part A, shows how this instruction works. ASLB shifts ACCB's bits one place to the left. ACCB's MS bit, B7, shifts to the C flag. A binary zero shifts into ACCB's LS bit.

Assume that ACCB contains $8F prior to ASLB execution. What do ACCB and the C flag contain after ASLB execution? The C flag sets, and ACCB = $1E. Figure 1–2, part B, shows the C flag and ACCB's contents both before and after ASLB execution. A substantial number of shift and rotate instructions include the C flag.

The condition code register's V bit sets to indicate that the result of an addition operation has an upset sign bit. Either an add or subtract instruction can generate such an addition result. The ALU executes two's complement subtraction. Two's complement subtraction finds the difference by adding the minuend to the two's complemented subtrahend.[7]

FIGURE 1–2
Reprinted with the permission of Motorola.

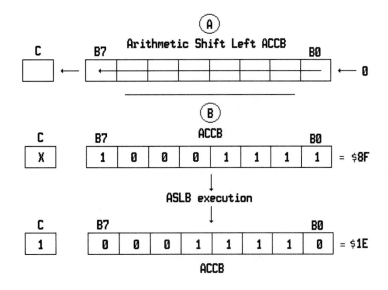

---

[6]Motorola uses the terms "higher" and "lower" to describe unsigned operands. Motorola uses the terms "greater" and "lesser" to describe signed operands.
[7]Find the two's complement of a number by inverting all of its binary bits, then adding binary one to the result. An alternative procedure is to start with the LS binary bit and move leftwards. Up to and including the first binary one encountered, do not invert these bits. Invert all bits to the left of the first binary one. This procedure results in a two's complement.

> **Negative and Positive Operands**
>
> A signed operand's MS bit indicates its polarity. For example, assume that ACCA contains $8C. If the programmer assumes this datum to be signed, $8C is a negative number. $8C is negative because its MS bit is a binary one:
>
> ```
> ACCA = %10001100 = $8C
> ```
>
> If the datum's MS bit = 0, the number is positive. If ACCA contains $7F, for example, this would be a positive number, because its MS bit is a binary zero:
>
> ```
> ACCA = %01111111 = $7F
> ```
>
> If ACCA holds the signed operand $7F, this value equates to decimal +127:
>
> ```
> ACCA = %01111111 = $7F = +127
> ```
>
> If ACCA holds the signed operand $8C, this value equates to decimal -116:
>
> ```
> ACCA = %10001100 = $8C = -116
> ```
>
> Negative hex or binary numbers are two's complement values. That is, a true hex value, e.g., $74 (116) is negated by two's complementing it. Thus, the two's complement of $74 is $8C:
>
> ```
> %01110100 = $74 =  116 (true value)
> %10001100 = $8C = -116 (negated value)
> ```

Assume, for example, that the ALU adds the contents of ACCB to the contents of a memory location and returns the sum to ACCB. If ACCB = $8B and the memory location contains $A3, the sum should be negative, since both $8B and $A3 are negative. However, the ALU adds $8B to $A3 and returns the sum ($2E) to ACCB. The sign bit of $2E is a zero, indicating a positive result. The V flag sets to indicate an upset in the sign bit. An overflow of the sum into the sign bit (i.e., a one carry from bit 6 to bit 7) causes such an upset. Of course, ACCB and the C flag contain a valid sum : $12E. The C flag has become the sign bit.

The V flag mechanism exclusive ORs the carry into the bit 7 addition, with the carry out of bit 7 addition, to generate the V flag state. The ALU includes the V bit in conditional branch tests involving signed operands.

The CCR's Z bit sets to indicate that a CPU operation generated a zero result. For example, assume that the D register contains $7ABC, and that memory locations $0000–$0001 contain $8544. The CPU executes an instruction that adds these two operands and returns the sum to the D register. The ALU generates this result:

`$7ABC + $8544 = $(1)0000.`

The ALU returns $0000 to the D register. The C flag sets to indicate that the sum exceeds the D register's capacity. The Z flag sets because the D register's contents equal $0000; i.e., all D register bits are zeros.

The CCR's N bit is simply a copy of the result register's MS (sign) bit. Thus, the N flag sets to indicate a negative result. For example, assume that the CPU subtracts the contents

> **Sign Bit Upset Sets the V Flag**
>
> ```
>  Addend (ACCB)         $8B =    %10001011 = -117
> +Augend (memory)       $A3 =    %10100011 =  -93
>                    $(1)2E = %(1)00101110 = -210
>
> V flag = 1, C flag = 1, ACCB = $2E
> ```

of a memory location from the contents of ACCB, and returns the difference to ACCB. If ACCB = $1A and the memory location contains $3A, the ALU returns the difference ($E0) to ACCB. This return sets the N flag because ACCB's MS bit = 1. The result is negative because the ALU subtracted greater positive number $3A from lesser positive number $1A.

Finally, the CCR's H (half-carry) bit sets when an addition operation causes a one carry from bit 3 to bit 4. A single M68HC11 instruction uses the H flag. The DAA (decimal adjust ACCA) instruction uses the H flag when it decimalizes a BCD (binary-coded decimal) sum returned by the ALU to ACCA.

To summarize, the CPU's data handling function includes the ALU, accumulators/D register, and CCR. The ALU performs a variety of logical and mathematical operations on data. The ALU also executes a limited number of arithmetic operations on address information. These address-oriented operations update four address pointers and address latch. The ALU also performs logical tests on CCR bits to determine whether the CPU should execute conditional program branches.

The accumulators/D register hold data to be manipulated by the ALU, and the results of that manipulation. These registers also form the termini for data reads and writes to memory or I/O. ACCA and ACCB can operate as independent 8-bit registers, or can combine to form the D register. ACCA forms the MS byte of the D register, and ACCB forms the LS byte.

The 8-bit CCR incorporates three kinds of bits. The X and I bits mask sixteen different interrupt types. The S bit controls the M68HC11's response to a STOP instruction. The H, N, Z, V, and C bits flag conditions resulting from a CPU operation. The C flag indicates result register overflow, or the subtraction of a higher from a lower value operand. The C flag also forms a register or memory extension bit during shift and rotate execution. The V flag indicates an upset in the sign-bit system during an addition operation. The Z flag indicates when a CPU operation returns zero to a result register. The N flag is a copy of the result register's MS (sign) bit. The H flag indicates the value of a carry from bit three to bit four during the addition of BCD operands. The DAA (Decimal Adjust ACCA) instruction uses the H flag when it decimalizes a BCD sum.

Program instructions generate address and data handling activities. The CPU's instruction handling function decodes instruction opcodes and generates execution control signals and enables. The following section describes the CPU's instruction handling function.

M68HC11 CPU Instruction Handling

Refer again to Figure 1–1. This figure's leftmost dotted line box surrounds the CPU's instruction handling function. The instruction handler decodes instruction opcodes and generates control signals and enables that are necessary to execute these instructions.

Read data and instruction opcodes reach the CPU via the internal data bus (C). An accumulator (F) latches read data. The instruction decoder/controller/sequencer (I) latches instruction opcodes. This decoder translates an opcode into sequenced enables. An instruction opcode generally consists of 1 byte. However, some seventy-five opcodes include a prebyte. Motorola's addition of the Y and D registers to the CPU architecture necessitated these prebytes.

The instruction handler's controller-sequencer generates read/write and address strobe control signals, along with sequenced enable signals. The instruction handler distributes sequenced enables throughout the microcontroller. Enables cause ALU components to execute their functions—such as addition or a logic shift. Above all, these sequenced enables generate information transfers—e.g., from address latch to address bus, or from memory to accumulator. Consider the sequenced enables necessary to execute a simple read instruction such as `LDAA A` (load ACCA from address $000A):

1. Enable transfer of $000A to the address latch. These enables clear the address latch's MS byte and transfer a program byte ($0A) from memory to the address latch's LS byte.
2. Cause the address latch to drive the address bus with $000A.
3. Generate an address strobe.
4. Drive the read/write (R/$\overline{W}$) control line high.
5. Enable the transfer of a data byte from memory to data bus, and from data bus to ACCA.

9

The reader should appreciate that the CPU accomplishes all of these transfers in one microsecond (1μs).[8]

Although an instruction may generate complex ALU data manipulations, the majority of sequencer outputs enable data and address transfers from one M68HC11 component to another. To make these information transfers, the sequencer enables transmission gates or three-state buffers.

The foregoing discussion deals with instruction execution. But before it can execute an instruction, the CPU must fetch the instruction's opcode from a program byte in memory. An opcode fetch involves these controller/sequencer actions:

1. Enable an address word transfer (16 bits) from the program counter (B) to the address latch (A).
2. Cause the program counter to increment, so that it points at the next program byte.
3. Generate an address strobe and drive the R/$\overline{W}$ control line high.
4. Enable transfer of the instruction opcode from memory to data bus (C).
5. Enable transfer of the opcode from data bus to instruction decoder (I).
6. Decode the opcode.

Thus, an instruction fetch consists of the above six steps. The M68HC11 on board an EVBU or EVB consumes 500 nanoseconds in an instruction fetch. This 500-nanosecond period constitutes one machine cycle. Motorola's instruction set information always lists the number of machine cycles consumed by an instruction fetch and execute. An instruction fetch is automatic. When the CPU completes its execution of an instruction, the controller/sequencer automatically generates the six-step fetch of the next program opcode.

In summary, the instruction handler decodes an instruction opcode received from program memory via the data bus. The decoder/controller/sequencer translates an opcode into sequenced enables and control bus signals. Sequenced enables cause information transfers throughout the microcontroller. Sequenced enables also cause the ALU to perform mathematical or logical operations on data—as required by the instruction. The majority of instruction execution actions involve information transfers. The controller/sequencer automatically causes the CPU to fetch an opcode from memory. An instruction fetch consists of an opcode transfer from memory to data bus, to instruction decoder, then decoding of the opcode. This opcode fetch (a bus-read cycle) constitutes one machine cycle. Motorola's instruction set listing measures instruction fetch and execute times in terms of machine cycles.

M68HC11 Programmer's Model

Refer to Figure 1–1. The M68HC11 Programmer's Model[9] incorporates these eight key registers:

  ACCA (A)
  ACCB (B)
  D register (D)
  X Index Register (IX)
  Y Index Register (IY)
  Stack Pointer (SP)
  Program Counter (PC)
  Condition Code Register (CCR)

Additionally, this Programmer's Model breaks the Condition Code Register (CCR) into its component bits (from MS to LS):

  STOP Disable (S)
  X-Interrupt Mask (X)
  Half Carry (H)
  I-Interrupt Mask (I)

---

[8]The author assumes that the M68HC11 runs with 2MHz internal clocks.
[9]Motorola Inc., *M68HC11 Reference Manual* (1990), pp. 1-2.

Negative Flag (N)
Zero Flag (Z)
Overflow from bit 6 to bit 7 (Sign Upset, V)
Carry/Borrow from MSB (C)

Every M68HC11 instruction involves at least one of these registers or bits. Motorola's instruction set listing provides operational descriptions and Boolean expressions for all instructions. Each description and Boolean expression explains an instruction in terms of the eight key registers or CCR bits.

For example, consider this instruction mnemonic:

`INS`

Motorola describes INS as:

`Increment Stack Pointer`

INS's Boolean expression is:

`SP+1—>SP`

As another example, consider this mnemonic:

`LDY`

Motorola describes LDY as:

`Load Index Register Y`

LDY's Boolean expression is:

`M:M+1—>Y`

where M:M+1 stands for the contents of a memory address (M), and the next higher address (M+1).

Thus, Motorola uses the Programmer's Model to explain instructions to the user.

1.4 BUFFALO ROM SYSTEM COMMANDS

The BUFFALO ROM Monitor Program manages overall EVB or EVBU system operations. In addition to its assembler/disassembler and utility subroutines, BUFFALO includes nineteen system-level commands. These commands differ in minor ways between EVBU and EVB.[10] This section discusses ten BUFFALO system commands which the EVB and EVBU share. The reader shall use these ten commands while doing this book's hands-on exercises. This chapter's exercises give the reader opportunity to apply these commands.

Memory Display (MD) Command

The MD (Memory Display) command allows the user to display the hex contents of any M68HC11 address on the terminal screen. Here are some examples of MD syntax and use.

`MD`   <return>

Use the MD command by itself, and BUFFALO assumes you wish to examine 144 bytes of memory starting at address $0000. BUFFALO displays nine lines of memory, 16 bytes to a line. Thus, BUFFALO displays the contents of the following memory addresses:

$0000–$000F
$0010–$001F
$0020–$002F

---

[10]The EVBU's BUFFALO ROM includes a STOPAT command. STOPAT executes on programs in ROM, where the BR (breakpoint) system will not. The EVB's BUFFALO ROM does not include STOPAT. The EVB uses two VERIFY commands, one for each of its serial I/O ports. Because the EVBU has only one serial I/O port, it uses only one VERIFY command.

```
            $0030–$003F
            $0040–$004F
            $0050–$005F
            $0060–$006F
            $0070–$007F
            $0080–$008F
```

```
MD B604                                                              <return>
```

Here, BUFFALO again assumes you wish to examine 144 bytes of memory, starting with a 16-byte line which includes address $B604. BUFFALO displays the contents of these addresses:

```
    $B600–$B60F
    $B610–$B61F
    $B620–$B62F
    $B630–$B63F
    $B640–$B64F
    $B650–$B65F
    $B660–$B66F
    $B670–$B67F
    $B680–$B68F
```

```
MD B735 B757                                                         <return>
```

From this command BUFFALO assumes you wish to display inclusive, 16-byte lines of memory—from the line which contains address $B735 to the line which contains address $B757. BUFFALO displays these addresses:

```
    $B730–$B73F
    $B740–$B74F
    $B750–$B75F
```

```
MD B647 B76A                                                         <return>
```

This command line tells BUFFALO to scroll through multiple lines of 16 bytes each—from the line which contains address $B647 to the line which contains address $B76A.

Thus, the MD command causes BUFFALO to display the hex contents of memory locations, in lines of 16 bytes each. A sixteen-byte line would appear as shown in this example:

```
MD 14 1A                                                             <return>
0010 4D 79 AA 44 6F 67 FF 48 61 73 20 46 6C 65 61 73 My Dog Has Fleas
```

The terminal display shows one 16-byte line which includes addresses $0014 and $001A. Note that BUFFALO displays the translation of any ASCII codes in the line. BUFFALO displays this translation at the end of the 16-byte line.

The BUFFALO MD (Memory Display) command allows the user to examine the contents of all M68HC11 internal memory and all directly addressable external memory locations.

Memory Modify (MM) Command

The MM (Memory Modify) command allows the user to initialize operand or data bytes in RAM or EEPROM. A user can also key in machine code program bytes with this command.

To modify the contents of a RAM or EEPROM location (e.g., address $002A), use this command line:

```
MM 2A                                                                <return>
```

In response, BUFFALO displays the address, its contents, and a prompt:

```
002A FF _
```

where $002A is the address, $FF is the current contents of this address, and the underline is a prompt.

If you do not wish to change the contents of this address, simply press the Return key to return to the BUFFALO system prompt. If you wish to change the contents of this address, enter the new, two-digit hex value (e.g., $97) and then the Return key:

002A FF 97_   <return>

BUFFALO has now stored $97 to address $002A. To confirm this write, one can use the MM command again:

MM 2A   <return>

BUFFALO should now show that $97 was stored to address $002A:

002A 97 _

Using MM and the keyboard space bar, the user can store a string of data/program bytes to consecutive RAM/EEPROM addresses. For example, assume that you wish to store the following data to addresses $B606–$B60A:

$B606 = $AA
$B607 = $BB
$B608 = $CC
$B609 = $DD
$B60A = $EE

To store this data, use the following command line:

MM B606   <return>

BUFFALO responds by displaying:

B606 FF _

where FF is the current contents of address $B606. (Let us assume that addresses $B606–$B60A all contain $FF.)

Enter the first datum and press the space bar:

B606 FF AA _   <space bar>

BUFFALO responds by storing $AA to address B606, then displaying the contents of the next location (B607):

B606 FF AA FF _

Enter the next datum and press the space bar:

B606 FF AA FF BB_   <space bar>

BUFFALO stores $BB to address B607, then displays the contents of address B608:

B606 FF AA FF BB FF _

Continue to enter data bytes and press the space bar until the last datum has been entered. Then press the return key:

B606 FF AA FF BB FF CC FF DD FF EE _   <return>

Use the BUFFALO MD command to confirm that the data were entered correctly:

MD B606 B60A   <return>
B600 FF FF FF FF FF FF AA BB CC DD EE FF FF FF FF FF

Thus, the BUFFALO MM (Memory Modify) command can display and modify a single byte or a string of consecutive bytes in RAM or EEPROM. In accomplishing this book's hands-on exercises, the reader shall often use this command to initialize data bytes.

Block Fill (BF) Command

Use the BF (Block Fill) command to load a string of memory addresses with a constant value. Here is an example:

```
BF B60A B71D A1                                                          <return>
```

This command loads addresses $B60A, $B71D, and all intervening locations with the value $A1. The first number ($B60A) following the "BF" command specifies the low address to be filled. The next number ($B71D) specifies the final address to be filled. The final number ($A1) in the command line specifies the hex value to be stored to all locations, from starting address to final address. An EVBU user can use the BF command to write a value to RAM and EEPROM locations. Using BF, an EVB user can write only to RAM locations.

MOVE Command

Use the MOVE command to copy the contents of a block of addresses to RAM or EEPROM. For example, assume that you wish to copy a program, which presently resides in RAM, to EEPROM in order to save it after power down. Assume further that this program presently resides in RAM addresses $0000–$002A, and that you wish to copy this program to EEPROM, starting at address $B700:

```
MOVE 0 2A B700                                                           <return>
```

BUFFALO responds to this command line by copying the contents of $0000–$002A to addresses $B700–$B72A. Thus, the number following the MOVE command is the starting address of the block to be copied. The next number is the last address of the block to be copied. The final number is the destination's starting address.

Go (G) Command

The G (Go) command causes the M68HC11 to execute a program. Assume that you wish the microcontroller to execute a program which resides in memory block $B710–$B7A5. Use this command line to execute the program:

```
G B710                                                                   <return>
```

In response, BUFFALO loads the M68HC11's program counter with $B710. The microcontroller then fetches and executes program instructions starting at address $B710.

When doing this book's hands-on exercises, the reader shall use the G (Go) command to execute programs.

Assembler/Disassembler (ASM) Command

The ASM (Assembler/Disassembler) command invokes the BUFFALO one-line assembler/disassembler. The assembler compiles machine code from source code lines entered by the user. The disassembler translates machine code back into BUFFALO assembly code.

Refer to Figure 1–3. Assume that you wish to assemble the program shown in this figure. Also assume that you have just powered up an EVBU, and that all RAM locations contain $FF as the result of this power up. Invoke the assembler/disassembler with this command:

```
ASM 100                                                                  <return>
```

The number 100 following the ASM mnemonic indicates the hex address at which you wish to begin assembling the program. In response to this command, BUFFALO displays the following on the terminal screen:

```
0100 STX $FFFF >_
```

FIGURE 1–3

$ADDRESS	SOURCE CODE	COMMENTS
0100	LDX #100	Initialize X register
0103	LDAA #2	Load ACCA with constant $02
0105	ADDA 2	Add contents of address $0002
0107	STAA 9	Store sum to address $0009
0109	STAA C,X	Store sum to address $010C
010B	WAI	Halt

BUFFALO has disassembled the "power up" $FFs, to show that the first three $FF bytes mean "store X register's contents to address $FFFF." Of course, we know that these three bytes of $FF are coincidental with power up. We have no intention of retaining this STX $FFFF instruction. The prompt to the right of the disassembled line means that you can now assemble a replacement instruction. Enter the first source code line from Figure 1–3:

```
0100 STX $FFFF >LDX #100                                              <return>
```

In response, BUFFALO assembles your source code, and displays the result below the assembly code line. BUFFALO also displays the next assembly code line:

```
0100 STX $FFFF >LDX #100
     CE 0100
0103 STX $FFFF >_
```

Enter the second source code line from Figure 1–3:

```
0100 STX $FFFF >LDX #100
     CE 100
0103 STX $FFFF >LDAA #2                                               <return>
```

BUFFALO assembles and displays the second instruction. BUFFALO also disassembles the next instruction:

```
0100 STX $FFFF >LDX #100
     CE 0100
0103 STX $FFFF >LDAA #2
     86 02
0105 STX $FFFF >_
```

To assemble the program's remaining source code lines, enter them one at a time, pressing the return key after each line entry. BUFFALO responds to each <return> by assembling the line you just entered, and disassembling the next instruction. Follow this procedure until you enter all source code lines. Then, exit the assembler/disassembler by holding the control key down, and pressing the "A" key. This (CTRL)A returns you to a BUFFALO system prompt.

To check program entry, reinvoke the assembler/disassembler at the program's starting address:

```
ASM 100                                                               <return>
```

BUFFALO responds by disassembling the program's first line:

```
0100 LDX #$0100 >_
```

Check all program lines by pressing and re-pressing the return key. Correct any line by entering new source code at the prompt for that line.

You can also disassemble and correct any program instruction. To disassemble the program's fifth instruction, use this command from the BUFFALO system prompt:

```
ASM 109                                                               <return>
```

BUFFALO responds by disassembling the instruction at address $0109, and prompting the user to make a change:

```
0109 STAA $0C,X >_
```

When you have finished checking and correcting the program, use (CTRL)A to return to the BUFFALO system prompt.

The procedures discussed above work for both EVB and EVBU. However, one important difference between these boards exists: an EVBU user can assemble program directly to the M68HC11E9's on-chip EEPROM. The EVB board's BUFFALO assembler/disassembler cannot assemble program directly to the M68HC11A1's EEPROM. An EVB user must assemble the program in RAM, then use the MOVE command to copy it to EEPROM.

Using the BUFFALO assembler/disassembler is simple and straightforward. The assembler uses no labels or assembler directives. It does not save comments. It assembles or disassembles one instruction at a time. The assembler/disassembler calculates relative addresses. It has a limited repertoire of error messages. Usually, an entry error causes BUFFALO to display an error message, and to prompt the user to re-enter the source code line.

Register Modify (RM) Command

A review of Figure 1–1 shows that the CPU uses seven important registers. The RM (Register Modify) command allows a user to examine the current contents of these registers, and change their contents from the terminal keyboard. For example, enter this command:

RM   <return>

In response, BUFFALO displays the contents of the seven registers. For example:

```
P-0105 Y-FFFF X-B600 A-02 B-FF C-C0 S-0047
P-105 _
```

This display indicates the following registers and their contents:

Program counter = $0105
Y index register = $FFFF
X index register = $B600
ACCA = $02
ACCB = $FF
Condition code register = $C0
Stack pointer = $0047

This display's second line prompts the user to change the program counter's contents. To leave the PC's contents as is, simply press the return key, and return to the BUFFALO system prompt. To change the PC contents, enter the new value. For example:

```
P-0105 Y-FFFF X-FFFF A-02 B-FF C-C0 S-0047
P-105 0100                                                                              
```
<return>

BUFFALO loads the program counter with $0100 and returns to its system prompt. If the user presses the space bar instead of the return key, BUFFALO prompts the user to change the Y index register. The user can continue to change the contents of the registers, pressing the space bar after each change. At any time, the user can press the return key and return to the BUFFALO system prompt.

To change the contents of a single register—ACCB, for example—simply enter this command:

RM B   <return>

BUFFALO responds by displaying the contents of all eight registers, and prompting the user for a change to ACCB:

```
B-FF _
```

The user can enter a new ACCB value, then press the return key or the space bar. Again, the space bar allows the user to change other register values. The return key causes BUFFALO to exit the RM command and return to a system prompt.

The reader shall find the RM command particularly useful during system analysis and debug.

Trace (T) Command

The T (Trace) command causes BUFFALO to display current CPU status each time it fetches and executes a program instruction. After each fetch and execute, BUFFALO displays the assembly code line just executed (EVBU) or the opcode just executed (EVB), along with the resulting contents of the seven CPU registers. To use the T command, the reader must first use the BUFFALO RM command to point the program counter at the first instruction to be executed and traced. After initializing the PC, enter this command:

```
T                                                                <return>
```

In response, the CPU fetches and executes an instruction, then displays the instruction just executed and CPU status. A user can trace succeeding instructions by simply pressing the return key after each fetch and execute.

A user can trace up to 255 instructions by entering a suffix to the T command. To trace twenty consecutive instructions, for example, use this command:

```
T 14                                                             <return>
```

where "14" is $14 = 20. In response, BUFFALO displays CPU status after it executes each of the twenty instructions.

To return to full-speed operation after tracing an instruction, simply use the G (Go) command:

```
G                                                                <return>
```

In response, the CPU executes remaining program instructions without traces, and at full speed.

Breakpoint Set (BR) Command

Breakpoints allow the user to execute a program—using the G (Go) command—up to a desired breakpoint. At this breakpoint, fetch and execute stops, and BUFFALO displays current CPU status (the contents of the seven key CPU registers). A breakpoint must be an instruction opcode address, i.e., the beginning address of an instruction. The BR command allows a user to establish up to four breakpoint addresses. Assume, for example, that you wish to establish breakpoints at addresses $0105 and $0109. Enter this command line:

```
BR 105 109                                                       <return>
```

BUFFALO responds by establishing breakpoints at addresses $0105 and $0109,[11] then displaying the breakpoint table's contents:

```
BR 105 109
0105 0109 0000 0000
```

The user can then start program operation with the BUFFALO G command. The CPU executes the program up to the first breakpoint, then halts. BUFFALO displays current CPU status. To continue program execution to the next breakpoint, use the BUFFALO P command. See the next section for a discussion on the P command.

Use a minus sign to clear breakpoints from the breakpoint table. For example, assume that you wish to remove address $0109 from the breakpoint table:

```
BR -109                                                          <return>
```

---

[11] BUFFALO establishes breakpoints by temporarily substituting software interrupts (SWIs) at the locations listed in the breakpoint table.

In response, BUFFALO removes $0109 from the table and displays the new table contents:

```
BR -109
0105 0000 0000 0000
```

To erase all breakpoint addresses, use the minus sign as follows:

BR -   <return>

BUFFALO clears the breakpoint table and displays:

```
0000 0000 0000 0000
```

An EVBU user can establish breakpoints in either RAM or EEPROM. EVB users can establish breakpoints in RAM only.

Breakpoints facilitate system analysis and debugging. They obviate the need for repetitive instruction tracing, and can also be used with interrupt service routines. A programmer who is debugging or analyzing program loops finds breakpoints particularly useful. With the BR command, the user finds it easy to establish and remove breakpoints.

Proceed/Continue (P) Command

Use the P (Proceed/Continue) command in conjunction with breakpoints. After program execution halts at a breakpoint, use the P instruction to continue program execution from that breakpoint. Simply enter the following, and program execution will continue:

P   <return>

Thus, the P instruction allows the user to analyze an entire program, stopping at up to four breakpoints along the way.

Section 1.4 has discussed ten BUFFALO operating system commands. In doing this book's hands-on exercises, the reader shall use these commands frequently. Exercise 1.1 gives the reader an opportunity to use all ten of these BUFFALO system commands.

EXERCISE 1.1 Use BUFFALO System Commands to Load, Run, and Analyze Programs

REQUIRED EQUIPMENT: Motorola M68HC11EVB or EVBU, and terminal

GENERAL INSTRUCTIONS: This hands-on laboratory exercise gives the reader an opportunity to use BUFFALO ROM system commands presented in Section 1.4. Use these commands to load, run, move, analyze, and modify a DEMO program. In Part I of this exercise, connect a terminal to the EVBU/EVB and get a BUFFALO system prompt. In Part II, use the MM and MD commands to load and check the DEMO program, in machine code. In Part III, use the MM, G, RM, and T commands to initialize the program's augend, run it, and analyze its operation. In Part IV, MOVE the DEMO program to EEPROM, then use the BF command to write over the program's original locations in RAM. In Part V, use the BUFFALO assembler/disassembler to re-enter the DEMO program. In Part VI, use the assembler/disassembler to modify the DEMO program. Then use the BR, G, and P commands to analyze its operation. Exercise 1.1 amply demonstrates BUFFALO system commands' utility and ease of use.

PART I: Connect the EVBU/EVB to a Terminal and Obtain a System Prompt

Leave your EVBU/EVB unconnected from the terminal. Power up the terminal only and configure it as follows:

   Full-duplex
   Eight data bits, no parity

One stop bit
Automatic wrap-around
Jump scroll
9600 baud, both receive and transmit
English set-up
North American keyboard

Power down the terminal. Use a twenty-five conductor ribbon cable to connect your terminal to the EVBU or EVB board:

- EVBU—connect the ribbon cable between the terminal's RS-232 connector and P2 (see Appendix B).
- EVB—connect the ribbon cable between the terminal's RS-232 connector and P2 (see Appendix A).

Make appropriate power supply connections to your EVBU/EVB board:

- EVBU—make +5VDC and ground connections.
- EVB—make +5VDC, +12VDC, –12VDC and ground connections.

Power up the terminal.

Power up the EVBU/EVB board. If you have made proper connections, your screen should now display:

```
BUFFALO 3.x (int) - Bit User Fast Friendly Aid to Logical Operation_
```

A prompt should be flashing at the end of this display line.

If you do not get this display, press and release the EVBU/EVB reset button. If you still get no "BUFFALO . . ." display, seek help from the instructor. Follow these steps if no instructor is available:

1. Power down the EVBU/EVB board and terminal.
2. Recheck your terminal setup.
3. Assure that the ribbon cable connects to the proper RS-232 I/O connectors at the terminal and EVBU/EVB board.
4. Recheck power supply output voltage(s) and ground connection to the EVBU/EVB. Assure that power supply and EVBU/EVB share a common ground.

Assure that you obtain the "BUFFALO . . ." message after each EVBU/EVB push-button reset. Do not proceed until you can consistently get this message on the terminal screen.

Reset the EVBU/EVB to get a new "BUFFALO . . ." message, then press the return key on the terminal keyboard. You should now see the BUFFALO system prompt on the line below the "BUFFALO . . ." message:

```
>_
```

Show this prompt to the instructor.

_____ *Instructor Check*

In the remaining parts of this exercise, enter BUFFALO system commands from this prompt. Following sections call this prompt the "BUFFALO system prompt."

PART II: Load and Check DEMO Machine Code

Review pages 11–13. These sections explain BUFFALO system commands used in Part II.

Refer to Figures 1–4 and 1–5, which illustrate a simple DEMO program. This program adds two numbers and stores the sum—twice—to separate RAM memory locations. The addend is an immediate-mode operand. This operand is a constant, and therefore an integral part of the program itself. The augend is a byte stored in RAM. Overall, the DEMO program uses immediate, indexed, and direct (page zero) addressing modes. The reader should bear in mind that this DEMO program has no significant purpose, save to demonstrate CPU operations in several addressing modes. Therefore, DEMO uses indexed

FIGURE 1–4

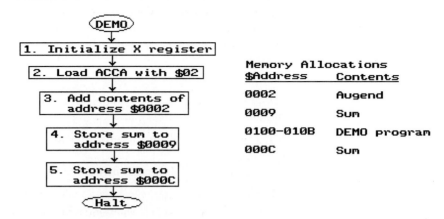

addressing mode simply to demonstrate its use, not to enhance program efficiency. Throughout this book, the reader shall have ample opportunity to apply indexed mode to appropriate problems. In the present case, the objective is to keep the program as simple as possible—to enhance reader insight.

Refer to Figure 1–4. This figure presents a flowchart which summarizes DEMO program execution. This figure also allocates memory space to program and data bytes. In block 1, the CPU uses immediate addressing mode to initialize the X index register's contents. Immediate mode values are part of the program itself. Therefore, the CPU reads two *program* bytes to the X register. Immediate mode allows the CPU to access constant values quickly and efficiently.

In block 2, the CPU again uses immediate mode to read the addend ($02) to Accumulator A. Of course, this immediate mode operand is a program byte. In block 3, the CPU uses direct addressing mode to read the *augend* value from address $0002 to the ALU. Simultaneously, the CPU copies the *addend* value from ACCA to the ALU. The ALU adds addend to augend, then returns the sum to ACCA. Thus, the sum replaces the addend in ACCA.

In block 4, the CPU uses direct addressing mode to copy the sum from ACCA to RAM address $0009. In block 5, the CPU uses indexed addressing mode to copy the sum from ACCA to RAM address $000C. After storing the sum to address $000C, the CPU halts.

Now refer to Figure 1–5. This figure lists a program which implements the Figure 1–3 flowchart. Actually, Figure 1–5 contains two lists of the same program. The list on the left is BUFFALO assembly source code. The list on the right is machine code. If a user enters

FIGURE 1–5

DEMO Program

ASSEMBLY CODE		MACHINE CODE		
$ADDRESS	SOURCE CODE	$ADDRESS	$CONTENTS	COMMENTS
0100	LDX #FFFF	0100	CE	Initialize X register
		0101	FF	" " "
		0102	FF	" " "
0103	LDAA #2	0103	86	Load ACCA with
		0104	02	constant $02
0105	ADDA 2	0105	9B	Add contents of
		0106	02	address $0002
0107	STAA 9	0107	97	Store sum to address
		0108	09	$0009
0109	STAA D,X	0109	A7	Store sum to address
		010A	0D	$000C
010B	WAI	010B	3E	Halt

source code, BUFFALO assembles these entries into corresponding machine code. However, a user can enter machine code directly, via the BUFFALO MM command. Thus, these two lists relate closely to each other. For example, if the reader enters the first line of source code at address $0100, then the following display should appear:

```
LDX #FFFF
```

BUFFALO assembles this source code into three machine-code program bytes:

$ADDRESS	$CONTENTS
0100	CE
0101	FF
0102	FF

The first byte is the opcode $CE. This opcode tells the CPU to transfer the next two program bytes ($FF and $FF) to the X index register. Thus, the CPU initializes its X index register's contents to $FFFF.

The reader enters the second line of source code at address $0103:

```
LDAA #2
```

BUFFALO assembles this source code into:

$ADDRESS	$CONTENTS
0103	86
0104	02

Again, the first byte is the opcode. It instructs the CPU to read the next program byte ($02) to ACCA. Thus, the CPU loads ACCA with the addend, $02.

Note that both of the foregoing source code lines include a pound sign (#) preceding the values to be transferred. A pound sign forces the BUFFALO assembler to use immediate addressing mode.

The reader enters the third line of assembly code at address $0105:

```
ADDA 2
```

BUFFALO assembles this machine code:

$ADDRESS	$CONTENTS
0105	9B
0106	02

The opcode $9B instructs the CPU to take four actions:

1. Transfer the addend from ACCA to the ALU.
2. Transfer the augend from the specified memory location to the ALU.
3. Add addend to augend.
4. Transfer the sum to ACCA.

Note that the source code uses no pound sign. Absence of a pound sign means that the number accompanying the mnemonic is an address, not an operand. BUFFALO sees this address as $0002; it resides on memory page zero, i.e., the address' high byte equals $00. Therefore, BUFFALO assembles the instruction using direct mode addressing. Direct mode is often called "page zero" addressing. Direct mode assumes that the effective address's high byte is $00 (page zero). Since the MS byte = $00, direct mode does not include it in the program. The machine code DEMO program expresses this address simply as $02.

The reader enters the fourth source code line at $0107:

```
STAA 9
```

BUFFALO assembles this source code as:

$ADDRESS	$CONTENTS
0107	97
0108	09

Because there is no pound sign, BUFFALO interprets the "09" as effective address $0009. Since this address resides on page zero, BUFFALO assembles a direct mode instruction. Opcode $97 instructs the CPU to store the contents of ACCA to the effective address, $0009.

The reader enters the fifth line of source code at address $0109:

```
STAA D,X
```

The comma and X in this source code force BUFFALO to assemble the machine code in indexed mode using the X register. BUFFALO interprets the hex number preceding the comma ($D) as the indexed mode offset value. BUFFALO assembles this 1-byte offset value in a program byte following the opcode:

$ADDRESS	$CONTENTS
0109	A7
010A	0D

The opcode instructs the CPU to take two actions: First, derive the instruction's effective address by adding the offset to the contents of the X register. Second, store ACCA's contents to this effective address.

$$\begin{aligned}\text{Effective address} &= \text{X register} + \text{offset}\\ &= \$FFFF + \$0D\\ &= \$000C\end{aligned}$$

Thus, the CPU stores the contents of ACCA to address $000C.

Finally, the reader enters this source code at address $010B:

```
WAI
```

BUFFALO assembles this source code as inherent mode opcode $3E. Opcode $3E causes the CPU to do certain housekeeping chores, then assume a wait state, pending a push-button reset by the user.

To complete this part of the lab exercise, enter the DEMO program in *machine code*. In Part V, you will have the opportunity to use the assembler/disassembler.

Refer again to Figure 1–5. You shall use the BUFFALO MM command to load DEMO in machine-code form. If you use an EVBU, this program will load as listed, into RAM locations $0100–$010B. However, if you use an EVB, addresses $0100–$010B are not available. The EVB locates user RAM at addresses $C000–$DFFF. Therefore, an EVB user must locate the Figure 1–5 program in the available RAM block. If you use an EVB, rewrite the program's machine code listing in the blanks below. Start with address $C000. The first four entries are already completed to get you started. If you use an EVBU, skip this task.

DEMO Program
MACHINE CODE

$ADDRESS	$CONTENTS	COMMENTS
C000	CE	Initialize X register
C001	FF	" " "
C002	FF	" " "
C003	86	Load ACCA with constant $02
____	__	_____
____	__	_____
____	__	_____
____	__	_____
____	__	_____
____	__	_____
____	__	_____
____	__	_____

Now, use the BUFFALO MM command to load the machine code program into RAM. From the BUFFALO system prompt, enter:

`MM xxxx`   <return>

where xxxx is the program's beginning address (0100 for the EVBU; C000 for the EVB). Enter program bytes, pressing the space bar between bytes. When you have entered all program bytes, press the return key to return to the BUFFALO system prompt.

Next, use the MD command to check program contents:

`MD xxxx`   <return>

where xxxx is the program's beginning address. Your machine code program bytes should now appear on the first line of the memory display. Check these entries against your machine code list to assure their accuracy. If you made any entry mistakes, use the MM command to correct them. You can change the contents of any RAM location by entering:

`MM yyyy`   <return>

where yyyy is the RAM address holding the byte to be corrected. You can use the MM command and space bar to examine all program memory addresses if you wish. As long as you refrain from entering new data, the contents of RAM remain unchanged when you use the MM command. Use the MD command to display the corrected machine code program. Show your terminal display to the instructor.

_____ *Instructor Check*

PART III: Initialize the Program Augend, Run the Program, Analyze Program Operation

Review the BUFFALO RM and T commands. DEMO program's augend resides at address $0002. The reader should note that both EVB and EVBU make a small amount of user RAM available at addresses $0000–$0035. BUFFALO uses RAM addresses $0036–$00FF, and will likely write over any data you have stored in this address block.

In the blank, predict the value of the sum if you initialize location $0002 to $04.

Sum = $_____

In the blanks, list the addresses where the DEMO program stores this sum.

$_____
$_____

Use the BUFFALO MM command to initialize the augend to $04.
Use the MM command to store $FF in the two sum locations.
Now, use the BUFFALO G command to run the program:

`G xxxx`   <return>

where xxxx is the beginning address of the program.

Reset the EVBU/EVB board (push-button). Use the MD command to examine the sum locations. They should both contain $06.

Demonstrate DEMO program operation to the instructor. Use the MM command to write $FF to each of the sum locations. Initialize the augend to a value specified by the instructor.

_____ *Instructor Check*

Use the MM and MD commands to re-initialize the sum locations to $FF.

Next, use the BUFFALO RM command to initialize the program counter (PC) to DEMO's beginning address. Use this command to access the PC:

`RM`   <return>

Initialize the PC to DEMO program's starting address. Press the return key to exit the RM command and obtain the BUFFALO system prompt.

Use the BUFFALO T command to fetch and execute the DEMO program's first instruction:

```
T                                                                        <return>
```

Fill in the blanks with data from the terminal screen.

```
X   = $_____
PC  = $_____
```

Press the return key to fetch and execute DEMO's second instruction. Fill in the blanks.

```
X    = $_____
PC   = $_____
ACCA = $_____
```

Press the return key to fetch and execute DEMO's third instruction. Record screen data in the blanks.

```
X    = $_____
PC   = $_____
ACCA = $_____
```

Press <return> to execute the fourth instruction. Record screen data.

```
X    = $_____
PC   = $_____
ACCA = $_____
```

Do not reset the EVBU/EVB board. Use the MD command to display the sum storage locations. To which location has the M68HC11 CPU stored the sum?

$_____

Use the T <return> command to fetch and execute the DEMO program's fifth instruction. Record screen data.

```
X    = $_____
PC   = $_____
ACCA = $_____
```

Use the MD command to display the sum storage locations. To which location did the fifth instruction store the sum?

$_____

Use the T command to fetch and execute the DEMO program's sixth instruction. Record screen data.

```
X    = $_____
PC   = $_____
ACCA = $_____
```

Reset the EVBU/EVB board.

This part of Exercise 1.1 demonstrates the ease with which the user can run, trace, and analyze an application program. Remember that after the initial fetch and execute, BUFFALO will trace an instruction if you simply press the return key. However, if you employ an MD command after a trace, use the T command to fetch and execute the next program instruction.

PART IV: MOVE the DEMO Program and Block Fill

Review the BF and MOVE commands.

Use the BUFFALO MOVE command to copy the DEMO program from RAM to EEPROM—starting at address $B700. Reset the EVBU/EVB board, then enter this command line:

```
MOVE xxxx yyyy B700                                                      <return>
```

where xxxx is the first RAM program address, yyyy is the final RAM program address, and B700 is the beginning address of EEPROM storage.

Use the MD command to display the contents of $B700 and following addresses. Assure that the MOVE command successfully copied the program from RAM to EEPROM.

Use the BUFFALO BF command to write $AA over each RAM DEMO program location. Use this command line:

```
BF xxxx yyyy AA                                                    <return>
```

where xxxx is the first RAM program address, yyyy is the final RAM program address, and $AA is the data to be written to the specified addresses.

Use MD commands to display both the new and the old program locations on the terminal screen. The old program locations should now contain $AA.

Show your display to the instructor.

_____ *Instructor Check*

PART V: Use the BUFFALO Assembler/Disassembler to Enter a Program

Carefully review pages 14–16, on the BUFFALO one-line assembler/disassembler.

Refer to the assembly source code listing in Figure 1–5. If you use an EVBU, you can enter DEMO program source code as listed. However, if you use an EVB, you must change the addresses accompanying the source codes to reflect available RAM. The DEMO program should reside in EVB RAM address $C000–$C00B. If you use an EVB, make a new assembly source code listing in the blanks below. Make address entries appropriate to DEMO's location in address block $C000–$C00B. The first two entries are completed to get you started. If you use an EVBU, skip this step.

DEMO Program
ASSEMBLY CODE

$ADDRESS	SOURCE CODE
C000	LDX #FFFF
C003	LDAA #2
____	_____
____	_____
____	_____
____	_____

Use the BUFFALO ASM command to invoke the assembler/disassembler. From the BUFFALO system prompt, enter:

```
ASM xxxx                                                           <return>
```

where xxxx is the beginning address of the program (0100 for the EVBU, C000 for the EVB). Enter all source code lines. After each line entry, press the return key to assemble the instruction. After entering the last source code line and pressing the return key, use (CTRL)A to return to the BUFFALO system prompt.

Use this command to reinvoke the assembler/disassembler at DEMO program's starting address:

```
ASM xxxx                                                           <return>
```

Use the return key to disassemble the DEMO program line by line, and check its accuracy. Use the assembler/disassembler to correct any program errors. When confident that the program is correct, show the instructor your complete disassembled program listing on the terminal screen.

_____ *Instructor Check*

Return to the BUFFALO system prompt. Use the MD command to display the contents of $B700–$B70B and $0100–$010B (EVBU) or $C000–$C00B (EVB). Note that the contents of these two memory blocks are identical. BUFFALO assembled machine code identical to the machine code which you entered earlier in this exercise.

Use the MM command to initialize the augend to $0A, and the two sum locations to $FF.

Use a G command to run the program in RAM. Use an MD command to check the sum locations, to assure that the program ran properly. Show the instructor the terminal display showing the sum locations.

_____ *Instructor Check*

PART VI: DEMO Program Modification and Analysis

Figure 1–6 flowcharts a modified DEMO program. This figure also provides memory allocations and an assembly source code listing.

This program initializes ACCA to $02, then adds an augend five times. Each addition generates a new sum, which the CPU stores to an assigned location (SUM1, SUM2, SUM3, etc.). After generating the fifth sum, the CPU halts. If the augend equals $04, for example, the program stores these SUMx values:

$ADDRESS	LABEL	$CONTENTS
0001	SUM1	06
0002	SUM2	0A
0003	SUM3	0E
0004	SUM4	12
0005	SUM5	16

Refer to the Figure 1–6 flowchart and its accompanying program list. In block 1 (program address $0100 or $C000), the CPU points its X register at the first sum location. To accomplish this, the CPU loads its X register with $0001, SUM1's address. In block 2 (program address $0103 or $C003), the CPU initializes ACCA to $02. In block 3 (program address $0105 or $C005), the CPU uses direct addressing mode to add the augend value (contents of $0000) to the contents of ACCA. In block 4 (program address $0107 or $C007), the CPU stores the sum to an effective address. This effective address is formed by the sum of the index register, plus the offset value accompanying the instruction. Since this offset value equals zero, the CPU stores the sum to an address pointed at by the X register.

FIGURE 1–6

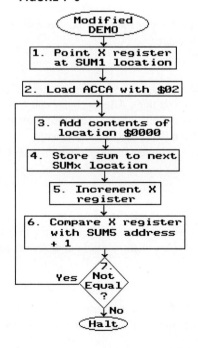

In block 5 (program address $0109 or $C009), the CPU increments (adds $0001) to the X register's contents. This points the X register at the next SUMx location. The instruction's mnemonic—INX—stands for "Increment X Register." If the CPU has generated five sums, stored them, and incremented the X register each time, the block 5 increment points the X register at the SUM5 location plus one. Therefore, in block 6 (program address $010A or $C00A), the CPU compares the incremented X register value with $0006, the address of SUM5 + 1. The CPX mnemonic stands for "Compare X Register." The number accompanying the CPX is the comparison value $0006. If this comparison shows that the X register contents does not equal $0006, the CPU branches back from Block 7 to Block 3, where it adds the augend again. If the X register equals $0006, the CPU has generated all five sums. The CPU does not branch back from block 7, but halts. The instruction that implements block 7 (program address $0100 or $C00D) uses the BNE (Branch if Not Equal) mnemonic. Note that the number accompanying the BNE mnemonic is the address ($0105 or $C005) to which the CPU branches back—if the block 6 comparison shows that the X register's contents do not equal the comparison value.

From the BUFFALO system prompt, invoke the assembler/disassembler. Use the assembler to compile and check Figure 1–6's Modified DEMO program.

Given an augend value of $0B, predict SUM1–SUM5 results. Enter these predicted values in the blanks:

SUM1 = $_____
SUM2 = $_____
SUM3 = $_____
SUM4 = $_____
SUM5 = $_____

Use the MM command to initialize the augend to $0B. Use the G command to run the Modified DEMO program. Reset the EVBU/EVB board. Use the MD command to display the SUM1–SUM5 results. These results should match your predictions. Show your displayed SUM1–SUM5 values to the instructor.

_____ *Instructor Check*

Review the BUFFALO breakpoint feature.

* Use the BR command to enter the following breakpoints into the BUFFALO breakpoint table:

```
EVBU:0103 0107 010A 010F
EVB: C003 C007 C00A C00F
```

Do not reset the EVBU or EVB board.

Analyze these breakpoint addresses, with reference to Figure 1–6. If the M68HC11 CPU executes the complete Modified DEMO program one time, how many times will execution halt at each of these breakpoints? Enter your answers in the blanks:

| Breakpoint | | Number |
EVBU	EVB	of Halts
0103	C003	_____
0107	C007	_____
010A	C00A	_____
010F	C00F	_____

Use the MM and MD commands to initialize the augend to $0B and all SUMx locations to $FF.

Use the BR command to assure that the breakpoint addresses are still in the breakpoint table.

Use the G command to start program execution:

```
G 100  (EVBU)                                                           <return>
G C000 (EVB)                                                            <return>
```

At the first breakpoint, record the relevant information in the blanks provided:

```
Flowchart Block # = _____
PC   = $_____
X    = $_____
ACCA = $_____
```

Use the P command to make the CPU fetch and execute to the next breakpoint:

P   <return>

Again, record pertinent information at this breakpoint:

```
Flowchart Block # = _____
PC   = $_____
X    = $_____
ACCA = $_____
```

Use the `P <return>` command to advance program execution to each breakpoint. At each breakpoint, record pertinent information in the blanks.

Note: Avoid an EVBU/EVB board reset while doing this procedure. If you must (or inadvertently) reset the EVBU/EVB board, start the entire procedure over again. An asterisk, following the previous instructor check above, indicates the restart point.

_____

```
Flowchart block # = _____
PC   = $_____
X    = $_____
ACCA = $_____
```
_____

```
Flowchart block # = _____
PC   = $_____
X    = $_____
ACCA = $_____
```
_____

```
Flowchart block # = _____
PC   = $_____
X    = $_____
ACCA = $_____
```
_____

```
Flowchart block # = _____
PC   = $_____
X    = $_____
ACCA = $_____
```
_____

```
Flowchart block # = _____
PC   = $_____
X    = $_____
ACCA = $_____
```
_____

```
Flowchart block # = _____
PC   = $_____
X    = $_____
ACCA = $_____
```
_____

```
Flowchart block # = _____
PC   = $_____
X    = $_____
ACCA = $_____
```
_____

```
Flowchart block # = _____
PC = $_____
X = $_____
ACCA = $_____
```
_____

```
Flowchart block # = _____
PC = $_____
X = $_____
ACCA = $_____
```
_____

```
Flowchart block # = _____
PC = $_____
X = $_____
ACCA = $_____
```
_____

Do not reset the EVBU/EVB board.

After entering your final data entries above, and before executing the final WAI instruction, take the following actions. These actions erase the breakpoint table contents, and restore all program instructions. Enter the command:

BR —   <return>

Enter the command

G xxxx   <return>

where xxxx is the address of the WAI instruction.

This procedure completes program execution. It is important to remember that if you were using breakpoints, take these two actions before resetting the EVBU/EVB board. Otherwise, the BUFFALO breakpoint utility will leave your program in a corrupted condition. If the program is corrupted, you must re-assemble the program instructions located at breakpoint addresses.

Reset the EVBU/EVB board. Use the BR command to examine the breakpoint table. What are this table's contents? What action caused BUFFALO to erase the breakpoint table's contents?

Invoke the assembler/disassembler; disassemble and examine the program. Is the program intact? Show your terminal display with the intact disassembled program to the instructor.

_____ *Instructor Check*

Exercise 1.1 has given the reader some practical experience with ten BUFFALO system commands. The remainder of this book gives the reader many occasions to use these commands.

1.5 M68HC11 ADDRESSING MODES

This section focuses on the M68HC11's six addressing modes. Exercise 1.1's DEMO program illustrates how immediate, direct, and indexed modes work. The exercise's Modified DEMO program adds two other modes—inherent and relative. Since the reader understands the Modified DEMO program, the author shall use its instructions as addressing mode examples, where possible.

Inherent Addressing Mode

Inherent mode instructions never address memory locations. A number of inherent mode instructions use no operand at all, and therefore have no need to address memory. Other inherent mode instructions use an implied, constant operand such as zero or one. The remaining inherent mode instructions manipulate values that reside in the CPU's eight key registers. Modified DEMO uses the inherent mode INX (Increment X Register) instruction. INX implies a constant operand $0001, which the CPU adds to the contents of the X register:

```
X reg. + 1 --> X reg.
```

Modified DEMO's WAI instruction also uses inherent mode. WAI uses no operand. To execute a WAI instruction, the M68HC11 stacks current register data, and assumes a wait state.

Note that INX and WAI use only one byte of program space. An inherent mode instruction requires only an opcode, since it addresses no memory location. For example, the INX instruction uses opcode $08. A wide majority of inherent mode instructions require only one byte of program space. However, eight Y index register associated, inherent mode instructions use 2-byte opcodes. For example, the Increment Y Register (INY) instruction uses the same opcode as INX—$08—but requires a 1-byte prefix, $18. Therefore, INY (inherent) requires a 2-byte opcode: $18 08.

Thus, inherent mode instructions are generally single-byte instructions (with eight exceptions), that use no operand, imply a constant operand, or manipulate CPU register data. The M68HC11 uses some 73 inherent mode instructions.

Immediate Addressing Mode

Immediate addressing mode uses constant-value operands, and includes these values in the program itself. Recall that assembly code indicates immediate mode operands with a pound sign (#). Modified DEMO uses three immediate mode instructions:

```
LDX  #1
LDAA #2
CPX  #6
```

The first instruction loads the X register with immediate value $0001. The `LDAA #2` instruction loads ACCA with immediate value $02. The third instruction compares the X register's contents with immediate value $0006. Twenty-nine M68HC11 instructions can use immediate mode addressing.

Extended Addressing Mode

Extended mode instructions address operands that reside at memory addresses. Consider a "Store ACCA" instruction: `STAA C06A`. This instruction causes the CPU to copy the contents of ACCA to the memory address accompanying the instruction mnemonic—in this example, location $C06A. The assembler compiles this source code into a 3-byte instruction: `$B7 C0 6A`. The first byte is the opcode; the second and third bytes specify the data transfer's destination address.

Consider another source code example: `LDX 10B`. To execute this instruction, the CPU loads its X register with the contents of address $010B (MS byte) and $010C (LS byte). The assembler compiles a 3-byte instruction: `$FE 01 0B`. To execute this instruction, the CPU transfers the contents of the specified address ($010B) and the following address, because the X register holds 2 bytes.

Thus, an extended mode instruction incorporates the operand's full 2-byte address. Forty-nine instructions can use extended mode.

Direct Addressing Mode

Direct mode addresses reside on memory page zero. That is, these addresses' most significant bytes are always $00. Since this high byte is zero, it can be eliminated from the instruction. Thus, a direct mode memory access saves a byte of program space, as compared with extended mode. Only the effective address's low byte accompanies the opcode to form an instruction. The great majority of direct mode instructions consist of two bytes. Only four out of forty direct mode instructions use opcode prebytes. Such prebytes make them 3-byte instructions. These 3-byte instructions read or write direct mode data to the D or Y register.

The Modified DEMO program uses this direct mode instruction:

```
ADDA 0
```

The instruction causes the CPU to add the contents of location $0000 to the contents of ACCA, and return the sum to ACCA. The 2-byte machine code instruction is $9B 00.

Thus, direct mode instructions access page zero memory addresses. Since the effective address's high byte is zero, it can be omitted from the instruction. The BUFFALO assembler uses direct mode for page-zero accesses whenever possible.

Indexed Addressing Mode

A review of Figure 1–1 reminds the reader that the CPU incorporates two index registers, X and Y. The CPU forms an indexed mode effective address by adding an unsigned program byte to the contents of one of these registers. The program byte is called the indexed mode offset. This byte-length offset resides at the program address following the opcode.

The original DEMO program (Figure 1–4) contains this indexed mode instruction:

```
STAA D,X
```

This instruction causes the CPU to add eight-bit offset $0D to the sixteen-bit contents of the X register (previously initialized by the DEMO program to $FFFF) and to form the sixteen-bit effective address, $000C:

```
$FFFF + $0D = $000C
```

The CPU then writes the contents of ACCA to this effective address.

The Modified DEMO program uses the same instruction, but with $00 offset:

```
STAA 0,X
```

In this case, the CPU forms the effective address by adding the offset ($00) to the contents of the X register. Because of the $00 offset, the effective address equals the X register's contents. In each Modified DEMO program loop, the CPU increments the X register. Thus, the CPU stores sums to a string of consecutive storage locations. Modified DEMO illustrates how indexed mode addressing can shorten and simplify repetitive program operations. Without indexed mode, Modified DEMO's task—generating and storing five sums—would require a program five times longer.

Indexed mode instructions using the Y index register require an opcode prebyte ($18) to differentiate them from X-indexed instructions. When possible, use X-indexed mode to keep programs as short as possible. Fifty-three M68HC11 instructions can use either X- or Y-indexed mode to address operands in memory.

Relative Addressing Mode

Relative mode instructions control program flow, rather than access operands. Figure 1–6's Modified DEMO program illustrates this program control concept. This program's BNE (Branch if Not Equal) instruction operates in relative mode. The BNE instruction references a preceding comparison of the X register with $0006. This comparison results in two possibilities: X register and $0006 are equal—or they are not. The BNE instruction bases program flow upon these two possibilities. If the two values are not equal, then program flow branches back to execute another program loop. If the two values are equal, program flow does not branch back, but continues to the next instruction in the program list. BNE is called a conditional branching instruction, because program flow depends upon meeting a condition. The M68HC11 instruction set includes twenty-one relative mode instructions. Of these twenty-one, eighteen are conditional branches. Thus, a programmer can control program flow based upon eighteen different kinds of conditions. The remaining three relative mode instructions are unconditional branches. An unconditional branching instruction alters program flow, regardless of prevailing conditions. For example, a program may operate continuously, executing and re-executing a program loop. An unconditional branching instruction—BRA (Branch Always) at the bottom of the loop—branches program flow to the top of the loop each time the CPU encounters it.

The reader should recall that the program counter (PC) controls program flow (see page 5). Each time the CPU fetches a program byte, it increments the program counter so that it points at the next byte in the program list. A relative mode instruction alters program flow by sign extending a 1-byte offset, called the relative address, then adding the 2-byte, sign-extended value to the program counter's contents. In the case of a conditional branch, the CPU adds the offset (relative address) if a condition is met; the CPU does not add the offset if the condition is not met. To branch forward (down) in the program, the machine code programmer uses a positive offset. To branch back (as in Modified DEMO), the machine code programmer uses a negative offset. The BUFFALO assembler calculates the relative address for an assembly code programmer. When entering assembly source code, simply enter the program flow's destination address. For example, Modified DEMO's BNE source code is

BNE XXXX   <return>

where XXXX is the destination address ($0105 for the EVBU; $C005 for the EVB). The assembler calculates the relative address, and assembles machine code instruction

$26 F6

where $26 is the opcode for BNE and $F6 is the relative address. In this example, the relative address ($F6) equals decimal negative ten. Thus, the CPU must add negative ten to the program counter contents to form the destination address, i.e., the new contents of the program counter. The CPU increments the program counter right after transferring its contents to the address latch. The relative address itself resides at program address $010E (EVBU) or $C00E (EVB) in the Modified DEMO program. Thus, by the time the CPU fetches this relative address, the PC contains $010F (EVBU) or $C00F (EVB). If the BNE conditions are met, the CPU sign extends the relative address to sixteen bits, then adds it to the PC's contents:

$$
\begin{array}{rl}
\text{EVBU:} & \text{program counter's contents} = \$010\text{F} \\
& + \underline{\text{sign-extended relative address} = \$\text{FFF6}} \\
& \text{16-bit sum returned to the PC} = \$0105
\end{array}
$$

$$
\begin{array}{rl}
\text{EVB:} & \text{program counter's contents} = \$\text{C00F} \\
& + \underline{\text{sign-extended relative address} = \$\text{FFF6}} \\
& \text{16-bit sum returned to the PC} = \$\text{C005}
\end{array}
$$

The CPU returns the 16-bit sum to the PC. Thus, the CPU will fetch the next program opcode from address $0105 (EVBU) or $C005 (EVB).

Because the relative address is a 1-byte offset, the CPU can branch up to 128 addresses back (relative address = $80), or branch up to 127 addresses forward (relative address = $7F). An attempt to assemble a branching instruction with a greater relative address generates an error message from the BUFFALO assembler.

Relative addressing enhances program portability. Although assembly source code specifies an address, the assembled relative address will work properly even if the program is moved to another memory block. The reader shall use relative addressing extensively to simplify programs and make them responsive to changing conditions.

1.6 M68HC11 INSTRUCTIONS

This section and its subsections consider most of the M68HC11's instruction set. Subsections categorize these instructions into seven functional groups. The author includes hands-on exercises, giving the reader opportunities to apply some of the instructions.

The reader must distinguish BUFFALO system commands from M68HC11 instructions. BUFFALO system commands allow the user to load, check, run, debug, and analyze

M68HC11 programs. The programs themselves consist of M68HC11 instructions, as compiled by the BUFFALO assembler. In fact, BUFFALO system commands simply call ROM program routines. These ROM routines—made up of M68HC11 instructions—execute the command's task, such as display memory contents on the terminal screen, or change a memory location's contents.

Data and Address Handling Instructions

Recall from Section 1.3 that the CPU incorporates eight key registers. The CPU's address handling function includes the program counter (PC), X and Y registers, and stack pointer (SP). The CPU's data handling function includes ACCA, ACCB, the D register, and condition code register (CCR). Future sections will discuss CCR-associated instructions and PC-associated instructions. The present section covers instructions that move information into and out of these remaining six key registers: ACCA, ACCB, D register, X register, Y register, and SP.

The instructions under discussion in this section use mnemonics which start with the letters LD, ST, T, and XG. The LD stands for "load." There is a load instruction for each of the six registers listed above:

LDAA—load accumulator A
LDAB—load accumulator B
LDD—load D register
LDX—load Y register
LDY—load Y register
LDS—load stack pointer

Refer to Figure 1–6 for two load instruction examples: LDX #1 loads the X register with the immediate-mode value $0001. LDAA #2 loads ACCA with the immediate-mode value $02.

Each of these six instructions can load values using immediate, direct, extended, or indexed addressing modes.

A "store" instruction's mnemonic starts with ST. All six registers have an associated store instruction:

STAA—store accumulator A
STAB—store accumulator B
STD—store D register
STX—store X register
STY—store Y register
STS—store stack pointer

Figure 1–5 includes two store instructions. The STAA 9 instruction stores the contents of ACCA to direct-mode address $0009. The STAA D,X instruction stores the contents of ACCA to address $000C, using indexed mode. Recall that the CPU forms the effective address by adding the contents of the index register ($FFFF in Figure 1–5) to the offset value accompanying the instruction ($0D in Figure 1–5). These six store instructions can use direct, extended, or indexed addressing.

If an instruction starts with T, it transfers the contents of one register to another:

TAB—transfers the contents of ACCA to ACCB.
TBA—transfers the contents of ACCB to ACCA.
TSX—transfers the contents of the stack pointer + 1 to the X register.
TSY—transfers the contents of the stack pointer + 1 to the Y register.
TXS—transfers the contents of the X register – 1 to the stack pointer.
TYS—transfers the contents of the Y register – 1 to the stack pointer.

These instructions merit further discussion. A transfer accumulator instruction leaves the same value in both accumulators. A TAB instruction leaves the original contents of ACCA in both accumulators. A TBA instruction leaves the original contents of ACCB in both accumulators.

The reader should note that the TSX and TSY instructions increment the stack pointer value before transferring it to the X or Y register. The CPU increments the SP value so that this value actually points at the last data pushed onto the stack. At any point in time, the SP points at the next location to receive stacked data. As soon as the CPU pushes a byte onto the stack, it decrements the SP. Motorola presumes that the programmer transfers the stack pointer to an index register so that a subsequent instruction can use indexed mode to access a byte on the stack itself. Since the transferred SP+1 value points directly at the top of the stack, the CPU can use indexed mode with zero offset to access this top byte. To compensate for the SP increment by TSX and TSY, the TXS and TYS instructions decrement the contents of the X or Y register, before transferring its contents to the SP.

A transfer instruction's destination is always another CPU register. Therefore, all transfer instructions use inherent addressing mode.

If an instruction's mnemonic begins with XG, it exchanges data between the D register and an index register:

   XGDX—X register contents to D register; D register contents to X register.
   XGDY—Y register contents to D register; D register contents to Y register.

The XGDX and XGDY instructions use inherent addressing mode.

Push and Pull Instructions

Push and pull instructions transfer the contents of ACCA, ACCB, and the index registers to and from the stack.

A programmer establishes a stack by taking two actions. First, allocate a block of user RAM as a stack. Second, insert an LDS instruction in the program. This LDS instruction must load the stack pointer with the highest address of the allocated RAM block (stack).

Because a stack grows from higher to lower addresses, the reader can view the initial SP value as the bottom of the stack. As the CPU pushes data bytes onto the stack, the SP decrements to lower and lower addresses within the RAM block. When the CPU pulls data bytes from the stack, the SP increments to higher RAM addresses. Thus, the stack grows or shrinks, depending upon whether the CPU pushes data onto the stack or pulls data from it.

Push instructions transfer the contents of ACCA, ACCB, or an index register onto the stack:

- PSHA—Transfer the contents of ACCA to the address pointed at by the stack pointer; then, decrement the stack pointer (ACCA $\rightarrow M_{SP}$; SP–1 $\rightarrow$ SP).
- PSHB—Transfer the contents of ACCB to the address pointed at by the stack pointer; then, decrement the stack pointer (ACCB $\rightarrow M_{SP}$; SP–1 $\rightarrow$ SP).
- PSHX—Transfer the contents of the X register's LS byte to the address pointed at by the SP. Decrement the SP. Then transfer the contents of the X register's MS byte to the address pointed at by the SP. Decrement the SP again ($X_{LS} \rightarrow M_{SP}$, SP–1 $\rightarrow$ SP; $X_{MS} \rightarrow M_{SP}$, SP–1 $\rightarrow$ SP).
- PSHY—Transfer the contents of the Y register's LS byte to the address pointed at by the SP. Decrement the SP. Then transfer the contents of the Y register's MS byte to the address pointed at by the SP. Decrement the SP again ($Y_{LS} \rightarrow M_{SP}$, SP–1 $\rightarrow$ SP; $Y_{MS} \rightarrow M_{SP}$, SP–1 $\rightarrow$ SP).

Pull instructions take just the opposite actions. Pull instructions transfer bytes from the stack to ACCA, ACCB or an index register.

- PULA—Increment the stack pointer. Transfer the contents of the address pointed at by the SP, to ACCA (SP+1 $\rightarrow$ SP; $M_{SP} \rightarrow$ ACCA).
- PULB—Increment the stack pointer. Transfer the contents of the address pointed at by the SP, to ACCB (SP+1 $\rightarrow$ SP; $M_{SP} \rightarrow$ ACCB).
- PULX—Increment the stack pointer. Transfer the contents of the address, pointed at by the stack pointer, to the X register's MS byte. Increment the stack pointer again. Transfer the contents of the address, pointed at by the stack pointer, to the X register's LS byte (SP+1 $\rightarrow$ SP, $M_{SP} \rightarrow X_{MS}$; SP+1 $\rightarrow$ SP, $M_{SP} \rightarrow X_{LS}$).
- PULY–Increment the stack pointer. Transfer the contents of the address, pointed at by the stack pointer, to the Y register's MS byte. Increment the stack pointer again. Transfer the contents of the address, pointed at by the stack pointer, to the Y register's LS byte (SP+1 $\rightarrow$ SP, $M_{SP} \rightarrow Y_{MS}$; SP+1 $\rightarrow$ SP, $M_{SP} \rightarrow Y_{LS}$).

Push and pull instructions transfer information between CPU registers and RAM locations pointed at by the stack pointer. Push and pull instructions use no offsets. Therefore, these instructions use inherent mode addressing only.

Programmers use the stack for temporary information storage, and to pass parameters from one program routine to another. Certain other M68HC11 instructions also make use of the stack. Jump to Subroutine (JSR) and Branch to Subroutine (BSR) instructions automatically stack the program counter contents. The Return from Subroutine (RTS) instruction automatically pulls the stacked PC value. The WAI and SWI (Software Interrupt) instructions automatically stack the processor context, i.e., the contents of the PC, X and Y registers, ACCA, ACCB, and the CCR. All M68HC11 hardware interrupt mechanisms also stack the processor context. The Return from Interrupt (RTI) instruction automatically pulls the processor context and returns it to the proper CPU registers. Exercise 1.2 gives the reader experience with some load, store, push, and pull instructions.

EXERCISE 1.2 Load, Store, Push, and Pull Instructions

REQUIRED EQUIPMENT: Motorola M68HC11EVB or EVBU, and terminal

GENERAL INSTRUCTIONS: This laboratory exercise gives the reader a chance to develop a DEMO2 program. This program uses several load, store, push, and pull instructions. Overall, DEMO2 accomplishes the following tasks:

1. Load the stack pointer, X register, and Y register, with initializing values.
2. Load ACCA with a series of ten data bytes from RAM addresses $0000–$0009.
3. After each data byte load, store it to RAM block $000A–$0013, in the same order as RAM block $0000–$0009.
4. After each data byte load from $0000–$0009 and store to $000A–$0013, push the data byte onto the stack.
5. After loading, storing, and pushing the data bytes from $0000–$0009, re-initialize the X register.
6. Pull the stacked data bytes to ACCB. After each pull, store the data byte back to RAM block $0000–$0009. Store these data bytes in reverse order from original (if a given value originated from address $0000, it should end up in address $0009; if a value originated from address $0001, it should end up in address $0008, etc.).
7. Before halting, re-initialize the stack pointer so that the WAI instruction does not stack processor context over data on the DEMO2 stack.

This program should increase the reader's insight into load and store operations and how the stack works. DEMO2 also uses the Compare X Register (CPX) and BNE (Branch if Not Equal) instructions, as used in Part VI of Exercise 1.1.

PART I: Analyze the DEMO2 Program Flow Chart

Refer to Figure 1–7 and its accompanying memory allocations list. You shall use the BUFFALO MM command to initialize the ten data bytes at $0000–$0009. The author suggests that you initialize these bytes to $00, $01, $02, etc., to facilitate program checking. In flowchart block 1, the CPU establishes the bottom of the stack at address $01FF (EVBU) or $C0FF (EVB). To accomplish this, the CPU simply loads the stack pointer (SP) with the bottom-of-stack address.

In blocks 2 and 3, the CPU loads its X and Y registers with initial values. DEMO2 uses the X register to manage loads and stores to the $0000–$0009 data block. The Y register manages writes to the $000A–$0013 data block. Therefore, the CPU initializes the X or Y register by pointing it at the initial address of its assigned block. To "point" the index register, the CPU loads it with the appropriate initial address.

In block 4, the CPU reads a byte, pointed at by the X register, to ACCA using indexed addressing mode. In block 5, the CPU writes this data byte from ACCA to an address

FIGURE 1-7

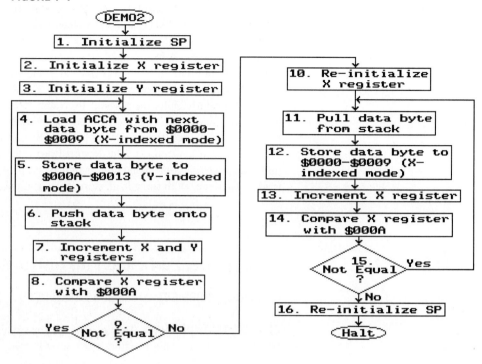

DEMO2 Memory Allocations

EVBU $ADDRESS	EVB $ADDRESS	CONTENTS
0000–0009	0000–0009	Data bytes
000A–0013	000A–0013	Write data bytes
0047	0047	Re-initialized SP
0100–0127	C000–C027	DEMO2 program
01FF	C0FF	Initial SP

pointed at by the Y register, using indexed addressing mode. In block 6, the CPU pushes this same data byte from ACCA onto the stack.

In block 7, the CPU uses the inherent mode INX and INY instructions to advance the X and Y index registers by one. These registers now point at either (1) the next address in their assigned RAM data blocks; or (2) the first address following their assigned data blocks. In the latter case, all data from $0000–$0009 have been loaded to ACCA, stored to $000A–$0013, and pushed onto the stack. The X register points at $000A (address following $0000–$0009), and the Y register points at $0014 (address following $000A–$0013).

In block 8, the CPU checks to see whether it has loaded, stored, and pushed all $0000–$0009 data bytes. The CPU accomplishes this by comparing the X register's contents with $000A. Recall from the paragraph above that after all $0000–$0009 data have been handled, the X register contains $000A. In block 9, the CPU executes a BNE (Branch if Not Equal) instruction. This conditional branching instruction uses relative addressing to return program flow to block 4—if the CPU has not loaded, stored, and pushed all $0000–$0009 data. If the CPU has loaded, stored, and pushed all $0000–$0009 data, the X register contents matches the Block 8 comparison value, and the CPU executes the next DEMO2 program instruction at block 10 rather than branching back.

By taking the "no" path from block 9, the CPU advances to the second part of the DEMO2 program. In this part, the CPU pulls and stores data back to $0000–$0009 in the reverse order from original. This task requires no extraordinary procedures. Recall that the stack operates in a "last on/first off" manner. The last byte pushed, in block 6, is the contents of address $0009. The first pull should therefore transfer this same data byte from the stack to ACCB. The CPU must store this byte to address $0000, to reverse the original order. Therefore, as the CPU pulls data from the stack, it must store these data to $0000, $0001, $0002, etc. DEMO2 uses the X register to manage these stores. Therefore, in block 10, the CPU must re-initialize its X register so that it points at the first storage location, $0000.

In blocks 11–15, the CPU executes a pull and store loop until it has pulled and stored all ten data bytes. In block 11, the CPU pulls a data byte from the stack to ACCB. In block 12, the CPU stores this data byte to the next $0000–$0009 address, using indexed addressing mode. In block 13, the CPU increments its X register to point at either (1) the next $0000–$0009 address; or (2) the address following the $0000–$0009 block, $000A.

Again, in block 14, the CPU compares its X register's contents with $000A. If X = $000A, all data bytes have been pulled and stored. If the X register does not contain $000A, the CPU branches back from block 15 ("yes" branch) to block 11. Block 15 uses a BNE instruction.

If the CPU takes the "no" path from block 15, it re-points the stack pointer at address $0047 (block 16). When the EVBU or EVB board powers up, BUFFALO establishes the stack at this address. In executing system commands, BUFFALO uses the stack. Be sure to examine the DEMO2 stack (using the MD command) after program execution. The CPU must re-initialize the stack pointer so that neither the WAI instruction (following block 16) nor BUFFALO writes over the DEMO2 stack data. Therefore, in block 16, the CPU loads the stack pointer with $0047. The CPU then halts by executing the WAI instruction.

PART II: Develop DEMO2 Source Code

Review Exercise 1.1, and the previous two sections. Develop source code that implements Figure 1–7's flowchart and adheres to its accompanying memory allocations. Enter your source code and appropriate addresses in the blanks provided below. There are comments to help you.

DEMO2

$ADDRESS	SOURCE CODE	COMMENTS
_____	_____	Initialize SP
_____	_____	Initialize X register
_____	_____	Initialize Y register
_____	_____	Load ACCA with next data byte from $0000–$0009
_____	_____	Store data byte to $000A–$0013
_____	_____	Push data byte onto stack
_____	_____	Increment X register
_____	_____	Increment Y register
_____	_____	Compare X register with $000A
_____	_____	If X register ≠ $000A, execute another load, store push loop
_____	_____	If X = $000A, re-initialize X register
_____	_____	Pull next data byte to ACCB
_____	_____	Store data byte to $0000–$0009
_____	_____	Increment X register
_____	_____	Compare X register with $000A
_____	_____	If X register ≠ $000A, execute another pull, store loop
_____	_____	Re-initialize SP
_____	_____	Halt

PART III: Load, Debug, and Demonstrate DEMO2

Use the BUFFALO assembler/disassembler to load and check the DEMO2 program. Use the MM command to initialize addresses $0000–$0009 with $00, $01, $02, etc. Test and debug DEMO2 program operation. When sure that DEMO2 runs properly, demonstrate its operation to the instructor.

_____ *Instructor Check*

Use the MM command to re-initialize locations $0000–$0009 to $00, $01, $02, etc. Run the program one time. Use the MD command to display addresses $C0F0–$C0FF (EVB) or $01F0–$01FF (EVBU). In the blanks, record the stack addresses and their contents.

$ADDRESS	$CONTENTS
_____	____
_____	____
_____	____
_____	____
_____	____
_____	____
_____	____
_____	____
_____	____
_____	____

Explain why the stack contains data in reverse order from RAM data block $000A–$0013.

PART IV: Laboratory Software Listing

DEMO2—EVB

$ADDRESS	SOURCE CODE	COMMENTS
C000	LDS #C0FF	Initialize SP
C003	LDX #0	Initialize X register
C006	LDY #A	Initialize Y register
C00A	LDAA 0,X	Load ACCA with next data byte from $0000–$0009
C00C	STAA 0,Y	Store data byte to $000A–$0013
C00F	PSHA	Push data byte onto stack
C010	INX	Increment X register
C011	INY	Increment Y register
C013	CPX #A	Compare X register with $000A
C016	BNE C00A	If X register ≠ $000A, execute another load, store push loop
C018	LDX #0	If X = $000A, re-initialize X register
C01B	PULB	Pull next data byte to ACCB
C01C	STAB 0,X	Store data byte to $0000–$0009
C01E	INX	Increment X register
C01F	CPX #A	Compare X register with $000A
C022	BNE C01B	If X register ≠ $000A, execute another pull, store loop
C024	LDS #47	Re-initialize SP
C027	WAI	Halt

DEMO2—EVBU

$ADDRESS	SOURCE CODE	COMMENTS
0100	LDS #1FF	Initialize SP
0103	LDX #0	Initialize X register
0106	LDY #A	Initialize Y register

010A	LDAA 0,X	Load ACCA with next data byte from $0000–$0009
010C	STAA 0,Y	Store data byte to $000A–$0013
010F	PSHA	Push data byte onto stack
0110	INX	Increment X register
0111	INY	Increment Y register
0113	CPX #A	Compare X register with $000A
0116	BNE 10A	If X register ≠ $000A, execute another load, store push loop
0118	LDX #0	If X = $000A, re-initialize X register
011B	PULB	Pull next data byte to ACCB
011C	STAB 0,X	Store data byte to $0000–$0009
011E	INX	Increment X register
011F	CPX #A	Compare X register with $000A
0122	BNE 11B	If X register ≠ $000A, execute another pull, store loop
0124	LDS #47	Re-initialize SP
0127	WAI	Halt

Clear, Increment, Decrement, Bit Set, and Bit Clear Instructions

The M68HC11's clear, increment, and decrement instructions can write zeros, add one, or subtract one from four of the CPU's key registers. These instructions can also affect memory bytes or bits.

Clear (CLR) instructions change the contents of an accumulator or memory to $00:

CLR—write $00 to a memory location.
CLRA—change ACCA's contents to $00.
CLRB—change ACCB's contents to $00.

The CLRA and CLRB instructions use inherent mode only.

Since the CLR instruction addresses a memory location, its execution requires extended or indexed mode. Consider this source-code line:

CLR A

The BUFFALO assembler compiles this machine-code instruction:

$7F 00 0A

where $7F is the CLR (extended) opcode, and $00 0A is the effective address. The M68HC11 executes this instruction by writing $00 to address $000A.

Decrement instruction mnemonics start with the letters DE. These instructions can operate on the following:

Memory (DEC)
ACCA (DECA)
ACCB (DECB)
X register (DEX)
Y register (DEY)
Stack pointer (DES)

A decrement instruction causes the CPU to subtract one from the contents of a specified CPU register or memory location. Assume, for example, that address $000A holds $FC. A DEC A instruction causes the CPU to reduce location $000A's contents from $FC to $FB. If ACCB holds $08, for example, an inherent mode DECB instruction would reduce ACCB's contents to $07. If the Y register holds $01B6, a DEY instruction reduces the Y register's contents to $01B5.

Because the DEC instruction addresses memory, it must use either extended or indexed addressing modes. Remaining decrement instructions subtract one from CPU registers. These instructions therefore use inherent mode.

Increment instructions bear a close resemblance to decrement instructions. Of course, increment instructions add one to a register or memory contents, rather than subtract one. Increment instruction mnemonics start with IN. These instructions can operate on the same memory or registers as decrement instructions. Likewise, increment instructions use the same addressing modes. Increment instructions include:

> INC—add one to the contents of a memory location.
> INCA—add one to ACCA's contents.
> INCB—add one to ACCB's contents.
> INX—add one to the X register's contents.
> INY—add one to the Y register's contents.
> INS—add one to the stack pointer's contents.

Exercise 1.2 has already made the reader familiar with the INX and INY instructions. The reader will find clear, increment, and decrement instructions particularly useful in implementing timing loops and software loop counters.

This section also considers two unique instructions, which clear or set selected bits in a memory location. These instructions are called BCLR (Bit Clear) and BSET (Bit Set). The programmer accompanies a BCLR or BSET mnemonic with a mask byte. This mask specifies which individual bits to clear or set. For example, assume that a memory byte contains $FF, and that the programmer wishes to clear bits 0, 2, 4, and 6. To clear these bits, the programmer accompanies the BCLR mnemonic with mask $55:

```
%01010101.
```

The mask's binary ones correspond to bits 0, 2, 4, and 6 of the memory byte. When the CPU executes this BCLR instruction, it clears corresponding memory bits to zero. Thus, the memory byte's contents change from $FF to:

```
%10101010 = $AA.
```

Remember that ones in the BCLR mask specify bits to be cleared.

Ones in the BSET mask byte specify memory bits to be set. For example, assume that a memory byte contains $B6:

```
%10110110
```

and that the programmer wishes to set bits 3 and 6. To accomplish this, the programmer accompanies a BSET instruction with mask $48:

```
%01001000
```

When the CPU executes this BSET instruction, it sets memory bits corresponding to ones in the mask byte. Thus, the CPU changes the memory location's contents from $B6 to:

```
%11111110 = $FE
```

Remember that ones in a BSET mask byte specify bits to be set.

BCLR and BSET instructions use direct or indexed modes to address the target memory address. For example, assume these conditions:

```
X = $0100.
Address $010C contains $CC.
The programmer wishes to change address $010C's contents to $44.
```

Given these conditions, the programmer uses this source code:

```
BCLR C,X 88
```

The "C,X" part of the source code addresses memory location $010C, using indexed mode. Mask byte $88 clears bits 3 and 7 of location $010C. After the CPU executes this instruction, address $010C contains $44:

$$\begin{aligned}\text{original contents of } \$010C &= \%11001100\\ \text{mask byte} &= \%10001000\\ \text{result} &= \%01000100\end{aligned}$$

The M68HC11 uses a substantial number of on-board control registers. These registers configure the M68HC11's internal devices and keep track of their status. BSET and BCLR allow a user to set or clear individual control register bits, without affecting other bits in the register.

FIGURE 1–8
Reprinted with the permission of Motorola.

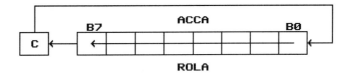

Shift, Rotate, and Logic Instructions

Shift and rotate instructions manipulate the contents of a memory location or an accumulator. A rotate instruction circulates bits right or left through the target accumulator or memory byte, into the C flag, and back to the target accumulator or memory. A single rotate instruction shifts bits only one place right or left. Consider a ROLA instruction: the letters RO stand for "rotate"; the letter L stands for "left," the direction of the rotate; and the A stands for ACCA, the target accumulator. Figure 1–8 diagrams the ROLA instruction. This figure shows that ACCA's MS bit shifts to the C flag, the C flag's content shifts into ACCA's LS bit, and ACCA bits 0 through 6 shift one bit left.

For example, assume that prior to ROLA execution, ACCA = $66, and the C flag is set. ROLA changes ACCA's contents to $CD, and resets the C flag.

A rotate right shifts an accumulator or memory location's LS bit into the C flag, the C flag into the accumulator or memory location's MS bit, and bits 7 through 1, one place right. Figure 1–9 illustrates this rotate right pattern.

As an ROR example, assume that ACCB contains $77, and that the C flag is set, prior to RORB execution. What are the contents of ACCB and the C flag after RORB execution? ACCB = $BB and the C flag is set. RORB shifts all ACCB bits one place right, the binary one in the C flag shifts into bit 7 of ACCB, and the binary one in bit 0 shifts to the C flag.

Rotate instructions can rotate bits in either accumulator, or in a memory location:

ROLA/RORA—rotate ACCA bits.
ROLB/RORB—rotate ACCB bits.
ROL/ROR—rotate bits in a memory location.

Rotate accumulator instructions use inherent mode.

ROL/ROR instructions use extended or indexed addressing mode. For example:

```
ROR C1A7
```

rotates address $C1A7's contents right, using extended mode.

Arithmetic and logical shift instructions also shift bits one place right or left in an accumulator, memory location or D register. These instructions also shift a bit into the C flag. However, an arithmetic or logical shift does not rotate the original C flag state into the target register or memory location.

Corresponding arithmetic and logical shifts left generate an identical response from the CPU. In fact, arithmetic and logical shift left instructions share the same opcodes:

- ASL/LSL—Shift a memory byte left. Opcode depends upon addressing mode.
- ASLA/LSLA—Shift ACCA left. Opcode = $48 (inherent).
- ASLB/LSLB—Shift ACCB left. Opcode = $58 (inherent).
- ASLD/LSLD—Shift the D register left. Opcode = $05 (inherent).

The reader can discern from this list that the letters ASL in the mnemonics stand for "Arithmetic Shift Left." The letters LSL stand for "Logical Shift Left."

All of these instructions shift a binary zero into the target register/memory byte's LS bit. Figure 1–10 diagrams arithmetic/logical shifts left. Since ASL/LSL instructions operate on memory locations, they use either extended or indexed addressing modes. The remaining shift left instructions use inherent mode, because they operate on CPU registers.

FIGURE 1–9
Reprinted with the permission of Motorola.

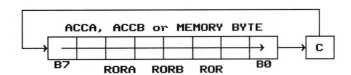

41

FIGURE 1–10
Reprinted with the permission of Motorola.

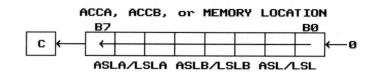

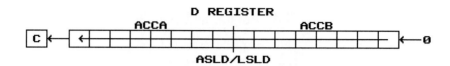

A user must distinguish between logical and arithmetic shifts right. All of these instructions shift bit 0 into the C flag. However, Arithmetic Shift Right (ASR) instructions retain the target register/memory's sign bit. Thus, ASR instructions do not shift a zero into the sign (MS) bit. Rather, this bit remains unchanged. Figure 1–11, part A diagrams ASR action. ASR instructions can do shifts on accumulators or memory locations:

ASR—arithmetically shift a memory byte right.
ASRA—arithmetically shift ACCA right.
ASRB—arithmetically shift ACCB right.

Logical Shift Right (LSR) instructions shift a zero into the target register/memory's MS bit, as shown in Figure 1–11, part B. LSR instructions can shift bits in a memory location, either accumulator, or the D register:

LSR—logic shift a memory byte right.
LSRA—logic shift ACCA right.
LSRB—logic shift ACCB right.
LSRD—logic shift the D register right.

Arithmetic and logical shifts right use distinct opcodes because of their differing actions. ASR and LSR instructions use extended or indexed modes to address a target memory location. All other shift right instructions operate on CPU registers, and therefore use inherent addressing mode.

Our discussion now turns to the M68HC11's logic instructions. These instructions fall in two categories: (1) instructions that find the one's or two's complement of a memory location or accumulator; and (2) instructions that AND, OR, or XOR the contents of an accumulator with a memory byte.

FIGURE 1–11
Reprinted with the permission of Motorola.

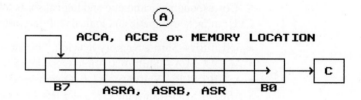

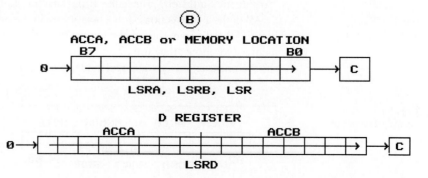

42

Complement (COM) instructions invert (one's complement) all bits in an accumulator or memory location:

    COM—invert a memory location's contents.
    COMA—invert the contents of ACCA.
    COMB—invert the contents of ACCB.

For example, assume that ACCB contains $AA:

```
ACCB = %10101010
```

A COMB instruction inverts these bits to $55:

```
ACCB = %01010101
```

A NEG instruction two's complements the contents of an accumulator or memory location:

    NEG—two's complement the contents of a memory byte.
    NEGA—two's complement ACCA's contents.
    NEGB—two's complement ACCB's contents.

For example, assume that ACCA contains $76:

```
ACCA = %01110110
```

A NEGA instruction changes ACCA's contents to $8A:

```
ACCA = %10001010
```

COM and NEG instructions use either extended or indexed mode to operate on memory bytes. COMA/NEGA and COMB/NEGB instructions use inherent mode to operate on accumulators.

ORAA and ORAB stand for "OR Accumulator A with memory" and "OR Accumulator B with memory." These instructions inclusive OR an accumulator with a memory location. For example, assume that ACCB contains operand $B5, and location $C005 contains $28. This source code uses extended addressing to OR the contents of ACCB and $C005:

```
ORAB C005
```

The CPU executes this instruction as follows:

$$\begin{aligned}
\text{OR ACCB} &= \%10110101 \\
\text{with location \$C005} &= \underline{\%00101000} \\
\text{return result to ACCB} &= \%10111101
\end{aligned}$$

Thus, the CPU returns $BD to ACCB.

ANDA and ANDB stand for "AND Accumulator A with memory" and "AND Accumulator B with memory." For example, assume that ACCB contains $B5, and location $C005 contains $28. This source code uses extended addressing mode to AND the contents of ACCB and $C005:

```
ANDB C005
```

The CPU executes this instruction as follows:

$$\begin{aligned}
\text{AND ACCB} &= \%10110101 \\
\text{with location \$C005} &= \underline{\%00101000} \\
\text{return result to ACCB} &= \%00100000
\end{aligned}$$

Thus, the CPU returns $20 to ACCB.

EORA and EORB stand for "Exclusive OR ACCA with memory" and Exclusive OR ACCB with memory." As an example, again assume that ACCB contains $B5 and location $C005 contains $28. This source code uses extended mode to exclusive OR the contents of ACCB with $C005:

```
EORB C005
```

The CPU executes this instruction as follows:

$$\begin{aligned}
\text{Exclusive OR ACCB} &= \%10110101 \\
\text{with location \$C005} &= \underline{\%00101000} \\
\text{return result to ACCB} &= \%10011101
\end{aligned}$$

Thus, the CPU returns $9D to ACCB.

Inclusive OR, AND, and Exclusive OR instructions can use immediate, direct, extended, or indexed modes to address memory bytes.

This section has introduced the reader to a variety of shift, rotate, and logic instructions. Exercise 1.3 gives the reader experience with a number of instructions introduced in this section as well as the previous one.

EXERCISE 1.3 — Bit Clear, Bit Shift, and Logic Instructions

REQUIRED EQUIPMENT: Motorola M68HC11EVB or EVBU, and terminal

GENERAL INSTRUCTIONS: Laboratory Exercise 1.3 gives the reader an opportunity to implement combinational logic in software. This implementation—called DEMO3—gives the reader experience with instructions covered in the previous two sections. The reader shall apply BCLR, COM, ANDA, ORAA, ASLA, and LSRA instructions to the combinational logic software. Part I of this exercise analyzes DEMO3's flowchart. In Part II, predict logic circuit output values for all sixteen input combinations. In Part III, develop DEMO3 source code. In Part IV, debug and demonstrate DEMO3 operation.

PART I: Analyze the DEMO3 Flowchart

DEMO3 software implements the sum-of-products equation and logic diagram shown at lower right of Figure 1–12. The user loads binary values for inputs A, B, C, and D into address $0000 (INPUT). Use the BUFFALO MM command to initialize INPUT. Input variables A, B, C, and D result in sixteen possible INPUT combinations, $00 through $0F. The user can initialize INPUT to any of these combinations. INPUT bit correspondence is:

$$\text{Input A} = \text{INPUT bit 3}$$
$$\text{Input B} = \text{INPUT bit 2}$$
$$\text{Input C} = \text{INPUT bit 1}$$
$$\text{Input D} = \text{INPUT bit 0}$$

DEMO3 emulates Figure 1–12's combinational logic to derive X. Of course, X equals binary one or binary zero, depending upon INPUT bit values. DEMO3 returns this X output value to address $0004, bit zero. Locations $0002 and $0003 function as temporary locations for the logic diagram's level-two values. Use an MM command to initialize

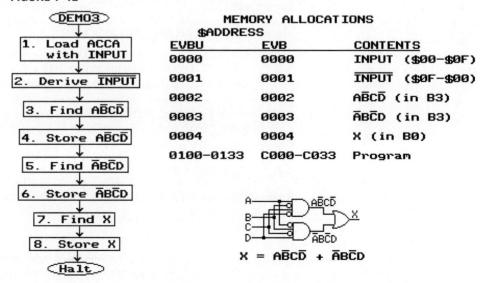

FIGURE 1–12

INPUT (address $0000) with desired input values. Use a G command to run DEMO3. Then use the MD command to examine the X output (address $0004).

Refer to Figure 1–12 flowchart blocks 1 and 2. In block 1, the CPU reads INPUT to ACCA from RAM location $0000. In block 2, the CPU takes three actions. First, the CPU stores the INPUT value from ACCA to $\overline{\text{INPUT}}$ location $0001. Second, the CPU executes a COM instruction (extended mode) to invert the value just stored in location $0001. This COM instruction inverts A, B, C, and D input variables to $\overline{A}$, $\overline{B}$, $\overline{C}$, and $\overline{D}$. These NOT-ed values reside in the following address $0001 bits:

$$\overline{A} = \text{bit 3}$$
$$\overline{B} = \text{bit 2}$$
$$\overline{C} = \text{bit 1}$$
$$\overline{D} = \text{bit 0}$$

Third, the CPU executes a BCLR instruction, to clear address $0001's high nibble. Use direct mode to access and clear bits 4–7:

```
BCLR 1 F0
```

Thus, in block 2, the CPU establishes inputs, as NOT-ed by the input inverters in Figure 1-12's logic diagram. These inputs now reside in location $0001 ($\overline{\text{INPUT}}$).

In flowchart block 3, the CPU derives an output from the top AND gate in Figure 1–12's logic diagram. This gate ANDs inputs A, $\overline{B}$, C, and $\overline{D}$. First, the CPU reads $\overline{\text{INPUT}}$ to ACCA from location $0001. To execute block 3's AND operation, the CPU must align A with $\overline{B}$, and C with $\overline{D}$. These pairs must correspond in ACCA ($\overline{B}$,$\overline{D}$) and location $0000 (A,C). To make this alignment, the CPU executes an ASLA instruction. After this ASLA, C and $\overline{D}$ reside in bit 1 of $0000 and ACCA respectively. A and $\overline{B}$ reside in bit 3. The CPU then ANDs ACCA with location $0000:

```
ANDA 0
```

To generate an output from the top AND gate, the CPU must now align the A/$\overline{B}$ result with the C/$\overline{D}$ result. To accomplish this, the CPU stores ACCA's contents to address $0002. After this store, copies of the first ANDA result reside in ACCA and location $0002. The CPU then executes two arithmetic shifts left (ASLAs) to align A/$\overline{B}$ with C/$\overline{D}$, in bit 3. Next, the CPU ANDs ACCA with the contents of address $0002, to derive A/$\overline{B}$/C/$\overline{D}$. This result represents the output from the top AND gate in Figure 1–12's logic diagram. This result resides in bit 3 of ACCA. All other ACCA bits are now irrelevant and should be cleared. In block 4, the CPU stores the result to location $0002, then executes a BCLR instruction to clear all address $0002 bits, except bit 3:

```
BCLR 2 F7
```

In blocks 5 and 6, the CPU derives an output from the bottom AND gate. The CPU loads ACCA with INPUT values from address $0000, then shifts ACCA left one time (ASLA) to align B with $\overline{A}$, and D with $\overline{C}$ (note that the inputs to the bottom AND gate are $\overline{A}$, B, $\overline{C}$, and D). The CPU then ANDs ACCA with location $0001, and returns the result to ACCA. This AND operation's significant results lie in ACCA's bit 1 ($\overline{C}$/D) and bit 3 ($\overline{A}$/B). The CPU stores ACCA's contents to address $0003. Next, the CPU shifts ACCA's contents left (ASLA) two times, to align $\overline{C}$/D in ACCA with $\overline{A}$/B in address $0003. The CPU then ANDs ACCA with location $0003, and returns the result to ACCA. The CPU has executed block 5. To execute block 6, the CPU stores ACCA's contents to address $0003, then clears all bits in address $0003, except bit 3. Bit 3 contains the $\overline{A}$/B/$\overline{C}$/D result. At this point, the output from the top AND gate resides in bit 3 of location $0002, and the output from the bottom AND resides in bit 3 of location $0003.

In block 7, the CPU performs a logical OR operation to derive the logic circuit's output value X. First, the CPU reads the contents of $0002 to ACCA. The CPU then ORs ACCA with address $0003:

```
ORAA 3
```

The X value now resides in ACCA, bit 3. The CPU executes three logic shifts right (LSRA) to move this X result from bit 3 to bit 0. The CPU then executes flowchart block 8 by storing the X value from ACCA to location $0004. Given INPUT values for A, B, C, and D, the CPU has derived output value X, and stored this result to bit 0 of address $0004.

PART II: Evaluate the Sum-of-Products Equation for all Input Combinations

Evaluate Figure 1–12's sum-of-products equation for all sixteen INPUT combinations. For each INPUT combination, enter output X's value in the truth table.

```
A B C D  X
0 0 0 0  __
0 0 0 1  __
0 0 1 0  __
0 0 1 1  __
0 1 0 0  __
0 1 0 1  __
0 1 1 0  __
0 1 1 1  __
1 0 0 0  __
1 0 0 1  __
1 0 1 0  __
1 0 1 1  __
1 1 0 0  __
1 1 0 1  __
1 1 1 0  __
1 1 1 1  __
```

PART III: Develop DEMO3 Source Code

Review the previous two sections. Develop source code that implements Figure 1–12's flowchart and adheres to its accompanying memory allocations. Enter your source code and appropriate addresses in the blanks provided below. Comments are included to help you along.

DEMO3

$ADDRESS	SOURCE CODE	COMMENTS
_____	_____	Load ACCA from INPUT
_____	_____	Store INPUT value to $0001
_____	_____	Invert INPUT to $\overline{\text{INPUT}}$
_____	_____	Clear $\overline{\text{INPUT}}$'s high nibble
_____	_____	Load ACCA from $\overline{\text{INPUT}}$
_____	_____	Align $\overline{B}$ with A, $\overline{D}$ with C
_____	_____	AND A with $\overline{B}$, C with $\overline{D}$
_____	_____	Store result to $0002
_____	_____	Align A/$\overline{B}$ with C/$\overline{D}$
_____	_____	" " " "
_____	_____	AND A/$\overline{B}$ with C/$\overline{D}$
_____	_____	Store A/$\overline{B}$/C/$\overline{D}$ to $0002
_____	_____	Clear all address $0002 bits, except B3
_____	_____	Load ACCA from INPUT
_____	_____	Align $\overline{A}$ with B, $\overline{C}$ with D
_____	_____	AND $\overline{A}$ with B, $\overline{C}$ with D
_____	_____	Store result to $0003
_____	_____	Align $\overline{A}$/B with $\overline{C}$/D
_____	_____	" " " "

		AND $\overline{A}$/B with $\overline{C}$/D
_____	_____	Store result to $0003
_____	_____	Clear all address $0003 bits, except B3
_____	_____	Load ACCA from $0002
_____	_____	OR with $0003
_____	_____	Shift X result to ACCA, bit 0
_____	_____	" " " " " " "
_____	_____	" " " " " " "
_____	_____	Store X result to $0004
_____	_____	Halt

PART IV: Load, Debug and Demonstrate DEMO3

Use BUFFALO's assembler/disassembler to load and check your DEMO3 program. Use MM commands to initialize INPUT with each of its sixteen combinations. For each INPUT value, run the program and check location $0004 to assure that the X result confirms your truth table prediction in Part II. When you're sure that DEMO3 runs properly, demonstrate its operation to the instructor.

_____ *Instructor Check*

PART V: Laboratory Software Listing

DEMO3—EVB

$ADDRESS	SOURCE CODE	COMMENTS
C000	LDAA 0	Load ACCA from INPUT
C002	STAA 1	Store INPUT value to $0001
C004	COM 1	Invert INPUT to $\overline{INPUT}$
C007	BCLR 1 F0	Clear $\overline{INPUT}$'s high nibble
C00A	LDAA 1	Load ACCA from $\overline{INPUT}$
C00C	ASLA	Align $\overline{B}$ with A, $\overline{D}$ with C
C00D	ANDA 0	AND A with $\overline{B}$, C with $\overline{D}$
C00F	STAA 2	Store result to $0002
C011	ASLA	Align A/$\overline{B}$ with C/$\overline{D}$
C012	ASLA	" " " "
C013	ANDA 2	AND A/$\overline{B}$ with C/$\overline{D}$
C015	STAA 2	Store A/$\overline{B}$/C/$\overline{D}$ to $0002
C017	BCLR 2 F7	Clear all address $0002 bits, except B3
C01A	LDAA 0	Load ACCA from INPUT
C01C	ASLA	Align $\overline{A}$ with B, $\overline{C}$ with D
C01D	ANDA 1	AND $\overline{A}$ with B, $\overline{C}$ with D
C01F	STAA 3	Store result to $0003
C021	ASLA	Align $\overline{A}$/B with $\overline{C}$/D
C022	ASLA	" " " "
C023	ANDA 3	AND $\overline{A}$/B with $\overline{C}$/D
C025	STAA 3	Store result to $0003
C027	BCLR 3 F7	Clear all address $0003 bits, except B3
C02A	LDAA 2	Load ACCA from $0002
C02C	ORAA 3	OR with $0003
C02E	LSRA	Shift X result to ACCA, bit 0
C02F	LSRA	" " " " " " "
C030	LSRA	" " " " " " "
C031	STAA 4	Store X result to $0004
C033	WAI	Halt

DEMO3—EVBU

$ADDRESS	SOURCE CODE	COMMENTS
0100	LDAA 0	Load ACCA from INPUT
0102	STAA 1	Store INPUT value to $0001
0104	COM 1	Invert INPUT to $\overline{INPUT}$
0107	BCLR 1 F0	Clear $\overline{INPUT}$'s high nibble

010A	LDAA 1		Load ACCA from $\overline{\text{INPUT}}$
010C	ASLA		Align $\overline{\text{B}}$ with A, $\overline{\text{D}}$ with C
010D	ANDA 0		AND A with $\overline{\text{B}}$, C with $\overline{\text{D}}$
010F	STAA 2		Store result to $0002
0111	ASLA		Align A/$\overline{\text{B}}$ with C/$\overline{\text{D}}$
0112	ASLA		" " " "
0113	ANDA 2		AND A/$\overline{\text{B}}$ with C/$\overline{\text{D}}$
0115	STAA 2		Store A/$\overline{\text{B}}$/C/$\overline{\text{D}}$ to $0002
0117	BCLR 2 F7		Clear all address $0002 bits, except B3
011A	LDAA 0		Load ACCA from INPUT
011C	ASLA		Align $\overline{\text{A}}$ with B, $\overline{\text{C}}$ with D
011D	ANDA 1		AND $\overline{\text{A}}$ with B, $\overline{\text{C}}$ with D
011F	STAA 3		Store result to $0003
0121	ASLA		Align $\overline{\text{A}}$/B with $\overline{\text{C}}$/D
0122	ASLA		" " " "
0123	ANDA 3		AND A$\overline{\text{A}}$/B with $\overline{\text{C}}$/D
0125	STAA 3		Store result to $0003
0127	BCLR 3 F7		Clear all address $0003 bits, except B3
012A	LDAA 2		Load ACCA from $0002
012C	ORAA 3		OR with $0003
012E	LSRA		Shift X result to ACCA, bit 0
012F	LSRA		" " " " " " "
0130	LSRA		" " " " " " "
0131	STAA 4		Store X result to $0004
0133	WAI		Halt

Arithmetic Instructions

M68HC11 arithmetic instructions provide for convenient, multiprecision programming in four functions—add, subtract, multiply, and divide. The M68HC11's instruction set includes these addition instructions:

- ABA—Add the contents of ACCA to the contents of ACCB. Return the 8-bit sum to ACCA. This is an inherent mode instruction.
- ADDA—Add the contents of a memory location to the contents of ACCA. Return the 8-bit sum to ACCA. This instruction can access the memory byte using immediate, direct, extended, or indexed mode.
- ADDB—Add the contents of a memory location to the contents of ACCB. Return the 8-bit sum to ACCB. ADDB uses immediate, direct, extended, or indexed mode to address the augend.
- ADDD—Add the contents of two memory locations (2 bytes) to the contents of the D register. Return the 16-bit sum to the D register. ADDD uses immediate, direct, extended, or indexed mode to access the 16-bit augend. Consider the following source code as an example:

ADDD 3

An assembler compiles this code into machine code instruction $D3 03. This instruction adds the contents of location $0004 to the D register's LS byte, and adds with carry the contents of $0003 to the D register's MS byte, then returns the sum to the D register. Consider another source code example:

ADDD #ABCD

From this code, the assembler compiles machine code instruction $C3 AB CD. This instruction adds immediate mode augend $ABCD to the contents of the D register, and returns the 16-bit sum to the D register.

Refer to the discussion of the condition code register's C flag. This discussion explains that the C flag sets when a sum overflows the result register. For example, the CPU adds $8F (in ACCA) to $A1 in memory. The sum, $130, overflows ACCA. The C flag sets to indicate this one carry from ACCA. The M68HC11's instruction set includes two add with carry instructions that facilitate multiprecision (multibyte) addition:

- ADCA—Add the contents of the C flag (zero or one), and the contents of a memory location to the contents of ACCA. Return the 8-bit sum to ACCA. Set the C flag if the sum overflows the accumulator. ADCA uses immediate, direct, extended, or indexed addressing modes.
- ADCB—Add the contents of the C flag, and the contents of a memory location, to ACCB. Return the 8-bit sum to ACCB. Set or reset the C flag to reflect the carry out of the accumulator. ADCB uses immediate, direct, extended, or indexed addressing modes.

Figure 1–13 flowcharts a program (DEMO4) that uses the ADDA, ADCA, and ROL instructions to do addition—with 3-byte precision.[12] Following paragraphs analyze this flowchart.

In block 1, the CPU clears the SUM MS location (address $0006). This location holds the sum's fourth and most significant byte. Given 3-byte addend and augend, this location holds a one if the addition of the addend and augend MS bytes generates a one carry. For example, this addition causes location $0006 to hold a one:

$$
\begin{aligned}
\text{3-byte addend} &= \$ABEEFF \\
\text{3-byte augend} &= \$9A3770 \\
\text{4-byte sum} &= \$0146266F
\end{aligned}
$$

In flowchart Block 11, the CPU will rotate the C flag state into address $0006, after it adds addend and augend MS bytes. In block 1, the CPU clears the contents of address $0006, to assure that it will hold no extraneous 1 bits after block 11's rotate.

In block 2, the CPU reads the addend's LS byte to ACCA from address $0002. In block 3, the CPU adds the augend's LS byte (from $0005). This ADDA returns the sum to ACCA. In block 4, the CPU stores the sum's LS byte from ACCA to location $0009 (SUM LS). This LS byte addition should use a standard ADDA instruction, rather than ADCA. This is the first addition operation, and no previously generated carry has any significance. However, all subsequent add operations in the program must use the ADCA instruction.

FIGURE 1–13

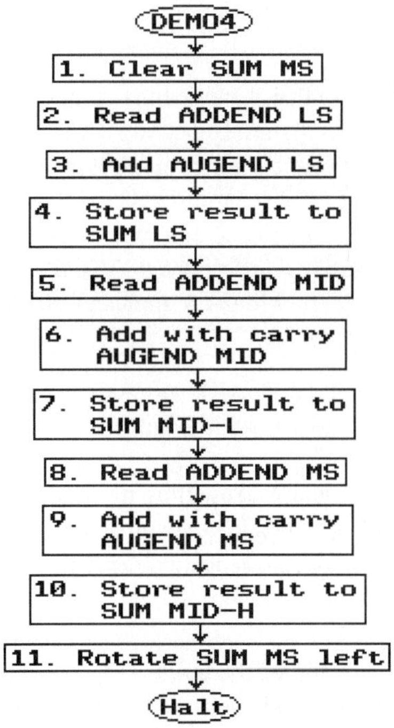

```
        DEMO4
          ↓
 1. Clear SUM MS
          ↓
 2. Read ADDEND LS
          ↓
 3. Add AUGEND LS
          ↓
 4. Store result to
    SUM LS
          ↓
 5. Read ADDEND MID
          ↓
 6. Add with carry
    AUGEND MID
          ↓
 7. Store result to
    SUM MID-L
          ↓
 8. Read ADDEND MS
          ↓
 9. Add with carry
    AUGEND MS
          ↓
10. Store result to
    SUM MID-H
          ↓
11. Rotate SUM MS left
          ↓
        Halt
```

MEMORY ALLOCATIONS $ADDRESS		
EVBU	EVB	CONTENTS
0000	0000	ADDEND MS
0001	0001	ADDEND MID
0002	0002	ADDEND LS
0003	0003	AUGEND MS
0004	0004	AUGEND MID
0005	0005	AUGEND LS
0006	0006	SUM MS
0007	0007	SUM MID-H
0008	0008	SUM MID-L
0009	0009	SUM LS
0100–0118	C000–C018	Program

---

[12]This section uses four programs (DEMO4, DEMO5, DEMO6, DEMO7) to demonstrate arithmetic instructions. The reader shall have an opportunity to develop and run these DEMO programs in Exercise 1.4.

In blocks 5, 6, and 7, the CPU loads, adds with carry, and stores the sum of the mid-significance addend and augend bytes. In block 5, the CPU loads ACCA from the ADDEND MID location ($0001). In block 6, the CPU adds with carry the contents of the AUGEND MID location ($0004). In block 7, the CPU stores the resulting sum from ACCA to the SUM MID-L location ($0008).

In blocks 8, 9, and 10, the CPU loads, adds with carry, and stores the sum of the MS addend and augend bytes. In block 10, the CPU stores the resulting sum from ACCA to the SUM MID-H location ($0007). This addition generates a zero or one carry, and the C flag holds this carry bit. In block 11, the CPU rotates this carry bit from the C flag into the SUM MS byte (location $0006). Use an ROL instruction and extended addressing to accomplish this block 11 task. After block 11 execution, the sum is complete, and the CPU halts.

The M68HC11's instruction set includes subtract instructions, which are counterparts to the adds:

- SBA—Subtract the contents of ACCB from the contents of ACCA. Return the difference to ACCA. Set the C flag if the subtrahend was higher than the minuend. SBA uses inherent addressing mode.
- SUBA—Subtract the contents of a memory location from the contents of ACCB. Return the difference to ACCA. Set the C flag if the subtrahend was higher than the minuend. SUBA can use immediate, direct, extended, or indexed addressing modes.
- SUBB—Subtract the contents of a memory location from the contents of ACCB. Return the difference to ACCB. Set the C flag if the subtrahend was higher than the minuend. SUBB can use immediate, direct, extended, or indexed addressing modes.
- SUBD—Subtract the contents of two memory locations (2 bytes) from the contents of the D register. Return the 16-bit difference to the D register. Set the C flag if the subtrahend was higher than the minuend. SUBD can use immediate, direct, extended, or indexed addressing modes.

Refer to the earlier discussion of the C flag. This discussion explains that the C flag sets when the ALU subtracts a higher number from a lower. To generate a valid difference, the CPU effectively "borrows" a binary one to complete the subtraction. In a multiprecision subtraction, the CPU must subtract such a borrow from the result of the next higher byte operation. The M68HC11's instruction set includes two subtract with carry instructions. These instructions subtract the C flag contents from their generated difference:

- SBCA—Subtract the C flag state and the contents of a memory location from the contents of ACCA. Return the difference to ACCA. Set the C flag if the combined C flag state and subtrahend were higher than the minuend. SBCA uses immediate, direct, extended, or indexed addressing modes.
- SBCB—Subtract the C flag state and the contents of a memory location from the contents of ACCB. Return the difference to ACCB. Set the C flag if the combined C flag state/subtrahend was higher than the minuend. SBCB uses immediate, direct, extended, or indexed addressing modes.

Figure 1–14 flowcharts a DEMO5 program. This program uses SUBB and SBCB instructions to subtract a 3-byte subtrahend from a 3-byte minuend. Following paragraphs analyze this flowchart.

Refer to blocks 1, 2, and 3. Executing these blocks, the CPU loads the minuend's LS byte, subtracts the subtrahend's LS byte, and stores the result in the LS byte of the difference. Block 2's subtract operation should use the SUBB command, because no relevant C flag condition exists from a previous subtract. Memory accesses can use direct addressing mode.

Blocks 4, 5, and 6 load, subtract, and store the mid-significant bytes of the minuend, subtrahend, and difference. Here, the block 5 subtract must use an SBCB instruction, because a relevant C flag condition resulted from the low-byte subtraction (Blocks 1–3).

Blocks 7, 8, and 9 complete the 3-byte subtraction. The CPU loads, subtracts, and stores the most significant bytes of the minuend, subtrahend, and difference. Again, the block 8 subtract must use the SBCB instruction to obtain a valid result under all circumstances.

The M68HC11's multiply (MUL) instruction multiplies the contents of the two accumulators, and returns a 16-bit product to the D register. Thus, a programmer finds the MUL

FIGURE 1–14

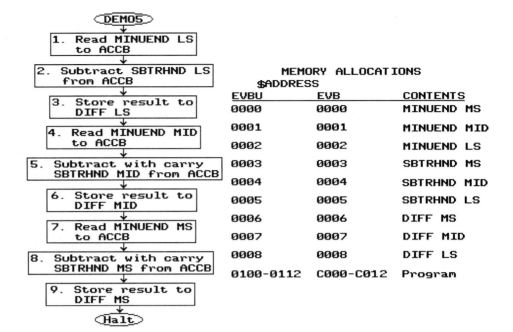

instruction easy to use: Load the A and B accumulators with multiplicand and multiplier, MUL, then store the 2-byte product from the D register to memory. MUL uses inherent addressing mode, since it accesses the contents of the two accumulators. The MUL instruction also makes multiprecision multiplication a relatively easy matter. Figures 1–15 and 1–16 illustrate a 3-byte by 1-byte multiplication program (DEMO6) which uses the MUL, ADCA, and STD instructions. Following paragraphs discuss this program.

Refer to Figure 1–15, part A, which illustrates DEMO6's strategy. This diagram shows that the CPU finds three intermediate products—P1, P2, and P3 (dotted-line boxes). P1 is the product of multiplier and multiplicand LS bytes. P2 is the product of multiplier and multiplicand mid-significance bytes. P3 is the product of multiplier and multiplicand MS bytes. Each of these products constitutes 2 bytes. The CPU stores P1's LS byte directly to the LS byte of the product (the oval on the far right). The CPU adds the MS byte of P1 to the LS byte of P2, and stores the result to the products's MID-L address (second oval from right). The CPU adds P2's MS byte, P3's LS byte, and the carry from the previous addition, then stores the result to the product's MID-H location (second oval from left). Finally, the CPU adds the carry from the previous addition to the MS byte of P3, and stores the resulting value in the product's MS location (far-left oval).

FIGURE 1–15

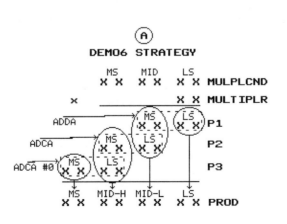

51

FIGURE 1–16

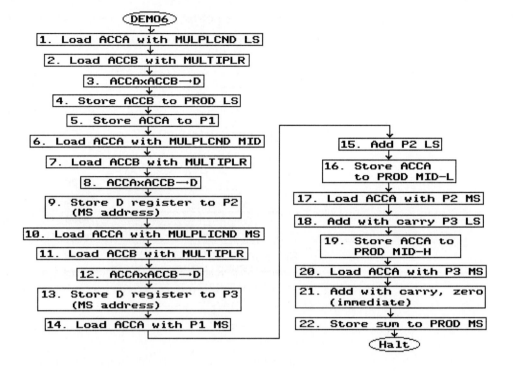

Figure 1–16 flowcharts the DEMO6 program. (Refer also to Figure 1–15, part B.) In flowchart blocks 1 through 5, the CPU generates intermediate product P1. In block 4, the CPU stores P1's LS byte directly from ACCB to address $0007, the LS byte of the product. In Block 6, the CPU stores P1's MS byte to address $000C. The CPU generates intermediate product P2 in flowchart blocks 7 through 9. In Block 9, the CPU stores P2 to its assigned locations: $000A and $000B.

In blocks 10 through 13, the CPU generates and stores intermediate product P3 to locations $0008 and $0009.

In blocks 14 through 22, the CPU sums and stores the final product (PROD) bytes. Note that block 15 adds P1's MS byte to P2's LS byte, using the ADDA instruction. Since the CPU stored P1's LS byte directly to the PROD LS address, this store generated no relevant C flag result. However, Block 18 uses the ADCA instruction since block 15's addition generates a relevant carry bit. Finally, note that block 21 should use this source code:

```
ADCA #0
```

which adds zero, with carry, to P3's MS byte. If block 18's addition caused the C flag to set, this ADCA instruction causes the CPU to add one to P3's MS byte, before storing it to the PROD MS location ($0004).

The M68HC11 instruction set includes two divide instructions:

- IDIV—Integer divide
- FDIV—Fractional divide

Both of these instructions divide the 16-bit contents of the D register (dividend/numerator) by the 16-bit contents of the X register (divisor/denominator), and return a 16-bit quotient to the X register. Both IDIV and FDIV return the 16-bit remainder to the D register. Both instructions use inherent mode. How then, do these instructions differ?

Use IDIV to divide whole integers. Use FDIV to carry the quotient to the right of the radix point. Consider this problem:

$$\$nnnn/\$dddd = \$qqqq.f_H f_H f_H f_H f_L f_L f_L f_L$$

where

$\$nnnn$ = a 16-bit numerator
$\$dddd$ = a 16-bit denominator
$\$qqqq$ = the 16-bit integer quotient
$f_H f_H f_H f_H f_L f_L f_L f_L$ = the 32-bit fractional quotient

Use IDIV and FDIV to solve the problem as follows:

1. Load the D register with $nnnn, and the X register with $dddd.
2. Execute IDIV. IDIV returns $qqqq to the X register and the remainder to the D register. Store the X register's contents as $qqqq. Leave the remainder in the D register.
3. Re-load the X register with $dddd.
4. Execute FDIV. This instruction divides the smaller remainder by the larger $dddd value and returns a fractional $f_H f_H f_H f_H$ value to the X register. FDIV also returns a new remainder to the D register. Store the X register's contents as $f_H f_H f_H f_H$. Leave the remainder in the D register.
5. Re-load the X register with $dddd.
6. Execute another FDIV. Store the X register's contents as $f_L f_L f_L f_L$.

The reader can see that a program could continue to re-load the X register with $dddd, and re-execute the FDIV instruction. Each FDIV execution would further generate 16 bits to the right of the radix point. Figure 1–17 flowcharts and allocates memory for the program just described. Figure 1–17 labels this program DEMO7.

The M68HC11's DAA (Decimal Adjust ACCA) instruction adjusts the sum of 2 packed BCD bytes. Recall that BCD stands for "binary coded decimal." BCD digits code decimal digits zero through nine as 4-bit words:

Decimal Digit	BCD Digit
0	0000
0	0001
0	0010
0	0011
0	0100
0	0101
0	0110
0	0111
0	1000
9	1001

Hence, ACCA or a memory location can contain two packed BCD digits. For example, ACCA holds the decimal value fifty-three as:

ACCA = BCD 01010011

FIGURE 1–17

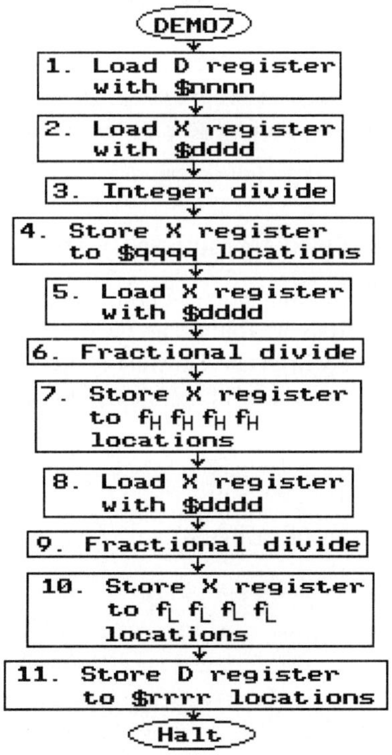

```
          ( DEMO7 )
             ↓
1. Load D register
   with $nnnn
             ↓
2. Load X register
   with $dddd
             ↓
3. Integer divide
             ↓
4. Store X register
   to $qqqq locations
             ↓
5. Load X register
   with $dddd
             ↓
6. Fractional divide
             ↓
7. Store X register
   to fH fH fH fH
   locations
             ↓
8. Load X register
   with $dddd
             ↓
9. Fractional divide
             ↓
10. Store X register
    to fL fL fL fL
    locations
             ↓
11. Store D register
    to $rrrr locations
             ↓
          ( Halt )
```

DEMO7
DATA STORAGE ALLOCATIONS

$ADDRESS	CONTENTS
0000	$nn MS (numerator)
0001	$nn LS (numerator)
0002	$dd MS (denominator)
0003	$dd LS (denominator)
0004	$qq MS (integer quot.)
0005	$qq LS (integer quot.)
0006	$fH fH MS (fractional quot.)
0007	$fH fH LS (fractional quot.)
0008	$fL fL MS (fractional quot.)
0009	$fL fL LS (fractional quot.)
000A	$rr MS (remainder)
000B	$rr LS (remainder)

The M68HC11 can add certain packed BCD values and get valid BCD results. For example:

$$\text{Addend in ACCA} = \text{BCD } 01010011 = 53$$
$$+ \text{Augend from memory} = \text{BCD } 00100100 = 24$$
$$\text{Sum returned to ACCA} = \text{BCD } 01110111 = 77$$

However, other BCD additions can generate invalid BCD results. For example:

$$\text{Addend in ACCA} = \text{BCD } 01010011 = 53$$
$$+ \text{Augend from memory} = \text{BCD } 10011000 = 98$$
$$\text{Sum returned to ACCA} = \text{BCD } 11101011 \text{ (invalid as BCD)}$$

In this case, the ALU has generated an invalid sum because both sum digits exceed 1001, the maximum BCD value.

A DAA instruction can rectify these invalid digits. In this case, DAA automatically adds 01100110 to the invalid sum:

$$\text{Invalid sum} = \quad 11101011$$
$$\text{DAA adds:} \quad 01100110$$
$$\text{Adjusted sum} = (C=1)\ 01000001 = 141$$

DAA adds 0110 to a BCD digit sum if its value exceeds 1001, or if the addition generates a one carry. The condition code register's H flag indicates a carry (called a "half carry") from the lower nibble sum to the higher nibble addition; the C flag indicates a carry from the higher-nibble sum.

The DAA instruction evaluates the BCD sum in ACCA and decimal adjusts it, if necessary. Use DAA after ABA, ADDA, or ADCA instructions only. Since DAA adjusts ACCA's contents, this instruction uses inherent addressing mode.

Arithmetic instructions facilitate multiprecision add, subtract, multiply, and divide operations. The DAA instruction rectifies invalid BCD sums in ACCA. Exercise 1.4 gives the reader a chance to develop and demonstrate the programs discussed here.

EXERCISE 1.4

Arithmetic Instructions

REQUIRED EQUIPMENT: Motorola M68HC11EVB or EVBU, and terminal

GENERAL INSTRUCTIONS: This laboratory exercise gives the reader an opportunity to develop, debug, and demonstrate programs DEMO4, DEMO5, DEMO6, and DEMO7 as discussed in the previous section. These programs demonstrate the power of M68HC11's arithmetic instructions.

PART I: Develop and Demonstrate DEMO4

Develop source code that implements Figure 1–13's flowchart and adheres to its accompanying memory allocations. If you use an EVBU, locate your DEMO4 program at address $0100. If you use an EVB, locate DEMO4 at address $C000. Enter implementing source code and appropriate addresses in the blanks provided below. Comments are included to help you along.

DEMO4

$ADDRESS	SOURCE CODE	COMMENTS
_____	_____	Clear SUM MS
_____	_____	Read ADDEND LS
_____	_____	Add AUGEND LS
_____	_____	Store result to SUM LS
_____	_____	Read ADDEND MID

_____	_____	Add with carry AUGEND MID-H
_____	_____	Store result to SUM MID-L
_____	_____	Read ADDEND MS
_____	_____	Add with carry AUGEND MS
_____	_____	Store result to SUM MID-H
_____	_____	Rotate SUM MS left
_____	_____	Halt

Load and debug your DEMO4 program. When confident that the program runs properly, demonstrate it to the instructor.

_____ *Instructor Check*

PART II: Develop and Demonstrate DEMO5

Develop source code that implements Figure 1–14's flowchart and adheres to its accompanying memory allocations. Enter implementing source code and appropriate addresses in the blanks provided below.

DEMO5

$ADDRESS	SOURCE CODE	COMMENTS
_____	_____	Load ACCB with MINUEND LS
_____	_____	Subtract SBTRHND LS
_____	_____	Store result to DIFF LS
_____	_____	Load ACCB with MINUEND MID
_____	_____	Subtract with carry SBTRHND MID
_____	_____	Store result to DIFF MID
_____	_____	Load ACCB with MINUEND MS
_____	_____	Subtract with carry SBTRHND MS
_____	_____	Store result to DIFF MS
_____	_____	Halt

Load and debug DEMO5. When sure that it runs properly, demonstrate its operation to the instructor.

_____ *Instructor Check*

PART III: Develop and Demonstrate DEMO6

Refer to Figures 1–15 and 1–16. Develop source code that implements Figure 1–16's flowchart and adheres to Figure 1–15's memory allocations. If you use an EVBU, locate your DEMO6 program at address $0100. If you use an EVB, locate DEMO6 at address $C000. Enter implementing source code and appropriate addresses in the blanks below.

DEMO6

$ADDRESS	SOURCE CODE	COMMENTS
_____	_____	Read MULPLCND LS to ACCA
_____	_____	Read MULTIPLR to ACCB
_____	_____	ACCA x ACCB --> D
_____	_____	Store ACCB to PROD LS
_____	_____	Store ACCA to P1 MS
_____	_____	Read MULPLCND MID to ACCA
_____	_____	Read MULTIPLR to ACCB
_____	_____	ACCA x ACCB --> D
_____	_____	Store D register to P2

$ADDRESS	SOURCE CODE	COMMENTS
_____	_____	Read MULPLCND MS to ACCA
_____	_____	Read MULTIPLR to ACCB
_____	_____	ACCA x ACCB --> D
_____	_____	Store D register to P3
_____	_____	Read P1 MS to ACCA
_____	_____	Add P2 LS
_____	_____	Store sum to PROD MID-L
_____	_____	Read P2 MS to ACCA
_____	_____	Add with carry P3 LS
_____	_____	Store sum to PROD MID-H
_____	_____	Read P3 MS to ACCA
_____	_____	Add with carry, zero
_____	_____	Store sum to PROD MS
_____	_____	Halt

Load and debug DEMO6. When sure that DEMO6 runs properly, demonstrate its operation to the instructor.

_____ *Instructor Check*

PART IV: Develop and Demonstrate DEMO7

Develop source code that implements Figure 1–17's flowchart, and adheres to its accompanying memory allocations. If you use an EVBU, locate your DEMO7 program at address $0100. If you use an EVB, locate DEMO7 at address $C000. Enter implementing source code and appropriate addresses in the blanks below.

DEMO7

$ADDRESS	SOURCE CODE	COMMENTS
_____	_____	Load $nnnn
_____	_____	Load $dddd
_____	_____	Integer divide
_____	_____	Store $qqqq
_____	_____	Load $dddd
_____	_____	Fractional divide
_____	_____	Store $f_H f_H f_H f_H$
_____	_____	Load $dddd
_____	_____	Fractional divide
_____	_____	Store $f_L f_L f_L f_L$
_____	_____	Store $rrrr
_____	_____	Halt

Load and debug DEMO7. When sure that DEMO7 runs properly, demonstrate its operation to the instructor.

_____ *Instructor Check*

PART V: Laboratory Software Listing

DEMO4—EVB

$ADDRESS	SOURCE CODE	COMMENTS
C000	CLR 6	Clear SUM MS
C003	LDAA 2	Read ADDEND LS
C005	ADDA 5	Add AUGEND LS
C007	STAA 9	Store result to SUM LS
C009	LDAA 1	Read ADDEND MID

$ADDRESS	SOURCE CODE	COMMENTS
C00B	ADCA 4	Add with carry AUGEND MID
C00D	STAA 8	Store result to SUM MID-L
C00F	LDAA 0	Read ADDEND MS
C011	ADCA 3	Add with carry AUGEND MS
C013	STAA 7	Store result to SUM MID-H
C015	ROL 6	Rotate SUM MS left
C018	WAI	Halt

DEMO4—EVBU

$ADDRESS	SOURCE CODE	COMMENTS
0100	CLR 6	Clear SUM MS
0103	LDAA 2	Read ADDEND LS
0105	ADDA 5	Add AUGEND LS
0107	STAA 9	Store result to SUM LS
0109	LDAA 1	Read ADDEND MID
010B	ADCA 4	Add with carry AUGEND MID
010D	STAA 8	Store result to SUM MID-L
010F	LDAA 0	Read ADDEND MS
0111	ADCA 3	Add with carry AUGEND MS
0113	STAA 7	Store result to SUM MID-H
0115	ROL 6	Rotate SUM MS left
0118	WAI	Halt

DEMO5—EVB

$ADDRESS	SOURCE CODE	COMMENTS
C000	LDAB 2	Load ACCB with MINUEND LS
C002	SUBB 5	Subtract SBTRHND LS
C004	STAB 8	Store result to DIFF LS
C006	LDAB 1	Load ACCB with MINUEND MID
C008	SBCB 4	Subtract with carry SBTRHND MID
C00A	STAB 7	Store result to DIFF MID
C00C	LDAB 0	Load ACCB with MINUEND MS
C00E	SBCB 3	Subtract with carry SBTRHND MS
C010	STAB 6	Store result to DIFF MS
C012	WAI	Halt

DEMO5—EVBU

$ADDRESS	SOURCE CODE	COMMENTS
0100	LDAB 2	Load ACCB with MINUEND LS
0102	SUBB 5	Subtract SBTRHND LS
0104	STAB 8	Store result to DIFF LS
0106	LDAB 1	Load ACCB with MINUEND MID
0108	SBCB 4	Subtract with carry SBTRHND MID
010A	STAB 7	Store result to DIFF MID
010C	LDAB 0	Load ACCB with MINUEND MS
010E	SBCB 3	Subtract with carry SBTRHND MS
0110	STAB 6	Store result to DIFF MS
0112	WAI	Halt

DEMO6—EVB

$ADDRESS	SOURCE CODE	COMMENTS
C000	LDAA 2	Read MULPLCND LS to ACCA
C002	LDAB 3	Read MULTIPLR to ACCB
C004	MUL	ACCA x ACCB --> D
C005	STAB 7	Store ACCB to PROD LS
C007	STAA C	Store ACCA to P1 MS
C009	LDAA 1	Read MULPLCND MID to ACCA
C00B	LDAB 3	Read MULTIPLR to ACCB
C00D	MUL	ACCA x ACCB --> D
C00E	STD A	Store D register to P2
C010	LDAA 0	Read MULPLCND MS to ACCA
C012	LDAB 3	Read MULTIPLR to ACCB
C014	MUL	ACCA x ACCB --> D
C015	STD 8	Store D register to P3
C017	LDAA C	Read P1 MS to ACCA

$ADDRESS	SOURCE CODE	COMMENTS
C019	ADDA B	Add P2 LS
C01B	STAA 6	Store sum to PROD MID-L
C01D	LDAA A	Read P2 MS to ACCA
C01F	ADCA 9	Add with carry P3 LS
C021	STAA 5	Store sum to PROD MID-H
C023	LDAA 8	Read P3 MS to ACCA
C025	ADCA #0	Add with carry, zero
C027	STAA 4	Store sum to PROD MS
C029	WAI	Halt

DEMO6—EVBU

$ADDRESS	SOURCE CODE	COMMENTS
0100	LDAA 2	Read MULPLCND LS to ACCA
0102	LDAB 3	Read MULTIPLR to ACCB
0104	MUL	ACCA x ACCB --> D
0105	STAB 7	Store ACCB to PROD LS
0107	STAA C	Store ACCA to P1 MS
0109	LDAA 1	Read MULPLCND MID to ACCA
010B	LDAB 3	Read MULTIPLR to ACCB
010D	MUL	ACCA x ACCB --> D
010E	STD A	Store D register to P2
0110	LDAA 0	Read MULPLCND MS to ACCA
0112	LDAB 3	Read MULTIPLR to ACCB
0114	MUL	ACCA x ACCB --> D
0115	STD 8	Store D register to P3
0117	LDAA C	Read P1 MS to ACCA
0119	ADDA B	Add P2 LS
011B	STAA 6	Store sum to PROD MID-L
011D	LDAA A	Read P2 MS to ACCA
011F	ADCA 9	Add with carry P3 LS
0121	STAA 5	Store sum to PROD MID-H
0123	LDAA 8	Read P3 MS to ACCA
0125	ADCA #0	Add with carry, zero
0127	STAA 4	Store sum to PROD MS
0129	WAI	Halt

DEMO7—EVB

$ADDRESS	SOURCE CODE	COMMENTS
C000	LDD 0	Load $nnnn
C002	LDX 2	Load $dddd
C004	IDIV	Integer divide
C005	STX 4	Store $qqqq
C007	LDX 2	Load $dddd
C009	FDIV	Fractional divide
C00A	STX 6	Store $f_H f_H f_H f_H$
C00C	LDX 2	Load $dddd
C00E	FDIV	Fractional divide
C00F	STX 8	Store $f_L f_L f_L f_L$
C011	STD A	Store $rrrr
C013	WAI	Halt

DEMO7—EVB

$ADDRESS	SOURCE CODE	COMMENTS
0100	LDD 0	Load $nnnn
0102	LDX 2	Load $dddd
0104	IDIV	Integer divide
0105	STX 4	Store $qqqq
0107	LDX 2	Load $dddd
0109	FDIV	Fractional divide
010A	STX 6	Store $f_H f_H f_H f_H$
010C	LDX 2	Load $dddd
010E	FDIV	Fractional divide
010F	STX 8	Store $f_L f_L f_L f_L$
0111	STD A	Store $rrrr
0113	WAI	Halt

Condition Code Register Instructions

The reader should review the discussion on the condition code register (CCR) on pages 6–9. This discussion describes each CCR bit and its purpose. Eight M68HC11 instructions manipulate bits in the CCR. All use inherent addressing mode. Two of these instructions transfer 8 bits between ACCA and the CCR:

- TAP—Transfer the contents of ACCA to the CCR.
- TPA—Transfer the contents of the CCR to ACCA.

Recall that the X bit masks certain interrupts, and that the S bit controls the M68HC11's response to the STOP instruction.[13] The M68HC11 has no special instructions to set/reset the S and X bits. Therefore, the programmer must use the TAP instruction to clear the X bit,[14] and set or reset the S bit. By using a TPA then TAP, the programmer can manipulate X and S, without disturbing the other CCR bits. First, use TPA to transfer the CCR's contents to ACCA. Second, OR ACCA's contents with mask:

```
%bb000000
```

The two "b"s in the mask represent desired S and X states. Mask bit 7 corresponds to the S bit, and mask bit 6 to the X bit. The mask saves the contents of the H, I, N, Z, V, and C bits. Third, use TAP to transfer the contents of ACCA back to the CCR.

The remaining six CCR-oriented instructions simply set or reset the C, I, and V bits:

- SEC—Set the C flag.
- CLC—Clear the C flag.
- SEV—Set the V flag.
- CLV—Clear the V flag.
- SEI—Set the I mask.
- CLI—Clear the I mask.

Thus, the programmer has a selection of eight M68HC11 instructions to manipulate bits in the condition code register.

Test and Compare Instructions

The M68HC11 can test the contents of ACCA or ACCB and set or reset the CCR's N and Z flags accordingly:

- TSTA—Test ACCA's contents for zero or minus.
- TSTB—Test ACCB's contents for zero or minus.

These instructions use inherent addressing mode. A TSTA or TSTB instruction automatically subtracts zero from the contents of the subject accumulator, then sets or resets the N and Z flags accordingly. A TSTA or TSTB leaves the target accumulator with its contents intact. Thus, a TSTA or TSTB instruction sets the Z flag if the accumulator contains $00, and sets the N flag if the accumulator's bit 7 is set. TSTA or TSTB automatically clears the V and C flags, and leaves the S, X, H, and I bits unchanged.

Two bit-test instructions allow the M68HC11 to test individual bits in a memory byte or accumulator. These instructions AND the contents of an accumulator with a memory byte, and set the N and Z flags accordingly. Both instructions leave the original contents of accumulator and memory byte intact:

- BITA—AND the contents of ACCA with the contents of a memory byte. Set or reset the N and Z flags accordingly. The original contents of ACCA and the memory byte are unaffected.
- BITB—AND the contents of ACCB with a memory byte. Set or reset the N and Z flags accordingly. The original contents of ACCB and memory are unaffected.

---

[13]Chapter 2 covers the X, S, and I bits in detail.
[14]Once a TAP clears the X bit, this bit cannot be set again, even by another TAP execution. Only an M68HC11 chip reset will set the X bit.

BITA or BITB is most useful for testing a single bit in memory or an accumulator. A subsequent conditional branching instruction can control program flow according to the state of the tested bit. For example, assume that the programmer wishes to test bit 4 of ACCA. The following instruction would test this bit and set the Z and N flags accordingly:

```
BITA #10
```

The "#10" part of the source code indicates that immediate mode program byte %00010000 is AND-ed with the contents of ACCA. Thus, $10 can be considered a mask byte. If ACCA bit 4 = 1, BITA resets the Z flag. Assume that ACCA contains $7A, for example:

$$\begin{aligned} \text{ACCA} &= \%01111010 \\ \text{Mask } \$10 &= \underline{\%00010000} \\ \text{BITA result} &= \%00010000 \\ Z&=0, N=0 \end{aligned}$$

If ACCA bit 4 = 0, BITA sets the Z flag. Assume that ACCA = $6A:

$$\begin{aligned} \text{ACCA} &= \%01101010 \\ \text{Mask } \$10 &= \underline{\%00010000} \\ \text{BITA result} &= \%00000000 \\ Z&=1, N=0 \end{aligned}$$

Remember that BITA and BITB leave the original contents of both accumulator and memory intact.

BITA and BITB can access a memory byte using immediate, direct, extended, or indexed addressing modes. The programmer can test a bit or bits in a memory byte by loading the accumulator with the mask, then accessing the memory byte to be tested using direct, extended, or indexed mode. For example, assume that address $C01A contains $A5, and that the programmer wishes to test bits 3 and 7 of this memory byte and set or reset the Z and N flags accordingly. These instructions would execute the test:

```
LDAB #88
BITB C01A
```

In this example, the BITB instruction ANDs the contents of ACCB and address $C01A:

$$\begin{aligned} \text{Mask in ACCB} &= \%10001000 \\ \$C01A &= \underline{\%10100101} \\ \text{BITB result} &= \%10000000 \\ Z&=0, N=1 \end{aligned}$$

Of course, the CPU leaves the original contents of ACCB and address $C01A intact.

BITA and BITB instructions automatically reset the V flag. BITA and BITB leave all CCR bits unchanged, except for the N, Z, and V bits.

Thus, BITA and BITB instructions allow the CPU to test individual bits in either memory or an accumulator, and set or reset the N and Z flags accordingly.

Part VI of Exercise 1.1 (Figure 1–6) introduced the CPX instruction. In this Modified DEMO program, the CPX instruction compares an immediate-mode, 16-bit value with the contents of the X register. If the comparison value and X register contents match, a conditional branching instruction (BNE) routes program flow to the WAI instruction, halting the CPU. The M68HC11's instruction set includes six such compare instructions, including CPX. Each of these instructions makes its comparison by executing a test subtraction of the comparison value from the target CPU register. This test subtraction sets or clears the N, Z, V, and C flags.[15] Once the flags have set or cleared, the CPU restores the target register and comparison value to their original values. The CPU does not save the difference generated by this comparison.

---

[15]Recall that the CPU executes two's complement subtraction. To execute a two's complement subtraction, the CPU two's complements the subtrahend, then adds this two's complemented value to the minuend. The V flag sets to indicate an upset in the sign bit during this addition operation. The C flag state is an inversion of the carry out from bit 7's addition.

Five of the six compare instructions use comparison values residing in memory. These instructions can use immediate, direct, extended, or indexed addressing mode to access the comparison value. These five instructions are:

- CMPA—Compare the contents of ACCA with a byte in memory.
- CMPB—Compare the contents of ACCB with a byte in memory.
- CPD—Compare the contents of the D register with a 2-byte value in memory.
- CPX—Compare the contents of the X register with a 2-byte value in memory.
- CPY—Compare the contents of the Y register with a 2-byte value in memory.

The sixth compare instruction uses inherent addressing mode. CBA compares the contents of ACCA with ACCB, and sets or clears the N, Z, V, and C flags accordingly.

In most instances, a programmer uses test and compare instructions prior to a conditional branching instruction. Recall from pages 31–32 that conditional branching instructions direct program flow, based upon prevailing conditions. The following section discusses instructions that control program flow.

Program Control Instructions

Program control instructions alter the program counter's contents, and thereby cause program flow to jump or branch. The reader observed the operation of the BNE (Branch if Not Equal) instruction in the Modified DEMO and DEMO2 programs.[16]

Program control instructions fall into four categories: (1) unconditional branching instructions; (2) conditional branching instructions; (3) program jump instructions; and (4) return instructions.

The M68HC11's instruction set includes two unconditional branching instructions: BRA (Branch Always) and BSR (Branch to Subroutine). To execute a BRA, the CPU sign extends its accompanying relative address byte, then adds this 16-bit, sign-extended offset to the contents of the program counter. The CPU then fetches its next program instruction from the address pointed at by the new PC value.

The BSR instruction causes the CPU to take three actions. First, the CPU fetches the relative address accompanying the BSR opcode. Second, the CPU stacks the program count. At this point, the program counter contains the address of the next instruction following the BSR. Third, the CPU sign extends the relative address and adds it to the contents of the PC. The CPU then fetches its next program instruction from the address pointed at by the new PC value.

As stated above, BSR stands for "branch to subroutine." A subroutine is a program routine which resides in a block of memory separate from the main program. Usually, a subroutine contains program instructions which the CPU executes more than one time during the course of main program execution, and at different points in the main program. If the programmer integrated this routine into the main program, it would have to reside at two or more places in the main program's memory space. This would waste memory. If this routine becomes a subroutine, the CPU can branch to it (BSR) from two or more places in the main program. Since BSR stacks the address of the next main program instruction, the CPU can always find its way back from the subroutine by pulling the stacked return address, and loading the PC with it.

The M68HC11 instruction set includes eighteen conditional branching instructions. Recall that a conditional program branch depends upon meeting a condition. If the condition is not met, program flow remains unaltered; execution continues with the next instruction in the program.

Branch conditions consist of specified CCR status flag values. Some conditional branching instructions execute a branch if a single CCR flag meets a specified state requirement. For the remaining conditional branch instructions, branch execution depends upon combinations of CCR flag states.

---

[16]See Exercise 1.1, Part VI, Relative Addressing Mode (pp. 31–32), and Exercise 1.2.

Single-flag branching instructions come in pairs—one for flag set, the other for flag reset:

- BCC—Branch if Carry Clear. The CPU branches if the previous instruction's execution causes the C flag to reset.
- BCS—Branch if Carry Set. The CPU branches if the previous instruction's execution causes the C flag to set.
- BNE—Branch if Not Equal.[17] Branch if the previous instruction's execution caused the Z flag to reset.
- BEQ—Branch if Equal. Branch if the previous instruction's execution caused the Z flag to set.
- BPL—Branch if Plus. Branch if the previous instruction's execution caused the N flag to reset.
- BMI—Branch if Minus. Branch if the previous instruction's execution caused the N flag to set.
- BVC—Branch if V Flag Clear. Branch if the previous instruction's execution caused the V flag to reset.
- BVS—Branch if V Flag Set. Branch if the previous instruction's execution caused the V flag to set.

Four branching instructions specify conditions caused by subtract or compare instructions operating on unsigned operands.[18] These instructions are as follows:

- BHI—Branch if Higher. Branch if the unsigned minuend was higher than the unsigned subtrahend. To make the branch, both the C and Z flags must be reset (C + Z = 0).
- BHS—Branch if Higher or Same. Branch if the unsigned minuend was higher than, or the same as, the unsigned subtrahend. To make the branch, the C flag must be reset. Note that this branching instruction is the same as BCC. Both mnemonics use the same opcode: $24.
- BLO—Branch if Lower. Branch if the unsigned minuend was lower than the unsigned subtrahend. To make the branch, the C flag must be set. Note that this instruction is the same as BCS. Both BLO and BCS use opcode $25.
- BLS—Branch if Lower or Same. Branch if the unsigned minuend was the same as, or lower than, the unsigned subtrahend. To make the branch, either the C flag or the Z flag must be set (C + Z = 1).

These four unsigned branching instructions have counterparts for signed minuends and subtrahends. Because these branching instructions presume the subtraction of signed operands, they bring the N and Z flags into play.

It is important to clarify the meaning of the terms "greater" and "lesser" in the context of signed operands. In general, the more positive (less negative) a number, the greater it is. For example, which is greater: $33 or $F6?

$33 is greater because it equals +51. $F6 equals –10.[19]

Which is lesser: $7A or $8B?

$8B is lesser because it equals –117. $7A equals +122.

The four signed operand-oriented conditional branching instructions are:

- BGT—Branch if Greater Than. Branch if the signed minuend was greater than the signed subtrahend. To make the branch, the Z flag must be reset, and the N and V flags equal $[Z \cdot (N \oplus V) = 0]$.
- BGE—Branch if Greater or Equal. Branch if the signed minuend was equal to or greater than the signed subtrahend. To make the branch, the N and V flags must be equal $(N \oplus V = 0)$.
- BLT—Branch if Less Than. Branch if the signed minuend was less than the signed subtrahend. To make the branch, the N and V flags must be unequal $(N \oplus V = 1)$.

---

[17]Recall that Modified DEMO and DEMO2 use this conditional branching instruction.
[18]See Arithmetic Instructions (pp. 48–54) and Test and Compare Instructions (pp. 59–61).
[19]See the side bar accompanying M68HC11 CPU Data Handling on page 8.

- BLE—Branch if Less or Equal. Branch if the signed minuend was less than or equal to the signed subtrahend. To make the branch, either the Z flag must be set, or the N and Z flags unequal $[Z + (N \oplus V) = 1]$.

You have learned about eight conditional branching instructions to be used after a subtract or compare. Four of these instructions presume unsigned minuend and subtrahend. The other four instructions presume signed operands. Exercise 1–5 includes DEMO8, a number sort program which uses most of these conditional branching instructions.

Two final branching instructions base their actions on the states of specific bits in a memory byte:

- BRCLR—Branch if Bits Clear.
- BRSET—Branch if Bits Set.

A BRCLR or BRSET instruction includes a mask byte which accompanies the instruction opcode. This mask specifies the target memory bits to be tested. BRCLR and BRSET actually use two addressing modes. Relative mode specifies the branch destination. Either direct or extended mode accesses the memory byte to be tested. Here is a source code example:

```
BRCLR 2A 88 128
```

$2A is the direct mode address to be tested ($002A), and $88 is the mask: %10001000. $0128 is the branch destination address. To execute this BRCLR instruction, the CPU reads the contents of memory location $002A, and tests bits 3 and 7, as specified by the mask. If these two bits are both clear, the CPU executes the branch to destination address $0128.

Here is another source code example:

```
LDX #0
BRSET A,X 77 C000
```

To execute this BRSET instruction, the CPU uses indexed addressing mode to read the contents of address $000A. The CPU tests bits 0, 1, 2, 4, 5, and 6 of this read byte. If all of these test bits are set, the CPU branches program flow to the instruction at address $C000.

Finally, this section's discussion ends with three instructions that simply store a value to the program counter. These instructions are:

- JMP—Jump. Store a new, specified address to the program counter.
- JSR—Jump to Subroutine. Push the address of the next program instruction onto the stack. Store a new, specified address to the program counter.
- RTS—Return from Subroutine. Pull an address from the stack and restore it to the program counter.

As stated above, the JMP instruction simply stores a specified address to the PC. This address can be either an extended or indexed mode value. For example:

```
JMP C050
```

To execute this instruction, the CPU stores $C050 to the PC. Therefore, the CPU will next fetch and execute the instruction at address $C050.

Another example assumes that the X register contains $0100:

```
JMP 1A,X
```

In this example, the CPU executes the JMP instruction by adding the offset ($1A) to the contents of the X register ($0100), and storing the sum ($011A) to the PC. Therefore, the CPU will next fetch and execute the instruction at address $011A.

The JSR instruction also stores a specified value to the PC. In addition, JSR causes the CPU to stack the return address. This return address is the address of the instruction following the JSR. Thus, BSR and JSR instructions operate in very much the same manner. However, there are two differences between them:

1. BSR uses relative mode. The CPU adds the relative address accompanying the opcode, to the PC contents. On the other hand, JSR uses direct, extended, or indexed mode values. For example:

```
JSR 2A
```

stores the direct mode value $002A to the PC. Another example:

```
JSR FFC7
```

stores the extended mode value $FFC7 to the PC. For a third example, assume that the Y register contains $C000:

```
JSR A,Y
```

Here, the CPU stores $C00A to the PC.

2. Because it uses a 1-byte relative address, the BSR instruction can redirect program flow up to 128 addresses back or 127 addresses forward. A JSR instruction can redirect program flow to any address on the M68HC11's memory map.

Remember that both JSR and BSR stack a return address before redirecting program flow.

The third instruction, RTS, assumes that the CPU has executed a JSR or BSR instruction, and has fetched and executed a subroutine. The programmer should make the final subroutine instruction an RTS—Return from Subroutine. By executing an RTS instruction, the CPU completes its execution of a subroutine by pulling the return address from the stack, and re-loading the PC with this value. Thus, the CPU returns to the next main program instruction.

Because a BSR or JSR stacks the return address, and an RTS pulls this address to the PC, a programmer can nest subroutines. Nested subroutines are subroutines called from subroutines. A programmer can nest subroutines because any JSR or BSR instruction stacks the return address on the top of the stack. Remember that the stack is a "last on/first off" arrangement. Thus, an RTS at the end of a nested subroutine will return program flow to the next instruction in the subroutine—which called the nested subroutine. This arrangement allows the programmer to nest as many subroutines as memory space allows. Exercise 1.5 includes DEMO8, a program which nests two subroutines.

Section 1.7 has discussed some 134 instructions! However, Chapter 2 will discuss several more. These remaining instructions are largely associated with M68HC11 interrupts.

EXERCISE 1.5 Compare and Program Control Instructions

REQUIRED EQUIPMENT: Motorola M68HC11EVB or EVBU, and terminal

GENERAL INSTRUCTIONS: This laboratory presents DEMO8, a program that exercises compare and program control instructions. These instructions include conditional branches and JSRs. DEMO8 includes a MAIN program and nine subroutines. Two of these subroutines are nested. Overall, DEMO8 analyzes a set of five, 8-bit test operands and classifies this set into five different subsets:

   Higher than $64 (unsigned)
   Lower or same as $4B (unsigned)
   Greater or equal to $E7 (signed)
   Less than $F6 (signed)
   Greater than $1E (signed)

A single test operand can belong to more than one subset. Take operand $F3, for example. This operand belongs to three subsets:

1. Unsigned operands higher than $64
2. Signed operands greater or equal to $E7
3. Signed operands less than $F6

Therefore, DEMO8 reads each test operand in turn, and assigns it to appropriate subsets.

DEMO8 uses six RAM memory blocks of five locations each. Test operands and each of the five subsets are assigned one of these RAM blocks. Figure 1–18 lists these RAM blocks and their labels. If a test operand fails to qualify for a subset, DEMO8 fills its place in the subset's RAM block with $00.

FIGURE 1–18

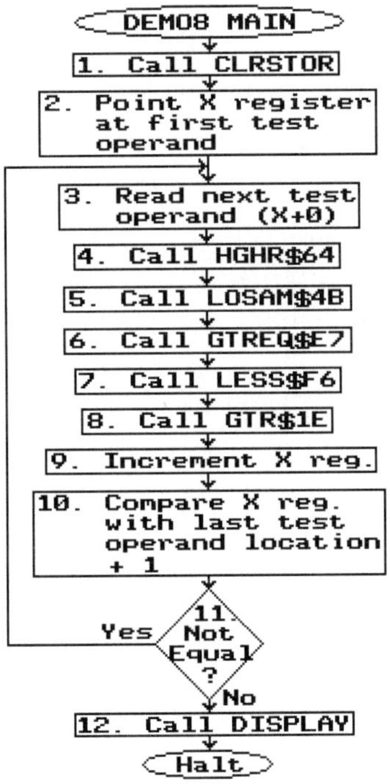

```
        ┌─DEMO8 MAIN─┐
               ↓
        1. Call CLRSTOR
               ↓
        2. Point X register
           at first test
           operand
               ↓
     →  3. Read next test
           operand (X+0)
               ↓
        4. Call HGHR$64
               ↓
        5. Call LOSAM$4B
               ↓
        6. Call GTREQ$E7
               ↓
        7. Call LESS$F6
               ↓
        8. Call GTR$1E
               ↓
        9. Increment X reg.
               ↓
       10. Compare X reg.
           with last test
           operand location
           + 1
               ↓
              /11\
         Yes/ Not \
        ───<  Equal >
            \  ?  /
             \   /No
               ↓
       12. Call DISPLAY
               ↓
           ┌─Halt─┐
```

```
MEMORY ALLOCATIONS
   $ADDRESS
EVBU        EVB         CONTENTS
0000-0004   0000-0004   Test Operands
0005-0009   0005-0009   Operands high-
                        er than $64
000A-000E   000A-000E   Operands low-
                        er or same as
                        $4B
000F-0013   000F-0013   Operands great-
                        er or equal to
                        $E7
0014-0018   0014-0018   Operands less
                        than $F6
0019-001D   0019-001D   Operands great-
                        er than $1E
0100        C000        DEMO8 MAIN
0125        C025        HGHR$64 sub-
                        routine
0130        C030        LOSAM$4B sub-
                        routine
013A        C03A        GTREQ$E7 sub-
                        routine
0145        C045        LESS$F6 sub-
                        routine
0150        C050        GRTR$1E sub-
                        routine
0160        C060        DISPLAY sub-
                        routine
0180        C080        CLRSTOR sub-
                        routine
```

DEMO8 displays the test operand set and subsets on six lines of the terminal screen. On the first line, DEMO8 displays the set of test operands. On succeeding lines, DEMO8 displays the five subsets. To execute this display, DEMO8 uses two BUFFALO utility subroutines, .OUTCRL and .OUT1BS. .OUTCRL sends an ASCII line feed and carriage return to the terminal. Using .OUTCRL, DEMO8 can display the test operand set and each subset on its own display line. .OUT1BS converts a byte pointed at by the X register to two ASCII characters, and writes them to the terminal screen. .OUT1BS then increments the X register. Thus, DEMO8 uses .OUT1BS to write set and subset data to the terminal screen. DEMO8 calls these BUFFALO routines as nested subroutines.

In Part I of this exercise, the reader shall analyze the DEMO8 software. Part II is concerned with developing DEMO8 source code. In Part III, the reader will predict DEMO8 results, then debug and demonstrate program operation.

PART I: Analyze DEMO8 Software

Refer to Figure 1–18. This figure flowcharts DEMO8's MAIN program. MAIN uses a modular approach. That is, MAIN manages test operand accesses, and calls modules, (i.e., subroutines), which classify and display the set and subsets. Figure 1–18's flowchart blocks 2, 3, 9, 10, and 11 manage test operand accesses. In block 2, the CPU loads its X register with the address of the first test operand ($0000). In block 3, the CPU reads a test operand using indexed mode. After analyzing and classifying this operand (blocks 4–8), the CPU increments its X register to point at the next test operand (block 9). However, if all five test operands have been analyzed, block 9's increment points the X register at the last test operand location, plus one ($0005). To test for this condition, the CPU compares the incremented X value with $0005 (block 10). If the X register's contents do not equal $0005, the CPU takes the "yes" branch from block 11 (a BNE instruction) back to Block 3, where it reads the next test operand. If the X register's contents equal $0005, the CPU has classified all test operands into subsets. Therefore, the CPU takes the "no" path from block 11 to display the analysis results (block 12) and halt.

MAIN program's other blocks consist of subroutine calls, which the CPU executes with JSR instructions. Block 1's CLRSTOR subroutine clears all subset RAM locations ($0005–$001D). Blocks 4–8 call the subroutines which qualify each test operand for subset memberships. HGHR$64 (block 4) qualifies and stores test operands in the "higher than $64" subset. LOSAM$4B (block 5) qualifies and stores test operands in the "lower or same

FIGURE 1-19

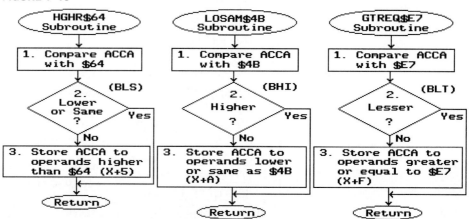

as $4B" subset. Block 6 calls GTREQ$E7 subroutine, which qualifies and stores test operands in the "greater or equal to $E7" subset. In block 7, the CPU calls the LESS$F6 subroutine, which qualifies operands into the "less than $F6" subset. The CPU calls GTR$1E in block 8. This subroutine qualifies test operands for the "greater the $1E" subset.

In block 12, the CPU calls the DISPLAY subroutine. This subroutine displays six lines of data on the terminal screen. The first line is the set of five test operands. The remaining lines display subset data. To accomplish this task, DISPLAY nests two subroutines, .OUTCRL and .OUT1BS.

Refer to Figures 1–19 and 1–20, which flowchart the subset categorization subroutines. These five subroutines have a close similarity, and the following description treats them generically.

The reader should understand that the MAIN program loads ACCA with the test operand (Figure 1–18, block 3). Note that block 1 of each categorization subroutine compares this test operand with a criterion value. For example, the "higher than $64" (HGHR$64) subroutine compares the test operand with $64. Recall that such a comparison sets or resets the N, Z, V, and C status flags.

In block 2, each routine takes a "yes" branch or a "no" path based upon the state of the status flag(s). The CPU takes the "yes" branch if the test operand meets a condition opposite to the subroutine's criterion. Note the recommended conditional branch mnemonic accompanying each block 2. The CPU branches around block 3 if the test operand does not meet the subroutine's general criterion. As an example, consider the HGHR$64 subroutine. A test operand of $13 is lower than the subroutine's criterion value ($64). If the CPU executes HGHR$64, and evaluates $13, it takes block 2's "yes" branch, bypassing flowchart block 3. If the test operand is $F3, the CPU takes the "no" path, and executes block 3. Thus, if the test operand meets the subroutine's general criterion (e.g., higher than $64), the CPU takes the "no" path from Block 2, and executes block 3.

FIGURE 1-20

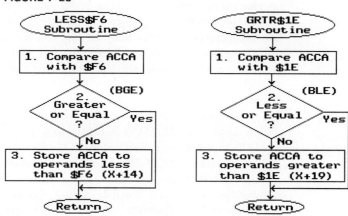

In block 3, the CPU stores the test operand to an appropriate subset RAM storage block. Subset RAM blocks contain five addresses each. The MAIN program points the X register at an address in the $0000–$0004 RAM block. The first subset resides at addresses $0005–$0009, the second subset at $000A–$000E, and so on. For its block 3 store, the evaluation subroutine can use an appropriate indexed mode offset, as noted in block 3.

After the CPU either bypasses or executes block 3, it returns to the MAIN program. To return, the CPU executes an RTS instruction. After evaluating all five test operands, DEMO8 calls the DISPLAY subroutine. Figure 1–21 flowcharts this subroutine. DISPLAY causes the M68HC11 to show the operand set and each subset on a separate terminal screen line. Thus, the terminal screen shows six lines of five operands each. To display these operands, the CPU calls the BUFFALO .OUT1BS subroutine. Recall that .OUT1BS displays a byte pointed at by the X register and, after displaying this byte, .OUT1BS increments the X register.

DISPLAY must show the contents of RAM addresses $0000–$001D on the terminal screen in lines of five bytes each. Therefore, in Figure 1–21, block 1, the CPU points its X register at address $0000. In block 2, the CPU calls (JSR) the .OUTCRL subroutine, at address $FFC4. .OUTCRL sends a carriage return and line feed to the terminal, so that the data display starts on a new line. DISPLAY shows data bytes in groups of five. Therefore, in block 3, the CPU loads ACCB with $05. ACCB functions as a display loop counter.

In block 4, the CPU calls (JSR) the .OUT1BS subroutine, at address $FFBE. This subroutine displays a byte and advances the X register to the next storage location. In block 5, the CPU decrements the loop count. If this decrement reduces the loop count to zero, a group of operands (set or subset) has been displayed. If the counter (ACCB) does not contain zero, the CPU must display more bytes to complete the set or subset. Therefore, if ACCB does not contain $00, the CPU takes the "yes" branch from block 6 (BNE), to block 4, where it displays another data byte. If ACCB contains $00, the CPU takes the "no" path to block 7. If the CPU has displayed all six groups of five, the last .OUT1BS execution incremented the X register to $001E. Therefore, in block 7, the CPU compares the X register's contents with this terminal value. If the X register's contents do not equal $001E, the CPU takes the "yes" branch from block 8 (BNE), back to block 2, where it displays a new line of five data bytes. If the X register's contents equal $001E, the CPU takes the "no" path from block 8, and returns from the DISPLAY subroutine (RTS) to the MAIN program.

Thus, the DISPLAY subroutine nests .OUTCRL and .OUT1BS subroutines to display test operands (set) and categorized results (subsets) in six lines of 5 bytes each.

FIGURE 1–21

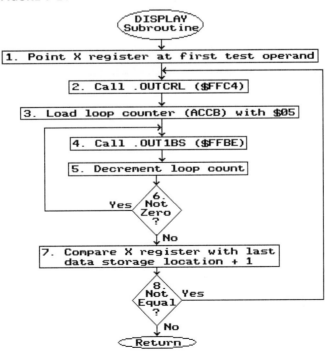

FIGURE 1-22

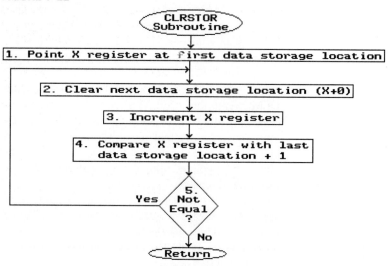

As its first action, DEMO8's MAIN program calls the CLRSTOR subroutine. CLRSTOR write $00 to all subset RAM locations—from address $0005 through address $001D. Figure 1–22 flow charts the CLRSTOR subroutine. In block 1, the CPU points its X register at the first subset location: $0005. In block 2, the CPU clears the subset location, using indexed mode with a $00 offset. In block 3, the CPU increments the X register to point at the next subset location. In block 4, the CPU compares the incremented X register value with the address of the last subset byte, plus one ($001E). If the X register contains $001E, the CPU has written $00 to all subset locations. The CPU therefore takes the "no" path from block 5, and returns from the subroutine (RTS). If the X register does not contain $001E, the CPU must clear another subset location. Therefore, the CPU takes the "yes" branch from block 5 (BNE) to block 2, where it clears the next subset location.

Having analyzed DEMO8 program flow, the reader can develop this program's source code.

PART II: Develop DEMO8 Source Code

Using Section 1.6 and Figures 1–18 through 1–22 for reference, develop DEMO8 source code. Enter your source code and appropriate addresses in the blanks below. Comments are included to help you.

DEMO8—MAIN

$ADDRESS	SOURCE CODE	COMMENTS
_____	_____	Call CLRSTOR
_____	_____	Initialize X register
_____	_____	Read next test operand
_____	_____	Call HGHR$64
_____	_____	Call LOSAM$4B
_____	_____	Call GTREQ$E7
_____	_____	Call LESS$F6
_____	_____	Call GTR$1E
_____	_____	Increment X register
_____	_____	Compare X register with last test operand location + 1
_____	_____	If not equal, test another operand
_____	_____	Call DISPLAY
_____	_____	Halt

HGHR$64 Subroutine

$ADDRESS	SOURCE CODE	COMMENTS
_____	_____	Compare test operand with $64
_____	_____	If lower or same, return
_____	_____	If higher, store test operand
_____	_____	Return

LOSAM$4B Subroutine

$ADDRESS	SOURCE CODE	COMMENTS
_____	_____	Compare test operand with $4B
_____	_____	If higher, return
_____	_____	If lower or same, store test operand
_____	_____	Return

GTREQ$E7 Subroutine

$ADDRESS	SOURCE CODE	COMMENTS
_____	_____	Compare test operand with $E7
_____	_____	If less, return
_____	_____	If greater or equal, store test operand
_____	_____	Return

LESS$F6 Subroutine

$ADDRESS	SOURCE CODE	COMMENTS
_____	_____	Compare test operand with $F6
_____	_____	If greater or equal, return
_____	_____	If less, store test operand
_____	_____	Return

GRTR$1E Subroutine

$ADDRESS	SOURCE CODE	COMMENTS
_____	_____	Compare test operand with $1E
_____	_____	If less or equal, return
_____	_____	If greater, store test operand
_____	_____	Return

DISPLAY Subroutine

$ADDRESS	SOURCE CODE	COMMENTS
_____	_____	Initialize X register
_____	_____	Call .OUTCRL
_____	_____	Initialize loop counter
_____	_____	Call .OUT1BS
_____	_____	Decrement loop count
_____	_____	If loop count not zero, display another byte
_____	_____	If loop count zero, compare X register with last data storage location + 1
_____	_____	If not equal, display five more data bytes
_____	_____	If equal, return

CLRSTOR Subroutine

$ADDRESS	SOURCE CODE	COMMENTS
_____	_____	Initialize X register
_____	_____	Clear next data storage location
_____	_____	Increment X register
_____	_____	Compare X register with last data storage location + 1
_____	_____	If not equal, clear next location
_____	_____	If equal, return

PART III: Predict DEMO8 Results; Debug and Demonstrate DEMO8 Operation

You shall use the BUFFALO MM command to store a set of five test operands to locations $0000–$0004. Predict DEMO8 results (subset memberships) based upon these test operands:

$0000 = $8C
$0001 = $13
$0002 = $03
$0003 = $D1
$0004 = $2B

Enter your predicted values in the blanks below, as DEMO8 would write them to the terminal screen. Enter 00 in each blank which does not contain a subset member. The first line of test operands is entered to get you started.

8C	13	03	D1	2B
__	__	__	__	__
__	__	__	__	__
__	__	__	__	__
__	__	__	__	__

Have the instructor check your predictions.

_____ *Instructor Check*

Use the BUFFALO assembler/disassembler to load and check your DEMO8 software. Use an MM command to enter the test operands specified above. Debug DEMO8 operation. When you are confident that DEMO8 runs properly, demonstrate its operation to the instructor.

_____ *Instructor Check*

PART IV: Laboratory Software Listing

DEMO8—MAIN—EVB

$ADDRESS	SOURCE CODE	COMMENTS
C000	JSR C080	Call CLRSTOR
C003	LDX #0	Initialize X register
C006	LDAA 0,X	Read next test operand
C008	JSR C025	Call HGHR$64
C00B	JSR C030	Call LOSAM$4B
C00E	JSR C03A	Call GTREQ$E7
C011	JSR C045	Call LESS$F6
C014	JSR C050	Call GTR$1E
C017	INX	Increment X register
C018	CPX #5	Compare X register with last test operand location + 1
C01B	BNE C006	If not equal, test another operand
C01D	JSR C060	Call DISPLAY
C020	WAI	Halt

DEMO8—MAIN—EVBU

$ADDRESS	SOURCE CODE	COMMENTS
0100	JSR 180	Call CLRSTOR
0103	LDX #0	Initialize X register
0106	LDAA 0,X	Read next test operand
0108	JSR 125	Call HGHR$64
010B	JSR 130	Call LOSAM$4B
010E	JSR 13A	Call GTREQ$E7
0111	JSR 145	Call LESS$F6
0114	JSR 150	Call GTR$1E
0117	INX	Increment X register
0118	CPX #5	Compare X register with last test operand location + 1
011B	BNE 106	If not equal, test another operand
011D	JSR 160	Call DISPLAY
0120	WAI	Halt

HGHR$64 SUBROUTINE—EVB

$ADDRESS	SOURCE CODE	COMMENTS
C025	CMPA #64	Compare test operand with $64
C027	BLS C02B	If lower or same, return
C029	STAA 5,X	If higher, store test operand
C02B	RTS	Return

HGHR$64 SUBROUTINE—EVBU

$ADDRESS	SOURCE CODE	COMMENTS
0125	CMPA #64	Compare test operand with $64
0127	BLS 12B	If lower or same, return
0129	STAA 5,X	If higher, store test operand
012B	RTS	Return

LOSAM$4B SUBROUTINE—EVB

$ADDRESS	SOURCE CODE	COMMENTS
C030	CMPA #4B	Compare test operand with $4B
C032	BHI C036	If higher, return
C034	STAA A,X	If lower or same, store test operand
C036	RTS	Return

LOSAM$4B SUBROUTINE—EVBU

$ADDRESS	SOURCE CODE	COMMENTS
0130	CMPA #4B	Compare test operand with $4B
0132	BHI 136	If higher, return
0134	STAA A,X	If lower or same, store test operand
0136	RTS	Return

GTREQ$E7 SUBROUTINE—EVB

$ADDRESS	SOURCE CODE	COMMENTS
C03A	CMPA #E7	Compare test operand with $E7
C03C	BLT C040	If less, return
C03E	STAA F,X	If greater or equal, store test operand
C040	RTS	Return

GTREQ$E7 SUBROUTINE—EVBU

$ADDRESS	SOURCE CODE	COMMENTS
013A	CMPA #E7	Compare test operand with $E7
013C	BLT 140	If less, return
013E	STAA F,X	If greater or equal, store test operand
0140	RTS	Return

LESS$F6 SUBROUTINE—EVB

$ADDRESS	SOURCE CODE	COMMENTS
C045	CMPA #F6	Compare test operand with $F6
C047	BGE C04B	If greater or equal, return
C049	STAA 14,X	If less, store test operand
C04B	RTS	Return

LESS$F6 SUBROUTINE—EVBU

$ADDRESS	SOURCE CODE	COMMENTS
0145	CMPA #F6	Compare test operand with $F6
0147	BGE 14B	If greater or equal, return
0149	STAA 14,X	If less, store test operand
014B	RTS	Return

GRTR$1E SUBROUTINE—EVB

$ADDRESS	SOURCE CODE	COMMENTS
C050	CMPA #1E	Compare test operand with $1E
C052	BLE C056	If less or equal, return
C054	STAA 19,X	If greater, store test operand
C056	RTS	Return

GRTR$1E SUBROUTINE—EVBU

$ADDRESS	SOURCE CODE	COMMENTS
0150	CMPA #1E	Compare test operand with $1E
0152	BLE 156	If less or equal, return
0154	STAA 19,X	If greater, store test operand
0156	RTS	Return

DISPLAY SUBROUTINE—EVB

$ADDRESS	SOURCE CODE	COMMENTS
C060	LDX #0	Initialize X register
C063	JSR FFC4	Call .OUTCRL
C066	LDAB #5	Initialize loop counter
C068	JSR FFBE	Call .OUT1BS
C06B	DECB	Decrement loop count
C06C	BNE C068	If loop count not zero, display another byte
C06E	CPX #1E	If loop count zero, compare X register with last data storage location + 1
C071	BNE C063	If not equal, display five more data bytes
C073	RTS	If equal, return

SUBROUTINE—EVBU

$ADDRESS	SOURCE CODE	COMMENTS
0160	LDX #0	Initialize X register
0163	JSR FFC4	Call .OUTCRL
0166	LDAB #5	Initialize loop counter
0168	JSR FFBE	Call .OUT1BS
016B	DECB	Decrement loop count
016C	BNE 168	If loop count not zero, display another byte
016E	CPX #1E	If loop count zero, compare X register with last data storage location + 1
0171	BNE 163	If not equal, display five more data bytes
0173	RTS	If equal, return

CLRSTOR SUBROUTINE—EVB

$ADDRESS	SOURCE CODE	COMMENTS
C080	LDX #5	Initialize X register
C083	CLR 0,X	Clear next data storage location
C085	INX	Increment X register

C086		CPX #1E	Compare X register with last data storage location + 1
C089		BNE C083	If not equal, clear next location
C08B		RTS	If equal, return

CLRSTOR SUBROUTINE—EVBU

$ADDRESS	SOURCE CODE	COMMENTS
0180	LDX #5	Initialize X register
0183	CLR 0,X	Clear next data storage location
0185	INX	Increment X register
0186	CPX #1E	Compare X register with last data storage location + 1
0189	BNE 183	If not equal, clear next location
018B	RTS	If equal, return

1.7 CHAPTER SUMMARY

Chapter 1 discusses the M68HC11 microcontroller, its architecture, and its instruction set.

The M68HC11E9 incorporates a variety of useful, built-in devices, and surprisingly large amounts of on-chip memory. This memory include 512 bytes of RAM, 12K bytes of ROM, and 512 bytes of EEPROM. Other built-in M68HC11E9 features include:

　　Two bidirectional parallel ports
　　One input-only parallel port
　　One output-only parallel port
　　One mixed, input-only, output-only, and I/O port
　　A full-duplex UART
　　A synchronous serial interface
　　An A-to-D converter
　　A pulse accumulator
　　A sophisticated, 16-bit timer system.

A central processing unit (CPU) provides the M68HC11's computing power. Figure 1–1 shows that the CPU divides into three functions—address handling, data handling, and instruction handling. The CPU's address handling section consists of an address latch and four address pointer registers. The address latch drives the address bus with an address to be accessed (effective address). The address latch gets its address information from the program counter—to access program bytes, from the data bus or from an address pointer. Address information may originate from program bytes (direct and extended addressing modes). Address information may be the sum of an index register's contents and a program byte (indexed addressing mode). Address information may come from the stack pointer, if the CPU is pushing a byte onto, or pulling a byte off, the stack.

The CPU's data handling function includes ALU, accumulators/D register, and CCR. The ALU performs a wide variety of logical and mathematical operations on data. The ALU also executes a limited number of arithmetic operations on address information. Finally, the ALU performs logical tests on CCR bits to determine whether the CPU should execute conditional program branches.

The accumulators/D register hold data to be manipulated by the ALU, and the results of that manipulation. These registers also form termini for data reads and writes to memory or I/O devices. ACCA and ACCB can operate as independent 8-bit registers, or can combine to form the D register. ACCA forms the D register's MS byte, and ACCB forms its LS byte.

The 8-bit CCR incorporates three kinds of bits. X and I bits mask sixteen different interrupt types. The S bit controls the M68HC11's response to the STOP instruction (see Chapter 2). The H, N, Z, V, and C bits flag conditions resulting from a CPU operation. The C flag indicates result register overflow, or the subtraction of a higher from a lower value operand. The C flag also forms a register or memory extension bit during shift and rotate executions. The V flag indicates an upset in the sign bit system during an addition operation. The Z flag indicates when a CPU operation returns zero to the result register. The N flag is a copy of the result register's MS (sign) bit. The H flag indicates the value of a carry from bit 3 to bit 4 during the addition of BCD operands. A DAA instruction uses the H flag when it decimalizes a BCD sum.

CPU instruction handling decodes an instruction received from program memory via the data bus. The instruction handler translates an opcode into sequenced enables and control bus signals. Sequenced enables cause information transfers throughout the microcontroller. Sequenced enables also cause the ALU to perform mathematical or logical operations on data, as required by the instruction. The instruction handler automatically causes the CPU to fetch an opcode from memory. An instruction fetch consists of an opcode transfer from memory to the data bus, to instruction decoder, then decoding of the opcode. This opcode fetch (a bus read cycle) constitutes one machine cycle. Motorola's instruction set defines instruction fetch and execution times in terms of machine cycles.

BUFFALO ROM's monitor program includes nineteen system-level commands. Section 1.4 discusses ten of these commands, which will be used throughout this book's remaining exercises.

The BUFFALO MD (Memory Display) command allows a user to display the contents of memory addresses at the terminal. The MM (Memory Modify) command can display and modify a single byte or a string of consecutive bytes in RAM or EEPROM. The BF (Block Fill) command loads a string of memory addresses with a constant, specified value. The BUFFALO MOVE command copies the contents of an address block to RAM or EEPROM. The G (Go) command causes the M68HC11 to execute a program.

The ASM (Assembler/Disassembler) command invokes the BUFFALO one-line assembler/disassembler. The assembler compiles machine code from source code lines entered by the user. The disassembler translates machine code back into BUFFALO assembly code. The BUFFALO assembler uses no labels or assembler directives, and it does not save comments. BUFFALO assembles or disassembles one instruction at a time. This assembler calculates relative addresses, and has a limited error message repertoire.

The BUFFALO RM (Register Modify) command allows a user to examine the current contents of the CPU's seven key registers, and use the terminal keyboard to change their contents.

The T (Trace) command causes BUFFALO to display current CPU status each time it fetches and executes a program instruction. After each fetch and execute, BUFFALO displays the assembly code line or opcode just executed, along with the resulting contents of the seven key CPU registers. To use the T command, the reader must first use the RM command to point the program counter at the first instruction to be executed and traced.

Breakpoints allow a user to execute a program—using the G command—up to a desired breakpoint. At this breakpoint, fetch and execute stops, and BUFFALO displays current CPU status (the contents of the seven key CPU registers). A breakpoint must be an instruction opcode address. The BR command allows a user to establish up to four breakpoint addresses. After program execution halts at a breakpoint, a BUFFALO P (Proceed) instruction is used to continue program execution from the breakpoint. The P instruction allows a user to analyze an entire program, stopping at up to four breakpoints along the way. Breakpoints facilitate system analysis and debugging. They eliminate the need for repetitive instruction tracing, and can be used with interrupt service routines.

The M68HC11 uses six addressing modes. Inherent mode instructions are generally 1 byte in length (with eight exceptions). Inherent mode instructions use no operand, imply a constant operand, or manipulate CPU register data. Immediate addressing uses constant-value operands, and includes these values in the program itself. Assembly code indicates immediate mode operands with a pound sign (#). Extended mode instructions address operands residing at memory addressees. An extended mode instruction incorporates the operand's full, 16-bit address. Direct mode addresses reside on memory page zero. Since the address' high byte is zero, direct mode instructions omit it from the machine code. Thus, a direct mode instruction saves a byte of program space, as compared with its extended mode counterpart.

The M68HC11's CPU incorporates two index registers: X and Y. The CPU forms an indexed mode effective address by adding an unsigned program byte to the contents of an index register. The program byte is called the offset. This byte-length offset resides in the program at the address following the indexed-mode opcode.

Relative mode instructions control program flow rather than access operands or manipulate bits. Some relative mode instructions are conditional branching instructions because program flow depends upon meeting a condition. Unconditional branching instructions alter program flow regardless of prevailing conditions. A relative mode instruction alters program flow by sign extending a 1-byte offset, called the relative address, then adding the 2-byte, sign-extended value to the program counter's contents. The offset byte resides at the program address following the relative mode opcode. In the case of a conditional branch, the CPU adds the offset if a condition is met. To branch forward, the machine code programmer uses a positive offset. To branch back in the program, the programmer uses a negative offset. Relative addressing enhances program portability. A relative address will work properly, even if the program is moved to another memory block.

Section 1.6 discusses the M68HC11's instruction set. Data and address handling instruction mnemonics start with the letters LD, ST, T, or XG. LD stands for "load." A load instruction reads information from memory to a CPU register. There are load instructions for ACCA, ACCB, the D register, the index registers, and the stack pointer. ST stands for "store." A store instruction writes information from a CPU register to memory. ACCA, ACCB, the D register, the index registers, and the stack pointer all have associated store instructions. If an instruction starts with T, it transfers the contents of one CPU register to another. These instructions transfer data between accumulators or between an index register and stack pointer. If an instruction's mnemonic begins with XG, it exchanges data between the D register and an index register.

Push and pull instructions transfer the contents of ACCA, ACCB, and the index registers to and from the stack. As the CPU pushes data bytes onto the stack, the stack pointer (SP) decrements to lower addresses within the stack's RAM block. When the CPU pulls data from the stack, the SP increments to higher RAM addresses. Thus, the stack grows or shrinks, depending upon whether the CPU pushes data onto the stack or pulls data from it. Push and pull instructions use inherent addressing mode.

Clear instructions change the contents of an accumulator or memory to $00. A decrement instruction causes the CPU to subtract one from the contents of a specified CPU register or memory location. Increment instructions add one to a register or memory contents. The bit clear (BCLR) instruction uses a mask byte to specify specific bits to be cleared within a memory byte. This mask resides at the program address following the opcode and address bytes. Ones in the mask byte specify corresponding memory bits to be cleared. The bit set (BSET) instruction likewise uses a mask to specify memory bits to be set.

Shift and rotate instructions manipulate the contents of a memory byte, an accumulator, or the D register. A rotate instruction circulates bits right or left through the target accumulator or memory byte, into the C flag, and from the C flag back to the target accumulator or memory byte. A single rotate instruction shifts bits one place right or left.

Arithmetic and logical shift instructions shift bits one place right or left in an accumulator, memory location, or D register. These instructions also shift a bit into the C flag. However, arithmetic or logical shifts do not rotate the C flag into the target memory byte or register. Arithmetic and logical shifts left take identical actions and share the same opcodes. These instructions shift binary zero into the target register or memory location's LS bit. The MS bit shifts to the C flag. However, the reader must distinguish between logical and arithmetic shifts right. All shift right instructions shift the target register or memory byte's LS bit into the C flag. However, arithmetic shift right instructions retain the MS (sign) bit. Logical shift right instructions shift a binary zero into the MS bit.

M68HC11 logic instructions fall into two categories:

1. Instructions which find the one's or two's complement of a memory byte or accumulator.
2. Instructions which AND, OR, or XOR the contents of an accumulator with a memory byte, and return the result to the accumulator.

M68HC11 arithmetic instructions provide for convenient, multiprecision programming in four functions: add, subtract, multiply, and divide. Addition instructions can add the contents of memory to ACCA, ACCB, or the D register. The ABA instruction adds the contents of the two accumulators. Add with carry instructions add the C flag state, along with the contents of a memory byte, to ACCA or ACCB.

The M68HC11's instruction set includes subtract instructions which are counterparts to the adds. These instructions can subtract memory bytes from ACCA, ACCB, or the D register. The SBA instruction subtracts the contents of ACCB from ACCA. Subtract with carry instructions subtract the C flag's content, along with a memory byte from ACCA or ACCB.

The M68HC11's multiply (MUL) instruction multiplies the contents of the two accumulators and returns a 16-bit product to the D register.

Both integer divide and fractional divide instructions divide the D register's contents by the contents of the X register. These instructions return the quotient to the X register, and the remainder to the D register. The integer divide (IDIV) instruction divides whole integers. The fractional divide (FDIV) carries the quotient to the right of the hex point.

The M68HC11's DAA (Decimal Adjust ACCA) instruction evaluates a BCD sum in ACCA and decimal adjusts it as necessary. Use DAA only after an ABA, ADDA, or ADCA operation on packed BCD digits.

Eight M68HC11 instructions manipulate bits in the condition code register (CCR). Two of these instructions (TAP and TPA) transfer 8 bits between the CCR and ACCA. A programmer must use the TAP instruction to clear the X bit, and set or reset the S bit. The remaining six CCR-oriented instructions set or reset the C, I, or V bits.

Four instructions allow the CPU to test the contents of an accumulator or memory byte, and set or reset the N and Z flags accordingly. TSTA and TSTB test an accumulator's contents. BITA and BITB allow the CPU to test individual bits in a memory byte or accumulator. These bit test instructions AND the contents of the target accumulator with a memory byte, then set or reset the N and Z flags according to the AND-ed result. BITA and BITB do not save the AND-ed result, but use it for test purposes only.

The M68HC11's instruction set includes six compare instructions. Each of these instructions makes its comparison by subtracting the comparison value from a target CPU register, then setting or resetting the N, Z, V, and C flags according to the result. Target registers include ACCA, ACCB, and the D, X, and Y registers. Compare instructions do not save test subtraction results.

Program control instructions alter the program counter's contents, and thereby cause program flow to jump or branch. These instructions fall into four categories: (1) unconditional branching instructions, (2) conditional branching instructions, (3) program jump instructions, and (4) return instructions.

An unconditional branch instruction sign extends its accompanying relative address byte to 16 bits, then adds this sign-extended offset to the program counter's contents. The CPU thus fetches its next program instruction from the new address pointed at by the PC. One unconditional branch instruction, BSR, also stacks the current PC value (i.e., the next sequential program address). BSR allows the CPU to branch to a subroutine, then return from it by pulling the stacked value back to the PC.

A conditional branch instruction may also sign extend a relative address and add it to the PC's contents. However, this action depends upon meeting a condition. If the condition is not met, program flow remains unaltered. Branch conditions consist of specified CCR status flag values, as generated by the execution of a previous program instruction.

Program jump instructions simply store a new value to the program counter. One jump instruction, JSR, also stacks the current PC value. JSR allows the CPU to jump to a subroutine, then return by pulling the stacked PC value.

Return instructions, RTS and RTI, assume that the CPU has executed a subroutine or interrupt service routine, and must now return to the next instruction in the program. Chapter 2 covers the RTI (Return from Interrupt) instruction in detail. The CPU executes an RTS (Return from Subroutine) instruction at the end of a subroutine. An RTS causes the CPU to pull the return address from the stack and re-load the PC with this value. Thus, the CPU returns to the next program instruction. Because a BSR or JSR instruction stacks the return address, and an RTS pulls this address from the top of the stack, a programmer can nest subroutines, i.e., call subroutines from subroutines.

EXERCISE 1.6 Chapter Review

Answer the questions below by neatly entering your responses in the appropriate blanks.

1. An M68HC11A1 incorporates _____ EPROM, _____ EEPROM as its nonvolatile program memory. (Enter an X in the appropriate blank).
2. An M68HC11E9 holds the BUFFALO operating system in its _____ (RAM/ROM/EEPROM).
3. M68HC11's PORTC provides _____ (how many?) parallel I/O lines, and can operate with or without _____. PORTD provides _____ (how many?) parallel I/O lines.
4. A user can configure eight PORT _____ (which port?) lines as general-purpose inputs. PORT _____ (which port?) is output only. PORT _____ incorporates input-only, output-only, and I/O lines.
5. Describe the purpose of the M68HC11's SCI.

   The SCI operates _____ synchronously _____ asynchronously. (Enter an X in the appropriate blank.) The SPI provides _____ synchronous _____ asynchronous serial communications over PORT _____ lines.
6. The M68HC11's A-to-D converter uses PORT _____ pins as analog inputs. This ADC can scale up to _____ (how many?) inputs simultaneously.
7. Describe the purpose of the M68HC11's pulse accumulator.

8. Describe the purpose of the M68HC11's input capture feature.

   The input capture mechanism uses PORT _____ pins.
9. Describe the purpose of the output compare function.

   The output compare mechanism uses PORT _____ pins.
10. List the components of the M68HC11 CPU's address handling function.

    A.

    B.

    C.

    D.

    E.

11. When the CPU fetches a program byte from memory, the address latch gets its address information from the _____.
12. When the CPU uses extended or direct addressing to access memory, the address latch gets its address information from _____.
13. When using indexed addressing to access memory, the CPU loads the address latch with the sum of the _____ and a _____.

14. When the CPU pushes or pulls stacked information, the address latch's contents come from the _____.
15. List the components which make up the CPU's data handling function.

    A.

    B.

    C.

16. Summarize the tasks performed by the ALU.

17. List the tasks performed by the accumulators/D register.

    A.

    B.

    C.

18. List the three types of bits contained by the CCR and summarize the functions of these bits.

    A.

    B.

    C.

19. Assume that ACCA holds $19 and memory location $0004 contains $7A. In the blanks, enter the states of the C, N, V, and Z flags after the CPU executes an ADDA 4 instruction: C = _____, N = _____, V = _____, Z = _____.
20. Assume that ACCB contains $19 and memory location $0004 contains $7A. Enter the states of the C, N, V, and Z flags after the CPU executes a SUBB 4 instruction: C = _____, N = _____, V = _____, Z = _____.
21. Assume that the D register contains $ABCD, and memory locations $0000 and $0001 contain $ABCD. Enter the states of the C, N, V, and Z flags after the CPU executes a SUBD 0 instruction: C = _____, N = _____, V = _____, Z = _____.
22. Explain why some M68HC11 opcodes consist of two bytes: a prebyte, and an opcode byte.

23. Summarize the tasks performed by the CPU's instruction handling function.

24. Define a CPU machine cycle.

25. Although an instruction may generate complex ALU data manipulations, the majority of controller/sequencer outputs enable

26. List controller/sequencer actions involved in an opcode fetch.

    Step 1.

    Step 2.

    Step 3.

    Step 4.

    Step 5.

    Step 6.

27. Refer to Figure 1–5. This figure actually contains _____ (how many?) program list(s). How many programs does Figure 1–5 illustrate? _____
28. Explain how assembly source code differs from machine code.

29. Hand assemble machine code from this source code line: LDX #1. Enter the appropriate machine code bytes in the blanks:
    $_____
    $_____
    $_____
30. Hand disassemble source code from this machine code instruction: $A7 14. Enter the appropriate source code line in the blank: _____
    If the X register contains $C000, to what address would this instruction store ACCA's contents? $_____
31. The BUFFALO MD C004 C014 command displays _____ (how many?) bytes, from address $_____ to address $_____.
32. The BUFFALO MM 0 command allows the user to key in _____ starting at address $_____.
33. The BUFFALO BF 0 33 CC command writes $_____ to all addresses from $_____ to $_____.
34. The BUFFALO MOVE C000 C03A B700 command copies the contents of address $_____ through $_____ to addresses $_____ through $_____.

35. The BUFFALO G B780 command causes the M68HC11 to _____, starting at address $_____.
36. The BUFFALO ASM 120 command causes the M68HC11 to _____ and _____ program, starting at address $_____.
37. Define the term "assemble" in the context of the BUFFALO assembler/disassembler.

Define the term "disassemble."

38. The BUFFALO RM A instruction allows the user to display the contents of all key _____ and modify the contents of _____. By using the _____, the user can modify the contents of the other key CPU registers.
39. To trace fifteen instructions, starting at address $0100, the user first enters this BUFFALO command: _____. The user then loads the _____ with $_____. Next, the user enters this BUFFALO command: _____. In response, BUFFALO displays the _____ just executed and the contents of _____, each time it fetches and executes one of the fifteen instructions.
40. Assume that you wish to analyze a program located at address $C00A. To establish breakpoints at addresses $C014 and $C036, use this BUFFALO system command: _____. Next, use the BUFFALO _____ command to fetch and execute program instructions up to the first breakpoint. To continue fetch and execute from this breakpoint, use the BUFFALO _____ command.
41. The INCB instruction uses _____ addressing mode.
42. The LDD #ABCD instruction uses _____ addressing mode.
43. The STD 1B6 instruction uses _____ addressing mode.
44. The STD A instruction uses _____ addressing mode.
45. The LDD 6,Y instruction uses _____ addressing mode. If the Y register contains $C055, this instruction accesses data from addresses $_____ and $_____.
46. The BNE 10A instruction uses _____ addressing mode. If this instruction resides at address $0125, the relative address is $_____.
47. The LDD 9 instruction uses _____ addressing mode to _____ (action) the _____ register with the contents of addresses $_____ and $_____.
48. The STS C0AB instruction uses _____ addressing mode to _____ the contents of the _____ to addresses $_____ and $_____.
49. The TSY instruction uses _____ addressing mode to _____ the contents of _____ to the _____.
50. The XGDY instruction uses _____ addressing mode to _____ the contents of the _____ with the contents of the _____.
51. In each blank, enter Boolean notation which explains the listed push or pull instruction.

   PSHA:

   PSHB:

   PSHX:

   PSHY:

   PULA:

PULB:

PULX:

PULY:

52. List the mnemonics for six instructions (other than push and pull instructions) which use the stack.

53. The CLR 9 instruction uses _____ addressing mode to write $_____ to address $_____.
54. If ACCB = $00, and the CPU executes the DECB instruction, the new contents of ACCB are $_____. DECB uses _____ addressing mode.
55. If address $010F contains $FF, and the CPU executes an INC 10F instruction, the new contents of $010F are $_____. The INC 10F instruction uses _____ addressing mode.
56. If address $C026 contains $BC, and the CPU executes a BSET C026 41 instruction, the new contents of address $C026 are $_____. This instruction uses _____ addressing mode.
57. Assume that the C flag is set and the D register contains $ABCD. After the CPU executes an ROLA instruction, the C flag is _____ (set/reset), and the D register contains $_____.
58. Assume that the C flag is reset and location $0120 contains $B7. After the CPU executes an ROR 120 instruction, the C flag is _____, and location $0120 contains $_____.
59. Address $0026 contains $A6. After the CPU executes an LSL 26 instruction, the C flag is _____, and address $0026 contains $_____. In the blank, enter source code which uses a different mnemonic to generate the identical action and results: _____.
60. The D register contains $F173. After the CPU executes an LSRD instruction, the C flag is _____, and the D register contains $_____.
61. ACCA contains $A4. After the CPU executes an ASRA instruction, the C flag is _____, and ACCA contains $_____.
62. Address $0123 contains $9B. After the CPU executes a COM 123 instruction, address $0123 contains $_____.
63. ACCB contains $9B. After the CPU executes a NEGB instruction, ACCB contains $_____.
64. ACCA contains $67 and address $C00B contains $98. After the CPU executes an EORA C00B instruction, ACCA contains $_____ and address $C00B contains $_____.
65. ACCB contains $67. After the CPU executes an ANDB #98 instruction, ACCB contains $_____.
66. Address $0008 contains $66, and ACCA contains $55. After the CPU executes an ORAA 8 instruction, address $0008 contains $_____ and ACCA contains $_____.
67. Memory address $0009 contains $CE, and memory address $000A contains $87. The D register contains $3706. After the CPU executes an ADDD 9 instruction, the D register contains $_____, and the C flag is _____ (set/reset).
68. Assume that the C flag is set, and ACCB contains $89. After the CPU executes an ADCB #A4 instruction, ACCB contains $_____ and the C flag is _____ (set/reset).
69. The D register contains $012A. Location $0013 contains $76 and location $0014 contains $01. After the CPU executes a SUBD 13 instruction, the D register contains $_____ and the C flag is _____.
70. ACCA contains $C7, memory address $0000 contains $67, and the C flag is set. After the CPU executes an SBCA 0 instruction, ACCA contains $_____ and the C flag is _____.

71. Assume that ACCA = $9B and ACCB = $87. After the CPU executes a MUL instruction, the D register contains $_____.
72. The X register contains $099A, and the D register contains $107D. The CPU divides the numerator by the denominator, using one IDIV and one FDIV instruction. The CPU stores the IDIV quotient to addresses $000A–$000B, and the FDIV quotient to addresses $000C–$000D. Enter the final contents of these memory locations: $000A = $_____, $000B = $_____, $000C = $_____, $000D = $_____.
73. Assume that ACCA contains packed BCD digits 01011001_{BCD}, and ACCB contains 00110111_{BCD}. If the CPU executes an ABA instruction, it returns the number $_____ to ACCA. The CPU then executes a DAA instruction. DAA causes the CPU to add the number $_____ to the contents of ACCA. After DAA execution, the C flag is _____ (set/reset), and ACCA contains _____$_{BCD}$.
74. If the CPU executes a LDAA #80 instruction, then TAP, it _____ (sets/resets) the S bit and _____ (set/resets) the X bit.
75. The SEV instruction _____ (sets/resets) the _____ flag in the CCR.
76. Assume that ACCA contains $AB, and the CPU executes the TSTA instruction. After TSTA execution, ACCA contains $_____, the N flag = _____ (1/0), the Z flag = _____, the V flag = _____, and the C flag = _____.
77. ACCB contains $9D, the Y register contains $C000, and memory location $C00F contains $A6. After the CPU executes the BITB F,Y instruction ACCB contains $_____, the N flag = _____, the Z flag = _____, and the V flag = _____.
78. The D register contains $ABCD, memory location $0000 contains $DC, and memory location $0001 contains $BA. After the CPU executes the CPD 0 instruction, the D register contains $_____, the N flag = _____, the Z flag = _____, the V flag = _____, and the C flag = _____.
79. List the three actions taken by the CPU when it executes a BSR instruction.

    A.

    B.

    C.

80. If a BSR instruction resides at address $C025, and the subroutine to be called resides at address $C065, the relative address must be $_____.
81. The CPU subtracts $35 from $82, then executes a BVS instruction. The CPU (_____ will, _____ will not) branch.
82. The CPU subtracts $35 from $82, then executes a BHI instruction. The CPU (_____ will, _____ will not) branch.
83. The CPU subtracts $35 from $82, then executes a BGT instruction. The CPU (_____ will, _____ will not) branch.
84. The BRSET 35 32 150 instruction tests bits _____ (enter bit numbers) of the data byte at address $_____. If this memory byte contains $A9, the CPU (_____ will, _____ will not) branch. If this memory byte contains $B6, the CPU (_____ will, _____ will not) branch.
85. Describe the major difference between the JMP and JSR instructions.

86. Describe the actions taken by the CPU when it executes an RTS instruction.

87. Define the term "nested subroutine."

88. Explain why nested subroutines are possible.

2 M68HC11 Resets And Interrupts

2.1 GOAL AND OBJECTIVES

Resets and interrupts are similar in two ways. First, resets and interrupts are extraordinary events. They cause the microcontroller to deviate from its normal fetch and execute. Second, a reset or interrupt causes the microcontroller to copy a special address to its program counter, and then execute the routine which resides at that address.

But resets and interrupts have important differences. Resets cause the microcontroller to abort its normal fetch and execute, then prepare to start again "from scratch." A reset causes the microcontroller to lose the current contents of its accumulators and other key registers. Resets occur because the user wants to start program execution over again, or because something has gone wrong. A reset of any kind causes the M68HC11 to turn off many of its internal devices, and mask (disable) most interrupts. Thus, when routine fetch and execute restarts, the CPU must re-initialize these internal devices, and re-enable appropriate interrupts.

In contrast, interrupts cause the CPU to push its current context (accumulators, index registers, program counter, and condition code register) onto the stack for safekeeping, while it services the interrupt. After servicing the interrupt, the CPU can pull all of this stacked data back to its registers, and resume routine fetch and execute from where it left off. An interrupt does nothing to turn off internal devices. Although an interrupt may temporarily keep other interrupts from occurring, it does nothing to disable them. Interrupts, then, assume that the CPU will return to normal fetch and execute. Resets, on the other hand, make it necessary to re-start program execution from the beginning.

Chapter 2 examines resets, interrupts, their associated control registers, and relevant instructions. This chapter's goal is to help the reader understand how reset and interrupt mechanisms work, and how to use resets and interrupts most advantageously. By carefully studying this chapter and completing its exercises, the reader should achieve these objectives:

Measure the voltage at which power-on resets (POR) occur.
Measure the voltage at which an undervoltage detector generates an external circuit reset.
Develop software which demonstrates computer operating properly (COP) resets. Predict COP timeout, and compare with actual COP performance.
Develop software and connect hardware which demonstrate IRQ and XIRQ interrupts.
Develop software which demonstrates Illegal Opcode Trap and SWI interrupts.

Resets and interrupts play a key role in virtually all aspects of M68HC11 operation. To properly understand the remainder of this text, the reader must grasp the concepts presented in this chapter. Fortunately, the reader shall find M68HC11 resets and interrupts both impressive and interesting.

2.2 M68HC11 RESETS

M68HC11 resets have an important purpose: to establish or re-establish basic starting conditions for system operation. The M68HC11 may establish starting conditions because a user just powered it up. The microcontroller may re-establish starting conditions because the user elected to do so, or because something went wrong during system operation. In any case, the M68HC11 takes reset actions because its $\overline{\text{RESET}}$ pin (pin 17) has been driven low, then high. Either on-chip circuits or external circuits can drive the M68HC11's $\overline{\text{RESET}}$ pin low, then high.

In all instances, the M68HC11 takes some reset-associated actions when its $\overline{\text{RESET}}$ pin is first driven low. The M68HC11 takes further actions when its $\overline{\text{RESET}}$ pin is then driven high. Four situations can affect $\overline{\text{RESET}}$ pin levels:

1. An external circuit drives the $\overline{\text{RESET}}$ pin low, then high.
2. An internal circuit on board the M68HC11 detects a chip power up, and drives the $\overline{\text{RESET}}$ pin low, then high.
3. An internal circuit detects slow or stopped M68HC11 clocks. If clocks are slow, but not stopped, the internal circuit drives the $\overline{\text{RESET}}$ pin low, then high. If clocks are stopped, the internal circuit drives the $\overline{\text{RESET}}$ pin low, but may not be able to drive it back high.
4. Internal circuitry detects that the CPU appears to be lost or hung up. This internal circuitry drives the $\overline{\text{RESET}}$ pin low, then high.

When an internal or external circuit drives $\overline{\text{RESET}}$ low, the M68HC11 automatically establishes its memory map, and disables most of its interrupts and internal hardware functions. When the $\overline{\text{RESET}}$ pin rises from low to high, the CPU vectors to a reset service routine provided by the system designer. If the first or second situation (above) causes the reset, the CPU vectors to an initialization routine. This routine prepares internal and external devices for system operation. The third and fourth situations cause the CPU to vector to special emergency routines. These routines may attempt to find the problem, warn the user, or take corrective action.

This section examines how resets are generated, and how the M68HC11 responds to them.

Reset Generation

As described above, resets are generated in four ways. Following paragraphs examine each of these four reset causes in more detail.

External circuits generate a reset by driving the M68HC11's $\overline{\text{RESET}}$ pin low, then high. To generate this external hardware reset, the $\overline{\text{RESET}}$ pin must remain low for six E clock cycles[1] before rising back high. Commonly, a system designer connects the $\overline{\text{RESET}}$ pin to both a push-button pull-up circuit, and an undervoltage sense circuit. This combination generates resets in two ways: First, a system user can reset the M68HC11 by pressing the push-button, then releasing it.[2] Second, an undervoltage sensor drives the $\overline{\text{RESET}}$ pin low when it senses a significant drop in V_{DD} (chip power). Figure 2–1 shows how an undervoltage sensor and push-button pull-up circuit connect to the M68HC11 on board the EVBU.[3]

---

[1] The M68HC11's E clock runs at one-fourth of the external clock source frequency. This external clock source connects to M68HC11's XTAL and EXTAL pins and functions as the frequency determining device of a Pierce oscillator. Both EVB and EVBU use an external 8 MHz clock source. Thus, the E clock frequency is 2 MHz. The M68HC11 makes the E clock available at its E clock pin (pin 5). System designers use the E clock to synchronize M68HC11 I/O operations.

[2] The user must press the push-button for at least six E clock cycles. Six E clock cycles equal 1.5 microseconds. The user cannot avoid pressing the push-button for a time much longer than 1.5 microseconds.

[3] Motorola Inc., *M68HC11EVBU Universal Evaluation Board User's Manual* (Motorola, Inc.: 1990), p. 6-15. Adapted with permission.

FIGURE 2–1
Reprinted with the permission of Motorola.

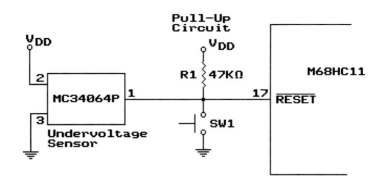

If the M68HC11 chip powers up (i.e., its V_{DD} pin rises from 0V to about 5V), the M68HC11 automatically resets. Motorola calls this kind of reset a power-on reset (POR). A POR circuit on-board the M68HC11 detects V_{DD} pin's rise from 0V, and drives the microcontroller's $\overline{RESET}$ pin low for 4064 E clock cycles. This internal POR circuit drives the $\overline{RESET}$ pin low only if V_{DD} rises from 0V.

If M68HC11 clocks slow down, an internal clock monitor reset circuit can drive the $\overline{RESET}$ pin low for four E clock cycles, then back high. A user enables or disables this clock monitor by writing a bit to the Systems Configuration Options Register (OPTION). The OPTION register resides at address $1039 in the M68HC11's control register block.[4] Figure 2–2 shows the CME bit, within the OPTION register. CME (Clock Monitor Enable, bit 3) enables or disables clock monitor detect circuitry. Enable clock monitor resets by writing a binary one to CME, and disable these resets by writing a binary zero. Any chip reset clears the CME bit, disabling the clock monitor function. Thus, only program instructions can enable the clock monitor's detect circuitry.

The clock monitor's internal circuitry uses an RC timeout to detect a slow or stopped clock. If the clock monitor detects no internal clock edge during the RC timeout period, it drives the $\overline{RESET}$ pin low for four clock cycles. If internal clocks have completely stopped, the four cycles never elapse, and the $\overline{RESET}$ pin remains low. In this case, the M68HC11 can initialize control register bits, but cannot establish pin levels or vector to an appropriate reset service routine because these actions require a clock. In such a case, the user must intervene to correct the problem.

Manufacturing variations cause actual RC timeouts to vary from one chip production lot to the next. Motorola guarantees that the clock monitor will detect internal clocks running at frequencies below 10 KHz. Further, Motorola states that the clock monitor will not react to clock frequencies above 200 KHz.[5]

During the sixth E clock cycle after the $\overline{RESET}$ pin was driven low, the M68HC11 checks it again. If $\overline{RESET}$ is still low, the M68HC11 assumes that a POR (power-on) or external circuit caused the reset. Recall that the internal POR circuitry drives $\overline{RESET}$ low for 4064 E clock cycles. Recall also that an external circuit must drive $\overline{RESET}$ low for *at least* six full clock cycles. If the M68HC11 finds that $\overline{RESET}$ has returned high before six E clock cycles have elapsed, it assumes that the clock monitor has detected a slow clock, or that the CPU is lost. In this case, the M68HC11 makes a further check to determine whether the clock monitor generated the reset. If the clock monitor caused the reset, the CPU vectors to an appropriate reset response routine.

FIGURE 2–2
Reprinted with the permission of Motorola.

OPTION $1039

B7		B5	B4	B3		B1	B0
		IRQE	DLY	CME		CR1	CR0

---

[4] As the reader shall see in this and subsequent chapters, sixty-four control and status registers associate with the M68HC11's internal devices. These registers reside in a block of memory addresses, starting at $1000.
[5] Motorola Inc., *M68HC11 Reference Manual* (1990), p. 5-9.

FIGURE 2–3
Reprinted with the permission of Motorola.

CONFIG $103F

B7					B2		B0
					NOCOP		

The computer operating properly (COP) system forms the fourth reset source. The COP system detects a "hung-up" or lost M68HC11 CPU. The CPU can get hung up if it transmits a handshake signal to a peripheral and goes into a program polling loop, waiting for the peripheral to respond. If the peripheral fails to respond, the CPU gets hung up, i.e., stuck in the polling loop indefinitely. Programming bugs can cause the CPU to get lost. The reader has likely experienced a situation where a program bug has sent the CPU "off into the weeds." The CPU may wander in this software "wilderness" indefinitely. COP circuitry detects problems such as these and responds by driving the $\overline{RESET}$ pin low for four E clock cycles, then driving this pin back high.

The COP system uses a watchdog timer. This watchdog timer operates when the COP system is enabled, and the CPU executes program instructions. With its COP system enabled, the CPU must write $55 then $AA to a control register before the watchdog timer times out. If the CPU fails to execute these two writes within the timeout period, the COP system drives $\overline{RESET}$ low, then high. Writing $55 then $AA to the control register causes the watchdog timer to clear and begin a new timeout. A system designer can select from four watchdog timeout periods. Thus, a hung or lost CPU fails to return to the $55/$AA write instructions, and the COP system resets the M68HC11.

The COP system uses bits in three control registers: CONFIG, OPTION, and COPRST. A user enables the COP system by clearing a bit in the Configuration Control Register (CONFIG, control register block address $103F). Refer to Figure 2–3, which shows the NOCOP control bit (CONFIG, bit 2). This bit's name (NOCOP) means that setting it disables the COP system. Thus, to enable the COP system, the user must reset the NOCOP bit. The CONFIG register actually resides in chip EEPROM. The reader cannot program the CPU to write a zero to NOCOP, and thereby turn on the COP system. Rather, MOVE a control byte from RAM to CONFIG, using the BUFFALO MOVE command. Other CONFIG register bits help establish the M68HC11's memory map. A user must avoid changes to memory map bits when moving a control byte to CONFIG. Exercise 2.1 provides detailed instructions on this procedure. Further, as long as NOCOP remains reset, the COP system operates every time the M68HC11 executes a program. Assure that NOCOP = 1 before executing programs which fail to provide for COP operation. To re-configure the CONFIG register and disable the COP system, the reader can use the BUFFALO BULKALL command, which writes $0F to CONFIG. This value disables the COP system and properly establishes the M68HC11's memory map.

Two bits in the OPTION register give the system designer a choice of four watchdog timeout periods. Motorola calls these bits the COP timer rate select bits, CR1 and CR0. Refer to Figure 2–2, which illustrates the OPTION register. The four COP timeouts derive from the M68HC11's E clock. Internal clock divider chains divide the E clock frequency by 2^{15}. This $E/2^{15}$ frequency feeds the COP timer. CR1/CR0 bit settings cause the COP timer to divide the $E/2^{15}$ frequency by 1, 4, 16, or 64:

CR1	CR0	$E/2^{15}$ divided by
0	0	1
0	1	4
1	0	16
1	1	64

EVBU or EVB E clocks run at 2 MHz. Therefore, the base COP timer rate is 2 MHz/2^{15} = 61.035 Hz. CR1 and CR0 cause this base rate to be divided by 1, 4, 16, or 64. For example, if CR1/CR0 = %10, COP's timer divides the 61.035 Hz base rate by 16:

$$61.035 \text{ Hz}/16 = 3.8146875 \text{ Hz}$$

FIGURE 2-4

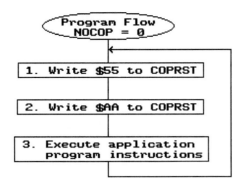

Reciprocate 3.815 Hz to find timeout in seconds:

$$1/3.815 \text{ Hz} = 262.14\text{ms}$$

Timeout values for all four CR1/CR0 combinations are as follows:

CR1	CR0	$E/2^{15}$ divided by	Timeout
0	0	1	16.384ms
0	1	4	65.536ms
1	0	16	262.14ms
1	1	64	1.049s

The M68HC11 time protects OPTION register's CR1/CR0 bits. The CPU must write to time-protected bits within sixty-four bus cycles after a chip reset. Thus, the M68HC11 must initialize CR1/CR0 as a part of its POR/external circuit reset service routine. BUFFALO's reset routine initializes CR1/CR0 to %11. An EVBU/EVB user cannot change these values.

Recall that the CPU must write $55, then $AA to a control register, to prevent a COP reset and start a new timeout period. This control register is the Arm/Reset COP Timer Circuitry Register (COPRST, control register block address $103A). The CPU writes $55 to COPRST to arm the circuitry which clears the COP timer. The CPU writes $AA to COPRST to actually clear the COP timer. The CPU can execute other instructions between its $55 and $AA writes. Of course, the CPU must execute its second write ($AA) before COP timeout elapses.

Figure 2–4 illustrates program flow with COP enabled. After writing $AA to COPRST (block 2), the M68HC11 must execute application program instructions (block 3), return, and execute blocks 1 and 2, before COP timeout. The following routine can efficiently write $55, the $AA to COPRST:

```
LDD #55AA
STAA 103A
STAB 103A
```

Reset Actions

Section 2.2 explains that M68HC11 resets establish or re-establish basic starting conditions for system operation. The previous section discussed four conditions which drive the $\overline{\text{RESET}}$ pin low, then high. This section describes actions taken by the M68HC11 in response to changes in $\overline{\text{RESET}}$ pin levels.

Regardless of reset source, the M68HC11 takes these actions when its $\overline{\text{RESET}}$ pin is driven low:

1. The M68HC11 completes establishment of its memory map by initializing the RAM and I/O Mapping Register (INIT, control register block address $103D) to $01. Refer to Figure 2–5, which shows the INIT register.

    INIT register bits B7–B4 (RAM3–RAM0) establish the upper hex digit of M68HC11 internal RAM addresses. When $\overline{\text{RESET}}$ goes low, the M68HC11 forces this digit to

FIGURE 2–5
Reprinted with the permission of Motorola.

INIT $103D

	B7	B6	B5	B4	B3	B2	B1	B0
	RAM3	RAM2	RAM1	RAM0	REG3	REG2	REG1	REG0
RESET =	0	0	0	0	0	0	0	1

zero. This action locates internal RAM at address block $0XXX. The M68HC11E9 on board the EVBU provides two, 256-byte pages of internal RAM. This internal RAM resides at addresses $0000–$01FF. The M68HC11A1 on board the EVB provides one page of internal RAM, at addresses $0000–$00FF.[6]

INIT register bits B3–B0 (REG3–REG0) establish the upper hex digit of control register block addresses. When RESET goes low, the M68HC11 forces REG3–REG0 to %0001. Thus, the M68HC11's internal control and status registers reside at memory locations $1000–$103F.

2. The M68HC11 forces the contents of its PIOC (Parallel I/O Control Register, control register block address $1002) to %00000X11. This value disables all parallel I/O-associated interrupts. Chapters 3, 4, and 11 describe parallel I/O systems.
3. The M68HC11 incorporates an extensive and sophisticated timer system. Most of this system's timer functions are based on a free-running counter. When RESET goes low, the M68HC11 initializes this counter to $0000, and disables all timer-associated functions and interrupts. Chapters 8 and 9 deal with timer system functions.
4. Depending upon the state of the NOCOP bit in the CONFIG register, the M68HC11 enables or disables the COP system (see pages 88–89).
5. The M68HC11 disables serial I/O systems and their associated interrupts. Chapters 5 and 6 deal with serial I/O.
6. The M68HC11 disables its on-board analog-to-digital converter. Chapter 10 describes this A-to-D converter.
7. The M68HC11 sets interrupt parameters and priorities, but disables all interrupt mechanisms, except for illegal opcode traps.
8. The M68HC11 disables the clock monitor system (see page 87).

Overall, the M68HC11 initializes bits in twenty-four control registers when its RESET pin is driven low. These bits establish most of the conditions listed above.

When the RESET pin rises from low to high, the CPU copies the address of an appropriate reset service routine to its program counter (PC), then executes this routine. For EVB and EVBU, these reset routines are located at addresses $E000, $00FA, and $00FD. If POR or external circuitry caused the reset, the CPU vectors to a routine at ROM address $E000. If the clock monitor system caused the reset, the CPU vectors to a routine at RAM address $00FD. If the COP system generated the reset, the CPU vectors to a routine at RAM address $00FA. To vector to one of these routines, the M68HC11 copies the routine's address—$E000 or $00FA or $00FD—to its PC from storage locations in ROM. Three pairs of ROM locations contain reset routine starting addresses:

1. ROM locations $FFFE and $FFFF contain the POR/external circuit routine's address ($E000).
2. ROM Locations $FFFC and $FFFD contain the clock monitor routine's address ($00FD).
3. ROM locations $FFFA and $FFFB contain the COP routine's address ($00FA).

Thus, depending upon the source of the reset, the M68HC11 loads its PC from $FFFE/$FFFF, $FFFC/$FFFD, or $FFFA/$FFFB.

How does the M68HC11 know which reset routine address to copy to its PC? POR and external pin resets hold the RESET pin low for at least six E clock cycles. The clock

---

[6]To a user, the EVB appears to have much more than 256 bytes of internal RAM. In fact, the EVB has as much as 16K bytes of additional RAM. However, all of this added RAM is external, consisting of MCM6264 8K × 1-byte memory chips.

monitor and COP systems drive $\overline{\text{RESET}}$ low for only four E clock cycles. During the sixth E clock cycle, the CPU checks the state of the $\overline{\text{RESET}}$ pin. If $\overline{\text{RESET}}$ is still low, the M68HC11 knows that POR or an external circuit generated the reset. In this case, the M68HC11 loads its PC from ROM locations $FFFE/$FFFF. If the $\overline{\text{RESET}}$ pin has returned high after five clock cycles, the M68HC11 knows that the source is either the clock monitor or COP system. In this case, the M68HC11 checks the status of these two systems to determine which one caused the reset. If the clock monitor caused the reset, the M68HC11 loads its PC from ROM locations $FFFC/$FFFD. If the COP system generated the reset, the M68HC11 loads its PC from $FFFA/$FFFB.

In the case of clock monitor and COP resets, the BUFFALO operating system uses pseudovectors. A pseudovector is a JMP instruction which re-vectors the CPU to the reset service routine's actual starting address. The reader should invoke BUFFALO's assembler/disassembler, and disassemble the clock monitor reset service "routine" at address $00FD. Disassembly shows that this "routine" is actually a pseudovector: JMP $E35A. Thus, if the clock monitor generates the reset, the M68HC11 loads its PC with $00FD (copied from addresses $FFFC/$FFFD). The CPU starts fetching and executing at address $00FD. The instruction at $00FD (a pseudovector) causes the CPU to jump to address $E35A, then fetch and execute from there. The CPU pseudovectored to the routine at $E35A. The COP service routine at $00FA also contains a pseudovector.

BUFFALO operating system's default clock monitor and COP pseudovectors cause the CPU to jump to the same routine—at address $E35A. This routine simply STOPs the CPU.[7] However, the pseudovectors themselves reside in RAM (addresses $00FA and $00FD). Therefore, the programmer can easily change these pseudovectors so that the CPU jumps to a custom routine. For example, using the assembler or MM instruction, the user could change the COP pseudovector to:

```
JMP 100.
```

The user can then assemble a custom COP service routine at RAM address $0100. A COP reset would cause the M68HC11 to vector, then pseudovector to this routine and execute it.

To summarize, M68HC11 resets can be generated from four sources. In all four of these cases, the M68HC11 establishes initial operating conditions when its $\overline{\text{RESET}}$ pin is driven low. The microcontroller establishes these operating conditions by initializing control bits in twenty-four of its control registers. When the $\overline{\text{RESET}}$ pin rises from low to high, the M68HC11 CPU vectors to one of three reset service routines:

> If POR or an external circuit generated the reset, the CPU vectors to a service routine at $E000 (in ROM).
> If clock monitor generated the reset, the CPU vectors to RAM location $00FD.
> If COP generated the reset, the CPU vectors to RAM location $00FA.

BUFFALO automatically establishes pseudovectors at the COP and clock monitor routine addresses. The user can easily modify these pseudovectors to provide for custom COP and clock monitor service routines.

EXERCISE 2.1   M68HC11 Resets

REQUIRED MATERIAL: 1 Potentiometer, 20K Ohms

REQUIRED EQUIPMENT: Motorola M68HC11 EVBU or EVB, and terminal
Variable power supply, 0–5VDC, 500mA
Voltmeter, 0–10VDC
Oscilloscope and probe

---

[7] Refer to pages 103–106 for analysis of the STOP instruction.

FIGURE 2-6

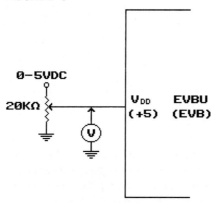

GENERAL INSTRUCTIONS: In this laboratory exercise, the reader shall demonstrate, test, and analyze three M68HC11 resets. Parts I and II test and analyze power-on resets (PORs). Part III analyzes external circuit resets generated by undervoltage sensors and the EVBU/EVB board's reset switch. Part IV demonstrates and analyzes the computer operating properly (COP) mechanism. Carefully read and execute this exercise's detailed procedures and instructions.

PART I: Power-On Resets

Power down the EVB or EVBU. Assure that you have a variable 0–5VDC power supply with at least 500mA output current capability. If you use an EVB, connect this board's +12VDC and –12VDC terminals to appropriate power supply terminals—powered down. Assure that all voltage supplies share a common ground. Make the EVB or EVBU's V_{DD}/+5V power supply connection as shown in Figure 2–6. Connect a voltmeter to the EVBU or EVB's +5V/V_{DD} terminal as shown in the figure.

Turn on the power supplies and adjust the EVBU/EVB +5VDC input to zero. Turn off the power supplies, and turn them back on. Observe the terminal screen as you use the potentiometer to *slowly* increase the voltage at the EVBU/EVB's +5VDC input. Increase this voltage until the terminal screen indicates an M68HC11 reset. Note the voltage at which this power-on reset occurs (V_{POR}). Repeat this procedure several times to assure that you have measured V_{POR} as accurately as possible. Enter your measured value in the blank.

V_{POR} = _____ V

Leave all connections in place, and turn off all power supplies.

PART II: Laboratory Analysis, POR Resets

Answer the questions below by neatly entering your responses in the appropriate blanks.

1. Describe how the M68HC11 indicates that it has executed a power-on reset (POR).

2. The M68HC11 copies the POR service routine's starting address to its PC from ROM addresses $_____. The contents of these ROM addresses are $_____. This service routine is also used by _____ resets.
3. Explain why it was necessary to increase V_{DD} from zero, in Step 2.

PART III: External Circuit Resets, Demonstration and Analysis

Refer to Figure 2–1, which depicts the EVBU's external circuit $\overline{RESET}$ pin connections. Note from this figure that the EVBU uses an MC34064P undervoltage sensor to drive the M68HC11E9's $\overline{RESET}$ pin low, when V_{DD} decreases. The EVB also uses an undervoltage sensor. This sensor consists of a Schmitt trigger gate interposed between the $\overline{RESET}$ pin's push-button pull-up circuit and the $\overline{RESET}$ pin itself. When V_{DD} decreases to approximately 4.5V, the Schmitt trigger switches states and drives $\overline{RESET}$ low. Use the following procedure to observe undervoltage sensor operation.

Step 1. Assure that the EVBU or EVB is powered down. Check your PART I V_{DD} connections. Make sure they are intact. Connect an oscilloscope to the M68HC11's $\overline{RESET}$ pin (pin 17). Set the oscilloscope's volts per division to 1V/Div. Adjust the oscilloscope so that its zero-volt reference is on the first or second horizontal graticule line from the bottom of the scope display. As configured, the oscilloscope indicates whether the $\overline{RESET}$ pin is low (at zero volts) or high (up five divisions from zero reference). Turn on the +5VDC power supply and adjust the potentiometer for +5VDC at the EVBU/EVB's V_{DD} connector (Figure 2–6). Assure that the M68HC11 executes a POR reset. If the microcontroller will not reset with the scope connected, disconnect the probe and try again. If the EVBU/EVB now resets, the oscilloscope's input resistance is too small. Try again with a ×10 probe, or use another scope with a higher input resistance.

Step 2. Assure that the M68HC11 is reset, and V_{DD} = +5VDC. Watch the oscilloscope display. Use the potentiometer to *gradually* reduce V_{DD} until the undervoltage sensor circuit drives the $\overline{RESET}$ pin low. When $\overline{RESET}$ goes low, leave the V_{DD} voltage as is. Record the V_{DD} value at which the undervoltage sensor drives the $\overline{RESET}$ pin low.

V_{DD} undervoltage = _____ V

Note that even though V_{DD} is still at a relatively high value (approximately 4.5V), the undervoltage sensor keeps the $\overline{RESET}$ pin low.

Step 3. Gradually increase V_{DD} until the undervoltage sensor drives the $\overline{RESET}$ pin back high. Record the V_{DD} value at which this transition occurs.

V_{RESET} = _____ V

Note that this transition causes the M68HC11 to vector to the external circuit reset service routine. Note also that your recorded voltage is less than the nominal V_{DD} value: +5VDC.

Return V_{DD} to its nominal +5VDC level. Observe the oscilloscope display and terminal screen as you *slowly* press and release the EVBU/EVB reset button. In the blank, describe what happens when you press the reset button, and what happens when you release it.

Press reset button:

Release reset button:

Explain why the M68HC11 writes the "BUFFALO 2.5 . . ." message to the terminal screen only when you release the push-button.

PART IV: COP Resets, Demonstration and Analysis

Step 1. Refer to the COP watchdog timeout chart on page 88. Use this chart to determine the M68HC11's timeout—when OPTION bits CR1/CR0 = %11. Enter this timeout value in the blank.

COP Timeout = _____ s

FIGURE 2-7

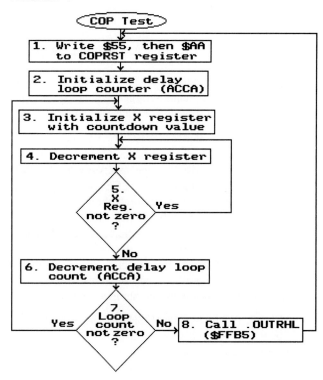

Step 2. Develop a test program which facilitates COP function analysis. This software accomplishes the following tasks: resets the watchdog timer, executes a variable delay, displays a zero to the terminal screen at the end of the delay, and repeats.

Locate your COP test program at address $0100 (EVBU) or $C000 (EVB). Figure 2-7 flowcharts this program. .OUTRHL (Figure 2-7, Block 8) is a BUFFALO ROM utility subroutine. It converts ACCA's least significant nibble to ASCII, and outputs this code to the terminal. The terminal displays this value on its screen. The COP Test program calls .OUTRHL only after ACCA has decremented to zero. Thus, .OUTRHL should cause a zero to appear at the terminal display. If the delay routine is short enough, the CPU can reset the watchdog timer, execute the delay, then display a zero and branch back to reset the watchdog timer—all before the watchdog timer times out and the COP mechanism generates a reset. On the other hand, if the CPU cannot do the delay, display a zero, and branch back in time, then the COP mechanism generates a chip reset and STOPs the CPU. The STOP-ed CPU writes no more zeros to the terminal display.

Your COP Test program should initialize the delay loop counter to $06, and the initial X register countdown value to $D600. List your COP Test program source code in the blanks below.

COP Test

$ADDRESS	SOURCE CODE	COMMENTS
_____	_____	_____
_____	_____	_____
_____	_____	_____
_____	_____	_____
_____	_____	_____
_____	_____	_____
_____	_____	_____
_____	_____	_____
_____	_____	_____

_____	_____	_____
_____	_____	_____
_____	_____	_____
_____	_____	_____
_____	_____	_____

Step 3. Assemble and run your COP Test program. If it operates properly, a new zero should appear on the terminal screen every second. Demonstrate your properly running software to the instructor.

_____ *Instructor Check*

Step 4. Enable the COP timeout mechanism by carefully accomplishing the following tasks. Check off each task as you do it.

_____ Task 1: Erase the contents of the M68HC11's EEPROM with this BUFFALO command:

`BULKALL`   <enter>

_____ Task 2: Push the EVBU/EVB's reset button. Use a BUFFALO MD command to display the CONFIG register contents. CONFIG resides at address $103F. assure that CONFIG contains the value $0F.

_____ Task 3: Use an MD command to display the OPTION register's contents. OPTION resides at address $1039. Assure that this register contains $93. This means that CR1/CR0 = %11, and that the COP timeout equals 1.049 seconds.

_____ Task 4: Use an MM command to store $0B in location $0000. Use an MD command to assure that location $0000 now contains $0B.

_____ Task 5: Use the BUFFALO MOVE command to copy the contents of location $0000 to the CONFIG register:

`MOVE 0 0 103F`   <enter>

_____ Task 6: Push the EVBU/EVB's reset button. Then use an MD command to display the CONFIG register's contents. Assure that CONFIG now contains $0B. If CONFIG = $0B, then NOCOP = 0, and the COP mechanism is enabled.

You can now test the COP timeout mechanism.

Step 5. Run your COP test software. Does it execute the delay and display a zero before COP timeout expires?

_____ Yes _____ No

Step 6. Increase the COP Test delay to two seconds: modify Figure 2–7, block 2's instruction so that it initializes the delay loop counter to $0C rather than $06. Run the COP Test software. Does it execute the delay and display a zero before COP timeout?

_____ Yes _____ No

Step 7. Try various delay loop counter values. Determine the *greatest* loop count value that consistently prevents COP timeout for a complete line of zeros on the terminal display. Add one to this loop count value and record in the blank.

Loop Count (max) + 1 = _____

Step 8. Change Figure 2–7, block 2's instruction so that it initializes the loop counter to the value entered in the blank above.

Step 9. Refer to Figure 2–7, block 3. Systematically test various X register initial values. Find the greatest X register initial value that consistently prevents COP timeout for a complete line of zeros on the terminal display. Enter this value in the blank.

Greatest X register initial value = $_____

Step 10. Determine the number of E clock cycles required by the COP Test routine to decrement the index register from your entered step 9 value, the number of times indicated in step 7. First, multiply the step 7 value by the step 9 value:

X register value (step 9) × loop count value (step 7) = $_____

Now multiply this product by the number of E clock cycles required to execute Figure 2–7, blocks 4 and 5 (these two blocks consume six E clock cycles). This new product represents the approximate number of E clock cycles in the COP Test program's delay routine.[8] Enter this new hex product (E clock total) in the blank.

E clock total = $_____

Convert this hex value to decimal:

E clock total = _____

The EVBU/EVB's E clock runs at 2 MHz. Therefore, each E clock cycle requires 500ns. Calculate the COP Test program's approximate total delay time by multiplying the E clock total by 500ns:

E clock total × 500ns = _____ s

Is this calculated value close to the COP timeout value established by the OPTION register's CR1/CR0 bits?

_____ Yes _____ No

Important: Use the following procedure to turn off the COP mechanism.

1. Execute the BULKALL command:

   BULKALL   <enter>

2. Push the EVBU/EVB reset button.
3. Use an MD command to display the CONFIG register's contents. Assure that CONFIG now contains $0F.

PART V: Laboratory Software Listing

COP TEST PROGRAM—EVB

$ADDRESS	SOURCE CODE	COMMENTS
C000	LDD #55AA	Write $55 then $AA to COPRST register
C003	STAA 103A	" " " " " " "
C006	STAB 103A	" " " " " " "
C009	LDAA #6	Initialize delay loop counter (ACCA)
C00B	LDX #D600	Initialize X register countdown value
C00E	DEX	Decrement X register
C00F	BNE C00E	If X register not zero, decrement again
C011	DECA	If X register zero, decrement loop count (ACCA)
C012	BNE C00B	If loop count not zero, do another countdown
C014	JSR FFB5	If loop count zero, call .OUTRHL subroutine
C017	BRA C000	Branch back and clear watchdog timer

COP TEST PROGRAM—EVBU

$ADDRESS	SOURCE CODE	COMMENTS
0100	LDD #55AA	Write $55 then $AA to COPRST register
0103	STAA 103A	" " " " " " "
0106	STAB 103A	" " " " " " "
0109	LDAA #6	Initialize delay loop counter (ACCA)
010B	LDX #D600	Initialize X register countdown value
010E	DEX	Decrement X register
010F	BNE 10E	If X register not zero, decrement again
0111	DECA	If X register zero, decrement loop count (ACCA)

---

[8]This product is approximate because it ignores blocks 2, 3, 6, and 7. We can ignore this software overhead because the CPU executes these blocks infrequently—as compared with blocks 4 and 5.

0112		BNE 10B	If loop count not zero, do another countdown
0114		JSR FFB5	If loop count zero, call .OUTRHL subroutine
0117		BRA 100	Branch back and clear watchdog timer

2.3 M68HC11 INTERRUPTS

The reader can understand a microcontroller interrupt by comparing it with a typical, real-life situation:

Let us suppose that you are sitting at your desk working on an assignment. You have been working on this assignment for some time. Your work on this assignment equates to the microcontroller's routine application program execution. The phone rings. Your mother is on the line, and explains that she has an emergency problem. Mother's phone call is an interrupt. This phone call equates with an internal or external device's interrupt request. You want to assist your mother, so you carefully set aside your homework, keeping it in good order so that you can efficiently return to it after you have responded to your mother's problem. Carefully setting aside your homework equates to pushing the present microcontroller context (condition codes, index register values, accumulator values, and program counter values) onto the stack. You get out a piece of paper and write down the information about the problem as your mother relates it to you. Her car has a dead battery and is stalled at Fifth and Main. You write down the address, turn on the phone recorder, and go to Fifth and Main. Getting the address, tuning on the phone recorder, and going to the address equate to three microcontroller actions: First, the microcontroller sets flags which prevent further interrupts from occurring until it returns from the interrupt (turning on the phone recorder). Second, the microcontroller fetches the address of an interrupt software routine (writing down the address). Third, the microcontroller vectors to the interrupt service routine (going to Fifth and Main). When you arrive at Fifth and Main, you take care of your mother's problem. This equates to the microcontroller executing the interrupt service routine. After solving Mother's problem, you return home to your desk and check the phone recorder for any other pending interrupt requests. If there are no requests, you move your homework materials back to their original positions on your desk and continue working where you left off. Your return to the homework equates with several microcontroller actions. After executing the interrupt service routine, the microcontroller recovers its original context from the stack (moving the homework materials back in front of you). If there are no pending interrupts (checking the phone recorder), the microcontroller continues routine program execution from where it left off.

This analogy permits one more comparison between you and the microcontroller. Perhaps you are working against a critical deadline, and you do not want any interruptions. In that case, you turn down the sound on the phone recorder and turn off the ringer so you will not be interrupted. Likewise, the microcontroller can set certain condition code bits and control register flags to keep interrupts from occurring while it executes critical business.

This analogy may give the reader a sense of the logic and flow of microcontroller interrupts. Following sections consider interrupt types, vectors and pseudovectors, interrupt masks, interrupt-associated instructions, and four kinds of interrupts. M68HC11 interrupt mechanisms are logical and sensible. By carefully studying the following sections, one can gain a thorough and practical understanding of M68HC11 interrupts.

Types of Interrupts

The M68HC11 uses twelve types of interrupt. Seven of these interrupt types have unique special purposes. The remaining five are general-purpose types. This section briefly introduces the reader to all twelve interrupt types.

1. The M68HC11's Serial Communications Interface (SCI) uses SCI Serial System Interrupts. The SCI provides asynchronous, two-way serial communication between the M68HC11 and other microcontrollers (or terminal or host computer). The M68HC11 uses SCI Serial System Interrupts for a variety of control or error detection purposes. Control register bits allow the user to select which purpose(s) SCI interrupts serve.

2. The M68HC11's Serial Peripheral Interface (SPI) facilitates synchronous, two-way data exchanges between microcontrollers and external devices. The SPI uses SPI Serial Transfer Complete Interrupts to control data transmission over SPI lines.
3. The M68HC11's Pulse Accumulator counts the number of times an external device applies specified rising and/or falling edges to an input pin. A Pulse Accumulator Input Edge Interrupt occurs each time the external device applies a specified edge to the pulse accumulator's input pin.
4. The Pulse Accumulator maintains its input edge count in a special 8-bit register. Pulse Accumulator Overflow Interrupts can interrupt the CPU each time the count register's 8-bit contents roll over from $FF to $00.
5. The M68HC11 incorporates a flexible and sophisticated timer section. This timer section bases a wide variety of functions on a 16-bit, free-running counter. The Timer Overflow Interrupt can interrupt the CPU each time the free-running count rolls over from $FFFF to $0000.
6. The M68HC11's timer section includes an Output Compare feature. Output compare mechanisms compare 16-bit values, established by the user, with the free-running count each time this count increments. When a match occurs, the output compare mechanism drives an assigned output pin to a user-specified level (high, low, or toggle). The user can enable up to five Output Compare Interrupts, one for each output compare mechanism. When enabled, an output compare mechanism interrupts the CPU upon a successful compare.
7. The M68HC11's timer section also includes Input Capture mechanisms. Each input capture mechanism monitors an assigned M68HC11 input pin. When a peripheral drives this pin with a specified edge (rising, falling, or toggle), the input capture mechanism records the free-running count. The EVB's M68HC11A1 incorporates three input capture mechanisms. The EVBU's M68HC11E9 includes three or four input capture mechanisms.[9] Each input capture mechanism includes an Input Capture Interrupt. When enabled, an input capture mechanism interrupts the CPU when it detects a specified edge at its assigned pin.

Interrupt types 1 through 7 (above) have unique special purposes, as discussed. The remaining five interrupt types have general purposes, i.e., they can be used for a variety of applications, depending upon the system's designer's needs and creativity.

8. The M68HC11 generates an interrupt when the CPU fetches an illegal opcode. The CPU's first step in the execution of any instruction is to fetch and decode that instruction's opcode. If the CPU decodes the opcode, and finds that it is not included in the M68HC11's instruction set, the M68HC11 generates an Illegal Opcode Trap Interrupt. All interrupts cause the CPU to vector to service routines. Each interrupt type has a unique interrupt vector destination. The user can load a custom routine at the illegal opcode destination. The user can then include illegal opcodes in a program. Each time the CPU encounters one of these illegal opcodes, it vectors to the custom routine. Thus, the user has developed what amounts to a customized "instruction."
9. The Real Time Interrupt (RTI) mechanism generates recurring interrupts at a periodic rate selected by the user. When interrupted, the CPU vectors to a user-developed, RTI interrupt service routine. A user can select from four periodic RTI rates, each derived from the E clock frequency.
10. The SWI (Software Interrupt) instruction causes the CPU to vector to a user-supplied custom routine. SWIs provide a useful alternative to a subroutine call (JSR, BSR) where it is important to save the entire processor context on the stack. Recall that an interrupt causes the M68HC11 to stack its condition codes, index register values, accumulator values, and program count. A subroutine call stacks only the program count. Thus, the programmer can use the SWI instruction where the main program and service routine use different processor context values.
11. M68HC11 pin 19 is labelled "$\overline{\text{IRQ}}$" (Interrupt Request). By driving this pin low, an external device can generate an IRQ Interrupt. The CPU responds to the interrupt by stacking its context and vectoring to a user-supplied IRQ interrupt service routine. Parallel I/O handshake signals can also generate an IRQ interrupt. Pages 106–109 describe pin-generated IRQ interrupts in more detail. Chapter 4 describes IRQ interrupts generated by parallel I/O handshakes.

---

[9] A user can assign the fourth pin to either an output compare or input capture mechanism.

12. M68HC11 pin 18 is labelled "$\overline{\text{XIRQ}}$" (Pseudo Nonmaskable Interrupt Request). By driving this pin low, an external device can cause the CPU to stack its context and vector to a custom XIRQ service routine. Pages 110–112 cover $\overline{\text{XIRQ}}$ interrupts in detail.

This section provides a brief description of the M68HC11's twelve interrupt types. M68HC11 internal devices generate seven of these interrupt types. Each of these seven interrupts has a special purpose: to make the CPU attend to the internal device's immediate needs. The five remaining interrupts serve more general purposes. The M68HC11 generates three of these general-purpose types internally: Program instructions generate SWIs and illegal opcode interrupts; the M68HC11's timer section generates periodic RTIs. The two remaining general-purpose interrupts service the needs of external devices. An external device requests service by driving an interrupt-associated M68HC11 pin ($\overline{\text{IRQ}}$ or $\overline{\text{XIRQ}}$) low. The M68HC11 can enable or inhibit ten out of these twelve interrupts. All ten can be enabled or inhibited by condition code register bits (global control). Nine of the ten can also be inhibited or enabled by local control register bits. The next section describes global and local interrupt masks.

Global and Local Interrupt Masks

A user cannot disable illegal opcode traps.[10] The user generates software interrupts by writing an SWI instruction into program routines. Of the remaining ten interrupt types, all are maskable at the global (CCR) level, and nine are also masked locally by bits in control registers.

The condition code register's X bit masks XIRQs globally. Any M68HC11 reset causes the X bit to set, and X = 1 inhibits XIRQ interrupts. Software can enable XIRQ interrupts by changing the X bit from one to zero. This software can change the X bit from one to zero, but not from zero to one. Once XIRQs have been enabled, the user cannot disable them, except by resetting the microcontroller. Motorola calls XIRQs "Pseudo-Nonmaskable Interrupts," because once unmasked, software cannot re-mask them. Refer to Figure 2–8, which shows the M68HC11's condition code register bits. The X mask is CCR bit 6.

A programmer uses the TAP instruction (Transfer from ACCA to CCR) to reset the X bit. For example:

```
CLRA
TAP
```

The TAP instruction copies ACCA's contents ($00) to the condition code register, clearing the X bit.

The M68HC11 masks the following interrupts at both the global (CCR) and local (control register) levels:

> SCI Serial System Interrupts
> SPI Serial Transfer Complete Interrupts
> Pulse Accumulator Input Edge Interrupts
> Pulse Accumulator Overflow Interrupts
> Timer Overflow Interrupts
> Timer Output Compare Interrupts
> Timer Input Capture Interrupts
> Real-Time Interrupts (RTI)
> Interrupt Requests (IRQ).

FIGURE 2–8
Reprinted with the permission of Motorola.

Condition Code Register (CCR)

B7	B6	B5	B4	B3	B2	B1	B0
S	X	H	I	N	Z	V	C

---

[10]See the discussion on illegal opcode traps on pages 112–116.

The condition code register's I bit (bit 4, Figure 2–8) enables and disables these interrupts at the global level. Clearing the I mask enables these interrupts, and setting the I mask disables them.

A user can program the CPU to set or clear the I mask at any time by using two instructions:

    SEI = Set I mask
    CLI = Clear I mask

IRQ interrupts constitute a special case. IRQ interrupts can be generated in two ways: (1) by a low level at the $\overline{\text{IRQ}}$ pin, or (2) by a parallel I/O handshake signal. Only the I bit can mask $\overline{\text{IRQ}}$ pin-generated interrupts. IRQs generated by the parallel I/O mechanism are masked locally and by the global I mask. Thus, $\overline{\text{IRQ}}$ pin-generated interrupts are masked globally (I mask), and parallel I/O-generated IRQs are masked both globally and locally.

Pages 106–109 discuss $\overline{\text{IRQ}}$ pin-generated interrupt requests. Chapters 4 through 10 discuss interrupts that must be enabled at both the global and local levels. When interrupted, the M68HC11 vectors to a special interrupt service routine. The next section discusses interrupt vectors and pseudovectors.

Interrupt Vectors and BUFFALO ROM Pseudovectors

The M68HC11 responds to an unmasked interrupt request by vectoring to an appropriate interrupt service routine, then executing it. Each of the twelve interrupt types has an assigned service routine starting address. The M68HC11 vectors to the appropriate service routine by copying this routine's starting address to its program counter. The M68HC11 then begins fetching and executing at the service routine's starting address. The M68HC11 copies service routine starting addresses to its PC from storage locations $FFD6–$FFF9 in ROM. Table 2–1 lists these ROM locations, and the interrupt service routine starting addresses which they contain.

TABLE 2–1

Interrupt Type	Interrupt Service Routine Starting Address	ROM Locations Holding Interrupt Service Routine Starting Address
SCI Serial System	$00C4	$FFD6–$FFD7
SPI Serial Transfer Complete	$00C7	$FFD8–$FFD9
Pulse Accumulator Input Edge	$00CA	$FFDA–$FFDB
Pulse Accumulator Overflow	$00CD	$FFDC–$FFDD
Timer Overflow	$00D0	$FFDE–$FFDF
Input Capture 4/Output Compare 5	$00D3	$FFE0–$FFE1
Output Compare 4	$00D6	$FFE2–$FFE3
Output Compare 3	$00D9	$FFE4–$FFE5
Output Compare 2	$00DC	$FFE6–$FFE7
Output Compare 1	$00DF	$FFE8–$FF89
Input Capture 3	$00E2	$FFEA–$FFEB
Input Capture 2	$00E5	$FFEC–$FFED
Input Capture 1	$00E8	$FFEE –$FFEF
Real Time Interrupt	$00EB	$FFF0–$FFF1
IRQ	$00EE	$FFF2–$FFF3
XIRQ	$00F1	$FFF4–$FFF5
SWI	$00F4	$FFF6–$FFF7
Illegal Opcode Trap	$00F7	$FFF8–$FFF9

Upon an EVB/EVBU POR or push-button reset, the BUFFALO operating system stores default pseudovectors to all interrupt service routine starting addresses listed in Table 2–1. Recall that a pseudovector is a JMP instruction which re-vectors the CPU to the interrupt service routine's actual starting address. Thus, an interrupted CPU stacks its processor context, loads its PC with an interrupt address from ROM storage, and begins fetching and executing at this address. At this interrupt address, the CPU fetches the pseudovector JMP instruction and executes it by loading its PC with the beginning address of the actual interrupt service routine. The CPU pseudovectors to the actual interrupt service routine and fetches and executes from there.

Since all interrupt pseudovectors reside in RAM, the user can easily modify them to suit a particular application. The first thing an interrupted CPU does is stack its processor context. This aspect of interrupt execution is discussed below.

The Interrupt Stack and RTI Instruction

All twelve interrupt types cause the CPU to stack its context, i.e., the contents of its program counter, index registers, accumulators, and CCR. When the CPU returns from interrupt, it pulls these data from the stack and returns them to their original registers. Thus, the CPU can resume fetching and executing from the place where it was interrupted. Assume, for example, that the stack pointer contains $0048 at the time of an interrupt request. The CPU stacks its contents as shown in Figure 2–9. After each context byte is stacked, the stack pointer (SP) decrements. After all context bytes have been stacked, SP's contents equal the CCR's stack location minus one.

Consideration of this automatic stacking process prompts an interesting question: What happens to the stack if a nested interrupt occurs? A nested interrupt occurs while the CPU executes a previous interrupt's service routine. A nested interrupt therefore causes the CPU to stack the previous interrupt service routine's context. This automatic stacking process causes the stack to grow to twice its original size. In the case of Figure 2–9, for example, the SP's contents would decrement to $0035, after stacking of the interrupt routine's context data. A series of nested interrupts causes the stack to grow excessively, perhaps writing processor context bytes over program or data in RAM. For this reason, good programming practice discourages nested interrupts. The M68HC11 includes two features that help prevent unintentional nested interrupts. In the case of all interrupts, the M68HC11 automatically sets the I mask as soon as the CPU has finished stacking processor context. To permit nested I-maskable interrupts, the CPU must execute a CLI instruction as part of the interrupt service routine itself. However, during I-maskable interrupts, the M68HC11 does not automatically set the X mask. Therefore, a peripheral could generate an XIRQ interrupt while the CPU executes an I-maskable interrupt service routine. In the case of XIRQ interrupts, the M68HC11 automatically sets both the I and X masks as soon as the CPU has finished stacking processor context.

A programmer should make the interrupt service routine's final instruction an RTI (Return from Interrupt). An RTI instruction causes the CPU to pull its original context from the stack, and restore all registers to their original condition—at the time the interrupt was

FIGURE 2–9

SP after CPU context stacked →	$003F	----
	$0040	CCR
	$0041	ACCB
	$0042	ACCA
	$0043	X Register – MS
	$0044	X Register – LS
	$0045	Y Register – MS
	$0046	Y Register – LS
	$0047	PC – MS
SP at time of interrupt →	$0048	PC – LS

generated. Refer again to Figure 2–9. When pulling data from the stack, the M68HC11 increments the SP, then pulls a byte from the stack. To execute an RTI instruction, the M68HC11 would increment the SP to $0040F, then pull the CCR byte. The M68HC11 would then increment the SP to $0041 and pull the ACCB byte. The M68HC11 continues this process until it has pulled the PC's LS byte. At this point, the CCR, accumulators, index registers and PC have all returned to their original condition, and the SP = $0048. The CPU then fetches and executes from where it left off at the time of the interrupt.

I-Maskable Interrupt Priorities

The M68HC11 prioritizes I-maskable interrupts. If two or more I-maskable interrupts are requested simultaneously, the M68HC11 services the interrupt with the highest priority first, then services the others in priority order. Of course, illegal opcode, SWI, and XIRQ (X = 0) interrupts have supreme priority, since they are nonmaskable. Upon a POR or pushbutton reset, the M68HC11 establishes default I-maskable interrupt priorities as shown in the second column of Table 2–2.

The user can program the CPU to promote any Table 2–2 interrupt to highest I-maskable priority. The promotion routine must set the I mask, write promotion bits to the HPRIO (Highest Priority Interrupt and Misc.) register, then reset the I mask to permit interrupts. Such a routine promotes one interrupt to highest I-maskable priority, but leaves the others in default priority order.

Refer to Figure 2–10, which illustrates the HPRIO register (control register block address $103C). To promote a particular interrupt to top priority, the CPU writes a unique bit combination to HPRIO Priority Select bits PSEL3–PSEL0. Table 2–2 shows these unique bit combinations. For example, to promote Timer Overflow Interrupts to top priority, the CPU clears all PSEL bits:

```
SEI          Set I mask
CLRA         $00 --> ACCA
STAA 103C    Write $00 to HPRIO
CLI          Clear I mask
```

To promote Output Compare 2 Interrupts, the CPU writes $0C to HPRIO:

```
SEI          Set I mask
LDAA #C      $0C --> ACCA
STAA 103C    Write $0C to HPRIO
CLI          Clear I flag
```

TABLE 2–2

Interrupt Type	Default Priority	HPRIO Promote to #1 priority			
		PSEL3	PSEL2	PSEL1	PSEL0
IRQ	1	0	1	1(0)	0(1)
RTI	2	0	1	1	1
Input Capture 1	3	1	0	0	0
Input Capture 2	4	1	0	0	1
Input Capture 3	5	1	0	1	0
Output Compare 1	6	1	0	1	1
Output Compare 2	7	1	1	0	0
Output Compare 3	8	1	1	0	1
Output Compare 4	9	1	1	1	0
Input Capture 4/Output Compare 5	10	1	1	1	1
Timer Overflow	11	0	0	0	0
Pulse Accumulator Overflow	12	0	0	0	1
Pulse Accumulator Input Edge	13	0	0	1	0
SPI Serial Transfer Complete	14	0	0	1	1
SCI Serial System	15	0	1	0	0

FIGURE 2-10

HPRIO $103C

B7				B3	B2	B1	B0
				PSEL3	PSEL2	PSEL1	PSEL0

In each of these short routines, the first instruction sets the I mask. The PSEL bits are not writable unless I = 1, inhibiting I-maskable interrupts. Clearing the I mask after promoting an interrupt is not mandatory. Of course, all I-maskable interrupts are inhibited until the CPU clears the I mask. Remember that promoting an I-maskable interrupt does not give it priority over nonmaskable interrupts—illegal opcode trap, SWI, and XIRQ (X = 0).

Programmers often use WAI and STOP instructions in conjunction with interrupts. The following section discusses these instructions and how they relate to M68HC11 interrupts.

M68HC11 WAI and STOP Instructions

As a general rule, application designers use interrupts for two reasons: First, interrupts can notify the CPU of a system malfunction. For example, designers use XIRQs to notify the CPU of potentially catastrophic situations, such as a bus contention (two output drivers attempting to drive a single bus line simultaneously). Second, external or internal devices use interrupts to request service from the CPU. Interrupts allow the CPU to execute routine business without having to check periodically for system problems or whether a device needs service. Thus, interrupts enhance operational efficiency.

However, in some applications the microcontroller has no routine business to execute. The CPU has nothing better to do than wait for a device to request service. In such a circumstance, the programmer has two choices: a program polling loop or waiting for an interrupt. In a polling loop, the microcontroller does not use interrupts. Rather, the CPU checks and rechecks external or internal devices to see if they need service. In a polling loop, the CPU constantly fetches and executes instructions that cause it to check the devices, then branch back and check them again. A programmer who finds a polling loop undesirable can select from two instructions that cause the CPU to stop fetching and executing, and simply wait for an interrupt. These instructions are WAI (wait for interrupt) and STOP (stop processing). Both instructions cause the microcontroller to reduce power consumption while it waits. Thus, designers of battery-powered applications find WAI and STOP particularly useful. Each instruction has unique features and advantages. Following paragraphs describe the WAI and STOP instructions.

Refer to Figure 2–11, which flowcharts M68HC11's execution of a WAI instruction. "WAI" means "wait for interrupt." Interrupts cause the CPU to stack its processor context,

FIGURE 2-11

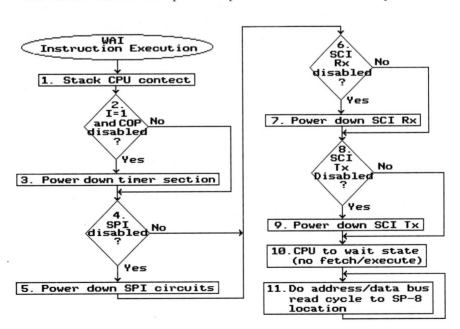

FIGURE 2–12

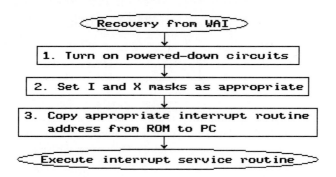

then vector to an interrupt service routine. The WAI instruction anticipates an interrupt. WAI execution causes the CPU to pre-stack its context (Figure 2–11, block 1). Then, when an interrupt occurs, the CPU can skip context stacking. Thus, the CPU vectors to the interrupt service sooner and more efficiently.

In block 2, the M68HC11 checks its I mask and COP system. If I = 1 *and* NOCOP = 1, the microcontroller turns off its timer section (block 3) to reduce WAI state power consumption. In Blocks 4, 6, and 8, the M68HC11 checks three internal devices to see whether they are enabled or disabled. If an SPI or SCI circuit is disabled, it follows that the CPU must be waiting for an interrupt unassociated with it. Therefore, if any of three SPI/SCI circuits are disabled, the M68HC11 powers them down to further reduce power consumption (blocks 5, 7, and 9).

In block 10, the CPU enters a wait state. Now refer to block 11; while the CPU waits, it executes continuous read bus cycles to the CCR location on the stack. The CPU continuously drives its internal address bus with the CCR's stack location. Memory applies the stacked CCR contents to the M68HC11's internal data bus, but the CPU does not latch it.

Figure 2–12 flowcharts tasks executed by the CPU when an unmasked interrupt causes recovery from the M68HC11's reduced-power WAI state. In block 1, the CPU powers up the circuits that it powered down upon entry into the WAI state. In block 2, the CPU sets appropriate CCR interrupt mask(s). When *any* interrupt drives it from WAI, the CPU sets the I mask. If an XIRQ causes the CPU to recover from WAI, it sets both the I and X masks. In block 3, the CPU vectors to the appropriate interrupt service routine, by copying that routine's starting address from ROM storage to its PC. The CPU then begins fetching and executing the interrupt service routine.

Motorola designed the WAI instruction to save time and reduce power when the CPU has nothing better to do than wait for an interrupt. WAI saves recovery time by pre-stacking the processor's context. WAI saves power by shutting down unused timer, SPI, and SCI circuits.

The STOP instruction can act as an NOP,[11] or it can literally stop the microcontroller in its tracks, causing a drastic reduction in power consumption. If STOP acts as an NOP, it merely consumes two machine cycles and processing proceeds to the next program instruction. If properly enabled, STOP causes processing to cease by halting all internal M68HC11 clocks. When processing STOPs in this way, power consumption decreases to .5mW–1.5mW, as compared with normal fetch and execute levels as high as 150mW.

The condition code register's S bit determines whether a STOP instruction acts as an NOP or actually halts internal clocks. Refer to Figure 2–8. The S bit is the CCR's most significant bit. If S = 1, the STOP instruction operates as an NOP. If S = 0, STOP halts M68HC11 processing. A programmer can use the TAP instruction to set or reset the S bit at any point in program execution. The M68HC11 recovers from a STOP when an external circuit drives the $\overline{\text{RESET}}$ or $\overline{\text{XIRQ}}$ pin low. The M68HC11 also recovers from STOP if the I mask is reset, and an external device drives the $\overline{\text{IRQ}}$ pin low.

Figure 2–13 summarizes STOP execution. The M68HC11 evaluates the CCR's S bit. If S = 1, the CPU resumes fetch and execute with the next program instruction. If S = 0, the M68HC11 stops all internal clocks, reducing chip power consumption to a minimum level. If an external circuit drives the M68HC11's $\overline{\text{RESET}}$ pin low, the microcontroller resumes operation by establishing processor starting conditions, and vectoring to the POR/external circuit reset routine (see pages 89–91).

---

[11] NOP stands for "no operation" or "no op."

FIGURE 2–13

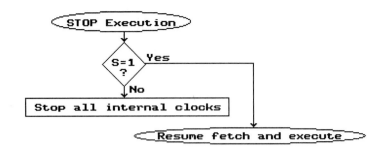

The M68HC11 can also recover from a STOP if either (1) an external device drives the $\overline{\text{XIRQ}}$ pin low; or (2) I = 0, and an external device drives the $\overline{\text{IRQ}}$ pin low. Figure 2–14 flowcharts M68HC11 response under these two conditions.

Refer to Figure 2–14, block 1, and Figure 2–2. OPTION register bit 4 is labelled DLY (Enable Oscillator Start-Up Delay—on exit from STOP). Any M68HC11 reset establishes the DLY bit's default condition as zero. However, the M68HC11 might have to execute critical timing sequences immediately upon its recovery from STOP. In such a case, the M68HC11's POR/external circuit reset service routine should set the DLY bit.[12] As its first act (block 1, Figure 2–14), the CPU checks the DLY bit. If DLY = 1, the CPU executes block 2 by delaying for 4064 E clock cycles, to give its on-board Pierce oscillator a chance to stabilize. All internal clocks derive their frequencies from this Pierce oscillator. If oscillator frequency stabilizes, then critical timing sequences will be exact. If DLY = 0, the CPU takes the "yes" branch to block 3, without executing block 2.

In Figure 2–14, block 3, the CPU determines whether a low level at the $\overline{\text{XIRQ}}$ pin caused the recovery from STOP. If $\overline{\text{XIRQ}}$ is high, an external device has caused the exit from STOP by driving the IRQ pin low (where I = 0), generating an $\overline{\text{IRQ}}$ interrupt. If an unmasked, pin-generated IRQ interrupt caused STOP recovery, the CPU takes the "no" path from block 3 and processes this interrupt. If the $\overline{\text{XIRQ}}$ pin is low, the CPU takes the "yes" branch to block 4, where it checks the state of the X bit in the CCR. If X = 0, the CPU takes the "no" path and processes the XIRQ interrupt. If X = 1, the CPU takes the "yes" branch from block 4, and re-commences fetching and executing at the next program instruction.

Thus, a STOP instruction can cause two possible courses of action. If S = 1, a STOP instruction causes the processor to continue fetching and executing at the next program instruction. In this case, STOP functions as an NOP instruction. If S = 0, STOP causes the M68HC11 to halt all internal clocks, and assume a low-power state. The M68HC11 can recover from this state if an external device drives its $\overline{\text{XIRQ}}$ or $\overline{\text{RESET}}$ pin low. If an external

FIGURE 2–14

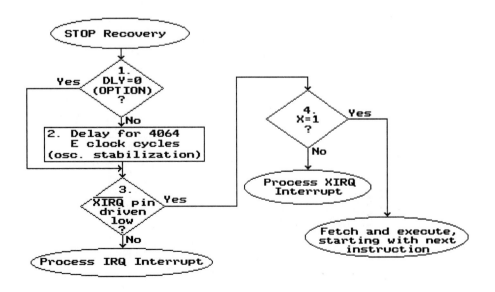

---

[12] The reset service routine must set the DLY bit because it is time protected. Recall that the CPU must write to a time-protected bit within sixty-four bus cycles after a reset.

105

device drives its $\overline{\text{XIRQ}}$ pin low, the M68HC11 will execute an XIRQ service if X = 0. If X = 1, the M68HC11 resumes fetching and executing at the next instruction. The M68HC11 also recovers from STOP if I = 0, and an external device drives its $\overline{\text{IRQ}}$ pin low. In this case, the M68HC11 executes an IRQ interrupt service.

A programmer can use a WAI or STOP instruction when the M68HC11 has nothing better to do than wait for a peripheral to generate an interrupt. Each of these instructions has unique advantages. The WAI instruction anticipates an interrupt by pre-stacking processor context. This saves time when a device generates the anticipated interrupt. Any unmasked internal or external interrupt causes recovery from the WAI state. Likewise, any reset source can cause a recovery from WAI. Before entering the WAI state, the CPU powers down disabled SPI, SCI, and timer circuits to reduce power consumption. Power down reduces WAI state power consumption to 15mW—150mW, depending upon which circuits actually power down.

A STOP instruction can halt the M68HC11 in its tracks—or function as an NOP, depending upon the state of the CCR's S bit. If a STOP instruction halts the processor, it reduces power consumption to the .5mW–1.5mW range.[13] An external circuit can cause recovery from STOP by driving either the $\overline{\text{RESET}}$ or $\overline{\text{XIRQ}}$ pin low. If its $\overline{\text{RESET}}$ pin is driven low, the M68HC11 executes a normal POR/external circuit reset. If its $\overline{\text{XIRQ}}$ pin is driven low, the M68HC11 executes an XIRQ interrupt service (if its CCR X bit = 0). If X = 1, the CPU begins fetching and executing at the next program location. An external circuit can also cause the M68HC11 to exit STOP by generating an unmasked (I = 0) IRQ interrupt. In this case, the M68HC11 responds by servicing the interrupt.

The following section analyzes $\overline{\text{IRQ}}$ pin-generated interrupts, and presents an IRQ test routine for use in Exercise 2.2.

$\overline{\text{IRQ}}$ Pin-Generated Interrupts

Recall that parallel I/O handshakes, or high-to-low $\overline{\text{IRQ}}$ pin transitions, can generate an IRQ. This section deals with $\overline{\text{IRQ}}$ pin-generated interrupts. First, it examines $\overline{\text{IRQ}}$ pin input circuitry. Second, this section presents an IRQ test configuration and software, which the reader shall execute in Exercise 2.2.

Figure 2–15 diagrams $\overline{\text{IRQ}}$ pin input circuitry.[14] As with many M68HC11 pins, this $\overline{\text{IRQ}}$ input serves two purposes. During factory testing, Motorola uses this pin as a power input (V_{PPBULK} = +20VDC) to facilitate parallel bulk EEPROM programming. N-channel input transistors (A) apply this +20V power to the M68HC11's EEPROM. Of course, a system designer uses the $\overline{\text{IRQ}}$ pin to generate interrupt requests. When an external device is not requesting an IRQ interrupt, it applies a high to the $\overline{\text{IRQ}}$ pin. This high turns on the N-channel transistor in IRQ Input Buffer B. This high also turns off input buffer B's P-channel transistor. Therefore, the buffer outputs a logic low to IRQ interrupt control logic. To generate an IRQ interrupt, the external peripheral drives the $\overline{\text{IRQ}}$ pin low. This falling edge turns on input buffer B's P-channel transistor, and turns off the N-channel device. The buffer outputs a high to the IRQ logic.

FIGURE 2–15
Reprinted with the permission of Motorola.

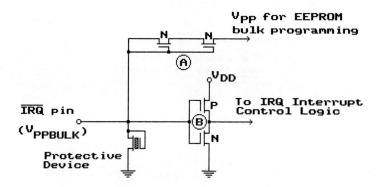

---

[13]The EVBU's M68HC11E9 consumes .5mW when STOP-ed, because it operates in single-chip mode. The EVB's M68HC11A1 consumes 1.5mW, because it operates in expanded multiplexed mode.
[14]Motorola Inc., *M68HC11 Reference Manual*, p. 2-30. Adapted with permission.

FIGURE 2–16

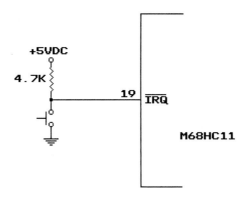

A programmer has the option of making IRQ interrupts level-sensitive or edge-sensitive. To make IRQs edge-sensitive, the CPU sets the IRQE (IRQ Edge) bit in the OPTION register. Refer to Figure 2–2, which shows the IRQE bit (bit 5). To make IRQs level sensitive, the IRQE bit must be clear. All M68HC11 resets clear the IRQE bit. Thus, by default, a low level at the $\overline{\text{IRQ}}$ pin generates an IRQ interrupt. The M68HC11 time-protects the IRQE bit. Therefore, to make IRQs edge-sensitive, the POR/external circuit reset routine must set the IRQE bit within sixty-four bus cycles after reset. If IRQE = 1, making IRQs edge-sensitive, IRQ interrupt control logic latches falling edges at the $\overline{\text{IRQ}}$ pin. IRQ control logic remains in this latched state (IRQ pending) until the M68HC11 actually services the interrupt. If IRQE = 0, making IRQs level-sensitive, an external peripheral must hold the $\overline{\text{IRQ}}$ pin low until the M68HC11 responds. Otherwise, the M68HC11 will never recognize the interrupt request.

Following paragraphs describe an IRQ test configuration and software, which the reader shall use in Exercise 2.2. This configuration allows the user to request IRQ interrupts using a push-button switch. This switch connects to the $\overline{\text{IRQ}}$ pin and a 4.7K Ohm pull-up resistor. Figure 2–16 diagrams this switch configuration. By pressing the push-button, a user grounds the $\overline{\text{IRQ}}$ pin. With the switch closed, current flows between ground and the +5VDC supply. The 4.7K Ohm pull-up resistor drops 5V. With the push-button switch open, the voltage supply applies 5V to the $\overline{\text{IRQ}}$ pin through the pull-up resistor.

Supporting software uses IRQ interrupts to record either (1) the number of times that the user pressed the push-button for less than a second, or (2) the number of seconds that the user pressed the push-button continuously. If the user presses the push-button for more than a second, the software records key-press duration, in seconds. If the user presses the push-button for less than a second, the software records this event as one key-press. Software maintains the key-press count in RAM address $0000. To read the key-press count, reset the EVBU or EVB, then use a BUFFALO MD (Memory Display) command to display the contents of location $0000. Supporting software consists of three parts: (1) MAIN program, (2) IRQ Interrupt Service routine, and (3) DELAY subroutine. Figure 2–17 and its accompanying program listing illustrate the MAIN program and IRQ Interrupt Service routine. Refer to block 1 of this figure, and program addresses $0100–$0102. In this block, the CPU establishes the bottom of the stack at RAM address $01FF, well away from the supporting software. Under certain operating conditions, this stack will grow to twelve bytes.[15]

Refer to Figure 2–17, block 2, and program addresses $0103–$0107. In block 2, the CPU initializes the IRQ pseudovector to `JMP 120`. Recall from Table 2–1 that the IRQ pseudovector resides at RAM locations $00EE–$00F0. The POR/external circuit reset routine has already loaded RAM address $00EE with a JMP opcode. Therefore, in block 2, the CPU simply stores the IRQ Interrupt Service routine's address ($0120) to locations $00EF and $00F0.

In block 3, program addresses $0108–$010A, the CPU initializes the key-press count (address $0000) to $00. The CPU then clears the I mask to enable IRQ interrupts (block 4, program address $010B).

---

[15]The stack grows to twelve bytes when the M68HC11 CPU calls the DELAY subroutine from the IRQ Interrupt Service routine.

FIGURE 2-17

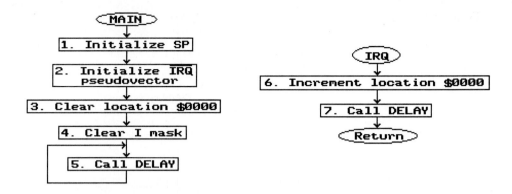

IRQ Demonstration, MAIN Program, and IRQ Interrupt Service Routine

$ADDRESS	MAIN SOURCE CODE	COMMENTS
0100	LDS #1FF	Initialize SP
0103	LDD #120	Initialize IRQ pseudovector
0106	STD EF	" " "
0108	CLR 0	Clear keypress count
010B	CLI	Clear I mask
010C	JSR 130	Call DELAY subroutine
010F	BRA 10C	Call DELAY again

$ADDRESS	IRQ INTERRUPT SERVICE ROUTINE SOURCE CODE	COMMENTS
0120	INC 0	Increment keypress count
0123	JSR 130	Call DELAY subroutine
0126	RTI	Return

After clearing the I mask, the CPU enters a continuous loop where it calls a one-second DELAY subroutine, then branches back to call a DELAY again. Refer to Figure 2–17, block 5, and program addresses $010C–$0110. These instructions constitute "routine program execution," and keep the CPU busy fetching and executing while it waits for an IRQ interrupt. In Exercise 2.2, the reader will connect an ammeter in series with the EVBU/EVB's V_{DD} connection. As the microcontroller executes this loop, the reader will record current draw and calculate board power consumption. The reader shall compare this "fetch and execute" power consumption with M68HC11 power use in the WAI and STOP modes.

Figure 2–17, blocks 6 and 7 (program addresses $0120–$0126) constitute the IRQ Interrupt Service routine. Upon an IRQ interrupt, the M68HC11 automatically sets its I mask before vectoring to this service routine. Therefore, the microcontroller will not recognize another interrupt request until it returns (RTI) from the IRQ interrupt. After vectoring and pseudovectoring to this routine, the CPU executes block 6, in which it simply increments the key-press count at address $0000. In block 7, the CPU calls the DELAY subroutine. After one second elapses, the CPU returns from the DELAY subroutine and executes the RTI instruction (program address $0126). This RTI returns the CPU to where it left off in the MAIN program or the DELAY subroutine called from the MAIN program. Recall that an RTI instruction restores the CCR's contents with its I mask cleared. If the user has never released the push-button, another IRQ interrupt occurs immediately. As long as the user continues to press the push-button, the M68HC11 executes and re-executes IRQ interrupts every second, incrementing the key-press count each time. When the user finally releases the push-button, the CPU returns from interrupt, to continue MAIN program or

FIGURE 2–18

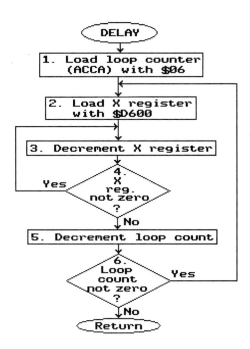

IRQ Demonstration, DELAY Subroutine

$ADDRESS	DELAY SUBROUTINE SOURCE CODE	COMMENTS
0130	LDAA #6	Initialize loop counter (ACCA)
0132	LDX #D600	Load X register with countdown value
0135	DEX	Decrement X register
0136	BNE 135	If X register not zero, decrement again
0138	DECA	If X register zero, decrement loop count
0139	BNE 132	If loop count not zero, count down again
013B	RTS	If loop count zero, return

subroutine execution from where it left off originally. If the user presses the push-button for less than one second, the CPU increments the key-press count once and returns from the interrupt after a second, to continue routine MAIN or DELAY execution from where it left off. Thus, the CPU records the number of times the push-button was pressed (if less than one second) or the number of seconds that the button was pressed.

Figure 2–18 and its accompanying program listing illustrate the DELAY subroutine.[16] This subroutine consumes one second by loading the M68HC11's X register with $D600 and decrementing it to $0000, six times.

If the CPU calls DELAY from the MAIN program, program flow returns to the MAIN program. If the CPU calls DELAY from the IRQ Interrupt Service routine, program flow returns to this interrupt routine. If the CPU calls the DELAY from the MAIN program, the I mask is reset during DELAY execution. An IRQ interrupt can occur at any time. If the CPU calls the DELAY from the IRQ service routine, I = 1, preventing nested IRQ interrupts.

Figures 2–16 through 2–18 illustrate an IRQ test configuration and support software, which the reader shall implement in Exercise 2.2. This configuration also facilitates a comparison of M68HC11 power consumption under routine fetch/execute, WAI, and STOP conditions. The section that follows presents an XIRQ test configuration and software for use in Exercise 2.2.

---

[16]This subroutine uses the same strategy as the delay in the COP test program, Figure 2–7.

XIRQ Interrupts

Motorola developed XIRQs to replace NMI interrupts, which it uses in a number of 6800 family processors. "NMI" stands for "nonmaskable interrupt." Indeed, NMIs are never maskable; a peripheral can generate an NMI any time. Motorola replaced NMIs with XIRQs to solve two problems, which are discussed below.

Problem 1. In most applications, the POR/external pin reset service routine initializes the stack pointer. SP initialization locates the stack in a block of RAM addresses well away from critical program or data bytes. One cannot predict the SP's contents prior to this initialization; the SP might be pointing at important data or program information. With NMIs, a peripheral can inadvertently generate an NMI interrupt prior to SP initialization. The processor responds by stacking its context over important data or program bytes. XIRQs prevent this problem because the M68HC11 masks them as a part of its initial reset actions. Thus, XIRQs are inhibited until software clears the X mask.

Problem 2. Because NMIs are nonmaskable, it is easy to inadvertently nest them. If the peripheral's NMI request bounces—three times for example—it generates three nested NMI interrupts. These three nested interrupts make the stack three times its normal size, perhaps writing over vital information. XIRQs prevent this problem because the M68HC11 automatically sets the X mask upon its entry into the XIRQ service, preventing nested interrupts.[17]

By making XIRQ's X bit maskable, Motorola solved these two problems. However, XIRQs still have the benefits of NMIs. Once software clears the X bit, software cannot set it again. Recall that only a reset or an XIRQ interrupt itself causes the X bit to set. In the case of an XIRQ interrupt, RTI execution at the end of the XIRQ routine clears the X mask. Therefore, XIRQs are unmasked again as soon as the CPU returns from interrupt.

M68HC11 pin 18 serves a single purpose—as the XIRQ interrupt input. Figure 2–19 shows that $\overline{XIRQ}$ pin logic consists solely of an inverting input buffer, similar to the $\overline{IRQ}$ pin's.[18] A low level at the $\overline{XIRQ}$ pin turns on the P-channel transistor and turns off the N-channel, causing a high output to XIRQ interrupt control logic. This high generates an XIRQ interrupt, if X = 0.

In Exercise 2.2, the reader shall again use a push-button switch to demonstrate XIRQ interrupts. Connect this switch as for the IRQ demonstration. To change this circuitry from its IRQ to XIRQ configurations, merely re-connect the pull-up output from the $\overline{IRQ}$ pin to the $\overline{XIRQ}$ pin. Figure 2–20 shows how the pull-up circuit connects to the M68HC11's $\overline{XIRQ}$ pin.

XIRQ supporting software uses interrupts to record either the number of times or the number of seconds that the user presses the push-button. XIRQ support software maintains this count in RAM address $0000. However, two alternative MAIN programs allow the reader to evaluate the WAI and STOP instructions. Refer to Figure 2–21 and its accompanying program lists. This figure flowcharts the MAIN (WAI) and MAIN (STOP) programs. Except for interchangeable use of WAI and STOP, these two MAIN programs are identical. The following description discusses and compares both of these programs.

Blocks 1, 2, and 3 of both Figure 2-21 flowcharts are identical, as is implementing source code. Block 1 (program addresses $C000/$C020) establish a stack area safely separated from program locations. In Block 2 (program addresses $C003/$C023), the CPU

FIGURE 2–19
Reprinted with the permission of Motorola.

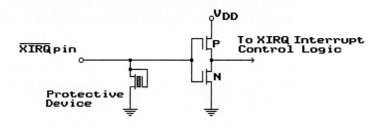

[17]Motorola Inc., *M68HC11 Reference Manual*, p. 5-20.
[18]Motorola Inc., *M68HC11 Reference Manual*, p. 2-24.

FIGURE 2-20

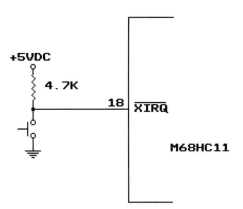

FIGURE 2-21

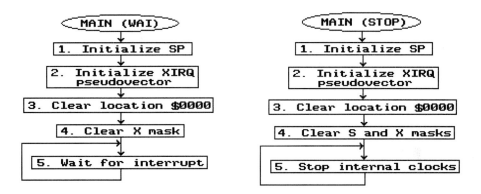

XIRQ Demonstration Software, Alternative MAIN Programs

$ADDRESS	MAIN (WAI) SOURCE CODE	COMMENTS
C000	LDS #C0FF	Initialize stack pointer
C003	LDD #C040	Initialize XIRQ pseudovector
C006	STD F2	" " "
C008	CLRA	Clear keypress count
C009	STAA 0	" " "
C00B	TAP	Clear X mask
C00C	WAI	Wait for interrupt
C00D	BRA C00C	Wait again

MAIN (STOP) $ADDRESS	SOURCE CODE	COMMENTS
C020	LDS #C0FF	Initialize stack pointer
C023	LDD #C040	Initialize XIRQ pseudovector
C026	STD F2	" " "
C028	CLRA	Clear keypress count
C029	STAA 0	" " "
C02B	TAP	Clear S bit and X mask
C02C	STOP	Stop internal clocks
C02D	BRA C02C	Stop clocks again

initializes the M68HC11's BUFFALO XIRQ pseudovector. This pseudovector directs XIRQ interrupt program flow to address $C040. Block 3 (program addresses $C008/$C028) clears accumulator A and store this value ($00) to key-press count location $0000.

Refer to block 4 of both MAIN programs and to the source code that implements them (program addresses $C00B–$C02B). In both cases, the CPU merely transfers the contents of ACCA ($00) to the CCR. In both cases, this transfer clears both the S and X masks. The

FIGURE 2-22

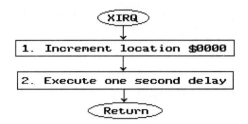

XIRQ Demonstration Software, XIRQ Interrupt Service Routine

$ADDRESS	XIRQ INTERRUPT SERVICE ROUTINE SOURCE CODE	COMMENTS
C040	INC 0	Increment keypress count
C043	LDAA #6	Execute one-second delay
C045	LDX #D600	" " "
C048	DEX	" " "
C049	BNE C048	" " "
C04B	DECA	" " "
C04C	BNE C045	" " "
C04E	RTI	Return

S mask is unrelated to WAI execution. However, the CPU will STOP internal clocks only if S = 0. Of course, the M68HC11 honors XIRQ requests if X = 0.

Refer to blocks 5 of the MAIN programs. These blocks execute the WAI and STOP instructions. At program address $C00C of the MAIN (WAI) program, the CPU executes the WAI instruction. The CPU executes WAI by stacking its context—powering down unused timer, SPI, and SCI circuits—and entering a wait state. When an XIRQ request causes recovery from WAI, the CPU sets the X and I masks, then vectors to the XIRQ service routine. At program address $C02C of the MAIN (STOP) program, the M68HC11 executes a STOP by turning off all of its internal clocks. Recall that the CPU does not stack its context when it executes the STOP instruction. When an XIRQ request causes recovery from STOP, the CPU stacks its context, sets the X and I masks, then vectors to the XIRQ interrupt service.

Refer to Figure 2–22 and its accompanying program list. Since only the XIRQ service uses a one-second delay, it integrates the delay routine rather than calling it as a subroutine. Integrating a one-second delay into the XIRQ routine saves program space. In flowchart block 1 (program addresses $C040–$C042), the CPU increments the key-press count. In block 2 (program addresses $C043–$C04D), the CPU consumes one second by decrementing its X register from $D600 to zero, six times. After executing this one-second delay, the CPU returns from the interrupt, back to the MAIN program from which it originated (Figure 2–21). Upon its return from interrupt, the CPU branches back to the WAI or STOP instruction, to re-enter the WAI state or again STOP its internal clocks. In Exercise 2.2, the reader shall connect an ammeter in series with the EVBU/EVB's V_{DD} connection. This configuration allows the reader to evaluate M68HC11 power consumption during WAI and STOP execution.

Illegal Opcode Interrupts

The M68HC11 stacks its context and vectors to a special routine when it fetches an illegal opcode. What factors make an opcode "illegal"? First, the CPU must fetch what it expects to be an opcode byte. The great majority of opcodes consist of 1 byte. However, a significant number of Y and D register-associated instructions consist of 2 bytes. Therefore, the CPU may be fetching the first byte or the second byte of an instruction opcode, when it finds this byte to be "illegal." The following values constitute a legal prebyte, in a 2-byte opcode:

$18, $1A, $CD

For example, the legal opcode assembled from source code LDAA 0,Y is $18 A6.

An *illegal* opcode (either 1 or 2 bytes) has no associated mnemonic. For example, $42 is an illegal value because it has no relationship to any M68HC11 mnemonic. Thus, these 1- and 2-byte opcodes are all illegal:

$42
$1842
$1A42
$CD42

The reader shall use the value $42 to generate illegal opcode traps in Exercise 2.2. Following paragraphs describe illegal opcode test software, for use in this exercise.

Refer to Figure 2–23 and its accompanying program list. Overall, this MAIN program executes a routine task, writing a zero to the terminal screen at one-second intervals. Note,

FIGURE 2–23

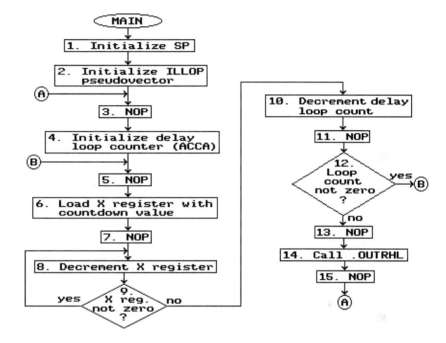

Illegal Opcode Trap Demonstration Software, MAIN Program

$ADDRESS	MAIN PROGRAM SOURCE CODE	COMMENTS
0100	LDS #1FF	Initialize stack pointer
0103	LDD #130	Initialize ILLOP pseudovector
0106	STD F8	" " "
0108	NOP	No op
0109	LDAA #6	Initialize delay loop count
010B	NOP	No op
010C	LDX #D600	Load X register with countdown value
010F	NOP	No op
0110	DEX	Decrement X register
0111	BNE 110	If X register not zero, decrement again
0113	DECA	If X register zero, decrement loop count
0114	NOP	No op
0115	BNE 10B	If loop count not zero, count down again
0117	NOP	No op
0118	JSR FFB5	Call .OUTRHL
011B	NOP	No op
011C	BRA 108	Do another delay

however, that this program includes no operation (NOP) instructions throughout. These single-byte instructions provide convenient places to substitute an illegal opcode. In Exercise 2.2, the reader shall substitute an illegal opcode ($42) for each NOP ($01), and observe the results.

Refer to blocks 1 and 2 of Figure 2–23 (program addresses $0100–$0107). In block 1, the CPU establishes the bottom of the stack at location $01FF, well away from program areas. In block 2, the CPU changes the illegal opcode trap pseudovector to: JMP 130. This pseudovector establishes the illegal opcode interrupt service routine at address $0130.

Blocks 4 through 12 (program addresses $0109–$0116) constitute the one-second delay. This routine uses ACCA as a delay loop counter. The CPU initializes this counter to $06 in Block 4. Each time the CPU decrements its X register from $D600 to zero, it reduces the ACCA value by one. After the sixth X register decrement cycle, the CPU decrements the ACCA count value to zero. When ACCA = $00, the CPU escapes the one-second delay loop, and executes block 14 (program addresses $0118–$011A). In block 14, the CPU calls the BUFFALO .OUTRHL utility subroutine. This subroutine converts ACCA's low nibble ($0) to ASCII, and writes it to the terminal display. After displaying a zero, the CPU branches back to block 3, where it executes another one-second delay.

If it encounters an illegal opcode while executing the MAIN program, the CPU vectors to the ILLOP Interrupt Service routine at address $0130. Refer to Figure 2–24 and its accompanying program list. ILLOP causes the CPU to display this message on the terminal screen:

```
ILLOP @ $XXXX
```

where $XXXX is the program address of the illegal opcode.

How can the CPU determine the illegal opcode's address ($XXXX) for display at the terminal? Recall that any interrupt causes the CPU to stack its context, including the program count. For all interrupts *except illegal opcode traps,* the CPU stacks PC values that point at the instruction to be executed next—upon return from the interrupt. However, when the CPU encounters an illegal opcode, it stacks a PC value pointing directly at the illegal opcode location. Thus, the CPU stacks the illegal opcode's actual address as the PC value. To determine the value of "$XXXX," the CPU accesses the stacked PC contents. Refer to Figure 2–8. Note that the PC-MS location lies eight address locations away from the SP value after context stacking. Thus, the PC-MS byte is located at SP + 8. To access the stacked PC value, the CPU transfers the contents of SP + 1 to an index register.[19] The CPU can then read the stacked PC value to its D register, using indexed mode, with an offset of $07. Having read the PC value to its D register, the CPU can then display it at the terminal, as the illegal opcode's location ($XXXX).

After displaying the $XXXX value on the terminal screen, the CPU must return from the interrupt (RTI) to continue MAIN program execution. To execute an RTI, the CPU pulls its stacked context to ACCA, ACCB, CCR, index registers, and the PC. Recall that the stacked PC value is the illegal opcode's address. Thus, this PC value would return the CPU back to the illegal opcode, and immediately generate another illegal opcode trap. The CPU would be caught in a continuous string of interrupts. To avoid this problem, the CPU must increment the stacked PC value to point at the next MAIN program instruction. Furthermore, the CPU must increment this PC value before it encounters the RTI instruction. To do this, the CPU reads the stacked PC value, using indexed mode with an offset of $07. The CPU then adds one to this $XXXX value so that it points at the next MAIN program instruction. Then, the CPU stores this value back to the stack, using indexed mode with an offset of $07. The CPU can then execute an RTI, which returns it to the next MAIN program instruction.

Refer again to Figure 2–24 and its accompanying program list, which illustrate the ILLOP Interrupt Service routine. Recall that ILLOP causes the M68HC11 to display the error message:

```
ILLOP @ $XXXX.
```

---

[19] A TSX or TSY instruction, which transfers the SP value to an index register, causes the CPU to automatically add one to the SP value before transferring it. Therefore, the CPU transfers SP + 1.

FIGURE 2-24

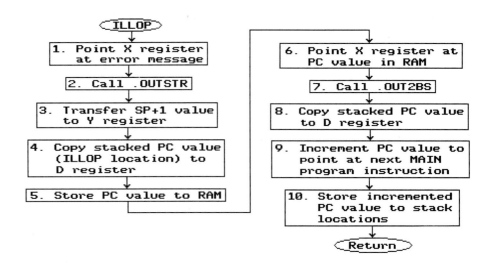

Illegal Opcode Demonstration Software, ILLOP Interrupt Service Routine, and ILLOP Message ASCII Data

	ILLOP INTERRUPT SERVICE ROUTINE	
$ADDRESS	SOURCE CODE	COMMENTS
0130	LDX #0	Point X register at ASCII message data
0133	JSR FFC7	Call .OUTSTR
0136	TSY	Transfer SP+1 to Y register
0138	LDD 7,Y	Copy stacked PC value to D register
013B	STD A	Store PC value to RAM
013D	LDX #A	Point X register at PC value in RAM
0140	JSR FFC1	Call .OUT2BS
0143	LDD 7,Y	Copy stacked PC value to D register
0146	ADDD #1	Increment PC value to point at next MAIN program instruction
0149	STD 7,Y	Store incremented PC value back to stack
014C	RTI	Return

	ILLOP MESSAGE ASCII DATA	
$ADDRESS	$CONTENTS	CHARACTER
0000	49	I
0001	4C	L
0002	4C	L
0003	4F	O
0004	50	P
0005	20	<SP>
0006	40	@
0007	20	<SP>
0008	24	$
0009	04	<EOT>

The ILLOP routine uses the BUFFALO .OUTSTR utility subroutine to display the following parts of the error message:

ILLOP<SP>@<SP>$<EOT>.

This message's ASCII codes reside at RAM addresses $0000–$0009. In block 1 (program addresses $0130–$0132), the CPU points its X register at the first ASCII message address ($0000). The CPU then calls the .OUTSTR subroutine (block 2, program addresses $0133–$0135). This subroutine causes the CPU to write successive ASCII bytes (pointed at

115

by the X register) from RAM to the terminal screen, until it encounters the "end of transmission" (EOT, $04) code. The screen now displays:

```
ILLOP @ $
```

Next, the CPU must access the stacked PC value, which points at the illegal opcode address. In block 3 (program address $0136), the CPU executes the TSY instruction. This instruction causes the CPU to read the SP's contents, add one to this value, then transfer it to the Y register. This operation leaves the SP's contents unchanged, and stores value SP + 1 to the Y register. Note that the stacked PC value resides at Y + $07.

In block 4 (program addresses $0138–$013A), the CPU reads the stacked PC value to its D register. To do this, the CPU uses Y indexed mode with an offset of $07. The CPU must now write this PC value to the terminal screen. To accomplish this, the CPU will use the BUFFALO .OUT2BS utility subroutine. This subroutine assumes that the write value is in memory locations pointed at by the X register. Therefore, the CPU must transfer the PC value, now in the D register, to RAM.

Refer to block 5 (program addresses $013B–$013C). The CPU writes its D register's contents (the PC value) to RAM addresses $000A and $000B. In block 6 (program addresses $013D–$013F), the CPU prepares for .OUT2BS execution by pointing its X register at RAM location $000A.

In block 7 (program addresses $0140–$0142), the CPU calls .OUT2BS. This routine writes the contents ($XXXX) of locations $000A and $000B to the terminal screen. The screen's message now reads:

```
ILLOP @ $XXXX
```

where $XXXX = the illegal opcode's program address.

Before returning from the ILLOP routine, the CPU must increment the stacked PC value. .OUT2BS execution writes over the contents of ACCA. Therefore, the CPU must re-read the PC's contents from the stack. In block 8 (program addresses $0143–$0145), the CPU re-reads the stacked PC value to its D register, using indexed mode with an offset of $07. In block 9 (program addresses $0146–$0148), the CPU adds one to this value. In block 10 (program addresses $0149–$014B), the CPU stores this incremented value to stack locations SP + 8 (PC-MS) and SP + 9 (PC-LS). The CPU can now RTI to the MAIN program instruction following the illegal opcode.

In Exercise 2.2, the reader shall substitute an illegal opcode for each MAIN program NOP and note the effects. The reader will see that the CPU can unerringly display the location of an illegal opcode.

A Software Interrupt (SWI) Demonstration

Software interrupts (SWIs) become an attractive alternative to subroutine calls when the MAIN program and routine to be called use different processor contexts. This section presents an SWI application that uses two sets of index register values. An SWI causes the CPU to stack the MAIN program's X and Y register values automatically. An RTI, at the end of the SWI service routine, restores MAIN program's X and Y register values. An SWI enhances program efficiency, as compared with JSR or BSR instructions. If the MAIN program used a JSR or BSR, it would have to use program instructions to store its X and Y registers to temporary storage locations. MAIN would also have to use instructions to retrieve these X and Y register values from temporary storage. The temporary storage locations, along with the extra load and store instructions, would waste program space.

The SWI interrupt routine itself does not benefit from automatic stacking. Therefore, this interrupt routine must maintain its Y register values in temporary storage.

This section's SWI program demonstrates the use of SWIs in efficiently saving MAIN program register values when the routine being called requires a different context.

Overall, the SWI demonstration program calculates totals for four operand groups—five operands in each group. The MAIN program transfers each operand group in turn to five memory locations labelled "operands to be added." After transferring an operand group, the MAIN program calls an SWI. The SWI routine finds the total of the "operands to be added," and stores this total to an assigned memory location. The MAIN program also manages program operations to assure that the CPU halts after it has calculated and stored totals for all four operand groups.

Figure 2–25 and its accompanying software lists illustrate the SWI demonstration's MAIN program and memory allocations. In block 1 through block 5 of Figure 2–25, the CPU initializes its stack pointer, index register values, and SWI pseudovector. The CPU establishes its stack in block 1, program addresses $0100–$0102. Note that the stack resides well away from program and data areas in RAM. In block 2, program addresses

FIGURE 2–25

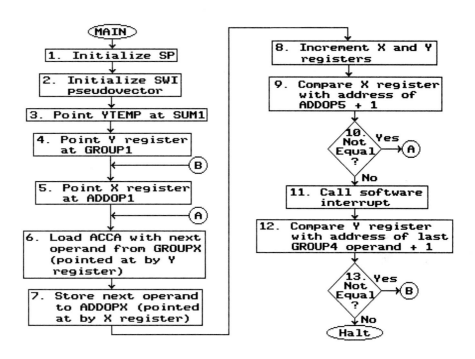

SWI Demonstration

MEMORY ALLOCATIONS	
$ADDRESS	COMMENTS
	SUMS
0000	SUM 1
0001	SUM2
0002	SUM3
0003	SUM4
	OPERANDS TO BE ADDED
0004	ADDOP1
0005	ADDOP2
0006	ADDOP3
0007	ADDOP4
0008	ADDOP5
	OPERAND GROUPS
0009-000D	GROUP1
000E-0012	GROUP2
0013-0017	GROUP3
0018-001C	GROUP4
0020-0021	YTEMP
0100	MAIN Program
130	SWI Interrupt Service Routine
01FF	Initial SP Contents

MAIN PROGRAM		
$ADDRESS	SOURCE CODE	COMMENTS
0100	LDS #1FF	Initialize stack pointer
0103	LDD #130	Initialize SWI pseudovector
0106	STD F5	" " "
0108	LDD #0	Point YTEMP at SUM1
010B	STD 20	" " "
010D	LDY #9	Point Y register at GROUP1
0111	LDX #4	Point X register at ADDOP1
0114	LDAA 0,Y	Read next GROUPx operand
0117	STAA 0,X	Write operand to next ADDOPx
0119	INY	Increment Y register
011B	INX	Increment X register
011C	CPX #9	Not all GROUPx operands transferred?
011F	BNE 114	If yes, transfer another operand
0121	SWI	If no, call software interrupt
0122	CPY #1D	Not all groups totalled?
0126	BNE 111	If yes, transfer another group
0128	WAI	If no, halt

$0103–$0107, the CPU initializes its SWI pseudovector, which resides at RAM addresses $00F4–$00F6. After block 2 initialization, the pseudovector is:

```
JMP 130.
```

Next, the CPU initializes a Y register value for use by the SWI Interrupt Service routine. The SWI routine maintains this Y register value in two RAM locations ($0020–$0021) labelled YTEMP. In block 3 (program addresses $0108–$010C), the CPU writes this initial Y register value ($0000) to the YTEMP locations. In Block 4 (program addresses $010D–$0110), the CPU initializes the MAIN program's Y register by pointing it at the first GROUP1 address (see Memory Allocations list). Finally, the CPU completes its initialization phase by pointing its X register at the first operand to be added: ADDOP1 (block 5, program addresses $0111–$0113).

In Figure 2–25, block 6 through block 10, the CPU transfers five operands (GROUPx) from operand group storage to "operands to be added" locations, ADDOP1–ADDOP5 ($0004–$0008). Each time the CPU executes block 6 (program addresses $0114–$0116), it loads ACCA with an operand from group storage, pointed at by the Y register. The CPU then stores this operand from ACCA to the "operands to be added" location, pointed at by the X register (block 7, program addresses $0117–$0118). In block 8 (program addresses $0119–$011B), the CPU increments both the Y and X registers so that they point at the next addresses in their respective data blocks. In blocks 9 and 10 (program addresses $011C–$0120), the CPU determines whether it has transferred a complete group of five operands from GROUPx to ADDOP1–ADDOP5. After transferring the last GROUPx operand to ADDOP5, the CPU increments its X register to $0009 (block 8). Therefore, in block 9, the CPU compares its X register value with $0009. If the X register has not reached $0009, the CPU takes the "yes" branch from block 10 to block 6, where it completes another data-byte transfer loop. If the X register does contain the value $0009, the CPU escapes the data transfer loop; it has transferred a group of five operands (GROUPx) from the operand group storage area to the "operands to be added" area in RAM.

Refer to Figure 2–25, block 11 (program address $0121). Having transferred five operands into the "operands to be added" RAM block, the CPU calls a software interrupt. The SWI Interrupt Service routine adds the five transferred operands and stores the total to an appropriate SUMx location (i.e., the CPU stores the GROUP1 sum to SUM1, the GROUP2 sum to SUM2, and so on). Upon its return from the SWI routine, the CPU must determine whether it has added and stored totals for all operand groups (GROUP1–GROUP4).

In completing the entire transfer and addition process, the CPU increments its MAIN program Y register value through all of the GROUPx storage location values. After the last group transfer and addition, MAIN program's Y register should contain $001D—the last GROUP4 storage location + 1. Therefore, in block 12 (program addresses $0122–$0125), the CPU compares its Y register's contents with $001D. If the Y register and comparison value are unequal, the CPU branches back from block 13 (program addresses $0126–$0127) to block 5, where it transfers and adds another group of operands. If Y register's contents equal the comparison value, the CPU takes the "no" path from block 13 and halts. Following paragraphs explain how the SWI Interrupt Service routine totals the ADDOPx values and stores this sum to an appropriate SUMx location in RAM.

Figure 2–26 and its accompanying program list illustrate the SWI Interrupt Service routine. Recall that the MAIN program's initialization phase establishes SWI routine's initial Y register value and stores it to YTEMP. This initial value points at the SUM1 location ($0000). In Figure 2–25, block 1 (program addresses $0130–$0132), the CPU reads YTEMP to its Y register. On the first SWI routine execution, YTEMP contains the SUM1 address. The MAIN program and SWI routine use different Y register values. The MAIN program's Y register points at GROUPx operands. The SWI routine's Y register points at SUMx locations. An SWI causes the CPU to automatically stack the MAIN program's Y register contents. An RTI instruction causes the CPU to automatically recover this Y register value from the stack. The SWI routine itself has no automatic stacking to depend upon. Therefore, the SWI routine must store and recover its Y register from YTEMP.

FIGURE 2-26

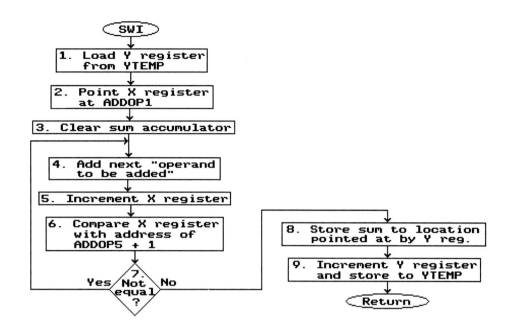

SWI Demonstration, SWI Interrupt Service Routine

$ADDRESS	SWI INTERRUPT SERVICE ROUTINE SOURCE CODE	COMMENTS
0130	LDY 20	Copy YTEMP to Y register
0133	LDX #4	Point X register at ADDOP1
0136	CLRA	Clear sum accumulator
0137	ADDA 0,X	Add next ADDOPx
0139	INX	Increment X register
013A	CPX #9	Not all ADDOPx values added?
013D	BNE 137	If yes add another ADDOPx
013F	STAA 0,Y	If no, store sum to SUMx
0142	INY	Increment Y register
0144	STY 20	Copy Y register to YTEMP
0147	RTI	Return

Both the MAIN program and SWI routine use the X register to point at ADDOPx locations ("operands to be added"). However, by the time the CPU executes the SWI instruction, the MAIN program's X register points at ADDOP5 + 1. The SWI routine must re-point the X register at ADDOP1 so that addition can commence. Therefore, in Figure 2–26, block 2 (program addresses $0133–$0135), the CPU points its X register at the ADDOP1 location.

In block 3, program address $0136, the CPU clears ACCA. The SWI routine accumulates the sum in ACCA. The CPU must therefore clear this register to assure an accurate total. In block 4 (program addresses $0137–$0138), the CPU adds the next ADDOPx value to ACCA, and returns the sum to ACCA. The CPU then increments its X register to point at the next ADDOPx location (block 5, program address $0139). In block 6 (program addresses $013A–$013C), the CPU determines whether it has added all five ADDOPx values to the accumulating ACCA total. If all five ADDOPx values have been added, the block 5 increment points the X register at ADDOP5 + 1. Therefore, in block 6, the CPU compares its X register's contents with the ADDOP5 + 1 location: $0009. If X register contents do not equal $0009, the CPU takes the "yes" branch from block 7 (program addresses $013D–$013E), back to block 4, where it adds another ADDOPx value to ACCA. If its X register does contain $0009, the CPU escapes the addition loop and executes block 8.

Recall that in block 1, the CPU transferred an appropriate SUMx address from YTEMP to the Y register. In block 8 (program addresses $013F–$0141), the CPU stores the sum from ACCA to the SUMx location pointed at by the Y register. The CPU then increments the Y register so that it points at the next SUMx location, and transfers this value back to YTEMP (block 9, program addresses $0142–$0146). The RTI instruction at program address $0147 causes the CPU to recover its MAIN program context from the stack, and return to MAIN program, block 12.

This SWI software demonstrates how software interrupts can increase program efficiency when the MAIN program and service routine(s) use different processor contexts. An SWI instruction causes the CPU to automatically stack its context. However, the SWI service routine proper does not benefit from automatic stacking. Therefore, the SWI routine must save its own processor context before executing an RTI instruction. Of course, an RTI causes the CPU to automatically recover its MAIN program context from the stack.

Exercise 2.2 allows the reader to actually demonstrate IRQs, XIRQs, Illegal Opcode Traps, and SWIs. The reader shall also have the opportunity to compare microcontroller power consumption in WAI and STOP states.

EXERCISE 2.2 M68HC11 Interrupts

REQUIRED MATERIALS: 1 Push-button switch, push-to-close
1 Resistor, 4.7K Ohms

REQUIRED EQUIPMENT: Motorola M68HC11 EVBU or EVB, and terminal
Voltmeter, 0–10VDC
Milliammeter, 0–200mA

GENERAL INSTRUCTIONS: Exercise 2.2 gives the reader an opportunity to observe operating characteristics of IRQ, XIRQ, ILLOP, and SWI interrupts. Also, this laboratory exercise analyzes EVBU/EVB power consumption in the fetch and execute, WAI, and STOP modes. Parts I, III, V, and VII use software and hardware configurations presented in the previous four sections. By carefully following the specific instructions given in each part of this exercise, the reader can demonstrate the power and flexibility of four general-purpose interrupts.

PART I: IRQ Demonstration

1. Power down the EVBU or EVB. Connect a voltmeter across the EVBU or EVB's +5VDC power supply connections so that you can accurately measure the actual voltage applied by the power supply to the EVB or EVBU board.
2. Connect an ammeter in series with the EVB or EVBU board's connection to the +5VDC power supply.
3. Refer to Figure 2–16. Connect push-button switch circuitry as shown in this figure. Assure that you connect the switch circuitry to pin 19 ($\overline{\text{IRQ}}$) of the M68HC11.
4. Refer to the software lists accompanying Figure 2–17 and Figure 2–18. If you use an EVBU, power it up and assemble the software as listed. If you use an EVB, modify the listed software for compatibility with available RAM. If you use an EVB, allocate memory as follows:

```
Key-press count—$0000
MAIN program—$C000
IRQ Interrupt Service routine—$C020
DELAY subroutine—$C030
Initial stack pointer—$C0FF.
```

If you use an EVB, modify listed subroutine call, branching, SP initialization, and pseudovector initialization instructions to reflect these memory allocations.

5. Run the software, debug both hardware and software. Assure that circuitry and software operate properly. Recall that location $0000 should contain (1) the total number of seconds that you pressed the push-buttons, and/or (2) the number of times you pressed the push-buttons for less than a second. When the hardware and software operate properly, demonstrate them to the instructor.

_____ *Instructor Check*

6. Start the MAIN program again. Do *not* press the push-button. Read the voltmeter and ammeter while the M68HC11 executes the Figure 2–17, block 5 delay loop. Enter your voltage and current readings in the appropriate blanks.

V_{+5VDC} supply = _____ V
$I_{fetch\ and\ execute}$ = _____ mA

Calculate EVBU/EVB power consumption by multiplying the measured voltage and current values. Enter this power consumption value (to four significant figures) in the blank.

$P_{fetch\ and\ execute}$ = _____ mW

PART II: Analysis, IRQ Demonstration

Answer the questions below by neatly entering your responses in the appropriate blanks.

1. Explain why software establishes the stack at $01FF ($C0FF).

2. List source code (per Figure 2–17, block 2) which would establish the IRQ Interrupt Service routine at address $0150 (instead of $0130).

$ADDRESS	SOURCE CODE
0103	_____
____	_____

3. Explain the necessity of Figure 2–17, block 4.

4. Describe the purpose of the MAIN program's one-second delay loop.

5. By pressing the push-button for four seconds, the user generates ____ (how many?) IRQ interrupts. These interrupts increment location $0000 to $ ____. These interrupts are ____ nested ____ unnested? Explain why this is the case.

Which block in Figure 2–17's IRQ Interrupt Service flowchart resets the I mask? _____ .

6. Describe the purpose of the IRQ routine's one-second delay.

7. Motorola specifies M68HC11 $P_{\text{fetch and execute}}$ as 85mW for the M68HC11E9 on the EVBU, and 150mW for the M68HC11A1 on the EVB.[20] Compare this specified value with your measured value. Explain the difference between specified and measured power consumption figures.

PART III: XIRQ Demonstration and WAI vs. STOP Comparison

1. Leave the voltmeter and ammeter connected to the +5VDC power supply in the same manner as PART I. Power down the EVBU/EVB. Refer to Figure 2-20. Re-connect the pull-up circuit output to the M68HC11's XIRQ pin (pin 18), as shown in this figure.
2. Refer to the program lists accompanying Figures 2–21 and 2–22. If you use an EVB, assemble the software as listed. If you use an EVBU, modify the listed software for compatibility with available RAM. If you use an EVBU, allocate memory as follows:

```
Key-press count—$0000
MAIN (WAI) program—$0100
MAIN (STOP) program—$0120
XIRQ Interrupt Service routine—$0140
Initial stack pointer—$01FF.
```

If you use an EVBU, modify branching, SP initialization, and pseudovector initialization instructions to reflect these memory allocations.

3. Run and debug system software and hardware. Assure that both MAIN programs run properly, and that location $0000 accurately records the number or duration of key-presses.
4. Start the MAIN (WAI) program. Do *not* press the push-button. Read the voltmeter and ammeter while the M68HC11 is in the WAI state. Enter your readings in the appropriate blanks.

$V_{\text{+5VDC supply}}$ = _____ V
I_{WAI} = _____ mA

Calculate EVB/EVBU board power consumption by multiplying measured voltage and current values. Enter calculated power consumption (four significant figures) in the blank.

P_{WAI} = _____ mW

5. Start the MAIN (STOP) program. Do *not* press the push-button. Read the voltmeter and ammeter while the M68HC11 is STOP-ed. Enter your readings in the appropriate blanks.

$V_{\text{+5VDC supply}}$ = _____ V
I_{STOP} = _____ mA

Calculate power consumption in the STOP state. Enter your calculated value in the blank.

P_{STOP} = _____ mW

6. Demonstrate system operation to the instructor. Demonstrate both the MAIN (WAI) and the MAIN (STOP) programs. Also show the instructor your measured and calculated values.

_____ *Instructor Check*

---

[20]Motorola Inc., *MC68HC11E9, HCMOS Single-Chip Microcontroller* (Phoenix, AZ: Motorola Inc., 1988), p. 11-2. These figures assume that V_{DD} = +5VDC, and that all M68HC11 ports are configured as inputs. The EVB's microcontroller consumes more power because it operates in expanded multiplexed mode (see Chapter 11). The EVBU's microcontroller runs in single-chip mode.

PART IV: Laboratory Analysis, XIRQ Demonstration

Answer the questions below by neatly entering your responses in the appropriate blanks.

1. Review M68HC11 WAI and STOP Instructions (pp. 103–106) and Part II of this laboratory (question 7). Enter predicted chip M68HC11 power consumption in the blanks:

   $P_{\text{fetch and execute}}$ = _____ mW
   P_{WAIT} = _____ mW
   P_{STOP} = _____ mW

   Now, re-record your measured EVBU/EVB board power consumption figures from Parts I and III of this laboratory exercise.

   $P_{\text{fetch and execute}}$ = _____ mW
   P_{WAIT} = _____ mW
   P_{STOP} = _____ mW

2. Compare your predicted and measured values. Explain why your measured values generally exceed predicted values.

3. Subtract your predicted P_{STOP} from measured P_{STOP}. This difference should represent power consumed by other EVB or EVBU components. Enter this difference in this blank.

   P_{DIFF} = _____ mW

4. Refer to Figure 2–11, which depicts WAI instruction power-down actions. From this flowchart list the devices that you would expect the M68HC11 to power down, upon execution of the XIRQ demonstration program's WAI instruction.

   A. _____

   B. _____

   C. _____

   Which Figure 2–11 circuits would the M68HC11 not power down when it executes XIRQ demo's WAI instruction? _____. Why? _____.

5. Estimate M68HC11 chip WAI-state power consumption by subtracting P_{DIFF} (question 3) from measured P_{WAI} (question 1). Enter this figure in the blank. P_{WAI} (M68HC11) = _____ mW

   Is estimated M68HC11 chip WAI state power consumption reasonable in light of your predicted value in question 1 above? ____

6. The MAIN (WAI) routine resets the I mask before executing the WAI instruction. Because I = 0, the WAI instruction leaves the M68HC11's timer section powered up. Follow these steps to determine timer section power consumption:

   Step 1. Use the BUFFALO assembler to change the MAIN (WAI)'s CLI instruction to NOP.

   Step 2. Start the program; measure and record I_{WAI}.

   I_{WAI} = _____ mA

   Step 3. Multiply your step 2 figure by $V_{\text{+5VDC supply}}$, to derive P_{WAI}.

   P_{WAI} (I = 1) = _____ mW

   Subtract P_{DIFF} (question 3, above) from the step 3 product to find M68HC11 chip power consumption in the WAI state, when I = 1.

   $P_{\text{WAI}} - P_{\text{DIFF}} = P_{\text{WAI}}$ (M68HC11) = _____ mW

   Compare question 5 and question 6 P_{WAI} (M68HC11) values. By turning off its timer section, the M68HC11 saves a further _____ mW of power.

PART V: Illegal Opcode Trap Demonstration

1. Refer to the program lists accompanying Figure 2–23 and Figure 2–24. If you use an EVBU, assemble the software as listed. If you use an EVB, modify the listed software for compatibility with available RAM. If you use an EVB, allocate memory as follows:

$0000–$0009	ILLOP message ASCII data
$C000	MAIN program
$C030	ILLOP Interrupt Service routine
$C0FF	Initial stack pointer

   If you use an EVB, modify branching, SP initialization, and pseudovector initialization instructions to reflect the above memory allocations.

2. Test and debug the ILLOP demonstration software. As assembled, the MAIN program should write zeros to the terminal screen at one-second intervals. Refer to Figure 2–23. Use the BUFFALO MM command to substitute an illegal opcode ($42) for block 3's NOP opcode ($01). Test the software again. Assure that the M68HC11 now writes the following information to the terminal screen at one-second intervals:

   ILLOP @ $XXXX 0

   where $XXXX = the address of the illegal opcode. When the software runs properly, demonstrate it to the instructor.

   _____ *Instructor Check*

PART VI: Laboratory Analysis, ILLOP Demonstration Software

Refer to the Figure 2–23 flowchart and Table 2–3 below. For each Figure 2–23 NOP block, in turn, execute the following steps:

Step 1. Refer to your program listing and note the NOP instruction's program address. Enter this address in the appropriate blank under opcode $Address in Table 2–3.

Step 2. Use a BUFFALO MM command to change the contents of Step 1's program address from $01 to $42 (an illegal opcode).

Step 3. Start the ILLOP demonstration program, and carefully observe the M68HC11's response. Put an "X" in the appropriate Table 2–3, "yes" or no" blank, to indicate whether or not the ILLOP interrupt screen message correctly records the illegal opcode's location.

Step 4. Briefly and accurately describe the M68HC11's terminal screen response to the illegal opcode. Enter your description in Table 2–3 under M68CH11 Screen Response.

Step 5. Reset the EVBU/EVB. Use the BUFFALO assembler or MM command to change the illegal opcode back to an NOP. Go to the next NOP and do steps 1–5 again.

Table 2–3 already contains information for Figure 2–23, block 3, to provide an example.

TABLE 2–3

Fig. 2-23 Block	Opcode $Address	Correct ILLOP Address Identified? Yes	No	M68HC11 Screen Response
3	C008	X		ILLOP message, then zero, each second.
5	_____	__	__	_____
7	_____	__	__	_____
11	_____	__	__	_____
13	_____	__	__	_____
15	_____	__	__	_____

PART VII: SWI Demonstration Program

1. Refer to the program lists accompanying Figure 2–25 and Figure 2–26. If you use an EVBU, assemble the software as listed. If you use an EVB, modify listed software for compatibility with available RAM. If you use an EVB, allocate memory as follows:

   ```
   $0000–$0003      SUM1–SUM4
   $0004–$0008      ADDOP1–ADDOP5
   $0009–$001C      GROUP1–GROUP4
   $0020–$0021      YTEMP
   $C000            MAIN program
   $C030            SWI Interrupt Service routine
   $C0FF            Initial stack pointer
   ```

   If you use an EVB, modify branching, SP initialization, and pseudovector initialization instructions to reflect the above memory allocations.

2. Test and debug the SWI demonstration software. When the software runs properly, demonstrate it to the instructor.

   _____ *Instructor Check*

PART VIII: Laboratory Analysis, SWI Demonstration Software

Answer the questions below by neatly entering your responses in the appropriate blanks.

1. Refer to Figure 2–25. Does the MAIN program reset the I mask? Why is this the case?

2. Refer to Figures 2–25 and 2–26. Describe the purpose of YTEMP.

3. Explain why SWIs are sometimes more useful and efficient than JSRs or BSRs.

4. Refer to Figure 2–25, block 9. Explain why the CPU compares its X register's contents with the address of ADDOP5 + 1.

5. Refer to Figure 2–25, block 12. Explain why the CPU compares its Y register with the last GROUP4 address + 1.

6. Refer to Figure 2–26, block 6. Explain why the CPU compares its X register with the ADDOP5 address + 1.

PART IX: Laboratory Software Listing

IRQ DEMONSTRATION—MAIN PROGRAM—EVB

$ADDRESS	SOURCE CODE	COMMENTS
C000	LDS #C0FF	Initialize stack pointer
C003	LDD #C020	Initialize IRQ pseudovector
C006	STD EF	" " "

C008	CLR 0	Clear keypress count
C00B	CLI	Clear I mask
C00C	JSR C030	Call DELAY subroutine
C00F	BRA C00C	Call DELAY again

IRQ DEMONSTRATION—MAIN PROGRAM—EVBU

(See program list accompanying Figure 2–17.)

IRQ INTERRUPT SERVICE ROUTINE—EVB

$ADDRESS	SOURCE CODE	COMMENTS
C020	INC 0	Increment keypress count
C023	JSR C030	Call DELAY subroutine
C026	RTI	Return

IRQ INTERRUPT SERVICE ROUTINE—EVBU

(See program list accompanying Figure 2–17.)

IRQ DEMONSTRATION—DELAY SUBROUTINE—EVB

$ADDRESS	SOURCE CODE	COMMENTS
C030	LDAA #6	Initialize loop counter (ACCA)
C032	LDX #D600	Load X register with countdown value
C035	DEX	Decrement X register
C036	BNE C035	If X register not zero, decrement again
C038	DECA	If X register zero, decrement loop count
C039	BNE C032	If loop count not zero, count down again
C03B	RTS	Return

IRQ DEMONSTRATION—DELAY SUBROUTINE—EVBU

(See program list accompanying Figure 2–18.)

XIRQ DEMONSTRATION—MAIN (WAI) PROGRAM—EVB

(See program list accompanying Figure 2–21.)

XIRQ DEMONSTRATION—MAIN (WAI) PROGRAM—EVBU

$ADDRESS	SOURCE CODE	COMMENTS
0100	LDS #1FF	Initialize stack pointer
0103	LDD #140	Initialize XIRQ pseudovector
0106	STD F2	" " "
0108	CLRA	Clear keypress count
0109	STAA 0	" " "
010B	TAP	Clear X mask
010C	WAI	Wait for interrupt
010D	BRA 10C	Wait again

XIRQ DEMONSTRATION—MAIN (STOP) PROGRAM—EVB

(See program list accompanying Figure 2–21.)

XIRQ DEMONSTRATION—MAIN (STOP) PROGRAM—EVBU

$ADDRESS	SOURCE CODE	COMMENTS
0120	LDS #1FF	Initialize stack pointer
0123	LDD #140	Initialize XIRQ pseudovector
0126	STD F2	" " "
0128	CLRA	Clear keypress count
0129	STAA 0	" " "
012B	TAP	Clear S bit and X mask
012C	STOP	Stop internal clocks
012D	BRA 12C	Wait again

XIRQ INTERRUPT SERVICE ROUTINE—EVB

(See program list accompanying Figure 2–22.)

XIRQ INTERRUPT SERVICE ROUTINE—EVBU

$ADDRESS	SOURCE CODE	COMMENTS
0140	INC 0	Increment keypress count
0143	LDAA #6	Execute one-second delay
0145	LDX #D600	" " "
0148	DEX	" " "
0149	BNE 148	" " "
014B	DECA	" " "
014C	BNE 145	" " "
014E	RTI	Return

ILLOP DEMONSTRATION—MAIN PROGRAM—EVB

$ADDRESS	SOURCE CODE	COMMENTS
C000	LDS #C0FF	Initialize stack pointer
C003	LDD #C030	Initialize ILLOP pseudovector
C006	STD F8	" " "
C008	NOP	No op
C009	LDAA #6	Initialize delay loop counter
C00B	NOP	No op
C00C	LDX #D600	Load X register with countdown value
C00F	NOP	No op
C010	DEX	Decrement X register
C011	BNE C010	If X register not zero, decrement again
C013	DECA	If X register zero, decrement loop count
C014	NOP	No op
C015	BNE C00B	If loop count not zero, count down again
C017	NOP	No op
C018	JSR FFB5	Call .OUTRHL
C01B	NOP	No op
C01C	BRA C008	Do another delay

ILLOP DEMONSTRATION—MAIN PROGRAM—EVBU

(See program list accompanying Figure 2–23.)

ILLOP INTERRUPT SERVICE ROUTINE—EVB

$ADDRESS	SOURCE CODE	COMMENTS
C030	LDX #0	Point X register at ASCII message data
C033	JSR FFC7	Call .OUTSTR
C036	TSY	Transfer SP+1 to Y register
C038	LDD 7,Y	Copy stacked PC value to D register
C03B	STD A	Store PC value to RAM
C03D	LDX #A	Point X register at PC value in RAM
C040	JSR FFC1	Call .OUT2BS
C043	LDD 7,Y	Copy stacked PC value to D register
C046	ADDD #1	Increment PC value to point at next MAIN program instruction
C049	STD 7,Y	Store incremented PC value back to stack
C04C	RTI	Return

ILLOP INTERRUPT SERVICE ROUTINE—EVBU

(See program list accompanying Figure 2–24)

ILLOP MESSAGE ASCII DATA—EVB AND EVBU

(See list accompanying Figure 2–24.)

SWI DEMONSTRATION—MAIN PROGRAM—EVB

$ADDRESS	SOURCE CODE	COMMENTS
C000	LDS #C0FF	Initialize stack pointer
C003	LDD #C030	Initialize SWI pseudovector
C006	STD F5	" " "
C008	LDD #0	Point YTEMP at SUM1
C00B	STD 20	" " "
C00D	LDY #9	Point Y register at GROUP1
C011	LDX #4	Point X register at ADDOP1
C014	LDAA 0,Y	Read next GROUPx operand
C017	STAA 0,X	Write operand to next ADDOPx
C019	INY	Increment Y register
C01B	INX	Increment X register
C01C	CPX #9	Not all GROUPx operands transferred?
C01F	BNE C014	If yes, transfer another operand
C021	SWI	If no, call software interrupt
C022	CPY #1D	Not all groups totalled?
C026	BNE C011	If yes, transfer another group
C028	WAI	If no, halt

SWI DEMONSTRATION—MAIN PROGRAM—EVBU

(See program list accompanying Figure 2–25.)

SWI INTERRUPT SERVICE ROUTINE—EVB

$ADDRESS	SOURCE CODE	COMMENTS
C030	LDY 20	Copy YTEMP to Y register
C033	LDX #4	Point X register at ADDOP1
C036	CLRA	Clear sum accumulator
C037	ADDA 0,X	Add next ADDOPx
C039	INX	Increment X register
C03A	CPX #9	Not all ADDOPx values added?
C03D	BNE C037	If yes, add another ADDOPx
C03F	STAA 0,Y	If no, store sum to SUMx
C042	INY	Increment Y register
C044	STY 20	Copy Y register to YTEMP
C047	RTI	Return

SWI INTERRUPT SERVICE ROUTINE—EVBU

(See program list accompanying Figure 2–26.)

2.4 CHAPTER SUMMARY

M68HC11 resets establish or re-establish basic starting conditions for system operation. When an external or internal circuit drives its $\overline{\text{RESET}}$ pin low, then high, the M68HC11 executes a full reset sequence. The microcontroller executes part of this sequence when the $\overline{\text{RESET}}$ pin goes low, and completes the sequence when $\overline{\text{RESET}}$ returns high.

M68HC11 resets are generated in four ways:

1. External circuits generate a reset by driving the $\overline{\text{RESET}}$ pin low, then back high after at least six E clock cycles have elapsed.
2. If its V_{DD} pin rises from 0V to about 5V, the M68HC11 executes a power-on reset (POR). Internal POR detect circuitry drives the $\overline{\text{RESET}}$ pin low for 4064 E clock cycles, then back high.
3. If it detects slow M68HC11 clocks, an internal clock monitor circuit can drive the $\overline{\text{RESET}}$ pin low for four E clock cycles, then back high. The CPU enables the clock monitor by setting the CME bit in the OPTION register.
4. The M68HC11's COP system detects a hung or lost CPU, and responds by driving the $\overline{\text{RESET}}$ pin low for four E clock cycles, then back high. A user enables the COP system

by resetting the CONFIG register's NOCOP bit. The COP system uses a watchdog timer. To prevent a chip reset, the CPU must periodically write $55 then $AA to the COPRST register before the watchdog timer times out. If the CPU fails to execute these writes within the timeout period, the COP system resets the M68HC11. By writing $55 then $AA to COPRST, the CPU resets the watchdog timer and starts a new timeout. Thus, a hung or lost CPU fails to return to the $55/$AA write instructions, and the COP system resets the M68HC11. The CPU selects from four timeout periods, by writing to the CR1/CR0 bits in the OPTION register.

When internal or external circuits drive $\overline{\text{RESET}}$ low, the M68HC11 takes these actions:

1. Establishes the internal RAM and register block memory maps by initializing the INIT register.
2. Disables all parallel I/O-associated interrupts.
3. Initializes the timer system's free-running counter to $0000. Disables all timer-associated functions and interrupts.
4. Enables or disables the COP system, depending upon the state of the CONFIG register's NOCOP bit.
5. Disables serial I/O systems and their associated interrupts.
6. Disables on-chip A-to-D.
7. Establishes interrupt priorities by initializing the HPRIO register. Disables all interrupts but illegal opcode traps.
8. Disables the clock monitor.

When an internal or external circuit drives the $\overline{\text{RESET}}$ pin back high, the M68HC11 takes these actions: (1) Copies the address of an appropriate reset service routine from ROM storage to the program counter, and (2) fetches and executes the reset service routine. The POR mechanism and external circuits hold the $\overline{\text{RESET}}$ pin low for at least six E clock cycles. The COP and clock monitor systems drive the $\overline{\text{RESET}}$ pin low for only four E clock cycles. During the sixth E clock cycle, the CPU checks the $\overline{\text{RESET}}$ pin's state. If $\overline{\text{RESET}}$ = 0, the CPU copies the POR/external circuit reset service routine's address to the PC. If $\overline{\text{RESET}}$ = 1, the CPU checks clock monitor and COP status to determine which of these systems generated the reset. The CPU then copies the appropriate service routine's starting address from ROM to the PC.

The BUFFALO operating system establishes clock monitor and COP pseudovectors in RAM. A pseudovector is a JMP instruction which re-vectors the CPU to the reset service routine's actual starting address. The user can easily modify these pseudovectors to provide for custom COP and clock monitor reset service routines.

The M68HC11 uses twelve interrupt types. Internal devices can generate seven of these interrupt types. Each of these seven interrupts has a special purpose—to make the CPU attend to an internal device's unique, immediate needs. The five remaining interrupts serve more general purposes. The M68HC11 generates three of these general-purpose interrupts internally. Program instructions generate SWIs and illegal opcode traps. The M68HC11's timer section generates periodic RTIs (real-time interrupts). The two remaining general-purpose interrupts service the needs of external devices. An external device requests service by driving an interrupt-associated pin ($\overline{\text{IRQ}}$ or $\overline{\text{XIRQ}}$) low.

The M68HC11 responds to an unmasked interrupt request by vectoring to an appropriate interrupt service routine, then executing it. The M68HC11 vectors to the appropriate service routine by copying this routine's starting address to its program counter. Upon a POR or push-button reset, the BUFFALO operating system stores default pseudovectors to all interrupt service routine starting addresses. These pseudovectors cause the M68HC11 to JMP to the actual service routine's starting address. Thus, the CPU vectors, then pseudovectors to an appropriate interrupt service routine.

All twelve interrupt types cause the CPU to stack its context—PC, X/Y registers, ACCA/ACCB, and CCR. The programmer should make the final interrupt service routine instruction an RTI. RTI causes the CPU to pull its original context from the stack, and restore all its registers to their original condition at the time of the interrupt. The CPU then fetches and executes from where it left off at the time of the interrupt.

The M68HC11 prioritizes I-maskable interrupts. If two or more I-maskable interrupts are requested simultaneously, the M68HC11 services the interrupt with the highest priority first, then services the others in priority order. No I-maskable interrupt has priority over illegal opcode, SWI, and XIRQ (X = 0) interrupts, since they are unmaskable. By writing appropriate PSEL3–PSEL0 bits to the HPRIO register, the CPU can promote any I-maskable interrupt to top I-maskable priority. The HPRIO register's PSEL3–PSEL0 bits are writable only if I = 1. The CPU must set the I mask before writing to PSEL3–PSEL0, then reset the I mask after this write.

Interrupts allow the CPU to execute routine business without having to periodically check to see if there is a system problem or a device needs service. However, in some applications the microcontroller has no routine business to execute. In this situation, a programmer can select from two instructions that cause the CPU to stop fetching and executing, and simply wait for an interrupt. These instructions are WAI and STOP. Both instructions cause the M68HC11 to reduce power consumption while it waits.

The WAI instruction saves both time and power. A WAI instruction anticipates an interrupt by pre-stacking processor context. This saves time when a device generates the anticipated interrupt. Any unmasked internal or external interrupt causes recovery from the WAI state. Likewise, any reset source can cause a recovery from WAI. Before entering the WAI state, the CPU powers down disabled SPI, SCI, and timer circuits to reduce power consumption. This power down reduces WAI state power consumption to 15mW–150mW, depending upon which circuits are turned off.

The STOP instruction can halt the M68HC11 "in its tracks" or function as an NOP, depending upon the state of the CCR's S bit. If S = 0, a STOP causes processing to cease by halting all internal M68HC11 clocks. If S = 1, the STOP instruction acts as an NOP that consumes two machine cycles. If the STOP instruction halts processing, it reduces chip power consumption to the .5mW-1.5mW range. An external circuit can cause recovery from STOP by driving either the $\overline{RESET}$ or $\overline{XIRQ}$ pin low. If its $\overline{RESET}$ pin is driven low, the M68HC11 executes a normal POR/external circuit reset. If its $\overline{XIRQ}$ pin is driven low, the M68HC11 executes an XIRQ interrupt service, if X = 0. If X = 1, the CPU begins fetching and executing at the next program location. An external circuit can also cause the M68HC11 to exit STOP by generating an unmasked (I = 0) IRQ interrupt. In this case, the CPU responds by servicing the interrupt.

Figure 2–15 diagrams $\overline{IRQ}$ pin input circuitry. During factory testing, Motorola uses this pin as a power input to facilitate bulk EEPROM programming. Of course, system designers use the $\overline{IRQ}$ pin to generate interrupts. $\overline{IRQ}$ pin logic uses a totem-pole buffer to transfer inverted $\overline{IRQ}$ pin levels to IRQ interrupt control logic.

A programmer has the option of making IRQ interrupts level- or edge-sensitive. To make IRQs edge-sensitive, the CPU sets the IRQE bit in the OPTION register. To make IRQs level-sensitive, the IRQE bit must be clear. All M68HC11 resets clear IRQE. The M68HC11 time protects this bit. An edge-sensitive IRQ remains pending until the M68HC11 services it. If IRQs are level sensitive, an external device must hold the $\overline{IRQ}$ pin low until the M68HC11 responds.

Figures 2–16 through 2–18 illustrate IRQ test configuration and software, which are implemented in Exercise 2.2. This test configuration allows the user to request IRQ interrupts by pressing a push-button switch (see Figure 2–16). Supporting software uses IRQ interrupts to record either the number of times the user pressed the push-button for less than a second, or the number of seconds that the user pressed the push-button continuously.

IRQ test software consists of three parts: MAIN program, IRQ Interrupt Service routine, and DELAY subroutine. The MAIN program initializes the stack pointer and IRQ pseudovector, and clears the I mask. The MAIN program then enters a continuous loop where it calls the one-second DELAY subroutine, then branches back to call DELAY again.

The IRQ Interrupt Service routine increments the key-press counter (location $0000), then calls the DELAY subroutine. After one second elapses, the CPU returns from the DELAY subroutine and executes an RTI. This RTI returns the CPU to where it left off in the MAIN program or DELAY subroutine. If a user presses the push-button continuously, the M68HC11 executes and re-executes IRQ interrupts every second, incrementing the key-press count each time. When the user finally releases the push-button, the CPU returns from interrupt, to continue MAIN program or DELAY subroutine execution from where it left off. If the user presses a push-button for less than one second, the CPU increments the key-press count once and returns from the interrupt after a second to continue MAIN/DELAY execution. The DELAY subroutine consumes one second by loading the M68HC11's X register with $D600 and decrementing it to $0000, six times.

Motorola replaced NMIs (nonmaskable interrupts) with XIRQs (pseudo-nonmaskable interrupts) to solve two problems: First, inadvertent NMIs could precede SP initialization and cause the CPU to stack its context over vital program or data. Second, a bouncing NMI request nests NMI interrupts, causing abnormal stack growth and possible program disruption. XIRQs solve the first problem by automatically setting the X mask upon any chip reset. XIRQs cannot occur until software properly initializes the SP, then clears the X mask. XIRQs solve the second problem by setting the X flag upon entry into an XIRQ service, thus preventing nested interrupts.

Pages 110–112 present an XIRQ test configuration and software. This application again uses a switch connected to the $\overline{\text{XIRQ}}$ pin. XIRQ supporting software again uses interrupts to record either the number of times or the number of seconds that the user presses the push-button. Two alternative MAIN programs allow the user to evaluate the WAI and STOP instructions. Except for interchangeable use of WAI and STOP, these two MAIN programs are identical. They initialize the SP, initialize the XIRQ pseudovector, clear the keypress duration counter, clear appropriate S and X bits, then either STOP processing or enter the WAI state. The XIRQ Interrupt Service increments the keypress counter, then executes a one-second delay, before returning to the MAIN program from which it originated.

The next section presents Illegal Opcode Trap demonstration software. The M68HC11 stacks its context and vectors to a special routine when it fetches an illegal opcode. An illegal opcode has no associated mnemonic, although the BUFFALO operating system disassembles an illegal opcode as an "ILLOP."

The illegal opcode demonstration's MAIN program executes a routine task—writing a zero to the terminal screen at one-second intervals. However, this program distributes NOP instructions throughout. These NOPs provide convenient places to substitute an illegal opcode.

If it encounters an illegal opcode while executing the MAIN program, the CPU vectors to the ILLOP Interrupt Service routine. ILLOP causes the CPU to display this message:

```
ILLOP @ $XXXX
```

where $XXXX is the illegal opcode's address.

The CPU can easily determine the $XXXX value. All interrupts cause the CPU to stack its context, including the program count. For all interrupts except illegal opcode traps, the CPU stacks a PC value that points at the next instruction to be executed, upon return from the interrupt. However, an illegal opcode trap causes the CPU to stack a PC value pointing directly at the illegal opcode location. To access this stacked PC value, the CPU transfers the SP + 1 value to an index register. The CPU can then read the stacked PC value to its D register, using indexed mode with an appropriate offset. The CPU displays this value on the terminal screen as "$XXXX."

After displaying "$XXXX," the CPU must return (RTI) from the ILLOP service routine. To execute an RTI, the CPU pulls its stacked context, including the PC. This stacked PC value points at the illegal opcode. Therefore, the CPU must increment the stacked PC so that it points at the next MAIN program instruction. Otherwise, the CPU would return to the illegal opcode and immediately generate another illegal opcode. The microcontroller would execute a continuous string of interrupts. To increment the stacked PC value, the CPU reads it using indexed mode. The CPU adds one to this value, and stores it back to the stack using indexed mode.

The last section of this chapter presents Software Interrupt (SWI) demonstration software. SWIs become an attractive alternative to subroutine calls when the MAIN program and routine to be called use different processor contexts. The SWI demonstration software uses two sets of index register values. Overall, this software calculates totals for four operand groups with five operands in each group. The MAIN program transfers each operand group in turn to five memory locations, which are labeled "operands to be added." After transferring an operand group, the MAIN program calls an SWI. The SWI routine finds the total of the "operands to be added" and stores this total to an assigned memory location. The MAIN program manages program operations to assure that the CPU halts after it has calculated and stored totals for all four operand groups. This SWI software demonstrates how SWIs can efficiently save MAIN program register values when the routine being called requires a different context.

EXERCISE 2.3   Chapter Review

Answer the questions below by neatly entering your responses in the appropriate blanks.

1. List the four conditions that affect $\overline{RESET}$ pin levels.

    1.

    2.

    3.

    4.

2. Summarize actions taken by the M68HC11 when an external or internal circuit drives its $\overline{RESET}$ pin low.

3. Describe actions taken by the M68HC11 when its $\overline{RESET}$ pin rises from low to high.

4. To generate an external circuit reset, the external circuit must drive $\overline{RESET}$ _____ (high/low) for at least _____ (how many?) E clock cycles.
5. If the M68HC11's V_{DD} pin rises to about 5V from _____ V, internal _____ (name) circuitry drives the $\overline{RESET}$ pin _____ (high/low) for _____ (how many?) E clock cycles.
6. To enable clock monitor resets, the CPU _____ (sets/resets) the bit in the _____ register. Clock monitor circuitry drives the $\overline{RESET}$ pin _____ (low/high) for _____ E clock cycles, when its detects _____.
7. To enable COP resets, the user must _____ (set/reset) the _____ bit in the _____ register. The COP system detects a _____ or _____ CPU, and drives the $\overline{RESET}$ pin _____ (low/high) for _____ E clock cycles. To avoid a COP reset, the CPU must write _____ to the _____ register, before _____ times out.
8. To select a 65.536ms COP timeout, the CPU writes %_____ to the _____ bits in the _____ register.
9. INIT register bits RAM3–RAM0 establish the _____ of the M68HC11's internal _____.
10. INIT register bits REG3–REG0 establish the _____ of the M68HC11's internal _____ addresses.
11. If NOCOP = 1, the M68HC11 _____ (disables/enables) the computer operating properly system, when the $\overline{RESET}$ pin is driven low.
12. When an internal or external circuit drives its $\overline{RESET}$ pin low, the M68HC11 gives _____ interrupts top I-maskable priority.
13. After an external circuit drives the $\overline{RESET}$ pin from low to high, the M68HC11 copies $_____ from ROM addresses $_____ to its program counter.
14. After the COP system drives the $\overline{RESET}$ pin from low to high, the M68HC11 copies $_____ from ROM addresses $_____ to its PC.

15. Explain how the M68HC11 determines that it should copy the POR/external circuit reset routine's address to its PC.

16. Explain how the M68HC11 determines whether it should copy the COP or clock monitor service routine address to its PC.

17. Define "reset pseudovector."

18. To pseudovector the CPU to a COP reset service routine at $0150, locate a _____ (source code) instruction at addresses $_____.

19. List the seven interrupt types generated by internal devices.
    1.
    2.
    3.
    4.
    5.
    6.
    7.

20. List the five interrupt types classified by the author as "general purpose."
    1.
    2.
    3.
    4.
    5.

21. List the two interrupt types generated by program instructions.
    1.
    2.

22. List the two general-purpose interrupt types generated by external devices.
    1.
    2.

133

23. List the nine interrupt types masked by local control bits.

    1.

    2.

    3.

    4.

    5.

    6.

    7.

    8.

    9.

24. List the three interrupt types which cannot be masked by the global I bit.

    1.

    2.

    3.

25. List the two interrupt types which are *not* masked either globally or locally.

    1.

    2.

26. XIRQ interrupts are masked by the _____ bit in the _____ register.
27. IRQ pin-generated interrupts are enabled by _____.
28. IRQs generated by the parallel I/O mechanism are enabled by both _____ and _____.

29. You wish to pseudovector the CPU to an Input Capture 3 interrupt service routine at location $C040. In the blanks, enter the pseudovector's address and source code.

    $ADDRESS   **SOURCE CODE**
    _____   _____

30. You wish to pseudovector the CPU to a parallel I/O interrupt service routine at location $0155. In the blanks, enter the pseudovector's address and source code.

    $ADDRESS   **SOURCE CODE**
    _____   _____

31. During an I-maskable interrupt, the M68HC11 automatically sets the _____ mask(s) after stacking the processor context. In the blanks, list the interrupt types that can nest within an I-maskable interrupt routine.

    1.

    2.

    3.

32. During an XIRQ interrupt, the M68HC11 automatically sets the _____ mask(s) after stacking the processor context. In the blanks, list the interrupt types which can nest within an XIRQ routine.

    1.

    2.

    3.

33. Explain why an interrupt service routine's final instruction should be an RTI.

34. In the blanks, enter source code for a routine that promotes Real-Time Interrupts to highest I-maskable priority, then enables I-maskable interrupts.

$ADDRESS	SOURCE CODE
0125	_____
_____	_____
_____	_____
_____	_____

35. The _____ and _____ instructions cause the microcontroller to stop fetching and executing and simply to wait for an interrupt.

36. The _____ instruction saves time by pre-stacking the processor context before entering a _____.

37. A WAI instruction causes the M68HC11 to power down _____ circuits if they will not be involved in the anticipated interrupt.

38. The CPU sets the _____ when any interrupt drives it from the WAI state. The CPU sets the I and X masks if an _____ causes it to recover from WAI.

39. While in the WAI state, the CPU executes continuous _____ to the _____ location on the stack.

40. Assume that the M68HC11 is in the WAI state and I = 1. How does the M68HC11 respond when an external peripheral drives its IRQ pin low?

41. Enter an "X" in the blank preceding the M68HC11 state that consumes the least power.

    _____ Normal fetch and execute

    _____ WAI, I = 0, NOCOP = 1

    _____ WAI, I = 1, NOCOP = 1

    _____ WAI, I = 1, NOCOP = 0

42. Describe the M68HC11's response to a STOP instruction when S = 1.

43. Describe the M68HC11's response to a STOP instruction when S = 0.

44. If the STOP instruction halts processing, it reduces power consumption to the _____ mW range.

45. Assume that the M68HC11 is STOP-ed and X = 1. How does the M68HC11 respond when an external circuit drives its $\overline{\text{XIRQ}}$ pin low?

46. Assume that the M68HC11 is STOP-ed, and I = 1. How does the M68HC11 respond when an external circuit drives its $\overline{\text{IRQ}}$ pin low?

   How does the STOP-ed M68HC11 respond if I = 0 and an external circuit drives its $\overline{\text{IRQ}}$ pin low?

47. Assume that the M68HC11 is STOP-ed, I = 1, and X = 0. How does the M68HC11 respond if an external circuit drives its $\overline{\text{RESET}}$ pin low, then high?

48. Enter an "X" in the blank preceding the M68HC11 state that consumes the least power.

   _____ Normal fetch and execute

   _____ WAI, I = 0, NOCOP = 1

   _____ WAI, I = 1, NOCOP = 1

   _____ WAI, I = 0, NOCOP = 0

   _____ WAI, I = 1, NOCOP = 0

   _____ STOP, S = 1

   _____ STOP, S = 0

   Which of the listed states consume the same amount of power? _____
   _____ Why? _____

49. Refer to Figure 2–15. Describe the purpose of input transistors A.

50. Refer to Figure 2–15. Which component sends an IRQ request to IRQ interrupt control logic? _____

51. To make IRQs edge sensitive, the _____ reset routine must write binary _____ to the _____ bit of the_____ register, within _____ bus cycles after reset.

52. Refer to Figures 2–16, 2–17, and 2–18. Assume that you wish to configure this application to measure *only* the number of key presses, regardless of duration. To make this change, you would _____ (set/reset) the IRQE bit. Explain why you would take this action.

53. Refer to Figure 2–17. Describe how you would modify the block 2 instruction to locate the IRQ routine at address $0150.

54. Refer to Figure 2–17. Explain why the MAIN program calls the DELAY subroutine in block 5.

55. Refer to Figure 2–17. Explain why the CPU increments location $0000 (block 6).

56. Refer to Figure 2–17. Explain why the CPU calls the DELAY routine in block 7.

57. Refer to Figure 2–18. Summarize the process used by the DELAY subroutine to consume a second.

58. Motorola developed XIRQs to replace _____ interrupts. Describe the difference between these two interrupt types.

59. Describe how XIRQs keep inadvertent interrupts from stacking processor context bytes over vital program or data.

60. Describe how XIRQs prevent nested interrupts and uncontrolled stack growth due to XIRQ request bounce.

61. Refer to Figure 2–21. Explain why this figure shows two MAIN programs.

62. Refer to blocks 4 of Figure 2–21. Explain why identical source code can be used to execute either of these blocks.

63. Compare Figures 2–17 and 2–21. Explain why Figure 2–21's programs do not call a one-second delay.

64. Refer to Figure 2–22. Explain why the XIRQ Interrupt Service routine executes a one-second delay.

65. Define the term *illegal opcode*.

66. Refer to Figure 2–23. Explain why this flowchart contains so many NOP blocks.

67. Refer to Figure 2–23. Summarize the task accomplished by the MAIN program.

68. Refer to Figure 2–24. Describe the task performed in block 2 of this flowchart.

69. Refer to Figure 2–24. Explain why the CPU executes block 3 of this flowchart.

70. Refer to Figure 2–24. Explain why the CPU stores the accessed PC value to RAM (block 5).

71. Refer to Figure 2–24. Compare flowchart blocks 4 and 8. Explain why the CPU copies the stacked PC value to the D register again in block 8.

72. Refer to Figure 2–24, blocks 8, 9, and 10. Explain why the CPU increments the stacked PC value before executing an RTI.

73. Explain why SWIs are an attractive alternative to JSR/BSR instructions.

74. Refer to Figure 2–25. Explain the purpose of YTEMP (block 3).

75. Refer to Figure 2–25 and the memory allocations list that accompanies it. From what RAM address will the CPU read an operand when it executes block 6 for the twelfth time? $_____. To what RAM address will the CPU write this operand when it executes block 7 for the twelfth time? $_____. To which SUMx will this operand contribute? _____.
76. Refer to Figure 2–25. The block 9 comparison value is $_____. Explain why block 9 specifies this value.

77. Refer to Figure 2–25. The block 12 comparison value is $_____. Explain why block 12 specifies this value.

78. Refer to Figure 2–26. Explain why the CPU executes block 9.

79. Refer to Figure 2–26, blocks 4 through 9. Explain the X register's purpose in these flowchart blocks.

80. Refer to Figure 2–26, block 8. To what address will the CPU store the sum when it executes block 8 for the third time? $_____.

3 M68HC11 Parallel I/O

3.1 GOAL AND OBJECTIVES

A microcontroller can use serial communication or parallel communication to exchange data with a peripheral. Each technique has its advantages and disadvantages.

To write the contents of its accumulator to a serial peripheral, the M68HC11 must multiplex the 8 data bits—along with some special control and error detection bits called "framing bits"—and send them one at a time over a single data communications line. The major advantage of this technique is physical simplicity. Serial communications require only one or two data lines. This simplicity assumes greater importance when the peripheral is some distance from the microcontroller. But serial communications have a down side: it takes time to multiplex the data and framing bits and then send them over a serial communications line.[1] In turn, the serial peripheral must de-multiplex the serial data, check for errors, and strip the framing bits to make this data usable.

To write the contents of an accumulator to a parallel peripheral, the M68HC11 simply sends the 8 data bits simultaneously over eight parallel data lines, one line for each bit. Thus, parallel communications are much faster than serial communications. Additionally, parallel communications require no multiplexing and de-multiplexing. However, parallel communications require greater physical complication. For parallel communications with a remote peripheral, the M68HC11 requires eight data lines instead of one, and eight times the number of buffers, amplifiers, etc., needed to maintain signal quality and integrity over a long distance.

Chapters 3 and 4 present M68HC11 parallel I/O. Chapter 3 presents basic, general-purpose parallel I/O techniques and hardware. Chapter 4 deals with more advanced aspects of M68HC11 parallel I/O. The goal of Chapter 3 is to help the reader understand how to implement basic parallel data communications hardware and software.

After studying this chapter and completing its exercises and hands-on activities, you will be able to:

> Complete a wiring diagram which depicts specified connections between the M68HC11 and a parallel peripheral.
> Connect and demonstrate the operation of interfaces between the M68HC11 and parallel peripherals.

---

[1] The fastest the M68HC11 can send serial bits is 9600 bits/second (baud).

3.2 ON-CHIP RESOURCES USED BY THE M68HC11 FOR PARALLEL I/O

The M68HC11 includes these resources for implementing parallel I/O:

- Port B Data Register (PORTB, control register block address $1004).[2] PORTB latches data written to it by the M68HC11's CPU. Having latched this write data, PORTB drives output pins PB0–PB7 (pins 42–35) with the write data. If, for example, the CPU writes the contents of ACCA to PORTB, PB0 is driven with ACCA's D0 data bit, PB1 with D1, and so on. PORTB is an output port only. If the CPU reads PORTB, this port returns the last data written to it.
- Port C Data Register (PORTC, control register block address $1003). A programmer can configure PORTC pins as either inputs or outputs. CPU writes to PORTC cause write data to appear on PORTC pins configured as outputs. A write to PORTC bits configured as inputs causes PORTC to latch and retain these data bits. PORTC does not drive its pins configured as inputs with write data. If program instructions change a PORTC pin from input to output, that pin is immediately driven with the last data bit written to it and retained (latched). If the CPU reads PORTC, this port returns the states of PORTC pins configured as inputs, along with the last data written to pins configured as outputs.
- Data Direction Register for Port C (DDRC, control register block address $1007). Writing bits to DDRC configures corresponding PORTC pins as either inputs or outputs. Writing binary one to a DDRC bit configures its corresponding PORTC pin as an output. Writing a zero configures as an input. With a system reset, all DDRC bits reset and configure all PORTC pins as inputs.

To summarize:

> PORTB is a dedicated parallel output (write) port.
> DDRC configures corresponding PORTC bits as either inputs or outputs.
> A PORTB read returns the last data stored to it.
> Reading PORTC bits configured as outputs also returns the last data stored to them.
> Storing data to PORTC bits configured as inputs causes these data to be latched.
> Changing a PORTC pin from input to output causes it to be driven with the last bit written to it, before the configuration change.

A system designer can also use PORTA and PORTD pins for general purpose parallel I/O. However, these ports have other primary purposes.[3] Therefore, this chapter limits its discussion to PORTB and PORTC, which the M68HC11 dedicates to general purpose parallel I/O.

3.3 PARALLEL I/O PROGRAM FLOW

Since the M68HC11 can read or write to either PORTB or PORTC, four possible operations result:

> Read PORTB
> Write PORTB
> Read PORTC
> Write PORTC

Following sections cover each of these possibilities.

Read PORTB

The Figure 3–1 flowchart and its accompanying program segment example show how simple it is to read PORTB. Recall, however, that a PORTB read returns data previously written to this register since its pins are outputs only.

---

[2] Default addresses $1000–$103F contain the M68HC11's control register block. A system designer can move this register block to any 4K page boundary on the M68HC11's 64K direct addressable memory map. To make such a move, the designer changes bits 3 through 0 of the INIT register (RAM and I/O Mapping Register) to the most significant hex digit of the register block's new beginning address. The M68HC11 time protects its INIT register.
[3] The M68HC11 uses PORTA mainly for timer system functions. Chapters 7, 8, and 9 cover this subject. Chapters 5 and 6 cover PORTD. The M68HC11 uses PORTD for serial communications.

FIGURE 3-1

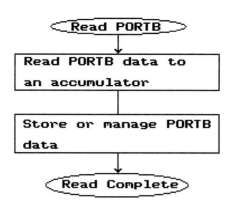

Read PORTB

$ADDRESS	SOURCE CODE	COMMENTS
010A	LDAA 1004	Read PORTB
010D	STAA A	Store read data to address $000A
010F	----	(next operation)

Write to PORTB

Writing to PORTB is equally simple. Refer to Figure 3-2 and the accompanying sample program segment. A PORTB write drives all of its pins with write data.

Read PORTC

A PORTC read requires proper DDRC initialization. Configure PORTC pins connected to a read peripheral as inputs by writing zeros to corresponding DDRC bits. The Figure 3-3 flowchart and its accompanying program segment assume the following:

> PORTC pins PC3–PC0 connect to a read peripheral.
> PORTC pins PC7–PC4 connect to a write peripheral.
> A PORTC read returns previous write data from pins PC7–PC4. We wish to zero out these data bits, after a PORTC read.

FIGURE 3-2

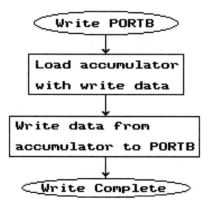

Write PORTB

$ADDRESS	SOURCE CODE	COMMENTS
010A	LDAA A	Load accumulator with write data from address $000A
C00C	STAA 1004	Write data to PORTB
C00F	----	(next operation)

FIGURE 3-3

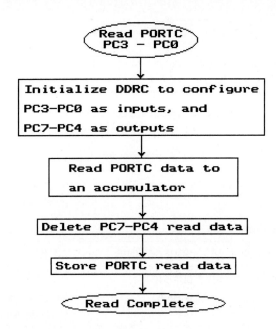

Read PORTC (PC3–PC0)

ADDRESS	SOURCE CODE	COMMENTS
0100	LDAA #F0	Load DDRC control word
0102	STAA 1007	Store control word to DDRC (PC3–PC0 as inputs; PC7–PC4 as outputs)
0105	LDAA 1003	Read PORTC
0108	ANDA #F	Mask $0F makes all high nibble bits = 0 (PC7–PC4 data deleted)
010A	STAA 0	Store PORTC read data
010C	----	(next operation)

Refer to Figure 3–3 and its accompanying program segment. The instruction at address $0108 ANDs mask $0F with the read data byte, to delete all ones from the high nibble. As an example, assume that the CPU reads data byte $A9 from PORTC. ANDing this byte with mask $0F returns $09:

```
%10101001 = $A9    PORTC read data
%00001111 = $0F    Logical AND with mask $0F
%00001001 = $09    Result
```

Write to PORTC

PORTC writes also require DDRC initialization. Writing data to PORTC pins configured as inputs causes PORTC to latch these data. Remember, however, that PORTC pins configured as inputs cannot be driven with write data. The Figure 3–4 flowchart and its accompanying program segment make the same assumptions as the PORTC read example above, with one addition:

- PORTC pins PC3–PC0 connect to a read peripheral.
- PORTC pins PC7–PC4 connect to a write peripheral.
- A PORTC read returns previous write data from pins PC7–PC4. We want the CPU to delete these data bits after a PORTC read.
- A PORTC data write routine immediately follows the PORTC read.

From the above, the reader can correctly generalize that programming for general-purpose parallel I/O is simple and straightforward.

FIGURE 3-4

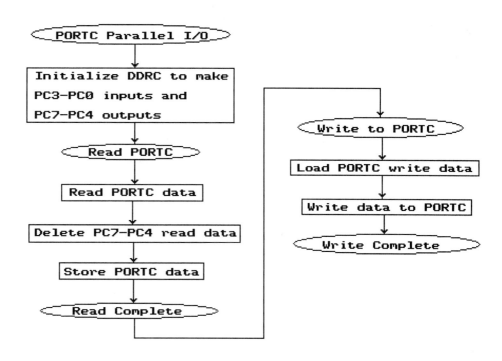

Read and Write PORTC

$ADDRESS	SOURCE CODE	COMMENTS
0100	LDAA #F0	Initialize DDRC
0102	STAA 1007	" " "
0105	LDAA 1003	Read PORTC
0108	ANDA #F	Delete ones in high nibble
010A	STAA 0	Store read data
010C	LDAA A	Load PORTC write data
010E	STAA 1003	Write to PORTC
0111	----	(next operation)

3.4 PORTB PIN LOGIC AND TIMING

PORTB functions as an output port. However, a PORTB read returns the last data written. Figure 3–5 contains a simplified schematic of PORTB pin logic.[4]

M68HC11 Reset and PORTB

Refer to Figure 3–5.[5] When the M68HC11 operates in single-chip mode,[6] the Mode A (MDA) input is low. This low from the MDA input disables AND gate 3 (i.e., causes it to output a constant low), but enables AND gates 2 and 4. Reset assertion causes flip-flop 7 to set, making its Q output high. This output high is buffered (9) and applied to totem-pole output 10. This high turns on the N-channel transistor, and turns off the P-channel device. Flip-flop 7's set state (Q=1) holds the PB pin low, after the reset input negates.

---

[4]Motorola, Inc., *M68HC11 Reference Manual* (1990), p. 7-12.
[5]This PORTB description makes reference to circled numbers accompanying gates and components in the diagram. Thus, "AND gate 2" refers to the gate labeled with the circled number 2. Figure 3–5 shows pin logic for *one* of the eight PORTB pins. The other seven PORTB pins duplicate this logic.
[6]This chapter's discussion assumes that the M68HC11 operates in single-chip mode. Chapter 11 deals with expanded multiplexed mode.

FIGURE 3–5
Reprinted with the permission of Motorola.

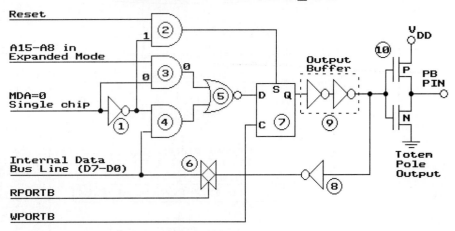

Write to PORTB

Refer to Figures 3–5 and 3–6.[7] To initiate a PORTB write, the M68HC11 asserts WPORTB (write PORTB). Assume that a high also appears on the internal data bus line. This high data bit drives AND gate 4's output high, and NOR gate 5's output low. Flip-flop 7 latches this low. This low is buffered (9) and applied to the gates of the totem-pole output 10, turning the P-channel device on and the N-channel device off, driving the PORTB output pin high. Thus, PORTB logic translates a high on the internal data bus line to a high at the output pin. Of course, an internal bus line low would be latched as Q=1 at flip-flop 7, and translated to a PORTB output pin low, by turning on the totem-pole's N-channel transistor, and turning the P-channel transistor off.

Read PORTB

To initiate a PORTB read, the M68HC11 asserts the RPORTB (read PORTB) line. RPORTB assertion enables transmission gate 6. The latched state of flip-flop 7 is buffered (9), inverted by gate 8 and applied to the internal data bus line. Thus, PORTB pin logic drives the internal data bus line with its output pin's logic level.

FIGURE 3–6
Reprinted with the permission of Motorola.

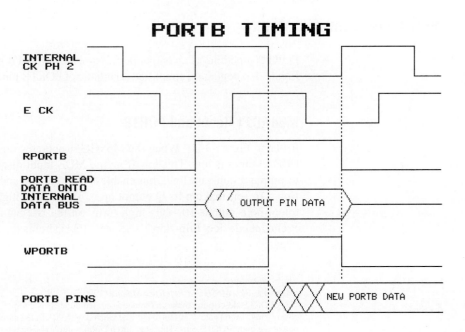

---

[7]Motorola Inc., *M68HC11 Reference Manual*, p. 7-13.

3.5 PORTC PIN LOGIC AND TIMING

Refer to Figure 3–7.[8] A write bit has a single data path from internal data bus line to the PORTC pin: PORTC flip-flop 3 latches this write bit and applies it to the gates of the totem-pole output transistors (17) via NOR gate/buffer 15 and NAND gate/buffer 16. The totem-pole drives the PORTC pin.

A read bit, however, has two possible data paths—depending upon where it originates. If the read bit originates at the PORTC pin (because the corresponding DDRC bit = 0), it is buffered (10), and passes via transmission gate 9 to the internal data bus line.

If the read bit originates at the $\overline{Q}$ output of PORTC flip-flop 3 (because the corresponding DDRC bit = 1), then logic gate (12) inverts it and transmission gate 11 applies it to the internal data bus line.

Following sections cover PORTC reads and writes as they relate to PORTC pin configurations.

Write to PORTC; DDRC Bit = 1

Recall that when the CPU writes a binary one to a DDRC bit, this one configures a corresponding PORTC pin as an output. DDRC consists of eight flip-flops, one for each PORTC pin (see Figure 3–7, flip-flop 2). Refer to Figures 3–7 and 3–8.[9] DDRC flip-flop 2 controls read and write data paths. The CPU writes a binary one to DDRC flip-flop 2 by asserting WDDRC (write DDRC) and storing a high from the internal data bus line. The resulting high from the Q output of DDRC flip-flop 2 causes NOR gate 14 to output a low, enabling NOR gate/buffer 15. The low from NOR gate 14 is also inverted by gate 13 and enables NAND gate/buffer 16. Thus, a write bit from PORTC flip-flop 3 can now be applied to the gates of totem-pole output (17).

Assume that DDRC flip-flop 2 latches a binary one. Assume also that the CPU then writes a binary one to PORTC. WPORTC (write PORTC) assertion clocks this write bit from the internal data bus line into PORTC flip-flop 3. As a result, PORTC flip-flop 3's $\overline{Q}$ output switches low. NAND gate/buffer 16 applies this low to the gate of the totem-pole

FIGURE 3–7
Reprinted with the permission of Motorola.

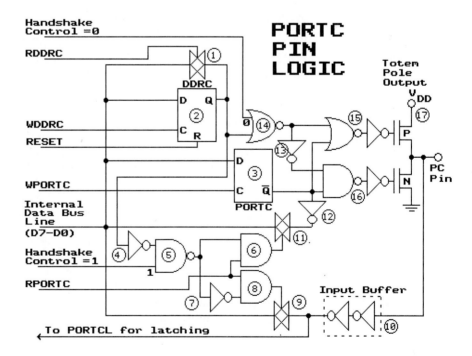

---

[8]Motorola Inc., *M68HC11 Reference Manual*, p. 7-19. Remember that this diagram depicts the pin logic for only one of the eight PORTC bits.
[9]Motorola Inc., *M68HC11 Reference Manual*, p. 7-21.

FIGURE 3–8
Reprinted with the permission of Motorola.

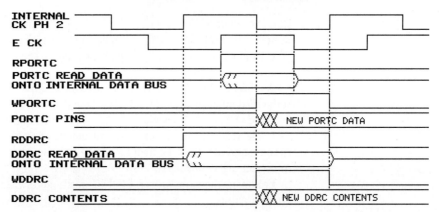

N-channel transistor. This N-channel device turns off. The low at flip-flop 3's $\overline{Q}$ output is also applied to the gate of the totem-pole P-channel transistor via NOR gate/buffer 15. The P-channel transistor turns on, causing an output high at the PORTC pin.

Read PORTC; DDRC Bit = 1

If DDRC flip-flop 2 stores a one, logic gate 4 inverts DDRC's Q output from high to low. Logic gate 4 applies this low to logic gate 5. Logic gate 5's output switches high. Gate 5 applies its output high to gate 6 and inverter 7. Inverter 7 applies its output low to gate 8. Gate 6 outputs a high, and gate 8, a low when they are strobed by a RPORTC (read PORTC) pulse. The low from gate 8 disables transmission gate 9. The high output from gate 6 enables transmission gate 11. Thus, when the DDRC bit = 1, a PORTC read causes the state of the $\overline{Q}$ output of PORTC flip-flop 3 to be inverted by gate 12 and applied via transmission gate 11 to the internal data bus line.

Write to PORTC; DDRC Bit = 0

When the CPU writes binary zero to a DDRC bit, this zero configures its corresponding PORTC pin as an input. Writing a binary zero to DDRC flip-flop 2 switches its Q output low. This low, along with a low from the handshake control line, causes gate 14's output to switch high. NOR gate buffer 15 applies this high to the gate of the totem-pole P-channel transistor. The P-channel transistor switches off. The high from gate 14 is inverted to a low by gate 13 and applied to the gate of the totem-pole N-channel transistor via NAND gate/buffer 16. This low turns the N-channel transistor off. Under these conditions, PORTC flip-flop 3 latches a data bit from the internal data bus. However, PORTC pin logic does not couple this bit to the PORTC pin.

Read PORTC; DDRC Bit = 0

If the DDRC bit = 0, then DDRC flip-flop 2's Q output = 0. This low is inverted (4) and applied to gate 5, along with a high from handshake control. Consequently, gate 5's output switches low. This low causes gate 6 to output a low, disabling transmission gate 11. The low from gate 5 is inverted by gate 7, and applied to gate 8. When strobed by the RPORTC (read PORTC) signal, gate 8 outputs a high, enabling transmission gate 9. Thus, buffer 10 and transmission gate 9 couple the state of the PORTC pin to the internal data bus line.

Read DDRC

A DDRC read asserts RDDRC (read DDRC). RDDRC assertion enables transmission gate 1. Transmission gate 1 couples DDRC flip-flop 2's Q output to the internal data bus line.

PORTC and PORTCL

PORTC reads (DDRC bit = 0) couple PORTC pin levels directly to the internal data bus via input buffer 10 and transmission gate 9. Thus, PORTC does not latch its pin states where DDRC bits = 0. However, PORTCL (PORTC Latched Data Register, control register block address $1005) may latch the states of the PORTC pins. PORTCL asynchronously latches PORTC pin levels during special handshake operations. Chapter 4 covers parallel I/O using handshakes.

EXERCISE 3.1 — M68HC11 PORTB and PORTC Pin Logic

AIM: Analyze the operation of PORTB and PORTC pin logic.

INTRODUCTORY INFORMATION: By completing this exercise, you will get practice and experience in analyzing PORTB and PORTC pin logic, and thereby gain insight into the operation of these parallel ports.

PART I: PORTB Pin Logic

Refer to Figure 3–9. The inputs and outputs of diagram components are labeled with lower case letters. For example, the reset input to AND gate 2 is labeled with an "a." The problems below refer to this input as "2a." Each problem below specifies a particular kind of operation, such as "write a zero to PORTB." Given the operation specified, enter an "X" in the appropriate blank.

EXAMPLE:
OPERATION: Reset PORTB

Component/Pin	Logic State Zero	Logic State One	Device State On	Device State Off
Reset (pulse)		X		
2a		X		
2b		X		
2c		X		
7Q		X		
10a		X		
10b		X		
P-channel transistor				X
N-channel transistor			X	
PB pin	X			

1. Write a zero to PORTB

Component/Pin	Logic State Zero	Logic State One	Device State On	Device State Off
2a				
2b				
2c				
3b				
3c				
4a				
4b				

FIGURE 3–9
Reprinted with the permission of Motorola.

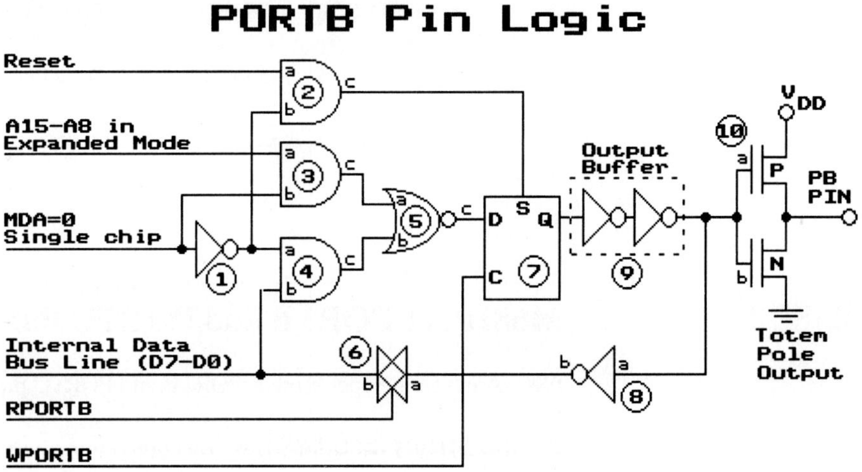

PORTB Pin Logic

Component/Pin	Logic State		Device State	
	Zero	One	On	Off
4c				
5a				
5b				
5c				
WPORTB (pulse)				
7Q				
10a				
10b				
P-channel transistor				
N-channel transistor				
PB pin				

2. Write a one to PORTB

Component/Pin	Logic State		Device State	
	Zero	One	On	Off
2a				
2b				
2c				
3b				
3c				
4a				
4b				
4c				
5a				
5b				
5c				
WPORTB (pulse)				
7Q				
10a				
10b				
P-channel transistor				
N-channel transistor				
PB pin				

3. Read PORTB; 7Q = 1

Component/Pin	Logic State	
	Zero	One
2a	___	___
2b	___	___
2c	___	___
3b	___	___
3c	___	___
4a	___	___
4b	___	___
4c	___	___
5a	___	___
5b	___	___
5c	___	___
WPORTB (pulse)	___	___
8a	___	___
8b	___	___
RPORTB (pulse)	___	___
6a	___	___
6b	___	___
Internal data bus line	___	___

4. Read PORTB; 7Q = 0

Component/Pin	Logic State	
	Zero	One
2a	___	___
2b	___	___
2c	___	___
3b	___	___
3c	___	___
4a	___	___
4b	___	___
4c	___	___
5a	___	___
5b	___	___
5c	___	___
WPORTB (pulse)	___	___
8a	___	___
8b	___	___
RPORTB (pulse)	___	___
6a	___	___
6b	___	___
Internal data bus line	___	___

PART II: PORTC Pin Logic

Refer to Figure 3–10. Components are labeled in the same way as in Part I of this exercise. Given each specified operation and conditions, enter an "X" in the appropriate blank.

FIGURE 3–10
Reprinted with the permission of Motorola.

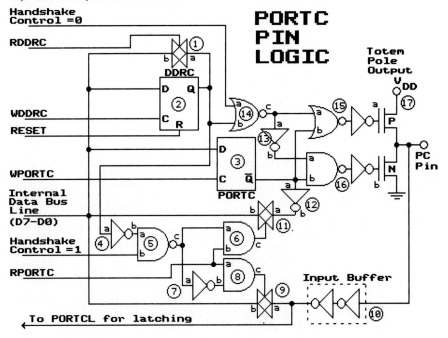

EXAMPLE:
OPERATION: Reset DDRC

Component/Pin	Logic State		Device State	
	Zero	One	On	Off
Reset (pulse)		X		
2Q	X			
4a	X			
4b		X		
5a		X		
5b		X		
5c	X			
6a	X			
7a	X			
7b		X		
8b		X		
RPORTC (pulse)	X			
6c	X			
8c	X			
Gate 11				X
Gate 9				X

1. Write a one to PORTC; DDRC bit = 1

Component/Pin	Logic State		Device State	
	Zero	One	On	Off
2Q				
14a				

152

Component/Pin	Logic State		Device State	
	Zero	One	On	Off
14b	___	___		
14c	___	___		
13a	___	___		
13b	___	___		
WPORTC (pulse)	___	___		
3D	___	___		
$3\overline{Q}$	___	___		
15a	___	___		
15b	___	___		
16a	___	___		
16b	___	___		
17a	___	___		
17b	___	___		
P-channel transistor			___	___
N-channel transistor			___	___
PC pin	___	___		

2. Write a zero to PORTC; DDRC bit = 0

Component/Pin	Logic State		Device State	
	Zero	One	On	Off
2Q	___	___		
14a	___	___		
14b	___	___		
14c	___	___		
13a	___	___		
13b	___	___		
WPORTC (pulse)	___	___		
3D	___	___		
$3\overline{Q}$	___	___		
15a	___	___		
15b	___	___		
16a	___	___		
16b	___	___		
17a	___	___		
17b	___	___		
P-channel transistor			___	___
N-channel transistor			___	___

3. Read PORTC; $3\overline{Q} = 1$; DDRC bit = 1

Component/Pin	Logic State		Device State	
	Zero	One	On	Off
2Q	___	___		
15a	___	___		
16a	___	___		
15b	___	___		
16b	___	___		
17a	___	___		

Component/Pin	Logic State		Device State	
	Zero	One	On	Off
17b	——	——		
P-channel transistor			——	——
N-channel transistor			——	——
PC pin	——	——		
4a	——	——		
4b	——	——		
5a	——	——		
5b	——	——		
5c	——	——		
7a	——	——		
7b	——	——		
6a	——	——		
RPORTC (pulse)	——	——		
6c	——	——		
8b	——	——		
8c	——	——		
Gate 11			——	——
Gate 9			——	——
12a	——	——		
11a	——	——		
Internal data bus line	——	——		

4. Read PORTC; PC pin = 1; $3\overline{Q} = 0$; DDRC bit = 0

Component/Pin	Logic State		Device State	
	Zero	One	On	Off
2Q	——	——		
15a	——	——		
16a	——	——		
15b	——	——		
16b	——	——		
17a	——	——		
17b	——	——		
P-channel transistor			——	——
N-channel transistor			——	——
4a	——	——		
4b	——	——		
5a	——	——		
5b	——	——		
5c	——	——		
7a	——	——		
7b	——	——		
6a	——	——		
RPORTC (pulse)	——	——		
6c	——	——		
8b	——	——		

8c	__ __
Gate 11	__ __
Gate 9	__ __
Internal data bus line	__ __

3.6 THE DL-1414 ALPHANUMERIC DISPLAY

A DL-1414 alphanumeric display serves as the write peripheral for hands-on laboratory exercises in this chapter and Chapter 4. This device is small, durable, and compatible with the protoboard specified in the following laboratory exercises.

The DL-1414 has these features:

- Each of four seventeen-segment LEDs has its own address.
- A built-in RAM latches display addresses and ASCII display data words.
- An integral character generator converts stored ASCII data words to displayed characters. This generator converts ASCII codes $20 through $5F. All other ASCII codes cause a blank display.
- Built-in LED drivers provide consistently bright and clear character displays.
- Inputs are TTL compatible.
- The DL-1414 requires only a +5VDC power supply.

Figure 3–11 shows the arrangement and addresses of the DL-1414's four, seventeen-segment LEDs.[10] Note that LED #0 (address $00) shows the disposition of its seventeen segments. Each seventeen-segment LED can display a character.

Figure 3–12 depicts the DL-1414 pin arrangement and designations. D6–D0 input ASCII data words to the DL-1414. In the laboratory exercises, connect these pins to PORTB. A1–A0 input LED address bits to the DL-1414. In the laboratory exercises, connect these pins to PORTC. A1–A0 address the DL-1414's LEDs as follows:

A1	A0	LED #
0	0	0
0	1	1
1	0	2
1	1	3

The $\overline{WR}$ (write enable) pin is also an input pin. Asserting this pin causes the DL-1414 RAM to latch D6–D0 and A1–A0 data. This pin is low-active.

FIGURE 3–11
Courtesy of Siemens Components Inc.
Adapted with permission.

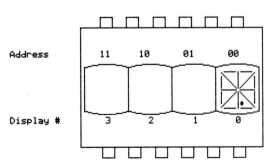

---

[10]Siemens Components Inc., *DL1414T Data Sheet* (Cupertino, CA: Siemens Components, Inc., Optoelectronics Division, August 1989).

FIGURE 3-12
Courtesy of Siemens Components Inc. Adapted with permission.

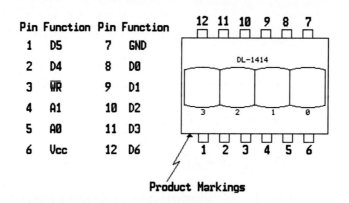

The DL-1414 Display Process

The Figure 3-13 flowchart shows how to display an ASCII character on one of the four DL-1414 LEDs. PORTB and PORTC writes accomplish the actions depicted in this figure.

DL-1414 Timing Characteristics

Figure 3-14 shows DL-1414 timing minimums.[11] These minimums are specified as follows:

T_W = 325ns: Assert $\overline{WR}$ for at least this long.

T_{DS} = 250ns: $\overline{WR}$ assertion and D6–D0 data must coincide for at least this long.

T_{WD} = 75ns: The DL-1414 requires this period to recognize $\overline{WR}$ assertion.

T_{AS} = 400ns: Apply address data to pins A1-A0 for at least this long, before negating $\overline{WR}$.

T_{AH} = 50ns: After negating $\overline{WR}$, wait at least this long before applying new address bits to pins A1–A0.

T_{DH} = 50ns: Apply ASCII data bits to pins D6-D0 for at least this long after negation of the *shortest possible* $\overline{WR}$ (T_W = 325ns).

FIGURE 3-13

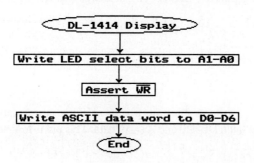

FIGURE 3-14
Courtesy of Siemens Components Inc. Adapted with permission.

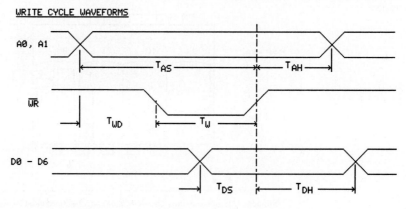

---

[11]Siemens Components Inc., *DL1414T Data Sheet*.

EXERCISE 3.2   Interfacing the DL-1414 Alphanumeric Display

AIM: Given bit identifications for the DL-1414 and M68HC11, complete a wiring diagram by adding pin numbers and lines to indicate interface wiring connections.

INTRODUCTORY INFORMATION: By completing this exercise, you will learn how to connect the interface for laboratory Exercise 3.3.

INSTRUCTIONS:

1. Study this chart. It describes the purposes of the various PORTB and PORTC bits in an interface with the DL-1414.

PORT	M68HC11 PIN#	BIT	PURPOSE
C	9	0	LED Address (A0)
C	10	1	LED Address (A1)
B	42	0	ASCII Data Bit
B	41	1	"
B	40	2	"
B	39	3	"
B	38	4	"
B	37	5	"
B	36	6	"
B	35	7	Assert $\overline{\text{WR}}$

2. Neatly complete the diagram below. Be sure that you comply with these requirements:
   A. The grid on the top of the diagram represents M68HC11 pins.
   B. The symbol on the bottom of the diagram represents the DL-1414. Neatly write all pin numbers on the appropriate DL-1414 pins.
   C. Neatly complete the diagram below by drawing lines to represent wires between the M68HC11 and DL-1414 pins. These lines must represent the following: (1) All required PORTB connections; (2) All required PORTC connections; (3) DL-1414 ground connection; and (4) DL-1414 Vcc connection. Draw lines with square corners. Draw these lines vertically and/or horizontally. Draw lines so that they are routed around (not across) the grid and DL-1414 symbol. Finally, account for all pins on the DL-1414.

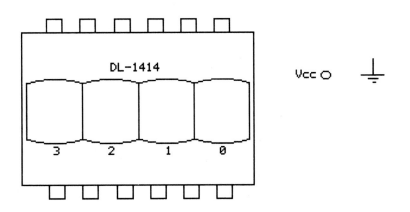

157

3.7 AN EXAMPLE OF PARALLEL OUTPUT INTERFACING

In Exercise 3.2, you completed a diagram depicting an interface between the M68HC11 and a write peripheral—the DL-1414 alphanumeric display. Figure 3–15 and its accompanying program list illustrate supporting software for this interface. This program causes the M68HC11 to display the word "TAXI" on the DL-1414.

Refer to block 1 of Figure 3–15, and program addresses $0000 and $0002. Configure PORTC as an output port by writing ones to all DDRC bits.

Refer to block 2 and program address $0005. By clearing PORTC, the CPU writes zeros to all PORTC output pins. PORTC bits 0 and 1 connect to DL-1414 address pins A0 and A1. Therefore, the CPU addresses the right-most display digit.

Refer to block 3 and program addresses $0008 and $000A. An ASCII "I" = $49. When the CPU loads ACCA with $49, ACCA's most significant bit (B7) becomes binary zero. Bits 0–6 of PORTB connect to the D0–D6 pins of the DL-1414. Bit 7 connects to the DL-1414 $\overline{WR}$ pin, which is low active. Therefore, writing $49 from ACCA to PORTB sends the ASCII data word to DL-1414 pins D0-D6. At the same time, this write asserts $\overline{WR}$. Thus, the DL-1414 displays "I" on its right-most LED.

Refer to block 4 and program address $000D. By incrementing address $1003, the M68HC11 changes the contents of PORTC from $00 to $01 (%00000001). This action

FIGURE 3–15

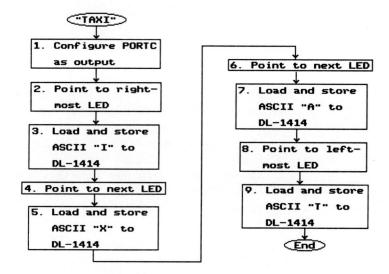

DL-1414 "TAXI" Display

$ADDRESS	SOURCE CODE	COMMENTS
0000	LDAA #FF	Configure PORTC as an output
0002	STAA 1007	" " " "
0005	CLR 1003	Address the right-most LED (#0)
0008	LDAA #49	Load ASCII "I"
000A	STAA 1004	Write ASCII "I" and write enable bit to PORTB
000D	INC 1003	Address the next LED (#1)
0010	LDAA #58	Load ASCII "X"
0012	STAA 1004	Write ASCII "X" and write enable bit to PORTB
0015	INC 1003	Address the next LED (#2)
0018	LDAA #41	Load ASCII "A"
001A	STAA 1004	Write ASCII "A" and write enable bit to PORTB
001D	INC 1003	Address the left-most LED (#3)
0020	LDAA #54	Load ASCII "T"
0022	STAA 1004	Write ASCII "T" and write enable bit to PORTB
0025	WAI	Halt

changes the DL-1414 A0 pin to a one, thus addressing the second display digit from the right.

Refer to block 5 and program addresses $0010 and $0012. The CPU loads the ASCII data word for "X" and writes it to PORTB. This write drives DL-1414 pins D6–D0 with the ASCII code for "X" and asserts $\overline{WR}$.

Refer to blocks 6 and 8, and program addresses $0015 and $001D. Prior to ASCII data word writes, the CPU increments DL-1414's A1–A0 pins to point at display #2, then display #3.

Refer to blocks 7 and 9, and program addresses $0018/$001A and $0020/$0022. After each A1–A0 increment, the CPU loads and writes an ASCII data word. The DL-1414 displays an "A" and then a "T." Since the DL-1414 latches ASCII data, it displays "TAXI" until the interface is powered down, or until new PORTC and PORTB writes change the data.

EXERCISE 3.3 Parallel Output Interfacing Laboratory

REQUIRED MATERIALS: 1 DL-1414 Alphanumeric Display

REQUIRED EQUIPMENT: Motorola M68HC11EVBU or EVB, and terminal

GENERAL INSTRUCTIONS: In this laboratory, interface the M68HC11 with the DL-1414. Then, develop software that demonstrates interface operation. *Carefully* execute the detailed instructions given below.

PART I: Connect the Interface

1. Power down the EVBU (EVB) and terminal. Insert the DL-1414 into your protoboard. Refer to Figure 3–16 and connect the pins of the DL-1414 to the pins of the M68HC11 via the header on the EVBU (EVB).[12]
2. Make your connections per these requirements.

PORT	BIT	PURPOSE
C	0	A0 LED Select
C	1	A1 LED Select
B	0–6	ASCII Data Word
B	7	DL-1414 $\overline{WR}$

3. Do not forget to connect ground and Vcc to the DL-1414. **Leave the EVBU (EVB) and DL-1414 powered down.**
4. Show your connections to the instructor.

_____ *Instructor Check*

FIGURE 3–16

EVBU (EVB) HEADER

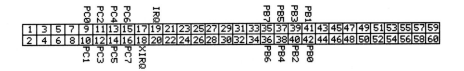

---

[12]Refer also to Exercise 3.2, in which you diagrammed the required connections for this laboratory exercise.

PART II: Display "TAXI" on the DL-1414.

1. Refer to the Figure 3–15 flowchart and its accompanying source code listing.
2. Power up the EVBU (EVB), interface, and terminal. Load, debug, and demonstrate the "TAXI" software.

_____ *Instructor Check*

PART III: Display a 0–1000 Decimal Count on the DL-1414.

1. Develop software that causes the M68HC11 to count in decimal between 0 and 1000 in one-second increments. Display each count on the DL-1414. Refer to the flowcharts in Figures 3–17 through 3–20 while developing your software. Allocate memory as shown in the Figure 3–17 chart.
2. Load, debug, and demonstrate the operation of your software to the instructor.

_____ *Instructor Check*

Note: If you plan to do laboratory Exercise 3.4, use a BUFFALO MOVE command to save your PART III software to the M68HC11's EEPROM.

PART IV: Laboratory Analysis

Answer the questions below by neatly entering your responses in the appropriate blanks.

1. Describe the purpose of the PORTC connections in this exercise.

2. Describe the two functions of the PORTB connections in this exercise.

3. In displaying the word "TAXI" on the DL-1414, the M68HC11 initializes the A1–A0 address bits by _____, then _____ this port to point at succeeding display digits.
4. In what order does the M68HC11 write the ASCII characters for "TAXI"?

FIGURE 3–17

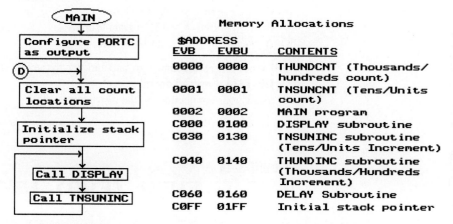

Programming Hints

1. Addresses $0000 and $0001 contain the current decimal count digits. Initialize these locations by clearing them.
2. Do subroutine calls (JSRs) to the DISPLAY subroutine, then the INSUNINC subroutine.
3. After the two subroutine calls, branch back (BRA) to the DISPLAY subroutine call.

FIGURE 3-18

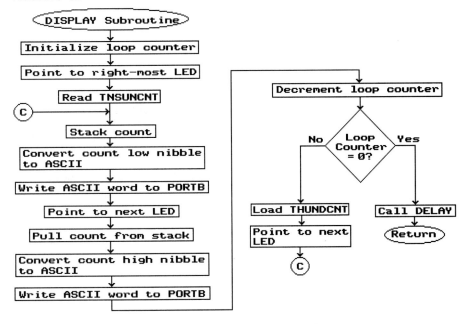

Programming Hints

1. Because the DISPLAY subroutine displays the tens/units count, then the thousands/hundreds count, initialize the loop counter to $02.

2. Stack (PSHA) the count to save the high nibble. Then, convert the count's low nibble to ASCII.

3. Convert the count's low nibble to ASCII by ANDing with mask $0F to clear all high nibble bits, and then ORing with mask $30.

4. After pulling the count from the stack (PULA), use four LSRA instructions to shift the high nibble to the lower four bits. Then, OR the result with mask $30 to convert to ASCII.

5. After converting and displaying both the tens/units and thousands/hundreds digits, call the DELAY subroutine. DELAY causes the DL-1414 to display the count for one second.

FIGURE 3-19

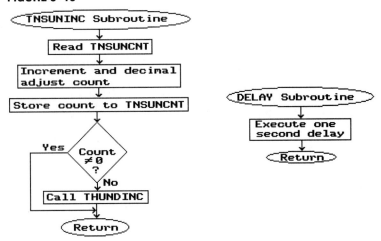

Programming Hints

1. This software counts in decimal. Therefore, increment TNSUNCNT by adding one (do not use INCA). After adding one, use the DAA instruction to decimal adjust the new count. DAA works with ACCA only, so use ACCA for this operation.

2. If adding one and decimal adjusting causes TNSUNCNT to roll over from 99 to 00, then call the THUNDINC subroutine.

FIGURE 3–20

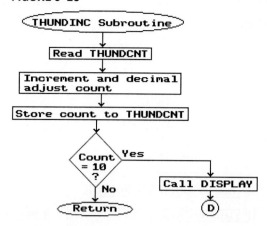

Programming Hints

1. Again, add one to the thousands/hundreds count, then decimal adjust the result (DAA).
2. Compare the count with 10. If the count equals 10, then call the DISPLAY subroutine, because the overall count equals 1000. The M68HC11 must display it—before returning to start the count from 0000 once more. Use the JMP instruction to return via flowchart terminal D. Use RTS for the return where the count does not equal 10.

5. The 0–1000 Decimal Count program maintains the tens/units count at address $_____, and the thousands/hundreds count at address $_____.

6. Incrementing TNSUNCNT and THUNDCNT requires the use of the _____ (mnemonic) and _____ (mnemonic) instructions. These instructions are necessary because

7. Refer to Figure 3–18. Describe the two-step process that converts the count's low nibble to ASCII.

8. Refer to Figure 3–18. Describe the five-step process that converts the high nibble of the count to ASCII.

9. Refer to Figure 3–18. Explain why the loop counter is initialized to $02.

10. Refer to Figure 3–18. Explain why the CPU calls the DELAY subroutine.

11. Refer to Figure 3–19. Explain why the CPU tests the tens/units count to see if it is zero.

12. Refer to Figure 3–20. Explain why the CPU calls the DISPLAY subroutine when the thousands/hundreds count equals 10.

PART V: Laboratory Software Listing

0–1000 COUNT AND DISPLAY—MAIN PROGRAM—EVB

$ADDRESS	SOURCE CODE	COMMENTS
0002	LDAA #FF	Configure PORTC as output
0004	STAA 1007	" " " "
0007	CLR 0	Clear THUNDCNT
000A	CLR 1	Clear TNSUNCNT
000D	LDS #C0FF	Initialize stack pointer
0010	JSR C000	Call DISPLAY subroutine
0013	JSR C030	Call TNSUNINC subroutine
0016	BRA 10	Do another count

0–1000 COUNT AND DISPLAY—MAIN PROGRAM—EVBU

$ADDRESS	SOURCE CODE	COMMENTS
0002	LDAA #FF	Configure PORTC as output
0004	STAA 1007	" " " "
0007	CLR 0	Clear THUNDCNT
000A	CLR 1	Clear TNSUNCNT
000D	LDS #1FF	Initialize stack pointer
0010	JSR 100	Call DISPLAY subroutine
0013	JSR 130	Call TNSUNINC subroutine
0016	BRA 10	Do another count

DISPLAY SUBROUTINE—EVB

$ADDRESS	SOURCE CODE	COMMENTS
C000	LDAB #2	Initialize loop counter
C002	CLR 1003	Point to right-most LED
C005	LDAA 1	Load TNSUNCNT
C007	PSHA	Stack count
C008	ANDA #F	Clear high nibble of count
C00A	ORAA #30	Convert count to ASCII
C00C	STAA 1004	Write ASCII to PORTB
C00F	INC 1003	Point to next LED
C012	PULA	Pull count from stack
C013	LSRA	Move high nibble into low nibble
C014	LSRA	" " " " " "
C015	LSRA	" " " " " "
C016	LSRA	" " " " " "
C017	ORAA #30	Convert count to ASCII
C019	STAA 1004	Write ASCII to PORTB
C01C	DECB	Decrement loop counter
C01D	BEQ C026	Does loop counter = 0?
C01F	LDAA 0	Load THUNDCNT
C021	INC 1003	Point to next LED
C024	BRA C007	Go back and display THUNDCNT
C026	JSR C060	Call DELAY subroutine
C029	RTS	Return

DISPLAY SUBROUTINE—EVBU

$ADDRESS	SOURCE CODE	COMMENTS
0100	LDAB #2	Initialize loop counter
0102	CLR 1003	Point to right-most LED
0105	LDAA 1	Load TNSUNCNT
0107	PSHA	Stack count
0108	ANDA #F	Clear high nibble of count
010A	ORAA #30	Convert count to ASCII
010C	STAA 1004	Write ASCII to PORTB
010F	INC 1003	Point to next LED
0112	PULA	Pull count from stack
0113	LSRA	Move high nibble into low nibble
0114	LSRA	" " " " " "
0115	LSRA	" " " " " "
0116	LSRA	" " " " " "
0117	ORAA #30	Convert count to ASCII

$ADDRESS	SOURCE CODE	COMMENTS
0119	STAA 1004	Write ASCII to PORTB
011C	DECB	Decrement loop counter
011D	BEQ 126	Does loop counter = 0?
011F	LDAA 0	Load THUNDCNT
0121	INC 1003	Point to next LED
0124	BRA 107	Go back and display THUNDCNT
0126	JSR 160	Call DELAY subroutine
0129	RTS	Return

TNSUNINC SUBROUTINE—EVB

$ADDRESS	SOURCE CODE	COMMENTS
C030	LDAA 1	Load TNSUNCNT
C032	ADDA #1	Increment count
C034	DAA	Decimal adjust count
C035	STAA 1	Store to TNSUNCNT
C037	BNE C03C	If count is not = 0, return
C039	JSR C040	If count = 0, call the THUNDINC subroutine
C03C	RTS	Return

TNSUNINC—EVBU

$ADDRESS	SOURCE CODE	COMMENTS
0130	LDAA 1	Load TNSUNCNT
0132	ADDA #1	Increment count
0134	DAA	Decimal adjust count
0135	STAA 1	Store to TNSUNCNT
0137	BNE 13C	If count is not = 0, return
0139	JSR 0140	If count = 0, call the THUNDINC subroutine
013C	RTS	Return

THUNDINC SUBROUTINE—EVB

$ADDRESS	SOURCE CODE	COMMENTS
C040	LDAA 0	Load THUNDCNT
C042	ADDA #1	Increment count
C044	DAA	Decimal adjust count
C045	STAA 0	Store to THUNDCNT
C047	CMPA #10	Does count = 10?
C049	BEQ C04C	If yes, call DISPLAY subroutine
C04B	RTS	If no, return
C04C	JSR C000	Call DISPLAY subroutine
C04F	JMP 7	Go back and start count at 0000

THUNDINC SUBROUTINE—EVBU

$ADDRESS	SOURCE CODE	COMMENTS
0140	LDAA 0	Load THUNDCNT
0142	ADDA #1	Increment count
0144	DAA	Decimal adjust count
0145	STAA 0	Store to THUNDCNT
0147	CMPA #10	Does count = 10?
0149	BEQ 14C	If yes, call DISPLAY subroutine
014B	RTS	If no, return
014C	JSR 100	Call DISPLAY subroutine
014F	JMP 7	Go back and start count at 0000

DELAY SUBROUTINE—EVB

$ADDRESS	SOURCE CODE	COMMENTS
C060	LDAA #6	Initialize loop counter
C062	LDX #D600	Load count-down value
C065	DEX	Decrement count
C066	BNE C065	If count does not = 0, decrement again
C068	DECA	If count = 0, decrement loop counter
C069	BNE C062	If loop counter does not = 0, count down X register again
C06B	RTS	If loop counter = 0, return

DELAY SUBROUTINE—EVBU

$ADDRESS	SOURCE CODE	COMMENTS
0160	LDAA #6	Initialize loop counter
0162	LDX #D600	Load count-down value
0165	DEX	Decrement count
0166	BNE 165	If count does not = 0, decrement again
0168	DECA	If count = 0, decrement loop counter
0169	BNE 162	If loop counter does not = 0, count down X register again
016B	RTS	If loop counter = 0, return

EXERCISE 3.4 Parallel Output Interfacing II

REQUIRED MATERIALS: 1 DL-1414 Alphanumeric Display
1 Push-button Switch, push to close
1 Resistor, 5.1 KΩ

REQUIRED EQUIPMENT: Motorola M68HC11EVBU or EVB, and terminal

GENERAL INSTRUCTIONS: In this laboratory exercise, add an IRQ interrupt feature to the interface and software you developed for Exercise 3.3. Carefully review and execute the instructions given below.

PART I: Modify the Interface

1. Power down your Exercise 3.3 interface, EVBU (EVB), and terminal.
2. Connect a pull-up circuit (Figure 3–21) to the $\overline{IRQ}$ pin of the M68HC11. Connect the pull-up output to the microcontroller's $\overline{IRQ}$ pin (pin 19).
3. Show your connections to the instructor.

_____ *Instructor Check*

PART II: Modify the 0000–1000 Software

1. Develop an IRQ service routine which displays "INT" for one second on the alphanumeric display—when an operator presses the interrupt push-button. After one second, regular 0000–1000 counting routine must resume. Use the Figure 3–22 flowchart as a guide to developing your IRQ service routine. Locate this routine at address $0170 (EVBU) or $C070 (EVB).

FIGURE 3–21

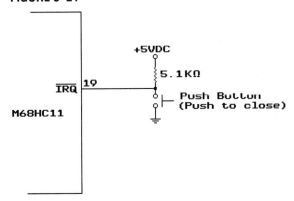

FIGURE 3-22

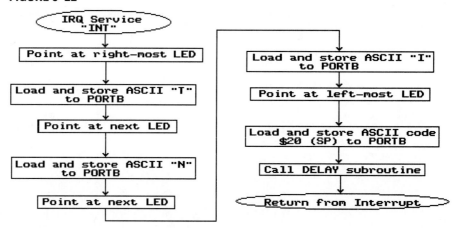

Note: In developing your software, don't forget that the MAIN program must clear the I mask. Otherwise, the processor will not recognize an IRQ interrupt. Also, remember the BUFFALO ROM IRQ pseudovector at location $00EE:

$00EE JMP 170

2. Load, test, and debug your software. When sure that it operates properly, demonstrate it to the instructor.

_____ *Instructor Check*

PART III: Laboratory Analysis

Answer the questions below by neatly entering your answers in the appropriate blanks.

1. Describe the purpose of the pull-up circuit connected to the M68HC11's $\overline{IRQ}$ pin.

2. Describe the purpose of the IRQ Interrupt Service routine.

3. Explain why the interrupt service routine calls a one-second delay.

4. Explain why the IRQ Interrupt Service routine writes ASCII code $20 to the left-most DL-1414 LED.

PART IV: Laboratory Software Listing

MODIFIED MAIN PROGRAM—EVB

$ADDRESS	SOURCE CODE	COMMENTS
0002	CLI	Clear I mask
0003	LDAA #FF	Initialize PORTC
0005	STAA 1007	" "
0008	CLR 0	Clear THUNDCNT
000B	CLR 1	Clear TNSUNCNT
000E	LDS #C0FF	Initialize stack pointer
0011	JSR C000	Call DISPLAY subroutine
0014	JSR C030	Call TNSUNINC subroutine
0017	BRA 11	Call DISPLAY subroutine again

MODIFIED MAIN PROGRAM—EVBU

$ADDRESS	SOURCE CODE	COMMENTS
0002	CLI	Clear I mask
0003	LDAA #FF	Initialize PORTC
0005	STAA 1007	" "
0008	CLR 0	Clear THUNDCNT
000B	CLR 1	Clear TNSUNCNT
000E	LDS #1FF	Initialize stack pointer
0011	JSR 100	Call DISPLAY subroutine
0014	JSR 130	Call TNSUNINC subroutine
0017	BRA 11	Call DISPLAY subroutine again

MODIFIED THUNDINC SUBROUTINE—EVB

$ADDRESS	SOURCE CODE	COMMENTS
C040	LDAA 0	Load THUNDCNT
C042	ADDA #1	Increment count
C044	DAA	Decimal adjust count
C045	STAA 0	Store THUNDCNT
C047	CMPA #10	Does count = 10?
C049	BEQ C04C	If yes, call DISPLAY subroutine
C04B	RTS	If no, return
C04C	JSR C000	Call DISPLAY subroutine
C04F	JMP 8	Go back and start count at 0000

MODIFIED THUNDINC SUBROUTINE—EVBU

$ADDRESS	SOURCE CODE	COMMENTS
0140	LDAA 0	Load THUNDCNT
0142	ADDA #1	Increment count
0144	DAA	Decimal adjust count
0145	STAA 0	Store THUNDINC
0147	CMPA #10	Does count = 10?
0149	BEQ 14C	If yes, call DISPLAY subroutine
014B	RTS	If no, return
014C	JSR 100	Call DISPLAY subroutine
014F	JMP 8	Go back and start count at 0000

IRQ SERVICE ROUTINE—EVB

$ADDRESS	SOURCE CODE	COMMENTS
C070	CLR 1003	Point at right-most LED
C073	LDAA #54	Load ASCII "T"
C075	STAA 1004	Store ASCII "T" to PORTB
C078	INC 1003	Point at next LED
C07B	LDAA #4E	Load ASCII "N"
C07D	STAA 1004	Store ASCII "N" to PORTB
C080	INC 1003	Point at next LED
C083	LDAA #49	Load ASCII "I"
C085	STAA 1004	Store ASCII "I" to PORTB
C088	INC 1003	Point at left-most LED
C08B	LDAA #20	Load ASCII SP character
C08D	STAA 1004	Store ASCII SP to PORTB
C090	JSR C060	Call DELAY subroutine
C093	RTI	Return from interrupt

IRQ SERVICE ROUTINE—EVBU

$ADDRESS	SOURCE CODE	COMMENTS
0170	CLR 1003	Point at right-most LED
0173	LDAA #54	Load ASCII "T"
0175	STAA 1004	Store ASCII "T" to PORTB
0178	INC 1003	Point at next LED
017B	LDAA #4E	Load ASCII "N"
017D	STAA 1004	Store ASCII "N" to PORTB
0180	INC 1003	Point at next LED
0183	LDAA #49	Load ASCII "I"

0185	STAA 1004	Store ASCII "I" to PORTB
0188	INC 1003	Point at left-most LED
018B	LDAA #20	Load ASCII SP character
018D	STAA 1004	Store ASCII SP to PORTB
0190	JSR 160	Call DELAY subroutine
0193	RTI	Return from interrupt

EXERCISE 3.5 Parallel Input/Output Interfacing

REQUIRED MATERIALS: 1 DL-1414 Alphanumeric Display
 1 DIP Switch (4 switches)
 4 Resistors, 5.1 KΩ

REQUIRED EQUIPMENT: Motorola M68HC11EVBU or EVB, and terminal

GENERAL INSTRUCTIONS: In this exercise, connect four DIP-mounted switches as pull-ups. Connect the pull-up outputs to PORTC pins PC3–PC0. Connect PORTB pins PB6–PB0 to the D6–D0 inputs on the DL-1414. Connect PORTB pin PB7 to the DL-1414 $\overline{WR}$ pin. Tie DL-1414 A1 and A0 pins high, to permanently address LED #3. With these connections, the DIP-based pull-up circuits function as a read peripheral, and the DL-1414 functions as a write peripheral. Develop supporting software that causes the M68HC11 to continuously read PORTC, convert the read data to ASCII, and write the ASCII data word to the DL-1414. This software causes the DL-1414 to continuously display a decimal number on its LED #3. The displayed number always reflects the state of the DIP-mounted switches—if they are set to %0000–%1001. The software ignores switch settings %1010–%1111.

PART I: Connect the Interface

1. Power down the EVBU (EVB) and terminal. Refer to Figure 3–23; connect interface components to the M68HC11, via the EVBU (EVB) header.
2. Show your connections to the instructor.

_____ *Instructor Check*

FIGURE 3–23

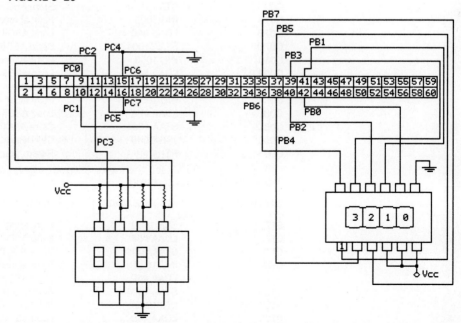

FIGURE 3-24

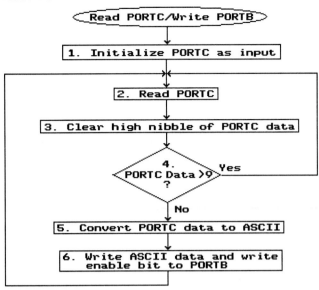

PART II: Software Development

Using the Figure 3-24 flowchart as a guide, develop supporting software for the interface. Locate your program at RAM address $0000.

Refer to Figure 3-24. After reading PORTC (flowchart block 2), clear the high nibble of the switch data by ANDing it with mask $0F (block 3). Accumulator bits 3-0 (low nibble) now contain the DIP switch data. Compare this data with 09. If switch data exceeds 09, branch back (block 4) and read PORTC again (block 2). Stay in this loop until the interface operator changes the DIP switch settings to %1001 or lower. In this way, the software never recognizes switch settings $A-$F.

If DIP switch data is %1001 or lower, convert it to ASCII by ORing with mask $30 (block 5). After switch data conversion, the accumulator's MS bit is binary zero. When the M68HC11 writes the ASCII data to PORTB, this particular bit asserts the DL-1414 $\overline{\text{WR}}$ pin.

PART III: Interface Debug and Demonstration

Power up the EVBU (EVB), terminal, and interface. Load your software. Debug the interface and supporting software. Demonstrate interface operation to the instructor.

_____ *Instructor Check*

PART IV: Laboratory Analysis

Answer the questions below by neatly entering your answers in the appropriate blanks.

1. List the range of DIP switch combinations that the write peripheral can display.
2. In the spaces provided, enter a modified program that handles all input switch data from $0 through $F. HINT: Subtracting $9 from any of the hex digits $A-$F returns the correct low nibble of the ASCII data word. ORing the resultant low nibble with a proper mask returns correct ASCII bits 6 through 4.

$ADDRESS	SOURCE CODE	COMMENTS
_____	_____	_____
_____	_____	_____
_____	_____	_____
_____	_____	_____

3. Load, debug, and demonstrate your modified software to the instructor.

_____ *Instructor Check*

PART V: Laboratory Software Listing

READ PORTC/WRITE PORTB—EVB AND EVBU

$ADDRESS	SOURCE CODE	COMMENTS
0000	CLR 1007	Initialize PORTC as input
0003	LDAA 1003	Read PORTC switch data
0006	ANDA #F	Clear high nibble of switch data
0008	CMPA #9	Switch data > 9?
000A	BHI 3	If yes, do another PORTC read
000C	ORAA #30	If no, convert switch data to ASCII
000E	STAA 1004	Write ASCII and write enable to PORTB
0012	BRA 3	Do it again

SOFTWARE MODIFICATION FOR SWITCH DATA $0–$F—EVB AND EVBU

$ADDRESS	SOURCE CODE	COMMENTS
0000	CLR 1007	Initialize PORTC
0003	LDAA 1003	Read PORTC switch data
0006	ANDA #F	Clear high nibble
0008	CMPA #9	Is switch data > 9?
000A	BHI 13	If yes, branch to address $0013
000C	ORAA #30	If no, convert to ASCII 0–9
000E	STAA 1004	Store ASCII code and write enable to PORTB
0011	BRA 3	Do it again
0013	SUBA #9	If > 9, convert low nibble to ASCII
0015	ORAA #40	Convert high nibble to ASCII $A–$F
0017	STAA 1004	Store ASCII code and write enable to PORTB
001A	BRA 3	Do it again

3.8 CHAPTER SUMMARY

To communicate with its peripherals, the microcontroller uses two techniques—serial communication and parallel communication. Each of these techniques has a major advantage and disadvantage. Serial communication requires only a single data line or pair of data lines between microcontroller and peripheral. However, all binary bits must be transmitted over a data line, *one at a time.* Thus, serial communications are slow. Parallel communication is faster because each bit in a data word has its own data communications line between microcontroller and peripheral. But this increased speed necessitates the complexity and expense of multiple communications lines. This section covers M68HC11 general purpose parallel I/O.

General purpose parallel I/O uses several on-chip resources:

- PORTB is an output port that can be either read or written. PORTB reads return data previously written to this port. Writes to PORTB drive its pins with the write data.
- The M68HC11 can configure PORTC pins as either inputs or outputs. Writes to PORTC pins configured as outputs drives these pins with the write data. Reads of PORTC pins configured as inputs returns the states of these input pins. PORTC reads return the last data written to pins configured as outputs. PORTC writes to pins configured as inputs cause PORTC to latch and retain these data bits. PORTC pin logic will not drive its pins with write data, when these pins are configured as inputs.
- The M68HC11 uses DDRC bits to configure PORTC pins as inputs or outputs. Writing a one to a DDRC bit configures its equivalent PORTC pin as an output. Writing a zero configures the equivalent PORTC pin as an input.

This chapter's hands-on exercises use the DL-1414 alphanumeric display as the parallel write peripheral. This device has four, seventeen-segment LEDs, each individually addressable. The DL-1414 incorporates a RAM that latches input display data and address bits. The DL-1414 also has an on-board character generator that translates ASCII codes $20 through $5F into displayed alphanumeric characters and symbols. To display a single character on the DL-1414, the M68HC11 must do the following:

Write LED address bits to DL-1414 pins A0 and A1.
Assert the DL-1414's $\overline{WR}$ pin.
Write an ASCII data word ($20–$5F) to DL-1414 pins D0–D6.

This chapter's hands-on laboratory exercises involve these activities:

Write ASCII code for the word "TAXI" to the DL-1414 and display it.
Display a 0–1000 decimal count on the DL-1414. This count increments at one-second intervals, and returns automatically from count 1000 to count 0000.
Enhance the 0–1000 counting program with a push-button IRQ interrupt feature. This interrupt causes the DL-1414 to display "INT" for one second.
Read DIP switch data from PORTC, and write this switch data to PORTB for display by the DL-1414.

EXERCISE 3.6 Chapter Review

Answer the questions below by neatly entering your answers in the appropriate blanks.

1. List the two techniques generally used by the M68HC11 to exchange data with a peripheral.

2. Describe the main advantage of serial communications.

3. Describe the main disadvantage of serial communications.

4. Describe the main advantage of parallel communications.

5. Describe the main disadvantage of parallel communications.

6. PORTB is an _____ (input or output) port only.

7. Describe the data returned when the M68HC11 reads PORTB.

8. A programmer can configure PORTC pins as either _____ or _____.
9. Describe what happens when the CPU writes data to PORTC pins configured as inputs.

10. Describe the data returned when the M68HC11 reads PORTC pins configured as outputs.

11. Describe the data returned when the M68HC11 reads PORTC pins configured as inputs.

12. To configure PORTC pins 0, 2, 4, and 6 as inputs, and pins 1, 3, 5, and 7 as outputs, the M68HC11 CPU must write control word $_____ to _____ (name of register), at address $_____.
13. Describe the effect on PORTB pins, when the M68HC11 writes to PORTB.

14. PORTC pins connected to a read peripheral must be configured as _____, by writing _____ to corresponding _____ bits.

$ADDRESS	SOURCE CODE
C000	LDAA #AA
C002	STAA 1007
C005	LDAA 1003
C008	ANDA #_____
C00A	STAA 0

15. Refer to the program segment above. Describe the purpose of the instruction at address $C008.

16. Refer to the program segment above. What hexadecimal mask value belongs in the blank accompanying the ANDA instruction at address $C008? $_____
17. Explain the purpose of flip-flop 7 in the PORTB pin logic (Figure 3–5).

18. What happens to the data latched by this flip-flop (see question 17) when the M68HC11 reads PORTB?

Note: Questions 19–22 below refer to Figure 3–7. Answer these questions by entering the circled numbers from the Figure 3–7 diagram.

19. A write bit is latched by component _____ and applied to the gates of totem pole—component _____, via component _____ or component _____.
20. If the DDRC bit = 0, then the read bit originates at the _____ (specify this origin by name), is buffered by component _____, passed by component _____, and applied to the internal data bus line.
21. If the DDRC bit = 1, then a read bit originates at the _____ (specify this origin by name), is inverted by component _____, passed by component _____, and applied to the internal data bus line.
22. Component _____ controls the data paths for reads and writes.
23. Each DL-1414 LED contains _____ LED segments.
24. The DL-1414 character generator converts ASCII codes $_____ through $_____.
25. To address the DL-1414 LED second from the left, A1 must equal %_____ and A0 must equal %_____.

26. To display a "T" on a DL-1414's LED, input bits D6–D0 must be:

D6– – – – – –D0

$ADDRESS	SOURCE CODE
C000	LDAA #3
C002	STAA 1007
C005	LDAA #1
C007	STAA 1003
C00A	LDAA #56
C00C	STAA 1004
C00F	WAI

27. The program segment above displays the letter _____ at the second LED from the _____ (left) _____ (right) on the DL-1414.

28. Refer to the program segment above. Describe the purpose of the instructions at addresses $C000 and C002.

29. Refer to the program segment above. Describe the purpose of the instructions at addresses $C005 and $C007.

30. Refer to the program segment above. Describe the two purposes served by the instructions at addresses $C00A and $C00C.

4 Parallel I/O Using the Simple-Strobed and Full-Handshake Modes

4.1 GOALS AND OBJECTIVES

Many I/O applications use asynchronous parallel communications. "Asynchronous" means that the receiver and transmitter use their own separate clocks to strobe or enable communication operations.[1] When microcontroller and peripheral use asynchronous parallel communications, they need ways of sending "I've sent data" and "I've received data" messages to each other to coordinate these communications. These messages are called "handshakes." Handshake messages consist of binary signal levels, or toggles, sent over dedicated handshake lines.

M68HC11 simple-strobed and full-handshake modes furnish simple and efficient ways of sending these messages between microcontroller and peripherals. The goals of this chapter are to help the reader understand how to implement simple-strobed and full-handshake I/O, and to instill an appreciation for the versatility and flexibility of these M68HC11 handshake resources.

After studying this chapter and completing its exercises and hands-on activities you will be able to:

Identify microcontroller resources used for simple-strobed and full-handshake operations.
Derive control words that configure the microcontroller for simple-strobed and full-handshake operations.
Connect interfaces and develop supporting software for simple-strobed and full-handshake communications.
Determine whether descriptive phrases apply to simple-strobed I/O, to full-handshake I/O, or to both.

---

[1]In many cases, providing a synchronous clock signal to both transmitter and receiver becomes impractical. These devices may be physically remote from each other; thus, providing an undistorted and noise-free clock signal to both receiver and transmitter requires needless expense and complication. Exercise 4.2 furnishes another example where synchronous clocking is impractical. In this case, the transmitter consists of push-button switches. The human operator is incapable of synchronizing key-presses with the receiver's 2 MHz clock signal. One could devise a strobing scheme to synchronize key-press data, but this would entail needless complication.

4.2 SIMPLE-STROBED I/O

Use simple-strobed I/O for two-way parallel data transfers between the microcontroller and its peripherals, when you require a flexible handshaking scheme. Simple-strobed I/O is flexible for two reasons: First, it can provide two-way communications with a single read/write peripheral. Second, it permits communications between the M68HC11 and two peripherals—one a read and the other a write device.[2]

On-Chip Resources Used by the M68HC11 for Simple-Strobed I/O

Simple-strobed I/O uses these on-chip resources:

- Strobe A (STRA, M68HC11 pin 4). A read peripheral uses the STRA pin to inform the M68HC11 that it has driven parallel communication lines with read data. Thus, the peripheral uses STRA to tell the microcontroller "I've sent data."
- Strobe B (STRB, M68HC11 pin 6). The M68HC11 uses STRB to tell a write peripheral that it has driven parallel communications lines with write data. Thus, the M68HC11 uses STRB to tell a peripheral "I am writing data."
- Parallel I/O Control Register (PIOC, control register block address $1002). In simple-strobed I/O the PIOC has two purposes. First, it configures the microcontroller for parallel I/O operations. Second, a PIOC bit serves as a status flag during simple-strobed operations.
- Port C Data Register (PORTC), and Port C Latched Data Register (PORTCL). These registers reside in the control register block at addresses $1003 and $1005, respectively. PORTC/PORTCL receive read data from a peripheral. PORTCL consists of D flip-flops connected to PORTC pins. Asynchronous STRA assertion by the read peripheral causes PORTCL to latch PORTC input data so that the CPU can read it. A PORTCL read clears the PIOC status flag for the next data exchange.
- Data Direction Register for Port C (DDRC, control register block address $1007). For simple-strobed operations, DDRC initializes selected PORTC pins as data inputs.[3]
- Port B Data Register (PORTB, control register block address $1004). The microcontroller uses PORTB to send write data to a peripheral. Recall from Chapter 3 that PORTB functions as an output port only.

The Sequence of Events in a Simple-Strobed Data Transfer

Figure 4–1 illustrates the steps involved in a data exchange between microcontroller and peripheral. Remember that the microcontroller can read from one peripheral and write to another, or read and write to a single peripheral.

Refer to blocks 2 and 6 of the flowchart. The read peripheral sends a handshake message (STRA—block 2) which means "I've sent read data." The microcontroller sends a similar message (STRB—block 6) which means "I've sent the write data."

Refer to block 2. How does the M68HC11 know that the peripheral has asserted STRA? The M68HC11 knows because STRA assertion sets a status flag in the PIOC: the Strobe A Flag (STAF). The STAF is PIOC bit 7 (see Figure 4–2).

Programmers can choose from two techniques to find out whether STRA has asserted, and the STAF set. These two techniques are status polling and interrupt control.

---

[2]Exercise 4.2 uses both a read and a write peripheral. The read peripheral consists of a bank of push-buttons. The write peripheral displays key-press data.

[3]Simple-strobed I/O uses PORTC as an input port. Therefore, configure all PORTC pins connected to a read peripheral as inputs. Full-handshake I/O uses PORTC for both inputs and outputs. As shown in Chapter 3, PORTC pins can also function as general-purpose inputs or outputs. These pins can serve as general-purpose I/O pins concurrently with other PORTC pins used for simple-strobed operations.

FIGURE 4–1

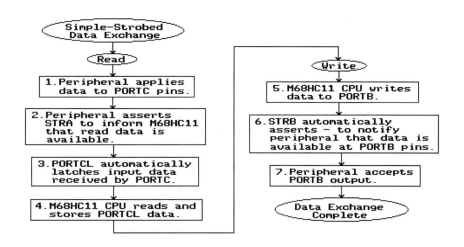

Status Polling, Simple-Strobed Mode

In status polling, the CPU periodically reads the PIOC and checks the STAF bit. If STAF = 1, the CPU reads input data from PORTCL. A PIOC read followed by a PORTCL read automatically resets the STAF. After reading PORTCL, the CPU can store the read data, then write output data to PORTB. This action automatically asserts STRB for two Phase Two (PH2) clock cycles.

The Figure 4–3 flowchart and program segment summarize a simple-strobed data exchange using status polling. Refer to block 1 of the Figure 4–3 flowchart and to program addresses $0100–$0105. By writing $03 to the PIOC register, the CPU configures for simple-strobed mode, with interrupts disabled.[4] By clearing all DDRC bits, the CPU configures all PORTC pins as inputs. Here we assume that the read peripheral is an 8-bit device. PORTB is an output port only and cannot be configured any other way.

Refer to block 2 and program address $0108. If data exchanges are relatively infrequent, the CPU can conduct other routine business between its recurring PIOC reads (block 3). For example, it could do mathematical operations on data read from PORTCL in previous data exchanges.

Refer to blocks 3 and 5, and program addresses $0120–$0125. The CPU must read PIOC then PORTCL to reset the STAF. Of course, the PORTCL read loads ACCA with data sent by the peripheral.

Refer to block 7 and program address $012A. Writing output data to PORTB causes STRB to assert for two clock cycles. STRB assertion starts with the first positive-going edge of the internal PH2 clock—after the PORTB write.

Interrupt Control, Simple-Strobed Mode

With interrupt control, STRA assertion by the peripheral causes the PIOC STAF bit to set. Also, this STRA assertion generates an internal IRQ interrupt within the microcontroller. Thus, the microcontroller can conduct routine operations without doing the periodic PIOC reads required by status polling. When the peripheral asserts STRA, this action generates an IRQ interrupt. The M68HC11 can then service this interrupt by reading PIOC and PORTCL. Reading PIOC and PORTCL captures the read data and resets the STAF bit.

FIGURE 4–2
Reprinted with permission of Motorola.

PIOC

STAF	STAI	CWOM	HNDS	OIN	PLS	EGA	INVB
B7							B0

---

[4]This write sets the EGA and INVB bits in the PIOC. These bits define the STRA active edge and STRB active level. See page 180 for a discussion of these bits.

FIGURE 4–3

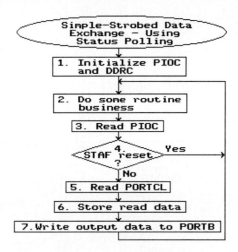

Simple-Strobed Data Exchange—Status Polling

$ADDRESS	SOURCE CODE	COMMENTS
0100	LDAA #3	Initialize PIOC
0102	STAA 1002	" "
0105	CLR 1007	Initialize DDRC
0108	(Do routine business)	
----		" " "
----		" " "
0120	LDAA 1002	Read PIOC
0123	BPL 108	If STAF=0, do more routine business
0125	LDAA 1005	If STAF=1, read PORTCL, capturing read data and resetting the STAF
0128	STAA 0	Store read data
012A	STAA 1004	Write this data to peripheral, via PORTB
012D	BRA 108	Do it again

In cases where data exchanges come at infrequent or unpredictable times, interrupt control becomes useful. For example, a peripheral collecting data on an occasional environmental phenomenon might well be interfaced to the M68HC11 using interrupt control.

Setting PIOC bit 6 establishes interrupt control (see Figure 4–4). This bit is called the STRA Interrupt Enable (STAI) bit.

The Figure 4–5 flowchart and accompanying program segments summarize a simple-strobed data exchange using interrupt control. Refer to block 1 and program addresses $0100–$010B. By writing $43 to the PIOC, the CPU initializes for simple-strobed mode with interrupts enabled (STAI = 1).[5] By clearing the DDRC register, the CPU configures all PORTC pins as inputs. Again, we assume the read peripheral to be an 8-bit device. PORTB functions only as an output. The instructions at $0108–$010B initialize the BUFFALO IRQ pseudovector to $0130.[6]

FIGURE 4–4
Reprinted with permission of Motorola.

PIOC

STAF	STAI	CWOM	HNDS	OIN	PLS	EGA	INVB
B7							B0

---

[5]This write also sets the EGA and INVB bits in the PIOC. These bits define the STRA active edge and STRB active level. See page 180.

[6]Thus, when an IRQ occurs, the M68HC11 vectors to address $00EE, and starts fetching and executing. The first instruction fetched is JMP xxxx (where xxxx = pseudovector address). The instructions at $0108–$010B make xxxx = $0130.

FIGURE 4–5

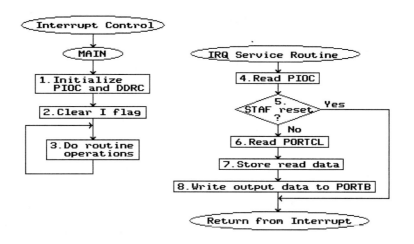

Simple-Strobed Data Exchange—Interrupt Control

$ADDRESS	SOURCE CODE	COMMENTS
	MAIN PROGRAM	
0100	LDAA #43	Initialize PIOC
0102	STAA 1002	" "
0105	CLR 1007	Initialize DDRC
0108	LDD #130	Initialize IRQ pseudovector
010B	STD EF	" " "
010D	CLI	Clear I flag
010E	(Do routine business)	

012E	BRA 10E	Keep doing routine business
	IRQ SERVICE ROUTINE	
0130	LDAA 1002	Read PIOC
0133	BPL 13D	If STAF=0, return
0135	LDAA 1005	If STAF=1, read PORTCL, to capture read data and reset STAF
0138	STAA 0	Store read data
013A	STAA 1004	Write this data to peripheral, via PORTB
013D	RTI	Return from interrupt

Refer to block 2 and program address $010D. An M68HC11 reset automatically sets the I mask in the condition code register. The I mask inhibits IRQ interrupts until the CPU clears it. When the M68HC11 vectors to the IRQ service routine, it again automatically sets the I mask for the duration of the interrupt routine to avoid nested interrupts. When the M68HC11 executes the RTI instruction, it automatically resets the I mask.

Refer to block 3 and program addresses $010E–$012E. The microcontroller can do other routine business until the peripheral initiates a data exchange by asserting STRA.

Refer to blocks 4 and 5, and program addresses $0130–$0133. In some cases another device—other than the communicating peripheral in our flowchart—uses the external $\overline{IRQ}$ pin to initiate interrupts for its own purposes. In such a case, an IRQ could be generated by either the communicating peripheral or the device connected to the external $\overline{IRQ}$ pin. The microcontroller uses the block 4 and block 5 actions to find out which device in fact initiated the IRQ. If a device other than the communicating peripheral initiated the IRQ, STAF = 0 because STRA was not asserted. In this case the M68HC11 must return from the interrupt (as shown in the flowchart) or pseudovector to another appropriate service routine.

FIGURE 4–6
Reprinted with permission of Motorola.

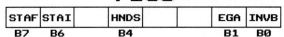

The PIOC and Simple-Strobed Mode

As explained above, the PIOC serves two purposes. First, it configures the M68HC11 for desired parallel I/O operations. Second, the STAF indicates when the communicating peripheral supplies read data. PIOC bits 0, 1, 4, 6, and 7 combine to serve these two functions (see Figure 4–6).

The STAF bit (B7) sets when the communicating peripheral asserts STRA. The M68HC11 CPU resets the STAF by reading the PIOC, then PORTCL.

The STAI (B6) enables the interrupt control function. Clearing the STAI bit disables interrupts, and setting this bit enables interrupts:

    STAI = 0   interrupts disabled
    STAI = 1   interrupts enabled

HNDS (B4) is the Handshake/Simple-Strobe Mode Select bit. Clearing the HNDS bit selects simple-strobed mode. Setting the HNDS bit selects the full-handshake mode:

    HNDS = 0   simple-strobed mode
    HNDS = 1   full-handshake mode

EGA (B1) is Active Edge Select for STRA Assertion. Clearing the EGA bit means that the peripheral asserts STRA by driving the STRA pin from high to low. Setting the EGA means that a low-to-high transition asserts STRA:

    EGA = 0   high-to-low transition asserts STRA
    EGA = 1   ow-to-high transition asserts STRA

INVB (B0) is Invert STRB Output. STRB assertion can be defined as either high or low. Clearing INVB makes STRB active low, and setting INVB makes STRB active high:

    INVB = 0   STRB active low
    INVB = 1   STRB active high

After an M68HC11 reset, the PIOC register contains either $03 or $07 (the state of PIOC bit 2 is unpredictable). In either case, a reset configures the microcontroller as follows:

    interrupt control disabled
    Simple-strobed I/O mode
    STRA—rising edge asserts
    STRB—active high asserts

EXERCISE 4.1 PIOC Control Words for Simple-Strobed Mode

AIM: Derive control words that configure the microcontroller for simple-strobed operation.

INTRODUCTORY INFORMATION: This exercise teaches you how to calculate PIOC control words. These control words configure the M68HC11 for simple-strobed I/O operations.

INSTRUCTIONS: Use the configuration descriptions and refer to Figure 4–6 while deriving appropriate PIOC control words. Enter these hex control words in the blanks provided.

Configuration	PIOC Control Word
EXAMPLE Interrupt disabled Simple-strobed I/O STRA—positive-going edge asserts STRB—active high	$___03___
Interrupt disabled Simple-strobed I/O STRA—negative-going edge asserts STRB—active low	$___00___
Interrupt enabled Simple-strobed I/O STRA—positive-going edge asserts STRB—active low	$___42___
Interrupt enabled Simple-strobed I/O STRA—positive-going edge asserts STRB—active high	$___43___
Interrupt disabled Simple-strobed I/O STRA—negative-going edge asserts STRB—active high	$___01___
Interrupt enabled Simple-strobed I/O STRA—negative-going edge asserts STRB—active high	$___41___
Interrupt enabled Simple-strobed I/O STRA—negative-going edge asserts STRB—active low	$___40___
Interrupt disabled Simple-strobed I/O STRA—positive-going edge asserts STRB—active low	$___02___

EXERCISE 4.2 Simple-Strobed I/O

REQUIRED MATERIALS: 1 DL-1414 Alphanumeric Display
4 Resistors, 5.1 KΩ
4 Push-button Switches, push-to-close
1 74HC20 Dual, Four-Input NAND

REQUIRED EQUIPMENT: Motorola M68HC11EVBU or EVB, and Terminal

GENERAL INSTRUCTIONS: Refer to Figure 4–7. In this laboratory, use simple-strobed I/O to detect push-button key-presses, then display key-press data on the alphanumeric display. A bank of four push-button switches serves as the input (read) peripheral and the alphanumeric display serves as the output (write) peripheral. Each push-button is a switch in a pull-up circuit. The output (active low) of each pull-up circuit connects to two points: (1) its respective PORTC pin, and (2) its assigned input to a NAND gate. The output from the

NAND gate connects to the M68HC11 STRA pin. Thus, pressing a push-button not only supplies input data to the microcontroller, but also asserts STRA. This read peripheral (made up of the four push-buttons) is only 4 bits wide, and connects to bits 4-7 of PORTC. Configure these pins as inputs. This leaves four PORTC pins (bits 3–0) available for general-purpose I/O. Use bits 0 and 1 of PORTC to supply address information to the alphanumeric display—that is, configure PORTC bits 0 and 1 as general-purpose outputs.

The microcontroller performs routine operations by counting between 0 and 1000 at one second intervals, displaying each count on the alphanumeric display. Refer to Chapter 3, Exercise 3.4, for detailed information on the DL-1414 and the zero to 1000 count software.[7]

PART I: Connect the Interface

Power down the EVBU (EVB) and terminal. Refer to Figure 4–7 while connecting the interface components to the pins of the M68HC11, via the EVBU (EVB) header.

Make your connections per these specifications:

PORT	BIT	PURPOSE
B	0–6	ASCII data words to DL-1414
C	0	A0 address bit to DL-1414 (output)
C	1	A1 address bit to DL-1414 (output)
C	2–3	Grounded—configured as inputs
C	4	Push-button 3 (input)
C	5	Push-button 2 (input)
C	6	Push-button 1 (input)
C	7	Push-button 0 (input)
(D)	6	STRA input
(D)	7	STRB output

Show your connections to the instructor.

_____ *Instructor Check*

PART II: Keypad Read and Display—Status Polling

Develop software which executes these tasks:

1. Count repeatedly from 0 to 1000 at one-second intervals, displaying each count on the alphanumeric display (routine operation). Refer to Chapter 3, Exercise 3.3 for detailed information on the counting software.

FIGURE 4–7

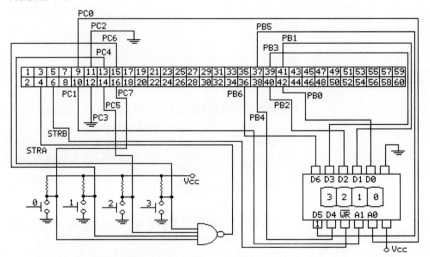

---

[7]Initialize the HNDS bit to select simple-strobed mode. In this configuration STRB asserts for two internal clock cycles coincident with a write to PORTB. STRB asserts even if there has been no previous STRA assertion or PORTCL read. Thus, we can use STRB to write enable the DL-1414 during routine counting operations.

2. Between each count, call the Display Clear subroutine (DISCLR). This subroutine extinguishes all segments on the alphanumeric display. Then, call the Key-press Detect and Display subroutine (KYDETDIS). KYDETDIS polls the PIOC to find out if the operator has pressed a key on the input peripheral. If a key has been pressed, KYDETDIS displays the push-button number for one second on the appropriate DL-1414 LED. For example, if the operator presses key number 2, then a "2" should appear on LED number 2 of the DL-1414.

KYDETDIS uses this algorithm:

1. Read the PIOC and check the STAF. If the STAF is reset, return from KYDETDIS. If the STAF is set, go to step 2.
2. Read key-press data from PORTCL.
3. Decode the key-press data by determining how many rotate left executions it takes to move the key-press bit into the C flag. The number of rotate executions equals the decoded key press value. For example, two rotate executions means that push-button number 2 was pressed.
4. Load both accumulators with decoded key press data (e.g., $02). Write ACCB to PORTC as address information for the alphanumeric display (recall that PORTC bits 0 and 1 are LED address outputs). Convert ACCA information to ASCII and write to PORTB (data output to the DL-1414).
5. Call a one-second delay (DELAY subroutine).
6. Return from KYDETDIS.

Use simple-strobed I/O with interrupts disabled. Use the memory allocation chart and the Figures 4–8, 4–9, and 4–10 as guides to software development.

FIGURE 4–8

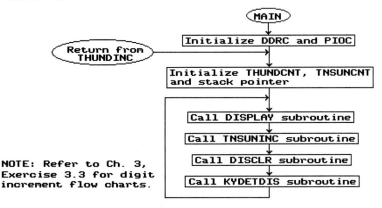

NOTE: Refer to Ch. 3, Exercise 3.3 for digit increment flow charts.

Programming Hints—MAIN Program

1. Initialize the PIOC for simple-strobed operation, with STRA low-to-high assertion, and STRB active low.
2. Configure PORTC bits 2–7 as inputs and bits 0–1 as outputs.

Memory Allocations

$ADDRESS		
EVBU	EVB	CONTENTS
0000	0000	THUNDCNT (Thousands/Hundreds Count)
0001	0001	TNSUNCNT (Tens/Units Count)
0002	0002	MAIN program
0100	C000	DISPLAY subroutine
0130	C030	TNSUNINC (Tens/Units Increment) subroutine
0140	C040	THUNDINC (Thousands/Hundreds Increment) subroutine
0160	C060	DELAY subroutine
0170	C070	KYDETDIS (Key-press Detect and Display) subroutine
01A0	C0A0	DISCLR (Display Clear) subroutine
01FF	C0FF	Initial Stack Pointer

FIGURE 4-9

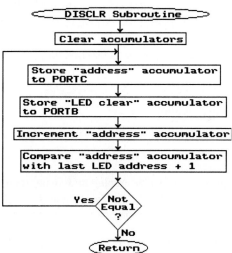

Programming Hints—DISCLR Subroutine

1. Use the "address" accumulator to address each LED in turn—$00, then $01, then $02, thn $03. Thus, the last LED + 1 = $04.
2. The contents of the "segment clear" accumulator remains as $00. This value extinguishes all LED segments.

FIGURE 4-10

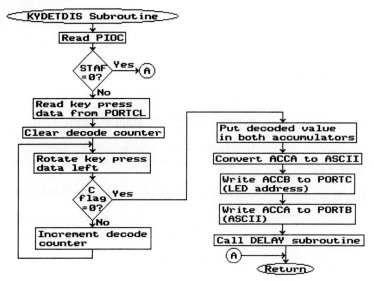

Programming Hints—KYETDIS Subroutine

1. To test for STAF=0, use the BPL (Branch if Plus) instruction. BPL checks bit 7 and branches if bit 7 = 0. The STAF is PIOC bit 7.
2. Make ACCB the "decode counter." Initialize ACCB to $00 and increment for each rotation (ROLA) of key-press data. Use TBA to transfer the decoded value into ACCA, leaving this decoded value in both accumulators.
3. When both accumulators contain the decoded key-press value, use ACCB's contents as the segment address. Convert ACCA's contents to ASCII. Use an ORAA instruction with mask $30 to convert ACCA's contents to ASCII:

   ORAA #30

4. The DELAY subroutine—located at $0160 ($C060)—should consume about one second. Count down the X register from $D600 to zero, six times.

Power up the EVBU (EVB) and interface. Load, debug, and demonstrate the operation of your software to the instructor.

_____ *Instructor Check*

PART III: Keypad Read and Display—Interrupt Control

Generally, the interface and supporting software execute the same tasks as in Part II. However, the software now uses interrupt control rather than status polling to detect and display a key-press. Make no hardware changes. Software changes are minimal:

1. PIOC initialization must enable interrupt control.
2. The MAIN program must clear the I mask.
3. Interrupt control uses IRQ vectors. The IRQ service routine is the Key-Press Detect and Display (KYDETDIS) routine at $0170 ($C070). Therefore, initialize the BUFFALO ROM IRQ pseudovector as follows:

   $00EE JMP 170 (EVBU) or $00EE JMP C070 (EVB)

4. Call the Display Clear (DISCLR) subroutine from the KYDETDIS interrupt service routine, rather than the MAIN program.

FIGURE 4-11

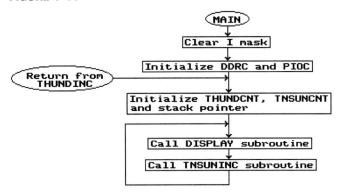

FIGURE 4-12

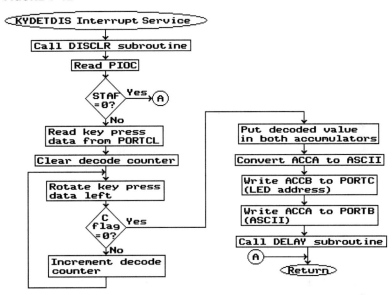

5. KYDETDIS is now an IRQ service routine. Return from KYDETDIS with an RTI rather than RTS.
6. Modify the JMP instruction in the THUNDINC subroutine. Assure that this JMP directs program flow to the CLR 0 instruction in the MAIN program (see Exercise 3.3 in Chapter 3). Assure that all jumps and branches reflect software modifications.
7. Review the Figure 4–11 and 4–12 flowcharts for changes in the MAIN program and KYDETDIS routine.

Load, debug, and demonstrate the operation of your software to the instructor.

_____ *Instructor Check*

PART IV: Laboratory Analysis

Answer the questions below by neatly entering your responses in the appropriate blanks.

1. Describe the read peripheral used in the laboratory.

2. The read peripheral connects to bits _____ (bit numbers) of the _____ register.
3. Where do the remaining bits of this register (see question 2) connect?

4. PORTC bits _____ (bit numbers) are configured as inputs and PORTC bits _____ are configured as outputs.
5. Describe the write peripheral used in the laboratory.

6. The write peripheral's data input pins connect to bits _____ of the _____ register.
7. STRB connects to the _____ pin of the _____ (device name).
8. The DISCLR subroutine functions to extinguish all _____. Explain why this routine is necessary.

9. The DISCLR subroutine writes $_____ to the _____ addresses.
10. Refer to the KYDETDIS routine. Enter ones and zeros in the blanks to indicate the contents of ACCA after a PORTCL read. Assume that the operator pressed key number 1. Enter question marks for unpredictable bits.

    D7— — — — — — —D0

11. Refer to the KYDETDIS routine. Given the situation in question 10, how many ROLA executions will it take to reset the C flag? _____ Explain why the decode counter contents would be only $01 when the C flag resets.

12. When you converted the interface and software to interrupt control, which routine became the IRQ service routine? _____
13. During conversion to interrupt control, the DISCLR subroutine call moved to the KYDETDIS routine. Why was this change necessary?

14. Discuss the relative advantages and disadvantages of status polling versus interrupt control—with reference to the interface and peripherals used in this laboratory. Which technique was most satisfactory? Why?

PART V: Laboratory Software Listing

MAIN PROGRAM—STATUS POLLING—EVB

$ADDRESS	SOURCE CODE	COMMENTS
0002	LDAA #3	Initialize DDRC
0004	STAA 1007	" "
0007	DECA	Initialize PIOC
0008	STAA 1002	" "
000B	CLR 0	Initialize counts
000E	CLR 1	" "
0011	LDS #C0FF	Initialize SP
0014	JSR C000	Call DISPLAY subroutine
0017	JSR C030	Call TNSUNINC subroutine
001A	JSR C0A0	Call DISCLR subroutine
001D	JSR C070	Call KYDETDIS subroutine
0020	BRA 14	Continue routine counting

MAIN PROGRAM—INTERRUPT CONTROL—EVB

$ADDRESS	SOURCE CODE	COMMENTS
0002	LDAA #42	Initialize PIOC
0004	STAA 1002	" "

$ADDRESS	SOURCE CODE	COMMENTS
0007	LDAA #3	Initialize DDRC
0009	STAA 1007	" "
000C	LDD #C070	Initialize IRQ pseudovector
000F	STD EF	" " "
0011	CLI	Clear interrupt mask
0012	CLR 0	Initialize counts
0015	CLR 1	" "
0018	LDS #C0FF	Initialize SP
001B	JSR C000	Call DISPLAY subroutine
001E	JSR C030	Call TNSUNINC subroutine
0021	BRA 1B	Continue routine counting

MAIN PROGRAM—STATUS POLLING—EVBU

$ADDRESS	SOURCE CODE	COMMENTS
0002	LDAA #3	Initialize DDRC
0004	STAA 1007	" "
0007	DECA	Initialize PIOC
0008	STAA 1002	" "
000B	CLR 0	Initialize counts
000E	CLR 1	" "
0011	LDS #1FF	Initialize SP
0014	JSR 100	Call DISPLAY subroutine
0017	JSR 130	Call TNSUNINC subroutine
001A	JSR 1A0	Call DISCLR subroutine
001D	JSR 170	Call KYDETDIS subroutine
0020	BRA 14	Continue routine counting

MAIN PROGRAM—INTERRUPT CONTROL—EVBU

$ADDRESS	SOURCE CODE	COMMENTS
0002	LDAA #42	Initialize PIOC
0004	STAA 1002	" "
0007	LDAA #3	Initialize DDRC
0009	STAA 1007	" "
000C	LDD #170	Initialize IRQ pseudovector
000F	STD EF	" " "
0011	CLI	Clear interrupt mask
0012	CLR 0	Initialize counts
0015	CLR 1	" "
0018	LDS #1FF	Initialize SP
001B	JSR 100	Call DISPLAY subroutine
001E	JSR 130	Call TNSUNINC subroutine
0021	BRA 1B	Continue routine counting

DISPLAY SUBROUTINE—EVB

$ADDRESS	SOURCE CODE	COMMENTS
C000	LDAB #2	Initialize loop counter
C002	CLR 1003	Point to right-most LED
C005	LDAA 1	Load TNSUNCNT
C007	PSHA	Stack count
C008	ANDA #F	Convert low nibble to ASCII
C00A	ORAA #30	" " " "
C00C	STAA 1004	Write ASCII word to PORTB
C00F	INC 1003	Point to next LED
C012	PULA	Pull count from stack
C013	LSRA	Convert high nibble to ASCII
C014	LSRA	" " " "
C015	LSRA	" " " "
C016	LSRA	" " " "
C017	ORAA #30	" " " "
C019	STAA 1004	Write ASCII word to PORTB
C01C	DECB	Decrement loop counter
C01D	BEQ C026	Loop counter = 0?
C01F	LDAA 0	If not, load THUNDCNT
C021	INC 1003	Point to next LED
C024	BRA C007	Encode and display THUNDCNT

$ADDRESS	SOURCE CODE	COMMENTS
C026	JSR C060	Call DELAY subroutine
C029	RTS	Return

DISPLAY SUBROUTINE—EVBU

$ADDRESS	SOURCE CODE	COMMENTS
0100	LDAB #2	Initialize loop counter
0102	CLR 1003	Point to right-most LED
0105	LDAA 1	Load TNSUNCNT
0107	PSHA	Stack count
0108	ANDA #F	Convert low nibble to ASCII
010A	ORAA #30	" " " "
010C	STAA 1004	Write ASCII word to PORTB
010F	INC 1003	Point to next LED
0112	PULA	Pull count from stack
0113	LSRA	Convert high nibble to ASCII
0114	LSRA	" " " "
0115	LSRA	" " " "
0116	LSRA	" " " "
0117	ORAA #30	" " " "
0119	STAA 1004	Write ASCII word to PORTB
011C	DECB	Decrement loop counter
011D	BEQ 126	Loop counter = 0?
011F	LDAA 0	If not, load THUNDCNT
0121	INC 1003	Point to next LED
0124	BRA 107	Encode and display THUNDCNT
0126	JSR 160	Call DELAY subroutine
C029	RTS	Return

TNSUNINC SUBROUTINE—EVB

$ADDRESS	SOURCE CODE	COMMENTS
C030	LDAA 1	Load TNSUNCNT
C032	ADDA #1	Increment count
C034	DAA	Decimal adjust count
C035	STAA 1	Store count
C037	BNE C03C	Count not equal to zero?
C039	JSR C040	If no, call THUNDINC subroutine
C03C	RTS	If yes, return

TNSUNINC SUBROUTINE—EVBU

$ADDRESS	SOURCE CODE	COMMENTS
0130	LDAA 1	Load TNSUNCNT
0132	ADDA #1	Increment count
0134	DAA	Decimal adjust count
0135	STAA 1	Store count
0137	BNE 13C	Count not equal to zero?
0139	JSR 140	If no, call THUNDINC subroutine
013C	RTS	If yes, return

THUNDINC SUBROUTINE—STATUS POLLING—EVB

$ADDRESS	SOURCE CODE	COMMENTS
C040	LDAA 0	Load THUNDCNT
C042	ADDA #1	Increment count
C044	DAA	Decimal adjust count
C045	STAA 0	Store count
C047	CMPA #10	Compare count with 10
C049	BEQ C04C	Count = 10?
C04B	RTS	If no, return
C04C	JSR C000	If yes, call DISPLAY subroutine
C04F	JMP B	Start counting at 0000 again

THUNDINC SUBROUTINE—INTERRUPT CONTROL—EVB

$ADDRESS	SOURCE CODE	COMMENTS
C040	LDAA 0	Load THUNDCNT
C042	ADDA #1	Increment count

$ADDRESS	SOURCE CODE	COMMENTS
C044	DAA	Decimal adjust count
C045	STAA 0	Store count
C047	CMPA #10	Compare count with 10
C049	BEQ C04C	Count = 10?
C04B	RTS	If no, return
C04C	JSR C000	If yes, call DISPLAY subroutine
C04F	JMP 12	Start counting at 0000 again

THUNDINC SUBROUTINE—STATUS POLLING—EVBU

$ADDRESS	SOURCE CODE	COMMENTS
0140	LDAA 0	Load THUNDCNT
0142	ADDA #1	Increment count
0144	DAA	Decimal adjust count
0145	STAA 0	Store count
0147	CMPA #10	Compare count with 10
0149	BEQ 14C	Count = 10?
014B	RTS	If no, return
014C	JSR 100	If yes, call DISPLAY subroutine
014F	JMP B	Start counting at 0000 again

THUNDINC SUBROUTINE—INTERRUPT CONTROL—EVBU

$ADDRESS	SOURCE CODE	COMMENTS
0140	LDAA 0	Load THUNDCNT
0142	ADDA #1	Increment count
0144	DAA	Decimal adjust count
0145	STAA 0	Store count
0147	CMPA #10	Compare count with 10
0149	BEQ 14C	Count = 10?
014B	RTS	If no, return
014C	JSR 100	If yes, call DISPLAY subroutine
014F	JMP 12	Start counting at 0000 again

DELAY SUBROUTINE—EVB

$ADDRESS	SOURCE CODE	COMMENTS
C060	LDAA #6	Load loop counter
C062	LDX #D600	Load X register with count-down value
C065	DEX	Decrement X register
C066	BNE C065	If X register ≠ 0, decrement again
C068	DECA	Decrement loop counter
C069	BNE C062	If loop counter ≠ 0, count down X register again
C06B	RTS	Return

DELAY SUBROUTINE—EVBU

$ADDRESS	SOURCE CODE	COMMENTS
0160	LDAA #6	Load loop counter
0162	LDX #D600	Load X register with count-down value
0165	DEX	Decrement X register
0166	BNE 165	If X register ≠ 0, decrement again
0168	DECA	Decrement loop counter
0169	BNE 162	If loop counter ≠ 0, count down X register again
016B	RTS	Return

KYDETDIS SUBROUTINE—STATUS POLLING—EVB

$ADDRESS	SOURCE CODE	COMMENTS
C070	LDAA 1002	Read PIOC
C073	BPL C08B	If STAF = 0, return
C075	LDAA 1005	Read PORTCL
C078	CLRB	Clear decode counter
C079	ROLA	Rotate keypress data left
C07A	BCC C07F	If C = 0, then decode counter = correct key-press data
C07C	INCB	If C ≠ 0, increment decode counter
C07D	BRA C079	Rotate key-press data again
C07F	TBA	Put decoded key-press data into both accumulators

$ADDRESS	SOURCE CODE	COMMENTS
C080	ORAA #30	Convert ACCA key-press data to ASCII
C082	STAB 1003	Write LED address (in ACCB) to PORTC
C085	STAA 1004	Write ASCII key-press data to PORTB
C088	JSR C060	Call DELAY subroutine
C08B	RTS	Return

KYDETDIS INTERRUPT SERVICE ROUTINE—EVB

$ADDRESS	SOURCE CODE	COMMENTS
C070	JSR C0A0	Call DISCLR subroutine
C073	LDAA 1002	Read PIOC
C076	BPL C08E	If STAF = 0, return
C078	LDAA 1005	Read PORTCL
C07B	CLRB	Clear decode counter
C07C	ROLA	Rotate key-press data left
C07D	BCC C082	If C = 0, then decode counter = correct key-press data
C07F	INCB	If C ≠ 0, increment decode counter
C080	BRA C07C	Rotate key-press data again
C082	TBA	Put decoded key-press data into both accumulators
C083	ORAA #30	Convert ACCA key-press data to ASCII
C085	STAB 1003	Write LED (in ACCB) to PORTC
C088	STAA 1004	Write ASCII key-press data to PORTB
C08B	JSR C060	Call DELAY subroutine
C08E	RTI	Return from interrupt

KYDETDIS SUBROUTINE—STATUS POLLING—EVBU

$ADDRESS	SOURCE CODE	COMMENTS
0170	LDAA 1002	Read PIOC
0173	BPL 18B	If STAF = 0, return
0175	LDAA 1005	Read PORTCL
0178	CLRB	Clear decode counter
0179	ROLA	Rotate key-press data left
017A	BCC 17F	If C = 0, then decode counter = correct key-press data
017C	INCB	If C ≠ 0, increment decode counter
017D	BRA 179	Rotate key-press data again
017F	TBA	Put decoded key-press data into both accumulators
0180	ORAA #30	Convert ACCA key-press data to ASCII
0182	STAB 1003	Write LED (in ACCB) to PORTC
0185	STAA 1004	Write ASCII key-press data to PORTB
0188	JSR 160	Call DELAY subroutine
018B	RTS	Return

KYDETDIS INTERRUPT SERVICE ROUTINE—EVBU

$ADDRESS	SOURCE CODE	COMMENTS
0170	JSR 1A0	Call DISCLR subroutine
0173	LDAA 1002	Read PIOC
0176	BPL 18E	If STAF = 0, return
0178	LDAA 1005	Read PORTCL
017B	CLRB	Clear decode counter
017C	ROLA	Rotate key-press data left
017D	BCC 182	If C = 0, then decode counter = correct key-press data
017F	INCB	If C ≠ 0, increment decode counter
0180	BRA 17C	Rotate key-press data again
0182	TBA	Put decoded key-press data into both accumulators
0183	ORAA #30	Convert ACCA key-press data to ASCII
0185	STAB 1003	Write LED (in ACCB) to PORTC
0188	STAA 1004	Write ASCII key-press data to PORTB
018B	JSR 160	Call DELAY subroutine
018E	RTI	Return from interrupt

DISCLR SUBROUTINE—EVB

$ADDRESS	SOURCE CODE	COMMENTS
C0A0	CLRA	Clear ACCA (Initial LED address)
C0A1	CLRB	Clear ACCB (LED data)

$ADDRESS	CONTENTS	COMMENTS
C0A2	STAA 1003	Store LED address to PORTC
C0A5	STAB 1004	Store LED clear data to PORTB
C0A8	INCA	Increment LED address
C0A9	CMPA #4	Compare address with last LED address + 1
C0AB	BNE C0A2	If not equal, clear another LED
C0AD	RTS	Return

DISCLR SUBROUTINE—EVBU

$ADDRESS	CONTENTS	COMMENTS
01A0	CLRA	Clear ACCA (Initial LED address)
01A1	CLRB	Clear ACCB (LED clear data)
01A2	STAA 1003	Store LED address to PORTC
01A5	STAB 1004	Store LED clear data to PORTB
01A8	INCA	Increment LED address
01A9	CMPA #4	Compare address with last LED address + 1
01AB	BNE 1A2	If not equal, clear another LED
01AD	RTS	Return

4.3 FULL-HANDSHAKE I/O

Although full-handshake I/O resembles simple-strobed I/O, it also differs in three important respects. First, full-handshake mode permits either input or output data transfers, but not both. Secondly, full-handshake mode does not use PORTB. Rather, this mode uses PORTC/PORTCL only, to handle data transfers. Therefore, full-handshake I/O frees PORTB for routine nonhandshake writes to another peripheral. Third, full-handshake mode can interlock the STRA and STRB handshakes so that the peripheral knows exactly when the M68HC11 is responding to STRA, and when it is ready for another data exchange.

On-Chip Resources Used by the M68HC11 for Full-Handshake I/O

Full-handshake I/O uses these on-chip resources:

- Strobe A (STRA, pin 4). The peripheral uses STRA to inform the M68HC11 that it has completed its part of the data transfer. STRA active edge assertion means that the peripheral has either applied read data to the PORTC pins of the M68HC11, or that it has latched write data from PORTC. In a special version of output full-handshakes, the peripheral can use a STRA active level to actually strobe the appearance of write data on the M68HC11's PORTC data output pins.
- Strobe B (STRB, pin 6). The M68HC11 uses STRB to tell the peripheral that it is ready for the next data transfer. STRB assertion means that the M68HC11 has either read data from PORTCL or written data to PORTCL.
- Parallel I/O Control Register (PIOC, register block address $1002). As in simple-strobed mode, the PIOC serves two functions. First, the PIOC configures the M68HC11 for full-handshake operations. This configuration can include interlocked STRA and STRB. Second, the PIOC STAF bit serves status flag functions similar to those in simple-strobed mode.
- Port C Data Register and Port C Latched Data Register (PORTC and PORTCL, register block addresses $1003 and $1005 respectively). The microcontroller must read or write to PORTCL for full-handshake data transfers. Read and write data are applied to PORTC pins even though the routing is via PORTCL. PORTCL consists of D flip-flops, which asynchronously latch read data appearing on PORTC pins. PORTCL simply routes output data to PORTC pins, via PORTC pin logic.
- Data Direction Register for PORT C (DDRC, register block address $1007). As in simple-strobed mode, DDRC configures PORTC pins as inputs or outputs. Use DDRC in conjunction with bits in the PIOC to select either full-handshake input or full-handshake output operation. In general, DDRC controls data flow direction of the PORTC pins, and

the DDRC is configured so that full-handshake operations work in harmony with the PIOC set-up. However, a system designer can use DDRC to set up full-handshake PORTC outputs so that they must be strobed by STRA active levels. With this arrangement the peripheral asserts the STRA active level to make write data appear on selected PORTC pins.[8] This special variation is discussed later in the chapter.

Remember that full-handshake mode does not use PORTB because communication is one-way. DDRC can configure PORTC/PORTCL for either reads or writes, since PORTC is bidirectional.

Interlocked and Pulsed Full Handshake

Full-handshake I/O comes in two modes—interlocked and pulsed. The difference between these modes lies in the duration of STRB. In interlocked operation, STRB remains asserted until the peripheral asserts the next STRA active edge, setting the PIOC's STAF bit. STRB negates only for the period between STAF assertion and a microcontroller CPU read or write to PORTCL. Thus, in interlocked operation STRB negation means that the CPU is busy checking status, managing data, or reading/writing PORTCL.

In pulsed operation, STRB asserts for only two internal clock cycles—after a CPU read or write to PORTCL. In pulsed mode, STRB negation does not necessarily mean that the microcontroller CPU is busy doing its part of the data transfer. In fact, STRB is negated most of the time.

In either interlocked or pulsed full handshake, the peripheral generally uses STRA to inform the M68HC11 that it has done its part of the data transfer. The exception to this rule is the special variation where the programmer uses DDRC to set up a special output handshake mode. This mode makes STRA an integral part of output data transfers. In this special output mode, the peripheral uses the STRA active level to actually strobe the appearance of write data on PORTC pins. In any of these modes, the M68HC11 always uses STRB to inform the peripheral that it has done its part, and that it is ready for another data transfer.

The M68HC11 can use either pulsed or interlocked handshakes. It can use these handshakes for either input or output operations. Four combinations result:

1. input - interlocked
2. output - interlocked
3. input - pulsed
4. output - pulsed

Following sections discuss these combinations.

Input-Interlocked Full-Handshake

Refer to Figures 4–13 and 4–14.[9] In Figure 4–13, block 2, the peripheral applies an active edge to the M68HC11's STRA pin. STRA active-edge assertion causes three responses from the M68HC11 (Figure 4–13, block 3):

- PORTCL latches PORTC input data.
- STAF sets.
- STRB negates.

The peripheral asserts STRA asynchronously with respect to the M68HC11's internal clock. Since the STRA active edge enables PORTCL, this register asynchronously latches PORTC data. However, the STAF sets and STRB negates synchronously with the M68HC11's PH2 clock. STAF set and STRB negation coincide with the first or second PH2 clock edge after STRA assertion, depending upon whether STRA allowed adequate set-up time prior to the first PH2 edge.

---

[8]Motorola Inc., *M68HC11 Reference Manual* (1990) pp. 7-20 and 7-42. Motorola calls this arrangement the "three-state variation of output handshakes."
[9]Motorola Inc., *M68HC11 Reference Manual*, p. 7-41.

FIGURE 4–13

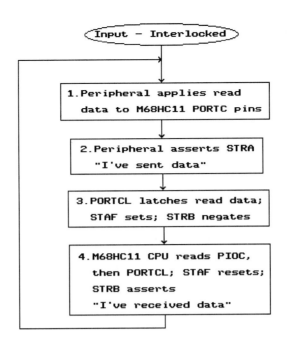

FIGURE 4–14
Reprinted with the permission of Motorola.

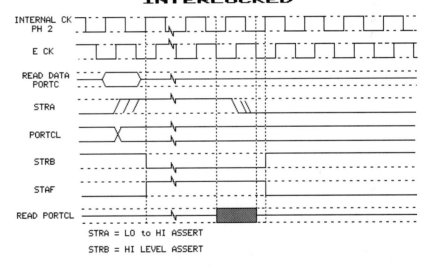

Refer to Figure 4–13, block 4. To clear the STAF, the M68HC11 CPU must read the PIOC, then read PORTCL. A PORTCL read causes STRB to assert, sending an "I've received data" message to the peripheral. Thus, the peripheral knows that it can send more read data to PORTC (see block 1 of flowchart).

Output-Interlocked Full-Handshake

Refer to block 1 of Figure 4–15 and to Figure 4–16.[10] To reset the STAF and assert STRB, the M68HC11 CPU reads the PIOC and then writes output data to PORTCL. The output data appears on the PORTC pins.

Refer to Figure 4–15, blocks 2 and 3. By asserting the STRA active edge, the peripheral causes the M68HC11's STAF to set and STRB to negate, thus sending an "I've received data" message to the microcontroller.

---

[10]Motorola Inc., *M68HC11 Reference Manual*, p. 7-42.

FIGURE 4–15

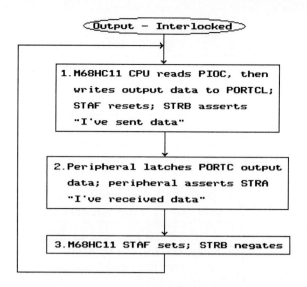

FIGURE 4–16
Reprinted with the permission of Motorola.

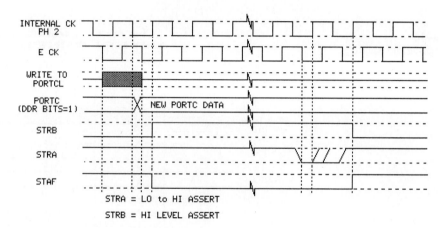

Input-Pulsed Full-Handshakes

Refer to blocks 2 and 3 of Figure 4–17 and to Figure 4–18.[11] As in input-interlocked mode, STRA active-edge assertion causes PORTCL to latch PORTC data asynchronously. STRA assertion also causes the STAF to set synchronously. However, in input-pulsed mode, STRB has already negated.

Refer to Figure 4–17, block 4. The CPU must read the PIOC, then PORTCL, in order to clear the STAF and assert STRB. Note that STRB asserts for only two PH2 clock cycles and negates automatically.

Output-Pulsed Full-Handshake

Refer to block 1 of Figure 4–19 and to Figure 4–20.[12] To reset the STAF and assert STRB, the M68HC11 must read the PIOC then write output data to PORTCL. This output data appears on PORTC pins. Note that STRB asserts for only two PH2 clock cycles, then negates automatically.

Refer to Figure 4–19, blocks 2 and 3. The peripheral latches the write data, then asserts the STRA active edge. STRA assertion causes the M68HC11's STAF to set. At this point, STRB has already negated (see block 1 of flowchart).

---

[11] Motorola Inc., *M68HC11 Reference Manual*, p. 7-41.
[12] Motorola Inc., *M68HC11 Reference Manual*, p. 7-42.

FIGURE 4–17

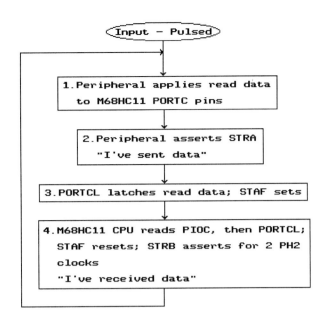

FIGURE 4–18
Reprinted with the permission of Motorola.

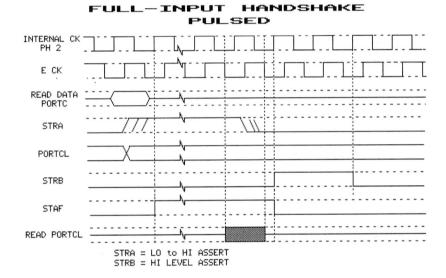

FIGURE 4–19

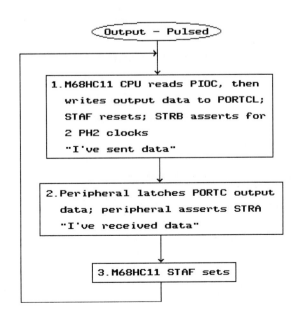

195

FIGURE 4–20
Reprinted with the permission of Motorola.

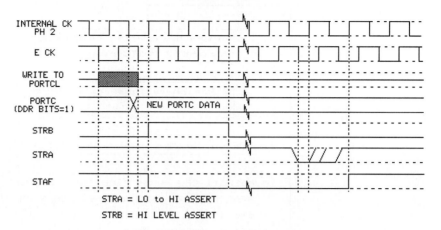

Three-State Variation of Output Full-Handshake

The M68HC11 provides a special type of output full-handshake, called the "three-state variation." This variation works with either output-pulsed or output-interlocked full-handshakes.

The programmer establishes the three-state variation by initializing relevant DDRC bits to zeros instead of ones. When so established, this variation changes the way write data appears on the selected PORTC pins. Recall that with conventional output full-handshakes, a PORTCL write causes the output data to also appear on the PORTC pins. With the three-state variation, a PORTCL write does not cause write data to appear on the PORTC pins. Rather, PORTCL data appears on the PORTC pins only when the peripheral brings STRA to its active level. The Figure 4–21 flow chart and Figure 4–22 timing diagram illustrate the sequence of events in the three-state variation of output full-handshakes.[13]

Refer to the timing diagram and block 1 of the three-state variation flowchart. By reading the PIOC and then writing to PORTCL, the M68HC11 resets the STAF and asserts STRB. However, this write data does not appear on the PORTC pins. STRB asserts normally—either for two PH2 clock cycles (pulsed) or until the STAF sets in block 5 (interlocked).

FIGURE 4–21

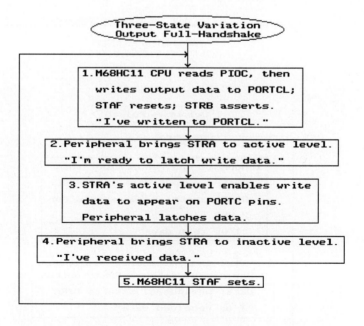

---

[13]Motorola Inc., *M68HC11 Reference Manual*, p. 7-42.

FIGURE 4-22
Reprinted with the permission of Motorola.

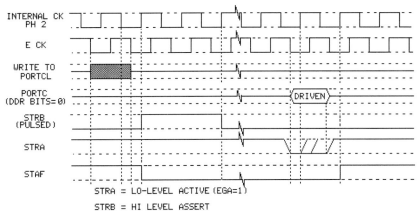

Refer to blocks 2 and 3. By driving STRA to its active level, the peripheral causes the write data to appear on the PORTC pins. The peripheral can then latch this write data.

Refer to blocks 4 and 5. The peripheral brings STRA to its inactive level by asserting the STRA active edge. The STAF sets (and STRB negates, if the M68HC11 is in interlocked mode).

Note that the three-state variation uses STRA active and inactive levels. All other full-handshake modes use only STRA active edges, as defined by the PIOC EGA bit:

EGA = 1   positive-going edge asserts
EGA = 0   negative-going edge asserts

The three-state variation STRA active level is also defined by the EGA bit:

EGA = 1   low-level active
EGA = 0   high-level active

Refer to the three-state variation timing diagram (Figure 4–22). Since EGA = 1, STRA's active edge is positive-going, and active level is low. Thus, write data appears on the PORTC pins when STRA is active low level. The STAF responds to STRA's positive-going active edge.

As in simple-strobed mode, programmers can use two methods to determine when the STAF has set: status polling and interrupt control.

Status Polling—Full-Handshake Mode

With status polling, the M68HC11's CPU periodically reads the PIOC and checks to see if the STAF = 1. If the CPU finds that STAF = 1, then it either reads or writes PORTCL, depending upon whether the M68HC11 is in input or output mode. Figures 4–23 and 4–24 flowcharts illustrate full-handshake status polling strategies for both input and output data.

Refer to block 1 of the input full-handshake status polling flowchart (Figure 4–23). The PIOC configures the M68HC11 for full-handshake mode (HNDS = 1). The DDRC configures PORTC as an input.

Refer to block 2. If data inputs are relatively infrequent, the CPU can execute other routine operations between its recurring PIOC reads (block 3).

Refer to blocks 3 and 5. The CPU must read the PIOC and then PORTCL to reset the STAF. Also, a PORTCL read causes STRB to assert for two PH2 clock cycles (pulsed mode) or until the peripheral asserts STRA (interlocked mode).

Refer to block 1 of the output full-handshake status polling flowchart (Figure 4–24). Again, the PIOC configures the M68HC11 for full-handshake operation (HNDS = 1). DDRC configures PORTC pins as outputs (DDRC bits = 1) or for the three-state variation (DDRC bits = 0).

FIGURE 4–23

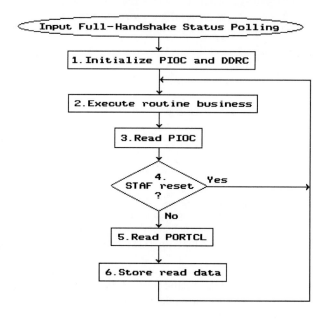

FIGURE 4–24

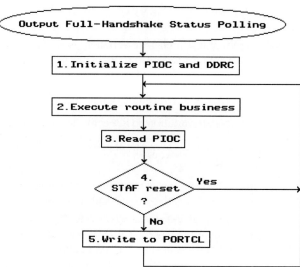

Refer to block 2. If the peripheral is relatively slow to latch output data and assert the STRA active edge, the M68HC11 can execute other routine operations while it is waiting for this response from the peripheral.

Refer to blocks 3, 4, and 5. Note that the M68HC11 will not write data until the peripheral asserts the STRA active edge, causing the STAF to set. When the peripheral does assert STRA, the M68HC11 CPU reads the PIOC then writes to PORTCL. These actions reset the STAF and assert STRB. STRB asserts for two PH2 clock cycles (pulsed mode) or until the STAF sets again (interlocked mode). When the CPU writes to PORTCL, the write data also appears on PORTC pins—where DDRC bits were initialized to ones. However, for bits where DDRC was initialized to zeros, write data appears on PORTC pins only when the peripheral asserts the STRA active level. Recall that with three-state variation, the STRA active edge sets the STAF. The STRA active edge occurs when the peripheral negates the STRA active level.

Interrupt Control—Full Handshake Mode

When the M68HC11 operates in full-handshake mode with interrupt control, STRA active-edge assertion by the peripheral causes the STAF to set and generates an IRQ interrupt within the microcontroller. With interrupt control the M68HC11 can do other routine

FIGURE 4–25

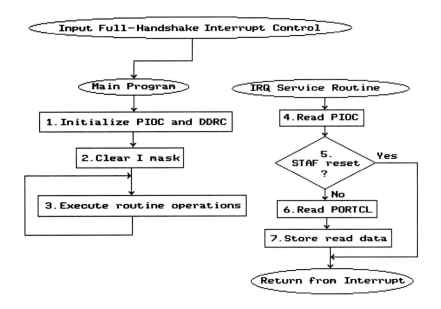

business without periodically reading the PIOC and checking the STAF. When interrupted, the M68HC11 vectors to an IRQ service routine in which it reads the PIOC and then reads or writes to PORTCL. The flowcharts in Figures 4–25 and 4–26 illustrate full-handshake interrupt control for input and output operations.

Refer to block 1 of the input full-handshake interrupt control flowchart (Figure 4–25). The PIOC configures the M68HC11 for full handshake mode (HNDS = 1) with interrupt enabled (STAI = 1). The DDRC configures PORTC as an input.

Refer to block 2. The M68HC11 must clear the I mask. This action enables IRQ interrupts. Recall that a system reset causes the I mask to set, inhibiting IRQ interrupts.

Refer to blocks 4, 5, and 6. To clear the STAF and assert STRB (sending the "I've received data" signal), the CPU must read the PIOC, then PORTCL. If external hardware IRQ interrupts are also being used, the CPU can test the STAF bit. If STAF = 0, the M68HC11 can return from the interrupt without reading PORTCL (as shown in the flowchart) or can pseudovector to an appropriate external IRQ service routine. If the full-handshake interrupt control function generates the IRQ interrupt, then the CPU reads the PIOC and finds that STAF = 1. The CPU then reads PORTCL, resetting the STAF and asserting STRB. STRB asserts for either two PH2 clock cycles (in pulsed mode) or until the peripheral asserts the STRA active edge (interlocked mode).

Refer to block 1 of the output full-handshake interrupt control flowchart (Figure 4–26). The PIOC configures the microcontroller for full-handshake mode (HNDS = 1) with

FIGURE 4–26

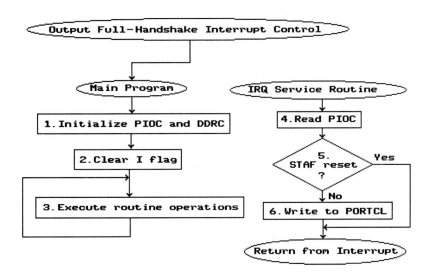

199

interrupt enabled (STAI = 1). The DDRC configures PORTC pins as outputs (DDRC bits = 1) or for the three-state variation (DDRC bits = 0).

Refer to blocks 4, 5, and 6. The M68HC11 CPU must read the PIOC and write to PORTCL to clear the STAF and assert STRB (sending the "I've written data" signal). If an IRQ is generated via the external $\overline{IRQ}$ pin, then the CPU finds the STAF reset (block 5) and returns from the interrupt as shown, or pseudovectors to an appropriate service routine. If STAF = 1, the CPU reads the PIOC and then writes to PORTCL. Accordingly, STRB asserts for two PH2 clock cycles (in pulsed mode) or until the peripheral asserts the STRA active edge (interlocked mode). The PORTCL write causes data to appear on PORTC pins where DDRC bits were initialized to ones. Where DDRC bits were initialized to zeros, write data appears on PORTC pins only when the peripheral asserts the STRA active level. Remember that the STRA active level will always be followed by the STRA active edge—setting the STAF.

The PIOC and Full-Handshake Mode

Refer to Figure 4–27. All PIOC bits, except bit 5, relate to full-handshake mode. The STAF (Strobe A Flag, bit 7) indicates that the communicating peripheral asserted a STRA active edge.

The STAI (Strobe A Interrupt Enable, bit 6) enables the interrupt control function.

STAI = 0	interrupt disabled
STAI = 1	interrupt enabled

HNDS (Handshake/Simple-Strobe Mode Select, bit 4) selects either full-handshake or simple-strobed mode.

HNDS = 0	Simple-strobed mode
HNDS = 1	Full-handshake mode

OIN (Output/Input Handshake Select, bit 3) selects full-input handshakes or full-output handshakes. The OIN bit becomes relevant only when HNDS = 1 (full-handshake mode selected).

OIN = 0	Full input handshakes
OIN = 1	Full output handshakes

PLS (Pulse Mode Select for STRB Output, bit 2) determines whether full handshakes operate in pulsed or interlocked mode.

PLS = 0	Interlocked mode
PLS = 1	Pulsed mode

EGA (Active-Edge Select for STRA, bit 1) selects the active edge for STRA assertion. EGA also selects the STRA active level. This level strobes write data onto the PORTC pins when the M68HC11 operates in the three-state variation.

EGA = 0	Negative-going transition is active edge. High is active level.
EGA = 1	Positive-going transition is active edge. Low is active level.

INVB (Invert STRB Output, bit 0) defines the STRB active level.

INVB = 0	STRB active low
INVB = 1	STRB active high

FIGURE 4–27
Reprinted with permission of Motorola.

PIOC

STAF	STAI	CWOM	HNDS	OIN	PLS	EGA	INVB
B7							B0

A note is in order about PIOC bit 5, which is CWOM (Port C Wired-OR Mode). This bit converts PORTC pins from CMOS push-pull outputs to N-channel open-drain outputs.

EXERCISE 4.3   **PIOC Control Words for Full-Handshake Mode**

AIM: Derive control words which configure the microcontroller for full-handshake operation.

INTRODUCTORY INFORMATION: By completing this exercise you will learn how to calculate PIOC control words. These words configure the M68HC11 for full-handshake data exchanges.

INSTRUCTIONS: Use Figure 4–27 and the configuration descriptions to derive PIOC control words. Enter each hex control word in its appropriate blank.

Note: Configure PORTC pins for push-pull operation (CWOM = 0) in all cases below.

Configuration	PIOC Control Word
EXAMPLE Interrupt disabled Output handshake Interlocked mode STRA—positive-going edge STRB—active-low	$ __1A__
Interrupt enabled Input handshake Interlocked mode STRA—negative-going edge STRB—active-high	$ _____
Interrupt enabled Output handshake Pulsed mode STRA—negative-going edge STRB—active-low	$ _____
Interrupt disabled Output handshake Interlocked mode STRA—negative-going edge STRB—active-high	$ _____
Interrupt enabled Output handshake Pulsed mode STRA—positive-going edge STRB—active-high	$ _____
Interrupt disabled Input handshake Interlocked mode STRA—negative-going edge STRB—active-low	$ _____

Interrupt enabled
Output handshake
Interlocked mode
STRA—positive-going edge
STRB—active-high $_____

Interrupt disabled
Input handshake
Pulsed mode
STRA—negative-going edge
STRB—active-high $_____

Interrupt disabled
Input handshake
Pulsed mode
STRA—negative-going edge
STRB—active-low $_____

Interrupt enabled
Input handshake
Pulsed mode
STRA—positive-going edge
STRB—active-low $_____

EXERCISE 4.4 Full-Handshake I/O Laboratory

REQUIRED MATERIALS:
1 Resistor, 5.1 KΩ
1 Push-button Switch, push to close
2 74HC241 Octal 3-State Driver

REQUIRED EQUIPMENT: 2 Motorola M68HC11EVBUs or M68HC11EVBs, and 2 terminals

GENERAL INSTRUCTIONS: To complete this exercise, you need a lab partner. Interface two EVBUs (or EVBs) to each other. One EVBU (EVB) functions as a data transmitter (TX) while the other is the receiver (RX). Cross-couple STRA and STRB lines between the two EVBUs. STRB of the transmitter becomes STRA for the receiver, and vice versa. Configure the transmitter for output handshakes in interlocked mode, with DDRC bits = 1. Configure the receiver for input handshakes in interlocked mode. Disable both transmitter and receiver interrupts. However, use the external $\overline{IRQ}$ pin on the transmitter to initiate 16-byte data transfers. Connect a pull-up circuit to this $\overline{IRQ}$ pin so that when the operator presses a push-button, it generates an interrupt. The IRQ interrupt service routine initiates the data transfer. The transfer itself uses status polling. When the EVBUs (EVBs) complete the 16-byte data transfer, they signal completion on their respective terminal screens. The transmitter writes "TX" to its screen, and the receiver writes "RX" to its screen.

PART I: Connecting the Interface

Power down the EVBUs (EVBs) and terminals. Use Figures 4–28 and 4–29 as guides to interface construction.
   Show your connections to the instructor.

_____ *Instructor Check*

FIGURE 4-28

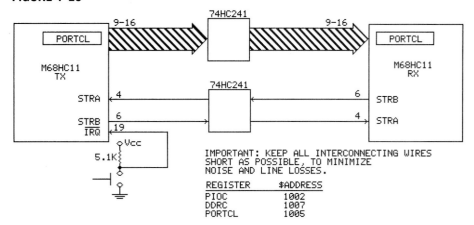

FIGURE 4-29

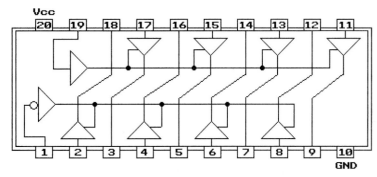

PART II: Software Development

The following is a general description of the software.

1. Both RX and TX initialize their stack pointers, PIOCs, and DDRCs. Additionally, the TX clears its I mask.
2. The RX initializes its X register to point at the RAM data storage block. The RX then enters a polling loop, where it checks and rechecks the STAF bit. When the STAF sets (handshake from the TX indicating that it has sent a byte), the RX reads PORTCL and stores the read data byte to memory (using indexed mode). Next, the RX increments its X register and checks to see if it has received and stored 16 bytes. If yes, the RX uses .OUTSTR to display "RX" on the terminal display, then halts. If no, the RX branches back to the STAF polling loop.
3. After clearing its I mask, the TX waits for an IRQ interrupt (push-button initiated). When interrupted, the TX points its X register at the RAM data storage block. The TX then reads a data byte from this block, and writes it to PORTCL. Next, the TX increments its X register and checks to see if it has transmitted 16 bytes. If yes, the TX calls a two-second delay, then returns from interrupt to its MAIN program. Here, the TX uses .OUTSTR to display "TX" on the terminal screen, and then halts. If the TX has not transmitted 16 bytes, it enters a polling loop. In this polling loop, the TX checks and rechecks the STAF until it sets. STAF = 1 flags a handshake from the RX. This handshake indicates that the RX has received the transmitted byte. When the STAF sets, TX reads the next byte from its data storage block and writes it to PORTCL.

Note: The $\overline{IRQ}$ is level sensitive. The TX calls a two-second delay so that a single key press will not generate multiple interrupts. Also, remember to initialize the transmitter's BUFFALO ROM IRQ pseudovector for the IRQ Service Routine at address $0130:

$00EE   JMP 130

Use the memory allocation chart, and the flowcharts in Figures 4–30 through 4–33, as guides to software development.

Memory Allocations

EVBU $ADDRESS	EVB $ADDRESS	CONTENTS—RX
0000–000F	0000–000F	Data storage
0010–0012	0010–0012	ASCII "RX" data
0100	C000	MAIN program
01FF	C0FF	Initial stack pointer

EVBU $ADDRESS	EVB $ADDRESS	CONTENTS—TX
0000–000F	0000–000F	Data storage
0010–0012	0010–0012	ASCII "TX" data
0100	C000	MAIN program
0130	C030	IRQ Interrupt Service routine
0160	C060	DELAY subroutine
01FF	C0FF	Initial stack pointer

FIGURE 4–30

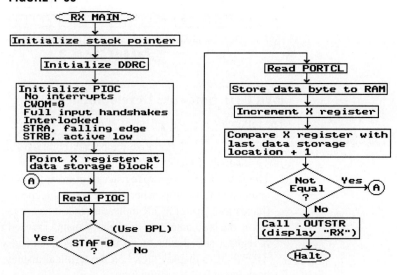

FIGURE 4–31

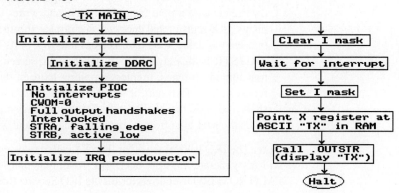

FIGURE 4-32

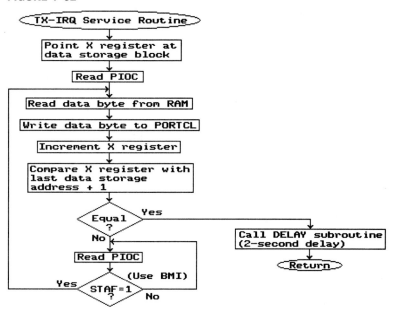

FIGURE 4-33

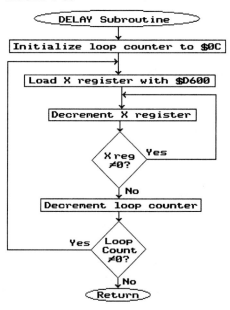

PART III: Interface Debug and Demonstration

Power up the EVBUs (EVBs), terminals, and interface. Load your software. Debug the interface and supporting software. Demonstrate interface operation to the instructor.

_____ *Instructor Check*

PART IV: Laboratory Analysis

Answer the questions below by neatly entering your answers in the appropriate blanks.

1. Explain why this laboratory exercise requires two EVBUs (EVBs).

2. How many bytes of data do the transmitter and receiver exchange, when the operator presses the push-button one time _____?
3. Explain how the interface operator initiates a data exchange. Describe the chain of events caused by this action.

4. How do the TX and RX indicate that the data exchange is complete?

5. Describe the purpose of the 74HC241 three-state drivers.

6. Describe the function of the IRQ interrupt.

7. For purposes of the data exchanges, which technique do the TX and RX use—interrupt control or status polling _____?
8. Do the TX/RX use interlocked or pulsed handshakes _____?
9. Explain why the TX IRQ service routine calls a two-second delay.

10. Evaluate the ability of your interface and supporting software to make reliable data exchanges between TX and RX.

PART V: Laboratory Software Listing

MAIN PROGRAM—RECEIVER—EVB

$ADDRESS	SOURCE CODE	COMMENTS
C000	LDS #C0FF	Initialize SP
C003	CLR 1007	Initialize DDRC
C006	LDAA #10	Initialize PIOC
C008	STAA 1002	" "
C00B	LDX #0	Point X register at data storage block
C00E	LDAA 1002	Read PIOC
C011	BPL C00E	If STAF = 0, re-read PIOC
C013	LDAA 1005	If STAF = 1, read PORTCL
C016	STAA 0,X	Store received byte to memory
C018	INX	Increment X register
C019	CPX #10	Compare X register with last data storage location + 1
C01C	BNE C00E	If not equal, read PIOC again
C01E	JSR FFC7	If equal, call .OUTSTR (display "RX")
C021	WAI	Halt

MAIN PROGRAM—RECEIVER—EVBU

$ADDRESS	SOURCE CODE	COMMENTS
0100	LDS #1FF	Initialize SP
0103	CLR 1007	Initialize DDRC
0106	LDAA #10	Initialize PIOC
0108	STAA 1002	" "

$ADDRESS	SOURCE CODE	COMMENTS
010B	LDX #0	Point X register at data storage block
010E	LDAA 1002	Read PIOC
0111	BPL 10E	If STAF = 0, re-read PIOC
0113	LDAA 1005	If STAF = 1, read PORTCL
0116	STAA 0,X	Store received byte to memory
0118	INX	Increment X register
0119	CPX #10	Compare X register with last data storage location + 1
011C	BNE 10E	If not equal, read PIOC again
011E	JSR FFC7	If equal, call .OUTSTR (display "RX")
0121	WAI	Halt

MAIN PROGRAM—TRANSMITTER—EVB

$ADDRESS	SOURCE CODE	COMMENTS
C000	LDS #C0FF	Initialize SP
C003	LDAA #FF	Initialize DDRC
C005	STAA 1007	" "
C008	LDAA #18	Initialize PIOC
C00A	STAA 1002	" "
C00D	LDD #C030	Initialize IRQ pseudovector
C010	STD EF	" " "
C012	CLI	Clear I mask
C013	WAI	Wait for interrupt
C014	SEI	Set I mask to prevent further interrupts
C015	LDX #10	Point X register at ASCII "TX"
C018	JSR FFC7	Call .OUTSTR (display "TX")
C01B	WAI	Halt

MAIN PROGRAM—TRANSMITTER—EVBU

$ADDRESS	SOURCE CODE	COMMENTS
0100	LDS #1FF	Initialize SP
0103	LDAA #FF	Initialize DDRC
0105	STAA 1007	" "
0108	LDAA #18	Initialize PIOC
010A	STAA 1002	" "
010D	LDD #130	Initialize IRQ pseudovector
0110	STD EF	" " "
0112	CLI	Clear I mask
0113	WAI	Wait for interrupt
0114	SEI	Set I mask to prevent further interrupts
0115	LDX #10	Point X register at ASCII "TX"
0118	JSR FFC7	Call .OUTSTR (display "TX")
011B	WAI	Halt

IRQ INTERRUPT SERVICE ROUTINE—TRANSMITTER—EVB

$ADDRESS	SOURCE CODE	COMMENTS
C030	LDX #0	Point X register at data storage block
C033	LDAA 1002	Read PIOC
C036	LDAA 0,X	Read data byte from memory
C038	STAA 1005	Write data byte to PORTCL
C03B	INX	Increment X register
C03C	CPX #10	Compare X register with last data storage location + 1
C03F	BEQ C048	If equal, call DELAY
C041	LDAA 1002	If not equal, read PIOC
C044	BMI C036	If STAF = 1, transmit another byte
C046	BRA C041	If STAF = 0, read PIOC again
C048	JSR C060	Call DELAY subroutine
C04B	RTI	Return

IRQ INTERRUPT SERVICE ROUTINE—TRANSMITTER—EVBU

$ADDRESS	SOURCE CODE	COMMENTS
0130	LDX #0	Point X register at data storage block
0133	LDAA 1002	Read PIOC
0136	LDAA 0,X	Read data byte from memory

0138	STAA 1005	Write data byte to PORTCL
013B	INX	Increment X register
013C	CPX #10	Compare X register with last data storage location + 1
013F	BEQ 148	If equal, call DELAY
0141	LDAA 1002	If not equal, read PIOC
0144	BMI 136	If STAF = 1, transmit another byte
0146	BRA 141	If STAF = 0, read PIOC again
0148	JSR 160	Call DELAY subroutine
014B	RTI	Return

DELAY SUBROUTINE—TRANSMITTER—EVB

$ADDRESS	SOURCE CODE	COMMENTS
C060	LDAA #C	Load loop counter
C062	LDX #D600	Load X register with count-down value
C065	DEX	Decrement X register
C066	BNE C065	If X register ≠ 0, decrement again
C068	DECA	Decrement loop counter
C069	BNE C062	If loop counter ≠ 0, count down X register again
C06B	RTS	Return

DELAY SUBROUTINE—TRANSMITTER—EVBU

$ADDRESS	SOURCE CODE	COMMENTS
0160	LDAA #C	Load loop counter
0162	LDX #D600	Load X register with count-down value
0165	DEX	Decrement X register
0166	BNE 165	If X register ≠ 0, decrement again
0168	DECA	Decrement loop counter
0169	BNE 162	If loop counter ≠ 0, count down X register again
016B	RTS	Return

EXERCISE 4.5 Full-Handshake/Three-State Variation

REQUIRED MATERIALS: 1 DL-1414 Alphanumeric Display
1 Resistor, 5.1 KΩ
1 Push-button Switch, push-to-close

REQUIRED EQUIPMENT: Motorola M68HC11EVBU or EVB, and terminal

GENERAL INSTRUCTIONS: Refer to Figure 4–34. In this exercise, use full-handshakes with three-state variation to detect push-button key-presses. The push-button pull-up circuit generates STRA. Pressing the push-button outputs a STRA active level (low); releasing the push-button outputs the STRA active edge. The STRA active edge sets the STAF bit and causes the microcontroller to write an ASCII character from its RAM data block to PORTCL. This write resets the STAF bit, and asserts STRB (in interlocked mode). STRB connects to the DL-1414's $\overline{WR}$ pin. A STRA active level (key press) causes the ASCII data word to appear on PORTC pins, and the DL-1414 to display the ASCII character. Push-button release generates a STRA active edge, three-states the PORTC pins, and initiates another ASCII data write cycle.

The M68HC11 uses status polling to detect STRA active edges. Between its periodic reads of the PIOC, the microcontroller performs routine operations by counting in hexadecimal at a one-second rate and displaying each count on the terminal display.

PART I: Connect the Interface

Power down the EVBU (EVB) and terminal. Refer to Figure 4–34. Connect interface components to M68HC11 pins via the EVBU (EVB) header.

FIGURE 4-34

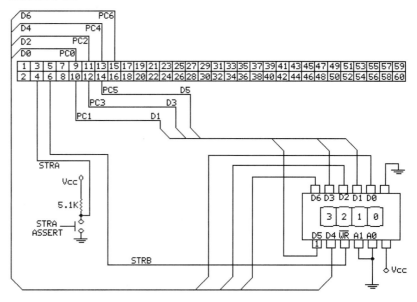

Show your connections to the instructor.

_____ *Instructor Check*

PART II: Software Development

Refer to block 1 of the MAIN program flowchart (Figure 4–35). Initialize the M68HC11 for full-handshake I/O with interlocked STRB and STRA three-state variation. A push-button press causes the STRA circuit to output a low. Therefore, initialize PIOC's EGA bit for STRA active low level, and positive-going active edge (push-button release). Disable interrupts because the software uses status polling. Since STRB asserts the DL-1414's $\overline{\text{WR}}$ pin, make it active low.

Refer to block 2. Since the three-state variation is used, clear DDRC (all DDRC bits = 0).

Refer to block 3. Use indexed addressing to access the sixteen ASCII data words (stored at memory addresses $0000–$000F). Therefore, point the X register at location $0000.

Refer to blocks 4 and 5. An initial push-button press causes a STRA active level. PORTCL data appears on PORTC pins. Therefore, initialize PORTCL with the first ASCII

FIGURE 4-35

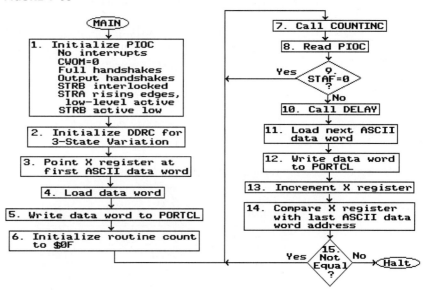

data word to be displayed by the DL-1414. After the first push-button release, the STRA active edge sets the STAF, and causes the microcontroller to access and write the next ASCII word to PORTCL.

Refer to block 6. Display a routine count between push-button presses. Increment and display this count on the terminal display at one-second intervals. Since the CPU will increment it before an initial display, initialize the count to –01 ($FF). Maintain this routine count in ACCA.

Refer to block 7. Call the COUNTINC subroutine, located at address $0150 ($C050). This subroutine increments the routine count and displays it on the terminal screen.

Refer to blocks 8 and 9. Load ACCB with the PIOC contents from address $1002. Test the STAF bit using the BPL instruction. If the STAF = 0, do another routine count.

Refer to blocks 10 through 12. If STAF = 1, call the DELAY subroutine. This delay negates the effects of push-button key bounce, which might otherwise cause several inadvertent STRA active levels and edges in rapid succession. After return from the one-second delay, load ACCB with the next ASCII data word - from the RAM data block ($0000–$000F). Use indexed addressing for this read, with an offset of $01. This read requires an $01 offset because the CPU read the original ASCII data word—at address $0000—during the initialization phase (see block 4). Write ACCB to PORTCL at address 1005.

Refer to blocks 13 through 15. Increment the X register so that the M68HC11 can access the next ASCII data word from the RAM data block. Compare the X register's contents with the address of the last ASCII data word ($000F). If the X register contents do not equal $000F, then branch back to display another routine count on the terminal screen. Conversely, if the X register contents equal $000F, the push-button has been pressed sixteen times. The M68HC11 has displayed all ASCII data characters on the DL-1414. Therefore, the M68HC11 halts.

Refer to block 1 of the COUNTINC (Routine Count Increment and Display subroutine) flowchart (Figure 4–36). Increment the routine count, which ACCA holds.

FIGURE 4–36

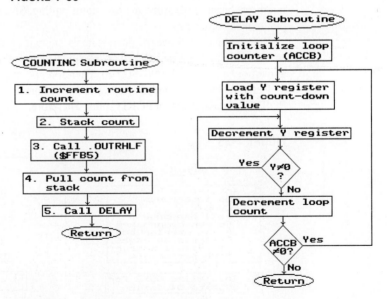

Memory Allocations

EVBU $ADDRESS	EVB $ADDRESS	CONTENTS
$0000–000F	$0000–000F	ASCII data words
$0100	$C000	MAIN program
$0140	$C040	DELAY subroutine
$0150	$C050	COUNTINC subroutine

Refer to block 2. Stack the routine count. The .OUTRHLF subroutine destroys $A through $F count values.

Refer to block 3. Display the routine count on the terminal screen by calling the .OUTRHLF utility subroutine, located at address $FFB5. This routine converts ACCA's low nibble to ASCII, and sends it to the terminal screen.

Refer to block 4. Restore the routine count to ACCA by pulling it from the stack.

Refer to block 5. Call the DELAY subroutine to display the routine count for one second on the terminal screen.

Locate the DELAY subroutine at address $0140 ($C040). Refer to the DELAY subroutine flowchart in Figure 4–36.

PART III: Interface Debug and Demonstration

Power up the EVBU (EVB), terminal and interface. Load your software. Load the RAM data block ($0000–$000F) with sixteen ASCII data words. Debug the interface and supporting software.

Demonstrate the operation of your interface and supporting software to the instructor.

_____ *Instructor Check*

PART IV: Laboratory Analysis

Answer the questions below by neatly entering your answers in the appropriate blanks.

1. Explain why the DL-1414 display goes blank one second after a push-button release.

2. What is the effect on the ASCII data word displays (DL-1414) if the PIOC PLS bit is changed from zero to one (PIOC initialized to $1B)? Why does this observed change occur?

3. What is the effect on the ASCII data word displays (DL-1414) if the DDRC is initialized to $FF instead of $00? Why does this change occur?

PART V: Laboratory Software Listing

MAIN PROGRAM—EVB

$ADDRESS	SOURCE CODE	COMMENTS
C000	LDAA #1A	Initialize PIOC
C002	STAA 1002	" "
C005	CLR 1007	Initialize DDRC
C008	LDX #0	Point X register at first ASCII data word
C00B	LDAA 0,X	Load first ASCII data word
C00D	STAA 1005	Write first ASCII data word to PORTCL
C010	LDAA #0F	Initialize routine count to $0F
C012	JSR C050	Call Routine Count Increment and Display
C015	LDAB 1002	Read PIOC
C018	BPL C012	STAF = 0?
C01A	JSR C040	If STAF = 1, call de-bounce delay
C01D	LDAB 1,X	Load next ASCII data word
C01F	STAB 1005	Write ASCII data word to PORTCL
C022	INX	Increment X register

$ADDRESS	SOURCE CODE	COMMENTS
C023	CPX #F	Compare X register with last ASCII data word address
C026	BNE C012	If not equal, display another routine count
C028	WAI	If equal, halt

MAIN PROGRAM—EVBU

$ADDRESS	SOURCE CODE	COMMENTS
0100	LDAA #1A	Initialize PIOC
0102	STAA 1002	" "
0105	CLR 1007	Initialize DDRC
0108	LDX #0	Point X register at first ASCII data word
010B	LDAA 0,X	Load first ASCII data word
010D	STAA 1005	Write first ASCII data word to PORTCL
0110	LDAA #0F	Initialize routine count to $0F
0112	JSR 150	Call Routine Count Increment and Display
0115	LDAB 1002	Read PIOC
0118	BPL 112	STAF = 0?
011A	JSR 140	If STAF = 1, call de-bounce delay
011D	LDAB 1,X	Load next ASCII data word
011F	STAB 1005	Write ASCII data word to PORTCL
0122	INX	Increment X register
0123	CPX #F	Compare X register with last ASCII data word address
0126	BNE 112	If not equal, display another routine count
0128	WAI	If equal, halt

DELAY SUBROUTINE—EVB

$ADDRESS	SOURCE CODE	COMMENTS
C040	LDAB #6	Load loop counter
C042	LDY #D600	Load Y register with count-down value
C046	DEY	Decrement Y register
C048	BNE C046	If Y register ≠ 0, decrement again
C04A	DECB	Decrement loop counter
C04B	BNE C042	If loop counter ≠ 0, count down Y register again
C04D	RTS	Return

DELAY SUBROUTINE—EVBU

$ADDRESS	SOURCE CODE	COMMENTS
0140	LDAB #6	Load loop counter
0142	LDY #D600	Load Y register with count-down value
0146	DEY	Decrement Y register
0148	BNE 146	If Y register ≠ 0, decrement again
014A	DECB	Decrement loop counter
014B	BNE 142	If loop counter ≠ 0, count down Y register again
014D	RTS	Return

COUNTINC SUBROUTINE—EVB

$ADDRESS	SOURCE CODE	COMMENTS
C050	INCA	Increment routine count
C051	PSHA	Stack count
C052	JSR FFB5	Call OUTRHLF
C055	PULA	Pull count from stack
C056	JSR C040	Call DELAY subroutine
C059	RTS	Return

COUNTINC SUBROUTINE—EVBU

$ADDRESS	SOURCE CODE	COMMENTS
0150	INCA	Increment routine count
0151	PSHA	Stack count
0152	JSR FFB5	Call OUTRHLF

0155	PULA		Pull count from stack
0156	JSR 140		Call DELAY subroutine
0159	RTS		Return

4.4 CHAPTER SUMMARY

Many applications use asynchronous parallel I/O. Asynchronous means that microcontroller and peripheral do not share a common clock signal. In such cases, microcontroller and peripheral must send "I've sent data" and "I've received data" messages to each other to coordinate their asynchronous data exchanges. These handshake messages consist of binary signal levels, or toggles, sent over dedicated lines.

The M68HC11 uses its STRA and STRB pins to receive and send handshakes. The peripheral sends its handshake message to the STRA pin. The microcontroller sends its handshake from STRB. The M68HC11 uses PORTB, PORTCL, and PORTC to exchange the actual parallel data with a peripheral.

The M68HC11 provides two kinds of handshakes: simple-strobed and full-handshake. The microcontroller also provides a three-state variation of output full-handshakes.

Simple-strobed handshakes allow the microcontroller to make bidirectional data transfers with a single peripheral, or to receive input data from one peripheral and send output data to another. In simple-strobed mode, the M68HC11 uses PORTCL and PORTC for read data, and PORTB for write data. PORTB writes automatically generate a STRB handshake signal. STRA assertion by a peripheral causes PORTCL flip-flops to latch read data from PORTC pins. STRA assertion also sets a status flag (STAF), informing the microcontroller that the read data is available from PORTCL.

Full-handshakes allow only unidirectional data exchanges with a peripheral. Full-handshake operations do not use PORTB. Rather, the M68HC11 uses PORTCL (and PORTC pins) for these read or write operations. A PORTCL read or write automatically asserts STRB. Full-handshake mode gives the M68HC11 user two ways to tailor STRB duration. In pulsed mode, the STRB signal lasts for two internal Phase Two clock cycles. In interlocked mode, the STRB signal lasts until the peripheral asserts a STRA active edge.

The three-state variation three-states PORTC output pins until the peripheral asserts a STRA active level. STRA active level causes the M68HC11 to drive its PORTC pins with data formerly written to PORTCL. Thus, the peripheral uses STRA to control the appearance of write data on PORTC pins.

The Parallel I/O Control Register (PIOC) configures the M68HC11 for simple-strobed or full-handshake modes. The PIOC also controls other handshake parameters:

- Whether full-handshakes apply to output or to input data.
- Whether full-handshakes use pulsed or interlocked STRB signals.
- Whether the STRA active edge is positive-going or negative-going, and its active level low or high.
- Whether STRB is active low or active high.

The PIOC also contains the STAF (PIOC bit 7). Once the STAF sets as the result of a STRA active edge, the M68HC11 CPU takes two actions to reset the STAF again:

1. Reads the PIOC.
2. Reads or writes PORTCL (full-handshake), or writes to PORTB (simple-strobed).

The second operation also automatically asserts the STRB signal.

The M68HC11 must respond when the STAF bit sets. A user can choose from two techniques for detecting a STAF set, and generating this response. The first technique is called status polling. Program the M68HC11 to make recurring PIOC reads, checking the STAF each time. If STAF = 1, the M68HC11 can take appropriate action to read or write PORTCL, or write to PORTB. The second technique is interrupt control. The STRA active edge generates an IRQ interrupt. The IRQ service routine causes the CPU to read or write

PORTCL, or write to PORTB. A PIOC control bit allows the user to enable or inhibit IRQ interrupt control. Thus, the M68HC11 provides simple and flexible means for handshaking during parallel data transfers.

EXERCISE 4.6 Chapter Review

PART I: In the blank preceding each descriptive phrase, indicate whether that phrase describes the simple-strobed mode (SS), the full-handshake mode (FH), or both the simple-strobed and full-handshake modes (B).

SS = Simple-Strobed
FH = Full-Handshake
B = Both Simple-Strobed and Full-Handshake

EXAMPLE:

___B___ Used for parallel I/O

_____ Uses both PORTB and PORTC

_____ Can use interrupt control

_____ Uses STRA to generate PORTCL reads. STRA cannot generate PORTCL writes.

_____ STRA pin is input; STRB pin is output

_____ Enabled by setting PIOC bit 4

_____ Normally used with only a single peripheral

_____ Uses PORTC for data inputs only

_____ Must read PIOC, then PORTCL to reset STAF

_____ Uses PORTC/PORTCL for all data transfers

_____ Uses the STAF bit for status polling

_____ Uses PORTD lines PD6 and PD7

_____ Uses STRA and STRB

_____ Uses interlocked or pulsed mode

_____ Convenient for use with two peripherals

_____ Enabled by clearing PIOC bit 4

_____ Write to PORTB causes STRB to assert for two PH2 clock cycles

_____ Convenient for alternating input and output operations

_____ Must read PIOC, then write to PORTCL to reset the STAF bit

_____ PORTCL asynchronously latches input data appearing on PORTCL pins

_____ Used for either data inputs or outputs, but not both

_____ Uses PORTB as the output register

PART II: Answer the questions below by neatly entering your answers in the appropriate blanks.

1. Describe the difference between synchronous and asynchronous data exchanges.

2. Define the term "handshake."

3. Explain how simple-strobed I/O uses STRA and STRB.

4. List the two purposes of the PIOC regarding simple-strobed I/O.

5. Explain how the M68HC11 uses PORTC and PORTCL during simple-strobed I/O operations.

6. Explain how the M68HC11 uses PORTB during simple-strobed I/O operations.

7. In a simple-strobed data exchange, which kind of data is exchanged first—read data or write data? _____ How does the receiving device know that this data has been transmitted?

8. List the two generally used programming techniques for determining whether the PIOC's STAF bit has set.

9. How does the M68HC11's CPU reset the STAF bit during simple-strobed operations?

10. Will the STAF reset if the M68HC11 CPU executes other instructions between its PIOC and PORTCL reads?

$ADDRESS	SOURCE CODE
010C	CLR 1002
010F	CLR 1007
0112	(Routine operations)

0130	LDAA 1002
0133	BPL 110
0135	LDAA 1005
0138	STAA A
013A	STAA 1004
013D	BRA 110

11. In the program above, the instruction at address $_____ initializes the PIOC.
12. The program above defines STRB assertion as active _____.
13. In the program above, the instruction at address $_____ tests the STAF bit.
14. In the program above, the M68HC11 CPU reads PORTCL by executing the instruction at address $_____.
15. In the program above, the M68HC11 writes data to PORTB by executing the instruction at address $_____.
16. Describe the action which causes STRB to assert during simple-strobed operations. How long does STRB assert?

17. Explain how simple-strobed interrupt control differs from status polling.

18. How does the programmer establish simple-strobed interrupt control?

$ADDRESS	SOURCE CODE
C000	LDAA #41
C002	STAA 1002
C005	CLR 1007
C008	CLI
C009	(Routine operations)

C01F	BRA C009
*****	
C050	LDAA 1002
C053	BPL C05D
C055	LDAA 1005
C058	STAA A
C05A	STAA 1004
C05D	RTI

19. The program above defines STRA assertion as a _____-going edge.
20. In the program above, interrupts cannot occur until the CPU executes the instruction at address $_____.
21. In the program above, the instruction at address $_____ tests the STAF bit. If STAF = 0, the CPU next executes the instruction at address $_____.
22. In the program above, the instruction at address $_____ establishes interrupt control.
23. In the program above, the instruction at address $_____ automatically resets the I mask (interrupt mask), in addition to its main function.
24. In the program above, the instruction at address $_____ identifies spurious or external IRQ interrupts.
25. Full-handshake I/O differs from simple-strobed I/O in three ways. List these three differences.

26. List the two possible messages conveyed by the peripheral when it asserts the STRA active edge during full-handshake operations.
27. In a special version of output full-handshakes, the peripheral uses the STRA active level to

28. During full-handshake operations, STRB assertion means that the M68HC11 has either _____ or _____.
29. List the PIOC's two purposes regarding full-handshake I/O.

30. Explain the difference between interlocked STRB and pulsed STRB.

31. During input-interlocked full-handshake operations, STRA active-edge assertion causes three responses from M68HC11. List these responses.

32. Which of the above responses (question 31) occur synchronously with the M68HC11 internal clock? _____
33. During input-interlocked full-handshake operations, the M68HC11 resets the STAF bit by _____.
34. During output-interlocked full-handshake operations, the STRA active edge causes two responses from the M68HC11. List these responses.

35. During output-interlocked full-handshake operations, the M68HC11 resets the STAF bit by _____.
36. During input-pulsed full-handshake operations, the STRA active edge causes two responses from the M68HC11. List these responses.

37. During output-pulsed full-handshake operations, the STRA active edge causes one response from the M68HC11. What is this response? _____
38. How does the programmer establish the output full-handshake "three-state variation?"

39. Describe how the "three-state variation" differs from normal full-handshake output operations.

$ADDRESS	SOURCE CODE
0100	LDAA #1F
0102	STAA 1002
0105	CLR 1007

40. Refer to the program segment above. Enter an X in the blank preceding each mode or characteristic enabled by this program segment.

    _____ Status polling

    _____ Interrupt control

    _____ Simple-strobed I/O

    _____ Full-input handshakes

    _____ Full-output handshakes

    _____ STRB pulsed

    _____ STRB interlocked

    _____ STRA low-level active

    _____ STRA high-level active

    _____ STRA positive-going edge asserts

    _____ STRA negative-going edge asserts

    _____ STRB active low

    _____ STRB active high

    _____ Three-state variation

41. In the three-state variation, the _____ causes the STAF to set.
42. In the three-state variation, the _____ causes the PORTC pins to be driven with write data.

$ADDRESS	SOURCE CODE
0120	LDAA #15
0122	STAA 1002
0125	CLR 1007
0128	(Routine operations)

0130	LDAA 1002
0133	BPL 128
0135	LDAA 1005
0138	STAA 0
013A	BRA 128

43. Refer to the program list above. Enter an X in the blank preceding each mode or characteristic enabled by the PIOC initialization.

    _____ Status polling

    _____ Interrupt control

    _____ Full-input handshakes

    _____ Full-output handshakes

    _____ STRB pulsed

    _____ STRB interlocked

    _____ Three-state variation

44. In the program above, the instruction at address $_____ asserts STRB.
45. In the program above, the instructions at addresses $_____ and $_____ reset the STAF.

$ADDRESS	SOURCE CODE
0100	LDAA #1B
0102	STAA 1002
0105	CLR 1007
0108	(Routine operations)

0110	LDAA 1002
0113	BPL 108
0115	LDAA 0
0117	STAA 1005
011A	BRA 108

46. Refer to the program list above. Enter an X in the blank preceding each mode or characteristic enabled by the PIOC initialization.

    _____ Status polling

    _____ Interrupt control

    _____ Full-input handshakes

    _____ Full-output handshakes

    _____ STRB pulsed

    _____ STRB interlocked

    _____ Three-state variation

47. In the program above, the instruction at address $_____ asserts STRB.
48. In the program above, the instructions at addresses $_____ and $_____ reset the STAF.
49. In the program above, the _____ negates STRB.

$ADDRESS	SOURCE CODE
0100	LDAA #57
0102	STAA 1002
0105	CLR 1007
0108	CLI
0109	(Routine operations)

0120	BRA 109
********	
0150	LDAA 1002

```
0153            BPL 15A
0155            LDAA 1005
0158            STAA 0
015A            RTI
```

50. Refer to the program list above. Enter an X in the blank preceding each mode or characteristic enabled by the PIOC initialization.

    _____ Status polling

    _____ Interrupt control

    _____ Full-input handshakes

    _____ Full-output handshakes

    _____ STRB pulsed

    _____ STRB interlocked

    _____ Three-state variation

51. In the program above, what possible events cause execution of the instruction at address $0150?

52. In the program above, a branch from address $0153 to address $015A indicates that

53. In the program above, the M68HC11 asserts STRB by executing the instruction at address $_____.

54. In the program above, the M68HC11 negates the STAF by executing the instructions at addresses $_____ and $_____.

$ADDRESS	SOURCE CODE
0000	LDAA #5F
0002	STAA 1002
0005	LDAA #FF
0007	STAA 1007
000A	CLI
000B	(Routine operations)
—	
—	
0020	BRA B
********	
0100	LDAA 1002
0103	BPL 10B
0105	LDAA 150
0108	STAA 1005
010B	RTI

55. Refer to the program list above. Enter an X in the blank preceding each mode or characteristic enabled by the PIOC initialization.

    _____ Status polling

    _____ Interrupt control

    _____ Full-input handshakes

    _____ Full-output handshakes

    _____ STRB pulsed

    _____ STRB interlocked

    _____ Three-state variation

56. In the program above, what causes the CPU to take the "no" path and execute the instruction at address $0105?

57. In the program above, what possible events cause the M68HC11 to execute the instruction at address $0100?

58. In the program above, the M68HC11 asserts STRB by executing the instruction at address $_____.

59. In the program above, the M68HC11 clears the STAF bit by executing the instructions at addresses $_____ and $_____.

5 The M68HC11 Serial Communications Interface (SCI)

5.1 GOAL AND OBJECTIVES

An EVBU or EVB uses asynchronous serial communications to exchange data with its terminal. Serial communication's simple data line connections make it easier to implement in many cases. Where parallel I/O requires multiple data lines, serial communications require only one or two data lines—depending upon the mode used.

Serial communications can use full-duplex mode or half-duplex mode. Full-duplex communications use two data lines. One line carries serial data from peripheral to microcontroller. The other line carries data from microcontroller to peripheral. Full-duplex operations speed up serial communications, because data can travel in both directions at the same time. Half-duplex mode uses a single communications line to send data in both directions. Of course, the single data line makes simultaneous bidirectional data transmission impossible.

Serial communications require more complex supporting software than parallel communications. One can illustrate this complexity by comparing the number of M68HC11 status and control registers used in parallel versus serial communications:

> Parallel I/O uses only one status/control register—the Parallel I/O Control Register (PIOC).
> Serial I/O uses four such registers—the SCI Baud Rate Control Register (BAUD), SCI Control Registers 1 and 2 (SCCR1/SCCR2), and the SCI Status Register (SCSR).

Note that serial I/O register names include the acronym "SCI." This acronym stands for Serial Communications Interface. The M68HC11's SCI generates asynchronous serial communications with one or more peripherals.[1] "Asynchronous" means that data transmitter and receiver do not share a common clock signal. To coordinate asynchronous serial communications, the data transmitter must send special code bits before and after the serial data bits. These "framing" bits inform the data receiver when the serial data begins and when it ends. Framing bits serve the same purpose as the capital letter at the beginning of this sentence and the period at its end.

This chapter's goal is to help the reader understand how to implement asynchronous serial communications between the M68HC11 and its peripherals, using the SCI.

---

[1]Chapter 6 presents the Serial Peripheral Interface, which provides for synchronous serial data exchanges.

After studying this chapter and completing its exercises and hands-on activities, you will be able to:

> Identify microcontroller resources used for asynchronous serial communications.
> Derive control words which configure the SCI for serial communications.
> Develop and demonstrate SCI software routines which take control of the EVBU (EVB) serial communications port, and communicate back and forth with the system terminal.

5.2 INTRODUCTION TO THE M68HC11 SCI

The SCI has six features, which make it flexible and relatively easy to use:

1. The M68HC11 SCI can use either full-duplex mode, or half-duplex mode. The SCI uses two PORTD pins: RxD—receive data, and TxD—transmit data. RxD is M68HC11 pin 20 (PD0), and TxD is pin 21 (PD1). The RxD pin connects internally to a serial in-parallel out shift register. The SCI shifts a serial data word into this register from the RxD line. The SCI then transfers the data word, in parallel, to a parallel in-parallel out register called the Receive Data Register Buffer. The M68HC11's CPU can then read the contents of this buffer. The TxD pin connects internally to a parallel in-serial out shift register. This shift register connects to a parallel in-parallel out register called the Transmit Data Register Buffer. To transmit data, the CPU writes to the Transmit Data Register Buffer. This buffer transfers the write data in parallel to the shift register. The shift register drives the output TxD pin serially with the write data.

    The RxD and TxD pins can connect to two separate data lines for full-duplex operation; or to a single data line for half-duplex operation. When the TxD and RxD pins connect to a single line, the SCI cannot transmit and receive at the same time. Full-duplex configuration allows simultaneous transmit and receive operations.

2. The M68HC11 SCI uses the NRZ (Nonreturn to Zero) format. This format calls binary ones "marks" and binary zeros "spaces." A serial data word consists of either 8 or 9 data bits, plus 2 framing bits. The SCI transmits the least significant (LS) bit first. A framing bit—called the start bit—precedes the LS data bit. This start bit is a "space," or binary zero. Another framing bit—called the stop bit—follows the most significant (MS) data bit. The stop bit is a "mark." Often, the ninth data bit functions as a parity bit. Section 5.3 covers framing and parity in more detail.

3. The SCI incorporates a software-programmable baud rate generator. Baud is the rate at which the serial data and framing bits appear on a data line. Measure baud rate in bits per second.

4. A user can configure the SCI for 8 or 9 data bits. The SCI cannot generate parity bits automatically. The user must program for parity bit generation.

5. The SCI can control serial communications with interrupts or with status polling. These conditions can generate interrupts:
    - Transmit Data Register Empty (SCI has transmitted a data word).
    - Receive Data Register Full (SCI has received a data word).
    - Transmit Complete (SCI has transmitted a multiword message).
    - Idle Line (RxD line is idling; the SCI has received a multiword message).

6. The SCI incorporates a "wake up" feature. Use this feature when the M68HC11 functions as a receiver on a multidrop serial communications line. This feature puts the SCI into a "dormant" state, when it has not been selected by the transmitter. The "wake up" feature brings the SCI out of dormancy for the beginning of each serial message. Upon wake up, the SCI checks special address bits to see whether the transmitter has selected it or not. If not, the SCI can return to the dormant state. This feature reduces software overhead at M68HC11's functioning as multidrop receivers.

This chapter's succeeding sections elaborate on the SCI's six features.

5.3 M68HC11 SCI DATA FORMAT

The M68HC11's SCI uses standard TTL voltage levels for its marks and spaces. The SCI drives the TxD line to 4.2V for a mark (binary one), and to .4V for a space (binary zero). The SCI recognizes 3.5V–5V on its RxD line as a mark, and 0V–1V as a space.[2] Thus, the serial communication lines toggle between these voltages to make binary ones and zeros.

When the serial data line idles, it stays at the binary one level. During breaks between serial messages or parts of messages, the serial data line goes to binary zero. An ASCII break character drives the data line low for the period of time it takes to send a data word and framing bits.

Refer to Figure 5–1.[3] This figure diagrams marks and spaces as they toggle the serial communications line. The left of the diagram shows an idle communications line, indicated by a continuous binary one. The serial line goes low for one bit time to indicate that data bits follow.[4] This low constitutes a framing bit—called the start bit.

The Figure 5–1 diagram then shows nine bit times, one for each data bit. The serial line goes low or high for the duration of each bit time, depending upon whether the data bit is binary zero or one. Recall that a serial data word (less framing bits) consists of either 8 or 9 data bits (a user option).

Following the data bits, the serial communications line goes high for one bit time. This high bit constitutes a framing bit—the stop bit. An idle line, break character, or another start bit can follow the stop bit. The Figure 5–1 diagram shows a start bit and the beginning of another data word.

The eight or nine data bits, with their start and stop bits, constitute a data frame. Thus, a data frame can consist of 10 or 11 bits. Use an 11-bit frame to send a standard 8-bit binary word, and a parity bit. Note in Figure 5–1 that the SCI transmits the most significant bit last. A programmer can use this most significant data bit (bit 8) as a parity bit. Use 10-bit frames for sending ASCII characters. ASCII code itself uses bits 0 through 6. A programmer can use the eighth bit (bit 7) for parity. Parity generation and checking provides an easily implemented, but only approximate check for transmission errors.

A communications system can use either odd or even parity. Both transmitter and receiver must agree whether parity is odd or even. A parity generator makes the parity bit high or low, depending upon the number of high data bits (not including framing bits). For example, assume a 9-bit data word, with bit 8 (MS) assigned to even parity:

```
The data word = %_D8 P 0 1 0 1 0 1 1 1 _D0
```

Since the data word proper contains five binary ones, the parity generator makes the parity bit high, to provide even parity:

```
The data word = %_D8 1 0 1 0 1 0 1 1 1 _D0
```

Now the data word contains an even number of binary ones. If this example used odd parity, then the parity generator would make its bit a zero—to retain an odd number of high bits.

FIGURE 5–1
Reprinted with the permission of Motorola.

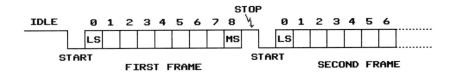

---

[2]Motorola Inc., *MC68HC11E9 HCMOS Single-Chip Microcontroller, Advance Information* (Motorola Inc.: 1988), p. 11-2. These figures assume $V_{DD} = 5VDC$ and $V_{SS} = 0V$.
[3]Motorola Semiconductor Products Sector, *MC68HC11 Course Notes* (Phoenix: Motorola Inc.: 1990), p. 128.
[4]If the serial communications data line operates at 9600 baud, each bit causes the line to be high or low for 104µs. Thus, at 9600 baud, one bit time = 104µs.

The data receiver employs a parity checker. The parity checker simply counts the number of binary ones in the received data word, less the framing bits but including the parity bit. With even parity, an even number of high bits causes the parity checker to assume that there are no errors in the data word. An odd number of high bits indicates one or an odd number of transmission errors. Thus, the receiver should ignore the data word, or request a retransmission. Of course, with odd parity, the parity checker looks for an odd number of high bits to indicate an error-free data word. The user can easily implement a software parity generator for 8-bit data words. Figure 5–2 flowcharts this parity generation program.

Refer to Figure 5–2 and its accompanying program list. This program generates even parity bits. However, the user can easily modify it for odd parity. The parity generator simply counts the number of binary-one bits in an 8-bit data word—less the parity bit. This program counts high bits by shifting the data word bit by bit into the C flag, and counting the number of times the C flag sets. The parity generator maintains a count of one bits in

FIGURE 5–2

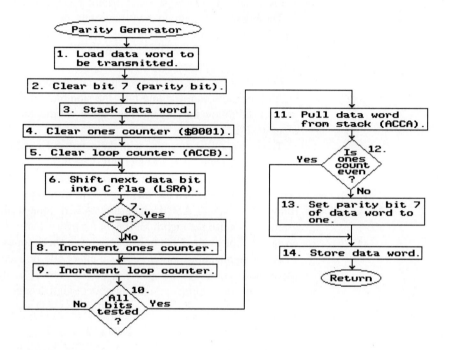

Parity Generator

$ADDRESS	SOURCE CODE	COMMENTS
0100	LDAA 0	Load data word
0102	ANDA #7F	Clear parity bit (D7)
0104	PSHA	Stack data word
0105	CLR 1	Clear ones counter
0108	CLRB	Clear loop counter
0109	LSRA	Shift data bit into C flag
010A	BCC 10F	C flag = 0?
010C	INC 1	If not, increment ones count
010F	INCB	Increment loop counter
0110	CMPB #7	All data bits tested?
0112	BNE 109	If not, test another bit
0114	PULA	If yes, pull data word from stack
0115	LSR 1	Is ones count even?
0118	BCC 11C	If yes, let D7 = 0
011A	ORAA #80	If not, make D7 = 1
011C	STAA 0	Store data word
011E	RTS	Return

address $0001. Bit shifts will destroy the data. Therefore, the CPU first stacks the data word to save it for later transmission.

The program must shift 7 data bits into the C flag. It uses a shift loop to accomplish this task. The parity generator executes the shift loop once for each data bit. ACCB holds a shift loop count. The parity generator escapes the shift loop after ACCB has incremented from zero to seven.

After escaping the shift loop, the parity generator restores the data word by pulling it from the stack. The parity generator then checks the binary ones count at address $0001. If it finds that bit zero of the ones count is binary one, the count must be odd. Therefore, the program sets the parity bit. The parity generator sets the parity bit by ORing the data word with mask $80. Modifying the parity generator for odd parity is simple: Change block 12 (Figure 5–2) to read "Is ones count odd?"

The programmer can also modify the parity generator to make it a parity checker:

- Change block 1 to read "Load received data word."
- Delete block 2.
- Make block 12 read "Is ones count even?" for even parity, or "Is ones count odd?" for odd parity.
- Change block 13 to read "Call error notification routine."

Implementing a parity generator/checker for 9-bit data words is somewhat more complex. The parity bit must reside in another register, separate from the data word proper. However, the programmer can apply the Figure 5–2 strategy with minimum added complication.

EXERCISE 5.1   M68HC11 SCI Data Format

AIM: Given data word length, parity specification, and message content, accurately diagram serial data frames.

INTRODUCTORY INFORMATION: By completing this exercise, reinforce your understanding of the NRZ message format. This exercise also illustrates how the SCI sends ASCII characters and other data over serial communication lines.

PART I: 10-Bit Frame, ASCII Data Words, Odd Parity

Neatly complete the diagrams below for the message "Howdy!". Each data frame diagram should include two idle line bit times, a start bit, ASCII data word, odd parity bit, a stop bit, and two more idle line bit times. The vertical dashed lines indicate bit widths. The horizontal dashed lines indicate binary ones and zeros. Format each data word as shown in the example for the letter "E."

PART II: 11-Bit Frame, Hexadecimal Data Words, Even Parity

Neatly complete the diagrams below for the hexadecimal data specified. Each data frame diagram should include two idle bit times, a start bit, 8 data bits (two hex digits), even parity bit, a stop bit and two more idle line bit times. The vertical dashed lines indicate bit times. The horizontal dashed lines indicate binary ones and zeros. Format each data word as shown in the example for data $1D.

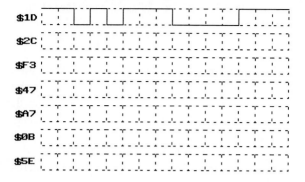

5.4 ON-CHIP RESOURCES USED BY THE M68HC11 SCI

The SCI uses five internal registers while sending and receiving serial data. Following sections cover these five registers.

SCI Data Register (SCDR, Control Register Block Address $102F)

To output a serial data word, the CPU writes it to the SCI Data Register (SCDR). This write causes the SCI to frame the data word, convert it to serial data, then drive the TxD pin with these serial bits.

Also, the CPU reads input serial data from the SCDR. For example, a peripheral sends a data frame to the M68HC11, over the RxD line. Receipt of this data causes the SCI to notify the CPU. The CPU can then read the data word (less framing bits) from the SCDR.

The M68HC11 CPU handles 8-bit data words by reading and writing the SCDR. But how does the M68HC11 handle 9-bit data words? The SCDR has only an 8-bit capacity. Another SCI register contains the ninth, most significant bit. SCCR1 (SCI Control Register 1) contains this ninth, MS bit.[5]

The CPU first writes the MSB of a 9-bit output data word to bit 6 of SCCR1, *then* writes the other 8 data bits to the SCDR. In response, the SCI frames the 9-bit data word and drives the TxD pin with it.

The CPU reads the input data word's MSB from SCCR1 bit 7 *before* reading the other 8 data bits from the SCDR. These two reads capture the entire 9-bit data word sent by the peripheral to the M68HC11's RxD pin.

The SCDR actually consists of four registers—two each for the transmit and receive functions. Figure 5–3 shows these four registers.[6]

Using two registers each for the transmit (TX) and receive (RX) functions double buffers the SCDR. Refer to the TX section of Figure 5–3. The CPU loads the TDR buffer by writing to bit 6 of SCCR1 (if the data word is 9 bits) and to the SCDR. Following discussions will call this buffer the TDR (Transmit Data Register). Thus, the TDR actually consists of 8 bits in the SCDR and 1 bit (labeled T8) in SCCR1. The 8-bit SCDR write transfers all 9 data bits to the shift register. These bits then appear serially on the TxD pin. As soon as the SCI accomplishes a parallel transfer to the shift register, it notifies the CPU that the "TDR is empty." While the shift register drives the TxD pin with the original data, the CPU can write new data to the TDR. "Double buffering" means, then, that the transmitter can handle two data words at a time: one shifting out serially over the TxD line, while the CPU writes the other to the TDR.

---

[5]This register also initializes the SCI for 8- or 9-bit operation and receiver wake up mode. SCCR1 is covered on pages 229–231.
[6]Motorola Semiconductor Products Sector, *MC68HC11 Course Notes*, p. 128.

FIGURE 5–3
Reprinted with the permission of Motorola.

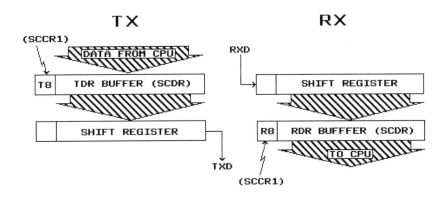

Refer to the RX section of Figure 5–3. The serial peripheral drives the M68HC11's RxD pin with a data frame. This data shifts serially into the receive shift register. The SCI then transfers this data word to the RDR buffer or "RDR" (Receive Data Register). The RDR consists of SCCR1 bit 7 (for the data word's MS bit) and the 8 bits of the SCDR. After data transfers to the RDR, the SCI notifies the CPU that the "RDR is full." The CPU then reads SCCR1 and SCDR to capture the 9-bit serial data word. Thus, the SCI receive section is also double-buffered. The CPU reads the RDR while the SCI shifts a new data frame into the receive shift register.

Finally, note that the SCI transmitter adds framing bits at the transmit shift register. The SCI receiver strips framing bits and checks for errors at the receive shift register. The SCI receiver checks for several kinds of errors: overrun errors (caused by too much data appearing at the RxD pin), noise errors, and framing errors. These errors are discussed in more detail on pages 231–233.

SCI Baud Rate Control Register (BAUD, Control Register Block Address $102B)

Refer to Figure 5–4 and its accompanying Baud Rate Table.[7] BAUD register bits 5, 4, and 2 through 0 determine the SCI baud rate.[8] Motorola labels bits 5 and 4 as the SCP1 and SCP0 prescaler bits. The baud rate prescaler section divides the E clock rate (2MHz for the EVBU or EVB) by 16. The prescaler then divides the resultant 125 KHz frequency by 1, 3, 4, or 13—depending upon the states of prescaler bits SCP1 and SCP0:

SCP1	SCP0	E/16 ÷ by
0	0	1
0	1	3
1	0	4
1	1	13

Assume, for example, that the user initializes SCP1 to binary one SCP0 to binary zero. The prescaler divides 125KHz by four, and the prescaler section outputs 31.25KHz.

The baud rate generator divides the prescaler output by 2^n, where n = the decimal value of baud rate select bits SCR2, SCR1, and SCR0:

SCR2	SCR1	SCR0	n	2^n
0	0	0	0	1
0	0	1	1	2
0	1	0	2	4
0	1	1	3	8
1	0	0	4	16
1	0	1	5	32
1	1	0	6	64
1	1	1	7	128

---

[7] Motorola Inc., *M68HC11 Reference Manual* (1990), p. 9-31. Extracted with permission. The Baud Rate Table is based on the EVBU (EVB) crystal frequency of 8MHz, and E clock frequency of 2MHz. Refer to *M68HC11 Reference Manual*, p. 9-31 for tables based on 8.4MHz, 4MHz, and 3.7MHz crystals.

[8] BAUD register bits 7 and 3 are used in M68HC11 special test mode for factory testing of microcontroller chips.

FIGURE 5-4
Reprinted with the permission of Motorola.

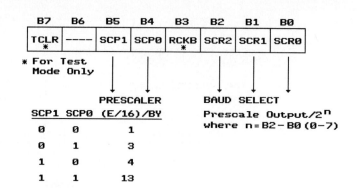

BAUD RATE TABLE (f_{XTAL} = 8MHz, E CK = 2MHz)

SCP1	SCP0	SCR2	SCR1	SCR0	BAUD	SCP1	SCP0	SCR2	SCR1	SCR0	BAUD
0	0	0	0	0	125K	1	0	0	0	0	31.25K
0	0	0	0	1	62.5K	1	0	0	0	1	15.625K
0	0	0	1	0	31.25K	1	0	0	1	0	7812.5
0	0	0	1	1	15.625K	1	0	0	1	1	3906
0	0	1	0	0	7812.5	1	0	1	0	0	1953
0	0	1	0	1	3906	1	0	1	0	1	977
0	0	1	1	0	1953	1	0	1	1	0	488
0	0	1	1	1	977	1	0	1	1	1	244
0	1	0	0	0	41.666K	1	1	0	0	0	9600
0	1	0	0	1	20.833K	1	1	0	0	1	4800
0	1	0	1	0	10.417K	1	1	0	1	0	2400
0	1	0	1	1	5.208K	1	1	0	1	1	1200
0	1	1	0	0	2.604K	1	1	1	0	0	600
0	1	1	0	1	1.302K	1	1	1	0	1	300
0	1	1	1	0	651	1	1	1	1	0	150
0	1	1	1	1	326	1	1	1	1	1	75

Continuing our example, assume that rate select bits SC2 through SC0 = 5. The baud rate generator divides the 31.25KHz prescaler output by 2^5 or 32. The selected baud rate then equals 977 baud. Thus, to select 977 baud, the user would initialize the BAUD register to $25:

$_{B7}$ 0 0 1 0 0 1 0 1 $_{B0}$ = $25

EXERCISE 5.2   M68HC11 SCI Baud Rate Generator

AIM: Given either BAUD register contents or baud rate generator data, derive corresponding counterpart values.

INTRODUCTORY INFORMATION: To complete this exercise, apply the information presented in this chapter on the BAUD register and baud rate generator. This application exercise clarifies and reinforces your understanding of this information.

INSTRUCTIONS: Refer to the Baud Rate Table and the information presented on pages 227–228 as you fill in the blanks below. In the first ten problems, derive baud rate generator data from the BAUD register contents given. Note the completed example for BAUD register = $27. For the second set of ten problems, determine the BAUD register contents from the baud rate generator information supplied. Note the completed example for BAUD register = $06.

$$f_{XTAL} = 8 \text{ MHz}; E\ CK = 2 \text{ MHz}$$

$BAUD Contents	Prescaler Output (Hz)	2^n	Baud Rate	
27	31.25KHz	128	244	EXAMPLE
06	_____	___	_____	
34	_____	___	_____	
15	_____	___	_____	
03	_____	___	_____	
22	_____	___	_____	
31	_____	___	_____	
00	_____	___	_____	
14	_____	___	_____	
33	_____	___	_____	
26	_____	___	_____	
06	125K	64	1953	EXAMPLE
___	31.25K	128	244	
___	41.667K	32	1.302K	
___	9.615K	4	2400	
___	31.25K	1	31.25K	
___	41.667K	16	2.604K	
___	9.615K	8	1200	
___	125K	2	62.5K	
___	41.667K	64	651	
___	9.615K	16	600	

SCI Control Register 1
(SCCR1, Control Register Block Address $012C)

SCCR1 has three functions:

1. Select SCI data word length—8 or 9 bits.
2. Provide storage for the ninth data bit in a serial data word.
3. Select the receiver wake up mechanism.

Refer to Figure 5–5. Motorola labels SCCR1 bit 4 the "M" bit. "M" stands for "Mode" or data frame format. Reset the M bit to format the SCI data frame for 1 start bit, 8 data bits, and 1 stop bit. Set the M bit to format for 1 start bit, 9 data bits, and 1 stop bit.

Recall that the M68HC11 CPU transmits or receives data bits by writing to the TDR (transmit data register) or reading the RDR (receive data register). The SCI Data Register (SCDR) contains 8 bits out of the 9-bit capacity of either TDR or RDR. Where the SCCR1 M bit = 0, the SCDR can hold the entire 8-bit data word. The CPU can receive or transmit simply by reading or writing to the SCDR. But if the M bit = 1, the SCDR has capacity for only 8 out of the 9 data bits. SCCR1 provides storage for the ninth bit (bit 8) of the data word. Figure 5–5 shows that SCCR1 bits 6 and 7 are labeled "T8" and "R8," respectively. The CPU writes a 9-bit data word's MS bit to T8. R8 latches a 9-bit receive data word's MS bit. Thus, R8 forms the ninth bit of the RDR, and T8 forms the ninth bit of the TDR.

System designers often use the ninth data bit for either parity or an additional stop bit. Motorola assigned R8 to SCCR1's MSB to facilitate checking of this ninth data bit. The CPU can use an N flag-oriented branching instruction (e.g., BPL) to check this bit.

The CPU must write to the SCCR1's T8 bit without disturbing the R8, M, or WAKE bits in this register. Therefore, the programmer should use the BSET or BCLR instruction to write a bit to T8. To write a one to T8 use:

```
LDX #1000      Point X register at control register block.
BSET 2C,X 40   Set the T8 bit.
```

FIGURE 5–5
Reprinted with the permission of Motorola.

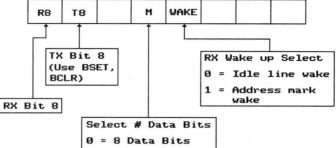

Write a zero to T8:

```
LDX #1000     Point X register at control register block.
BCLR 2C,X 40  Clear the T8 bit.
```

BSET or BCLR uses a mask to set or clear bits at the target address. Each 1 bit in the mask selects its target address counterpart for setting (BSET) or clearing (BCLR). Each zero in the mask leaves its counterpart unchanged. Mask $40 selects bit 6 to be set (BSET) or cleared (BCLR):

Mask $40 = %$_{B7}$ 0 1 0 0 0 0 0 0 $_{B0}$

The BSET and BCLR instructions use only direct and indexed addressing. Since SCCR1 resides on page $10, the programmer must use indexed mode.

Motorola calls SCCR1 bit 3 the "WAKE" bit. Recall that the M68HC11 uses the wake up feature when it is one of a number of receivers sharing a multidrop serial communications line. The receiver can go into a dormant state when not selected by the transmitter for communications via the multidrop line. The SCI wake up feature must bring the M68HC11 receiver out of dormancy for the beginning of each serial message—to check whether it has been selected or not. The SCCR1 WAKE bit determines how the M68HC11 receiver comes out of dormancy. To understand how the WAKE bit works, the reader must first understand more about M68HC11 receiver wake up operation.

To make the SCI dormant, the M68HC11 CPU sets a bit in another SCI control register—SCCR2. Setting this bit puts the receiver to sleep, and clearing the bit wakes the receiver up. Putting the receiver "to sleep" disables all SCI receiver status flags and interrupts. However, the receiver still shifts in all serial data. The transmitter puts addressing information in the first frame that it transmits. This addressing information identifies which multidrop receivers should receive the message to follow. Each multidrop receiver must read and check the addressing frame to see if the transmitter identified it as a message destination. If a particular SCI receiver is not addressed, it can put its SCI receiver back to sleep.

The M68HC11 receiver wake up function provides two ways of automatically waking the SCI receiver (i.e., enabling receiver status flags and interrupts):

   Idle line wake up
   Address mark wake up

With idle line wake up, the SCI automatically clears the SCCR2 wake up control bit when it detects that its RxD pin has remained high for one complete frame time. Thus, to wake up multidrop receivers between messages, the transmitter idles the serial communications line for at least one complete frame time.

With address mark wake up, the transmitter uses each serial data word's MSB as an "address bit."[9] The transmitter sets the MSB in the first frame of the message, and resets it in

---

[9] Where SCCR1's M bit = 0, SCDR bit 7 constitutes the "address bit." If SCCR1's M bit = 1, SCCR1's R8 bit constitutes the "address bit."

the remaining frames. When a dormant SCI receiver detects an "address mark" (a set MSB), it automatically clears the SCCR2 wake up control bit. The receiver wakes up and evaluates the addressing information in this initial frame to see whether it is selected.

Refer again to Figure 5–5. The SCCR1 WAKE bit (bit 3) determines how the M68HC11 SCI receiver wakes:

WAKE = 0 Idle line wake up
WAKE = 1 Address mark wake up

SCI Status Register (SCSR, Control Register Block Address $102E)

Refer to Figure 5–6. The M68HC11's SCSR contains four SCI status flags and three error flags:

TDRE (bit 7)—Transmit Data Register Empty
TC (bit 6)—Transmit Complete
RDRF (bit 5)—Receive Data Register Full
IDLE (bit 4)—Idle Line Detect
OR (bit 3)—Overrun Error
NF (bit 2)—Noise flag
FE (bit 1)—Framing Error

The M68HC11 CPU writes a 9-bit transmit data word to bit 6 of SCCR1 and the 8 bits of the SCDR. The CPU writes an 8-bit data word to the SCDR only. Either of these writes fills the TDR (Transmit Data Register). As soon as the SCI transfers this write data from the TDR to the shift register, the TDR becomes "empty." This transfer sets the TDRE flag (Transmit Data Register Empty) so that the CPU can write the next data word to the TDR.

Interpret the TDRE flag as follows:

TDRE = 1   The TDR is empty, and the M68HC11 CPU can now write another data word to SCCR1 bit 6 and the SCDR.
TDRE = 0   The TDR is not empty. Do not write to the SCDR or SCCR1.

The M68HC11 CPU clears the TDRE flag by (1) reading the SCSR, then (2) writing to the SCDR. After clearing the TDRE flag with these actions, the CPU can use status polling or interrupt control to determine when the TDRE sets again.

After the transmit shift register drives the TxD with a serial message's final framing bit, the TxD pin idles *and* the TC (Transmit Complete) flag sets. Either of these conditions sets the TC flag:

1. The SCI transmitter is enabled, the TxD pin idles, and TDRE = 1.
2. The transmit shift register has shifted out a complete frame to the TxD pin, and the SCI transmitter has been disabled.[10]

The TC bit seems to replicate the TDRE bit. In the case of a final message frame, however, the TDRE bit sets before the transmit shift register has transferred the final message bit to the TxD pin. In contrast, the TC flag sets only after the final message bit has driven the TxD pin. The M68HC11 MPU therefore uses the TC flag as an indication that it is safe to turn off a peripheral (e.g., a modem) connected to the TxD pin.

Interpret the TC flag as follows:

TC = 1   The SCI transmitter is idle and the transmit shift register has shifted out a complete frame to the TxD pin.
TC = 0   The SCI transmitter is still busy.

FIGURE 5–6
Reprinted with the permission of Motorola.

SCSR

7	6	5	4	3	2	1	0
TDRE	TC	RDRF	IDLE	OR	NF	FE	

---

[10]Motorola Inc., *MC68HC11E9 HCMOS Single-Chip Microcontroller, Advance Information* p. 5-9.

The M68HC11 MPU clears the TC flag by (1) reading the SCSR, when the TC flag is set; then (2) writing to the SCDR. The CPU can use either status polling or interrupt control to determine when the TC flag sets.

The M68HC11 captures a received 9-bit data word by reading bit 7 of SCCR1 (R8) and all 8 bits of the SCDR. The M68HC11 captures an 8-bit word by reading the SCDR. The CPU must not read this data until the SCI transfers it to the Receive Data Register (RDR). A data transfer from the receive shift register to the RDR sets the RDRF (Receive Data Register Full) status flag. RDRF = 1 informs the CPU that it can read the received data word. Thus, interpret the RDRF flag as follows:

RDRF = 1   The data word has been received and transferred to the RDR.
RDRF = 0   Data has not yet been transferred to the RDR; do not read.

The M68HC11 CPU clears the RDRF flag by (1) reading the SCSR, then (2) reading the SCDR. After clearing the RDRF flag with these actions, the CPU can use status polling or interrupt control to determine when the RDRF sets again.

The IDLE flag (Idle Line Detect) sets when the M68HC11's RxD line remains high for one full frame time (10 or 11 bits as defined by SCCR1's M bit). Interpret the IDLE flag as follows:

IDLE = 1   RxD pin has remained high for one full frame time.
IDLE = 0   The RxD line is receiving input data frames or break character(s).

The M68HC11 MPU clears the IDLE flag by (1) reading the SCSR, then, (2) reading the SCDR. Once cleared, the IDLE flag cannot set again until the RxD pin goes active, then idle. After clearing the IDLE flag, the CPU can use status polling or interrupt control to determine when the IDLE bit sets again.

The SCSR includes three flags which identify errors in the SCI receive process. As soon as the SCI shifts a complete data word into the receive shift register, it automatically transfers the data word to the RDR. However, the SCI makes this transfer only if the CPU has read the previous data transferred to the RDR. Thus, the CPU must read the RDR contents while a new data word shifts into the receive shift register. If the CPU does not complete the RDR read by the time the new data is completely shifted in, the new data is lost. The OR (Overrun Error) flag sets to indicate this loss. This arrangement gives the CPU at least one complete frame time to read the RDR.

Obviously, an OR flag indicates that the peripheral is sending data faster than the M68HC11 can process it. This situation can be caused by one of the following:

1. Baud rate too high
2. Excess M68HC11 software overhead

The M68HC11 CPU clears the OR flag by (1) reading the SCSR, then (2) reading the SCDR.

The SCI receiver takes continuous RxD pin samples—at a rate sixteen times the baud rate. Therefore, the SCI takes sixteen samples per bit. The SCI takes these multiple samples for three reasons:

1. Detect a start bit.
2. Determine if a data or stop bit is a one or zero.
3. Check for noise.

After it detects a stop bit, the SCI receiver begins looking for the next start bit. The SCI detects a possible start bit when a low sample follows three consecutive high samples. This low sample becomes the first of sixteen samples taken during the possible start bit time. The SCI receiver compares samples 3, 5, and 7. If two of these samples are low, the SCI qualifies the start bit, and begins sampling data bits.

For each data or stop bit in a frame, the SCI compares samples 8, 9, and 10. The SCI declares each data and stop bit as a one or a zero on the basis of these three samples. If two out of three samples are binary ones, the bit becomes a one; if two out of three are zeros, the bit becomes a zero. The exception to this rule is the start bit. The SCI uses samples 3, 5, and 7 to qualify the start bit. The SCI makes the qualified start bit a zero regardless of its 8, 9, and 10 samples.[11]

Samples 8, 9, and 10 must agree for each bit in a frame, or the SCI receiver sets the NF (Noise Flag) bit. This requirement applies to all bits in the frame, even the start bit. Interpret the NF bit as follows:

NF = 1   The SCI receiver detected noise in at least one bit time of the frame.
NF = 0   The SCI receiver detected no noise in the frame.

The M68HC11 CPU clears the NF bit by (1) reading the SCSR, when the NF flag is set, then (2) reading the SCDR.

SCSR bit 1 constitutes the Framing Error (FE) flag. The SCI receiver tests the incoming serial frame's stop bit (bit 10 or bit 11). If this stop bit is a one, the SCI resets the FE flag. If the stop bit is a zero, the SCI sets the FE flag to indicate a possible framing error. Interpret the FE flag as follows:

FE = 1   The SCI receiver detected a framing error or break character.
FE = 0   The SCR receiver did not detect a framing error.

The M68HC11 clears the FE flag by (1) reading the SCSR when the FE flag is set, then (2) reading the SCDR.

SCI Control Register 2
(SCCR2, Control Register Block Address $012D)

Refer to Figure 5–7. SCCR2 bits 7 through 4 enable interrupts. When corresponding SCSR flags set, they generate these interrupts.

- SCCR2 bit 7 is the TIE (Transmit Interrupt Enable) bit. If TIE = 1, the TDRE flag (SCSR, bit 7) generates an interrupt when it sets. Clearing the TIE bit disables these interrupts.
- SCCR2 bit 6 is the TCIE (Transmit Complete Interrupt Enable) bit. If TCIE = 1, the TC flag (SCSR, bit 6) generates an interrupt when it sets. Clearing the TCIE bit disables these interrupts.
- The RIE bit (Receive Interrupt Enable, SCCR2 bit 5) enables interrupts generated when the RDRF flag sets (SCSR bit 5). RIE = 1 enables these interrupts, and RIE = 0 disables them.
- The ILIE (Idle Line Interrupt Enable, SCCR2 bit 4) enables interrupts generated when the IDLE bit sets (SCSR bit 4). ILIE = 1 enables these interrupts and ILIE = 0 disables them.

Thus, SCCR2 bits 7 through 4 facilitate interrupt control associated with the TDRE, TC, RDRF, and IDLE flags.

SCCR2 bits 3 and 2 enable the SCI transmitter and receiver, respectively. Changing the TE (Transmit Enable) bit from a zero to a one enables the SCI transmitter, and also causes the transmitter to drive its TxD pin with a complete frame of logic ones (an idle character). Thus, the CPU can idle the TxD pin between messages by writing zero and then one to the

FIGURE 5–7
Reprinted with the permission of Motorola.

SCCR2

7	6	5	4	3	2	1	0
TIE	TCIE	RIE	ILIE	TE	RE	RWU	SBK

---

[11]Motorola Inc., *M68HC11 Reference Manual*, p. 9-21. Motorola explains that this arrangement avoids "an accidental wake up while using the idle-line variation of receiver wake up."

FIGURE 5–8

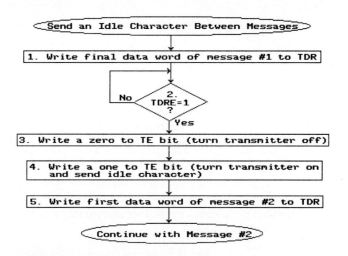

Send Idle Character Between Messages

$ADDRESS	SOURCE CODE	COMMENTS
----------------	--------------------------	
0120	LDAA A	Load data word (last, msg #1)
0122	STAA 2F,X	Write data word to SCDR
0124	LDAA 2E,X	Read SCSR
0126	BPL 124	TDRE = 1?
0128	BCLR 2D,X 8	If yes, turn TX off
012B	BSET 2D,X 8	Turn TX on and send idle character
012E	LDAA B	Load data word (first, msg #2)
0130	STAA 2F,X	Write data word to SCDR

TE bit. Figure 5–8 and its accompanying program segment illustrate this procedure. Please note that Figure 5–8 and the program segment make these assumptions:

- SCCR1's M bit = 1. Therefore, the data frame incorporates nine data bits. The frame's MS data bit serves as an additional stop bit. Thus, the tenth and eleventh bits constitute stop bits. The CPU writes a one to this additional stop bit (SCCR1 T8 bit) prior to execution of the Figure 5–8 program segment. T8 remains a one until changed by the CPU. Therefore, the CPU need not write a one to T8 for every frame transmitted.
- SCCR2's TIE and TCIE bits = 0. The program segment uses status polling to test for TDRE = 1.
- The X register points at the control register block (i.e., the X register contents = $1000).
- The final data word of message #1 originates at address $000A.
- The first data word of message #2 originates at address $000B.

Refer to block 1 of Figure 5–8 and the instructions at addresses $0120 and $0122. The CPU loads ACCA with the final data word of message #1, and then writes this data word to the SCDR.

Refer to block 2 and the instructions at addresses $0124 and $0126. The M68HC11 enters a polling loop in which it continually reads the SCSR and tests the TDRE flag. When the last data word from message #1 has transferred from the TDR to the transmit shift register, the TDRE flag sets. The CPU escapes the polling loop.

Refer to block 3 and the instruction at address $0128. The CPU turns off the transmitter by writing a zero to SCCR2's TE bit. This write clears the TE bit before the transmit shift register has time to drive the TxD pin with all bits of the final message #1 frame. Even though the TE bit has changed from one to zero, the SCI transmitter will not shut down until all current bits in the transmit shift register drive the TxD pin.

Refer to block 4 and the instruction at address $012B. The CPU writes a one to the SCCR2 TE bit. This write also occurs before the shift register has finished driving the TxD pin with the final message #1 data frame. Thus, both the block 3 and block 4 writes occur before the last bits of message #1 drive the TxD pin. Writing a one to the SCCR2 TE bit causes the SCI transmitter to send an idle character—a continuous high at the TxD pin for ten or eleven bit times. However, these binary ones will not appear on the TxD pin until the last message #1 bit has shifted out. Thus, the CPU inserts an idle character between message #1 and message #2.

Refer to block 5 and the instructions at addresses $012E and $0130. The CPU loads the first message #2 data word and writes it to the TDR. As soon as the final idle character bit has driven the TxD pin, the transmit shift register begins shifting out the initial message #2 data word. There has been no interruption between the idle character and the first frame of message #2.

SCCR2's RE bit (Receive Enable, bit 2) enables and disables the SCI receiver. Write a one to the RE bit to enable the receiver; write a zero to disable it. Disabling the SCI receiver leaves the receiver-oriented status flags (RDRF, IDLE, OR, NF, FE) as they were when the CPU reset the RE bit. If the receiver is disabled, reset receiver-oriented status flags cannot go set until the receiver is re-enabled.

SCCR2 bit 1 is the RWU (Receiver Wake Up) control bit. The previous discussion on receiver wake up refers to this RWU bit.

Set the RWU bit to make the SCI receiver "dormant." RWU = 1 disables all receiver-oriented status flags and interrupts. Recall that the SCI receiver shifts in RxD pin data even when asleep. The receiver shifts data in so that when it wakes, it can review addressing information embedded in an initial message frame.

Clear the RWU bit to wake the receiver. Normally, idle line wake up or address mark wake up writes this zero automatically.[12]

SCCR2's LSB is the SBK (Send Break) bit. Setting the SBK bit causes the SCI transmitter to drive its TxD pin with at least one frame time of binary zeros. The SCI transmitter applies this break character to the TxD pin as soon as the current data frame has shifted out. The SCI transmitter continues to drive its TxD pin with break characters until the CPU resets the SBK bit. When the CPU resets the SBK bit, the SCI transmitter will finish sending the current break character before it can send the next message.

A Note on PORTD and DDRD

The SCI uses PORTD pins PD0 and PD1 as RxD and TxD. The M68HC11 can also use these pins, along with PORTD pins PD2 through PD5, for general-purpose parallel I/O.[13] DDRD configures PORTD pins PD0–PD5 as general-purpose inputs or outputs.[14] In general, DDRD and PORTD operate in the same manner as DDRC and PORTC.[15] However, PD0 and PD1 have additional characteristics because the SCI uses them. If the SCI receiver is enabled (RE = 1), the CPU cannot drive PD0 with a general-purpose bit by writing to PORTD. Rather, PORTD latches this write bit. Where DDRD bit 0 is set, PD0 is configured as a general-purpose output. In this case, PORTD will drive PD0 with the latched write bit as soon as the SCI receiver is disabled. If DDRD bit 0 is reset, PD0 is configured as a general-purpose input. The latched write bit will not drive PD0 until the CPU disables the SCI receiver and changes DDRD bit 0 to binary one.

If DDRD bit 0 is reset, then reading PORTD returns the state of the PD0/RxD pin, whether the SCI receiver is enabled or not. If DDRD bit 0 is set, PORTD returns the bit it has latched.

If the SCI transmitter is enabled (TE = 1), or the transmitter is shifting out a frame to the TxD pin, the CPU cannot drive PD1 with a general-purpose bit by writing to PORTD. PORTD latches this write bit. If DDRD bit 1 is set, PORTD will drive PD1 with the latched bit as soon as the SCI transmitter has been disabled, and the SCI transmit shift register has driven the PD1/TxD pin with its last serial bit. If DDRD bit 1 is reset, PORTD will not drive PD1 with its latched bit until the CPU disables the SCI transmitter and changes DDRD bit 1 to binary one.

If DDRD bit 1 is reset, then reading PORTD returns the state of the PD1/TxD pin, whether the SCI transmitter is enabled or not. If the TE bit = 1, or the transmit shift register

---

[12]See the discussion on receiver wake up on pages 229–231.
[13]Port D Data Register (control register block address $1008).
[14]Data Direction Register for Port D (control register block address $1009).
[15]See Chapter 3 for a description of PORTC and DDRC operation.

is still shifting out serial data, a PORTD read returns the state of the TxD pin as it is sending a serial data bit. If DDRD bit 1 is set, a PORTD read returns the latched bit.

5.5 M68HC11 SCI INITIALIZATION

To initialize the SCI, the M68HC11 CPU must write to three registers:

- BAUD at address $102B
- SCCR1 at address $102C
- SCCR2 at address $102D

These registers reside at consecutive addresses in the control register block. The CPU can initialize two of these registers with a single write by loading a 2-byte register (D register, X register, or Y register) with two combined control words. The CPU can then write the 2-byte register to the lowest address of the SCI register pair. Of course, these two registers must reside at consecutive addresses.

Programmers customarily use indexed addressing to read and write the SCI registers. By using indexed addressing, the CPU can read and write these registers using 2-byte instructions, rather than the 3-byte instructions required by extended mode.

SCI programming often requires the use of BSET and BCLR commands when writing to the T8, TE, RE, RWU, and SBK bits in SCCR1 and SCCR2. Use BSET and BCLR to avoid disturbing the other bits in a register. The CPU *must* use indexed mode with BSET or BCLR instructions.

Usually the programmer initializes the X register to $1000 so that it points at the beginning address of the control register block. Therefore, reading or writing an SCI register using indexed mode entails an instruction offset equal to the low byte of that register's address.

Assume that the CPU must initialize its SCI as follows:

Baud = 9600
11-bit frames
Receiver interrupts enabled
Receiver enabled
Receiver asleep

Figure 5–9 and its accompanying program segment show how the CPU can initialize the SCI to meet these requirements.

FIGURE 5–9

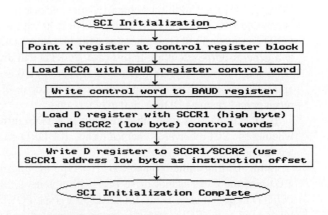

SCI Initialization

$ADDRESS	SOURCE CODE	COMMENTS
0100	LDX #1000	Initialize X register
0103	LDAA #30	Initialize BAUD register
0105	STAA 2B,X	" "
0107	LDD #1826	Initialize SCCR1/SCCR2
010A	STD 2C,X	" " "
010C	----------------------	----------------------------------

EXERCISE 5.3

SCI Initialization, and BSET/BCLR Source Code

AIM: Derive control words which configure the M68HC11 SCI. Develop BSET and BCLR source code that writes ones or zeros to the T8, TE, RE, RWU, and SBK bits of SCCR1 and SCCR2.

INTRODUCTORY INFORMATION: Part I of this exercise teaches you how to calculate SCCR1 and SCCR2 control words. These control words configure the SCI. Part II gives you practice at writing BSET and BCLR source code. This source code sets or clears the T8, TE, RE, RWU, and SBK bits of SCCR1 and SCCR2.

PART I: SCCR1 and SCCR2 Control Words

Use the configuration descriptions given below and refer to Figures 5–5 and 5–7, while deriving appropriate SCCR1 and SCCR2 control words. Enter these control words in the blanks provided. Make all nonrelevant SCCR1/SCCR2 bits binary zeros.

CONTROL WORDS

CONFIGURATION	SCCR1	SCCR2
EXAMPLE:		
11-bit frames Receiver interrupts enabled Receiver enabled Receiver asleep	$ 18	$ 26
10-bit frames Idle-line wake up Idle-line interrupts enabled Receiver enabled Receiver asleep	$_____	$_____
11-bit frames Transmit interrupts enabled Transmitter enabled	$_____	$_____
10-bit frames Transmit Complete interrupts enabled Receiver interrupts enabled Transmitter enabled Receiver enabled	$_____	$_____
11-bit frames Transmit interrupts enabled Transmit Complete interrupts enabled Transmitter enabled	$_____	$_____
10-bit frames Address-mark wake up Receiver interrupts enabled Receiver enabled Receiver awake	$_____	$_____

PART II: BSET and BCLR Masks

For each task listed below, develop complete and appropriate BSET or BCLR source code. Enter this code in the blank provided. Use indexed mode; assume that the X register points at address $1000.

TASK	SOURCE CODE
EXAMPLE:	
Write binary one to T8.	BSET 2C,X 40
Turn the receiver off.	
Turn the transmitter on.	
Turn the transmitter on and send break.	
Turn the transmitter off and stop sending breaks.	
Put the receiver to sleep.	
Enable the receiver and put it to sleep.	
Write binary zero to T8.	

5.6 TRANSMIT A MESSAGE—TYPICAL PROGRAM FLOW

Figures 5–10 and 5–11 and their accompanying program lists illustrate typical support software (MSGTX) for the SCI transmitter. This MSGTX software has some important characteristics:

- MSGTX uses interrupts to inform the CPU that the TDR is empty.
- MSGTX calls an MSGTRANS (Message Transfer) subroutine, located at address $0140. MSGTRANS checks to see if the message just transmitted is the final one of a multimessage string. If a final message has indeed been sent, MSGTRANS disables the transmitter and stops the M68HC11. A system reset causes recovery from this stop. After reset, the system operator sends the next message string by re-starting the MSGTX software from the beginning. If MSGTRANS finds that the message just transmitted is not the final one of the string, then it moves the next message into a block of RAM starting at address $0000. After executing this transfer, the CPU returns from the MSGTRANS subroutine. The MSGTX program then accesses the first data word of the next message and transmits it.
- MSGTX assumes that the final character in each message is ASCII "EOT" (End of Transmission), $04. The CPU checks each transmit data word to see if it is an EOT.
- MSGTX uses 11-bit frames. Bit 8 constitutes an additional stop bit. The CPU writes a binary one to the SCCR1 T8 bit only once, during SCCR1/SCCR2 initialization. This stop bit remains constant.

Figures 5–10 and 5–11 and their accompanying program list illustrate the MSGTX software. Refer to Figure 5–10, blocks 1 through 5, and program addresses $0100 through $0113. These instructions initialize the stack pointer, SCI interrupt pseudovector, X register, BAUD register, SCCR1, and SCCR2.

Refer to block 6 and program address $0114. The CPU points its Y register at a RAM block which holds the next message. The MSGTRANS subroutine moves each queued message into this RAM block.

Refer to blocks 7 and 8, and program addresses $0118 through $011B. The CPU reads a transmit data word from the RAM block to ACCB. Then, the CPU increments its Y register to point at the next transmit data word in the RAM block.

Refer to block 9 and program address $011D. The CPU clears the I mask to enable transmit interrupts.

Refer to Block 10 and program address $011E. The M68HC11 waits for a "TDR is empty" interrupt. After vectoring to the SCI interrupt service, the CPU writes the transmit

FIGURE 5-10

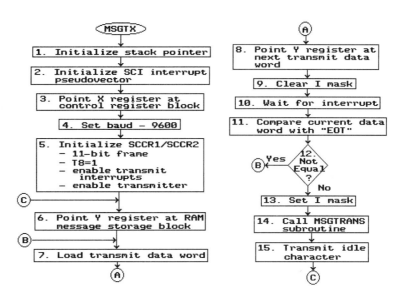

FIGURE 5-11

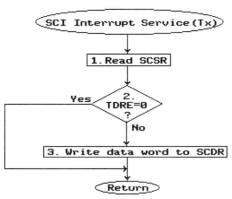

TX Software

$ADDRESS	MSGTX SOURCE CODE	COMMENTS
0100	LDS #1FF	Initialize stack pointer
0103	LDD #130	Initialize SCI interrupt pseudovector
0106	STD C5	" " " "
0108	LDX #1000	Initialize X register
010B	LDAA #30	Baud = 9600
010D	STAA 2B,X	" "
010F	LDD #5088	Initialize SCCR1/SCCR2
0112	STD 2C,X	" " " "
0114	LDY #0	Initialize Y register
0118	LDAB 0,Y	Load next data word
011B	INY	Increment Y register
011D	CLI	Clear I flag
011E	WAI	Wait for transmit interrupt

$ADDRESS	MSGTX SOURCE CODE	COMMENTS
011F	CMPB #4	Compare data word with EOT
0121	BNE 118	If not equal, transmit next data word
0123	SEI	Set I flag
0124	JSR 140	Call MSGTRANS subroutine
0127	BCLR 2D,X 8	Transmit idle character
012A	BSET 2D,X 8	" " " "
012D	BRA 114	Send another message

$ADDRESS	SCI INTERRUPT SERVICE (Tx) SOURCE CODE	COMMENTS
0130	LDAA 2E,X	Read SCSR
0132	BPL 136	If TDRE=0, return
0134	STAB 2F,X	If TDRE=1, write data word to SCDR
0136	RTI	Return

data word from ACCB to the SCDR. The SCI responds by driving its TxD pin with the data word and framing bits.

Refer to blocks 11 and 12, and program addresses $011F through $0122. After returning from the SCI Interrupt Service, the CPU executes block 11. Here, the CPU checks ACCB to see if it holds an ASCII EOT character. If ACCB does not hold an EOT, the CPU takes the "yes" branch from block 12 back to block 7. At block 7, the CPU accesses the next transmit data word. If ACCB = $04, the SCI has transmitted a complete message. Therefore, the CPU takes the "no" path from block 12 to block 13.

Refer to blocks 13 and 14, and program addresses $0123 through $0126. After inhibiting transmit interrupts (block 13), the CPU calls the MSGTRANS subroutine (block 14). MSGTRANS checks to see if any further messages are queued. If no more messages are queued, MSGTRANS uses the STOP instruction to halt the M68HC11. If another message is queued, MSGTRANS moves it to the RAM transmit data block at address $0000, then returns from subroutine. After returning from MSGTRANS, the CPU executes block 15.

Refer to block 15, and program addresses $0127 through $012E. In block 15, the M68HC11 transmits an idle character by clearing, then setting, the TE bit in SCCR2. Finally, the CPU branches back to block 6, where it re-initializes the Y register and transmits a new message.

Figure 5–11 flowcharts the SCI Interrupt Service routine. Refer to blocks 1 and 2, and program addresses $0130 through $0133. Recall that the CPU must read the SCSR and write to the SCDR—to reset the TDRE bit. Therefore, the CPU reads the SCSR and tests for a spurious interrupt with the BPL instruction. TDRE = 0 indicates a spurious interrupt, and the CPU branches to the RTI instruction.

Refer to block 3 and program address $0134. If TDRE = 1, the CPU transmits the data word by writing ACCB to the SCDR register.

EXERCISE 5.4 Transmitting a Message

AIM: Given a flowchart, write a program which uses the M68HC11 SCI to transmit a message made up of ASCII characters.

INTRODUCTORY INFORMATION: This exercise gives you experience in writing an SCI message-transmit program. This experience reinforces what you have learned about the SCI transmitter.

INSTRUCTIONS: Write a program that implements the flowchart shown in Figure 5–12. Enter your program source code and ASCII codes in the blanks provided below. Assure that your program meets these requirements:

1. Use 10-bit frames.
2. Use status polling to determine if the TDR is empty.
3. Baud = 2400.
4. Use indexed addressing (X register) to access the BAUD and SCI registers.
5. Use indexed addressing (Y register) to access the ASCII message characters.
6. Spell "Howdy!" with the ASCII message characters (refer to Exercise 5.1).
7. A RAM block at address $0000 contains the message's ASCII characters.
8. Make the last ASCII character in the message an "EOT."
9. Locate your transmit program at address $0100.

TRANSMIT "Howdy!" MESSAGE

$ADDRESS	SOURCE CODE	COMMENTS
_____	_____	_____
_____	_____	_____
_____	_____	_____
_____	_____	_____
_____	_____	_____
_____	_____	_____
_____	_____	_____

FIGURE 5-12

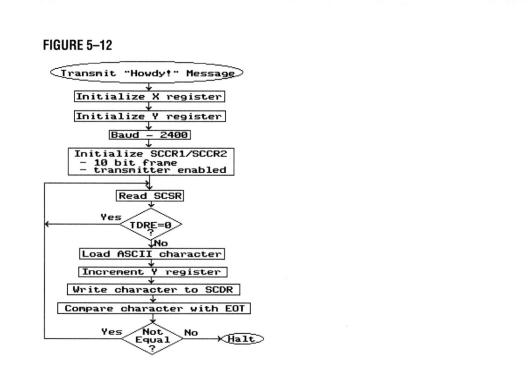

	ASCII CODES		
$ADDRESS	**$CONTENTS**	**CHARACTER**	
0000	48	H	**EXAMPLE**
0001	___	___	
0002	___	___	
0003	___	___	
0004	___	___	
0005	___	___	
0006	___	___	

5.7 RECEIVE A MESSAGE—TYPICAL PROGRAM FLOW

Figures 5–13 and 5–14 and their accompanying program list illustrate support software (MSGRX) for an M68HC11 connected as a receiver on a multidrop line. This software has several important characteristics:

- MSGRX uses interrupts to inform the CPU when its RDR is full.
- MSGRX presumes a MSGSVC (Message Service) subroutine, located at address $0140. As the CPU reads the incoming message data words from the SCDR, it stores them consecutively in a RAM address block beginning at address $0000. When an incoming EOT indicates the message end, the CPU calls MSGSVC. This subroutine copies the message from the RAM storage block ($0000) to dedicated memory. Thus, MSGSVC makes the RAM storage block at address $0000 available for the next incoming message.
- MSGRX assumes an 11-bit frame. Data bit 8 constitutes an "address" bit, to be used with address-mark wake up. Thus, bit 8 is binary one only in the first frame of a message. An

address mark causes the SCI receiver to wake. The receiver evaluates address information contained in this frame's remaining data bits. If these address bits equal $0A, the receiver stays awake. The CPU then reads message data words from the SCDR and stores them to the RAM storage block at address $0000.

Figures 5–13 and 5–14, along with their accompanying program list, illustrate the MSGRX software. Refer to Figure 5–13, blocks 1 through 6, and program addresses $0100 through $0117. These instructions initialize the SCI receiver at the beginning of a serial communication session. Note that block 6 execution enables the receiver, but also puts it to sleep. The receiver remains enabled and asleep throughout the serial communications session, *except* when it wakes briefly at the beginning of each message. If an initial message frame addresses this particular receiver, then it stays awake until it receives the complete message, and stores it in the RAM storage block ($0000). If this particular receiver is not addressed, it goes back to sleep until the initial frame of the next message.

Refer to block 7 and program address $0118. The CPU must clear the I mask to enable SCI interrupts.

Refer to block 8 and program address $0119. The CPU waits for an "RDR is full" interrupt. The SCI Interrupt Service routine causes the CPU to read a message data word from the SCDR, and store it. After reading and storing a data word, the CPU returns from interrupt to block 9.

Refer to block 9 and program address $011A. The CPU increments the Y register so that it points at the next available byte in the RAM data storage block.

Figure 5–14 flowcharts the SCI Interrupt Service routine. Refer to blocks 1, 2, and 3, and program addresses $0120 through $0129. When awakened by an address mark and interrupted by an "RDR is full" notification, the CPU reads the SCSR and SCDR (block 1). The CPU uses an LDD instruction to execute this read. LDD copies SCSR's contents to ACCA and SCDR's contents to ACCB. This read also resets the RDRF flag in the SCSR. However, the block 1 LDD copies SCSR to ACCA before this flag resets.

In Figure 5–14, block 2 the CPU confirms that RDRF equalled 1. In block 3, the CPU confirms that no SCSR error flags set. The CPU uses BITA commands to test the RDRF and error flags. A BITA instruction ANDs the byte under test with a mask byte. The ANDed result sets or resets the N and Z flags. BITA does not save this ANDed result. If RDRF = 1, a bit test with mask $20 resets the Z flag. If RDRF = 0, the Z flag sets. If the Z flag sets, the CPU branches (BEQ) from block 2 to the instruction that puts its receiver back to sleep (block 10). If BITA execution resets the Z flag, the CPU takes the "no" path to block 3. In block 3, the CPU uses a BITA instruction to test the OR, NF, and FE flags. This time, BITA ANDs the contents of ACCA with mask $0E. If any of the three error flags are set, this test causes the Z flag to reset. As a result, the CPU takes the "yes" branch from block 3 to block 10. If BITA execution sets the Z flag (no errors), the CPU takes the "no" path from block 3 to block 4.

Refer to block 4 and program addresses $012A through $012D. Recall that the received data word's MSB constitutes the address mark bit. SCCR1's R8 bit holds this address mark. In block 4, the CPU reads SCCR1 and uses the BPL instruction to see whether the R8 bit holds a mark or a space. If R8 = 1 (a mark), the CPU takes the "no" path from block 4 to block 5. If R8 = 0, the current message frame is not the initial one. Rather, this frame holds message data. The CPU has already evaluated the address frame during an earlier interrupt service. Therefore, the CPU takes the "yes" branch from block 4 to block 6.

Refer to block 5 and program addresses $012E through $0131. If the transmitter selects our M68HC11 receiver, the addressing frame contains $0A in its lower 8 data bits. The SCDR holds these 8 bits. In block 1, the M68HC11 reads the SCDR contents to ACCB. Thus, the CPU can find out if it is being addressed by comparing its ACCB contents with $0A. If ACCB does not hold $0A, the CPU takes the "yes" branch from block 5 to block 10. If ACCB = $0A, the CPU takes the "no" path from block 5 to block 6.

Refer to block 6 and program address $0132. Block 1 execution reads the contents of the SCDR to ACCB. The CPU has now confirmed that ACCB contains an error-free word, and that our M68HC11 has been selected as the message recipient. The CPU therefore stores this data from ACCB to its RAM data storage block.

FIGURE 5-13

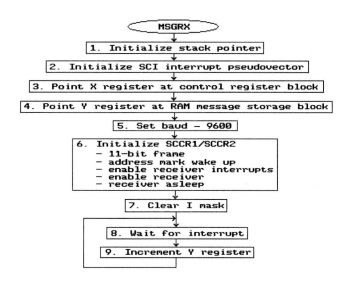

FIGURE 5-14

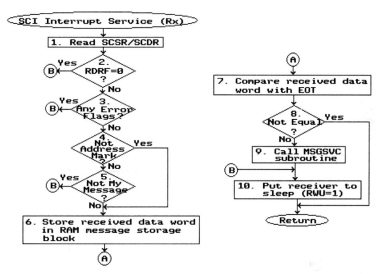

Rx Software

$ADDRESS	MSGRX SOURCE CODE	COMMENTS
0100	LDS #1FF	Initialize stack pointer
0103	LDD #120	Initialize SCI interrupt pseudovector
0106	STD C5	" " "
0108	LDX #1000	Initialize X register
010B	LDY #0	Initialize Y register
010F	LDAA #30	Baud = 9600
0111	STAA 2B,X	" "
0113	LDD #1826	Initialize SCCR1/SCCR2
0116	STD 2C,X	" " "
0118	CLI	Clear I flag
0119	WAI	Wait for interrupt caused by RDRF = 1
011A	INY	Increment Y register
011C	BRA 119	Wait for another interrupt

$ADDRESS	SCI INTERRUPT SERVICE (Rx) SOURCE CODE	COMMENTS
0120	LDD 2E,X	Read SCSR/SCDR
0122	BITA #20	RDRF = 0?
0124	BEQ 13C	If RDRF=0, put receiver to sleep and return
0126	BITA #E	Any error flags?
0128	BNE 13C	If yes, put receiver to sleep and return
012A	LDAA 2C,X	Read SCCR1
012C	BPL 132	Address mark?
012E	CMPB #A	If yes, is this my message?
0130	BNE 13C	If not my message, put receiver to sleep and return
0132	STAB 0,Y	Store received data word
0135	CMPB #4	Is data word an EOT?
0137	BNE 13F	If not EOT, return
0139	JSR 140	If EOT, call Message Service subroutine
013C	BSET 2D,X 2	Put receiver to sleep
013F	RTI	Return

243

Refer to blocks 7 through 10, and program addresses $0135 through $013F. In blocks 7 and 8, the CPU determines whether ACCB holds the final word of the message (EOT). If ACCB does not hold an EOT, the CPU takes the "yes" branch from block 8 to the RTI instruction. The CPU returns from interrupt without putting its receiver to sleep (block 10), because more message frames will be coming. If ACCB holds an EOT character, the CPU takes the "no" path from block 8 to block 9. In block 9, the CPU calls the MSGSVC subroutine. MSGSVC copies the complete message from the RAM storage block ($0000) to a memory block dedicated to this particular message. The CPU returns from MSGSVC to block 10. In block 10, the CPU puts its receiver to sleep, then returns from the interrupt.

EXERCISE 5.5

Receiving a Message

AIM: Given a flowchart, write a program which uses the M68HC11 SCI to receive a message made up of ten ASCII characters.

INTRODUCTORY INFORMATION: This exercise gives you experience at writing SCI receiver support software. This experience reinforces what you have learned about the M68HC11 SCI receiver.

INSTRUCTIONS: Write a program which implements the flowcharts shown in Figures 5–15 and 5–16. Enter your program in the blanks provided below. Assure that your program meets these requirements:

1. Use 10-bit frames.
2. Use interrupt control to determine if the RDR is full.
3. Baud = 9600.
4. Use indexed addressing (X register) to access the BAUD and SCI registers.
5. Use indexed addressing (Y register) to store received data words.
6. Store received data words in a block of RAM starting at address $0000.
7. When the RAM data storage block contains ten received characters, your software must set the I mask, disable the SCI receiver, then halt.
8. Locate your MAIN receive program at address $0100.
9. Locate your SCI Interrupt Service routine at address $0130.
10. Initialize the stack pointer to $01FF.

RECEIVE A TEN-CHARACTER MESSAGE

$ADDRESS	SOURCE CODE	COMMENTS

FIGURE 5-15

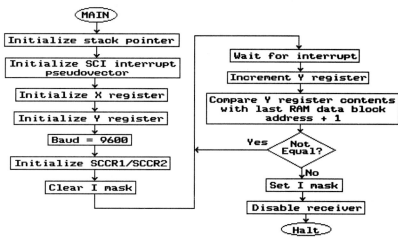

FIGURE 5-16

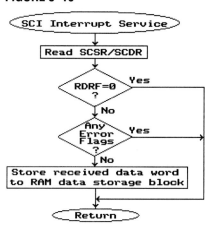

SCI INTERRUPT SERVICE

$ADDRESS	SOURCE CODE	COMMENTS

EXERCISE 5.6

Serial Communications Interface I (EVBU Only)

REQUIRED EQUIPMENT: Motorola M68HC11EVBU, and Terminal

GENERAL INSTRUCTIONS: The M68HC11E9 on board the EVBU uses its SCI to communicate with the terminal, via an MC145407 driver/receiver. This driver/receiver formats serial data for an EIA-232-D (RS-232C) bus connection between the EVBU and its terminal.

The EVB uses an entirely different approach. The M68HC11A1 on board the EVB communicates with its terminal via PORTC and an M68B50P ACIA (UART).[16] This UART serves essentially the same function as the SCI. The UART also formats serial data for an EIA-232-D connection between the EVB and its terminal.[17] Thus, this exercise requires the use of an EVBU and will not work with an EVB.

In this laboratory exercise, develop transmit and receive routines which take control of the EVBU's EIA-232 port, and communicate back and forth between the EVBU and terminal. Carefully execute the detailed instructions given below.

PART I: Software Development

Develop software which executes these tasks:

1. Transmit ASCII characters from the M68HC11 to the terminal. These ASCII characters appear on the terminal display as a prompt:

   PRESS 3 KEYS

2. Receive, check for errors, and store three key-press characters (ASCII) from the terminal. If the M68HC11 finds an error while receiving a character, it branches back to the transmit routine and again prompts the user to "PRESS 3 KEYS." In this case, the M68HC11 branches back without storing any key-press data.
3. After storing the valid key-press data, the CPU branches back to the transmit routine, and again prompts the user to "PRESS 3 KEYS."

Observe these program specifications:

1. Store user-prompt ASCII characters in a block of RAM starting at address $0000:

   PRESS<SP>3<SP>KEYS<EOT>

   The Transmit Routine (TX) uses the EOT as an indication that the SCI has transmitted all prompt message characters.

2. Locate your software starting at address $0100.
3. Concatenate the transmit and receive routines.
4. Store the three received ASCII key-press characters at RAM locations $000F through $0011.
5. Use SCSR status polling. Disable all SCI interrupt and wake up functions.
6. Use 10-bit serial data frames.

   Use the Figure 5–17 and Figure 5–18 flowcharts as guides to software development.

PART II: Software Debug and Demonstration

Load and debug your software. Demonstrate interface operation to the instructor.

_____ *Instructor Check*

PART III: Laboratory Analysis

Answer the questions below by neatly entering your answers in the appropriate blanks.

1. Describe the X register's purpose in the laboratory software.

---

[16]The M68HC11 on board the EVB actually operates in expanded multiplexed mode, but uses an M68HC24 PRU (Port Replacement Unit) to emulate single-chip mode.

[17]The EVB uses its M68HC11A1 SCI to download data from a host computer, using an RS-232 bus connection and the Motorola S-Record format.

FIGURE 5-17

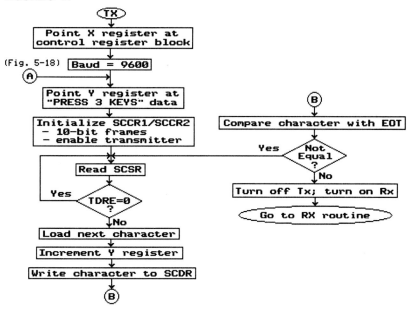

FIGURE 5-18

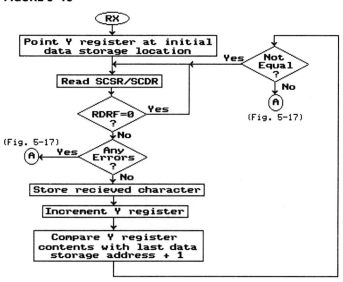

2. Explain why the software points the X register at the control register block (i.e., loads the X register with $1000).

3. Describe the Y register's two purposes in the laboratory software.

   (1)

   (2)

4. After three continuous rounds of prompts and key-press responses, how much key-press data is stored in the RAM data storage block?

5. List the software changes required to prompt the user to "PRESS 5 KEYS" and store the five key-press characters.

   (1)

   (2)

PART IV: Laboratory Software Listing

TRANSMIT ROUTINE (TX)—EVBU

$ADDRESS	SOURCE CODE	COMMENTS
0100	LDX #1000	Initialize X register
0103	LDAA #30	Baud = 9600
0105	STAA 2B,X	" "
0107	LDY #0	Initialize Y register
010B	LDD #8	Initialize SCCR1/SCCR2
010E	STD 2C,X	" " "
0110	LDAA 2E,X	Read SCSR
0112	BPL 110	If TDRE = 0, read SCSR again
0114	LDAA 0,Y	If TDRE = 1, load next character
0117	INY	Increment Y register
0119	STAA 2F,X	Write character to SCDR
011B	CMPA #4	Is character an EOT?
011D	BNE 110	If not EOT, read SCSR again
011F	LDAA #4	If EOT, disable Tx, and enable Rx
0121	STAA 2D,X	" " " " "

PROMPT MESSAGE —ASCII

$ADDRESS	$CONTENTS	CHARACTER
0000	50	"P"
0001	52	"R"
0002	45	"E"
0003	53	"S"
0004	53	"S"
0005	20	<SP>
0006	33	"3"
0007	20	<SP>
0008	4B	"K"
0009	45	"E"
000A	59	"Y"
000B	53	"S"
000C	04	<EOT>

RECEIVE ROUTINE (RX) —EVBU

$ADDRESS	SOURCE CODE	COMMENTS
0123	LDY #F	Initialize Y register (data storage)
0127	LDD 2E,X	Read SCSR/SCDR
0129	BITA #20	RDRF = 0?
012B	BEQ 127	If RDRF = 0 read SCSR/SCDR again
012D	BITA #E	If RDRF = 1, check for errors
012F	BNE 107	If error, branch to TX routine
0131	STAB 0,Y	If no errors, store received character
0134	INY	Increment Y register
0136	CPY #12	Has third character been stored?
013A	BNE 127	If not third character, read SCSR/SCDR again
013C	BRA 107	If third character, branch to TX routine

EXERCISE 5.7 Serial Communications Interface II (EVBU Only)

REQUIRED EQUIPMENT: Motorola M68HC11EVBU, and Terminal

GENERAL INSTRUCTIONS: In this laboratory exercise, develop transmit and receive routines which take control of the EVBU's EIA-232 port. Communicate back and forth between the EVBU and its terminal. As in Exercise 5.6, the M68HC11EVB cannot be used for this exercise.

PART I: Software Development

Your software must execute the tasks shown in the Figure 5–19 flowchart.
 Observe these program specifications:

1. Use both transmit and receive interrupts. Locate interrupt service routines at addresses $0160 and $0170:

   $0160—SCI Interrupt Service (Tx)
   $0170—SCI Interrupt Service (Rx)

2. Use the transmit routine twice: first, for a "PLEASE INPUT DATA" message; second, for a "MEMORY FULL" message. Use a transmit routine counter to keep track of which message is being transmitted. Clear this counter prior to transmitting the first message. Increment the counter prior to transmitting the second message. After transmitting the second message, test the counter. If counter contents equal one, the program halts. Locate this counter at address $002A.
3. Locate the MAIN program at address $0100. Concatenate the transmit and receive routines.
4. Store the ten received ASCII characters in a RAM data storage block, addresses $0000–$0009.
5. Locate ASCII transmit characters as follows:

   $000A–$001C: "PLEASE<SP>INPUT<SP>DATA(SP><EOT>"
   $001D–$0029: "MEMORY<SP>FULL<SP><EOT>"

FIGURE 5–19

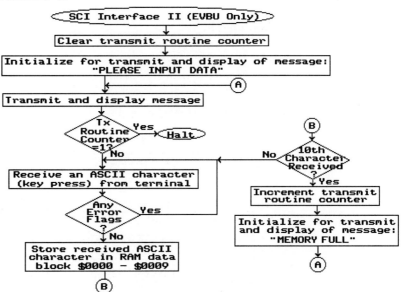

The transmit routine uses the EOT as an indication that it has transmitted all message characters.

6. Use 10-bit serial data frames.

Use the memory allocation chart and Figures 5–19 through 5–24 as guides to software development.

Memory Allocation

$ADDRESS	CONTENTS
$0000–$0009	Key-press data (ASCII)
$000A–$001C	ASCII "PLEASE<SP>INPUT <SP>DATA<SP><EOT>"
$001D–$0029	ASCII "MEMORY<SP>FULL<SP><EOT>"
$002A	Transmit routine counter
$0100	Main program
$0160	SCI Interrupt Service (Tx)
$0170	SCI Interrupt Service (Rx)
$01FF	Initial stack pointer

FIGURE 5–20

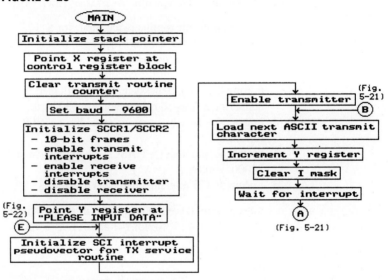

FIGURE 5–21

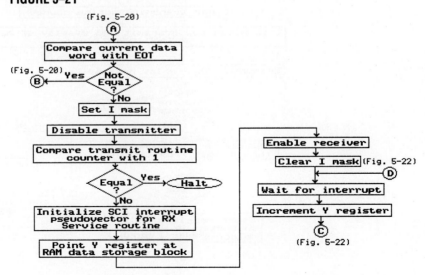

FIGURE 5-22

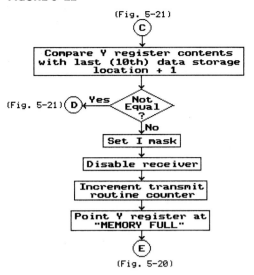

FIGURE 5-23

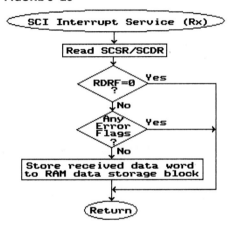

FIGURE 5-24

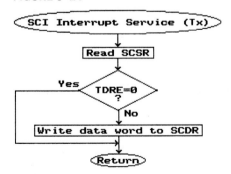

PART II: Software Debug and Demonstration

Load and debug your software. Demonstrate interface operation to the instructor.

_____ *Instructor Check*

PART III: Laboratory Analysis

Answer the questions below by neatly entering your answers in the appropriate blanks.

1. Describe the purpose of the transmit routine counter.

2. Explain why SCCR1/SCCR2 initialization disables both the SCI transmitter and receiver.

3. Explain why the SCI interrupt pseudovector must be initialized on three different occasions.

4. Explain why the Y register is incremented by the main program, rather than by the interrupt service routines.

5. Explain why each ASCII transmit character is loaded by the MAIN program rather than the TX Interrupt Service routine.

6. Explain why the MAIN program —rather than the TX Interrupt Service—tests each ASCII character to see whether it is an EOT.

7. Describe the X register's purpose in the laboratory software.

8. Describe the Y register's three purposes in the laboratory software.

   (1)

   (2)

   (3)

PART IV: Laboratory Software Listing

MAIN PROGRAM—EVBU

$ADDRESS	SOURCE CODE	COMMENTS
0100	LDS #1FF	Initialize stack pointer
0103	LDX #1000	Initialize X register
0106	CLR 2A	Clear transmit routine counter
0109	LDAA #30	Baud = 9600
010B	STAA 2B,X	" "
010D	LDD #A0	Initialize SCCR1/SCCR2
0110	STD 2C,X	" " "
0112	LDY #A	Point Y register at "PLEASE INPUT DATA"
0116	LDD #160	Initialize SCI pseudovector for TX Service routine
0119	STD C5	" " " " " " "
011B	BSET 2D,X 8	Enable transmitter
011E	LDAA 0,Y	Load next ASCII character
0121	INY	Increment Y register
0123	CLI	Clear I mask
0124	WAI	Wait for interrupt
0125	CMPA #4	Compare current data word with EOT
0127	BNE 11E	If not EOT, transmit another character
0129	SEI	If EOT, set I mask
012A	BCLR 2D,X 8	Disable transmitter
012D	LDAA 2A	Compare transmit routine counter with $01
013F	CMPA #1	" " " " " "
0131	BEQ 157	If equal, halt
0133	LDD #170	If not equal, initialize SCI interrupt pseudovector for RX Service routine
0136	STD C5	" " " " " " " "
0138	LDY #0	Point Y register at RAM data storage block
013C	BSET 2D,X 4	Enable receiver
013F	CLI	Clear I mask
0140	WAI	Wait for interrupt
0141	INY	Increment Y register
0143	CPY #A	Compare Y register with last data storage location +1
0147	BNE 140	If not equal, wait to receive another character
0149	SEI	If equal, set I mask
014A	BCLR 2D,X 4	Disable receiver
014D	INC 2A	Increment transmit routine counter
0150	LDY #1D	Point Y register at "MEMORY FULL"
0154	BRA 116	Transmit "MEMORY FULL"
0156	WAI	Halt

SCI INTERRUPT SERVICE (TX)—EVBU

$ADDRESS	SOURCE CODE	COMMENTS
0160	LDAB 2E,X	Read SCSR
0162	BPL 166	If TDRE = 0, return
0164	STAA 2F,X	Write ASCII character to SCDR
0166	RTI	Return

SCI INTERRUPT SERVICE (RX)—EVBU

$ADDRESS	SOURCE CODE	COMMENTS
0170	LDD 2E,X	Read SCSR/SCDR
0172	BITA #20	RDRF = 0?
0174	BEQ 17D	If yes, return
0176	BITA #E	Any error flags?
0178	BNE 17D	If yes, return
017A	STAB 0,Y	Store received data to RAM storage block
017D	RTI	Return

"PLEASE INPUT DATA"—ASCII

$ADDRESS	$CONTENTS	CHARACTER
000A	50	"P"
000B	4C	"L"
000C	45	"E"

$ADDRESS	$CONTENTS	CHARACTER
000D	41	"A"
000E	53	"S"
000F	45	"E"
0010	20	<SP>
0011	49	"I"
0012	4E	"N"
0013	50	"P"
0014	55	"U"
0015	54	"T"
0016	20	<20>
0017	44	"D"
0018	41	"A"
0019	54	"T"
001A	41	"A"
001B	20	<SP>
001C	04	<EOT>

"MEMORY FULL"—ASCII

$ADDRESS	$CONTENTS	CHARACTER
001D	4D	"M"
001E	45	"E"
001F	4D	"M"
0020	4F	"O"
0021	52	"R"
0022	59	"Y"
0023	20	<SP>
0024	46	"F"
0025	55	"U"
0026	4C	"L"
0027	4C	"L"
0028	20	<SP>
0029	04	<EOT>

5.8 CHAPTER SUMMARY

System designers use asynchronous serial communications to simplify physical connections between the M68HC11 and its peripherals. Although serial communication allows simpler connections, it requires more complex supporting software than does parallel I/O.

The M68HC11's Serial Communications Interface (SCI) facilitates asynchronous serial communication with peripherals. The SCI uses two M68HC11 pins: transmit data (TxD) and receive data (RxD). A designer can use these pins for either full-duplex or half-duplex operations.

The SCI formats data words to NRZ (Nonreturn to Zero) conventions. A user can configure the SCI for 8 or 9 data bits per frame. The SCI handles data frames consisting of 1 start bit, 8 or 9 data bits, and one stop bit.

The SCI incorporates a baud rate generator. A user can program for SCI baud rates from 75 bits/second to 125K bits/second.[18]

The SCI incorporates a "wake up" feature for use where the M68HC11 is a receiver on a multi-drop serial communications line. Wake-up mode allows the SCI receiver to "sleep" when not selected by the transmitter for communications via the multidrop line. The SCI wakes the receiver for the beginning of each serial message. Upon awaking, the receiver determines whether it is the message recipient or not. If not, the receiver can go back to sleep. This wake up feature reduces software overhead at each M68HC11 multidrop receiver.

---

[18]These baud rates assume that an 8MHz crystal is connected to the M68HC11's XTAL and EXTAL pins.

The SCI uses five internal registers while sending and receiving serial data:

- SCI Data Register (SCDR)
- SCI Baud Rate Control Register (BAUD)
- SCI Control Register 1 (SCCR1)
- SCI Status Register (SCSR)
- SCI Control Register 2 (SCCR2)

The CPU transmits a frame containing 8 data bits by writing them to the SCDR. The CPU captures a received 8-bit data word by reading the SCDR. The SCDR double buffers both the transmit and receive functions. The SCDR transmit function includes a TDR buffer and transmit shift register. This double buffer allows the CPU to write a new transmit data word to the SCDR (i.e., to the TDR buffer) as soon as the SCI has transferred the previous data word to the transmit shift register. The SCDR receive function includes a receive shift register and RDR buffer. Receiver double buffering allows the receive shift register to shift in new data from the RxD pin as soon as previously received data has transferred to the RDR buffer.

Five bits within the BAUD register control the SCI baud rate for both transmit and receive operations. BAUD register bits 5 and 4 constitute prescaler bits. The prescaler section divides the E clock rate by 16. The prescaler section then divides this result by 1, 3, 4, or 13 as controlled by the prescaler bits.

BAUD register bits 2, 1, and 0 are the baud rate select bits. The baud rate generator divides the prescaler output by 2^n where n = the value of these rate select bits.

The SCCR1 register has three functions. SCCR1's M bit selects serial data word length; either 8 or 9 bits. The WAKE bit determines how the SCI receiver awakes when it is connected to a multidrop line. If a user selects 9-bit data words, then SCCR1's R8 and T8 bits constitute the MSBs of the receive and transmit data words. Thus, R8 and T8 form the ninth (MS) data bits of the RDR and TDR, respectively.

When the SCI receiver's RxD pin connects to a multidrop communications line, SCCR1's WAKE bit determines how the receiver awakes to check addressing information in the first frame of a message. When WAKE = 0, an idle line character between messages wakes the receiver. If WAKE = 1, the SCI receiver wakes up when a received frame's most significant data bit is a binary one.

The M68HC11's SCSR contains four SCI status flags and three error flags. The TDRE (Transmit Data Register Empty) flag indicates when the latest TDR write data has transferred to the transmit shift register. The CPU clears the TDRE bit by reading the SCSR and then writing a new data word to the SCDR. The TC (Transmit Complete) flag indicates when the SCI transmitter has shifted out the final frame of a message to its TxD pin. The CPU uses the TC flag as an indication that it is safe to turn off a peripheral connected to the TxD pin. The CPU clears the TC bit by reading the SCSR, then writing to the SCDR. Thus, the TDRE and TC are transmit status flags.

The RDRF and IDLE bits are receiver status flags. The RDRF (Receive Data Register Full) flag indicates when a received data word has transferred from the receive shift register the RDR. The CPU clears the RDRF bit by reading the SCSR, then reading the SCDR. The IDLE (Idle Line Detect) flag indicates when the SCI's RxD line has been driven high for one full frame time, thus indicating the end of an incoming message. The CPU clears the IDLE bit by reading the SCSR, then reading the SCDR.

The SCSR's three error flags indicate problems with incoming serial data frames. The OR (Overrun Error) flag indicates that the receive shift register shifted in a complete new frame before the CPU could read the previous data word from the RDR. The CPU clears the OR bit by reading the SCSR, then reading the SCDR. The NF (Noise Flag) indicates that binary values of a received frame are varying within allotted bit times. These variations are most likely caused by noise. The CPU clears the NF flag by reading the SCSR, then reading the SCDR. The FE (Framing Error) bit indicates that the stop bit was a binary zero instead of a binary one. NRZ format calls for a stop bit = 1. The CPU clears the FE bit by reading the SCSR, then reading the SCDR.

SCCR2 bits 7 through 4 enable interrupts. Equivalent SCSR bits generate these interrupts when they set. The TIE (Transmit Interrupt Enable) bit enables interrupts when the TDRE bit sets. The TCIE (Transmit Complete Interrupt Enable) bit enables interrupts when the TC bit sets. The RIE (Receive Interrupt Enable) bit enables interrupts when the RDRF sets. The ILIE (Idle Line Interrupt Enable) bit enables interrupts when the IDLE flag sets. Thus, these interrupt enable bits facilitate interrupt control in testing the TDRE, TC, RDRF and IDLE flags.

SCCR2 bits 3 and 2 enable the SCI's transmitter and receiver. Changing the TE (Transmit Enable) bit from a zero to a one enables the SCI transmitter, and also causes the transmitter to send an idle character. The RE bit enables the SCI receiver.

SCCR2 bit 1 is the RWU (Receiver Wake Up) control bit. Setting this bit puts the receiver to sleep. The receiver wake up mechanism automatically writes a zero to this bit—to wake the receiver up.

SCCR2 bit 0 is the SBK (Send Break) bit. When the CPU writes a one to this bit, the SCI transmitter sends a break character. The transmitter continues to send break characters until the CPU writes a zero to the SBK bit.

EXERCISE 5.8 Chapter Review

Answer the questions below by neatly entering your answers in the appropriate blanks.

1. Describe the major advantage of asynchronous serial communications as compared with parallel I/O.

2. When compared with support software for parallel I/O, asynchronous serial communications software is more _____.
3. The two major serial communications modes are _____ and _____.
4. Full-duplex communications use _____ data line(s) between the microcontroller and its peripheral.
5. Half-duplex communications use _____ data line(s) between the microcontroller and its peripheral.
6. For fastest asynchronous serial communications, a system designer would choose _____-duplex mode.
7. The SCI uses two M68HC11 pins, called _____ and _____.
8. Each of these pins (question 7) connects internally to a _____.
9. The receive shift register is a _____-in, _____-out device.
10. The transmit shift register is a _____-in, _____-out device.
11. The transmit shift register connects in parallel to the _____.
    The CPU _____ (reads/writes to) this device to transmit a serial data word.
12. The receive shift register connects in parallel to the _____.
    The CPU _____ (reads/writes to) this device to transmit a serial data word.
13. Special bits called _____ bits inform the _____ (Tx/Rx) when a serial data word begins and ends.
14. An SCI data word can consist of _____ or _____ bits. An SCI data frame can consist of _____ or _____ bits. The SCI automatically generates (Tx) or checks (Rx) _____ (how many?) framing bits, called the _____ and _____ bits.
15. An SCI start bit is a binary _____ and a stop bit is a binary _____.
16. The first data bit of a frame is the _____ (MSB/LSB) of the data word.
17. The last data bit of a frame is the _____ of the data word.
18. The last bit of the frame is the _____ bit.
19. Define the term *baud rate*.

20. List the conditions that can generate an SCI interrupt.

    (1)

    (2)

    (3)

    (4)

21. The M68HC11 uses the wake up feature when its SCI receiver connects to a _____ _____.

22. The wake up feature brings the SCI receiver out of a dormant state for the beginning of each _____. The M68HC11 can then check _____ in the first of the message, to see whether its SCI receiver has been _____.

23. NRZ convention calls for a serial communications line to idle at the binary _____ level.

24. NRZ convention calls for a break character to consist of _____ (how many?) binary _____.

25. Both the transmitter and receiver must agree whether parity is _____ or _____.

```
  B10 P                     B0
% 1   0   0   1   1   0   1   1   0   1   0
```

26. The frame above uses _____ (odd/even) parity.

27. Refer to Figure 5–2 on page 224 and its accompanying program list. If the initial contents of address $0000 is ASCII code for the letter "E," what will be the contents of address $0001, just prior to execution of the instruction at address $0115? $_____ What will be the content of the C flag, just after execution of the instruction at address $0115? C = _____ What will be the contents of ACCA just after execution of the instruction at address $011A? ACCA = $_____ What will be the contents of address $0000, after execution of the program? $0000 = $_____

28. Refer to Figure 5–25, which illustrates a serial frame being transmitted. This 10-bit frame contains ASCII code $_____ for the letter _____. This frame uses _____ (odd/even) parity.

29. To transmit a 9-bit data word, the CPU writes data bit 8 to bit _____ of the _____ register, then writes data bits 0 through 7 to the _____ register.

30. To capture a received 9-bit data word, the CPU reads bit _____ of the _____ register, then reads the _____ register.

31. The SCDR transmit function incorporates both the TDR buffer and transmit shift register. This arrangement is called _____. Explain the advantage of this arrangement.

32. The SCDR receive function incorporates both an RDR buffer and receive shift register. This arrangement is called _____. Explain the advantage of this arrangement.

33. The M68HC11 SCI transmitter adds framing bits at the _____ (name the register).

FIGURE 5–25

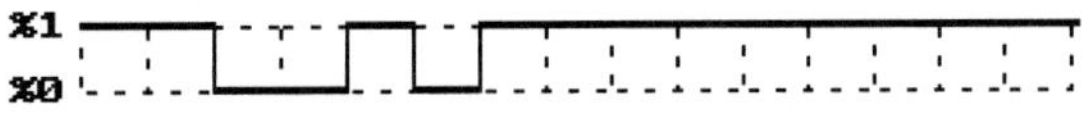

34. The M68HC11 SCI receiver strips and evaluates framing bits at the _____.
35. Bits _____ of the _____ register control the SCI baud rate.
36. Motorola calls BAUD register bits 5 and 4 the _____ bits. The prescaler divides the _____ clock rate by _____. The prescaler divides this result by _____, depending upon the states of the _____ bits.
37. If both prescaler bits = binary one, and the E clock rate = 2MHz, prescaler section output = _____Hz.
38. Motorola calls BAUD register bits 2, 1, and 0 the _____ bits. The baud rate generator divides the prescaler output by _____ where n = _____.
39. Given the prescaler section output from question 38, what is the SCI baud rate if all rate select bits are binary ones? _____
40. List the SCCR1 register's three functions.

    (1)

    (2)

    (3)

41. When SCCR1's M bit = 1, the data frame is _____ bits long.
42. To transmit the MSB of a 9-bit data word, the M68HC11 CPU writes it to the _____ bit of the _____ register.
43. To capture the MSB of a 9-bit data word, the M68HC11 CPU reads it from the _____ bit of the _____ register.

```
LDX #1000
BCLR 2C,X 40
```

44. The instructions shown above _____ (read/write) a binary _____ to the _____ bit of the _____ register.
45. Define SCI receiver "dormancy."

46. A dormant SCI receiver awakes so that it can read and evaluate the _____ of an incoming message.
47. Describe the wake up mechanism when WAKE = 0.

48. Describe the wake up mechanism when WAKE = 1.

49. List the two transmit-oriented status flags in the SCSR register.

    (1)

    (2)

50. The TDRE bit indicates that a transmit data word has transferred from the _____ to the _____. The CPU clears the TDRE flag by _____ _____.

51. List the two sets of conditions which cause the TC bit to set.

    (1)

    (2)

52. The M68HC11 CPU clears the TC flag by _____ _____.

53. List the two receiver-oriented status flags in the SCSR register.

    (1)

    (2)

54. The RDRF flag indicates that a received data word has transferred from the _____ _____ to the _____. The CPU clears the RDRF bit by _____.

55. The IDLE flag indicates that the SCI's _____ pin has been driven _____ (high/low) for _____ (duration). The CPU clears the IDLE bit by _____. Once cleared, the IDLE flag cannot set again until _____.

56. List the three error flags in the SCSR register.

    (1)

    (2)

    (3)

57. List the two conditions which cause the OR flag to set.

    (1)

    (2)

58. The CPU clears the OR bit by

59. The SCI samples RxD pin levels at a rate of _____ samples per bit time.
60. List the three reasons why the SCI samples RxD pin data levels.

    (1)

    (2)

    (3)

61. Describe how the SCI receiver detects a possible start bit.

62. The SCI receiver qualifies the start bit by checking sample numbers _____. If _____ (how many?) of these samples are _____ (high/low), the SCI qualifies the start bit.

63. Explain how the SCI receiver determines whether a given data bit is a one or a zero.

64. Describe the criteria used to reset the NF bit.

65. The CPU clears the NF bit by_____
   _____.

66. The FE flag sets if _____.
    The CPU clears the FE bit by _____
    _____.

67. SCCR2 bits 7 through 4 enable _____, which are generated when equivalent _____ register bits set. The CPU enables them by writing binary _____ to appropriate SCCR2 bits. Thus, these SCCR2 bits facilitate _____ in testing the _____ flags of the _____ register.

68. List the two functions served by SCCR2's TE bit.

    (1)

    (2)

Note: Questions 69–71 refer to Figure 5–8 on page 234 and its accompanying program segment.

69. Explain the purpose of the instructions at addresses $0124 and $0126.

70. The CPU executes the instructions at address $0128 and $012B before the transmit shift register has finished shifting out the _____ _____.

71. By executing the instructions at addresses $0128, $012B, $012E, and $0130 without delays or status polling, the CPU causes the SCI transmitter to drive its TxD pin with three end-to-end characters. Identify these characters.

    (1)

    (2)

    (3)

72. The CPU disables its SCI receiver by writing a binary _____ to the _____ bit in the _____ register. Disabling the receiver leaves the _____ flags as they were when _____.

73. The M68HC11 CPU puts its receiver to sleep by writing a binary _____ to the _____ bit in the _____ register. However, the receiver continues to _____ so that when it wakes, it can review _____ embedded in the _____.

74. The SCI transmitter sends a break character when the CPU writes a binary _____ to the _____ bit in the _____ register. The transmitter continues to send break characters until the CPU writes a binary _____ to the _____.

75. The SCI uses PORTD pin PD _____ as its RxD pin, and PD _____ as its TxD pin.

76. DDRD configures PORTD pins PD _____ through PD _____ as _____ inputs or outputs.

77. If the SCI receiver is enabled, the CPU cannot drive PORTD pin PD _____ with a _____ bit.

78. If DDRD bit 0 is a binary _____, a PORTD read returns the state of the _____ pin, whether the receiver is enabled or not.

79. The CPU cannot drive PD1 with a general purpose bit if the SCI transmitter is _____ or the SCI transmitter is _____. However, PORTD _____ this write bit.

80. If DDRD bit 1 is a binary _____, a PORTD read returns the state of the _____ pin, whether the SCI transmitter is enabled or not.

81. To initialize its SCI, the CPU must write to the _____, _____, and _____ registers.

82. Explain why programmers use indexed addressing to read and write to the SCI registers.

83. Usually the programmer initializes the X register to $_____ so that it points at _____ _____.

84. Explain why the programmer must often use the BSET and BCLR commands when writing to SCCR1 and SCCR2.

85. Refer to Figure 5–9 on page 236 and its accompanying program segment. The CPU initializes the BAUD register, using the instructions at addresses $_____ and $_____. The CPU initializes the _____ registers with the instructions at addresses $0107 and $010A.

Note: Questions 86–95 refer to Figures 5–10 and 5–11 on page 239, and their accompanying program list.

86. The instructions at addresses $0100–$0113 initialize the M68HC11 any time it is _____ by a _____.
87. The transmit message is stored in _____, starting at address $_____.
88. The CPU initializes the Y register to $_____. The Y register will be re-initialized to this value only after _____.
89. The CPU loads _____ (name the accumulator) with the transmit data word, because it will read the contents of the _____ register to accumulator _____ (see block 15).
90. Describe how the M68HC11 determines that it has transmitted the last frame of a message.

91. What two actions does the CPU take when it finds that its SCI has transmitted the final message frame?

    (1)

    (2)

92. MSGTRANS checks to see if _____. If another message must be sent, MSGTRANS moves it to address $_____. The CPU then _____.
93. After returning from the MSGTRANS subroutine, the CPU transmits _____ by executing the instructions at addresses $_____ and $_____.
94. List the two reasons why the CPU executes the instructions at addresses $0130 and $0132.

    (1)

    (2)

95. Explain the purpose of the instruction at address $0134.

Note: Questions 96–105 refer to Figures 5–13 and 5–14 on page 243, and their accompanying program lists.

96. Explain the function of bit 8 in the incoming serial data words.

97. The program is written from the point of view of a receiver connected to a _____ serial communications line. This receiver's "address" is $_____.
98. The instructions at program addresses $0100–$0117 serve to _____ the SCI receiver at the beginning of a serial communications session.
99. Define the phrase "serial communications session" as used in question 98.

100. After execution of the instruction at $0120, ACCA contains the contents of the _____ register and ACCB contains the contents of the _____ register.

101. The CPU uses the BITA command to test for two kinds of conditions. List these conditions.

    (1)

    (2)

102. After checking for errors, the CPU determines whether the received frame contains an _____. If yes, the CPU determines whether the received data word contains $_____.

103. If the initial message frame contains $0A, the CPU _____. Next, the CPU executes the instructions at addresses $_____ and $_____, to determine whether the received data word is an EOT character.

104. If the received character is an EOT, the M68HC11 calls the _____.

105. Describe the tasks performed by MSGSVC.

6 The M68HC11 Serial Peripheral Interface (SPI)

6.1 GOAL AND OBJECTIVES

The Serial Peripheral Interface (SPI) provides serial I/O. How does it differ from the Serial Communications Interface covered in the previous chapter?

1. The SPI provides synchronous serial communications between the M68HC11 and other devices. SPI communications use a common clock to time both transmitter and receiver.
2. The SPI presumes a system master and one or more slaves. The master selects a slave with which to communicate.
3. SPI data exchanges use no framing bits. The master initiates a data exchange by starting the synchronizing clock. The master ends the exchange after eight bit times by stopping the clock. Thus, the SPI requires no start or stop bits.
4. The SPI transmits and receives serial data in reverse order. The SPI exchanges the MS (most significant) bit of the data word first. In contrast, the SCI transmits the LS (least significant) bit first.
5. The SPI can exchange data at a greater bit rate (baud) than the SCI. An M68HC11 operating as a master can transmit SPI serial data up to 1M bits per second. An M68HC11 operating as a slave can handle SPI data at 2M bits per second.[1] Further, an SPI data exchange uses no framing bits, which makes it even more efficient.
6. SPI full-duplex data exchanges require three wires, instead of the two required by the SCI. These three wires are MOSI (Master Out Slave In), MISO (Master In Slave Out), and SCK (Serial Clock). The MOSI sends serial data from the master to the slave. The MISO sends serial data from the slave to the master. The SCK sends the synchronizing clock signal from the master to the slave. Bi-directional serial data flows simultaneously between master and slave.
7. An M68HC11 can function as either a master or a slave. An SPI communications network can include a master and multiple slaves. Therefore, the SPI facilitates data exchanges among multiple microcontrollers. Of course, the M68HC11 SPI also facilitates communications with a range of devices—from simple shift registers to display drivers and A/D converters.[2]

---

[1]Motorola Inc., *MC68HC11E9 HCMOS Single-Chip Microcontroller, Advance Information* (Motorola Inc.: 1988), p. 6-1. These bit rates presume that an 8MHz crystal connects between the M68HC11's XTAL and EXTAL pins, resulting in a 2MHz E clock rate.
[2]Motorola Inc., *M68HC11 Reference Manual* (1990), p. 8-1.

8. SPI programming is simpler than SCI programming (see Chapter 5). SPI programming involves four registers: the SPDR (SPI Data Register), SPCR (SPI Control Register), SPSR (SPI Status Register) and DDRD (Port D Data Direction Control Register).

This chapter's goal is to help the reader understand how to implement synchronous serial communications between an M68HC11 and its peripherals, using the SPI.

After studying this chapter and completing its exercises and hands-on activities, you will be able to:

> Derive initial SPI register values, given interface specifications.
> Use the SPI to interface an M68HC11 master with a slave shift register.
> Use SPIs to interface master and slave EVBUs (EVBs).

6.2 M68HC11 ON-CHIP RESOURCES USED BY THE SPI

The SPI uses four registers in exchanging data between master and slave. To interface and program an SPI, the reader must understand these registers.

SPI Data Register (SPDR, Control Register Block Address $102A)

The SPI master writes to its SPDR to initiate a data exchange with the slave. This SPDR write causes a chain of events:[3]

1. The master's SCK starts and is sent to the slave via the SCK line.
2. The master shifts its write data out to the slave over the MOSI (Master Out Slave In) line, while shifting in bits from the slave over the MISO (Master In Slave Out) line.
3. Simultaneously, the slave shifts data out to the master over the MISO line, while shifting in bits from the master over the MOSI line.
4. After eight bit times, the master's SCK automatically stops.
5. The stopped SCK halts the shifting of data by master and slave. The master and slave have synchronously exchanged serial data bytes.
6. Master and slave read their respective SPDRs to capture the exchanged data bytes.

Thus, the master's CPU initiates a data exchange by writing a byte to its SPDR. The data exchange ends automatically when the master's SCK stops after eight clock cycles. The SCK synchronizes MOSI and MISO data.

Refer to Figure 6–1.[4] An SPDR actually consists of two registers: (1) SPDR shift register, and (2) SPDR read data buffer. When the CPU writes to its SPDR, it actually writes to the SPDR shift register. When the CPU reads from its SPDR, it actually reads the SPDR read data buffer.

Note: The description that follows includes numbers in parentheses. These numbers refer to counterpart circled numbers in Figure 6–1.

The slave's CPU must write to its SPDR prior to a data exchange, because it must have a byte ready to transmit when the master starts the exchange. When the slave's CPU writes a byte to its SPDR, this byte transfers directly to the SPDR shift register (1). The master's CPU then writes a byte to its SPDR, and this byte also transfers directly to the master's SPDR shift register (2).

The master CPU's write causes the SCK to start (3). The SCK clocks the data out of the master's shift register, MSB first (4), to the MOSI line. The SCK from the master also clocks data out of the slave's SPDR shift register, MSB first, to the MISO data line (5).

As transmit data shifts out of each SPDR shift register (both master and slave), these shift registers also clock in received serial bits. The master's shift register clocks in bits from the MISO line (6). The slave shift register clocks in bits from the MOSI line (7).

---
[3]This chain of events assumes that both master and slave are M68HC11s.
[4]Motorola Inc., *MC68HC11E9 HCMOS Single-Chip Microcontroller, Advance Information*, p. 6-4.

FIGURE 6–1
Reprinted with the permission of Motorola.

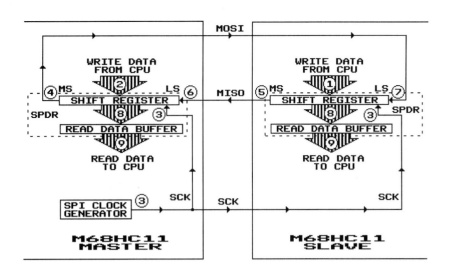

After eight SCK cycles, each SPDR shift register has clocked out 8 transmit bits and shifted in 8 receive bits. The SCK stops after these eight cycles. Master and slave SPDRs automatically transfer received bytes from their shift registers to their read data buffers (8). Master and slave CPUs can then read the received data bytes from their respective read data buffers (9).

The reader can see, then, that the SPDR single-buffers transmit (write) data, and double-buffers receive (read) data.

SPI Control Register (SPCR, Control Register Block Address $1028)

Refer to Figure 6–2. The M68HC11 configures its SPI by writing a control word to the SPCR register. The SPCR controls five SPI related functions:

- SPCR bits 0 through 3 control the SCK rate, polarity, and phase.
- Bit 4 controls whether the M68HC11 operates as master or slave.
- Bit 5 converts PORTD output pins from the totem-pole to the open-drain configuration.
- Bit 6 enables and disables the SPI.
- Bit 7 enables interrupts to facilitate interrupt control of SPI operations.

Motorola calls SPCR bits 1 (SPR1) and 0 (SPR0) the SPI Clock Rate Select bits. These bits control the SPI clock generator, which generates the SCK. The SPI clock generator divides the M68HC11 E clock rate by 2, 4, 16, or 32—depending upon the states of these two rate select bits:

SPR1	SPR0	E CK/by	SCK Rate (8MHz Xtal)[5]
0	0	2	1MHz
0	1	4	500KHz
1	0	16	125KHz
1	1	32	62.5KHz

FIGURE 6–2
Reprinted with the permission of Motorola.

SPCR

B7							B0
SPIE	SPE	DWOM	MSTR	CPOL	CPHA	SPR1	SPR0

---

[5]These SCK clock rates assume that an 8MHz crystal connects to the M68HC11's XTAL and EXTAL pins. Both EVB and EVBU use an 8MHz crystal.

FIGURE 6–3
Reprinted with the permission of Motorola.

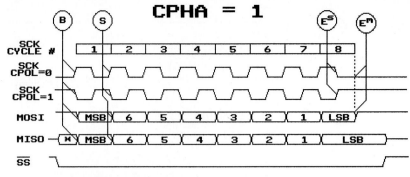

The master's SCK controls the bit transfer (baud) rate of SPI data exchanges. Therefore, the master CPU establishes this baud rate by initializing its SPCR clock rate select bits. The slave's SPR1/SPR0 bits have no control over the exchange, and their states need not match the master's.

SPCR bits 2 (CPHA, Clock Phase) and 3 (CPOL, Clock Polarity) combine to select an SPI data transfer format. SPI data transfer format relates to several questions about SPI timing:

1. When does the SPI begin a data exchange?
2. When does the SPI shift a bit out of its shift register—either to the MOSI or MISO line?
3. When is the end of a data exchange?

The previous sections have answered these questions in a general way. The data exchange begins when the master's CPU writes to its SPDR, and this write causes the SCK to start. A bit shifts out of the SPI shift register with each SCK cycle. The data exchange ends when the SCK stops after its eighth clock cycle. But these three questions must be answered more specifically. These specific answers vary, depending upon the states of the CPHA and CPOL bits in the SPCR. Further, these questions must sometimes be answered with reference to two additional factors: the master or slave's E clock and the state of the slave's $\overline{SS}$ (slave select) pin.

When does the data exchange begin? The CPHA and CPOL bits provide the answer. Refer to Figure 6–3.[6] If CPHA = 1, the data exchange begins with the initial edge of the SCK—the SCK's first transition from its idle state to its active level. This initial SCK edge is labeled "B" in Figure 6–3. When CPHA = 1, the CPOL bit defines the SCK's idle state and active level:

CPOL = 0   SCK's active level is high; SCK idles low.
CPOL = 1   SCK's active level is low; SCK idles high.

Thus, where CPHA = 1, both the master and slave drive the MOSI and MISO lines with the MSBs of their SPDR shift registers—when the SCK makes its first transition from idle state to its active level (Figure 6–3, point B). Remember that the master's CPU starts the clock with a write to its SPDR.

Refer to Figure 6–4.[7] If CPHA equals zero, then the question must be answered differently—with reference to the slave's $\overline{SS}$ (Slave Select) pin. When CPHA = 0, the data exchange begins when the master selects the slave by driving the slave's $\overline{SS}$ pin low. This point is labeled "B[1]" in Figure 6–4. This high-to-low transition causes the slave to drive the MISO line with the MS bit of its SPDR shift register. However, the master does not yet drive the MOSI with the MS bit of its shift register. After asserting the slave's $\overline{SS}$ pin, the master's CPU writes a transmit data word to its SPDR. This write causes the master's SPDR shift register to drive the MOSI line with its MSB. This point is labeled "B[2]" in

---
[6]Motorola Inc., *M68HC11 Reference Manual*, p. 8-3.
[7]Motorola Inc., *M68HC11 Reference Manual*, p. 8-2.

FIGURE 6–4
Reprinted with the permission of Motorola.

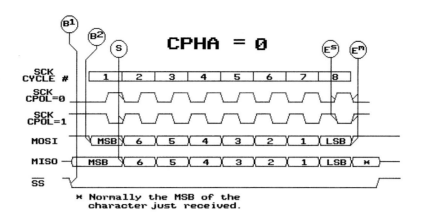

Figure 6–4. At point B² the SCK actually begins its first cycle. However, the SCK's first transition from idle occurs halfway through this cycle.

When does the SPI shift a bit out of its SPDR shift register? The short answer: when the SCK makes a transition. But the definition of SCK shift-causing transition depends upon the states of the CPOL and CPHA bits.

If CPHA = 1, then the SCK's shift-causing transition is from its inactive to active level (Point S in Figure 6–3). Of course, the CPOL bit defines SCK active and inactive levels:

CPHA = 1

CPOL State	Inactive Level	Active Level	Transition Causing a Shift Out
0	0	1	0 to 1
1	1	0	1 to 0

If CPHA = 0, the SCK's shift-causing transition is defined differently. The shift-causing transition is from the SCK's active level to inactive level (Point S in Figure 6–4). Again, the CPOL bit defines active and inactive levels:

CPHA = 0

CPOL State	Inactive Level	Active Level	Transition Causing a Shift Out
0	0	1	1 to 0
1	1	0	0 to 1

When is the end of the master/slave data exchange? Here, an overarching criterion helps define the data exchange's "end." We shall consider the data exchange ended when the SPI sets a status flag in its SPI Status Register. This flag is called the SPI Transfer Complete Flag (SPIF). Again, the answer depends upon the states of the CPHA and CPOL bits, *and* whether the M68HC11 in question is a master or a slave.

If the M68HC11 is a master, the data exchange ends (SPIF = 1) at the end of the eighth SCK cycle. Refer to Figure 6–4. If the master's CPHA control bit is binary zero, the end of the eighth cycle is also the final SCK edge. Figure 6–4 labels this final edge as "E^m." The final edge is a transition from the SCK's active to inactive levels.

Refer to Figure 6–3. If the master's CPHA control bit is binary one, the end of the eighth SCK cycle comes one half of the SCK's period after the final SCK edge. Figure 6–3 labels this point as "E^m." The final SCK edge is a transition from its active to inactive levels.

In both of these cases, the CPOL bit defines SCK's active and inactive levels:

CPOL State	Inactive Level	Active Level
0	0	1
1	1	0

If the M68HC11 is a slave, the end of the data exchange (SPIF = 1) is defined differently, depending upon the state of the slave's CPHA control bit. Refer to Figure 6–3. If the slave's

CPHA bit = 1, the end of the exchange occurs with the last SCK edge. This transition is from the SCK's active to inactive levels—as defined by the slave's CPOL control bit. The slave's end-of-exchange point is labeled "E^s" in Figure 6–3.

Refer to Figure 6–4. If the slave's CPHA = 0, the end of the exchange occurs with the second to the last SCK edge. This edge is labeled "E^s" in Figure 6–4. This transition is from the inactive to the active levels—as defined by the slave's CPOL control bit.

Data exchange "end" must be further defined with reference to the master and slave E clocks. These details will be discussed beginning on page 270.

Figure 6–2 shows that SPCR bit 4 constitutes the MSTR (Master/Slave Mode Select) control bit. The MSTR bit configures the SPI as a master or a slave:

    MSTR = 0   SPI is a slave.
    MSTR = 1   SPI is a master.

SPCR bit 5 is the DWOM (Port D Wired-OR Mode Select) control bit. This bit converts PORTD output pins from CMOS totem-pole to N-channel open-drain outputs. When the M68HC11's SPI is enabled, these outputs can include the SCK (PD4), MOSI (PD3), MISO (PD2), and $\overline{SS}$ (PD5). The DWOM control bit affects PORTD pins—configured as outputs—as follows:

    DWOM = 0   PORTD outputs are totem pole.
    DWOM = 1   PORTD outputs are N-channel open drain.

SPCR bit 6 is the SPE (SPI System Enable) control bit. The SPE bit enables and disables the SPI. If the SPE bit disables the SPI, then PORTD pins PD2–PD5 become general purpose I/O pins. If the SPI is enabled, PORTD pins PD2–PD5 assume their roles as MISO, MOSI, SCK, and $\overline{SS}$. The SPE control bit affects the SPI as follows:

    SPE = 0   SPI disabled
    SPE = 1   SPI enabled

Finally, SPCR bit 7 is called the SPIE (SPI Interrupt Enable) control bit. The previous discussion, defining the "end" of the SPI data exchange, refers to the SPIF (SPI Transfer Complete Flag). The SPI sets the SPIF at the end of the data exchange. SPI support software can use status polling to detect when the SPIF has set. However, the SPIE control bit facilitates interrupt control of the SPI data exchange process. The SPIE bit enables interrupts that the SPIF generates when it sets. SPIE affects the SPI as follows:

    SPIE = 0   Interrupts disabled; use status polling to detect SPIF = 1 condition.
    SPIE = 1   Enables interrupts generated when the SPIF sets.

The reader should note that the I mask disables SPI interrupts. The CPU must clear the I mask prior to SPI operations using interrupt control.

EXERCISE 6.1   SPCR Control Words

AIM: Derive control words which configure the SPI.

INTRODUCTORY INFORMATION: This exercise teaches you how to configure the SPI for baud rate, SCK phase and polarity, master or slave operation, and for status polling or interrupt control.

INSTRUCTIONS: The M68HC11 configures its SPI by writing a control word to the SPI Control Register (SPCR). Use the configuration descriptions given below, along with Figures 6–2, 6–3 and 6–4 to calculate appropriate SPCR control words. Enter your control words in the blanks provided. Assume that the M68HC11's E clock runs at 2 Mhz. Recall that slave control bits SPR1/SPR0 have no control over SCK rate. Therefore, initialize these bits to any combination from %00 to %11 in the slave's SPCR.

| Configuration | SPCR Control Word |

EXAMPLE

Baud = 125K
$\overline{SS}$ causes data exchange to begin
SCK active-high
SPI is slave
SPI outputs totem-pole
SPI disabled
Status polling $ 00–03

Baud = 1M
Data exchange begins with first SCK edge
SCK active-high
SPI is master
SPI outputs totem-pole
SPI enabled
Status polling $ _____

Baud = 500K
Exchange ends with final SCK edge
SCK idles high
SPI is slave
SPI outputs wired-OR
SPI disabled
Interrupt control $ _____

Baud = 62.5K
Exchange ends with final SCK edge
SCK idles low
SPI is master
SPI outputs wired-OR
SPI enabled
Interrupt control $ _____

Baud = 500K
Exchange ends 1µs after final SCK edge
SCK idles high
SPI is master
SPI outputs totem-pole
SPI enabled
Interrupt control $ _____

Baud = 1M
Exchange ends with last transition of SCK
    from inactive to active level
SCK active-high
SPI is slave
SPI outputs totem-pole
SPI disabled
Status polling $ _____

Baud = 62.5K
Exchange begins with first SCK edge
SCK active-low
SPI is master
SPI outputs wired-OR
SPI enabled
Interrupt control $ _____

Baud = 125K
$\overline{SS}$ causes data exchange to begin
SCK active-high
SPI is slave
SPI outputs totem-pole
SPI enabled
Interrupt control $_____

Baud = 500K
SPIF sets with final SCK edge
SCK idles low
SPI is slave
SPI outputs totem-pole
SPI enabled
Status polling $_____

Baud = 1M
SPIF sets with final SCK edge
SCK idles high
SPI is master
SPI outputs totem-pole
SPI enabled
Interrupt control $_____

SPI Status Register (SPSR, Control Register Block Address $1029)

Refer to Figure 6–5. The SPSR contains one status flag and two error flags. SPSR bit 7 is the SPI Transfer Complete Flag (SPIF). The SPIF sets to indicate the end of a data exchange. If the SPIE control bit = 1, the SPIF generates an interrupt when it sets. It has been established that data exchange "end" varies with the states of the MSTR, CPHA, and CPOL control bits in the SPCR. We must further define data exchange "end" with reference to the master and slave E clocks. Following paragraphs make this further distinction.

Refer to Figure 6–3. Recall that data exchange "end," for a master (E^m), comes at the end of the eighth SCK cycle. Figure 6–3 illustrates that this "E^m" point comes one half of the SCK's period after its final edge. Figure 6–6 magnifies this end-of-exchange point (CPHA = 1), and relates it to the master's E clock.[8] When CPHA = 1, the SPIF sets one half

FIGURE 6–5
Reprinted with the permission of Motorola.

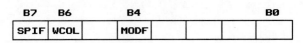

FIGURE 6–6
Reprinted with the permission of Motorola.

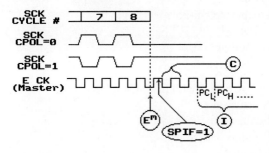

---

[8]Motorola Inc., *M68HC11 Reference Manual*, p. 8-13.

FIGURE 6–7

Reprinted with the permission of Motorola.

```
         CPHA = 0
         MSTR = 1
```

E clock period after the end of the final SCK cycle. The SPIF sets while the E clock is high. If SPI interrupts are disabled (SPIE = 0), the CPU fetches the next program opcode as soon as the E clock toggles low. This opcode fetch is labeled "C" in Figure 6–6. If the master's SPIE = 1, the CPU does not fetch a new op code. Rather, the CPU begins servicing the SPI interrupt by stacking the processor context. The CPU begins stacking in the third E clock cycle after the end of the final SCK cycle—point I in Figure 6–6.

Refer to Figure 6–4. This figure shows that when CPHA = 0, the master's data exchange ends with the final SCK edge (point E^m). Figure 6–7 magnifies the exchange's end, where CPHA = 0. Here, the SPIF sets one half E clock period after the final SCK edge. The SPIF sets while the E clock is high. If the CPU is using status polling, it fetches the next program opcode as soon as the E clock toggles low (point C in Figure 6–7). If the master's SPIE = 1, its CPU begins stacking the processor context in the third E clock cycle after the final SCK edge (Figure 6–7, point I).

Figures 6–6 and 6–7 show that the data exchange ends for the M68HC11 master with the end of its eighth SCK cycle. The SPI sets the SPIF during the first high E clock pulse, following the eighth SCK cycle. After the SPIF sets, the CPU fetches the next opcode as soon as the E clock toggles low (assuming that SPIE = 0). If SPIE = 1, the CPU begins stacking its context in the third E clock cycle after the end of the data exchange.

Figures 6–8 and 6–9 define the end of the master/slave data exchange when the SPI is a slave (MSTR = 0).[9] Refer to Figure 6–3. Recall that when CPHA = 1, the slave's end-of-exchange occurs with the final SCK edge. Figure 6–8 magnifies this end-of-exchange from the point of view of the slave's E clock. The master's SPI toggles the SCK. Therefore, the final SCK edge is asynchronous to the slave's E clock. The slave's SPI sets its SPIF synchronously with its own E clock. Because of the asynchronous relationship between the SCK and the slave's E clock, the slave will take from one to two E clock cycles to set its SPIF—following the final SCK edge from the master. If the slave's SPI interrupts are disabled (SPIE = 0), the slave's CPU fetches the next program opcode as soon as the E clock toggles low (after the SPIF sets). This opcode fetch is labeled "C" in Figure 6–8. If the

FIGURE 6–8

Reprinted with the permission of Motorola.

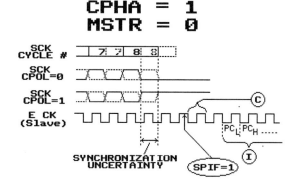

---

[9]Motorola Inc., *M68HC11 Reference Manual*, p. 8-13.

FIGURE 6–9
Reprinted with the permission of Motorola.

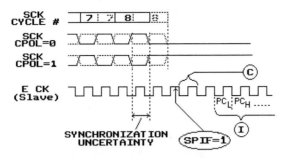

slave's SPIE = 1, its CPU does not fetch an opcode, but begins stacking processor context in the third E clock cycle after the SPIF sets. See point I in Figure 6–8.

Refer to Figure 6–4. Note that when CPHA = 0, the slave's end-of-exchange coincides with the second SCK edge from the last. Figure 6–9 magnifies the end-of-exchange where the M68HC11's CPHA and MSTR bits are both reset. Again, the slave requires one or two E clock cycles to respond to the asynchronous SCK and set its SPIF. If the slave's SPIE = 0, its CPU will fetch the next program opcode as soon as its E clock toggles low (after the SPIF sets). Refer to point C in Figure 6–9. If the slave's SPIE = 1, its CPU does not fetch an opcode, but begins stacking processor context in the third E clock cycle after the SPIF sets. See point I in Figure 6–9.

Thus, Figures 6–8 and 6–9 illustrate that an M68HC11 configured as an SPI slave requires one or two extra E clock cycles to respond to the asynchronous SCK and set its SPIF. The asynchronous relationship between master's SCK and slave's E clock causes this delay.

The SPSR includes two error flags, the WCOL and the MODF. WCOL detects errors committed by a slave SPI. The MODF mechanism detects contention by two masters for control of the MOSI, MISO, and SCK lines.

SPSR bit 6 is the WCOL (Write Collision Error Flag). A CPU write to its SPDR—during a data exchange—causes a write collision error. Recall that the SPDR is single-buffered. SPDR write data transfers directly to the SPI shift register. If the SPI allowed such a transfer during a data exchange, write data would corrupt the data being exchanged. However, the SPI disallows an inadvertent write. The SPI aborts the write, sets the WCOL flag, and completes the data exchange. The WCOL flag indicates that the write was aborted, and the offending write data was lost.

If the SPI is enabled, WCOL indicates a write collision error regardless of whether the offending M68HC11 is a master or slave. However, the master controls the beginning and end of the SPI data exchange and is therefore unlikely to commit a write collision error. The slave is usually the offender. The slave does not know when the master will start an SPI data exchange. Also, when the slave's CPHA = 0, it does not know when the master will officially end the data exchange.

To avoid write collision errors, the programmer must understand when an SPI data exchange begins and ends for write collision purposes. For an M68HC11 configured as a master, the data exchange begins when its CPU writes to the SPDR and ends when its SPIF sets. For an M68HC11 configured as a slave, the beginning and end of the exchange—for write collision purposes—depend upon the state of the slave's CPHA bit. If the slave's CPHA = 1, the data exchange begins with the first edge of the SCK, and ends when the slave's SPIF sets. If the slave's CPHA = 0, the data exchange begins when its $\overline{SS}$ pin is driven low, and ends when its $\overline{SS}$ pin is driven high. The master drives the slave's $\overline{SS}$ pin low and high. If the master is otherwise occupied after the end of a data exchange, it might delay returning the slave's $\overline{SS}$ pin high. The slave could then easily cause a write collision error, even though all bits have been shifted by its SPI, and its SPIF has set. Motorola suggests that, before writing to its SPDR, the slave should always read and check the state of PORTD, bit PD5. PORTD pin PD5 is the $\overline{SS}$ pin. Therefore, reading PORTD returns the state of the $\overline{SS}$ pin if it is configured as an input. Of course, the slave's CPU must not write to its SPDR until it reads PD5 and determines that this pin ($\overline{SS}$) is

FIGURE 6–10

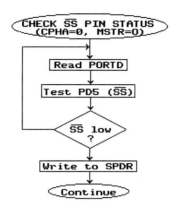

Check SS Pin Status (CPHA = 0, MSTR = 0)

$ADDRESS	SOURCE CODE	COMMENTS
C115	LDAA 8,X	Read PORTD
C117	BITA #20	Test PD5 (SS)
C119	BEQ C115	If SS is low, check again
C11B	LDAA A	If SS is high, load SPDR write data
C11D	STAA 2A,X	Write data to SPDR
C11F	-----------------	Continue

high.[10] Figure 6–10 and its accompanying program segment implement Motorola's suggested procedure. This software assumes that the slave's X register points at the beginning address of the control register block ($1000), and that SPDR write data resides at address $000A.

The CPU clears the WCOL bit by:

1. Reading the SPSR when WCOL = 1,
2. Then reading or writing to the SPDR.

Thus, the WCOL flag informs a slave that it has committed a write collision error. Such an error occurs when the slave attempts to write to its SPDR while an SPI data exchange is still in progress. When a write collision error occurs, the SPI aborts the SPDR write and sets the WCOL flag. A slave is more prone to write collision errors when its CPHA = 0. When its CPHA = 0, the slave should always check its SS pin (PD5) to assure that it is high before writing to the SPDR. The CPU clears the WCOL bit by reading its SPSR (while WCOL = 1), then accessing its SPDR.

SPSR's remaining bit is the Mode Fault Error Flag (MODF, bit 4). Mode fault errors happen only to M68HC11 masters. These errors can occur when there is more than one master in a multidrop SPI communications system. In such a case, the system designer must take great care to assure that only one master at a time has control of the SPI serial communications system. The reader can understand this imperative by considering the consequences of two masters driving an MOSI line—one master attempting to drive the line high, and the other attempting to drive it low! The MODF feature deals with this dire possibility.

To take advantage of the M68HC11's mode fault error feature, multiple masters must have their SS (Slave Select) pins interconnected and configured as inputs. A master detects a mode fault when another master contends for the common MOSI and MISO data lines. This other master signals its intention to use the SPI system data lines by driving the remaining masters' SS pins low. An SPI that is configured as a master calls a mode fault error when its SS pin is driven low.

When its SPI calls a mode fault error, the M68HC11 automatically does the following:

1. Resets the MSTR bit in its SPCR, re-configuring the SPI as a slave.
2. Resets the SPE bit in its SPCR, disabling the SPI.

---

[10]Motorola Inc., *M68HC11 Reference Manual*, p. 8-13.

3. Configures its PORTD pins PD2–PD5 as inputs by resetting all counterpart DDRD bits.
4. Sets the MODF bit in its SPSR.
5. Calls an SPI interrupt if SPIE = 1.

Figure 6–11 and its accompanying state table show a multiple-master system, arranged to take advantage of the SPI mode fault feature. This system has the following characteristics:

- The system consists of three M68HC11 masters and one M68HC11 slave.
- Masters can exchange data with the slave only.
- After initialization, each master continuously polls its PD1 pin. The system selects a particular master for a data exchange by driving the selectee's PD1 pin from high to low (a Request to Exchange, RTE).[11]
- An RTE causes a master M68HC11 to request the use of the MOSI, MISO, and SCK lines by writing a binary one to its PORTD, PD0 pin (a Slave Select Request, SR).
- The SR is sent from the master's PD0 pin to the Master Select Logic (MSL). The MSL grants this request by driving the $\overline{SS}$ pins low on the other two masters and the slave.
- The SPI MODF mechanisms disable the other two masters' SPIs; re-configure their MOSI, MISO, and SCK pins as inputs; reset their MSTR control bits; set their MODF bits; and generate SPI interrupts.
- In the case of simultaneous SRs by two or three masters, the MSL gives priority to master A, then master B. Master C has lowest priority (refer to the MSL State Table).
- After a data exchange, the selected master negates the $\overline{SS}$ signal to the slave and nonselected masters by writing a zero to the MSL (i.e., the selected master writes a binary zero to its PD0 pin).
- The slave and nonselected masters poll their PD5 ($\overline{SS}$) pins to determine when the data exchange is complete. When their $\overline{SS}$ pins go high, the nonselected masters and slave re-initialize for the next data exchange.

FIGURE 6–11

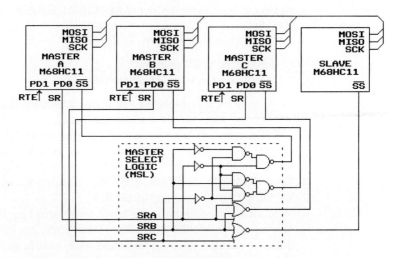

Master Select Logic (MSL) State Table

SRA	SRB	SRC	SSA	SSB	SSC	SS(Slave)	MASTER SELECTED
0	0	0	1	1	1	1	No requests
0	0	1	0	0	1	0	C
0	1	0	0	1	0	0	B
0	1	1	0	1	0	0	B
1	0	0	1	0	0	0	A
1	0	1	1	0	0	0	A
1	1	0	1	0	0	0	A
1	1	1	1	0	0	0	A

---

[11]The RTE could be initiated by an operator pressing a push-button in a pull-up circuit or by a system controller. Such a controller could be a fifth M68HC11.

FIGURE 6–12

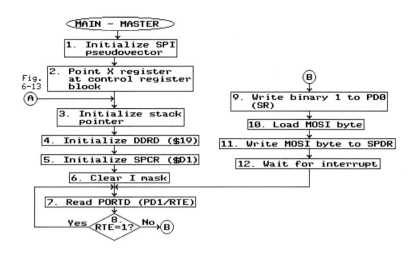

MAIN Program—Master

$ADDRESS	SOURCE CODE	COMMENTS
0100	LDD #130	Initialize SPI pseudovector
0103	STD C8	" " "
0105	LDX #1000	Initialize X register
0108	LDS #1FF	Initialize stack pointer
010B	LDAA #19	Initialize DDRD
010D	STAA 9,X	" "
010F	LDAA #D1	Initialize SPCR
0111	STAA 28,X	" "
0113	CLI	Clear I mask
0114	LDAA 8,X	Read PORTD (PD1/RTE)
0116	BITA #2	RTE = 1?
0118	BNE 114	If RTE = 1, read PD1/RTE again
011A	BSET 8,X 1	If RTE = 0, write 1 to PD0 (SR)
011D	LDAA 0	Load MOSI byte
011F	STAA 2A,X	Write MOSI byte to SPDR
0121	WAI	Wait for interrupt
0122	BRA 114	Look for another RTE

Figure 6–12 through Figure 6–15 flowcharts, along with their accompanying programs, illustrate supporting multimaster system software.

Refer to Figure 6–12—blocks 1, 2, and 3—and "MAIN–MASTER" program addresses $0100–$010A. For both EVBU and EVB, the SPI interrupt pseudovector resides in addresses $00C7–$00C9. Once the pseudovector (block 1) and X register (block 2) have been initialized, they need not be re-initialized after a data exchange. Note Terminal A, between blocks 2 and 3 in Figure 6–12. However, the stack pointer (block 3) must be re-initialized by the nonselected master, because its CPU branches to Terminal A rather than returning from interrupt. Re-initializing the stack pointer keeps the stack from growing to the point where it writes over program addresses.

Refer to Figure 6–12, blocks 4, 5, and 6, and "MAIN–MASTER" program addresses $010B–$0113. A master initializes its PORTD pins (block 4) as follows:

PORTD Pin	Pin Function	Input/Output	DDRD Bit Value
PD5	$\overline{SS}$	Input	0
PD4	SCK	Output	1
PD3	MOSI	Output	1
PD2	MISO	Input	0
PD1	RTE	Input	0
PD0	SR	Output	1

FIGURE 6-13

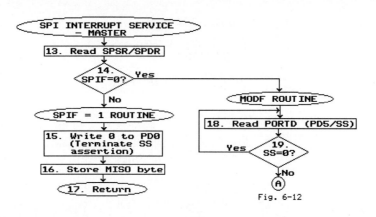

SPI Interrupt Service—Master

$ADDRESS	SOURCE CODE	COMMENTS
0130	LDD 29,X	Read SPSR/SPDR
0132	BPL 13A	If SPIF = 0, go to MODF routine
0134	BCLR 8,X 1	If SPIF = 1, terminate SR
0137	STAB 1	Store MISO byte
0139	RTI	Return

$ADDRESS	SOURCE CODE	COMMENTS
		(MODF ROUTINE)
013A	LDAA 8,X	Read PORTD (PD5/SS)
013C	BITA #20	SS = 0?
013E	BEQ 13A	If SS = 0, read PD5/SS again
0140	BRA 108	If SS = 1, re-initialize as master

FIGURE 6-14

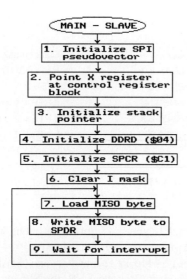

MAIN Program—Slave

$ADDRESS	SOURCE CODE	COMMENTS
0100	LDD #120	Initialize SPI pseudovector
0103	STD C8	" " " " "
0105	LDX #1000	Initialize X register
0108	LDS #1FF	Initialize stack pointer
010B	LDAA #4	Initialize DDRD
010D	STAA 9,X	" " "
010F	LDAA #C1	Initialize SPCR
0111	STAA 28,X	" " "
0113	CLI	Clear I mask
0114	LDAA 0	Load MISO byte
0116	STAA 2A,X	Write MISO byte to SPDR
0118	WAI	Wait for interrupt
0119	BRA 114	Load another MISO byte

FIGURE 6–15

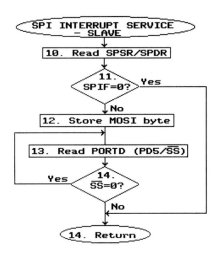

SPI Interrupt Service—Slave

$ADDRESS	SOURCE CODE	COMMENTS
0120	LDD 29,X	Read SPSR/SPDR
0122	BPL 12C	If SPIF = 0, return from interrupt
0124	STAB 1	If SPIF = 1, store MOSI byte
0126	LDAA 8,X	Read PORTD (PD5/SS)
0128	BITA #20	SS = 0?
012A	BEQ 126	If SS = 0, read PD5/SS again
012C	RTI	If SS = 1, return from interrupt

In block 5, a master initializes its SPCR as follows:

> SPI interrupts enabled
> SPI enabled
> PORTD outputs push-pull
> SPI is master
> SCK idles low
> Slave's data exchange begins when its $\overline{SS}$ pin is driven low by the MSL (see Figure 6–4).
> Baud = 500K

A master clears its I mask (block 6) to enable SPI interrupts.

Refer to Figure 6–12, blocks 7, 8, and 9, and "MAIN–MASTER" program addresses $0114–$011C. Here the master awaits a "Request to Exchange" (RTE) from the system controller by continuously reading its RTE pin (PORTD, pin PD1). When the system controller initiates an RTE by driving PD1 low, the master then requests use of the MOSI, MISO, and SCK lines by writing a binary one to its PD0 pin, thus initiating an SR (block 9). This SR is sent to the Master Select Logic (MSL). The MSL arbitrates SRs from the three masters. If a master is the first to issue an SR, the MSL selects it. In the case of a tie, the MSL selects the master with the highest priority. The MSL drives the $\overline{SS}$ pins low on the nonselected masters—causing their mode fault mechanisms to take over.

Refer to Figure 6–12, blocks 10, 11, and 12, and "MAIN–MASTER" program addresses $011D–$0123. If a master gets as far as block 10, the MSL has selected it and disabled the other two masters. Thus, the selected master loads the byte to be transmitted over the MOSI line (block 10). The selected master then initiates the master/slave data exchange by writing the transmit byte to its SPDR. The master/slave data exchange is interrupt driven so the master waits for the SPI interrupt to occur (block 12).

Refer to Figure 6–13, blocks 13 and 14, and "SPI INTERRUPT SERVICE–MASTER" program addresses $0130–$0133. An SPI interrupt is generated by SPIF = 1, or MODF = 1. The master must determine which of these bits has set so it can respond properly. To determine which status bit has set, the master loads its D register with SPSR data (to ACCA) and SPDR data (to ACCB).[12] The BPL command (block 14) causes a branch if bit 7 of ACCA is zero. ACCA bit 7 is the SPIF bit. Therefore, the program branches to the MODF routine if SPIF = 0. If SPIF = 1, the program doesn't branch, and continues with a SPIF = 1 routine.

Refer to Figure 6–13, blocks 15, 16, and 17, and "SPI INTERRUPT SERVICE–MASTER" program addresses $0134–$0139. The master begins its SPIF = 1 routine (block 15) by requesting that the MSL terminate the data exchange by driving $\overline{SS}$ pins high on the slave and nonselected masters. The master executes this step (block 15) by writing a zero to the MSL. The master then stores the received MISO byte (block 16). Recall that this MISO byte was transferred into ACCB by the SPSR/SPDR read (block 13). The master then returns from the interrupt (block 17).

Refer to Figure 6-13, blocks 18 and 19, and "SPI INTERRUPT SERVICE–MASTER program addresses $013A–$0141. If a master has gotten to block 18, it is a nonselected master. Its SPI has been disabled; its MSTR control bit has been reset; and its MOSI and SCK pins have been re-configured as inputs. The nonselected master has nothing to do at this point but wait for the end of the data exchange. Thus, in blocks 18 and 19, the nonselected master continuously reads the state of its $\overline{SS}$ pin (PD5). When its $\overline{SS}$ pin finally goes high, the nonselected master branches back to the MAIN program (Figure 6–12), to reconfigure itself as an eligible master. The nonselected master avoids a WCOL error by waiting to branch back until its $\overline{SS}$ pin returns high.

The slave's job is less complex, since it "doesn't care" which of the three masters it communicates with. Essentially, the slave loads its SPDR with a transmit byte (MISO); starts the data exchange when the MSL drives its $\overline{SS}$ low; ends the exchange when the MSL drives its $\overline{SS}$ pin high again; and stores the received byte before waiting for the next data exchange.

Refer to Figure 6–14, block 1 through block 6, and "MAIN–SLAVE" program addresses $0100–$0113. The slave's CPU initializes the SPI to be compatible with the system masters. In block 4, the slave configures its PORTD pins as follows:

PORTD Pin	Pin Function	Input/ Output	DDRD Bit Value
PD5	$\overline{SS}$	Input	0
PD4	SCK	Input	0
PD3	MOSI	Input	0
PD2	MISO	Output	1

In block 5, the slave initializes its SPCR as follows:

SPI interrupts enabled
SPI enabled
PORTD outputs push-pull
SPI is slave
SCK idles low
Data exchange begins when the MSL drives $\overline{SS}$ low (see Figure 6–4).

Refer to Figure 6–14, block 7 through block 9, and "MAIN–SLAVE" program addresses $0114–$0119. The slave prepares for a data exchange by writing a data byte (MISO) to its SPDR and waiting for an SPI interrupt. The SPIF generates this interrupt when it sets, indicating the end of the data exchange.

Refer to Figure 6–15, blocks 10 and 11, and "SPI INTERRUPT SERVICE–SLAVE" program addresses $0120–$0123. The slave CPU reads the contents of its SPSR to ACCA, and the contents of its SPDR to ACCB (block 10). The slave CPU then uses the BPL command to determine whether its SPIF has set (block 11). SPIF = 0 indicates a spurious interrupt. Therefore, the slave returns from the interrupt routine. If SPIF = 1, the slave continues to service the SPI interrupt.

Refer to Figure 6–15, block 12 through block 14, and "SPI INTERRUPT SERVICE–SLAVE" program addresses $0124–$012C. If the SPIF = 1, the slave stores the received

---

[12]Note that the block 13 read causes any set status bits (SPIF or MODF) to reset. However, this read transfers the SPSR bits to ACCA, before they reset.

FIGURE 6–16

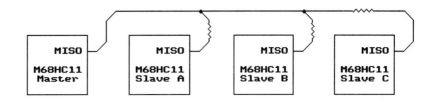

MOSI byte from ACCB to RAM address $0001 (block 12). The slave must now avoid a WCOL error by assuring that its $\overline{SS}$ pin has returned high. The slave therefore reads and re-reads the state of its $\overline{SS}$ pin until it returns high (blocks 13 and 14). When its $\overline{SS}$ pin finally goes high, the slave returns from the SPI interrupt.

Foregoing paragraphs describe a system which uses the M68HC11's mode fault feature to arbitrate among multiple SPI masters. Remember that the essential purpose of the MODF mechanism is to avoid a situation where two masters try to drive a serial data line simultaneously. Unfortunately, the MODF feature cannot protect against many other kinds of serial data line contention. Motorola cites an example where a master might inadvertently select more than one slave.[13] In such a case, two slaves simultaneously try to drive the MISO line, causing a catastrophic contention. Motorola suggests two ways of preventing hardware damage resulting from contention:

1. Connect a 1 KΩ to 10 KΩ resistance in series with each SPI driver (Figure 6–16). This resistance controls current in the case of data line contention. Motorola states that, in most cases, these series resistances will not adversely affect data exchanges.
2. Use a wired-OR option by setting the DWOM control bit in the SPCR register. Recall that this bit converts PORTD output pins from totem-pole to N-channel open-drain outputs. Connect open-drain outputs to a common external pull-up resistor, as shown in Figure 6–17.[14]

Port D Data Direction Register (DDRD, Control Register Block Address $1009)

DDRD configures PORTD pins PD0–PD5 as inputs or outputs. Writing a one to a DDRD bit configures its counterpart PORTD pin as an output. Writing a zero to a DDRD bit configures the counterpart PORTD pin as an input.

The SPI uses PORTD pins PD2–PD5. Configure these pins as inputs or outputs, depending upon whether the M68HC11 is master or slave. If an M68HC11 is a master (MSTR = 1), configure its MOSI and SCK pins as outputs by writing binary ones to DDRD bits 3 and 4. The SPI automatically configures a master's MISO pin as an input regardless of its counterpart DDRD bit. The master's $\overline{SS}$ pin can be configured as an input or an output by writing an appropriate control bit to its DDRD counterpart. In multiple-master systems, configure the $\overline{SS}$ pin as an input. In single-master systems, configure the master's $\overline{SS}$ pin as an output. The $\overline{SS}$ pin can then remain unconnected.

If an M68HC11 is a slave (MSTR = 0), configure its MISO pin as an output by writing a one to DDRD bit 2. The SPI automatically configures the slave's MOSI, SCK, and $\overline{SS}$ pins as inputs regardless of the states of their counterpart DDRD bits. The chart on the next page summarizes these SPI configurations.

FIGURE 6–17

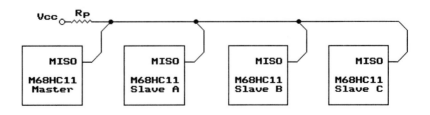

---

[13]Motorola Inc., *M68HC11 Reference Manual*, p. 8-10.
[14]Ibid.

PORTD Pin Configurations

PORTD Pin	M68HC11 Pin #	SPI Function	Master Input/Output	Slave Input/Output
PD2	22	MISO	Input*	Output
PD3	23	MOSI	Output	Input*
PD4	24	SCK	Output	Input*
PD5	25	$\overline{SS}$	**	Input*

*Configured automatically as input regardless of counterpart DDRD bit.
**Configure according to SPI system requirements.

EXERCISE 6.2 SPI Register Initialization

AIM: Given interface parameters, derive SPI register initial values, and list register addresses.

INTRODUCTORY INFORMATION: This assignment sheet gives you experience in working with relevant SPI registers.

INSTRUCTIONS:

1. Study the Figure 6–18 interface diagram. Complete the following chart. Base your entries on the interface diagram and these specifications:

   Configure the master's $\overline{SS}$ pin as an output.
   Slave uses status polling.
   Master uses an IRQ interrupt routine to initiate the data exchange.
   Make PORTD outputs totem-pole.
   SCK idles low.
   Slave's data exchange begins with a negative-going SCK edge.
   Baud rate = 500K

Note: Enter N/A where no value is relevant.

M/S	REGISTER	$ADDRESS	$INITIALIZING VALUE
M	DDRD	_____	_____
M	SPCR	_____	_____
M	SPDR	_____	_____
M	SPSR	_____	_____
S	DDRD	_____	_____
S	SPCR	_____	_____
S	SPDR	_____	_____
S	SPSR	_____	_____

2. Fill in the blanks in the following statement:
   If you want the slave to use interrupt control, but all other interface parameters remain the same, change the initializing value of the _____ (master/slave) _____ (name of register) to $_____ (initializing value).

FIGURE 6–18

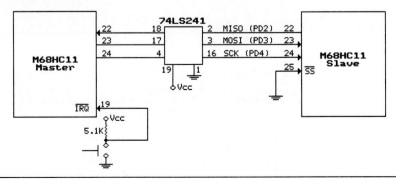

6.3 INTERFACING AN M68HC11 MASTER WITH A SLAVE SHIFT REGISTER

The SPI provides a simple means for interfacing the M68HC11 with a serial shift register. This section's example uses an SN74LS91N, 8-bit, serial in-serial out shift register as a slave. Refer to Figure 6–19, which illustrates the SN74LS91N's internal arrangement. Note that the serial S-R flip-flops are negative-edge triggered. However, serial data clocks into and out of the shift register with positive edges because of inverter/driver X. NAND gate Y provides two inputs (A and B)—one for input data, and the other as an active-high input enable. Inverter Z causes the serial S-R flip-flops to operate in the manner of D flip-flops.

Clocking in 8 serial bits requires eight positive-going clock edges. After these eight clock edges, the first bit clocked in drives the chip's Q_H output. Thus, it requires only seven more clock edges to drive Q_H with all 8 stored serial bits. The system designer must take this situation into account when interfacing the shift register's Q_H output with the M68HC11's MISO line. The M68HC11 master's SPI automatically clocks the slave shift register eight times, causing the high order bit to be lost. This high order bit shifts through the SPI's shift register in seven clocks and is lost on the eighth. Therefore, the system designer must connect a D flip-flop in series with the slave shift register's Q_H output and the master's MISO input. This D flip-flop causes the high bit to appear in the MSB of the master's shift register after eight SCK edges.

Figure 6–20 diagrams the interface between M68HC11 master and SN74LS91N slave. This interface has the following characteristics:

- The master transfers SPI serial data to the slave over the MOSI line.
- The SN74LS91N slave transfers data to the master via the D flip-flop. The slave shifts data out of its Q_H pin to the flip-flop. The flip-flop shifts data from its Q output to the master over the MISO line.
- The master enables serial data inputs to the slave by driving the slave's A input high. The master drives the A input high by writing a binary one to its PD0 pin.
- The interface operator initiates a master/slave data exchange by pressing the push-button. The low from the push-button pull-up circuit generates an IRQ interrupt at the master.
- The M68HC11's IRQ interrupt service routine causes the master to drive the slave's A input high, enabling data inputs to the slave. The interrupt routine then causes the master's CPU to write an MOSI data word to its SPDR, which starts the data exchange.

FIGURE 6–19

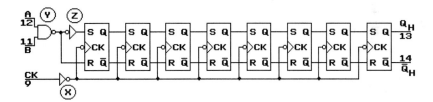

FIGURE 6–20

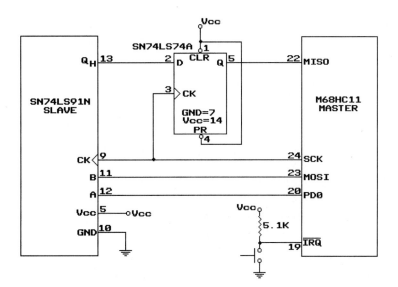

- The master's CPU controls the data exchange using status polling. When the 8-bit data exchange is complete, the master stores the MISO byte received from the slave. The CPU then disables the slave's data input by writing a binary zero to the PD0 pin.

Figures 6–21 and 6–22, along with their accompanying program list, illustrate the supporting software for the M68HC11–SN74LS91N interface. Refer to Figure 6–21, blocks 1 through 5, and MAIN program addresses $0100–$0112. The M68HC11 uses an IRQ service routine to initiate and control data exchanges. Therefore, the CPU initializes the IRQ pseudovector in block 1. In block 4, the CPU configures its MOSI, $\overline{SS}$, SCK, and PD0 pins

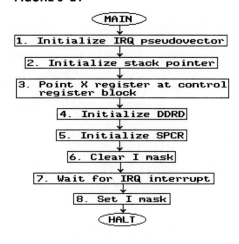

FIGURE 6–21

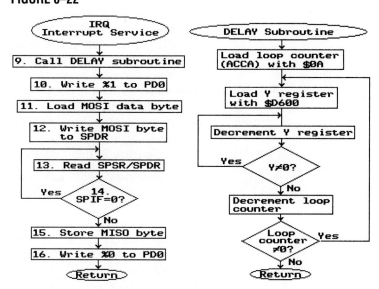

FIGURE 6–22

M68HC11—SN74LS91N Interface Software

MAIN

$ADDRESS	SOURCE CODE	COMMENTS
0100	LDD #150	Initialize IRQ pseudovector
0103	STD EF	" "
0105	LDS #1FF	Initialize stack pointer
0108	LDX #1000	Initialize X register
010B	LDAA #39	Initialize DDRD
010D	STAA 9,X	" "
010F	LDAA #5D	Initialize SPCR
0111	STAA 28,X	" "
0113	CLI	Clear I mask
0114	WAI	Wait for interrupt
0115	SEI	Set I mask
0116	WAI	Halt

IRQ Interrupt Service Routine

$ADDRESS	SOURCE CODE	COMMENTS
0150	JSR 190	Call DELAY subroutine
0153	LDAA #1	Write binary one to PD0
0155	STAA 8,X	" " " "
0157	LDAA 0	Load MOSI data byte
0159	STAA 2A,X	Write MOSI byte to SPDR
015B	LDD 29,X	Read SPSR/SPDR
015D	BPL 15B	If SPIF=0, read SPSR/SPDR again
015F	STAB 0	If SPIF=1, store MISO byte
0161	CLRA	Write binary zero to PD0
0162	STAA 8,X	" " " "
0164	RTI	Return

DELAY Subroutine

$ADDRESS	SOURCE CODE	COMMENTS
0190	LDAA #A	Initialize loop counter
0192	LDY #D600	Load Y register with count-down value
0196	DEY	Decrement Y register
0198	BNE 196	If Y register ≠ 0, decrement Y register again
019A	DECA	If Y register = 0, decrement loop counter
019B	BNE 192	If ACCA not ≠ 0, count down again
019D	RTS	If ACCA = 0, return

as outputs. Even though the $\overline{SS}$ pin remains open, the CPU configures it as an output to disable the mode fault error feature. In block 5, the CPU configures its SPI as follows:

> SPI interrupts disabled
> SPI enabled
> PORTD outputs totem-pole
> SPI is master
> SCK idles high
> Data exchange begins with SCK's initial edge (CPHA = 1)
> Baud = 500K

Refer to blocks 6 through 8 of Figure 6–21 and MAIN program addresses $0113–$0116. In block 6, the CPU clears its I mask to enable IRQ interrupts. In block 7, the CPU waits for an IRQ interrupt. The interface operator initiates a master/slave data exchange by pressing a push-button, which in turn generates an IRQ interrupt. After exchanging data with the slave, the CPU master returns from the IRQ Interrupt Service routine. The CPU then sets its I mask (block 8) and halts.

Refer to block 9 of Figure 6-22 and IRQ Interrupt Service routine address $0150. IRQ interrupts are level active. Any overlap between the active IRQ level (i.e., push-button press) and an attempt by the SPI to exchange data causes the SPI to malfunction. In block 9, the CPU calls the DELAY subroutine. This subroutine executes a two-second delay, which gives the interface operator time to release the push-button and avoid an overlap.

Refer to block 10 of Figure 6–22 and IRQ Interrupt Service routine addresses $0153–$0156. After returning from the DELAY subroutine, the CPU writes a binary one to its PD0 pin. This high enables the slave's input gate.

Refer to blocks 11 through 14 and IRQ Interrupt Service routine addresses $0157–$015E. The CPU causes the data exchange to begin by reading the MOSI byte from its RAM, and then writing this byte to the SPDR. The CPU then enters a polling loop where it keeps reading the SPSR to ACCA, and the SPDR to ACCB, until it finds that the SPIF (SPSR bit 7) has set.

Refer to blocks 15 and 16 and IRQ Interrupt Service routine addresses $015F–$0164. When its SPIF sets, the CPU knows that the data exchange is complete. The CPU then stores the MISO byte from ACCB to RAM address $0000. The CPU transferred the MISO byte from the SPDR to ACCB in block 13. Before returning from the IRQ interrupt and halting, the CPU writes a binary zero to its PD0 pin to disable the slave's data input (block 15).

EXERCISE 6.3 Interfacing an M68HC11 Master with an SN74LS91N Shift Register

REQUIRED MATERIALS: 1 SN74LS91N, 8-bit, Serial In-Serial Out Shift Register
1 SN74LS74A Dual, D Flip-flop
1 Resistor, 5.1 KΩ
1 Switch, push-button, push-to-close

REQUIRED EQUIPMENT: Motorola M68HC11EVBU (EVB), and terminal

GENERAL INSTRUCTIONS: Construct, operate, and analyze the operation of the microcontroller with a shift register interface presented in the previous section. Carefully execute the detailed procedures listed here.

PART I: Connect the Interface

Power down the EVBU (EVB) and terminal. Refer to Figure 6–20 and construct the interface. After double-checking all connections, show the instructor your connections.

_____ *Instructor Check*

PART II: Software Development

Refer to Figures 6–21 and 6–22, along with their accompanying program list. If you interface an EVBU, load the program as listed. If you interface an EVB, allocate memory as follows:

$ADDRESS	CONTENTS
0000	MOSI or MISO data byte
C000	MAIN program
C050	IRQ Interrupt Service routine
C090	DELAY subroutine
C0FF	Initial stack pointer

If you interface an EVB, modify listed software to match the above memory allocations. Modify these instructions:

    IRQ pseudovector initialization
    Stack pointer initialization
    Branch instruction destination addresses
    DELAY subroutine call

PART III: Interface Debug and Demonstration

Power up the EVBU (EVB), terminal, and interface. Load the support software. Debug the interface and software. When the interface functions properly, demonstrate it to the instructor.

_____ *Instructor Check*

PART IV: Laboratory Analysis

Answer the questions below by neatly entering your answers in the appropriate blanks.

1. Refer to MAIN program instructions located at addresses $010F–$0112 ($C00F–$C012). These instructions initialize CPOL to binary _____ and CPHA to binary _____.
2. Given SPCR initialization values established in question 1, the SPI shifts data from its shift register to the MOSI line on _____-going SCK edges. The SN74LS91N shift register shifts in data bits from the MOSI line on _____-going SCK edges.
3. The relationship referred to in question 2 allows _____ns of set-up time between the master's active edge and the slave's active edge.
4. Given SPCR initialization values established in question 1, the SN74LS91N shifts bits out to the MISO line on _____-going SCK edges. The M68HC11 SPI shifts in MISO data bits on _____-going SCK edges.
5. The relationship referred to in question 4 allows _____ns set-up time between the slave's active edge and the master's active edge.
6. An SN74LS74A flip-flop connects the _____ line between slave and master.
7. Explain the SN74LS74A's purpose.

8. The master enables serial data inputs to the slave by driving the slave's _____ high. The master does this by writing binary _____ to its _____ pin.
9. The interface operator initiates a data exchange by _____. This action generates an _____ at the master. The master's CPU controls the data exchange by using _____.
10. After vectoring to the IRQ Interrupt Service routine, the CPU calls the _____ subroutine.

11. Explain why the master calls this subroutine (question 10).

12. The CPU causes the data exchange to begin by writing the _____ to its _____.
    The CPU then enters a _____, which it escapes when _____.
13. Refer to Figure 6–22. The CPU reads the MISO byte in block _____ of the flowchart.
14. The support software initializes the master's SPCR so that CPOL = 1 and CPHA = 1. Empirically evaluate interface operation for all four CPOL/CPHA combinations. Record evaluation data in the chart below.

SPCR $CONTENTS	CPOL	CPHA	INTERFACE RELIABLE	UNRELIABLE
_____	0	0	_____	_____
_____	0	1	_____	_____
_____	1	0	_____	_____
_____	1	1	_____	_____

15. The support software initializes the master's SPCR for 500K baud. Empirically evaluate interface operation for all four SCK baud rates. Record evaluation data in the chart below. Initialize CPOL to binary one and CPHA to binary one for all evaluations.

SPCR $CONTENTS	BAUD	NTERFACE RELIABLE	UNRELIABLE
_____	1M	_____	_____
_____	500K	_____	_____
_____	125K	_____	_____
_____	62.5K	_____	_____

PART V: Laboratory Software Listing

MAIN PROGRAM—EVBU

(See program list accompanying Figures 6–21 and 6–22.)

IRQ INTERRUPT SERVICE ROUTINE—EVBU

(See program list accompanying Figures 6–21 and 6–22.)

DELAY SUBROUTINE—EVBU

(See program list accompanying Figures 6–21 and 6–22.)

MAIN PROGRAM—EVB

$ADDRESS	SOURCE CODE	COMMENTS
C000	LDD #C050	Initialize IRQ pseudovector
C003	STD EF	" " "
C005	LDS #C0FF	Initialize stack pointer
C008	LDX #1000	Initialize X register
C00B	LDAA #39	Initialize DDRD
C00D	STAA 9,X	" "
C00F	LDAA #5D	Initialize SPCR
C011	STAA 28,X	" "
C013	CLI	Clear I mask
C014	WAI	Wait for interrupt
C015	SEI	Set I mask
C016	WAI	Halt

IRQ INTERRUPT SERVICE ROUTINE—EVB

$ADDRESS	SOURCE CODE	COMMENTS
C050	JSR C090	Call DELAY subroutine
C053	LDAA #1	Write binary one to PD0

$ADDRESS	SOURCE CODE	COMMENTS
C055	STAA 8,X	" " "
C057	LDAA 0	Load MOSI data byte
C059	STAA 2A,X	Write MOSI byte to SPDR
C05B	LDD 29,X	Read SPSR/SPDR
C05D	BPL C05B	If SPIF = 0, read SPSR/SPDR again
C05F	STAB 0	If SPIF = 1, store MISO byte
C061	CLRA	Write binary zero to PD0
C062	STAA 8,X	" " "
C064	RTI	Return

DELAY SUBROUTINE—EVB

$ADDRESS	SOURCE CODE	COMMENTS
C090	LDAA #A	Initialize loop counter
C092	LDY #D600	Load count-down value
C096	DEY	Decrement Y register
C098	BNE C096	If Y register ≠ 0, decrement again
C09A	DECA	If Y register = 0, decrement loop counter
C09B	BNE C092	If loop counter ≠ 0, count down again
C09D	RTS	Return

6.4 INTERFACE AN SPI MASTER WITH AN SPI SLAVE

This section presents a simple interface between two M68HC11s—one a master and the other a slave. Refer to Figure 6–23, which illustrates interface connections.

The interface has the following characteristics:

- An SN74LS241 octal, three-state driver buffers the MOSI, MISO, and SCK lines between master and slave.
- Configure the master's $\overline{SS}$ pin as a general-purpose output. Leave this pin unconnected.
- The interface operator initiates a master/slave data exchange by pressing a push-button. The low from the push-button pull-up circuit generates an IRQ interrupt at the master.
- An IRQ interrupt service routine causes the master's CPU to write an MOSI data word to its SPDR, starting the data exchange.
- The master's CPU controls its part of the data exchange using status polling. After completion of an 8-bit data exchange, the master stores the byte received from the slave. The master CPU then returns from the IRQ service routine and displays "EXG" on its terminal display. "EXG" indicates completion of the data exchange.
- Tie the slave's $\overline{SS}$ pin low. Tying the $\overline{SS}$ pin low causes write collision errors *if* the slave's CPHA bit is reset (see the previous section). Recall that the slave must not write a new MISO byte to its SPDR until the current exchange is completed, or a write collision error results. If the slave's CPHA bit = 0, a data exchange does not end (for write collision purposes) until the slave's $\overline{SS}$ pin is returned high. However, if the slave's CPHA bit = 1, the slave's data exchange begins with the SCK's first active edge and ends when the slave's SPIF sets. A write collision will not occur as long as the slave's CPU waits until its SPIF bit has set—before writing a new data byte to its SPDR. Therefore, initialize the slave's CPHA bit to binary one.

FIGURE 6–23

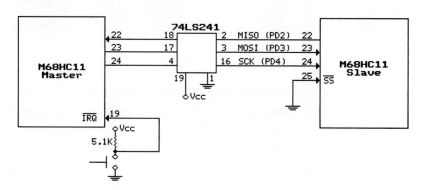

- The slave's CPU uses status polling to determine when its SPIF has set. The slave then stores the data byte received from the master and displays "EXG" on its terminal display. "EXG" indicates data exchange completion.

M68HC11 Master Software

Figures 6–24 and 6–25, along with their accompanying program list, illustrate supporting software for the master. Refer to Figure 6-24, blocks 1 through 5, and Master–MAIN program addresses $C000–$C012. The master uses an IRQ service routine to initiate and control its data exchange with the slave. Therefore, the master CPU must initialize its IRQ pseudovector. In block 4, the master CPU initializes its MOSI, $\overline{SS}$,

FIGURE 6-24

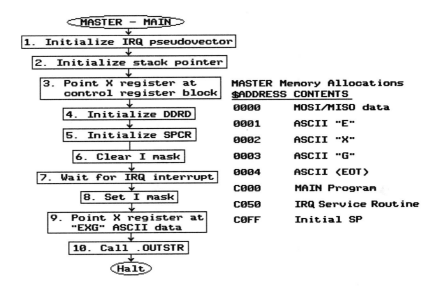

Master–Main Program

$ADDRESS	SOURCE CODE	COMMENTS
C000	LDD #C050	Initialize IRQ pseudovector
C003	STD EF	" " "
C005	LDS #C0FF	Initialize stack pointer
C008	LDX #1000	Initialize X register
C00B	LDAA #38	Initialize DDRD
C00D	STAA 9,X	" "
C00F	LDAA #55	Initialize SPCR
C011	STAA 28,X	" "
C013	CLI	Clear I mask
C014	WAI	Wait for interrupt
C015	SEI	Set I mask
C016	LDX #1	Point X register at "EXG" ASCII data
C019	JSR FFC7	Call .OUTSTR utility subroutine
C01C	WAI	Halt

Master–EXG" ASCII Data

$ADDRESS	SOURCE CODE	COMMENTS
0001	45	"E"
0002	58	"X"
0003	47	"G"
0004	04	<EOT>

FIGURE 6-25

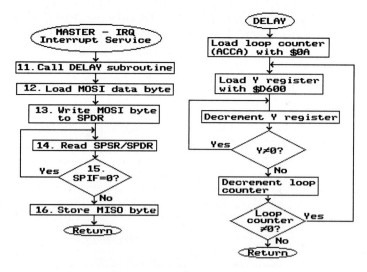

Master–IRQ Interrupt Service Routine

$ADDRESS	SOURCE CODE	COMMENTS
C050	JSR C090	Call DELAY subroutine
C053	LDAA 0	Load MOSI data byte
C055	STAA 2A,X	Write MOSI byte to SPDR
C057	LDD 29,X	Read SPSR/SPDR
C059	BPL C057	If SPIF = 0, read SPSR/SPDR again
C05B	STAB 0	If SPIF = 1, store MISO byte
C05D	RTI	Return

DELAY Subroutine

$ADDRESS	SOURCE CODE	COMMENTS
C090	LDAA #A	Initialize loop counter
C092	LDY #D600	Load Y register with count-down value
C096	DEY	Decrement Y register
C098	BNE C096	If Y register ≠ 0, decrement Y register again
C09A	DECA	If Y register = 0, decrement loop counter
C09B	BNE C092	If ACCA ≠ 0, count down again
C09D	RTS	If ACCA = 0, return

and SCK pins as outputs. Configure the master's $\overline{SS}$ pin as an output, and leave it unconnected. This disables the mode fault mechanism. In block 5, the master CPU configures its SPI as follows:

> SPI interrupts disabled
> SPI enabled
> PORTD outputs totem-pole
> SPI is a master
> SCK idles low
> Data exchange begins with SCK's initial edge (CPHA = 1)
> Baud = 500K

Refer to blocks 6 through 8 of Figure 6–24, and Master–MAIN program addresses $C013–$C015. After clearing its I mask (block 6), the master CPU waits for an IRQ interrupt. The interface operator initiates a data exchange by pressing the push-button, generating this IRQ interrupt. After exchanging data with the slave, the master returns from the IRQ interrupt and sets the I mask (block 8).

Refer to blocks 9 and 10 of Figure 6–24, and Master-MAIN program addresses $C016–$C01C. Before halting, the master CPU points its X register at an ASCII character list and calls the .OUTSTR utility subroutine. Recall that this subroutine displays ASCII characters from consecutive memory locations—pointed at by the X register. .OUTSTR stops displaying characters and returns when it encounters the ASCII EOT (End of Transmission) character.

Refer to block 11 of Figure 6–25, and Master–IRQ Interrupt Service routine address $C050. An overlap between the operator's IRQ push-button press and an attempt by the SPI to exchange data causes an SPI malfunction. Block 11 calls the DELAY subroutine. This subroutine gives the operator time to release the push-button and avoid the overlap.

Refer to blocks 12 through 16, and Master–IRQ Interrupt Service routine addresses $C053–$C05D. The master CPU causes the data exchange to begin by writing an MOSI data byte to its SPDR. The master CPU then goes into a polling loop where it repeatedly reads SPSR/SPDR (block 14) and tests the SPIF bit (block 15), until the SPIF sets. Of course, SPIF = 1 indicates data exchange completion. When the SPIF sets, the master CPU escapes the polling loop and stores the received MISO byte from ACCB to RAM address $0000 (block 16). The CPU then returns from the IRQ interrupt.

M68HC11 Slave Software

Figure 6–26 and its accompanying program list illustrate the supporting software for the slave. Refer to Figure 6–26, blocks 1 through 3 and Slave–MAIN program addresses $C000–$C00A. In block 2, the slave CPU initializes its MISO pin as an output, and all other

FIGURE 6–26

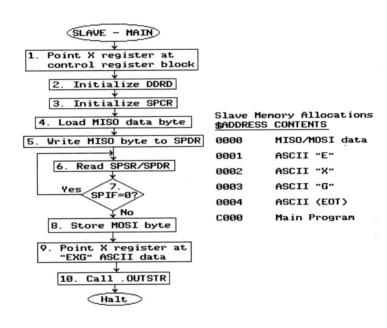

Slave–MAIN Program

$ADDRESS	SOURCE CODE	COMMENTS
C000	LDX #1000	Initialize X register
C003	LDAA #4	Initialize DDRD
C005	STAA 9,X	" "
C007	LDAA #45	Initialize SPCR
C009	STAA 28,X	" "
C00B	LDAA 0	Load MISO data byte
C00D	STAA 2A,X	Write MISO byte to SPDR
C00F	LDD 29,X	Read SPSR/SPDR
C011	BPL C00F	If SPIF = 0, read SPSR/SPDR again
C013	STAB 0	If SPIF = 1, store MOSI byte
C015	LDX #1	Point X register at "EXG" ASCII data
C018	JSR FFC7	Call .OUTSTR
C01B	WAI	Halt

Slave–"EXG" ASCII Data

$ADDRESS	$CONTENTS	CHARACTER
0001	45	"E"
0002	58	"X"
0003	47	"G"
0004	04	<EOT>

289

SPI pins as inputs. The slave's $\overline{SS}$ pin is tied low. When its $\overline{SS}$ pin is tied low, the slave CPU must initialize its CPHA to binary one in order to avoid write collision errors. In block 3, the slave CPU configures its SPI as follows:

SPI interrupts disabled
SPI enabled
PORTD outputs totem-pole
SPI is a slave
SCK idles low
Data exchange begins with SCK's initial edge (CPHA = 1)

Refer to Figure 6–26, blocks 4 through 8, and Slave–MAIN program addresses $C00B–$C014. In blocks 4 and 5, the slave CPU prepares for the data exchange by writing a MISO byte to its SPDR. The slave CPU must complete this write before the master initiates the data exchange. After writing the MISO byte, the slave CPU goes into a polling loop where it repeatedly reads SPSR/SPDR, and tests the SPIF bit (block 7) until the SPIF sets. When the SPIF sets, the slave CPU escapes the polling loop and stores the received MOSI byte from ACCB to RAM address $0000 (block 8).

Refer to Figure 6–26, blocks 9 and 10, and Slave–MAIN program addresses $C015–$C01B. After completing its part of the data exchange, the slave CPU displays "EXG" on its terminal display. The slave then halts.

EXERCISE 6.4   Interface an SPI Master with an SPI Slave

REQUIRED MATERIALS: 1 SN74LS241N Octal 3-State Driver
1 Resistor, 5.1 KΩ
1 Switch, push-button, push-to-close

REQUIRED EQUIPMENT: 2 Motorola M68HC11EVBs or EVBUs, and 2 terminals

GENERAL INSTRUCTIONS: Interface two M68HC11EVBs or EVBUs. One EVB (EVBU) functions as an SPI master and the other as an SPI slave. Connect a pull-up circuit to the $\overline{IRQ}$ pin of the master so that pressing the push-button generates an IRQ interrupt. The interrupt service routine initiates an SPI data transfer between master and slave.

PART I: Connect the Interface

Power down the EVBs (EVBUs) and terminals. Refer to Figure 6–23 and connect the interface. After double-checking all connections, show your interface to the instructor.

_____ *Instructor Check*

PART II: Software Development

Refer to Figures 6–24 and 6–25, along with their accompanying program lists. If you interface EVBs, load the program as listed. If you interface EVBUs, allocate memory as follows:

MASTER EVBU

$ADDRESS	CONTENTS
0000	MOSI or MISO data byte
0001–0004	ASCII characters
0100	MAIN program
0150	IRQ Interrupt Service routine
0190	DELAY subroutine
01FF	Initial stack pointer

SLAVE EVBU

$ADDRESS	CONTENTS
0000	MISO or MOSI data byte
0001–0004	ASCII characters
0100	MAIN program

If you interface an EVBU, modify listed software to match the above memory allocations. Modify these instructions:

    IRQ pseudovector initialization
    Stack pointer initialization
    Branch instruction destination addresses
    DELAY subroutine call

PART III: Interface Debug and Demonstration

Power up the EVBs (EVBUs), terminals, and interface. Load master and slave with support software. Debug the interface and software. When the interface functions properly, demonstrate it to the instructor.

                                                  _____ *Instructor Check*

PART IV: Laboratory Analysis

Answer the questions below by neatly entering your answers in the appropriate blanks.

1. Explain the SN74LS241A's purpose in the interface.

2. The interface operator initiates a data exchange by _____. This action generates an _____ at the master. The master's CPU controls the data exchange by using _____.

3. Explain why the master calls the DELAY subroutine.

4. Describe how the master indicates that it completed its part of the data exchange.

5. Explain why the slave's CPU initializes its CPHA bit to binary one.

6. The slave's CPU controls its part of the data exchange by using _____.

7. Describe how the slave indicates that it completed its part of the data exchange.

8. Both master and slave initialize their CPOL bits to binary _____ and CPHA bits to binary _____.

9. Empirically evaluate interface operation for all four CPOL/CPHA combinations. Record evaluation data in the chart below.

SPCR $CONTENTS				INTERFACE	
MASTER	SLAVE	CPOL	CPHA	RELIABLE	UNRELIABLE
_____	_____	0	0	_____	_____
_____	_____	0	1	_____	_____
_____	_____	1	0	_____	_____
_____	_____	1	1	_____	_____

10. Analyze the results recorded in the chart above. What factor made the interface unreliable? _____ Explain why this factor caused unreliable operation.

11. The laboratory software initializes the master's SPCR for 500K baud. Empirically evaluate interface operation for all four SCK baud rates. Record evaluation data in the chart below. **Note:** Initialize CPOL to binary zero and CPHA to binary one for all evaluations.

SPCR $CONTENTS			INTERFACE	
MASTER	SLAVE	BAUD	RELIABLE	UNRELIABLE
_____	_____	1M	_____	_____
_____	_____	500K	_____	_____
_____	_____	125K	_____	_____
_____	_____	62.5K	_____	_____

PART V: Laboratory Software Listing

MASTER–MAIN PROGRAM—EVB

(See program list accompanying Figures 6–24 and 6–25.)

MASTER–IRQ INTERRUPT SERVICE ROUTINE—EVB

(See program list accompanying Figures 6–24 and 6–25.)

DELAY SUBROUTINE—EVB

(See program list accompanying Figures 6–24 and 6–25.)

SLAVE–MAIN PROGRAM—EVB

(See program list accompanying Figure 6–26.)

MASTER–MAIN PROGRAM—EVBU

$ADDRESS	SOURCE CODE	COMMENTS
0100	LDD #150	Initialize IRQ pseudovector
0103	STD EF	" "
0105	LDS #1FF	Initialize stack pointer
0108	LDX #1000	Initialize X register
010B	LDAA #38	Initialize DDRD
010D	STAA 9,X	" "
010F	LDAA #55	Initialize SPCR
0111	STAA 28,X	" "
0113	CLI	Clear I mask
0114	WAI	Wait for IRQ interrupt
0115	SEI	Set I mask
0116	LDX #1	Point X register at "EXG" ASCII data
0119	JSR FFC7	Call .OUTSTR
011C	WAI	Halt

MASTER–IRQ INTERRUPT SERVICE ROUTINE—EVBU

$ADDRESS	SOURCE CODE	COMMENTS
0150	JSR 190	Call DELAY subroutine
0153	LDAA 0	Load MOSI data byte
0155	STAA 2A,X	Write MOSI byte to SPDR
0157	LDD 29,X	Read SPSR/SPDR
0159	BPL 157	If SPIF = 0, read SPSR/SPDR again
015B	STAB 0	If SPIF = 1, store MISO byte
015D	RTI	Return

DELAY SUBROUTINE—EVBU

$ADDRESS	SOURCE CODE	COMMENTS
0190	LDAA #A	Initialize loop counter
0192	LDY #D600	Load Y register with count-down value

$ADDRESS	SOURCE CODE	COMMENTS
0196	DEY	Decrement Y register
0198	BNE 196	If Y register ≠ 0, decrement again
019A	DECA	If Y register = 0, decrement loop counter
019B	BNE 192	If ACCA ≠ 0, count down again
019D	RTS	If ACCA = 0, return

SLAVE—MAIN PROGRAM—EVBU

$ADDRESS	SOURCE CODE	COMMENTS
0100	LDX #1000	Initialize X register
0103	LDAA #4	Initialize DDRD
0105	STAA 9,X	" "
0107	LDAA #45	Initialize SPCR
0109	STAA 28,X	" "
010B	LDAA 0	Load MISO data byte
010D	STAA 2A,X	Write MISO byte to SPDR
010F	LDD 29,X	Read SPSR/SPDR
0111	BPL 112	If SPIF = 0, read SPSR/SPDR again
0113	STAB 0	If SPIF = 1, store MOSI byte
0115	LDX #1	Point X register at "EXG" ASCII data
0118	JSR FFC7	Call .OUTSTR
011B	WAI	Halt

"EXG" ASCII DATA—MASTER AND SLAVE—EVB AND EVBU

$ADDRESS	$CONTENTS	CHARACTER
0001	45	"E"
0002	58	"X"
0003	47	"G"
0004	04	<EOT>

6.5 CHAPTER SUMMARY

The Serial Peripheral Interface (SPI) differs from the Serial Communications Interface (SCI) in several important ways:

- The SPI provides synchronous serial communications, where the SCI communicates asynchronously.
- SPI-based communication systems use a system master and one or more slaves. The master controls when an SPI data exchange starts, when it stops, and the baud rate at which the exchange occurs.
- Because SPI communications are synchronous, they require no framing bits.
- An SPI-based communications system can exchange data faster and more efficiently than an SCI system.

Master and slave M68HC11s exchange serial data simultaneously over full-duplex data communications lines. The MOSI (Master Out Slave In) line transfers data from master to slave. The MISO (Master In Slave Out) line transfers data from slave to master. These full-duplex lines originate and terminate at the master's and slave's SPDRs (SPI Data Registers).

The master initiates a data exchange by writing an MOSI byte to its SPDR. This write causes the synchronizing serial clock (SCK) to start. The SCK causes the master's SPDR to shift MOSI data bits to the slave's SPDR over the MISO line. The SCK also causes the slave's SPDR to shift MISO data bits to the master's SPDR over the MISO data line.

The SPDR actually consists of two registers: a shift register and a read data buffer. MOSI and MISO serial data shift simultaneously in and out of the master and slave shift registers. The master shifts MOSI data from its SPDR shift register to the slave's shift register. At the same time, the slave shifts MISO data from its shift register to the master's shift register. After 8 MOSI bits and 8 MISO data bits have exchanged, both master and slave automatically transfer the exchanged data from their respective SPDR shift registers to their SPDR read data buffers.

When the CPU writes to its SPDR, the write data transfers from an accumulator to the SPDR shift register. When the CPU reads from the SPDR, the read data transfers from the read data buffer to the accumulator.

The M68HC11 configures its SPI by setting or resetting bits in the SPCR (SPI Control Register):

- SPCR's SPR1/SPR0, CPOL, and CPHA bits control SCK rate, polarity, and phase.
- SPCR's MSTR bit controls whether the M68HC11 functions as a master or as a slave.
- The DWOM bit converts PORTD output pins from the totem-pole to the open-drain configuration.
- The SPE bit enables or disables the SPI.
- The SPIE bit determines whether the CPU uses interrupts or status polling to control SPI operations.

The SPCR's CPOL and CPHA control bits determine exactly when the master/slave data exchange begins. CPOL and CPHA define the exchange starting point as follows:

CPOL	CPHA	Exchange Start Definition
0	0	Exchange begins when the master drives the slaves's $\overline{SS}$ pin from high to low.
0	1	Exchange begins with the first low-to-high SCK edge.
1	0	Exchange begins when the master drives the slave's $\overline{SS}$ pin from high to low.
1	1	Exchange begins with the first high-to-low SCK edge.

CPOL and CPHA also determine how bit shifts synchronize with the SCK signal:

CPOL	CPHA	Shift Point Definition
0	0	SCK toggles from high to low.
0	1	SCK toggles from low to high.
1	0	SCK toggles from low to high.
1	1	SCK toggles from high to low.

Finally, the CPOL and CPHA control bits determine when an SPI data exchange ends. Here, we define data exchange "end" as the point where the SPIF (SPI Transfer Complete Flag) sets. The CPOL/CPHA bits make this determination differently for master and slave. CPOL/CPHA determine the end of the master's data exchange as follows:

SPI MASTER

CPOL	CPHA	Data Exchange End (SPIF=1)
0	0	Final SCK edge (high to low).
0	1	End of the eighth SCK cycle, i.e., one half of the SCK's period (T) after the final SCK edge (high to low).
1	0	Final SCK edge (low to high).
1	1	End of the eighth SCK cycle, i.e., one half of the SCK's period after the final SCK edge (low to high).

CPOL/CPHA determine the end of the slave's data exchange as follows:

SPI SLAVE

CPOL	CPHA	Data Exchange End (SPIF = 1)
0	0	Second-to-last SCK edge (low to high).
0	1	Final SCK edge (high to low).
1	0	Second-to-last SCK edge (high to low).
1	1	Final SCK edge (low to high).

The SPI Status Register (SPSR) contains one status flag (the SPIF) and two error flags (WCOL and MODF). The SPI Transfer Complete Flag (SPIF) sets to indicate the end of the master/slave data exchange. If SPI interrupts have been enabled (SPIE = 1 in the SPCR), the SPIF also generates an SPI interrupt when it sets.

A master's data exchange ends with the end of the eighth SCK cycle. The master's SPIF sets during the first high E clock pulse, following the eighth SCK cycle.

A slave's data exchange ends either at the final SCK edge (CPHA = 1) or the second SCK edge from the last (CPHA = 0). In either case, the master generates the SCK, and the SCK is therefore asynchronous with the slave's E clock. Thus, an SPI slave will require one or two extra E clock cycles to respond to the end-of-exchange SCK edge and to set its SPIF. The asynchronous relationship between the master's SCK and slave's E clock causes this delay.

SPSR's WCOL flag informs an M68HC11 that it has committed a write collision error. A slave usually commits this error when it attempts a write to its SPDR while a data exchange is still in progress. When a write collision error occurs, the SPI aborts the SPDR write that caused the error and sets the WCOL flag. A slave is most prone to write collision errors when its CPHA control bit is reset. When CPHA = 0, the data exchange is not complete (for write collision purposes) until the master has returned the slave's $\overline{SS}$ pin from low to high. When CPHA = 0, the slave can avoid a write collision error by checking the state of its $\overline{SS}$ pin before it writes to the SPDR. The CPU clears WCOL by reading its SPSR (while WCOL = 1), then accessing its SPDR.

The SPSR's MODF (Mode Fault) mechanism detects data line contention in a multiple-master SPI communications system. Since serial data line contention can be catastrophic, the multi-master system designer must take great care to assure that only one master at a time has control of a common MOSI line.

To take advantage of the MODF error detection feature, multiple masters must have their $\overline{SS}$ pins interconnected and configured as inputs. When one of the multiple masters intends to take control of the common MOSI data line, it drives the other masters' $\overline{SS}$ pins low. An SPI master—with its $\overline{SS}$ pin configured as an input—calls a mode fault error when its $\overline{SS}$ pin is driven low. A master responds to a mode fault error by automatically:

1. Resetting the MSTR bit in its SPCR
2. Resetting the SPE bit in its SPCR
3. Configuring its PORTD pins PD2–PD5 as inputs
4. Setting the MODF bit in its SPSR
5. Generating an SPI interrupt if the master's SPIE = 1

Figures 6–11 through 6–15, along with accompanying state table and program list, illustrate a multi-master SPI communications system. This system takes advantage of the mode fault detection feature.

Mode fault detection will not protect against other kinds of data line contention, such as a master's inadvertent selection of more than one slave. In such a case, these slaves would contend for the common MISO line. Motorola suggests using series 1 KΩ to 10 KΩ resistances, or the DWOM option, to protect slave MISO output drivers under such circumstances (see Figures 6–16 and 6–17).

The Port D Data Direction Register (DDRD) configures SPI pins as inputs or outputs. The SPI automatically configures the proper SPI pins as inputs regardless of the states of counterpart DDRD bits. However, the CPU must set counterpart DDRD bits to configure SPI output pins. The exception to this rule is the SPI master's $\overline{SS}$ pin. The master's CPU must configure this pin as either input or output.

Section 6.3 describes a simple interface between an M68HC11 master and an SN74LS91N shift register. The SN74LS91N functions as a slave. This interface requires a D flip-flop interposed between slave and master on the MISO data line. The interface requires this flip-flop because only seven SCK edges will drive the MISO line with all 8 serial bits from the slave. The master's SPI automatically clocks the slave shift register eight times. Without the D flip-flop, the MS bit from the slave would shift through the master's SPI shift register in seven SCK edges and be lost on the eighth edge.

The M68HC11–SN74LS91N interface includes these other important features:

- The master enables serial data inputs to the slave by driving the slave's A input high. When it drives this A input high, the master can clock MOSI data into the slave via the slave's B input.
- The interface operator initiates a master/slave data exchange by pressing a push-button, generating an IRQ interrupt at the master. The IRQ service routine causes the master to start the SPI data exchange.
- The master's CPU uses status polling to control the data exchange.

Section 6.4 describes an interface between two M68HC11s—SPI master and SPI slave. This interface has the following characteristics:

- An SN74LS241 three-state driver buffers the MOSI, MISO, and SCK lines between master and slave.
- The interface operator initiates a master/slave data exchange by pressing a push-button, generating an IRQ interrupt at the master. The IRQ service routine causes the master to start the data exchange.
- The master's CPU uses status polling to control its part of the data exchange. The master writes the message "EXG" to its terminal display to indicate that it has completed its part of the SPI data exchange.
- The slave's $\overline{SS}$ pin is tied low. Therefore, the slave's CPHA bit must be initialized to binary one to avoid write collision errors.
- The slave's CPU uses status polling to determine when its SPIF sets, indicating the end of the data exchange. The slave then writes the message "EXG" to its terminal display to indicate that its part of the SPI data exchange is complete.

EXERCISE 6.5

Chapter Review

Answer the questions below by entering your responses in the appropriate blanks.

1. List the four most important ways in which SPI communications differ from SCI communications.

    (1)

    (2)

    (3)

    (4)

2. A minimum of three lines must be used to connect an M68HC11 master with a slave. These lines are called the _____, _____, and _____.
3. The SPI master writes to its _____ to start a data exchange. This write causes the _____ to start. Bits shift from master to slave over the _____ line, and from slave to master over the _____ line.
4. After 8 bits have exchanged between master and slave, the master's _____ automatically stops, ending the _____.
5. Name the two registers which constitute the SPDR.

    (1)

    (2)

6. When the CPU writes to its SPDR, the write data transfers from the accumulator to the SPDR's _____. When the CPU reads its SPDR, the read data transfers from the _____ to the accumulator.

7. After eight SCK cycles, the SPI transfers the received serial data from its SPDR _____ to its SPDR _____.
8. The SPDR _____-buffers transmit (write) data and _____-buffers receive (read) data.
9. An 8MHz crystal connects to an M68HC11 master's XTAL and EXTAL pins. SPR1 = 1 and SPR0 = 0. The SPI baud rate is _____.
10. An 8MHz crystal connects to an M68HC11 slave's XTAL and EXTAL pins. SPR1 and SPR0 are both reset. The SPI baud rate is _____.
11. If all master and slave CPOL/CPHA control bits are reset, the SPI data exchange begins when _____.
12. If all master and slave CPOL/CPHA control bits are set, the SPI data exchange begins when _____.
13. If both master and slave have their CPOL bits set and their CPHA bits reset, serial data bits shift onto the MOSI and MISO data lines when the SCK toggles from _____.
14. If both master and slave have their CPOL bits reset and their CPHA control bits set, serial data bits shift onto the MOSI and MISO data lines when the SCK toggles from _____.
15. A master's CPOL and CPHA control bits are both set. When does its SPIF set?

16. A slave's CPOL and CPHA control bits are both set. When does the slave's SPIF set?

17. A CPU writes $A9 to its SPCR. Describe how this write configures the SPI.
    Baud = _____.
    Exchange begins when _____.
    SCK is active _____.
    SPI is a _____ (master/slave).
    SPI outputs are in the _____ configuration.
    SPI is _____ (enabled/disabled).
    CPU will use _____ (status polling/interrupt control).
18. A CPU initializes its SPCR to $5C. Describe how this write configures the SPI.
    Baud = _____.
    Exchange ends when _____.
    SCK idles _____.
    SPI is a _____ (master/slave).
    SPI outputs are in the _____ configuration.
    SPI is _____ (enabled/disabled).
    CPU will use _____ (status polling/interrupt control).
19. List the SPSR flags and their SPSR bit numbers.

20. The SPIF sets to indicate _____. The master's SPIF sets during the first _____, after the _____ SCK cycle.
21. An SPI slave requires _____ extra _____ to respond to the end-of-exchange SCK edge and set its SPIF. The _____ relationship between the master's _____ and the slave's _____ causes this delay.

$ADDRESS	SOURCE CODE
C115	LDAA 8,X
C117	BITA #20
C119	BEQ C115
C11B	LDAA A
C11D	STAA 2A,X
C11F	---------------

Notes:
(1) X Register contents = $1000
(2) Address $000A holds MISO data
(3) CPHA = 0

22. Study to the program segment above. Explain what this program segment does, and its purpose.

23. The M68HC11's mode fault mechanism detects _____ between multiple masters. To take advantage of this MODF error detection feature, the multiple masters have their _____ pins interconnected and configured as _____. When one master intends to take control of the common _____ data line, it drives the other masters' _____ low.

24. List the actions automatically taken by an M68HC11 master when its $\overline{SS}$ pin is driven low. Assume that this $\overline{SS}$ pin is configured as an input.

    (1)

    (2)

    (3)

    (4)

    (5)

25. Motorola suggests two possible ways to prevent catastrophic MISO data line contention between two or more SPI slaves. List these two ways.

    (1)

    (2)

26. The SPI automatically configures the proper pins (e.g., the slave's MOSI, SCK, and $\overline{SS}$ pins) as _____. However, both master and slave must configure appropriate SPI pins as outputs by _____.

27. What is the single exception to the rule stated in question 26?

28. Refer to Figure 6–20. Explain why this interface requires the SN74LS74A flip-flop.

29. Refer to Figure 6–20. How does the interface operator cause a master/slave data exchange to begin? _____ List the responses generated by this action.

30. Refer to Figure 6–22. Explain why the master calls the DELAY subroutine.

31. Refer to Figure 6–23. Explain the SN74LS241N's purpose in the interface.

32. Refer to Figure 6–23. Explain why the master's $\overline{SS}$ pin is unconnected.

33. Refer to Figure 6–23. How does the interface operator cause a master/slave data exchange to begin? _____. What response does the operator's action generate?

34. Refer to Figures 6–24 and 6–25. The master's CPU uses _____ to control its part of the SPI data exchange.

35. Refer to Figures 6–24 and 6–25. How does the master indicate that it has completed its part of the SPI data exchange?

36. Refer to Figure 6–25. Explain why the master calls the DELAY subroutine.

37. Refer to Figure 6–23. The slave's $\overline{SS}$ pin connects to _____. Given this connection, the slave must initialize its CPHA control bit to binary _____, in order to avoid _____.

38. Refer to Figure 6–26. The slave's CPU uses _____ to determine when the _____ sets, indicating that its part of the data exchange is complete.

39. Refer to Figure 6–26. How does the slave indicate that it has completed its part of the SPI data exchange?

$ADDRESS	SOURCE CODE
010A	LDD 29,X
010C	BPL 10A
010E	STAB 0
0110	--------------

40. Refer to the program segment above. The instruction at address $010A reads the _____ register to ACCA, and the _____ register to ACCB.
41. In the program segment above, the CPU uses _____ to control its part of the data exchange.
42. Refer to the program segment above. The instruction at address $010E causes the slave's CPU to store the _____ (MOSI/MISO) byte to RAM. This instruction would cause the master to store the _____ byte to RAM.

299

7 M68HC11 Free-Running Counter and Input Captures

7.1 GOAL AND OBJECTIVES

Chapters 7, 8, and 9 present M68HC11 timer functions, including input captures, output compares, and real-time (periodic) interrupts. Timer functions allow the M68HC11 to execute these tasks:

- Record the "times" at which up to four peripherals toggle timer input pins.[1]
- Use these recorded "times" to measure the pulse width, period, and frequency of an input square wave.
- Toggle up to five timer output pins at specified "times."
- Simultaneously generate up to four output square waves—each with a unique duty cycle and frequency.
- Generate interrupts at any of four periodic rates.

This chapter concentrates on two aspects of M68HC11 timer functions. First, it provides a detailed description of the timer section's heart—the free-running counter. Second, it describes the timer's input capture function and some of its applications.

Chapter 7's goal is to help the reader understand how to program and interface the M68HC11's free-running counter and input capture mechanisms.

After studying Chapter 7 and completing its exercises, you will be able to:

Develop software which uses the free-running counter to generate a specified time-out.

Develop software and an interface which uses the input captures to measure the time between external events.

Develop software and interfaces which use input capture pins for general purpose hardware interrupts.

7.2 THE M68HC11'S FREE-RUNNING COUNTER

The free-running counter is the heart of the M68HC11's timing section. The "time" of an input capture, output compare, or real time interrupt is actually a specific count.

---

[1]Some M68HC11 versions can track four input pins; some can track only three.

Upon an M68HC11 reset, the free-running counter begins counting at zero. For as long as the M68HC11 remains powered up, the free-running counter continues to count. When it reaches its terminal count, the free-running counter rolls the count over and starts counting at zero again. The only way to interrupt this counting is to reset or STOP the M68HC11. Otherwise, neither operator nor software can reset, modify, or interrupt the counting.

The free-running counter has a 16-bit capacity. It therefore counts from $0000 to $FFFF, then rolls the count over to $0000. Each time the free-running counter rolls its count over, the timer section sets a status flag. A user can program the M68HC11 so that it responds to this status flag by adding one to a count extension byte. In this manner, a programmer can extend the free-running counter's capacity to 24 bits. Using the D register, the programmer can easily extend the counter's capacity to 32 bits.

On-Chip Resources Used by the M68HC11 Free-Running Counter

The M68HC11 uses three registers to configure and operate the free-running counter:

1. The M68HC11's CPU accesses the current free-running count by reading the Timer Counter Register (TCNT, control register block addresses $100E–$100F). Since the free-running counter consists of 16 bits, the TCNT register holds 16 bits. Use a double-byte read instruction, such as LDD, to access the current free-running count value. The free-running counter continues its counting while the CPU reads the TCNT register. If the CPU accesses the TCNT with 2 single-byte reads, the high-byte read may come from an earlier count than the low-byte read. On the other hand, a 2-byte read (e.g., LDD or LDX) causes a special buffer to latch the count's low-byte value while the high byte is read. By latching the count's low byte, the timer assures that the TCNT register's 2 bytes form a coherent count value.

   Using a double-byte TCNT read, the CPU accesses the count, as of this read's second from the last MPU cycle.[2] For example, the CPU requires five MPU cycles to execute the LDD,X instruction:

MPU Cycle	CPU Action
1	Fetch and decode opcode.
2	Read indexed mode offset.
3	Add offset and X register to form target address.
4	Read target address contents.
5	Read target address + 1 contents.

   At cycle 4 the CPU reads the TCNT register's high byte. At this time, the timer section's special buffer latches the counter's low-byte value and retains it in the TCNT, until the CPU reads it during MPU cycle 5.

2. Refer to Figure 7–1. The CPU establishes the free-running count rate by initializing 2 control bits in Miscellaneous Timer Interrupt Mask Register 2 (TMSK2, control register block address $1024). Motorola calls these bits Timer Prescaler Select bits PR1 and PR0. An M68HC11 reset initializes these bits to binary zeros. When PR1 and PR0 are both reset, the free-running counter increments at the E clock rate (2MHz for the EVBU or EVB). Remaining PR1/PR0 bit combinations cause the count rate to equal the E clock frequency divided by 4, 8, or 16. The following chart summarizes the relationship between PR1/PR0 and free-running count rate:

TMSK2		Free-Running Count Rate
PR1	PR0	E Clock ÷ by
0	0	1
0	1	4
1	0	8
1	1	16

---

[2]Motorola Inc., *M68HC11 Reference Manual* (1990), p. 10-5.

FIGURE 7–1
Reprinted with the permission of Motorola.

TMSK 2

B7							B0
TOI	RTII	PAOVI	PAII			PR1	PR0

Of course, count rate determines the free-running counter's resolution and range. Counter resolution is the period of time represented by each count. For example, when the free-running count increments at 2MHz, the free-running counter stays at each count for 500ns. Therefore, resolution = 500ns.

Define counter range as the period of time it takes the count to increment from its initial to terminal values ($0000–$FFFF). For example, if PR1 = 1 and PR0 = 0, the free-running count increments at an E clock ÷ 8 rate. It therefore requires $FFFF x 4µs for the free-running counter to increment from $0000 to $FFFF. The following chart summarizes the relationship between PR1/PR0 and free-running counter resolution and range:

TMSK2		Free-Running Counter (E Clock = 2MHz)	
PR1	PR0	Resolution	Range
0	0	500ns	32.77ms
0	1	2µs	131.1ms
1	0	4µs	262.1ms
1	1	8µs	524.3ms

The M68HC11 time protects PR1 and PR0. The CPU must initialize time protected bits within 64 E clock cycles after a reset.

TMSK2, bit 7 constitutes the Timer Overflow Interrupt Enable (TOI) control. By setting TOI, the M68HC11 enables interrupts which occur each time the free-running count rolls over from $FFFF to $0000. This Timer Overflow Interrupt has a unique vector address. The CPU uses TOI interrupt control to extend the free-running counter's bit capacity and range.

3. Refer to Figure 7–2. A bit in the Miscellaneous Timer Interrupt Flag Register 2 (TFLG2, control register block address $1025) flags free-running count rollover. Motorola calls this bit (bit 7) the Timer Overflow Flag (TOF). The M68HC11 can use TOF status polling to extend the free-running counter's bit capacity and range.

All TFLG2 flags are Read/Modify/Write bits. After responding to a TFLG2 flag, the CPU must write a one to this bit in order to clear it. At the same time, the CPU must avoid writing a one to any TFLG2 bits that should remain unchanged.

There are two effective ways to reset a TFLG2 bit without disturbing other bits in the register:

1. Write a mask byte to TFLG2. The mask byte contains ones corresponding to any flags to be changed. To clear the TOF flag (TFLG2, bit 7), the programmer uses this source code:

```
LDAA #80
STAA 25,X
```

Note: X register contents = $1000

2. Use the BCLR instruction. The mask accompanying this instruction uses zeros to specify bits to be changed and ones to specify bits to be left unchanged. To clear the TOF bit, the programmer uses this source code:

```
BCLR 25,X 7F
```

Note: X register contents = $1000

FIGURE 7–2
Reprinted with the permission of Motorola.

TFLG2

B7							B0
TOF	RTIF	PAOVF	PAIF				

FIGURE 7–3

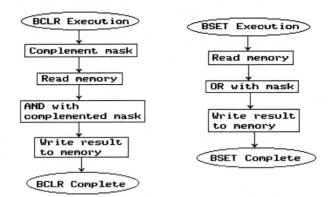

Avoid using the BSET instruction in an attempt to change read/modify/write bits. To understand why, refer to Figure 7–3. Assume that both TOF and PAOVF are set in the TFLG2 register. The CPU must reset the TOF bit only—leaving the PAOVF bit intact. If the CPU uses a `BCLR 25,X 7F` instruction, the following events occur:

1. Complement mask $7F to $80:

```
        %01111111
to      %10000000
```

2. Read TFLG2. AND this read data with the complemented mask:

```
            %10000000 (complemented mask)
AND with    %10100000 (TFLG2 read data)
result      %10000000
```

3. Write the ANDed result to TFLG2, clearing *only* the TOF bit.

If the CPU executes a `BSET 25,X 80` instruction, the following events occur:

1. Read TFLG2. OR this read data with the mask accompanying the BSET instruction:

```
            %10000000 (mask)
OR with     %10100000 (TFLG2 read data)
result      %10100000
```

2. Write the result to TFLG2, changing *both* the TOF and PAOVF flags from binary ones to binary zeros.

Thus, to clear TFLG2 flags, use either of these two methods:

1. Use extended or indexed mode to write a mask byte to TFLG2. The mask uses ones to specify the bits to be cleared, and zeros to specify the bits to be unchanged.
2. Use the BCLR instruction—with a mask that uses zeros to specify bits to be cleared, and ones to specify bits to remain unchanged.

Avoid the BSET instruction in an attempt to clear TFLG2's read/modify/write bits.

Extend the Free-Running Counter's Bit Capacity and Range

Figure 7–4 and its accompanying program list illustrate software that doubles the free-running counter's bit capacity. Doubling bit capacity multiplies the counter's range by $FFFF, extending it from 32.77ms to 35 minutes and 47.58 seconds.[3]

This software facilitates periodic task execution. For example, the M68HC11 could periodically capture and record temperatures from a number of remote sites. After counting for 35 minutes and 48 seconds, the CPU calls the Counter Service (CNTSVC) subroutine. This subroutine causes the M68HC11 to take and record temperature samples. After

---

[3] This figure assumes that the E clock rate = 2MHz, and PR1/PR0 are reset.

FIGURE 7-4

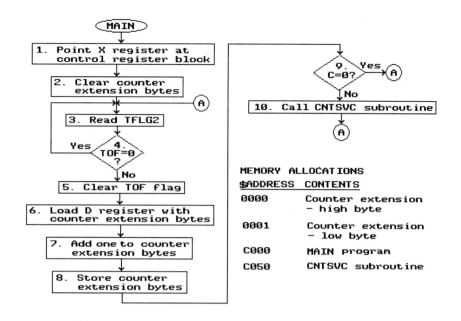

Free-Running Counter—2-Byte Extension

$ADDRESS	SOURCE CODE	COMMENTS
C000	LDX #1000	Initialize X register
C003	LDD #0	Clear counter extension high byte
C006	STD 0	Clear counter extension low byte
C008	LDAA 25,X	Read TFLG2
C00A	BPL C008	If TOF=0, read TFLG2 again
C00C	BCLR 25,X 7F	If TOF=1, clear TOF
C00F	LDD 0	Read counter extension word
C011	ADDD #1	Add one to counter extension bytes
C014	STD 0	Store counter extension word
C016	BCC C008	If C=0, no overflow; continue counting
C018	JSR C050	Call CNTSVC subroutine
C01B	BRA C008	Do another overflow cycle

returning from CNTSVC, the M68HC11 counts again for 35 minutes and 48 seconds, then re-calls CNTSVC. Thus, the M68HC11 captures temperature samples every 35 minutes, 48 seconds. Refer to blocks 1 and 2 of Figure 7-4, and program addresses $C000–$C007. Maintain free-running counter extension bytes in RAM locations $0000 and $0001. To make initial counter overflow as accurate as possible, the CPU initializes these extension bytes to $0000. View these extension bytes as a 16-bit counter extension word. The M68HC11 increments this word each time its free-running counter overflows.

Refer to blocks 3 through 5, and program addresses $C008–$C00E. The CPU polls its TFLG2 register to determine when the free-running counter overflows. This overflow causes the TOF flag (bit 7) to set in TFLG2. The CPU reads and re-reads TFLG2 until the TOF sets. When TOF = 1, the CPU escapes the polling loop. After escaping, the CPU uses a BCLR instruction to reset TOF.

Refer to blocks 6 through 8, and program addresses $C00F–$C015. The CPU reads the counter extension word from RAM and increments it. The CPU then writes this incremented extension word back to RAM.

Refer to blocks 9 and 10, and program addresses $C016–$C01C. Extension word increments (ADDD #1, block 7) eventually cause the extension word contents to roll over from $FFFF to $0000. This rollover causes the C flag to set. When the C flag sets, the free-running counter and its extension word have timed out 35 minutes and 47.58 seconds. If an extension word increment does not set the C flag, the CPU branches back from block 9 to block 3. If the C flag sets, complete 32-bit time-out has occurred, and the CPU calls the CNTSVC subroutine. This subroutine executes a task, such as temperature sampling. After returning from CNTSVC, the CPU branches back to block 3, to time-out another 35 minutes and 48 seconds.

Program the Free-Running Counter for a Specific Time-Out

Usually, a system designer must tailor free-running time-out to the task at hand. Assume, for example, that a user wants the M68HC11 to execute a subroutine (e.g., capture a number of temperature samples) every ten seconds. The designer makes some preliminary calculations and applies them to the software strategy illustrated in Figures 7–5 and 7–6.

A specific time-out strategy involves these steps:

1. Calculate the number of free-running time-outs (TOFs) in the specified period.
2. Determine how many additional free-running timer counts it takes to complete the specified period.
3. Develop software which calls a time-out service subroutine after the required number of TOFs and additional counts elapse.

Following paragraphs describe these procedures in more detail.

Step One: Calculate the number of TOFs in the specified time-out period. To find this figure, divide the desired time-out period by the free-running counter's range. For example, assume a desired time-out period of ten seconds. Also assume a free-running counter range of 32.77ms.

$$10s/32.77ms = 305.127 \text{ TOFs}$$

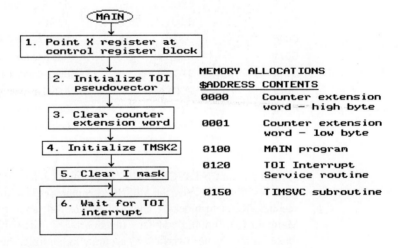

Ten-Second Time-Out—MAIN

$ADDRESS	SOURCE CODE	COMMENTS
0100	LDX #1000	Initialize X register
0103	LDD #120	Initialize TOI pseudovector
0106	STD D1	" " " " "
0108	LDD #0	Clear counter extension word
010B	STD 0	" " " " "
010D	LDAA #80	Initialize TMSK2
010F	STAA 24,X	" " "
0111	CLI	Clear I mask
0112	WAI	Wait for TOI interrupt
0113	BRA 112	Wait for interrupt again

FIGURE 7-6

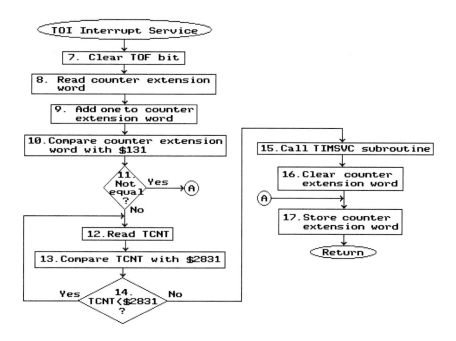

Ten-Second Time-Out—TOI Interrupt Service

$ADDRESS	SOURCE CODE	COMMENTS
0120	BCLR 25,X 7F	Clear TOF bit
0123	LDD 0	Load D register with counter extension word
0125	ADDD #1	Add one to counter extension word
0128	CPD #131	Compare counter extension word with $131
012C	BNE 13C	If not equal, store incremented counter extension word
012E	LDD E,X	If equal, read TCNT register
0130	CPD #2831	Compare TCNT with $283C
0134	BCS 12E	If C=1, read TCNT again
0136	JSR 150	If C=0, call TIMSVC subroutine
0139	LDD #0	Clear counter extension word
013C	STD 0	Store counter extension word
013E	RTI	Return

Step Two: Determine the number of additional free-running timer counts required to complete the specified period. To find this figure, multiply the fractional part of the previous step's result by the number of counts in the free-running counter's range:

$$.157 \times 65536 = 10289 \text{ counts}$$

Thus, the software should count 305 ($131) TOFs, then count an additional 10289 ($2831) free-running counts before calling the time-out service subroutine.

Figures 7-5 and 7-6, along with their accompanying program lists, illustrate the time-out software. This software has the following features:

- The software uses Timer Overflow Interrupts (TOIs). When the TOF sets, it generates a TOI interrupt. To enable these interrupts, the M68HC11 sets the TOI (Timer Overflow Interrupt Enable) control bit in TMSK2 and clears the global I mask.
- The TOI Interrupt Service routine increments a free-running counter extension word. This extension word records the number of times the TOF has set. The TOI Interrupt Service routine then compares the incremented extension word with the terminal number of TOFs ($131). A mismatch causes the CPU to return from the interrupt.

- If the extension word matches the terminal TOF count ($131), the CPU reads its TCNT register. The CPU reads and re-reads TCNT until its contents equal or exceed the number of additional counts required for specified time-out ($2831). The CPU then calls the Time-Out Service (TIMSVC) subroutine. After returning from this subroutine, the CPU clears the counter extension word and returns from the interrupt.
- Upon returning from the TOI Interrupt Service routine, the M68HC11 starts the time-out counting process again.

Refer to Figure 7–5, blocks 1 through 3, and MAIN program addresses $0100–$010C. In block 2, the CPU initializes the TOI pseudovector at addresses $00D1 and $00D2. In block 3, the CPU clears a 2-byte counter extension word at locations $0000 and $0001. This extension word uses 2 bytes to accommodate its $0131 terminal value.

Refer to blocks 4 through 6, and program addresses $010D–$0114. In block 4, the CPU enables TOI interrupts by writing a binary one to the TOI control bit. In block 5, the CPU clears the global I mask. The CPU then waits for a TOI interrupt. The CPU always returns from TOI interrupts to this wait state.

Refer to Figure 7–6, block 7, and TOI Interrupt Service program address $0120. Recall that the TOF flag is a read/modify/write bit. Therefore, the CPU uses the BCLR instruction to clear this flag.

Refer to blocks 8 and 9, and program addresses $0123–$0127. In block 8, the CPU reads the free-running counter extension word from RAM locations $0000–$0001. In block 9, the CPU increments the counter extension byte.

Refer to blocks 10 and 11, and program addresses $0128–$012D. In block 10, the CPU compares the incremented counter extension word with its final value: $0131. If the extension word has not yet incremented to this terminal value, the CPU takes the "yes" branch from block 11 to block 17. If the free-running counter extension word has incremented to $0131, the CPU takes the "no" path from block 11 to block 12. At block 12, the CPU begins monitoring the base free-running counter's progress toward its terminal count: $2831.

Refer to blocks 12 through 15, and program addresses $012E–$0138. The CPU monitors the free-running counter's progress by continuously reading TCNT and comparing its contents with the final free-running count: $2831. When TCNT finally equals or exceeds the comparison value, ten seconds have elapsed. The CPU escapes the TCNT read/compare loop and calls the TIMSVC subroutine (block 15). TIMSVC executes the periodic task.

Refer to blocks 16 and 17, and program addresses $0139–$013D. Upon returning from TIMSVC, the CPU clears the counter extension word to prepare it for another ten-second counting cycle. Note that if the block 10 comparison resulted in a mismatch, the CPU branched to block 17, where it stored the incremented free-running counter extension byte. After executing block 17, the CPU returns to the MAIN program to wait for another TOI interrupt.

EXERCISE 7.1 Fifteen-Second Time-Out and Display

REQUIRED EQUIPMENT: Motorola M68HC11EVB (EVBU), and terminal

GENERAL INSTRUCTIONS: Develop and demonstrate software that uses the M68HC11's free-running counter to time out fifteen-second intervals. A fifteen-second time-out entails a free-running counter extension word and software strategy as explained in the previous section. Carefully execute the following instructions.

PART I: Make Preliminary Calculations

Calculate the number of TOFs in a fifteen-second interval. Round the answer to six significant figures and enter in the blank:

#TOFs = _____

FIGURE 7-7

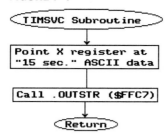

Fiffteen-Second Time-Out, Memory Allocations

EVBU $ADDRESS	EVB $ADDRESS	CONTENTS
0000	0000	Counter extension word—high byte
0001	0001	" " " " —low byte
0002	0002	"15<SP>sec.<EOT>" ASCII data
0100	C000	MAIN program
0120	C020	TOI Interrupt Service routine
0150	C050	TIMSVC subroutine

Calculate the number of additional free-running counts required to complete the fifteen-second time-out. Round the answer to five significant figures and enter in the blank:

Additional free-running counts = _____

Convert the "# TOFs" figure to three significant figures without rounding. Convert this integer to hexadecimal. Enter your hex result in the blank:

TOFs = $ _____

Convert the "additional free-running counts" figure to hexadecimal. Enter the hex result in the blank.

Additional free-running counts = $ _____

PART II: Software Development

Refer to the previous section and develop the fifteen-second time-out software. Allocate memory as shown in the chart above.

Refer to Figure 7-7. The Time-Out Service (TIMSVC) subroutine displays the message "15 sec."

PART III: Software Debug and Demonstration

Power up the EVB (EVBU) and terminal. Load and debug the software. When sure that the software runs properly, demonstrate its operation to the instructor.

_____ *Instructor Check*

PART IV: Laboratory Analysis

Answer the questions below by neatly entering your answers in the appropriate blanks.

1. Enter the appropriate hexadecimal values for specified time-outs listed on the next page. Determine and enter the minimum number of free-running counter extension bytes required for each specified time-out. Assume that the free-running counter's range equals 32.77ms.

309

$#TOFs (whole integers)	$Additional Free-Running Timer Counts (whole integers)	Minimum # Extension Bytes	Delay
_____	_____	_____	1 second
_____	_____	_____	5 seconds
_____	_____	_____	5 seconds
_____	_____	_____	1 minute
_____	_____	_____	3 minutes
_____	_____	_____	10 minutes

2. Refer to Figure 7–6, block 14. Explain why this block uses a BCS instruction rather than BNE.

3. Explain the effect that the question 2 situation has on time-out accuracy.

4. Given consideration of questions 2 and 3, how accurate is time-out? Base your calculation on an E clock rate of 2MHz. Within _____ µs.

5. Would the question 4 accuracy figure be valid for a TCNT comparison value of $0005? _____ Yes _____ No   Explain your answer.

6. Insert a breakpoint at the block 13 (compare TCNT) instruction. Determine minimum $TCNT value detectable within the accuracy specified in question 4. What is this minimum TCNT value? Explain how you calculated this figure.

PART IV: Laboratory Software Listing

MAIN PROGRAM—EVB

$ADDRESS	SOURCE CODE	COMMENTS
C000	LDX #1000	Initialize X register
C003	LDD #C020	Initialize TOI pseudovector
C006	STD D1	" " "
C008	LDD #0	Clear counter extension word
C00B	STD 0	" " " "
C00D	LDAA #80	Initialize TMSK2
C00F	STAA 24,X	" "
C011	CLI	Clear I mask
C012	WAI	Wait for TOI interrupt
C013	BRA C012	Wait for another TOI interrupt

MAIN PROGRAM—EVBU

$ADDRESS	SOURCE CODE	COMMENTS
0100	LDX #1000	Initialize X register
0103	LDD #120	Initialize TOI pseudovector
0106	STD D1	" " "
0108	LDD #0	Clear counter extension word

010B	STD 0	" " " "
010D	LDAA #80	Initialize TMSK2
010F	STAA 24,X	" "
0111	CLI	Clear I mask
0112	WAI	Wait for TOI interrupt
0113	BRA 112	Wait for another TOI interrupt

TOI INTERRUPT SERVICE—EVB

$ADDRESS	SOURCE CODE	COMMENTS
C020	BCLR 25,X 7F	Clear TOF bit
C023	LDD 0	Load D register with counter extension word
C025	ADDD #1	Add one to counter extension word
C028	CPD #1C9	Compare counter extension word with $1C9
C02C	BNE C03C	If not equal, store incremented counter extension word
C02E	LDD E,X	If equal, read TCNT register
C030	CPD #BC6A	Compare TCNT with $BC6A
C034	BCS C02E	If C=1, read TCNT again
C036	JSR C050	If C=0, call TIMSVC subroutine
C039	LDD #0	Clear D register (to clear counter extension word)
C03C	STD 0	Store D register to counter extension word
C03E	RTI	Return

TOI INTERRUPT SERVICE—EVBU

$ADDRESS	SOURCE CODE	COMMENTS
0120	BCLR 25,X 7F	Clear TOF bit
0123	LDD 0	Load D register with counter extension word
0125	ADDD #1	Add one to counter extension word
0128	CPD #1C9	Compare counter extension word with $1C9
012C	BNE 13C	If not equal, store incremented counter extension word
012E	LDD E,X	If equal, read TCNT register
0130	CPD #BC6A	Compare TCNT with $BC6A
0134	BCS 12E	If C=1, read TCNT again
0136	JSR 150	If C=0, call TIMSVC subroutine
0139	LDD #0	Clear D register (to clear counter extension word)
013C	STD 0	Store D register to counter extension word
013E	RTI	Return

TIMSVC SUBROUTINE—EVB

$ADDRESS	SOURCE CODE	COMMENTS
C050	LDX #2	Point X register at ASCII data
C053	JSR FFC7	Call .OUTSTR
C056	RTS	Return

TIMSVC SUBROUTINE—EVBU

$ADDRESS	SOURCE CODE	COMMENTS
0150	LDX #2	Point X register at ASCII data
0153	JSR FFC7	Call .OUTSTR
0156	RTS	Return

"15 sec." ASCII DATA—EVB AND EVBU

$ADDRESS	CONTENTS	CHARACTER
0002	31	"1"
0003	35	"5"
0004	20	\<SP\>
0005	73	"s"
0006	65	"e"
0007	63	"c"
0008	2E	"."
0009	04	\<EOT\>

FIGURE 7-8

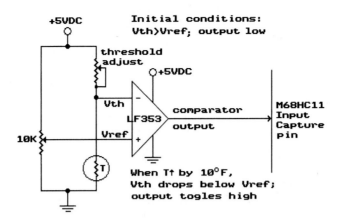

7.3 INTRODUCTION TO INPUT CAPTURES

Input captures allow the M68HC11 to record the times of external events. An external event must cause an input capture pin to toggle. For example, one can use a simple comparator circuit to determine when temperature rises by ten degrees from a reference level (see Figure 7–8). When temperature exceeds this ten-degree increase, the comparator drives an M68HC11 input capture pin from low to high. The M68HC11's input capture mechanism records the exact "time" that the input capture pin toggles. When temperature returns below the ten-degree threshold, the comparator output switches again—from high to low. The input capture mechanism can again record the exact "time" this toggle occurs.

An input capture function's recorded "time" consists of a free-running count value. The input capture mechanism records this free-running count in a special 16-bit Input Capture Register. At the system designer's option, an input capture can also generate an interrupt.

The M68HC11EVBU uses an M68HC11E9 microcontroller chip. This chip provides four input capture pins—and four associated input capture registers. Thus, the M68HC11E9 can handle up to four independent input capture tasks simultaneously.

The EVB uses an M68HC11A1. This chip provides three input capture pins and associated input capture registers. The M68HC11A1 can therefore handle three input capture tasks simultaneously.

7.4 ON-CHIP RESOURCES USED BY THE INPUT CAPTURE FUNCTION

The M68HC11E9 uses four pins, four data registers, one status register, and three control registers to operate its input capture mechanisms.[4] Following sections deal with these resources.

M68HC11E9 Input Capture Pins

Four PORTA pins share input capture (IC) and general purpose I/O functions:

PA0/IC3, pin 34
PA1/IC2, pin 33
PA2/IC1, pin 32
PA3/IC4/OC5, pin 31

Note that pin 31 has three functions:

PORTA 3 (general purpose I/O)
Input capture 4
Output compare 5 (see Chapter 8)

---

[4]Section 7.4 bases its discussion on the M68HC11E9 chip. Footnotes annotate differences between the E9 chip (used on the EVBU) and the A1 chip (used on the EVB).

FIGURE 7–9
Reprinted with the permission of Motorola.

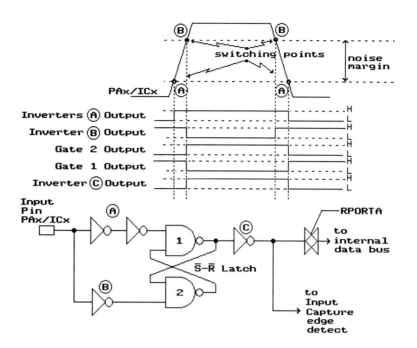

The CPU must take special actions to configure pin 31 as an input capture pin.[5] Pins 34, 33 and 32 are input-only pins. Pin 31 is bidirectional.

Each input capture pin uses an S-R latch to capture input toggles. Motorola makes this latch responsive to input edges but unresponsive to noise. The latch's input inverters have switch points tailored to increase noise margin.

Refer to Figure 7–9.[6] Inverter pair A switches at a lower than normal voltage level. Inverter B switches at a higher than normal level. A positive-going edge at the input pin causes the A inverters to switch at a relatively low voltage level, applying an early, inactive high level to NAND gate 1. Inverter B switches only after the positive-going edge reaches a relatively high voltage level. When inverter B switches, it applies an active low to NAND gate 2, driving its output high. This high cross-couples to NAND gate 1. NAND gate 1's inputs are now both high, which drives its output low. Inverter C converts this transition to a positive-going edge, and applies it to the input capture edge detect circuit. The S-R latch responds to noise only if its amplitude drops lower than the A inverters' switch point.

A negative-going edge at the input pin causes the B inverter to switch first, at a relatively high input voltage level. When it switches, inverter B applies an inactive high to NAND gate 2. Inverters A switch only after the input edge reaches a relatively low level. Upon switching, the A inverters apply a low to NAND gate 1, driving its output high. NAND gate 1's output high cross-couples to NAND gate 2, causing the S-R circuit to latch the transition. Inverter C converts NAND gate 1's high transition to a negative-going edge. Inverter C applies this edge to the Input Capture edge detect circuit. The S-R latch responds to noise only if its amplitude exceeds the B inverter's switch point.

Thus, Motorola designed input capture pins (IC1–IC4) for maximum noise margins. An external event generates an edge at an input capture pin. The pin's input capture mechanism records the "time" (free-running count) at which this edge occurs.

M68HC11E9 Timer Input Capture Registers

A peripheral toggles one of the M68HC11's input capture (IC) pins to indicate the occurrence of an external event. The M68HC11's input capture mechanism responds by recording the free-running count at the time the toggle occurred. A 16-bit Timer Input

---

[5]The M68HC11A1 does not use pin 31 for input captures.
[6]Motorola Inc., *M68HC11 Reference Manual*, p. 7-7.

Capture Register latches this free-running count. The M68HC11E9 incorporates an input capture register for each input capture pin:

IC Pin	Input Capture Register	Control Register Block Address
IC1	TIC1	$1010–$1011
IC2	TIC2	$1012–$1013
IC3	TIC3	$1014–$1015
IC4	TI4O5	$101E–$101F

Timer Input Capture/Output Compare Register (TI4O5) functions as pin IC4's counterpart register. Software can configure PA3/IC4/OC5 for either input captures or output compares. TI4O5 functions as an input capture register when PA3/IC4/OC5 operates as an input capture pin.[7] TIC1–TIC3 are read-only registers. When configured for input captures, TI4O5 is also a read-only register.

Thus, when a peripheral toggles an input capture pin, the pin's counterpart input capture register latches the free-running count at the time of the toggle. The CPU can then read this "time of occurrence" from the input capture register.

Pulse Accumulator Control Register (PACTL, Control Register Block Address $1026)

Previous discussion noted that M68HC11E9 pin PA3/IC4/OC5 can function as either an input capture or output compare pin. The M68HC11E9's CPU uses two PACTL register bits to configure this pin. Refer to Figure 7–10. DDRA3 constitutes the Data Direction Register Port A Bit 3 control. Setting DDRA3 configures the PA3/IC4/OC5 pin as an output. Clearing DDRA3 configures this pin as an input. Therefore, software usually clears the DDRA3 bit to configure the PA3/IC4/OC5 pin for input captures. However, the CPU can configure PA3/IC4/OC5 as an output, and enable input captures. This configuration causes the TI4O5 register to record the free-running count when the M68HC11E9 CPU writes to pin PA3.

PACTL bit I4/O5 constitutes the Input Capture 4/Output Compare 5 select bit. Clear this bit to configure the PA3/IC4/OC5 pin for output compares. Set the I4/O5 bit to configure for input captures. Setting the I4/O5 bit also configures the TI4O5 register for input captures. Thus, to configure pin PA3/IC4/OC5 for input captures, clear DDRA3 and set I4/O5 in the PACTL register.

Main Timer Interrupt Flag Register 1 (TFLG1, Control Register Block Address $1023)

Each input capture function has its own status flag. This flag sets when its counterpart input capture pin toggles with a specified edge. Refer to Figure 7–11. Input capture status flags reside in TFLG1 bits 3–0.

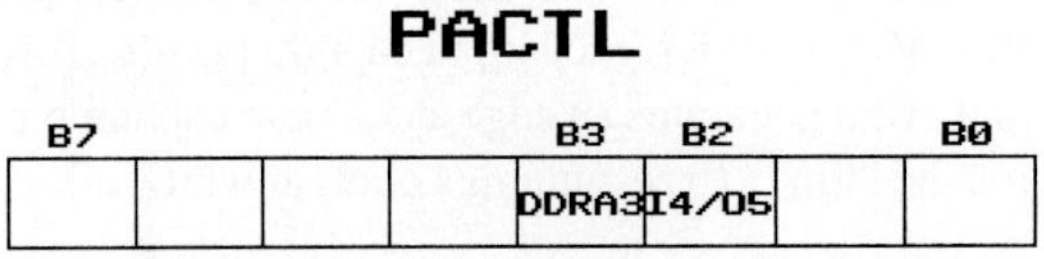

FIGURE 7–10
Reprinted with the permission of Motorola.

PACTL

B7			B3	B2		B0
			DDRA3	I4/O5		

FIGURE 7–11
Reprinted with the permission of Motorola.

TFLG1

B7			B3	B2	B1	B0
			I4O5F	IC1F	IC2F	IC3F

---

[7] The M68HC11A1 has no pin configurable as IC4. However, the A1's PA3 pin does function as Output Compare pin 5 (OC5).

Since PA3/IC4/OC5 can function as either an input capture or an output compare pin, the I4O5F flag provides a status bit for either function. The I4O5F bit becomes the IC4F bit if the input capture function is enabled. This bit becomes the OC5F flag if output compares are enabled.

As with TFLG2, all TFLG1 bits are read/modify/write. The M68HC11 CPU can change these bits only by writing binary ones to them. Refer to the discussion on pages 303–304 on how to clear selected read/modify/write bits in the TFLG1 and TFLG2 registers.

Main Timer Interrupt Mask Register 1
(TMSK1, Control Register Block Address $1022)

Each input capture function can generate interrupts, which are serviced at a unique vector address. A peripheral generates an input capture interrupt by toggling an IC pin with a specified edge.[8] Refer to Figure 7–12. The CPU enables IC interrupts by writing binary ones to TMSK1 bits 3–0. Write a zero to an ICxI bit to disable its corresponding interrupt. Note that TMSK1 arranges these interrupt enable bits in the same order as their counterpart flags in TFLG1. If PACTL bit I4/O5 = 1 (enabling IC4 functions), then setting TMSK1's I4O5I bit enables IC4 interrupts. If I4/O5 = 0, setting the I4O5I bit would enable Output Compare 5 interrupts.[9]

Thus, set bits in the TMSK1 register to enable IC interrupts. Specified IC pin toggles generate these interrupts.

Timer Control Register 2
(TCTL2, Control Register Block Address $1021)

A specified edge at an IC pin causes its associated TIC register to latch the free-running count. This specified edge also causes a counterpart IC flag to set—and may generate an interrupt (if enabled). What does "specified" mean? The TCTL2 register allots 2 bits to each IC pin. These 2 bits allow the user to choose from four edge-specification options:

EDGxB Bit	EDGxA Bit	Input Capture Edge Specification
0	0	Captures disabled.
0	1	Only positive-going edges trigger an IC response.
1	0	Only negative-going edges trigger an IC response.
1	1	Both positive- and negative-going edges trigger IC responses.

Refer to Figure 7–13. The CPU writes a control word to TCTL2 in order to specify active edges for any or all input capture pins.

FIGURE 7–12
Reprinted with the permission of Motorola.

TMSK1

B7				B3	B2	B1	B0
				I4O5I	IC1I	IC2I	IC3I

FIGURE 7–13
Reprinted with the permission of Motorola.

TCTL2

B7	B6	B5	B4	B3	B2	B1	B0
EDG4B	EDG4A	EDG1B	EDG1A	EDG2B	EDG2A	EDG3B	EDG3A

---

[8]If DDRA3 = 1 and I4/O5 =1 (in PACTL), the M68HC11E9 CPU can generate an IC interrupt by writing to PA3.
[9]M68HC11A1's TMSK1, bit 3 enables OC5 interrupts only.

EXERCISE 7.2   Initialize the M68HC11E9 for Input Captures

AIM: Derive control words which configure the microcontroller for input captures.

INTRODUCTORY INFORMATION: Complete this exercise to gain experience in calculating control words. These control words configure the M68HC11E9 for specified input capture operations.

INSTRUCTIONS: Use the specifications given, and refer to Sections 7.2 through 7.4 while formulating appropriate control words. The CPU writes these control words to the PACTL, TCTL2, TMSK1, and TMSK2 registers to initialize them for input captures. Enter hex control words in the blanks beside the addresses. Enter the control word which the CPU should write to that address—to configure the timer system according to the specifications. Enter "N/A" in a blank if a write to its corresponding address is unnecessary for either of two reasons: (1) a system reset initializes the register as specified, or (2) the register's contents have no relevance to the specifications. The CPU should write binary zeros to all nonapplicable control register bits.

Specifications	Address	Control Word

1. EXAMPLE

Use interrupt control to extend
   free-running counter range.
Use IC2 and IC4.
IC4 captures on falling edges.
IC2 captures on any edge.
IC4 uses interrupt control.
IC2 uses status polling.

	$1021	$ 8C
	$1022	$ 08
	$1024	$ 80
	$1026	$ 04

2. Do not extend free-running
   counter range.
Use IC2.
IC2 captures on rising edges.
IC2 uses status polling.

	$1021	$_____
	$1022	$_____
	$1024	$_____
	$1026	$_____

3. Use status polling to extend
   free-running counter range.
Use IC1, IC3, and IC4.
All captures on falling edges.
IC functions use interrupt control.

	$1021	$_____
	$1022	$_____
	$1024	$_____
	$1026	$_____

4. Use interrupt control to extend
   free-running counter range.
Use IC3 and IC4.
IC4 captures M68HC11 CPU writes
   to the IC4 pin.

IC4 captures on rising edges.
IC3 captures on falling edges.
IC3 uses interrupt control.
IC4 does not use interrupt control.

$1021	$_____
$1022	$_____
$1024	$_____
$1026	$_____

5. Use interrupt control to extend free-running counter range.
   Use all IC functions.
   All captures disabled.
   All captures use status polling.

$1021	$_____
$1022	$_____
$1024	$_____
$1026	$_____

6. Do not extend free-running counter range.
   Use IC1, IC2, and IC3.
   IC1 captures on rising edges.
   IC2 captures on falling edges.
   IC3 captures on any edge.
   IC1 and IC3 use interrupt control.
   IC2 uses status polling.

$1021	$_____
$1022	$_____
$1024	$_____
$1026	$_____

7. Use interrupt control to extend free-running counter range.
   Use all IC functions.
   IC4 and IC1 capture on any edge.
   IC2 captures on rising edges.
   IC3 captures on falling edges.
   IC4 and IC3 use interrupt control.
   IC1 and IC2 use status polling.

$1021	$_____
$1022	$_____
$1024	$_____
$1026	$_____

8. Do not extend free-running counter range.
   Use IC1 and IC2.
   IC captures on any edge.
   IC functions use status polling.

$1021	$_____
$1022	$_____
$1024	$_____
$1026	$_____

7.5 USING INPUT CAPTURES TO MEASURE TIME BETWEEN EVENTS

M68HC11 input captures can measure time between external events. When effectively programmed and interfaced, input captures can measure periods as small as 1μs.[10] By using free-running counter extension bytes, input captures can measure very long periods.

This section presents an input capture application that measures and displays periods between events. These periods can vary from one second to 35 minutes and 47 seconds. For purposes of this presentation, push-button presses constitute the external events. These key-presses represent any kind of external event that could toggle an IC pin.

Refer to Figure 7–14. An interface operator generates an external event by pressing a push-button in a pull-up circuit. A push-button press generates a negative-going edge at the M68HC11's IC3 pin.

Figures 7–15 through 7–19 introduce supporting software. This software measures and displays the number of minutes and seconds between push-button presses:

```
XX MIN. YY SEC.
```

Where XX = number of minutes in decimal, and YY = number of seconds in decimal.

The M68HC11 displays each succeeding measurement on a new line.

Refer to Figure 7–15. In MAIN program block 1, the CPU initializes its stack pointer, X register, interrupt pseudovectors, TCTL2, TMSK1, and TMSK2.

In block 2, the CPU clears the free-running counter's extension word (2 bytes). With this extension word, the free-running counter's range is 35 minutes and 47 seconds.

In block 3, the CPU initializes 4 bytes in RAM called "LSTCNT." LSTCNT's initial value consists of the free-running count, including extension bytes, at the time of initialization. After the first push-button press, LSTCNT contains the free-running count at the time of the previous push-button press.[11]

FIGURE 7–14

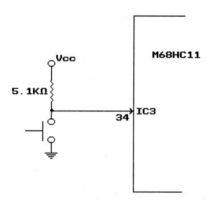

FIGURE 7–15

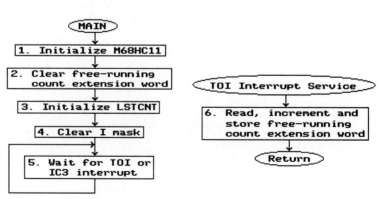

---

[10]Such an interface requires two IC pins to capture alternate edges. Internal IC response time and software overhead obviate the use of a single IC pin at high input frequencies.

[11]The M68HC11 CPU therefore makes its measurement from the time of LASTCNT initialization in Block 3, to the time of the first key-press.

FIGURE 7-16

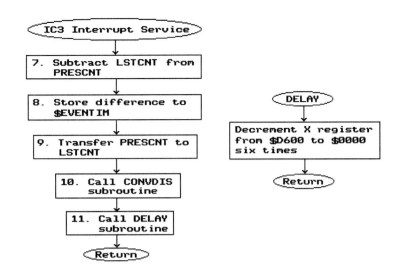

After clearing its I mask (block 4), the M68HC11 waits for either a TOI or IC3 interrupt (block 5). The timer section generates a TOI interrupt each time the free-running count rolls over from $FFFF to $0000. A push-button press generates an IC3 interrupt. The TOI Interrupt Service routine (block 6) simply increments the free-running count extension word.

Refer to Figure 7-16. An external event (push-button press) generates an IC3 interrupt. To measure the time between the previous and current events, the CPU must find the difference between the free-running counts at the times of these two key-presses. LSTCNT holds the count at the time of the previous event. PRESCNT (present free-running count) holds the time of the current key-press. Thus, to find the time between events, the CPU subtracts LSTCNT from PRESCNT (block 7). The difference constitutes the hexadecimal $Time Between Events ($EVENTIM). In block 8, the CPU stores $EVENTIM to its assigned 4-byte RAM block.

Looking ahead to the next key-press and measurement, the CPU must update LSTCNT to the time of the present key-press. In block 9, the CPU executes this update by transferring the PRESCNT value into LSTCNT.

$EVENTIM contains the number of free-running counts (at 500ns/count) between the previous and current key-presses. The M68HC11 must convert $EVENTIM into decimal minutes and seconds—and then display these values at the terminal. The Conversion and Display subroutine (CONVDIS) does this job. In block 10, the CPU calls the CONVDIS subroutine.

After returning from CONVDIS, the CPU calls the DELAY subroutine (block 11). DELAY executes a one-second delay, to de-bounce the external push-button.

Figures 7-17, 7-18, and 7-19 depict the CONVDIS (Conversion and Display) subroutine's three parts:

Part 1: Convert $EVENTIM to the decimal number of minutes between events (&MIN). See Figure 7-17.
Part 2: Convert $EVENTIM to the decimal number of seconds between events (&SEC). See Figure 7-18.
Part 3: Display &MIN and &SEC at the terminal display. See Figure 7-19.

Refer to Figure 7-17. In block 1, the CPU initializes the decimal number of minutes between events (&MIN) to zero. The CPU then subtracts $07270E00 from the $EVENTIM value (block 2). $07270E00 equals the number of free-running counts in one minute:

$$\$07270E00 = 120 \times 10^6$$
$$120 \times 10^6 \times 500ns = 60s$$

In block 3, the CPU determines whether the block 2 minuend was less than a minute (i.e., <$07270E00). $EVENTIM constitutes the minuend. If the minuend exceeded a minute's worth of counts, the CPU returns the subtraction result to $EVENTIM (block 4). Thus, the CPU reduces $EVENTIM by the number of free-running counts in one minute.

FIGURE 7-17

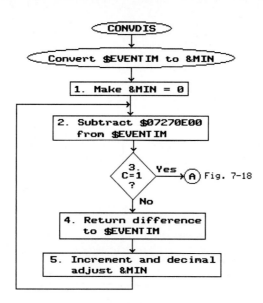

FIGURE 7-18

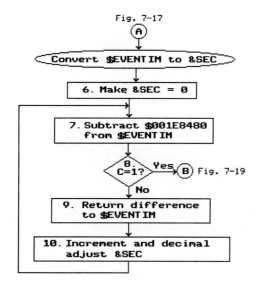

FIGURE 7-19

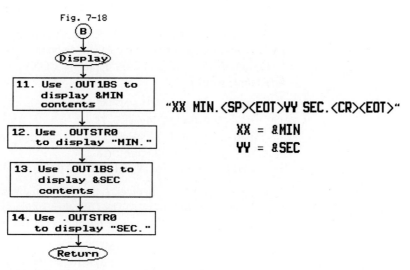

In block 5, the CPU increments the &MIN count. Thus, each time the CPU deducts a minute's worth of free-running counts from $EVENTIM, it increments $MIN. The CPU then branches from block 5 to block 2, where it subtracts another minute's worth of counts from $EVENTIM.

In block 3, the C flag sets if $07270E00 exceeded the contents of $EVENTIM. When the C flag sets, the CPU branches from block 3 before saving the subtraction's result. Thus, the CPU reduces $EVENTIM by as many minute values as possible. &MIN contains the number of successful subtractions, i.e., the number of minutes. At the point where the CPU branches from block 3, $EVENTIM contains the number of seconds remaining after subtraction of all possible minute values. The CPU branches from block 3 to Figure 7–18, block 6. Here, the CPU converts remaining $EVENTIM contents to decimal seconds.

Figure 7–18's strategy resembles that of Figure 7–17. The CPU deducts $001E8480 from $EVENTIM (block 7) until the C flag sets—indicating that $EVENTIM contains less than a second's worth of counts (block 8). Block 7's subtrahend equals the number of free-running counts in one second:

$$\$001E88480 = 2 \times 10^6$$
$$2 \times 10^6 \times 500ns = 1s$$

Each time the CPU makes a subtraction without setting the C flag, it saves the difference and returns it to $EVENTIM. The CPU increments &SEC (block 10) before branching back to make another subtraction.

When the C flag sets (block 8), $EVENTIM has been converted to &MIN and &SEC. The CPU must now display these decimal values on the terminal screen. The CPU branches from Figure 7–18, block 8 to Figure 7–19, block 11.

Refer to Figure 7–19. &MIN and &SEC reside at strategic locations in a display look-up table. In block 11, the CPU points its X register at &MIN and calls the .OUT1BS BUFFALO utility subroutine. .OUT1BS converts &MIN to ASCII and displays it. .OUT1BS returns an X register value of &MIN location plus one, which is the first ASCII code for "MIN.<SP><EOT>." In block 12, the CPU calls the .OUTSTR0 BUFFALO utility subroutine. This routine displays the "MIN><SP><EOT>" codes. Thus far, the terminal display shows:

XX MIN.

where XX = &MIN.

In block 13, the CPU points its X register at &SEC and calls .OUT1BS. The terminal display now shows:

XX MIN. YY

where YY = &SEC.

.OUT1BS returns an X register value which points at ASCII code for "SEC.<CR><EOT>." In block 14, the CPU calls .OUTSTR0 once more, to display the remainder of the message:

XX MIN. YY SEC.

The carriage return <CR> causes the CPU to display the next message one line down on the display.

To summarize, input captures allow the M68HC11 to measure the number of free-running counts between external events. Software can then convert the number of free-running counts to decimal minutes and seconds—and display these values on the terminal screen. The CPU finds the number of free-running counts between two successive events by subtracting the count recorded at the time of the previous event from the count recorded at the time of the current event.

EXERCISE 7.3

Use Input Captures to Measure Time Between Events

REQUIRED MATERIALS: 1 Resistor, 5.1 KΩ
1 Switch, push-button, push-to-close

REQUIRED EQUIPMENT: Motorola M68HC11EVBU (EVB), and terminal

GENERAL INSTRUCTIONS: For this laboratory exercise, connect a push-button/pull-up circuit to the M68HC11's IC3 pin. Develop and demonstrate software that uses input captures to measure and display times between key-presses. Carefully execute the following instructions.

PART I: Connect a Push-button/Pull-up Circuit

Connect a push-button/pull-up circuit to the M68HC11's IC3 pin, as shown in Figure 7–14. Show your connections to the instructor.

_____ *Instructor Check*

PART II: Software Development

Before proceeding, review Section 7.5 again. This section explains the software strategy used in this exercise. Refer to Figures 7–20 through 7–25 and develop time measurement software. Allocate memory as shown in the chart below.

PART III: Software Debug and Demonstration

Power up the EVBU (EVB) and terminal. Load and debug your software and interface. When sure that the interface runs properly, demonstrate its operation to the instructor.

_____ *Instructor Check*

Measure Time Between Events—Memory Allocations

EVBU $ADDRESS	EVB $ADDRESS	COMMENTS
0000–0001	0000–0001	Counter extension word (EXT)
0002–0003	0002–0003	LSTCNT—MS word
0004–0005	0004–0005	LSTCNT—LS word
0006–0007	0006–0007	$EVENTIM—MS Word
0008–0009	0008–0009	$EVENTIM—LS Word
000A–000B	000A–000B	Temporary Difference—LS Word (TEMP)
0010	0010	&Minutes Between Events (&MIN)
0011–0016	0011–0016	"MIN.<SP><EOT>" ASCII Data
0017	0017	&Seconds Between Events (&SEC)
0018–001D	0018–001D	"SEC<CR><EOT>" ASCII Data
0100	C000	MAIN program
0130	C030	TOI Interrupt Service routine (TOI)
0140	C040	IC3 Interrupt Service routine (IC3)
0162	C062	DELAY subroutine
0170	C070	CONVDIS subroutine
0170	C070	• Convert to &Minutes (CONV&MIN)
0195	C095	• Convert to &Seconds (CONV&SEC)
01BA	C0BA	• Display (DIS)
01FF	C0FF	Initial stack pointer

Note 1: One minute requires $07270E00 free-running counts. Divide this long-word value into two words:

```
$MIN-MS = $0727
$MIN-LS = $0E00
```

Note 2: One second requires $001E8480 free-running counts. Divide this long-word value into two words:

```
$SEC-MS = $001E
$SEC-LS = $8480
```

FIGURE 7-20

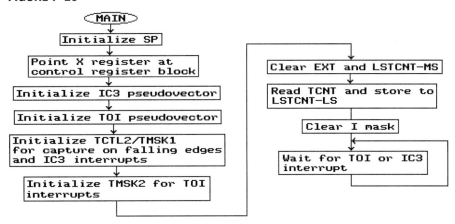

FIGURE 7-21

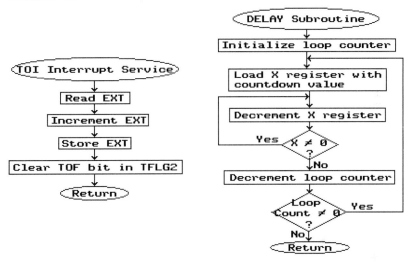

FIGURE 7-22

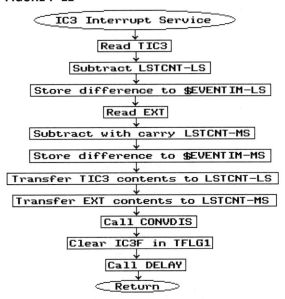

FIGURE 7-23

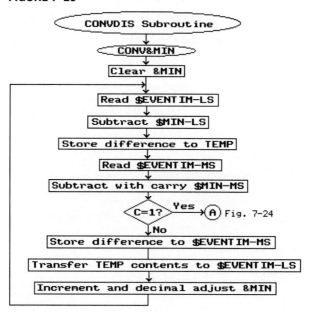

FIGURE 7-24

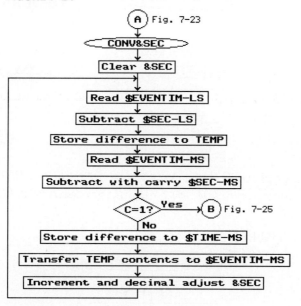

FIGURE 7-25

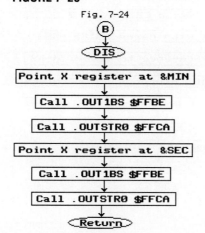

PART IV: Laboratory Analysis

Answer the questions below by neatly entering your answers in the appropriate blanks.

1. Modify your software to capture on IC3 rising edges. Observe and describe how this change affects interface operation.

2. Modify your software for capture on any IC3 edge. Observe and describe how this change affects interface operation.

3. Return your software to its original initialization values. Assure that the interface operates as it did in Part III of this exercise. Run the software. Press the push-button and hold it down continuously for ten seconds before releasing it. Describe the interface's response.

4. Explain why the M68HC11 behaved as you described in question 3.

5. Increase the interface's accuracy to hundredths of a second. The M68HC11 should now display measurement results as follows:

   `XX MIN. YY.HH SEC`

   where HH = decimal hundredths of a second.

   Add a CONV&HUNDSEC routine to the CONVDIS subroutine. Also, modify the CONVDIS subroutine's DIS section to display hundredths of a second. List your revised software in the blanks. **Programming Hint:** If the CPU initializes the stack pointer to $01FF ($C0FF), the stack will grow into the DIS routine memory space. Use BUFFALO's default stack pointer to avoid this problem.

CONV&HUNDSEC

$ADDRESS	SOURCE CODE	COMMENTS
_____	_____	_____
_____	_____	_____
_____	_____	_____
_____	_____	_____
_____	_____	_____
_____	_____	_____
_____	_____	_____
_____	_____	_____
_____	_____	_____
_____	_____	_____
_____	_____	_____
_____	_____	_____

		DIS	
$ADDRESS	SOURCE CODE		COMMENTS
_____	_____		_____
_____	_____		_____
_____	_____		_____
_____	_____		_____
_____	_____		_____
_____	_____		_____
_____	_____		_____
_____	_____		_____
_____	_____		_____

6. Load, debug, and demonstrate your modified software to the instructor.

_____ *Instructor Check*

PART V: Laboratory Software Listing

MAIN PROGRAM—EVB

$ADDRESS	SOURCE CODE	COMMENTS
C000	LDS #C0FF	Initialize stack pointer
C003	LDX #1000	Initialize X register
C006	LDD C040	Initialize IC3 pseudovector
C009	STD E3	" " "
C00B	LDD C030	Initialize TOI pseudovector
C00E	STD D1	" " "
C010	LDD #201	Initialize TCTL2/TMSK1
C013	STD 21,X	" " "
C015	LDAA #80	Initialize TMSK2
C017	STAA 24,X	" "
C019	LDD #0	Clear EXT and LSTCNT-MS
C01C	STD 0	Clear EXT
C01E	STD 2	Clear LSTCNT-MS
C020	LDD E,X	Read TCNT
C022	STD 4	Store TCNT to LSTCNT-LS
C024	CLI	Clear I mask
C025	WAI	Wait for TOI or IC3 interrupt
C027	BRA C025	Wait for another interrupt

MAIN PROGRAM—EVBU

$ADDRESS	SOURCE CODE	COMMENTS
0100	LDS #1FF	Initialize stack pointer
0103	LDX #1000	Initialize X register
0106	LDD 140	Initialize IC3 pseudovector
0109	STD E3	" " "
010B	LDD 130	Initialize TOI pseudovector
010E	STD D1	" " "
0110	LDD #201	Initialize TCTL2/TMSK1
0113	STD 21,X	" " "
0115	LDAA #80	Initialize TMSK2

$ADDRESS	SOURCE CODE	COMMENTS
0117	STAA 24,X	" "
0119	LDD #0	Clear EXT and LSTCNT-MS
011C	STD 0	Clear EXT
011E	STD 2	Clear LSTCNT-MS
0120	LDD E,X	Read TCNT
0122	STD 4	Store TCNT to LSTCNT-LS
0124	CLI	Clear I mask
0125	WAI	Wait for TOI or IC3 interrupt
0127	BRA 125	Wait for another interrupt

TOI INTERRUPT SERVICE ROUTINE—EVB

$ADDRESS	SOURCE CODE	COMMENTS
C030	LDD 0	Read EXT
C032	ADDD #1	Increment EXT
C035	STD 0	Store EXT
C037	BCLR 25,X 7F	Clear TOF bit
C03A	RTI	Return

TOI INTERRUPT SERVICE ROUTINE—EVBU

$ADDRESS	SOURCE CODE	COMMENTS
0130	LDD 0	Read EXT
0132	ADDD #1	Increment EXT
0135	STD 0	Store EXT
0137	BCLR 25,X 7F	Clear TOF bit
013A	RTI	Return

IC3 INTERRUPT SERVICE ROUTINE—EVB

$ADDRESS	SOURCE CODE	COMMENTS
C040	LDD 14,X	Read TIC3
C042	SUBD 4	Subtract LSTCNT-LS
C044	STD 8	Store difference to $EVENTIM-LS
C046	LDD 0	Read EXT
C048	SBCB 3	Subtract with carry LS byte of LSTCNT-MS
C04A	SBCA 2	Subtract with carry MS byte of LSTCNT-MS
C04C	STD 6	Store difference to $EVENTIM-MS
C04E	LDD 14,X	Transfer TIC3 contents to LSTCNT-LS
C050	STD 4	" " " " "
C052	LDD 0	Transfer EXT contents to LSTCNT-MS
C054	STD 2	" " " " "
C056	BSR C070	Call CONVDIS
C058	LDAA #1	Clear IC3F
C05A	STAA 1023	" "
C05D	BSR C062	Call DELAY
C05F	RTI	Return

IC3 INTERRUPT SERVICE ROUTINE—EVBU

$ADDRESS	SOURCE CODE	COMMENTS
0140	LDD 14,X	Read TIC3
0142	SUBD 4	Subtract LSTCNT-LS
0144	STD 8	Store difference to $EVENTIM-LS
0146	LDD 0	Read EXT
0148	SBCB 3	Subtract with carry LS byte of LSTCNT-MS
014A	SBCA 2	Subtract with carry MS byte of LSTCNT-MS
014C	STD 6	Store difference to $EVENTIM-MS
014E	LDD 14,X	Transfer TIC3 contents to LSTCNT-LS
0150	STD 4	" " " " "
0152	LDD 0	Transfer EXT contents to LSTCNT-MS
0154	STD 2	" " " " "
0156	BSR 170	Call CONVDIS
0158	LDAA #1	Clear IC3F
015A	STAA 1023	" "
015D	BSR 162	Call DELAY
015F	RTI	Return

DELAY SUBROUTINE—EVB

$ADDRESS	SOURCE CODE	COMMENTS
C062	LDAA #6	Initialize loop counter
C064	LDX #D600	Load X register with count-down value
C067	DEX	Decrement X register
C068	BNE C067	If X ≠ 0, decrement again
C06A	DECA	If X = 0, decrement loop counter
C06B	BNE C064	If ACCA ≠ 0, count down again
C06D	RTS	If ACCA = 0, return

DELAY SUBROUTINE—EVBU

$ADDRESS	SOURCE CODE	COMMENTS
0162	LDAA #6	Initialize loop counter
0164	LDX #D600	Load X register with count-down value
0167	DEX	Decrement X register
0168	BNE 167	If X ≠ 0, decrement again
016A	DECA	If X = 0, decrement loop counter
016B	BNE 164	If ACCA ≠ 0, count down again
016D	RTS	If ACCA = 0, return

CONVDIS SUBROUTINE—EVB

$ADDRESS	SOURCE CODE	COMMENTS
	CONV&MIN	
C070	CLR 10	Clear &MIN
C073	LDD 8	Read $EVENTIM-LS
C075	SUBD #E00	Subtract $MIN-LS
C078	STD A	Store difference to TEMP
C07A	LDD 6	Read $EVENTIM-MS
C07C	SBCB #27	Subtract with carry low byte of $MIN-MS
C07E	SBCA #7	" " high byte " "
C080	BCS C095	If subtrahend>minuend, CONV&MIN is complete; branch to CONV&SEC
C082	STD 6	If CONV&MIN is incomplete, store difference back to $EVENTIM-MS
C084	LDD A	Transfer TEMP contents to $EVENTIM-LS
C086	STD 8	" " " " "
C088	LDAA 10	Increment and decimal adjust &MIN
C08A	ADDA #1	" " " " "
C08C	DAA	" " " " "
C08D	STAA 10	" " " " "
C08F	BRA C073	Do another &MIN conversion loop
C091	NOP	
C092	NOP	
C093	NOP	
C094	NOP	
	CONV&SEC	
C095	CLR 17	Clear &SEC
C098	LDD 8	Read $EVENTIM-LS
C09A	SUBD #8480	Subtract $SEC-LS
C09D	STD A	Store difference to TEMP
C09F	LDD 6	Read $EVENTIM-MS
C0A1	SBCB #1E	Subtract with carry low byte of $SEC-MS
C0A3	SBCA #0	" " high byte " "
C0A5	BCS C0BA	If subtrahend>minuend, CONV&SEC is complete; branch to DIS
C0A7	STD 6	If CONV&SEC is incomplete, store difference back to $EVENTIM-MS
C0A9	LDD A	Transfer TEMP contents to $EVENTIM-LS
C0AB	STD 8	" " " " "
C0AD	LDAA 17	Increment and decimal adjust &SEC
C0AF	ADDA #1	" " " " "
C0B1	DAA	" " " " "
C0B2	STAA 17	" " " " "

$ADDRESS	SOURCE CODE	COMMENTS
C0B4	BRA C098	Do another &SEC conversion loop
C0B6	NOP	
C0B7	NOP	
C0B8	NOP	
C0B9	NOP	
	DIS	
C0BA	LDX #10	Point X register at $MIN
C0BD	JSR FFBE	Call .OUT1BS
C0C0	JSR FFCA	Call .OUTSTR0
C0C3	INX	Point X register at &SEC
C0C4	JSR FFBE	Call .OUT1BS
C0C7	JSR FFCA	Call .OUTSTR0
C0CA	RTS	Return

CONVDIS SUBROUTINE—EVBU

$ADDRESS	SOURCE CODE	COMMENTS
	CONV&MIN	
0170	CLR 10	Clear &MIN
0173	LDD 8	Read $EVENTIM-LS
0175	SUBD #E00	Subtract $MIN-LS
0178	STD A	Store difference to TEMP
017A	LDD 6	Read $EVENTIM-MS
017C	SBCB #27	Subtract with carry low byte of $MIN-MS
017E	SBCA #7	" " high byte " "
0180	BCS 195	If subtrahend>minuend, CONV&MIN is complete; branch to CONV&SEC
0182	STD 6	If CONV&MIN is incomplete, store difference back to $EVENTIM-MS
0184	LDD A	Transfer TEMP contents to $EVENTIM-LS
0186	STD 8	" " " " "
0188	LDAA 10	Increment and decimal adjust &MIN
018A	ADDA #1	" " " " "
018C	DAA	" " " " "
018D	STAA 10	" " " " "
018F	BRA 173	Do another &MIN conversion loop
0191	NOP	
0192	NOP	
0193	NOP	
0194	NOP	
	CONV&SEC	
0195	CLR 17	Clear &SEC
0198	LDD 8	Read $EVENTIM-LS
019A	SUBD #8480	Subtract $SEC-LS
019D	STD A	Store difference to TEMP
019F	LDD 6	Read $EVENTIM-MS
01A1	SBCB #1E	Subtract with carry low byte of $SEC-MS
01A3	SBCA #0	" " high byte " "
01A5	BCS 1BA	If subtrahend>minuend, CONV&SEC is complete; branch to DIS
01A7	STD 6	If CONV&SEC is incomplete, store difference back to $EVENTIM-MS
01A9	LDD A	Transfer TEMP contents to $EVENTIM-LS
01AB	STD 8	" " " " "
01AD	LDAA 17	Increment and decimal adjust &SEC
01AF	ADDA #1	" " " " "
01B1	DAA	" " " " "
01B2	STAA 17	" " " " "
01B4	BRA 198	Do another &SEC conversion loop
01B6	NOP	
01B7	NOP	
01B8	NOP	
01B9	NOP	
	DIS	
01BA	LDX #10	Point X register at $MIN
01BD	JSR FFBE	Call .OUT1BS
01C0	JSR FFCA	Call .OUTSTR0

$ADDRESS	SOURCE CODE	COMMENTS
01C3	INX	Point X register at &SEC
01C4	JSR FFBE	Call .OUT1BS
01C7	JSR FFCA	Call .OUTSTR0
01CA	RTS	Return

ASCII DATA—EVB AND EVBU

$ADDRESS	$CONTENTS	CHARACTER
0010	&MIN	
0011	4D	"M"
0012	49	"I"
0013	4E	"N"
0014	2E	"."
0015	20	<SP>
0016	04	<EOT>
0017	&SEC	
0018	53	"S"
0019	45	"E"
001A	43	"C"
001B	2E	"."
001C	0D	<CR>
001D	04	<EOT>

CONV&HUNDSEC ROUTINE AND MODIFIED DIS ROUTINE—EVB

$ADDRESS	SOURCE CODE	COMMENTS
C0BA	CLR 1A	Clear &HUNDSEC
C0BD	LDD 8	Read $EVENTIM-LS
C0BF	SUBD #4E20	Subtract $HUNDSEC-LS
C0C2	STD A	Store difference to TEMP
C0C4	LDD 6	Read $EVENTIM-MS
C0C6	SBCB #0	Subtract with carry $HUNDSEC-MS
C0C8	BCS C0D9	If subtrahend>minuend, branch to DIS
C0CA	STD 6	If CONV&HUNDSEC is complete, store difference back to $EVENTIM-MS
C0CC	LDD A	Transfer TEMP contents to $EVENTIM-LS
C0CE	STD 8	" " " " "
C0D0	LDAA 1A	Increment and decimal adjust &HUNDSEC
C0D2	ADDA #1	" " " " "
C0D4	DAA	" " " " "
C0D5	STAA 1A	" " " " "
C0D7	BRA C0BD	Do another &HUNDSEC conversion loop
	DIS	
C0D9	LDX #10	Point X register at &MIN
C0DC	JSR FFBE	Call .OUT1BS
C0DF	JSR FFCA	Call .OUTSTR0
C0E2	INX	Point X register at &SEC
C0E3	JSR FFBE	Call .OUT1BS
C0E6	JSR FFCA	Call .OUTSTR0
C0E9	INX	Point X register at &HUNDSEC
C0EA	JSR FFBE	Call .OUT1BS
C0ED	JSR FFCA	Call .OUTSTR0
C0F0	RTS	Return

CONV&HUNDSEC ROUTINE AND MODIFIED DIS ROUTINE—EVBU

$ADDRESS	SOURCE CODE	COMMENTS
01BA	CLR 1A	Clear &HUNDSEC
01BD	LDD 8	Read $EVENTIM-LS
01BF	SUBD #4E20	Subtract $HUNDSEC-LS
01C2	STD A	Store difference to TEMP
01C4	LDD 6	Read $EVENTIM-MS
01C6	SBCB #0	Subtract with carry $HUNDSEC-MS
01C8	BCS 1D9	If subtrahend>minuend, branch to DIS
01CA	STD 6	If CONV&HUNDSEC is complete, store difference back to $EVENTIM-MS

01CC	LDD A		Transfer TEMP contents to $EVENTIM-LS
01CE	STD 8		" " " " "
01D0	LDAA 1A		Increment and decimal adjust &HUNDSEC
01D2	ADDA #1		" " " " "
01D4	DAA		" " " " "
C0D5	STAA 1A		" " " " "
01D7	BRA 1BD		Do another &HUNDSEC conversion loop

DIS

01D9	LDX #10	Point X register at &MIN
01DC	JSR FFBE	Call .OUT1BS
01DF	JSR FFCA	Call .OUTSTR0
01E2	INX	Point X register at &SEC
01E3	JSR FFBE	Call .OUT1BS
01E6	JSR FFCA	Call .OUTSTR0
01E9	INX	Point X register at &HUNDSEC
01EA	JSR FFBE	Call .OUT1BS
01ED	JSR FFCA	Call .OUTSTR0
01F0	RTS	Return

MODIFIED ASCII DATA FOR CONV&HUNDSEC—EVB AND EVBU

$ADDRESS	$CONTENTS	CHARACTER
0010	&MIN	
0011	4D	"M"
0012	49	"I"
0013	4E	"N"
0014	2E	"."
0015	20	<SP>
0016	04	<EOT>
0017	&SEC	
0018	2E	"."
0019	04	<EOT>
001A	&HUNDSEC	
001B	53	"S"
001C	45	"E"
001D	43	"C"
001E	2E	"."
001F	0D	<CR>
0020	04	<EOT>

7.6 USE INPUT CAPTURE PINS FOR GENERAL-PURPOSE INTERRUPTS

A system designer can use three or four input capture pins for individual general-purpose interrupts.[12] An IC pin becomes a general-purpose interrupt pin if the CPU sets its counterpart ICxI bit in TMSK1. Each IC interrupt has a unique interrupt vector address, allowing the designer to develop a custom interrupt service routine for each IC pin used. EDGxB and EDGxA bits in TCTL2 allow the designer to specify individually the edges which generate interrupts. Finally, the ICxF bits in TFLG1 provide an additional measure of control over these interrupts.

Figures 7–26 through 7–28 illustrate the use of IC1, IC2, and IC3 as individual general-purpose interrupts. In this example, the MAIN program simulates routine business by counting at a two-second rate and displaying each count at the terminal. When generated by a specified edge at its ICx pin, each general-purpose interrupt displays a unique identifier on a new line of the terminal display. The interface uses three push-button/pull-up circuits to generate the interrupts. Figure 7–26 illustrates these interface connections.

---

[12]Recall that the M68HC11A1 (as used on the EVB) has three IC pins. The M68HC11E9 (EVBU) has three IC pins, plus a fourth which it can configure for either input captures or output compares.

FIGURE 7-26

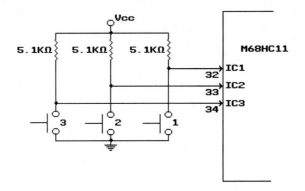

Refer to block 1 of Figure 7–27. First, the CPU initializes its stack pointer and points the X register at the control register block. Next, the CPU initializes pseudovectors for IC1, IC2, and IC3. Then, the CPU selects IC pin edges that generate the interrupts. To make these selections, the CPU sets appropriate bits in TCTL2. Finally, the CPU enables IC1, IC2, and IC3 interrupts at the local level by setting counterpart ICxI bits in TMSK1.

In block 2, the CPU clears the I mask to enable IC interrupts at the global level.

Refer to blocks 3 through 8 in Figure 7–27. To simulate normal business, the M68HC11 executes a simple count and display routine. ACCA functions as the counter register. In block 3, the CPU clears the counter register. In block 4, the CPU stacks the count. The .OUTRHLF routine (block 5) destroys count values $A through $F. Therefore, the CPU stacks the count to keep it intact. In block 5, the CPU calls the .OUTRHLF utility subroutine at address $FFB5. This subroutine converts ACCA's low nibble to ASCII and displays it at the terminal. After returning from .OUTRHLF, the CPU pulls the intact count value from the stack to ACCA (block 6). In block 7, the CPU increments the count. In block 8, the CPU calls the DELAY subroutine. This subroutine executes a two-second delay between counts. The CPU then branches back to block 4 to display the next count.

Figure 7–27, blocks 9 through 14 illustrate the software used by the three IC Interrupt Service routines. Each routine displays the following message on a new line of the terminal display:

ICxINT

The lower case "x" stands for "1," "2," or "3," depending upon which IC interrupt is generated. For example, if the interface operator presses push-button 2, the IC2 Interrupt Service routine displays this message:

IC2INT

Of course, each of these service routines must reside in a separate RAM block.

FIGURE 7-27

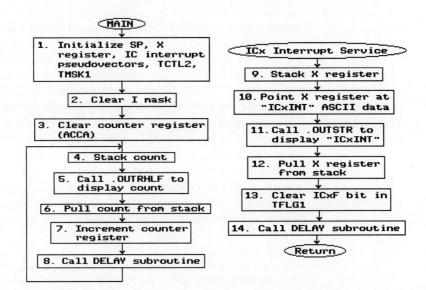

FIGURE 7-28

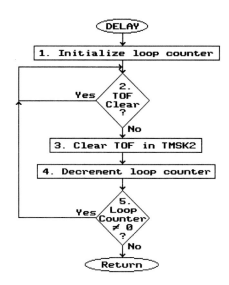

In block 9, the CPU stacks its X register (X = $1000), since it points the X register at the "ICxINT<CR><EOT>" ASCII data in block 10. In block 11, the CPU calls the .OUTSTR utility subroutine (address $FFC7). After returning from .OUTSTR, the CPU restores its X register's contents to $1000 by pulling this value from the stack (block 12). In block 13, the CPU uses the BCLR instruction to clear the ICxF bit in TFLG1. Recall that IC1F, IC2F, and IC3F are all read/modify/write bits.

In block 14, the CPU calls the DELAY subroutine, to de-bounce the push-button and give the interface operator time to release it.

Refer to Figure 7–28. To delay for two seconds, the CPU counts down sixty-one ($3D) free-running counter overflows. Recall that the free-running counter's range equals 32.77ms.[13] Sixty-one free-running counter rollovers equal:

$$61 \times 32.77\text{ms} = 1.99897\text{s}$$

In Figure 7–28, blocks 1 and 2, the CPU initializes a loop counter (ACCB) to $3D, then enters a polling loop, which it escapes when the TOF flag sets.[14] After escaping the polling loop, the CPU clears the TOF flag (block 3) and decrements the loop counter (block 4). If the loop counter has decremented to $00, two seconds have elapsed and the CPU returns from the DELAY subroutine (block 5). If the loop counter has not decremented to zero, the CPU branches from block 5 to block 2, where it enters another TOF polling loop.

Figures 7–26 through 7–28 illustrate how a system designer can use input capture pins for general-purpose interrupts. The M68HC11's I mask and ICxI bits provide two tiers of interrupt control. The TCTL2's EDGxB/EDGxA bits allow the designer to tailor these IC interrupts to the peripherals generating them. Each IC interrupt has its own vector address, and the ICxF bits in TFLG1 facilitate control over the individual interrupt service routines.

EXERCISE 7.4

Use Input Capture Pins for General-Purpose Interrupts

REQUIRED MATERIALS: 3 Resistors, 5.1 KΩ
3 Switches, push-button, push-to-close

REQUIRED EQUIPMENT: Motorola M68HC11EVBU (EVB), and terminal

---

[13]This figure assumes that an 8MHz crystal connects to the M68HC11 and that TMSK2's PR1/PR0 bits are both reset.
[14]Recall that the MAIN Program uses ACCA as the counter register.

GENERAL INSTRUCTIONS: In this laboratory exercise, connect separate push-button/pull-up circuits to the M68HC11's IC1, IC2, and IC3 pins. Develop software that uses these IC pins for general-purpose interrupts. Carefully execute the following procedures.

PART I: Connect Push-Button/Pull-Up Circuits

Refer to Figure 7–26. Connect push-button/pull-up circuits as shown. Show your connections to the instructor.

_____ *Instructor Check*

PART II: Software Development

Before proceeding, review Section 7.6. This section explains the software strategy used in this exercise. Refer to Figures 7–27 and 7–28. Develop interrupt software. Program the CPU to initialize TCTL2 so that falling edges generate all interrupts. Allocate memory as shown in the chart below.

PART III: Software Debug and Demonstration

Load and debug your software and interface. When confident that the interface operates properly, demonstrate it to the instructor.

_____ *Instructor Check*

PART IV: Laboratory Analysis

Answer the questions below by entering your answers in the appropriate blanks.

1. Describe how to modify the software so that rising edges generate IC1 interrupts.

2. Modify the software so that rising edges generate IC1 interrupts. Observe and describe how this change affects interface operation.

3. Describe how to modify the software so that any edge generates an IC3 interrupt.

Use IC Pins for General Purpose Interrupts—Memory Allocations

EVBU $ADDRESS	EVB $ADDRESS	CONTENTS
0000–0007	0000–0007	"IC1INT<CR><EOT>" ASCII data
0008–000F	0008–000F	"IC2INT<CR><EOT>" ASCII data
0010–0017	0010–0017	"IC3INT<CR><EOT>" ASCII data
0100	C000	MAIN program
0130	C030	IC1 Interrupt Service routine
0140	C040	IC2 Interrupt Service routine
0150	C050	IC3 Interrupt Service routine
0160	C060	DELAY subroutine
01FF	C0FF	Initial stack pointer

FIGURE 7-29

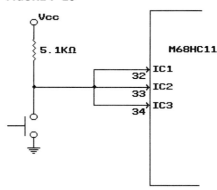

4. Modify the software so that any edge generates an IC3 interrupt. Observe and describe how this change affects software operations.

5. Return your software to its original initialization values. Assure that the interface operates as it did in Part III of this exercise. Move ASCII data and software to EEPROM. Power down the EVBU (EVB). Modify the interface so that a single push-button/pull-up circuit drives all three IC pins (IC1, IC2, IC3), as shown in Figure 7–29. Power up the EVBU (EVB) and interface. Move your ASCII data and software from EEPROM to its original locations in RAM. Observe and describe how this hardware change affects interface operation. Note particularly the order in which the M68HC11 executes the interrupt service routines when the push-button is pressed.

6. Explain why the M68HC11 executes the interrupt service routines in the observed order.

7. Recall that I-maskable interrupts can be promoted to highest I-maskable priority by writing an appropriate control word to the HPRI0 register. Promote IC3 interrupts with the following procedure:
   1. Press the EVBU (EVB) reset button.
   2. Use the BUFFALO MM (Memory Modify) command to write the appropriate promotion bits to PSEL3–PSEL0 in the HPRI0 register.
   3. **DO NOT RESET.** Run your software.

   Observe and describe how the HPRI0 register modification affects interface operation. particularly the order in which the M68HC11 executes the interrupt service routines.

8. Demonstrate the modified interrupt priorities (question 7) to the instructor.

   _____ *Instructor Check*

PART V: Laboratory Software Listing

MAIN PROGRAM—EVB

$ADDRESS	SOURCE CODE	COMMENTS
C000	LDS #C0FF	Initialize stack pointer
C003	LDX #1000	Initialize X register
C006	LDD #C030	Initialize IC1 interrupt pseudovector
C009	STD E9	" " " "

335

$ADDRESS	SOURCE CODE	COMMENTS
C00B	LDD #C040	Initialize IC2 interrupt pseudovector
C00E	STD E6	" " " "
C010	LDD #C050	Initialize IC3 interrupt pseudovector
C013	STD E3	" " " "
C015	LDD #2A07	Initialize TCTL2 and TMSK1
C018	STD 21,X	" " " "
C01A	CLI	Clear I mask
C01B	CLRA	Clear counter register
C01C	PSHA	Stack count
C01D	JSR FFB5	Call .OUTRHLF
C020	PULA	Pull count from stack
C021	INCA	Increment counter register
C022	JSR C060	Call DELAY subroutine
C025	BRA C01C	Display another count

MAIN PROGRAM—EVBU

$ADDRESS	SOURCE CODE	COMMENTS
0100	LDS #1FF	Initialize stack pointer
0103	LDX #1000	Initialize X register
0106	LDD #130	Initialize IC1 interrupt pseudovector
0109	STD E9	" " " "
010B	LDD #140	Initialize IC2 interrupt pseudovector
010E	STD E6	" " " "
0110	LDD #150	Initialize IC3 interrupt pseudovector
0113	STD E3	" " " "
0115	LDD #2A07	Initialize TCTL2 and TMSK1
0118	STD 21,X	" " " "
011A	CLI	Clear I mask
011B	CLRA	Clear counter register
011C	PSHA	Stack count
011D	JSR FFB5	Call .OUTRHLF
0120	PULA	Pull count from stack
0121	INCA	Increment counter register
0122	JSR 160	Call DELAY subroutine
0125	BRA 11C	Display another count

IC1 INTERRUPT SERVICE ROUTINE—EVB

$ADDRESS	SOURCE CODE	COMMENTS
C030	PSHX	Save X register contents
C031	LDX #0	Point X register at "IC1INT" ASCII data
C034	JSR FFC7	Call .OUTSTR
C037	PULX	Restore X register contents
C038	BCLR 23,X FB	Clear IC1F
C03B	JSR C060	Call DELAY subroutine
C03E	RTI	Return

IC1 INTERRUPT SERVICE ROUTINE—EVBU

$ADDRESS	SOURCE CODE	COMMENTS
0130	PSHX	Save X register contents
0131	LDX #0	Point X register at "IC1INT" ASCII data
0134	JSR FFC7	Call .OUTSTR
0137	PULX	Restore X register contents
0138	BCLR 23,X FB	Clear IC1F
013B	JSR 160	Call DELAY subroutine
013E	RTI	Return

IC2 INTERRUPT SERVICE ROUTINE—EVB

$ADDRESS	SOURCE CODE	COMMENTS
C040	PSHX	Save X register contents
C041	LDX #8	Point X register at "IC2INT" ASCII data
C044	JSR FFC7	Call .OUTSTR
C047	PULX	Restore X register contents
C048	BCLR 23,X FD	Clear IC2F
C04B	JSR C060	Call DELAY subroutine
C04E	RTI	Return

IC2 INTERRUPT SERVICE ROUTINE—EVBU

$ADDRESS	SOURCE CODE	COMMENTS
0140	PSHX	Save X register contents
0141	LDX #8	Point X register at "IC2INT" ASCII data
0144	JSR FFC7	Call .OUTSTR
0147	PULX	Restore X register contents
0148	BCLR 23,X FD	Clear IC2F
014B	JSR 160	Call DELAY subroutine
014E	RTI	Return

IC3 INTERRUPT SERVICE ROUTINE—EVB

$ADDRESS	SOURCE CODE	COMMENTS
C050	PSHX	Save X register contents
C051	LDX #10	Point X register at "IC3INT" ASCII data
C054	JSR FFC7	Call .OUTSTR
C057	PULX	Restore X register contents
C058	BCLR 23,X FE	Clear IC3F
C05B	JSR C060	Call DELAY subroutine
C05E	RTI	Return

IC3 INTERRUPT SERVICE ROUTINE—EVBU

$ADDRESS	SOURCE CODE	COMMENTS
0150	PSHX	Save X register contents
0151	LDX #10	Point X register at "IC3INT" ASCII data
0154	JSR FFC7	Call .OUTSTR
0157	PULX	Restore X register contents
0158	BCLR 23,X FE	Clear IC3F
015B	JSR 160	Call DELAY subroutine
015E	RTI	Return

DELAY SUBROUTINE—EVB

$ADDRESS	SOURCE CODE	COMMENTS
C060	LDAB #3D	Initialize loop counter
C062	BRCLR 25,X 80 C062	Test TOF until TOF = 1
C066	BCLR 25,X 7F	Clear TOF
C069	DECB	Decrement loop counter
C06A	BNE C062	If loop counter ≠ 0, test TOF again
C06C	RTS	If loop counter = 0, return

DELAY SUBROUTINE—EVBU

$ADDRESS	SOURCE CODE	COMMENTS
0160	LDAB #3D	Initialize loop counter
0162	BRCLR 25,X 80 162	Test TOF until TOF = 1
0166	BCLR 25,X 7F	Clear TOF
0169	DECB	Decrement loop counter
016A	BNE 162	If loop counter ≠ 0, test TOF again
016C	RTS	If loop counter = 0, return

"ICxINT" ASCII DATA—EVB AND EVBU

$ADDRESS	SOURCE CODE	COMMENTS
0000	49	"I"
0001	43	"C"
0002	31	"1"
0003	49	"I"
0004	4E	"N"
0005	54	"T"
0006	0D	\<CR\>
0007	04	\<EOT\>
0008	49	"I"
0009	43	"C"
000A	32	"2"
000B	49	"I"

000C	4E	"N"
000D	54	"T"
000E	0D	<CR>
000F	04	<EOT>
0010	49	"I"
0011	43	"C"
0012	33	"3"
0013	49	"I"
0014	4E	"N"
0015	54	"T"
0016	0D	<CR>
0017	04	<EOT>

7.7 CHAPTER SUMMARY

The M68HC11 bases its specialized timing functions upon a 16-bit, free-running count. This count provides "times" for periodic interrupts, input captures, or output compares. When the M68HC11 resets, the free-running counter begins counting at $0000 and continues to count up until the M68HC11 is either reset again or STOPed. When the free-running count rolls over from $FFFF to $0000, the timer section sets a status flag called the Timer Overflow Flag (TOF). A programmer can use TOFs to extend the free-running counter's range.

The M68HC11 CPU uses three registers to configure and operate the free-running counter:

- The CPU accesses the current free-running count by reading the TCNT (Timer Counter) register. To assure that it reads a coherent count value, the CPU must use a double-byte read, such as LDD.
- The M68HC11 writes to the TMSK2 (Timer Miscellaneous Interrupt Mask) register to select one of four free-running count rates. TMSK2's PR1/PR0 bits determine the count rate as E clock frequency divided by 1, 4, 8, or 16. Count rate determines the free-running counter's resolution and range.

    The CPU enables counter overflow interrupts (TOIs) by setting the TOI bit in TMSK2. When TOI = 1, the timer section generates an interrupt each time the free-running count rolls over from $FFFF to $0000.
- The TOF (Timer Overflow Flag), in the TFLG2 (Miscellaneous Timer Interrupt Flag) register, indicates free-running counter rollovers. A programmer uses this flag to extend the free-running counter's range.

All TFLG2 flags are read/modify/write bits. Clear TFLG2 bits with one of these techniques:

1. Use extended or indexed mode to write a mask byte to TFLG2. This mask uses ones to specify bits to be cleared and zeros to specify bits to be unchanged.
2. Use the BCLR instruction—with a mask that uses zeros to specify bits to be cleared, and ones to specify bits to be unchanged.

Extend the range of the free-running counter by establishing a counter extension word in RAM. Program the CPU to increment this word each time the TOF sets in TFLG2. This chapter presents a simple program that extends the free-running counter by 2 bytes, increasing the counter's bit width from 16 to 32 bits, and range from 32.77ms to 35 minutes and 48 seconds. Upon time-out, this program calls a Counter Overflow Service (CNTSVC) subroutine. This subroutine executes a measurement or control task. After returning from CNTSVC, the CPU times out another 35 minutes and 48 seconds, before recalling CNTSVC.

Usually, a system designer must tailor free-running counter time-out to a measurement or control application. Follow these steps to develop a specific time-out.

- Calculate the number of TOFs in the specified period.

- Determine the number of additional free-running counts (after the terminal TOF) required to complete the specified period.
- Develop software that counts the required number of TOFs, then calls a Time-Out Service (TIMSVC) subroutine after the required number of additional free-running counts. This software is illustrated on pages 306–308.

The time-out software uses Timer Overflow Interrupts (TOIs). The TOI interrupt service increments a counter extension word. After incrementing, the CPU compares the extension word's contents with its terminal value (number of TOFs in the specified period). A match causes the CPU to start reading and re-reading the TCNT register. The CPU executes this TCNT-read loop until TCNT contents exceed the number of required additional free-running counts. The CPU then calls the TIMSVC subroutine, which executes a measurement or control task.

The M68HC11 uses input captures to record the times of external events. Plan the interface so that the targeted event causes an input capture pin to toggle. When the IC pin toggles, the input capture mechanism records the free-running count in an Input Capture Register. This free-running count value constitutes the event's "time of occurrence."

Four M68HC11E9 PORTA pins share input capture and general purpose I/O duties. Pins PA0/IC3, PA2/IC2, and PA2/IC1 are input-only pins.[15] The PA3/IC4/OC5 pin is configurable as either input or output, and can be used for three functions: 1) general purpose I/O, 2) input captures, or 3) output compares.[16] The M68HC11E9 must specially configure this pin for input captures.

Each of these four input capture pins uses an S-R latch to capture input edges. Motorola designed this latch to be responsive to input toggles, but unresponsive to noise. The latch's input inverters have their switching points tailored for a maximum noise margin.

When the peripheral toggles an input capture pin, that pin's counterpart input capture register latches the free-running count. The CPU can then read the input capture register to access this "time-of-occurrence" data. All M68HC11E9 input capture registers are read-only registers, and are designated as TIC1, TIC2, TIC3, and TI4O5. The M68HC11E9 can configure the TI4O5 register as either a read-only input capture register or a readable and writable output compare register.[17]

The M68HC11E9 configures PORTA pin 3 for input captures by initializing 2 bits in the PACTL (Pulse Accumulator Control) register. PACTL's I4/O5 pin configures PA3 for either input captures or output compares:

    I4/O5 = 1   input captures
    I4/O5 = 0   output compares

PACTL's DDRA3 bit configures PA3 as an input or output pin:

    DDRA3 = 0   PA3 is an input
    DDRA3 = 1   PA3 is an output

Therefore, the M68HC11E9 CPU configures PA3 for input captures by clearing the DDRA3 bit and setting the I4/O5 bit in PACTL.[18]

The TFLG1 register (Main Timer Interrupt Flag Register 1) contains input capture flags IC1F–IC4F. These flags set when a peripheral applies specified edges to counterpart input capture pins. PA3/IC4/OC5 pin's flag becomes either input capture flag IC4F or output compare flag OC5F, depending upon how the M68HC11E9 configures this pin (using the I4/O5 bit in the PACTL register). ICxF bits are read/modify/write bits. The CPU clears an ICxF bit by writing a one to it.

The M68HC11E9 enables input capture interrupt control by writing ones to ICxI bits in the TMSK1 register (Main Timer Interrupt Mask Register 1). Specified IC pin toggles generate these interrupts. The I4O5I bit in TMSK1 becomes either IC4I or OC5I, depending upon the state of the I4/O5 control bit in PACTL.

---

[15]The M68HC11A1 also uses these pins for general purpose inputs and for input captures.
[16]The M68HC11A1 can use this pin for either output compares or general-purpose I/O, but not for input captures.
[17]The M68HC11A1 has a TOC5 register. This register functions for output compares only.
[18]The M68HC11A1's PACTL register does not use these 2 bits.

The CPU writes a control word to the TCTL2 register (Timer Control Register 2) to specify input capture pin active edges. TCTL2 allots two control bits (EDGxB and EDGxA) to each input capture pin. These bits allow the user to select from four IC edge specifications:

EDGxB	EDGxA	Edge Specification
0	0	Captures disabled
0	1	Positive-going edges
1	0	Negative-going edges
1	1	Both positive-going and negative-going edges

Input capture mechanisms can measure time between external events. When effectively programmed and interfaced, input captures can measure periods between events as small as 1μs, or as large as data and programming space permit. To make such measurements, the M68HC11 determines the number of free-running counts intervening between events. Software can then convert the number of free-running counts to decimal minutes, seconds, and fractions of seconds as small as 1μs.

Section 7.5 introduces software which measures and displays the period between external events—up to 35 minutes and 47 seconds. This software captures free-running counts at the times of the previous and latest events. By subtracting the previous count from the latest count, the M68HC11 CPU determines how many free-running counts intervened. The CPU converts this count difference into decimal minutes and seconds, and displays the results at the terminal.

Section 7.6 illustrates how a system designer can use input capture pins for general-purpose hardware interrupts. The M68HC11's I mask and ICxI bits (in TMSK1) provide two tiers of interrupt control. The TCTL2's EDGxB/EDGxA bits allow the designer to tailor IC pin responses to the peripherals generating the interrupts. Each IC interrupt has its own vector address, and the ICxF bits in TFLG1 facilitate control over the individual interrupt service routines.

EXERCISE 7.5 — Chapter Review

Answer the questions below by neatly entering your responses in the appropriate blanks.

1. The M68HC11's _____ provides "times" for input captures, output compares, and periodic interrupts.
2. The M68HC11's timer section sets a status flag each time the free-running count _____ from $_____ to $_____.
3. The M68HC11 accesses its current free-running count by reading the _____ register.
4. To assure that it reads a coherent free-running count value, the CPU must use a _____ read instruction.
5. The CPU selects the free-running count rate by writing to the _____ and _____ bits of the _____ register.
6. If the M68HC11 uses a 4MHz crystal, its CPU must write binary _____ to the PR1 bit, and binary _____ to the PR0 bit, to provide a free-running counter range of 524.28ms and resolution of 8μs.
7. To enable free-running counter overflow interrupts, the CPU writes binary _____ to bit #_____ of the _____ register.
8. The timer section can generate a TOI interrupt each time the _____ rolls over from $_____ to $_____.
9. In the blanks, enter source code which clears the PAOVF bit only, in TFLG2. Load a mask and store it using indexed addressing. Assume that the X register contains $1000.

   _____

   _____

10. In the blank, enter source code that clears the PAOVF bit only, in TFLG2. Use only one line of source code and indexed addressing. Assume that the X register contains $1000.

   _____

11. The _____ bit in the _____ register sets when the free-running count rolls over from $FFFF to $0000. A programmer uses this bit to extend the _____ and _____ of the _____.

12. An M68HC11 connects to a 4MHz crystal, and its PR1/PR0 bits are set. It runs software which extends its free-running counter to 24 bits. This software calls an overflow service subroutine when the free-running count rolls over from $FFFFFF to $000000. How often will the CPU call this subroutine?

    _____ minutes, _____ seconds.

13. An M68HC11 connects to a 4MHz crystal. Its PR1 bit = 1 and PR0 bit = 0. The CPU must execute a periodic measurement task every 30 seconds. Calculate the decimal number of TOFs, and the decimal number of additional free-running counts in this specified period.

    #TOFs = _____
    #Additional counts = _____

14. Given the results in question 13, how many free-running counter extension bytes does the support software require? _____ After counting the required number of TOFs, the support software calls a time-out service routine when the TCNT register's contents equal or exceed $_____.

15. If an M68HC11 uses _____ to record the times of external events, an external event must cause an _____ to toggle.

16. The M68HC11E9 can handle up to _____ (how many?) independent input capture tasks simultaneously. The M68HC11A1 can handle up to _____ such tasks.

17. Refer to Figure 7–9. After the S-R latch has responded to a positive-going edge, noise would have to drive the input _____ (higher than/lower than) the _____ (A/B) inverters' switch point to make the latch respond. After the S-R latch has responded to a negative-going edge, noise would have to drive the input _____ the _____ inverter's switch point, to make the latch respond.

18. When an external event causes the peripheral to toggle an IC pin, that pin's counterpart _____ latches the _____ at the time of the toggle. All TIC registers are _____-only. The _____ register is read-only when configured for _____. This register is readable or writable when configured for _____ purposes.

19. To configure the M68HC11E9's PA3 pin for input captures of external events, the CPU writes $_____ to address $_____.

20. If a peripheral toggles the IC3 pin with a specified edge, the TFLG1 register contains $_____ (assuming no other IC pins have toggled). Prepare TFLG1 for the next IC3 toggle with this source code (assume X = $1000):

    BCLR _____ , X _____

21. The M68HC11E9 CPU configures its IC4 and IC2 pins for interrupt control by writing $_____ to the _____ register.

22. To specify rising edges for IC4, falling edges for IC3, and disable input captures for IC1 and IC2, the M68HC11E9 CPU writes $_____ to the _____ register.

23. An M68HC11E9's timer section must be initialized per these specifications:
    Use interrupt control to extend free-running counter range.
    Use IC1 and IC4.
    IC4 captures on all edges.
    IC1 captures on rising edges.
    IC4 uses interrupt control.
    IC1 uses status polling.

Enter control words which the CPU must write to the following addresses.

```
$1021    $_____
$1022    $_____
$1024    $_____
$1026    $_____
```

24. Refer to Figure 7–15, block 2. The free-running count's extension word consists of _____ bytes.
25. Refer to Figure 7–15, block 3. Initially, LSTCNT holds the contents of the _____ and _____ at the time of initialization. After the first key-press, LSTCNT contains the _____ at the time of the _____.
26. Refer to Figure 7–15. Describe the purpose of the TOI Interrupt Service.

27. Refer to Figure 7–16. In which block does the CPU determine the number of free-running counts between the previous and most recent external events? _____
28. Refer to Figure 7–16. Explain the purpose of block 9.

29. Refer to Figure 7–17. Explain the relationship between blocks 2 and 5.

30. Refer to Figure 7–17. The CPU escapes the block 2 to block 5 loop when the block 2 subtrahend is _____ than the minuend. Loop escape indicates that the CPU has converted $EVENTIM to _____.
31. Refer to Figure 7–18. Explain why the CPU subtracts $001E8480 from $EVENTIM (block 7).

32. Refer to Figure 7–26. Describe the purpose of the three push-buttons.

33. Refer to Figure 7–27. Explain the purpose of blocks 3 through 8.

34. Refer to Figure 7–27. Explain why the CPU stacks the X register's contents in block 9 and pulls this value in block 12.

35. Refer to Figure 7–28. Given a free-running counter range of 32.77ms, the CPU initializes its loop counter to $_____ for a 1.5 second delay.
36. Figures 7–26 through 7–28 illustrate how a system designer can use IC pins for _____ _____.

8 M68HC11 Output Compare Functions

8.1 GOAL AND OBJECTIVES

Chapter 8 provides comprehensive coverage of M68HC11 output compare (OC) functions and their applications. These functions demonstrate impressive flexibility and utility. This chapter gives the reader an opportunity to develop and demonstrate software interfaces which

- Generate four simultaneous square waves, each with a unique frequency.
- Generate a square wave and measure its frequency.
- Turn output compare pins into a counter.
- Control the speed of a DC motor.

This chapter's goal is to help the reader understand how to program and interface the M68HC11's output compare mechanisms.

After studying Chapter 8 and completing its exercises, you will be able to

Derive control words that configure the M68HC11 for OC2–OC5 output compare operations.

Develop software that uses output compares to generate a square wave with a frequency range of 20 Hz–18 KHz and a 50% duty cycle.

Develop software that uses output compares to generate four simultaneous square waves with frequency ranges of 20 Hz–12 KHz and 50% duty cycles.

Develop software that uses output compares to generate a square wave, and uses input captures to measure that square wave's frequency.

Develop software that uses the M68HC11's five output compare pins as a five-stage, modulus thirty-two counter, with count rates from 30 to 46.5K counts per second.

Develop an interface and software that use output compares to control the speed of a DC motor.

8.2 ON-CHIP RESOURCES USED BY M68HC11E9 OC2–OC5 OUTPUT COMPARE FUNCTIONS

Each of the four M68HC11's output compare functions (OC2–OC5) has its own Output Compare Register (TOCx). An OCx output compare mechanism causes the M68HC11 to take programmed actions whenever the free-running count matches the contents of its TOC.[1] These actions can include the following:

1. Set, reset, or toggle a counterpart output compare pin.
2. Set a counterpart output compare status flag.
3. Generate a corresponding output compare interrupt.

Thus, output compares allow the M68HC11 to execute periodic tasks or to generate signals at OC pins. This section considers on-chip resources used by the OC2–OC5 functions. Output Compare 1 (OC1) represents a special case and is described in Section 8.5.

PORTA Pins PA3–PA6 (OC5/IC4–OC2)

When the M68HC11 successfully compares the free-running count with the contents of an output compare register, an output compare pin responds in one of four ways depending upon how it has been programmed:

1. The pin does nothing.
2. The pin sets.
3. The pin resets.
4. The pin toggles.

PORTA pins PA6–PA3 constitute these output compare pins. When not configured as OC pins, PA6–PA4 serve as general purpose outputs. When not configured for output compares, PA3 can serve as a general purpose I/O pin or input capture pin.[2] To understand how PA6–PA3 respond to these various requirements, we must review their pin logic.

Refer to Figure 8–1, which provides a block diagram of PA6–PA3 pin logic. The top block constitutes a latch for general-purpose inputs or input captures, which are used by the M68HC11E9's PA3 pin only. See pages 312–313 for a detailed discussion on this block. The following discussion treats PA6–PA3 as outputs only.[3]

The PAx Output Driver consists of N- and P-channel, enhancement mode MOS transistors in a totem-pole configuration. The output driver applies the "cheater" latch's contents to the output pin. Only one transmission gate (1, 2, or 3) operates at a time. The PORTA

FIGURE 8–1
Reprinted with the permission of Motorola.

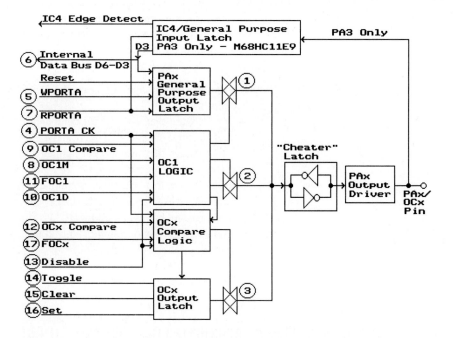

---

[1]See Chapter 7, Section 7.2 for a detailed description of the free-running counter.
[2]This statement applies to the M68HC11E9 (used on the EVBU). On the M68HC11A1 (EVB), PA3 functions identically to PA6–PA4. See the discussion in Section 7.4.
[3]Motorola Inc., *M68HC11 Reference Manual* (1990), p. 7-9. Adapted with permission.

clock (PORTA CK, 4), strobes the selected transmission gate. This gate couples a logic level to the cheater latch. PORTA CK 4 synchronizes output pin changes with the E clock.

Write PORTA (WPORTA) assertion (5) enables the PAx General-Purpose Output Latch. When thus enabled, the General-Purpose Output Latch captures a write bit from the Internal Data Bus line (6). If the CPU reads PORTA, then Read PORTA (RPORTA) assertion (7) causes the General-Purpose Output Latch to drive the Internal Data Bus line (6) with its captured state. OC1 Logic enables Transmission Gate 1 if neither OC1 Logic nor OCx logic (bottom two boxes) controls the PAx/OCx pin. If OC1 Logic enables Transmission Gate 1, the Cheater Latch captures the general-purpose latched state, and applies this level to the Output Driver. The Output Driver then drives the PAx pin.

When OC1M line 8 asserts (high), OC1 Logic takes precedence over all other PAx pin logic. When OC1M = 1, a successful OC1 compare causes OC1 Compare line 9 to assert. This assertion causes OC1 Logic to enable Transmission Gate 2. When enabled, Transmission Gate 2 couples the state of OC1D line 10 to the Cheater Latch. Thus, the OC1D state drives the PAx pin. PORTA CK 4 strobes this transfer operation.

The CPU can force an OC1 compare with FOC1 line 11. By asserting FOC1, the CPU causes OC1 logic to enable Transmission Gate 2, even though no successful OC1 compare has occurred.

The M68HC11 timer section asserts OCx Compare input 12 when the free-running count matches the contents of the OCx output compare register (TOCx). OCx Compare Logic responds to this assertion by enabling Transmission Gate 3. The state of the OCx Output Latch transfers to the Cheater Latch, then to the PAx Output Driver. Thus, a successful compare causes the PAx pin to assume the state of the OCx output latch.

The Disable (13), Toggle (14), Clear (15), and Set (16) lines determine what kind of state transfers from the OCx Output Latch to the PAx pin. Only one of these control lines asserts, while the others negate. If Disable line 13 asserts, Transmission Gate 3 will not be enabled, even when a successful compare occurs. If the Toggle line 14 asserts, the OCx Output Latch causes the PAx pin to toggle. If Clear line 15 asserts, the OCx Output Latch transfers a low to the PAx pin. If Set line 16 asserts, the OCx Output Latch transfers a high to the PAx pin. PORTA CK 4 strobes any state transfer from OCx latch to PAx output pin.

If OC1M line 8 asserts, and OC1 Compare 9 and OCx Compare 12 assert simultaneously, OC1 Logic takes priority. OC1 Logic enables Transmission Gate 2, and keeps the OCx Compare Logic from enabling Transmission Gate 3.

The CPU can force an OCx compare by asserting FOCx line 17. By asserting this line, the CPU causes OCx Compare Logic to enable Transmission Gate 3 and transfer a toggle, clear, or set to the PAx output pin. Of course, no such transfer occurs if Disable line 13 is asserted.

Thus, a PAx pin can function as an output compare (OCx) or a general-purpose output pin. The M68HC11E9's PA3 pin is a special case. The E9's PA3 pin can also function as an input capture or general-purpose input pin. When configured for output compares, PAx can respond to OC1 compares or OCx compares. OC1 output compares take priority. When configured for toggles, clears, or sets, OCx compares take precedence over general-purpose outputs. In fact, general-purpose outputs are inhibited. However, any time Disable line 13 asserts, the CPU can use the PAx pin for general-purpose outputs.[4] The M68HC11E9 can configure its PA3 pin for input captures or output compares but *not* both.

Refer to Figure 8–2.[5] This figure diagrams the OCx Compare Logic and OCx Output Latch blocks from Figure 8–1. Note that Figure 8–2 contains three main components—each surrounded by dotted lines.

- OCx Output Latch 19, which is an S-R latch.
- OCx Compare Logic 20, which consists of an OR gate, NAND gate, and inverters.
- Cheater Latch 21, which is distinct from the cheater latch that transfers states to the output driver (Figure 8–1). Note line 22, which connects to the output driver cheater latch.

In Figure 8–2, OCx Output Latch 19's Q output holds the next level for transfer to the output driver cheater latch. This level transfers via Transmission Gate 3 and line 22. PORTA CK 4 and OCx Compare Logic 20 strobe Transmission Gate 3.

---

[4] Or inputs, in the case of the M68HC11E9's PA3 pin.
[5] Ibid.

FIGURE 8-2
Reprinted with the permission of Motorola.

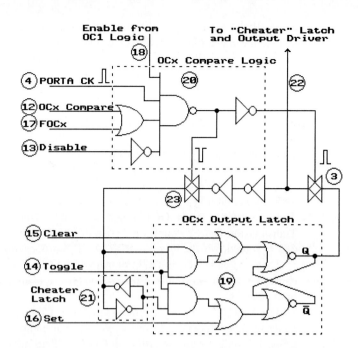

PORTA CK 4 and OCx Compare Logic strobe Transmission Gate 3 under these conditions:

1. OC1 Logic does not take precedence. When not taking precedence, OC1 Logic drives Enable line 18 high.
2. Either OCx Compare line 12 or FOCx line 17 asserts.
3. Disable line 13 is negated. As long as the Disable line = 1, OCx Compare Logic cannot enable Transmission Gate 3.

When these three conditions are met, OCx Compare Logic 20 strobes Transmission Gate 23 off. After latch 19's state has transferred to the pin's output driver, OCx Compare Logic 20 disables Transmission Gate 3 and enables Transmission Gate 23. Transmission Gate 23 couples the state of the output driver cheater latch from line 22 to Cheater Latch 21. Thus, Cheater Latch 21 captures the last state transferred from OCx Output Latch 19's Q output to the PAx pin. Note that this previous state transfers via Transmission Gate 23 to OCx Output Latch 19's top AND gate. Cheater Latch 21 inverts this previous state and applies it to the bottom AND gate. If Toggle line 14 is asserted, these previous state inputs cause OCx Output Latch 19 to toggle to a new state. Thus, Transmission Gate 23, Cheater Latch 21, and the OCx Output Latch 19's AND gates all function to make this latch toggle, *if* Toggle line 14 = 1.

If Clear line 15 = 1, OCx Output Latch 19's Q output is forced to binary zero.

If Set line 16 = 1, the OCx Output Latch 19's Q output is forced to binary one.

Timer Output Compare Registers (TOC2–TOC4 and TI4O5)

Each output compare function has its own Timer Output Compare Register (TOCx). The M68HC11 takes programmed actions when the free-running count matches the contents of a TOC. Each TOC contains 16 bits to match the free-running counter. TOC registers have the following register block addresses:

    TOC2 $1018–$1019
    TOC3 $101A–$101B
    TOC4 $101C–$101D
    TI4O5 $101E–$101F

The M68HC11E9's CPU configures the TI4O5 register as either TOC5 or TIC4 (Timer Input Capture Register 4) by writing control bits to the PACTL register. The next section reviews this initialization process.[6]

---
[6]This register is simply TOC5 on the M68HC11A1.

The CPU can either read or write to these TOC registers. The CPU writes to TOC registers to establish comparison values. Such a write must use a double-byte store instruction, such as STD, to avoid inadvertent comparisons. Such an inadvertent comparison may occur when the TOC register contains an unintended value. An unintended value results when the CPU uses two single-byte write instructions to initialize the TOC. After the first single-byte write, the TOC contains a value which is half old and half new: one byte contains part of the original TOC contents, and the other byte contains the value just written. If the free-running count happens to match this unintended value, a spurious comparison and response results. When the CPU writes to the MS byte of a TOC register, the timer mechanism prevents output compares for one bus cycle. This inhibit provides enough time to transfer the low byte to the TOC register. However, this inhibit provides enough time *only* if the CPU uses a double-byte write instruction.

Thus, the TOC registers contain values that the M68HC11's timer section compares with the free-running count. Matches generate output compares and programmed responses. The M68HC11 must use double-byte instructions when writing to TOC registers to avoid inadvertent comparisons.

Pulse Accumulator Control Register (PACTL, Control Register Block Address $1026)

Previous sections in Chapters 7 and 8 establish that the M68HC11E9 can configure PORTA pin 3 for input captures, output compares, or general-purpose I/O. The CPU uses two PACTL bits to configure the PA3 pin as either IC4 or OC5. These two bits also determine how other registers and bits operate. In fact, bits 2 and 3 of the M68HC11E9's PACTL register control the answers to these questions:[7]

1. Does pin PA3 function as an input capture pin (IC4) or as an output compare pin (OC5)?
2. Does the TI4O5 register function as an input capture register (TIC4) or as an output compare register (TOC5)?
3. Does the I4O5F bit in TFLG1 function as an input capture flag (IC4F) or as an output compare flag (OC5F)?
4. Does the I4O5I bit in TMSK1 enable Input Capture 4 Interrupts (IC4I) or Output Compare 5 Interrupts (OC5I)?
5. Should the M68HC11E9 write to TCTL2 to specify IC4 edges or to TCTL1 to specify OC5 actions?

Refer to Figure 8–3. PACTL bit 2 (I4/O5) determines whether the M68HC11E9 configures its PA3 pin for input captures or output compares:

    I4/O5 = 1   PA3 functions as Input Capture 4 (IC4) pin.
    I4/O5 = 0   PA3 functions as Output Compares 5 (OC5) pin.

PACTL bit 3 (DDRA3) determines whether the M68HC11E9 configures its PA3 pin as an input or output pin:

    DDRA3 = 1   PA3 functions as an output pin.
    DDRA3 = 0   PA3 functions as an input pin.

FIGURE 8–3
Reprinted with the permission of Motorola.

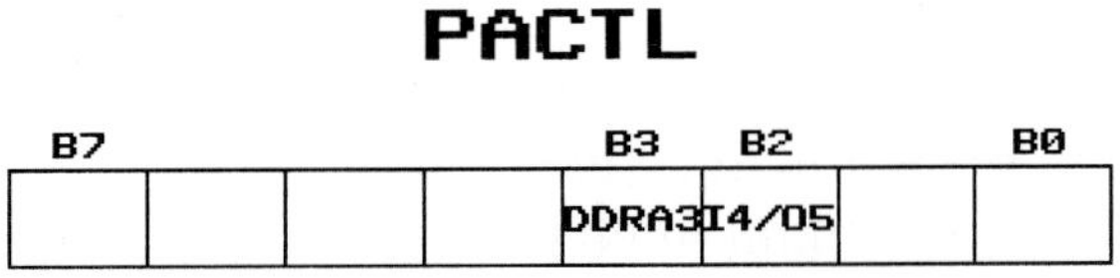

PACTL

B7			B3	B2		B0
			DDRA3	I4/O5		

---

[7]PACTL bits 2 and 3 are relevant only to the M68HC11E9 (EVBU). These bits are unused in the M68HC11A1's PACTL register. The M68HC11A1's PA3 pin can be configured only as an output compare pin (OC5) or a general-purpose output pin.

A chip reset clears both DDRA3 and I4/O5. Thus, a reset configures pin PA3 as an input, but also configures PA3 for output compares. If DDRA3 = 0, and I4/O5 = 0, the five questions are answered as follows:

1. Pin PA3 functions as an output compare pin (OC5).
2. TI4O5 functions as an output compare register (TOC5).
3. The I4O5F bit in TFLG1 functions as an output compare flag (OC5F).
4. The I4O5I bit in TMSK1 enables Output Compare 5 interrupts.
5. The M68HC11E9's CPU should write to TCTL1 to specify OC5 actions.

If the CPU writes to TCTL1 and specifies OC5 pin action (clear, set, or toggle), the output compare mechanism overrides the DDRA3 = 0 bit. Thus, the OC5/PA3 pin becomes an output so it can set, clear, or toggle as specified—even though DDRA3 = 0. If the CPU writes to TCTL1 and specifies no OC5 action, then pin PA3 reverts to a general-purpose input.

If the M68HC11E9 CPU clears the DDRA3 bit and sets the I4/O5 bit, the five questions are answered as follows:

1. Pin PA3 functions as an input capture pin (IC4).
2. TI4O5 functions as an input capture register (TIC4).
3. The I4O5F bit in TFLG1 functions as an input capture flag (IC4F).
4. The I4O5I bit in TMSK1 enables Input Capture 4 interrupts.
5. The M68HC11E9 CPU should write to TCTL2 to specify IC4 edges.

When the M68HC11E9's CPU sets the DDRA3 bit and clears the I4/O5 bit, it provides these answers to the five questions.

1. Pin PA3 functions as an output compare pin (OC5).
2. TI4O5 functions as an output compare register (TOC5).
3. The I4O5F bit in TFLG1 functions as an output compare flag (OC5F).
4. The I4O5I bit in TMSK1 enables Output Compare 5 interrupts.
5. The M68HC11E9 CPU should write to TCTL1 to specify OC5 actions. If the CPU writes to TCTL1 and specifies *no* pin action, then pin PA3 reverts to a general-purpose output.

When the M68HC11E9's CPU sets both DDRA3 and I4/O5, it answers the five questions as follows:

1. Pin PA3 functions as Input Capture 4 (IC4).
2. TI4O5 functions as an input capture register (TIC4).
3. The I4O5F bit in TFLG1 functions as an input capture flag (IC4F).
4. The I4O5I bit in TMSK1 enables Input Capture 4 interrupts.
5. Writes to either TCTL1 or TCTL2 are irrelevant. Writing to TCTL1 is irrelevant because PA3 is configured for input captures. A write to TCTL2 is irrelevant because PA3 is configured as an output.

If DDRA3 = 1 and I4/O5 = 1, the only way the M68HC11E9 CPU can generate a successful IC4 input capture is to write to PORTA bit 3.

Thus, the DDRA3 and I4/O5 bits in the M68HC11E9's PACTL register configure pin PA3 for input captures or output compares. These two bits also configure other status and control bits for input captures or output compares. Finally, DDRA3 and I4/O5 determine whether the TI4O5 register functions as an input capture register (TIC4) or an output compare register (TOC5).

Timer Control Register 1 (TCTL1, Control Register Block Address $1020)

A successful output compare can cause a counterpart OC pin to respond by setting, clearing, or toggling. Timer Control Register 1 (TCTL1) allows the user to program these OC2–OC5 responses. Refer to Figure 8–4. TCTL1 contains two control bits—OMx and OLx—for

FIGURE 8–4
Reprinted with the permission of Motorola.

```
                      TCTL1
  B7                                              B0
┌─────┬─────┬─────┬─────┬─────┬─────┬─────┬─────┐
│ OM2 │ OL2 │ OM3 │ OL3 │ OM4 │ OL4 │ OM5 │ OL5 │
└─────┴─────┴─────┴─────┴─────┴─────┴─────┴─────┘
```

each of the output compare pins OC2–OC5. These bits control the action taken by the OCx pin upon a successful compare:

OMx	OLx	OCx Pin Action
0	0	Disconnect the output compare mechanism from its OCx pin (no action).
0	1	Toggle the OCx pin.
1	0	Clear the OCx pin.
1	1	Set the OCx pin.

If the CPU clears counterpart OMx and OLx bits, the M68HC11E9's OC2-OC4 pins (PA6–PA4) can function as general-purpose outputs. If properly initialized, OC2–OC4 functions can still generate output compare flags and interrupts, even though their counterpart OC pins have been disconnected.

If the CPU clears OM5 and OL5, pin PA3/OC5 can be used as a general-purpose input *or* output, depending upon the state of the DDRA3 bit in the PACTL register. At the same time, the OC5 mechanism can still generate OC5 flags and associated interrupts.[8]

Main Timer Interrupt Mask Register 1
(TMSK1, Control Register Block Address $1022)

The CPU sets bits in the TMSK1 register to enable OC2–OC5 interrupts. When so enabled, output compare interrupts are generated by successful output compares. Each of these OC interrupts has its own vector address. The OC2 interrupt has a higher priority than OC3; OC3, higher than OC4; and OC4, higher than OC5. Of course, the I mask in the condition code register also enables and disables these interrupts. Refer to Figure 8–5. TMSK1 bits 6 (OC2I) through 3 (I4O5I) constitute the OC2–OC5 interrupt enable bits. TMSK1's I4O5I bit enables OC5 interrupts when the M68HC11E9's CPU has cleared the I4/O5 bit in its PACTL register.[9] Write a binary one to a TMSK1 OCxI bit to enable corresponding OCx interrupts. Clear the OCxI bit to disable corresponding OCx interrupts.

Main Timer Interrupt Flag Register 1
(TFLG1, Control Register Block Address $1023)

A match between the free-running count and the contents of a Timer Output Compare Register causes a corresponding flag to set in the Main Timer Interrupt Flag Register 1 (TFLG1). Figure 8–6 illustrates these OC2F, OC3F, OC4F, and I4O5F bits. TFLG1's I4O5F bit becomes the OC5F flag when the M68HC11E9's CPU resets the I4/O5 bit in its PACTL register.[10] Recall that all TFLG1 flags are read/modify/write bits. To reset an OCxF flag, the programmer must use either of these two methods:

1. Use extended or indexed mode to write a mask byte to TFLG1. This mask uses ones to specify OCxF flags to be cleared, and zeros to specify flags to remain unchanged.
2. Use the BCLR instruction with a mask that uses zeros to specify bits to be cleared and uses ones to specify bits to remain unchanged.[11]

---

[8]The M68HC11A1's PA3 pin is output only. The M68HC11A1's PACTL register contains no DDRA3 bit. Thus, the "A1's" PA3/OC5 pin functions identically to its PA6/OC2–PA4/OC4 pins.
[9]The M68HC11A1's TMSK1 register has no I4O5I bit. Rather, this bit is simply called OC5I. Likewise, the M68HC11A1's PACTL register has no I4/O5 bit.
[10]The M68HC11A1's TFLG1 register has no I4O5F bit. This bit is simply the OC5F flag.
[11]Refer to pages 303–304 for a detailed discussion on how to clear TFLG1 or TFLG2 bits.

FIGURE 8–5
Reprinted with the permission of Motorola.

TMSK1

B7				B3	B2	B1	B0
OC1I	OC2I	OC3I	OC4I	I4O5I			

FIGURE 8–6
Reprinted with the permission of Motorola.

TFLG1

B7							B0
OC1F	OC2F	OC3F	OC4F	I4O5F			

EXERCISE 8.1   Initialize the M68HC11E9 for OC2–OC5 Output Compares

AIM: Calculate control words that configure the microcontroller for OC2–OC5 output compare operations.

INTRODUCTORY INFORMATION: This exercise provides experience in calculating control words. These control words configure the M68HC11E9 for specified output compare operations.

INSTRUCTIONS: Use the specifications given, and refer to previous sections in this chapter while formulating appropriate control words. The M68HC11E9's CPU writes these control words to the PACTL, TCTL1, and TMSK1 registers to initialize them for OC2–OC5 output compares. Enter your hex control words in the blanks beside the three addresses. Enter control words that the CPU should write to each address in order to configure the timer system according to specifications. Enter "N/A" in a blank if a write to its corresponding address is unnecessary for either of two reasons: (1) a system reset initializes the register as specified, or (2) the contents of the register are not relevant to the specifications. The CPU should write zeros to all nonapplicable control register bits.

Specifications	Address	Control Word
EXAMPLE		
Use OC3		
Toggle OC3		
OC3 interrupt control		
	$1020	$ 10
	$1022	$ 20
	$1026	$ N/A
Use OC2 and OC4		
Set OC2		
Reset OC4		
OC2 status polling		
OC4 interrupt control		
	$1020	$_____
	$1022	$_____
	$1026	$_____
Use OC3 and OC5		
PA3 pin is output		
Set OC3		
Toggle OC5		
OC3 status polling		
OC5 interrupt control		

$1020	$_____
$1022	$_____
$1026	$_____

Use OC2, OC3, OC4
Toggle OC2
Toggle OC3
Reset OC4
OC2, OC4 use interrupt control
OC3 status polling

$1020	$_____
$1022	$_____
$1026	$_____

Use OC2, OC4, OC5
PA3 pin is input
Reset OC2
Toggle OC4
Set OC5
OC2 status polling
OC4, OC5 use interrupt control

$1020	$_____
$1022	$_____
$1026	$_____

Use OC2, OC3, OC4, OC5
PA3 pin is output
Toggle all pins (OC2–OC5)
All pins (OC2–OC5) use interrupt control

$1020	$_____
$1022	$_____
$1026	$_____

8.3 USE OUTPUT COMPARES TO GENERATE A SQUARE WAVE

Output compares OC2–OC5 make it easy to generate square waves. This section illustrates how a short and simple program can generate square waves with an approximate frequency range of 15 Hz to 20 Khz, and 50% duty cycles. This program uses OC5 and interrupt control.

To generate a particular frequency, the user must enter its $T/2$ value into RAM locations $0000–$0001. This value represents the hexadecimal number of free-running counts in the square wave's period (T), divided by two. For example, assume that you wish to generate a 1 KHz square wave:

$$T = 1/1 \text{ KHz} = 1\text{ms}$$
$$T/2 = 1\text{ms}/2 = 500\mu s$$

Recall that the free-running count increments at a 2 MHz rate. Therefore, each count = 500ns. Calculate the number of free-running counts in T/2:

$$(T/2)/500\text{ns} = 500\mu s/500\text{ns} = 1000$$

Thus, T/2 = 1000 timer counts. Convert this value to hexadecimal:

$$1000 = \$03E8$$

To generate a 1 KHz square wave, use the MM command to load locations $0000–$0001 with $03E8 and run the program.

Refer to Figure 8–7 and its accompanying program list. After initializing the stack pointer (block 1), the CPU stores the OC5 interrupt pseudovector to location $00D4 (block 2). In block 3, the CPU points its X register at the control register block.

351

FIGURE 8–7

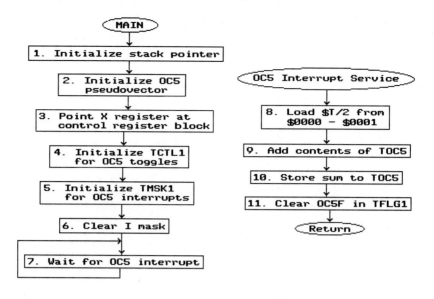

Square Wave Software

	MAIN	
$ADDRESS	SOURCE CODE	COMMENTS
0100	LDS #1FF	Initialize stack pointer
0103	LDD #120	Initialize OC5 pseudovector
0106	STD D4	" " "
0108	LDX #1000	Initialize X register
010B	LDAA #1	Initialize TCTL1 for OC5 toggles
010D	STAA 20,X	" " " "
010F	LDAA #8	Initialize TMSK1 for OC5 interrupts
0111	STAA 22,X	" " " "
0113	CLI	Clear I mask
0114	WAI	Wait for OC5 interrupt
0115	BRA 114	Wait for next OC5 interrupt

	OC5 Interrupt Service	
$ADDRESS	SOURCE CODE	COMMENTS
0120	LDD 0	Load $T/2
0122	ADDD 1E,X	Add contents of TOC5
0124	STD 1E,X	Store sum to TOC5
0126	BCLR 23,X F7	Clear OC5F in TFLG1
0129	RTI	Return

Memory Allocation

$ADDRESS	CONTENTS
$0000–$0001	$T/2

Refer to blocks 4 and 5, and MAIN program addresses $010B–$0112. Successful OC5 compares cause the OC5 pin to toggle if TCTL1 contains the proper value. TCTL1's OM5 bit must contain binary zero. The OL5 bit must contain a one. Block 4 initializes these bits. In block 5, the CPU enables OC5 interrupts by writing a binary one to the I4/O5 bit in TMSK1. OC5 is now initialized. This question arises: If this program initializes the M68HC11E9, why does it not initialize the DDRA3 and I4/O5 bits in the PACTL register?

This chapter has established that the M68HC11E9's OC5 pin can also be the IC4 pin. A user must properly initialize the PACTL register to configure this pin for output compares.

Before running the square wave generation program, the operator resets the microcontroller. This reset clears the DDRA3 and I4/O5 bits in the PACTL register. Thus, the I4/O5 bit configures the PA3/IC4/OC5 pin for output compares. The reset DDRA3 bit configures this pin as an input pin. Another question arises: If the OC5 pin is configured as an input pin, how can the M68HC11E9 generate an output square wave? This question, too, has been answered. If the CPU writes to TCTL1 and specifies OC5 pin action (clear, set, or toggle), the output compare mechanism overrides DDRA3.

Refer to blocks 6 and 7, and MAIN program addresses $0113–$0116. The condition code register's I mask enables and disables OC5 interrupts. Recall that a microcontroller reset causes the I mask to set. Therefore, the block 6 instruction clears the I mask. In block 7, the CPU waits for OC5 interrupts. Every time a T/2 period elapses, a successful OC5 compare sets the OC5F flag in TFLG1, generates an OC5 interrupt, and causes an automatic OC5 pin toggle. The OC5 Interrupt Service routine updates the TOC5 register for the next OC5 compare, and it clears the OC5F bit in TFLG1.

Refer to blocks 8, 9, and 10, and OC5 Interrupt Service addresses $0120–$0125. In these blocks, the CPU updates the TOC5 register with the next compare value. The OC5 pin has just toggled. It must toggle again after $T/2 free-running counts, i.e., it must toggle two times for each period of the square wave. Thus, the CPU loads the $T/2 value from RAM (block 8) and adds the current TOC5 value (block 9). The current TOC5 value generated the present OC5 interrupt. Therefore, if the CPU adds $T/2 to this current value, the sum will generate the next OC5 interrupt. The CPU writes this sum to TOC5 in block 10.

Refer to block 11 and OC5 Interrupt Service addresses $0126–$0128. An OC5 interrupt is generated when the OC5F bit sets. The M68HC11 must therefore clear this bit. Since TFLG1 bits are read/modify/write bits, the CPU uses the BCLR instruction. This instruction uses a mask that specifies the OC5F bit with a zero.[12]

The CPU then returns to the main program, block 7, to wait for the next interrupt, which occurs after $T/2 free-running counts elapse.

EXERCISE 8.2 Generate Square Waves with OC5

REQUIRED EQUIPMENT: Motorola M68HC11EVBU (EVB) and terminal
Oscilloscope and probe

GENERAL INSTRUCTIONS: In this laboratory exercise, implement the Section 8.3 software. Use an oscilloscope to verify and analyze OC5 output rectangular waveforms. Carefully execute the instructions below.

PART I: Connect the Oscilloscope

Connect oscilloscope channel 1 to pin 31 (OC5 pin) of the EVBU (EVB) header. Assure that the oscilloscope and EVBU (EVB) share a common ground. Show your connections to the instructor.

_____ *Instructor Check*

PART II: Software Development

Before proceeding, review Figure 8–7 and its accompanying program list. As listed, this software is compatible with the EVBU. If you use an EVB, modify the listed program for these memory allocations:

$ADDRESS	CONTENTS
0000–0001	$T/2
C000	MAIN program
C020	OC5 Interrupt Service routine
C0FF	Initial stack pointer

---

[12]Refer to pages 303–304 for a detailed discussion on TFLG1 and TFLG2 read/modify/write bits.

PART III: Debug and Demonstration

Calculate $T/2 for a 5 KHz square wave. Enter your calculated value in the blank:

$T/2 = $_____

Load and debug your program. Use your calculated value (above) as $T/2. When sure that the software runs properly, demonstrate its operation to the instructor. Show the instructor the 5 KHz output waveform on the oscilloscope.

_____ Instructor Check

PART IV: Laboratory Analysis

Step 1. Refer to the M68HC11 instruction set and calculate the number E clock cycles required to do the following:

1. Execute the WAI instruction at program address $0114 ($C014), including the interrupt vector fetch.
2. Execute the pseudovector JMP instruction.
3. Execute the OC5 Interrupt Service routine, including the RTI instruction.
4. Execute the BRA instruction at address $0115 ($C015).

Record the number of required E clock cycles in each blank below. Find the decimal Total of E clock cycles, and convert this Total to hexadecimal:

Execution	E CK Cycles
WAI + vector fetch (minimum)	_____
JMP (pseudovector)	_____
OC5 Interrupt Service	_____
BRA	_____
Decimal Total	_____
$Total	_____

$Total represents the minimum $T/2 value that will allow the M68HC11 to accurately produce an output square wave. Convert this minimum $T/2 value to predicted output maximum frequency (f):

$T/2_{MIN} \times 2 = $_____ = _____(decimal)

(decimal T/2 × 2) × 500ns = T = _____µs

$1/T = f_{MAX} = $_____KHz

You have just calculated maximum frequency that the OC5 function can reliably produce. Verify this prediction by observing and describing OC5 response for the following $T/2 values. Your descriptions should include a predicted decimal frequency based upon the given $T/2 value—compared with actual output performance as observed and measured using the oscilloscope.

$T/2 Value	Observations
32	_____
33	_____
34	_____

35

36

37

38

Step 2. The minimum frequency that OC5 can produce is based upon the maximum $T/2 value. What is maximum $T/2?

$T/2_{MAX}$ = $_____

Based upon this maximum $T/2, what is the predicted minimum decimal frequency that OC5 can produce?

f_{MIN} = $_____ Hz

Verify your prediction by observing and describing OC5 response for $T/2_{MAX}$. Enter your observations in the space below. Your observations should include f as measured using the oscilloscope.

Step 3. Increase OC5's output frequency range by converting your software from interrupt control to status polling. Refer to Figure 8–8 as a guide to software revision. List your revised software in the blanks below.

OC5 Output Square Wave—Status Polling

$ADDRESS	SOURCE CODE	COMMENTS

Step 4. Load, debug, and demonstrate your modified software to the instructor.

_____ *Instructor Check*

FIGURE 8-8

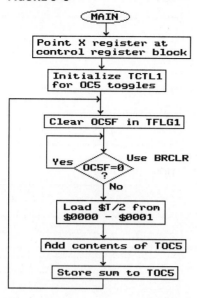

Step 5. Analyze the revised software. Use a systematic empirical method to determine the highest and lowest frequencies that OC5 can produce. Assure that the highest frequency is not an isolated one. That is, all greater $T/2 values must also generate reliable square waves. Enter your high and low values in the blanks:

	$T/2	Predicted f	Measured f
f_{HIGH}	_____	_____ KHz	_____ KHz
f_{LOW}	_____	_____ Hz	_____ Hz

PART V: Laboratory Software Listings

MAIN PROGRAM—EVB

$ADDRESS	SOURCE CODE	COMMENTS
C000	LDS #C0FF	Initialize stack pointer
C003	LDD #C020	Initialize OC5 pseudovector
C006	STD D4	" " "
C008	LDX #1000	Initialize X register
C00B	LDAA #1	Initialize TCTL1 for OC5 toggles
C00D	STAA 20,X	" " " "
C00F	LDAA #8	Initialize TMSK1 for OC5 interrupts
C011	STAA 22,X	" " " "
C013	CLI	Clear I mask
C014	WAI	Wait for OC5 interrupt
C015	BRA C014	Wait for next OC5 interrupt

MAIN PROGRAM—EVBU

(See program list accompanying Figure 8–7.)

OC5 INTERRUPT SERVICE ROUTINE—EVB

$ADDRESS	SOURCE CODE	COMMENTS
C020	LDD 0	Load $T/2
C022	ADDD 1E,X	Add contents of TOC5
C024	STD 1E,X	Store sum to TOC5
C026	BCLR 23,X F7	Clear OC5F in TFLG1
C029	RTI	Return

OC5 INTERRUPT SERVICE ROUTINE—EVBU

(See program list accompanying Figure 8–7.)

SOFTWARE MODIFICATION—EVB

$ADDRESS	SOURCE CODE	COMMENTS
C000	LDX #1000	Initialize X register
C003	LDAA #1	Initialize TCTL1 for OC5 toggles
C005	STAA 20,X	" " " "
C007	BCLR 23,X F7	Clear OC5F in TFLG1
C00A	BRCLR 23,X 8 C00A	Branch to self until OC5F = 1
C00E	LDD 0	Load $T/2
C010	ADDD 1E,X	Add contents of TOC5
C012	STD 1E,X	Store sum to TOC5
C014	BRA C007	Do another loop

SOFTWARE MODIFICATION—EVBU

$ADDRESS	SOURCE CODE	COMMENTS
0100	LDX #1000	Initialize X register
0103	LDAA #1	Initialize TCTL1 for OC5 toggles
0105	STAA 20,X	" " " "
0107	BCLR 23,X F7	Clear OC5F in TFLG1
010A	BRCLR 23,X 8 10A	Branch to self until OC5F = 1
010E	LDD 0	Load $T/2
0110	ADDD 1E,X	Add contents of TOC5
0112	STD 1E,X	Store sum to TOC5
0114	BRA 107	Do another loop

EXERCISE 8.3 — Generate Four Simultaneous Square Waves with OC2–OC5

REQUIRED EQUIPMENT: Motorola M68HC11EVBU (EVB) and terminal
Oscilloscope and probe

GENERAL INSTRUCTIONS: In this laboratory exercise, use the same software strategy as illustrated in Figure 8–7. However, use a single MAIN program, along with four OC Interrupt Service routines, to generate four simultaneous square waves. OC2–OC5 generate these simultaneous waveforms.

PART I: Software Development

Refer to Figure 8–7 and its accompanying program list. Also refer to the allocation chart at the top of page 358. Develop your software.

Calculate $T/2 values for OC2–OC5 based upon the following data. Enter your calculated values in the appropriate blanks.

Frequency	Output Pin	T	$T/2
10 KHz	OC2/PA6	____μs	____
3.3 KHz	OC3/PA5	____μs	____
1.667 KHz	OC4/PA4	____μs	____
833 Hz	OC5/PA3	____ms	____

PART II: Provide for Oscilloscope Connections

Make provisions to connect and disconnect the oscilloscope probes from the M68HC11's OC2–OC5 output pins.[13]

---

[13] If a multichannel oscilloscope or logic analyzer is available, use it to observe all OC2–OC5 outputs simultaneously.

Four Simultaneous Square Waves—Memory Allocations

EVBU $ADDRESS	EVB $ADDRESS	CONTENTS
0000–0001	0000–0001	$T/2–OC2
0002–0003	0002–0003	$T/2–OC3
0004–0005	0004–0005	$T/2–OC4
0006–0007	0006–0007	$T/2–OC5
00D3–00D5	00D3–00D5	OC5 pseudovector
00D6–00D8	00D6–00D8	OC4 pseudovector
00D9–00DB	00D9–00DB	OC3 pseudovector
00DC–00DE	00DC–00DE	OC2 pseudovector
0100	C000	MAIN program
0140	C040	OC2 Interrupt Service routine
0150	C050	OC3 Interrupt Service routine
0160	C060	OC4 Interrupt Service routine
0170	C070	OC5 Interrupt Service routine
01FF	C0FF	Initial Stack Pointer

PART III: Debug and Demonstration

Load and debug your software. When sure that it operates properly, demonstrate it to the instructor. Use $T/2 values from the chart above for your demonstration. Be prepared to reconnect the oscilloscope probes to any of the OC2–OC5 pins, as requested by the instructor.

_____ *Instructor Check*

PART IV: Laboratory Analysis

Answer the questions below by neatly entering your answers in the appropriate blanks.

1. Determine the maximum frequency (f_{MAX}) that OC2 can generate while OC3–OC5 generate their assigned frequencies, per Part I of this exercise. Use a systematic empirical test procedure to make this determination. Assure that this maximum frequency value is not isolated. That is, all greater $T/2 values should also produce reliable square waves. In the blanks, enter this maximum frequency's measured and calculated values along with its corresponding $T/2 value. Use empirically derived minimum $T/2 to find f_{MAX} (calculated).

    f_{MAX} (measured) = _____ KHz
    f_{MAX} (calculated) = _____ KHz
    $T/2_{MIN}$ = $_____

2. Compare the above results with the maximum frequency value obtained in question 1, part IV of Exercise 8.2. How does OC2's f_{MAX} compare with OC5's f_{MAX} obtained in Exercise 8.2?

    OC2's f_{MAX} was _____ (greater) _____ (lesser) than OC5's f_{MAX}.

    Explain the discrepancy between these two maximum frequencies.

3. Determine the minimum frequency (f_{MIN}) that OC2 can generate while OC3–OC5 generate their assigned frequencies. In the blanks, enter this minimum frequency's measured and calculated values along with its corresponding $T/2 value.

    f_{MIN} (measured) = _____ Hz
    f_{MIN} (calculated) = _____ Hz
    $T/2_{MAX}$ = $_____

4. Compare the above results with minimum frequency obtained for question 2, Part IV of Exercise 8.2. How does OC2's minimum compare with OC5's minimum in Exercise 8.2? _____

Explain why you obtained these results.

PART V: Laboratory Software Listing

MAIN—EVB

$ADDRESS	SOURCE CODE	COMMENTS
C000	LDS #C0FF	Initialize stack pointer
C003	LDD #C040	Initialize OC2–OC5 pseudovectors
C006	STD DD	" " " "
C008	ADDB #10	" " " "
C00A	STD DA	" " " "
C00C	ADDB #10	" " " "
C00E	STD D7	" " " "
C010	ADDB #10	" " " "
C012	STD D4	" " " "
C014	LDX #1000	Initialize X register
C017	LDAA #55	Initialize TCTL1 for OC2–OC5 toggles
C019	STAA 20,X	" " " " "
C01B	LDAA #78	Enable OC2–OC5 interrupts in TMSK1
C01D	STAA 22,X	" " " " "
C01F	CLI	Clear I mask
C020	BRA C020	Wait for OC2–OC5 interrupts

MAIN—EVBU

$ADDRESS	SOURCE CODE	COMMENTS
0100	LDS #1FF	Initialize stack pointer
0103	LDD #140	Initialize OC2–OC5 pseudovectors
0106	STD DD	" " " "
0108	ADDB #10	" " " "
010A	STD DA	" " " "
010C	ADDB #10	" " " "
010E	STD D7	" " " "
0110	ADDB #10	" " " "
0112	STD D4	" " " "
0114	LDX #1000	Initialize X register
0117	LDAA #55	Initialize TCTL1 for OC2–OC5 toggles
0119	STAA 20,X	" " " " "
011B	LDAA #78	Enable OC2–OC5 interrupts in TMSK1
011D	STAA 22,X	" " " " "
011F	CLI	Clear I mask
0120	BRA 120	Wait for OC2–OC5 interrupts

OC2 INTERRUPT SERVICE—EVB

$ADDRESS	SOURCE CODE	COMMENTS
C040	LDD 0	Load $T/2–OC2
C042	ADDD 18,X	Add TOC2 contents
C044	STD 18,X	Update TOC2
C046	BCLR 23,X BF	Clear OC2F in TFLG1
C049	RTI	Return

OC2 INTERRUPT SERVICE—EVBU

$ADDRESS	SOURCE CODE	COMMENTS
0140	LDD 0	Load $T/2–OC2
0142	ADDD 18,X	Add TOC2 contents
0144	STD 18,X	Update TOC2

| 0146 | BCLR 23,X BF | Clear OC2F in TFLG1 |
| 0149 | RTI | Return |

OC3 INTERRUPT SERVICE–EVB

$ADDRESS	SOURCE CODE	COMMENTS
C050	LDD 2	Load $T/2 - OC3
C052	ADDD 1A,X	Add TOC3 contents
C054	STD 1A,X	Update TOC3
C056	BCLR 23,X DF	Clear OC3F in TFLG1
C059	RTI	Return

OC3 INTERRUPT SERVICE—EVBU

$ADDRESS	SOURCE CODE	COMMENTS
0150	LDD 2	Load $T/2 - OC3
0152	ADDD 1A,X	Add TOC3 contents
0154	STD 1A,X	Update TOC3
0156	BCLR 23,X DF	Clear OC3F in TFLG1
0159	RTI	Return

OC4 INTERRUPT SERVICE—EVB

$ADDRESS	SOURCE CODE	COMMENTS
C060	LDD 4	Load $T/2–OC4
C062	ADDD 1C,X	Add TOC4 contents
C064	STD 1C,X	Update TOC4
C066	BCLR 23,X EF	Clear OC4F in TFLG1
C069	RTI	Return

OC4 INTERRUPT SERVICE—EVBU

$ADDRESS	SOURCE CODE	COMMENTS
0160	LDD 4	Load $T/2–OC4
0162	ADDD 1C,X	Add TOC4 contents
0164	STD 1C,X	Update TOC4
0166	BCLR 23,X EF	Clear OC4F in TFLG1
0169	RTI	Return

OC5 INTERRUPT SERVICE—EVB

$ADDRESS	SOURCE CODE	COMMENTS
C070	LDD 6	Load $T/2–OC5
C072	ADDD 1E,X	Add TOC5 contents
C074	STD 1E,X	Update TOC5
C076	BCLR 23,X F7	Clear OC5F in TFLG1
C079	RTI	Return

OC5 INTERRUPT SERVICE—EVBU

$ADDRESS	SOURCE CODE	COMMENTS
0170	LDD 6	Load $T/2–OC5
0172	ADDD 1E,X	Add TOC5 contents
0174	STD 1E,X	Update TOC5
0176	BCLR 23,X F7	Clear OC5F in TFLG1
0179	RTI	Return

8.4 INTEGRATE INPUT CAPTURES WITH OUTPUT COMPARES

Figures 8–9 through 8–17 and their accompanying program lists show how the M68HC11 can integrate output compare and input capture functions. This software causes the M68HC11 to generate a square wave at its OC2 (PA6) pin and to measure this output's frequency using IC2.

FIGURE 8–9

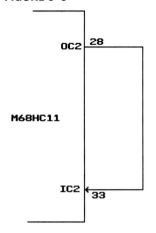

OC/IC Integration—Memory Allocations I

$ADDRESS	CONTENTS
0000	$T MS
0001	$T LS
0002	$T/2 MS
0003	$T/2 LS
0005	&f MS
0006	&f MID
0007	&f LS
0008	$f MS
0009	$f LS
000A	DIFF MS
000B	DIFF MID
000C	DIFF LS
000E	<CR> ASCII
000F	"I" ASCII
0010	"N" ASCII
0011	"P" ASCII
0012	"U" ASCII
0013	"T" ASCII
0014	<SP> ASCII
0015	"&" ASCII
0016	"f" ASCII
0017	<SP> ASCII
0018	<EOT> ASCII
0019	<CR> ASCII
001A	"&" ASCII
001B	"f" ASCII

$ADDRESS	CONTENTS
001C	<SP> ASCII
001D	"=" ASCII
001E	<SP> ASCII
001F	<EOT>
0020	BCD 10,000's
0021	BCD 1,000's
0022	BCD 100's
0023	BCD 10's
0024	BCD 1's
0025	TIC2#1 MS
0026	TIC2#1 LS

OC/IC Integration—Memory Allocations II

$ADDRESS	CONTENTS
C000	MAIN program
C01A	&fINPUT subroutine
C035	CK&fBNDS subroutine
C055	&fTO$f subroutine
C095	$fTOT/2 subroutine
C0C5	OC/IC subroutine
EEPROM	
B700	$TTO&f subroutine
B73A	DISP&f subroutine
B755	OC2 Interrupt Service routine

This integration software also makes frequency selection and measurement user friendly. Program operation can be summarized as follows:

1. The M68HC11 prompts the user to select a five-digit decimal frequency:

   `INPUT &f`

2. The user inputs a five-digit frequency request from the keyboard. The M68HC11 echoes back the requested value on the terminal display. The user must select a frequency in the range 00032–19999 Hz. The microcontroller responds to out-of-bounds entries by re-prompting the user to "INPUT &f."
3. Upon entry of the final decimal digit of an inbounds frequency value, the M68HC11 generates a square wave with a 50% duty cycle at the requested frequency.
4. The microcontroller measures the frequency of this output square wave and displays the measured value—in decimal—on the terminal display:

   `&f = XXXXX`

For necessary pin connections, refer to Figure 8–9. These connections entail a single wire between the OC2 (PA6) and IC2 (PA1) pins. A memory allocations list accompanies Figure 8–9.

Figure 8–10 and its accompanying program list depict the MAIN program. This software uses a modular strategy. The MAIN program consists mostly of subroutine calls. After initializing its stack pointer (block 1), the CPU calls the &fINPUT subroutine (block 2, program address $C003). &fINPUT prompts the user to enter a requested decimal frequency value. As the user enters digits from the keyboard, the CPU captures them, converts them from ASCII to BCD, and stores them in RAM locations $0020–$0024.

Refer to Figure 8–10, block 3 (program address $C006). After capturing and storing the requested frequency, the CPU calls the CK&fBNDS subroutine. This subroutine checks the

FIGURE 8-10

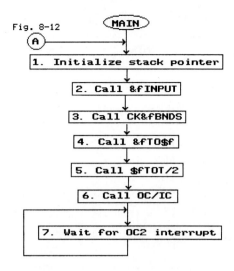

Main Program

$ADDRESS	SOURCE CODE	COMMENTS
C000	LDS #47	Initialize stack pointer
C003	JSR C01A	Call &fINPUT subroutine
C006	JSR C035	Call CK&fBNDS subroutine
C009	JSR C055	Call &fTO$f subroutine
C00C	JSR C095	Call $fTOT/2 subroutine
C00F	JSR C0C5	Call OC/IC subroutine
C012	WAI	Wait for OC2 interrupt
C013	BRA C012	Make another output edge

user's entry to determine whether it is inbounds. If the user entered an out-of-bounds value, the CPU returns to the MAIN program, block 1, where it re-initializes the stack pointer and again prompts the user to enter a frequency value. If the user enters a decimal value from 19999 to 00032, the CPU returns from the CK&fBNDS subroutine to block 4.

In Figure 8–10, block 4 (program address $C009), the CPU calls the &fTO$f subroutine. &fTO$f converts the BCD values stored at $0020–$0024 to hexadecimal. &fTO$f then stores the hexadecimal frequency value to RAM locations $0008 and $0009 (see Memory Allocations list).

Refer to Figure 8–10, block 5 (program address $C00C). After converting frequency from BCD to hexadecimal, the CPU calls the $fTOT/2 subroutine. To generate a square wave at the desired frequency, the CPU must convert hexadecimal frequency ($f) to its period value $T, then halve $T. The $fTOT/2 subroutine derives $T as the number of free-running counts—at 500ns per count—in the square wave's period. $fTOT/2 finds this value by "reciprocating" $f. To do this, the CPU divides the hex number of free-running counts in one second ($1E8480) by $f. The result equals $T, which is expressed in free-running counts. The CPU then divides $T by 2 to find $T/2. The CPU must add this value to the TOC2 register's contents after each successful output compare—to generate the requested frequency at a 50% duty cycle. The $fTOT/2 subroutine stores the $T value at RAM locations $0000–$0001. It stores $T/2 at RAM locations $0002–$0003.

Refer to Figure 8–10, block 6 (program address $C0CF). After deriving $T and $T/2, the M68HC11 calls the OC/IC subroutine. This subroutine generates initial output edges from the OC2 pin and makes a frequency measurement using IC2. After displaying measured frequency at the terminal, the CPU returns from the OC/IC subroutine where it waits for OC2 interrupts (block 7, program address $C012). The software uses OC2 interrupt control to generate the output square wave and uses IC2 status polling to make the frequency measurement.

FIGURE 8–11

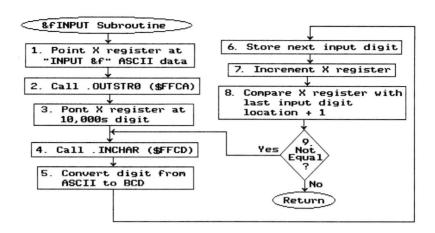

&fINPUT Subroutine

$ADDRESS	SOURCE CODE	COMMENTS
C01A	LDX #E	Point X register at "INPUT&f" ASCII data
C01D	JSR FFCA	Call .OUTSTR0
C020	LDX #20	Point X register at 10,000's digit
C023	JSR FFCD	Call .INCHAR
C026	ANDA #F	Convert input digit from ASCII to BCD
C028	STAA 0,X	Store input digit
C02A	INX	Point X register at next input digit storage location
C02B	CPX #25	Compare X register with last input digit storage location +1
C02E	BNE C023	If unequal, capture another input digit
C030	RTS	Return

Figure 8–11 and its accompanying program list illustrate the &fINPUT subroutine. This subroutine prompts the user to enter a requested frequency value. &fINPUT then captures the decimal digits entered by the user, converts them to BCD, and stores them in designated RAM locations.

The CPU begins &fINPUT execution by pointing its X register at the ASCII prompt data—located in RAM at addresses $000E–$0018 (see Figure 8–11, block 1 and program address $C01A). The CPU then calls the .OUTSTR0 utility subroutine (block 2). This subroutine writes the prompt "INPUT &f" to the terminal screen. Next, the CPU prepares to write key-press data to designated addresses in RAM: $0020–$0024. The CPU points its X register at the RAM location assigned to the BCD 10,000's digit (see Figure 8–11, Block 3 and program address $C020).

After prompting the user and initializing the X register for data storage, the CPU enters a block 4—block 9 loop. In this loop, the CPU captures key-presses, converts ASCII key-press data to BCD, and stores BCD digits to appropriate RAM locations (see Figure 8–11 and program addresses $C023–$C02F). In block 4, the CPU calls the .INCHAR utility subroutine. .INCHAR puts the CPU in a loop until the user presses a terminal key. .INCHAR then reads the ASCII key-press data to ACCA. In block 5, the CPU converts the ASCII key-press data to BCD by ANDing it with mask $0F. The CPU then stores the BCD digit to its appropriate location in RAM (block 6). In block 7, the CPU increments the index register so that it points at the next RAM BCD digit location. In block 8, the CPU compares this incremented X register value with the last BCD storage location + 1. If the X register value does not equal the last BCD location + 1, the CPU branches back from block 9 to block 4 to execute another loop. However, if the X register has incremented to $0025, the CPU has captured, converted, and stored all key-press data. Therefore, the CPU returns from &fINPUT.

Figure 8–12 and its accompanying software illustrate how the CPU checks BCD key-press data to assure that it is inbounds, i.e., within the range 19999 to 00032.

FIGURE 8-12

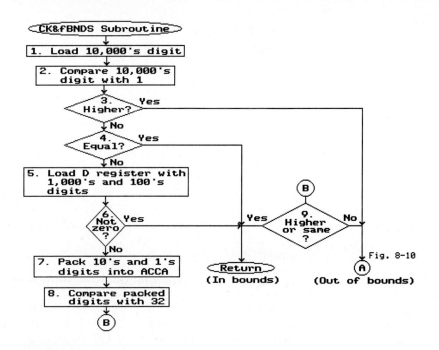

CK&fBNDS Subroutine

$ADDRESS	SOURCE CODE	COMMENTS
C035	LDAA 20	Load 10,000's digit
C037	CMPA #1	Compare digit with 1
C039	BHI C04D	If higher, &f is out-of-bounds
C03B	BEQ C050	If equal, &f is inbounds
C03D	LDD 21	Load 1,000's and 100's digits
C03F	BNE C050	If digits are not zero, &f is in bounds
C041	LDAA 23	Pack 10's and 1's digits
C043	ASLA	" " " "
C044	ASLA	" " " "
C045	ASLA	" " " "
C046	ASLA	" " " "
C047	ADDA 24	" " " "
C049	CMPA #32	Is &f equal to or higher than &32?
C04B	BCC C050	If yes, &f is inbounds
C04D	JMP C000	If &f is out-of-bounds, re-prompt user
C050	RTS	Return

In blocks 1 through 4 of Figure 8-12 (program addresses $C035-$C03C), the M68HC11 checks the BCD 10,000's digit to see whether it indicates an inbounds or out-of-bounds condition. After loading the 10,000's digit (block 1), the CPU compares it with one (block 2). In block 3, the CPU determines whether the 10,000's digit is higher than one. A 10,000's digit more than 1 indicates a keyboard entry of 2XXXX or more, an out-of-bounds condition. In such a case, the CPU branches back to the MAIN program to re-prompt the user for another request. In block 4, the CPU checks to see whether the 10,000's digit is equal to one. A 10,000's digit = 1 indicates a keyboard entry of 1XXXX, an in-bounds value. Therefore, the CPU branches to the RTS instruction (program address $C050) and returns to the MAIN program. If the 10,000's digit is not higher than one, and is not equal to one, the keyboard entry must be 0XXXX—which may or may not be in-bounds. The CPU must check the remaining BCD digits to ascertain whether requested frequency is too low.

Refer to Figure 8–12, blocks 5 and 6, and program addresses $C03D–$C040. In block 5, the CPU reads the BCD 1,000's and 100's digits to its D register in order to set/reset the Z flag. The CPU then checks to see if either of these digits is *not* a zero (block 6). If either or both of these digits are not zeros, the requested frequency is higher than 00032 and lower than 19999, i.e., inbounds. In this case, the CPU branches to the RTS instruction and returns to the MAIN program. However, if the 1,000's and 100's digits are both zeros, the requested frequency = 000XX. The CPU must now check the 10's and 1's BCD digits to assure that they are higher than 32.

Refer to Figure 8–12, blocks 7, 8, and 9, and program addresses $C041–$C04F. In block 7, the M68HC11 packs the BCD 10's and 1's digits into its A accumulator. The CPU can then check to see whether these packed digits are equal to or higher than 32, which is the minimum allowable value. In block 8, the CPU compares the packed BCD digits with 32. The CPU then determines whether these digits are higher or the same as 32. If they are, the requested frequency is inbounds, and the CPU branches to the RTS instruction. If the packed 10's and 1's digits are lower than 32, the requested frequency is out-of-bounds. In this case, the CPU jumps back to the beginning of the MAIN program to re-prompt the user for another request. If the CPU finds the requested frequency to be inbounds, it must convert this decimal value's BCD digits to hexadecimal.

Figure 8–13 and its accompanying program list show how the CPU converts five BCD digits—representing requested frequency—to hexadecimal. After pointing its X register at the BCD 10,000's digit (block 1), the CPU clears the hexadecimal frequency ($f) locations in RAM (block 2, program addresses $C058–$C05C). The CPU will accumulate its hex results in these $f locations.

In blocks 3, 4, and 5 of Figure 8–13 (program addressees $C05D–$C067), the CPU converts the BCD 10,000's digit to hexadecimal. Recall that this digit equals either one or zero. The CK&fBNDS subroutine disqualifies any higher value as out-of-bounds. In block 3, The CPU decrements the 10,000's digit. If block 3 execution decrements the 10,000's digit from one to zero, the N flag resets. If the 10,000's digit decrements from zero to $FF, the N flag sets. In block 4, the CPU tests the N flag. If N = 0, the CPU executes block 5. If N = 1, the CPU branches around Block 5. To execute Block 5, the CPU adds the hexadecimal equivalent of 10,000 to $f.

In blocks 6, 7, and 8 of Figure 8–13 (program addressees $C068 -$C074), the CPU converts the BCD 1,000's digit to hexadecimal. The 1,000's digit can be any value from one to nine. In block 6, the CPU decrements the 1,000's digit. If this decrement does not set the N flag (block 7), the CPU adds the hexadecimal equivalent of 1,000 to $f (block 8). The CPU stays in the block 6, 7, 8 loop until the BCD 1,000's digit decrements from zero to $FF, setting the N flag. When the N flag sets, $3E8 (1,000) has been added to $f as many times as indicated by the original BCD 1,000's digit.

The CPU uses the MUL instruction to convert the BCD 100's digit. Refer to Figure 8–13, blocks 9 through 12, and program addresses $C075–$C07D. The MUL instruction multiplies the contents of the A and B accumulators and returns a 16-bit product to the D register. Therefore, the CPU loads ACCA with the BCD 100's digit (block 9). The CPU then loads ACCB with the hexadecimal equivalent of 100 (block 10). In block 11, the CPU executes the MUL instruction. In block 12, it adds the product to $f.

The CPU also uses the MUL instruction to convert the BCD 10's digit (see blocks 13 through 16 and program addresses $C07E–$C086). The CPU multiplies the 10's digit by $A, the hexadecimal equivalent of 10. Finally, in block 17 (program addresses $C087–$C08D), the CPU adds the BCD 1's (unit's) digit to $f. This addition requires 2-byte precision because of possible carries into the high byte of the sum. The CPU has now converted decimal frequency (&f) to hexadecimal frequency ($f) and can return to the MAIN program.

The M68HC11 must generate a square wave at the requested frequency with a 50% duty cycle. The microcontroller uses output compares to generate this square wave. To do so, the CPU adds a hexadecimal value to its TOC2 register contents after each OC2 (PA6) pin toggle. This hexadecimal augend equals the number of free-running counts (at 500ns each) in the pulse width of the square wave, i.e., 50% of the period ($T). Therefore, the CPU must convert hexadecimal frequency ($f) to the number of free-running counts in that frequency's

FIGURE 8–13

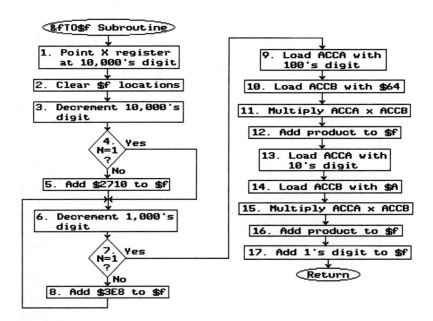

&fTo$f Subroutine

$ADDRESS	SOURCE CODE	COMMENTS
C055	LDX #20	Initialize X register
C058	LDD #0	Clear $f locations
C05B	STD 8	" "
C05D	DEC 0,X	Decrement 10,000's digit
C05F	BMI C068	If N=1, process 1,000's digit
C061	LDD 8	If N=0, add $2710 to $f
C063	ADDD #2710	" " " "
C066	STD 8	" " " "
C068	DEC 1,X	Decrement 1,000's digit
C06A	BMI C075	If N=1, process 100's digit
C06C	LDD 8	If N=0, add $3E8 to $f
C06E	ADDD #3E8	" " " "
C071	STD 8	" " " "
C073	BRA C068	Continue 1,000's processing
C075	LDAA 22	Load ACCA with 100's digit
C077	LDAB #64	Load ACCB with $64
CO79	MUL	ACCA x ACCB —> D register
C07A	ADDD 8	Add product to $f
C07C	STD 8	" " "
C07E	LDAA 23	Load ACCA with 10's digit
C080	LDAB #A	Load ACCB with $A
C082	MUL	ACCA x ACCB —> D register
C083	ADDD 8	Add product to $f
C085	STD 8	" " "
C087	LDAB 24	Add 1's digit to $f
C089	CLRA	" " "
C08A	ADDD 8	" " "
C08C	STD 8	" " "
C08E	RTS	Return

period ($T). It must then halve $T to find the hexadecimal value ($T/2) it will add to the TOC2 register's contents. To make this conversion, the CPU calls the $fTOT/2 subroutine. See Figure 8–14 and accompanying software list.

FIGURE 8-14

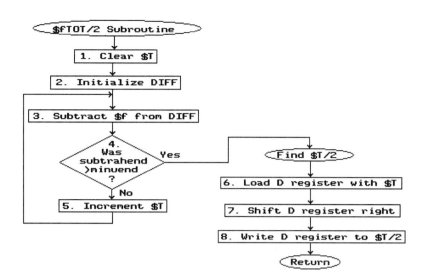

$fTOT/2 Subroutine

$ADDRESS	SOURCE CODE	COMMENTS
C095	LDD #0	Clear $T
C098	STD 0	" "
C09A	LDD #8480	Initialize DIFF to $1E8480
C09D	STD B	" " " " "
C09F	LDAA #1E	" " " " "
C0A1	STAA A	" " " " "
C0A3	LDD B	Subtract $f from DIFF
C0A5	SUBD 8	" " "
C0A7	STD B	" " "
C0A9	LDAA A	" " "
C0AB	SBCA #0	" " "
C0AD	STAA A	" " "
C0AF	BCS C0BA	If subtrahend>minuend, find $T/2
C0B1	LDD 0	If subtrahend not>minuend, increment $T
C0B3	ADDD #1	" " " " " " " "
C0B6	STD 0	" " " " " " " "
C0B8	BRA C0A3	Do another subtraction
C0BA	LDD 0	Find $T/2
C0BC	LSRD	" "
C0BD	STD 2	" "
C0BF	RTS	Return

The $fTOT/2 subroutine "reciprocates" the $f value and derives $T. The CPU can accomplish this "reciprocation" without floating point arithmetic if it interprets the numerator as the number of free-running counts in one second. The hexadecimal number of free-running counts in one second equals $1E8480 (at 500ns each). If the CPU divides this numerator by $f, the quotient equals $T, expressed as the number of free-running counts in the requested frequency's period. The CPU can then halve this value to derive $T/2—the TOC2 augend. The "reciprocation" process consumes the bulk of the $fTOT/2 subroutine.

Refer to blocks 1 and 2 of Figure 8–14, and program addresses $C095–$C0A2. In block 1, the CPU clears RAM locations where it will develop the $T value. In block 2, the CPU initializes the "reciprocation's" numerator to $1E8480. These 3 bytes in RAM are labeled DIFF, because this numerator will be reduced by successive subtractions.

Refer to blocks 3, 4, and 5 of Figure 8–14 (program addresses $C0A3–$C0B9). In block 3, the CPU subtracts $f from DIFF. This subtraction requires 3-byte precision. After each block 3 subtraction, the CPU tests the C flag to determine whether the subtraction returned a negative result (i.e., the subtrahend was larger than the minuend). If C = 1, the "reciprocation" is complete, and the CPU escapes the block 3, 4, 5 reciprocation loop. If C = 0, the CPU executes block 5 by adding one to the 2-byte $T value. After incrementing $T, the CPU branches back to block 3 to execute another subtraction of $f from DIFF. Thus, $T increments as many times as $f can be subtracted from DIFF without setting the C flag.

Refer to blocks 6, 7, and 8 of Figure 8–14 (program addresses $C0BA–$C0BE). In these blocks, the CPU divides $T by 2 to derive the TOC2 augend ($T/2). The CPU accomplishes this task by loading its D register with the $T value (block 6), and then executing a logic shift right (block 7). Shifting all D register bits right effectively divides its contents by 2, deriving $T/2. The CPU then stores the $T/2 value to its assigned RAM location (block 8). Having "reciprocated" $f and halved the result, the CPU returns to the MAIN program.

The M68HC11 can now generate and measure an OC2 output square wave. The CPU calls upon the OC/IC subroutine to execute these tasks. Figure 8–15 and its associated program list illustrate OC/IC.

In blocks 1, 2, and 3 of Figure 8–15 (program addresses $C0C5–$C0D3), the CPU initializes the necessary registers and pseudovector for OC2 square wave generation and IC2

FIGURE 8–15

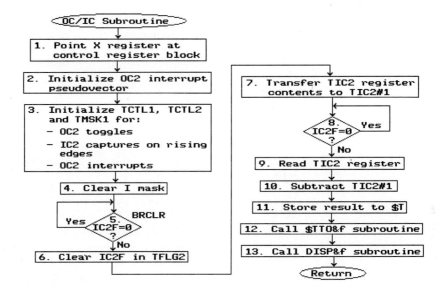

OC/IC Subroutine

$ADDRESS	SOURCE CODE	COMMENTS
C0C5	LDX #1000	Initialize X register
C0C8	LDD #B755	Initialize OC2 interrupt pseudovector
C0CB	STD DD	" " " " " "
C0CD	LDD #4004	Initialize TCTL1, TCTL2, TMSK1
C0D0	STD 20,X	" " " " " "
C0D2	STAA 22,X	" " " " " "
C0D4	CLI	Clear I mask
C0D5	BRCLR 23,X 2 C0D5	Branch to self until IC2F = 1
C0D9	BCLR 23,X FD	Clear IC2F
C0DC	LDD 12,X	Transfer TIC2 contents to TIC2#1
C0DE	STD 25	" " " " " "
C0E0	BRCLR 23,X 2 C0E0	Branch to self until IC2F = 1
C0E4	LDD 12,X	Read TIC2 register
C0E6	SUBD 25	Subtract TIC2#1
C0E8	STD 0	Store result to $T
C0EA	JSR B700	Call $TTO&f subroutine
C0ED	JSR B73A	Call DISP&f subroutine
C0F0	RTS	Return

frequency measurement. In block 1, the CPU initializes its X register. In block 2, the CPU initializes the pseudovector for an OC2 Interrupt Service routine, located at address $B755. The CPU uses interrupt control in generating the output square wave. In block 3, the CPU writes control words $40 to the TCTL1 register, $04 to the TCTL2 register, and $40 to the TMSK1 register. These control words generate OC2 toggles, enable input captures on rising edges, and enable OC2 interrupts.

In block 4 of Figure 8–15, the CPU begins square wave generation by clearing the global I mask. A successful output compare causes an OC2/PA6 pin toggle and generates an OC2 interrupt. The interrupt service routine causes the CPU to add its $T/2 value to TOC2—to set up for the next output edge and interrupt service.

Refer to block 5 through block 11 of Figure 8–15 (program addresses $C0D5–$C0E9). The M68HC11 uses input captures to measure the period ($T) of the generated square wave. The CPU expresses $T as the number of free-running counts in the square wave's period. To make this measurement, the CPU finds the difference (in free-running counts) between two successive positive-going edges of the square wave. To capture the free-running count at the time of the first rising edge, the CPU enters a polling loop (block 5). The CPU branches to itself until the IC2F flag in TFLG1 sets. IC2F = 1 indicates that IC2 has captured the initial edge and holds the corresponding free-running count in the TIC2 register. After clearing IC2F (block 6), the CPU reads TIC2, and stores this value to assigned locations in RAM, labeled TIC2#1 (block 7). Thus, in blocks 6 and 7, the CPU records the free-running count at the time of the first positive-going edge.

In block 8, the CPU again enters an IC2F polling loop. The CPU escapes this loop when the second rising edge sets IC2F. The TIC2 register now contains the free-running count at the time of the second edge. The CPU reads this count from TIC2 (block 9), then subtracts the TIC2#1 value from it (block 10). The difference equals $T: the number of free-running counts between edges. In block 11, the CPU stores this difference to the $T locations in RAM.

In block 12 of Figure 8–15, the CPU calls the $TTO&f subroutine. This subroutine converts $T (derived in blocks 5–11) to decimal frequency (&f). It stores the converted value, as packed BCD digits, in three designated RAM locations (labeled &f).

Finally, in block 13, the CPU calls the DISP&f subroutine. DISP&f causes the M68HC11 to display the measured decimal frequency on the terminal as "&f = XXXXX."

Upon return from the DISP&f subroutine, the CPU returns to the MAIN program where it continues generating the requested output square wave.

The above discussion notes that the OC/IC subroutine calls a nested subroutine known as $TTO&f. $TTO&f translates measured $T into its equivalent decimal frequency. This frequency value consists of five BCD digits packed into three RAM locations, which are labeled &f. $TTO&f makes this translation by "reciprocating" $T in a manner similar to the $fTOT/2 subroutine. The "reciprocation's" numerator equals the number of free-running counts in one second: $1E8480. The "reciprocation's" denominator is $T, also expressed in free-running counts. This calculation provides a decimal frequency value for display at the terminal. As $TTO&f increments the quotient, its decimal adjusts this value. In this way, $TTO&f provides a compact program for translating hexadecimal period into decimal frequency. However, this software is not dynamically efficient, as the reader will observe in Exercise 8.4. The translation of a relatively high frequency involves many iterations of the subtract/test/increment loop, which requires seconds to complete. We find such dynamic inefficiency acceptable since the M68HC11 makes only a single frequency measurement. In applications requiring continuous frequency measurements, $TTO&f's strategy would be unacceptable.[14]

Figure 8–16 and its accompanying software list illustrate the $TTO&f subroutine. In block 1, the CPU clears the RAM &f locations, where it will accumulate the decimal frequency value. In block 2, the CPU initializes the "reciprocation's" numerator, labeled DIFF, to $1E8480. The CPU then enters a subtract/test/increment loop (blocks 3, 4, and 5; program addresses $B710–$B733). In this loop, the CPU subtracts $T from DIFF (block 3),

---

[14]See Motorola Inc., *M68HC11 Reference Manual*, Section 10.6, Timer Example 10–1(b), pp. 10-42–10-44, 10-53, 10-54 for Motorola's $T to &f approach. Although this software takes up much more program space (less static efficiency), it requires less execution time (more dynamic efficiency).

FIGURE 8-16

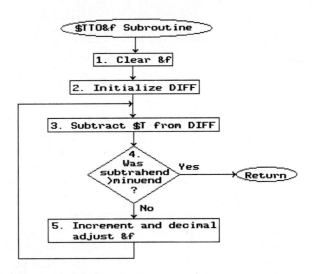

$TTO&f Subroutine

$ADDRESS	SOURCE CODE	COMMENTS
B700	LDD #0	Clear &f
B703	STD 5	" "
B705	STAA 7	" "
B707	LDD #8480	Initialize DIFF to $1E8480
B70A	STD B	" " " " "
B70C	LDAA #1E	" " " " "
B70E	STAA A	" " " " "
B710	LDD B	Subtract $T from DIFF
B712	SUBD 0	" " " " "
B714	STD B	" " " " "
B716	LDAA A	" " " " "
B718	SBCA #0	" " " " "
B71A	STAA A	" " " " "
B71C	BCS B735	If conversion is complete, return
B71E	LDAA 7	Increment and decimal adjust &f
B720	ADDA #1	" " " " " "
B722	DAA	" " " " " "
B723	STAA 7	" " " " " "
B725	LDAA 6	" " " " " "
B727	ADCA #0	" " " " " "
B729	DAA	" " " " " "
B72A	STAA 6	" " " " " "
B72C	LDAA 5	" " " " " "
B72E	ADCA #0	" " " " " "
B730	DAA	" " " " " "
B731	STAA 5	" " " " " "
B733	BRA B710	Do another conversion loop
B735	RTS	Return

until the subtraction causes a negative result, i.e., causes the C flag to set (block 4). When the C flag sets, $T has been subtracted &f times. Therefore, the CPU returns from the $TTO&f subroutine. If the CPU makes the block 3 subtraction without setting the C flag, it must increment and decimal adjust the frequency (block 5). The CPU uses the DAA instruction to decimal adjust the incremented frequency value. DAA can be used only after an 8-bit ADDA or ADCA instruction. Therefore, incrementing and adjusting the 3-byte frequency (&f) value requires twelve instructions (a major contributor to this software's dynamic inefficiency). Of course, after incrementing &f, the CPU must branch back to execute another subtract/test/increment loop.

Upon its return to OC/IC from the $TTO&f nested routine, the M68HC11 must display measured decimal frequency at the terminal. To do this, the CPU calls another nested subroutine: DISP&f. Figure 8–17 and its accompanying software list illustrate DISP&f.

Refer to Figure 8–17, blocks 1 through 3 and program addresses $B73A–$B73E. In block 1, the M68HC11 sets the global I mask to disable OC2 interrupts. Refer to the OC2 Interrupt Service routine to see why this step is necessary. The OC2 Interrupt Service assumes that the X register contains $1000. In Figure 8–17, block 2, the CPU stacks this X register value, and then repoints the X register at ASCII message data in RAM (block 3). If OC2 generated an interrupt, the new X register contents would cause an OC2 interrupt malfunction.

FIGURE 8–17

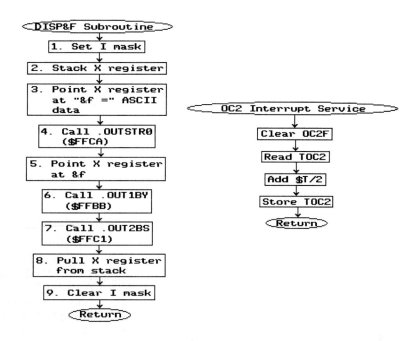

DISP&f Subroutine and OC2 Interrupt Service Routine

DISP&f Subroutine

$ADDRESS	SOURCE CODE	COMMENTS
B73A	SEI	Set I mask
B73B	PSHX	Stack X register
B73C	LDX #19	Point X register at "&f =" ASCII data
B73F	JSR FFCA	Call .OUTSTR0
B742	LDX #5	Point X register at &f
B745	JSR FFBB	Call .OUT1BY
B748	JSR FFC1	Call .OUT2BS
B74B	PULX	Restore X register
B74C	CLI	Clear I mask
B74D	RTS	Return

OC2 Interrupt Service Routine

$ADDRESS	SOURCE CODE	COMMENTS
B755	BCLR 23,X BF	Clear OC2F
B758	LDD 18,X	Read TOC2
B75A	ADDD 2	Add $T/2
B75C	STD 18,X	Write sum to TOC2
B75E	RTI	Return

Refer to blocks 4 through 7 and program addresses $B73F–$B74A. In these blocks the CPU uses the .OUTSTR0, .OUT1BY, and .OUT2BS utility subroutines to display the message

```
&f = XXXXX
```

where XXXXX equals the contents of the &f RAM locations, i.e., measured decimal frequency. Refer to the memory allocations accompanying Figure 8–9. The .OUT1BY routine displays the &f MS byte and returns an X register value that points at the &f MID location. The .OUT2BS routine then displays the &f MID and &f LS bytes.

Refer to blocks 8 and 9 of Figure 8–17 (program addresses $B74B–$B74D). In block 8, the CPU re-points its X register at the control register block by pulling the value it stacked (block 2). In block 9, the CPU clears the I mask, allowing OC2 interrupts to continue.[15] The CPU can then return to the MAIN program via the OC/IC subroutine.

As stated above, the OC2 Interrupt Service routine's main purpose is to update the TOC2 register for the next output toggle. Refer to Figure 8–17's OC2 Interrupt Service flowchart and its accompanying software list. In servicing the OC2 interrupt, the CPU has two tasks to accomplish: 1) clear the OC2F bit in TFLG1, and 2) update the TOC2 register for the next output edge. The CPU must clear the OC2F bit to prevent an inadvertent interrupt upon its return from the OC2 Interrupt Service routine. OC2 interrupts remain pending as long as OC2F = 1. After updating TOC2, the CPU returns to either the OC/IC subroutine or the MAIN program, depending upon the program count at the time of the interrupt.

EXERCISE 8.4   Integrate Input Captures with Output Compares

REQUIRED EQUIPMENT: Motorola M68HC11 EVB (EVBU) and terminal
Oscilloscope and probe

GENERAL INSTRUCTIONS: Load, debug, demonstrate, and analyze the software presented in Section 8.4. This software integrates output compares with inputs captures to generate and measure a square wave. Before proceeding, review Section 8.4, so that you fully understand the software strategy used in this exercise.

PART I: Connect the OC2 Pin, IC2 Pin, and Oscilloscope

Refer to Figure 8–9 and interconnect the OC2 and IC2 pins. Connect the oscilloscope to display and measure the OC2 output square wave. Show your connections to the instructor.

_____ *Instructor Check*

PART II: Software Development, Debug, and Demonstration

Refer to the software lists accompanying Figures 8–10 through 8–17. Source code entry procedures vary for EVB and EVBU users.

EVB users:

- Allocate memory as shown in the memory allocations list accompanying Figure 8–9.
- Assemble the $TTO&f, DISP&f, and OC2 Interrupt Service routines at locations $C000, $C03A, and $C055 in RAM. Then use the BUFFALO MOVE command to transfer these

---

[15] The nested .OUTSTR0, .OUT1BY, and .OUT2BS subroutines make DISP&f execution time quite long. Thus, it is possible that a pending interrupt might not be serviced in a timely manner while the I mask is set. By the time the CPU can vector to the OC2 Interrupt Service and update TOC2, the output square wave may have "skipped a beat." This "skipped beat" could be lengthy if the free-running count must roll over before returning to the updated TOC2 value. The program could eliminate this roll-over delay by modifying the OC2 Interrupt Service routine to read the TCNT register instead of TOC2. This modification would also minimize—but not eliminate—the "skipped beat" problem. However, reading TCNT instead of TOC2 introduces an additional inaccuracy into normal square wave generation. At the relatively low PRR values that OC2 can produce, this additional inaccuracy would be virtually undetectable and probably insignificant.

routines to their assigned EEPROM locations. The $TTO&f subroutine uses two branch instructions. When entering $TTO&F into RAM, modify the source code for these branches to BCS C035 and BRA C010. This source code makes the $TTO&f routine completely portable. MOVE it to address $B700 without any further modifications. Assemble the $TTO&f, DISP&f, and OC2 Interrupt Service routines *first*. MOVE them to EEPROM before assembling the rest of the software.

EVBU users:

- Allocate memory as shown in the memory allocations list accompanying Figure 8–9 with the following exceptions. Modify subroutine calls, jumps, and branches to reflect these allocations:

$ADDRESS	CONTENTS
0100	MAIN program
011A	&fINPUT subroutine
0135	CK&fBNDS subroutine
0155	&fTO$f subroutine
0195	&fTOT/2 subroutine
01C5	OC/IC subroutine

The EVBU's BUFFALO one-line assembler allows source code entry directly into specified EEPROM locations.

Load and debug the OC/IC integration software. When sure that the software operates properly, demonstrate its operation to the instructor. Show the instructor user-selected OC2 output square waves as displayed on the oscilloscope.

_____ *Instructor Check*

PART III: Laboratory Analysis

Answer the questions below by neatly entering your answers in the appropriate blanks.

1. Describe the purpose of each routine listed below.

    MAIN Program:

    &fINPUT:

    CK&fBNDS:

    &fTO$f:

    &fTOT/2:

    OC/IC:

    $TTO&f:

    DISP&f:

2. Explain why you made a connection between microcontroller pins 28 and 33.

3. Refer to Figure 8–10. Explain the reason for the block 7 loop to self.

4. Explain how the instruction at address $C026 ($0126) converts a digit from ASCII to BCD.

5. Refer to Figure 8–12, block 2. Why does the microcontroller compare the BCD 10,000's digit with one?

6. Refer to Figure 8–12, block 6. Explain why the CPU returns from the CK&fBNDS subroutine if either the 1,000's or 100's digit is other than zero.

7. Refer to Figure 8–12, block 7. Explain why the CPU packs the 10's and units BCD digits.

8. Refer to Figure 8–13. Describe the circumstances that cause the CPU to execute block 5.

9. Refer to Figure 8–13. What is the maximum number of times that the CPU could execute block 8? _____
10. Refer to Figure 8–13. What is the maximum hex product that the CPU could generate by executing block 11? $_____ What is the maximum hex product that the CPU could generate by executing block 15? $_____
11. Refer to Figure 8–14. To what hex value does the M68HC11 initialize DIFF? $_____ Why does the CPU initialize DIFF to this value?

12. Refer to Figure 8–14. Explain the reason for block 4.

13. Refer to Figure 8–14. Explain block 7's purpose and why its prescribed action achieves this purpose.

14. Refer to Figure 8–15. Explain why the CPU executes block 7.

15. Refer to Figure 8–16. To what hex value does the M68HC11 initialize DIFF? $_____
    Why does the CPU initialize DIFF to this value?

16. Refer to Figure 8–16. Why does it require twelve instructions to execute block 5?

17. Refer to Figure 8–17. Why does the DISP&f subroutine require blocks 2 and 8?

18. Refer to Figure 8–17. Why does the DISP&f subroutine require blocks 1 and 9?

19. Refer to Figure 8–17. List the OC2 Interrupt Service routine's two purposes.

20. Analyze the performance of the M68HC11 and OC/IC integration software. Observe its response over the range 32 Hz to 19.999 Khz. For each requested frequency, carefully and accurately measure the output square wave's frequency with the oscilloscope. Enter your oscilloscope measurements in the appropriate blank. Also enter the M68HC11's measured value, as displayed at the terminal, in the appropriate blank.

Requested Frequency	Measured Frequency (Oscilloscope)	Measured Frequency (IC2)
32 Hz	_____	_____
500 Hz	_____	_____
1 KHz	_____	_____

Frequency	Oscilloscope	IC2
1.5 KHz	_____	_____
2 KHz	_____	_____
2.5 KHz	_____	_____
3 KHz	_____	_____
3.5 KHz	_____	_____
4 KHz	_____	_____
4.5 KHz	_____	_____
5 KHz	_____	_____
5.5 KHz	_____	_____
6 KHz	_____	_____
6.5 KHz	_____	_____
7 KHz	_____	_____
7.5 KHz	_____	_____
8 KHz	_____	_____
8.5 KHz	_____	_____
9 KHz	_____	_____
9.5 KHz	_____	_____
10 KHz	_____	_____
10.5 KHz	_____	_____
11 KHz	_____	_____
11.5 KHz	_____	_____
12 KHz	_____	_____
12.5 KHz	_____	_____
13 KHz	_____	_____
13.5 KHz	_____	_____
14 KHz	_____	_____
14.5 KHz	_____	_____
15 KHz	_____	_____
15.5 KHz	_____	_____
16 KHz	_____	_____
16.5 KHz	_____	_____
17 KHz	_____	_____
17.5 KHz	_____	_____
18 KHz	_____	_____
18.5 Hz	_____	_____
19 KHz	_____	_____
19.5 KHz	_____	_____
19.999 KHz	_____	_____

Analyze your results. Assume that system response is adequate if it meets these criteria:

- Frequency, as measured with the oscilloscope, is within ±5% of the requested value.
- Frequency, as measured by IC2, is within ±5% of the counterpart oscilloscope measurement.

Given your recorded data and the above criteria, what is the acceptable frequency range of the M68HC11 and its OC/IC integration software?

_____Hz to _____Hz

PART IV: Laboratory Software Listing

MAIN PROGRAM—EVB

(See program list accompanying Figure 8–10.)

MAIN PROGRAM—EVBU

$ADDRESS	SOURCE CODE	COMMENTS
0100	LDS #47	Initialize stack pointer
0103	JSR 11A	Call &fINPUT subroutine
0106	JSR 135	Call CK&fBNDS subroutine
0109	JSR 155	Call &fTO$f subroutine
010C	JSR 195	Call $fTOT/2 subroutine
010F	JSR 1C5	Call OC/IC subroutine
0112	WAI	Wait for OC2 interrupt
0113	BRA 112	Make another output edge

&fINPUT SUBROUTINE—EVB

(See program list accompanying Figure 8–11.)

&fINPUT SUBROUTINE—EVBU

$ADDRESS	SOURCE CODE	COMMENTS
011A	LDX #E	Point X register at "INPUT &f" ASCII data
011D	JSR FFCA	Call .OUTSTR0
0120	LDX #20	Point X register at 10,000's digit
0123	JSR FFCD	Call .INCHAR
0126	ANDA #F	Convert input digit from ASCII to BCD
0128	STAA 0,X	Store input digit
012A	INX	Point X register at next input digit storage location
012B	CPX #25	Compare X register with last input digit storage location + 1
012E	BNE 123	If unequal, capture another input digit
0130	RTS	If equal, return

CK&fBNDS SUBROUTINE—EVB

(See program list accompanying Figure 8–12.)

CK&fBNDS SUBROUTINE—EVBU

$ADDRESS	SOURCE CODE	COMMENTS
0135	LDAA 20	Load 10,000's digit
0137	CMPA #1	Compare digit with 1
0139	BHI 14D	If higher, &f is out-of-bounds
013B	BEQ 150	If equal, &f is inbounds
013D	LDD 21	Load 1,000's and 100's digits
013F	BNE 150	If digits are not zero, &f is inbounds
0141	LDAA 23	Pack 10's and units digits
0143	ASLA	" " " " "
0144	ASLA	" " " " "
0145	ASLA	" " " " "
0146	ASLA	" " " " "
0147	ADDA 24	" " " " "
0149	CMPA #32	Is &f equal to or higher than &32?
014B	BCC 150	If yes, &f is inbounds
014D	JMP 100	If &f is out-of-bounds, re-prompt user
0150	RTS	Return

&fTO$f SUBROUTINE—EVB

(See program list accompanying Figure 8–13.)

&fTO$f SUBROUTINE—EVBU

$ADDRESS	SOURCE CODE	COMMENTS
0155	LDX #20	Initialize X register
0158	LDD #0	Clear $f locations
015B	STD 8	" " "
015D	DEC 0,X	Decrement 10,000's digit
015F	BMI 168	If N = 1, process 1,000's digit
0161	LDD 8	If N = 0, add $2710 to $f

$ADDRESS	SOURCE CODE	COMMENTS
0163	ADDD #2710	" " " "
0166	STD 8	" " " "
0168	DEC 1,X	Decrement 1,000's digit
016A	BMI 175	If N = 1, process 100's digit
016C	LDD 8	If N = 0, add $3E8 to $f
016E	ADDD #3E8	" " " "
0171	STD 8	" " " "
0173	BRA 168	Continue 1,000's digit processing
0175	LDAA 22	Load ACCA with 100's digit
0177	LDAB #64	Load ACCB with $64
0179	MUL	ACCA × ACCB —> D register
017A	ADDD 8	Add product to $f
017C	STD 8	" " "
017E	LDAA 23	Load ACCA with 10's digit
0180	LDAB #A	Load ACCB with $A
0182	MUL	ACCA × ACCB —> D register
0183	ADDD 8	Add product to $f
0185	STD 8	" " "
0187	LDAB 24	Add 1's digit to $f
0189	CLRA	" " "
018A	ADDD 8	" " "
018C	STD 8	" " "
018E	RTS	Return

$fTOT/2 SUBROUTINE—EVB

(See program list accompanying Figure 8–14.)

$fTOT/2 SUBROUTINE—EVBU

$ADDRESS	SOURCE CODE	COMMENTS
0195	LDD #0	Clear $T
0198	STD 0	" "
019A	LDD #8480	Initialize DIFF to $1E8480
019D	STD B	" " " " "
019F	LDAA #1E	" " " " "
01A1	STAA A	" " " " "
01A3	LDD B	Subtract $f from DIFF
01A5	SUBD 8	" " " "
01A7	STD B	" " " "
01A9	LDAA A	" " " "
01AB	SBCA #0	" " " "
01AD	STAA A	" " " "
01AF	BCS 1BA	If subtrahend > minuend, find $T/2
01B1	LDD 0	If subtrahend not > minuend, increment $T
01B3	ADDD #1	" " " "
01B6	STD 0	" " " "
01B8	BRA 1A3	Do another subtraction
01BA	LDD 0	Find $T/2
01BC	LSRD	" "
01BD	STD 2	" "
01BF	RTS	Return

OC/IC SUBROUTINE—EVB

(See program list accompanying Figure 8–15.)

OC/IC SUBROUTINE—EVBU

$ADDRESS	SOURCE CODE	COMMENTS
01C5	LDX #1000	Initialize X register
01C8	LDD #B755	Initialize OC2 interrupt pseudovector
01CB	STD DD	" " " " "
01CD	LDD #4004	Initialize TCTL1, TCTL2, TMSK1
01D0	STD 20,X	" " " " "
01D2	STAA 22,X	" " " " " "
01D4	CLI	Clear I mask
01D5	BRCLR 23,X 2 1D5	Branch to self until IC2F = 1
01D9	BCLR 23,X FD	Clear IC2F

01DC	LDD 12,X		Transfer TIC2 contents to TIC2#1
01DE	STD 25		" " " " "
01E0	BRCLR 23,X 2 1E0		Branch to self until IC2F = 1
01E4	LDD 12,X		Read TIC2 register
01E6	SUBD 25		Subtract TIC2#1
01E8	STD 0		Store result to $T
01EA	JSR B700		Call $TTO&f subroutine
01ED	JSR B73A		Call DISP&f subroutine
01F0	RTS		Return

$TTO&f SUBROUTINE—EVB AND EVBU

(See program list accompanying Figure 8–16.)

DISP&f SUBROUTINE—EVB AND EVBU

(See program list accompanying Figure 8–17.)

OC2 INTERRUPT SERVICE ROUTINE—EVB AND EVBU

(See program list accompanying Figure 8–17.)

DISPLAY ASCII DATA—EVB AND EVBU

$ADDRESS	CONTENTS	CHARACTER
000E	0D	<CR>
000F	49	"I"
0010	4E	"N"
0011	50	"P"
0012	55	"U"
0013	54	"T"
0014	20	<SP>
0015	26	"&"
0016	66	"f"
0017	20	<SP>
0018	04	<EOT>
0019	0D	<CR>
001A	26	"&"
001B	66	"f"
001C	20	<SP>
001D	3D	"="
001E	20	<SP>
001F	04	<EOT>

8.5 ON-CHIP RESOURCES USED BY THE OC1 OUTPUT COMPARE FUNCTION

This section considers Output Compare 1 (OC1) individually because it constitutes a special case. OC1 is special because it controls all five output compare pins (OC1–OC5). OC1 can assert this control even when individual OC pin functions (OC2–OC5) control their own pins. Thus, OC1 and individual OC pin mechanisms OC2–OC5 can jointly control OC pins. A system designer uses this joint control feature to vary the duty cycle of a square wave output from an OC2–OC5 pin. The designer can apply a variable duty cycle waveform to a number of important applications, such as DC voltage regulation or DC motor speed control.[16]

OC1 can control the actions of all five output compare pins (OC1–OC5). To exert such control, OC1 uses a number of on-chip resources:

    TOC1 register
    OC1M register
    OC1D register

---

[16]In Exercise 8.6, the reader will construct a motor speed control interface using OC1 and OC2.

PACTL register
PORTA pins PA7–PA3
TFLG1 register
TMSK1 register

This section considers each of these resources.

Timer Output Compare Register 1
(TOC1, Control Register Block Addresses $1016–$1017)

OC1 has its own Timer Output Compare Register (TOC1). The CPU writes comparison values to this register. When the free-running count matches the contents of TOC1, pins controlled by OC1 can either set or reset—as specified by the OC1M and OC1D registers.

Output Compare 1 Mask Register
(OC1M, Control Register Block Address $100C)

Refer to Figure 8–18. The CPU writes a mask byte to the OC1M register to assert OC1 control over output compare pins (OC1–OC5). By writing a binary one to an OC1M bit, the CPU enables OC1 control over a counterpart output compare pin. For example, consider a case where the CPU writes mask $A8 to the OC1M register. This mask asserts OC1 control over the following pins:

PORTA pin 7   OC1 pin
PORTA pin 1   OC3 pin
PORTA pin 3   OC5 pin

An OC1M mask byte does not necessarily give OC1 exclusive control over the OC2–OC5 pins. If the CPU also configures TCTL1 for OC pin toggles, sets, or resets, then OC1 and individual output compare mechanisms share joint control. However, if OC1 and an individual OC mechanism both respond to a match on the same free-running count, priority goes to OC1. PORTA pin 7 offers a special case. The CPU can configure this pin for general-purpose I/O, or it can assert exclusive OC1 control. OC1 can control pin PA7 only if it is configured as an output. See the following discussion on the PACTL register.

Output Compare 1 Data Register
(OC1D, Control Register Block Address $100D)

Refer to Figure 8–19. By writing a control word to OC1D, the CPU specifies pin actions that occur upon successful OC1 compares. However, these specified pin actions occur only if OC1M register bits enable OC1 control. If the CPU writes binary one to an OC1D bit, then successful OC1 compares cause a counterpart OCx pin to set. For example, if the CPU writes $A8 to OC1D, then a successful OC1 compare causes the OC1, OC3, and OC5 pins to set. If the CPU writes a zero to an OC1D register bit, successful OC1 compares cause the counterpart OC pin to reset.

FIGURE 8–18
Reprinted with the permission of Motorola.

OC1M

B7							B0
OC1M7	OC1M6	OC1M5	OC1M4	OC1M3			
(OC1)	(OC2)	(OC3)	(OC4)	(OC5)			

FIGURE 8–19
Reprinted with the permission of Motorola.

OC1D

B7							B0
OC1D7	OC1D6	OC1D5	OC1D4	OC1D3			
(OC1)	(OC2)	(OC3)	(OC4)	(OC5)			

FIGURE 8–20
Reprinted with the permission of Motorola.

Pulse Accumulator Control Register (PACTL, Control Register Block Address $1026)

Refer to Figure 8–20. PACTL register bit 7 contains the DDRA7 bit. The CPU must write a binary one to this bit to assert OC1 control over PORTA pin 7. Additionally, the CPU must write a binary one to bit 7 of the OC1M register. Of course, by writing a one to the DDRA7 bit, the CPU configures PORTA pin 7 as an output.

PORTA Pins PA7–PA3 (OC1–OC5)

Review pages 344–346 above for a detailed discussion on pins OC2–OC5. Following paragraphs describe only PORTA pin 7 (OC1).

Figure 8–21 provides a block diagram of PA7 (OC1) pin logic.[17] The Pulse Accumulator/General-Purpose Input Latch (top block) functions for general-purpose or pulse accumulator inputs.[18] This block consists of an S-R latch as discussed in Chapter 7. RPORTA (4) assertion enables Transmission Gate 3. The Pulse Accumulator/General-Purpose Input Latch can then drive Internal Data Bus Line D7 (7).

When enabled by WPORTA assertion (5), the General-Purpose Output Latch latches a write bit from Internal Data Bus Line D7 (7). OC1 Logic enables Transmission Gate 1 if its OC1M7 input (6) is negated (low). The PORTA CK (11) strobes this Transmission Gate 1 enable. The PORTA CK synchronizes PA7 pin changes with the M68HC11's E clock. When OC1 logic enables Transmission Gate 1, the "Cheater" Latch stores the state of the General-Purpose Output Latch. If DDRA7 = 1, the Output Driver inverts the bit stored in the "Cheater" Latch and applies it to the PA7 pin. Thus, if OC1 does not have control of pin PA7, the OC1M7 bit is low. In this case, OC1 Logic enables Transmission Gate 1. The PORTA clock strobes this enable. When enabled, Transmission Gate 1 applies the state of the General-Purpose Output Latch to the "Cheater" Latch and Output Driver. The Output Driver drives the PA7 pin *if* DDRA7 = 1.

As long as OC1M7 = 1, OC1 Logic will not enable Transmission Gate 1. Rather, OC1 Logic enables Transmission Gate 2, if either OC1 Compare (12) or FOC1 (9) asserts. The PORTA CK (11) strobes this enable. When enabled, Transmission Gate 2 applies the state

FIGURE 8–21
Reprinted with the permission of Motorola.

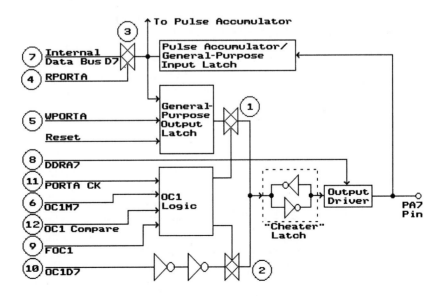

---

[17]Motorola Inc., *M68HC11 Reference Manual*, p. 7-10.
[18]Chapter 9 covers the M68HC11's pulse accumulator.

of the OC1D7 bit (10) to the "Cheater" Latch. If DDRA7 = 1, the Output Driver applies this bit to the PA7 pin. Thus, if OC1 has control of the PA7 pin, a successful OC1 compare causes the state of the OC1D7 bit (from the OC1D register) to be applied to pin PA7. However, pin PA7 assumes OC1D's state only if DDRA7 (in the PACTL register) is set.

A forced OC1 compare also drives pin PA7 with OC1D7's state. When the M68HC11 forces an OC1 compare, it asserts the FOC1 input (9). FOC1 assertion causes the OC1 Logic to enable Transmission Gate 2 (again strobed by PORTA CK). Transmission Gate 2 applies OC1D7's state to the "Cheater" Latch and Output Driver. The Output Driver drives pin PA7 pin with this state if DDRA7 = 1. Chapter 9 describes forced output compares in detail.

Main Timer Interrupt Flag Register 1 (TFLG1, Control Register Block Address $0123)

TFLG1 register's bit 7 constitutes the OC1F flag. This flag sets when the contents of the TOC1 register match the free-running count. The OC1F flag is a read/modify/write bit.

Main Timer Interrupt Mask Register 1 (TMSK1, Control Register Block Address $1022)

TMSK1 register's bit 7 constitutes the OC1I control bit. If the CPU sets this bit, successful OC1 compares generate OC1 interrupts.

8.6 CONTROLLING ALL OC PINS WITH OC1

OC1 can control output compare pins either exclusively or in conjunction with other OC functions. This section describes software that controls all five output compare pins, using OC1 exclusively. To assert exclusive OC1 control, disconnect pins OC2–OC5 from their individual output compare mechanisms. The CPU disconnects individual pin functions by writing zeros to relevant OMx/OLx bits in the TCTL1 register.

This section's software uses the five output compare pins as a five-stage, modulus thirty-two counter. These outputs count from %00000 to %11111. Pin PA7 (OC1) constitutes the most significant counter bit; PA3 (OC5) constitutes the least significant bit. The other OC pins output the three mid-significant bits as follows: PA6 (OC2) = 2^3, PA5 (OC3) = 2^2, PA4 (OC4) = 2^1. The software allows a user to select count rates from 30 to 46,500 counts per second. To maximize dynamic efficiency, OC1 uses status polling to control the five output compare pins.

Refer to Figure 8–22 and its accompanying program list. After initializing its X register (block 1), the CPU must initialize pins PA7 and PA3 as outputs (block 2). Pin PA7 is bidirectional on both the "A1" and "E9" versions of the M68HC11. Therefore the CPU must write binary one to PACTL's DDRA7 control bit. The M68HC11E9's PA3 pin is also bidirectional. Therefore, the "E9's" CPU must write binary one to the PACTL's DDRA3 bit. Of course, the M68HC11E9 must also reset PACTL's I4/O5 bit to configure PA3 for output compares. In contrast, the M68HC11A1's PA3 pin is an output-only OC pin. The "A1's" PACTL register has no DDRA3 bit.

Refer to program addresses $0103–$0109 and flow chart blocks 2 and 3. The CPU uses an LDD command to load its A register with the PACTL control word and its B register with the OC1M control word. The CPU then uses STAA and STAB commands to write these control words to the PACTL and OC1M addresses. The PACTL control word ($88) sets the DDRA7 and DDRA3 bits—and resets the I4/O5 bit. This control word also works for the "A1" version. The OC1M control word ($F8) sets all of the OC1Mx bits to assert OC1 control over the five output compare pins.

Refer to block 4 and program addresses $010A–$010B. The CPU loads a $T value from RAM addresses $0000–$0001. This $T value represents the number of free-running counts duration for each modulus thirty-two count. The operator uses the BUFFALO MM command to initialize $T before running the Modulus &32 Counter software.

FIGURE 8-22

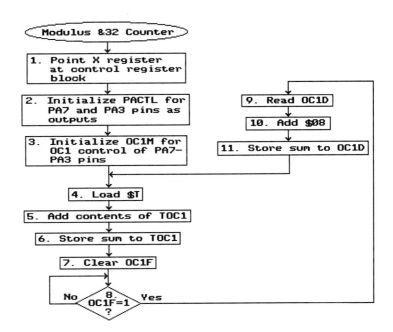

Modulus &32 Counter

$ADDRESS	SOURCE CODE	COMMENTS
0100	LDX #1000	Initialize X register
0103	LDD #88F8	Initialize PACTL and OC1M registers
0106	STAA 26,X	" " " " " " "
0108	STAB C,X	" " " " " " "
010A	LDD 0	Load $T
010C	ADDD 16,X	Add contents of TOC1
010E	STD 16,X	Store sum to TOC1
0110	BCLR 23,X 7F	Clear OC1F bit in TFLG1
0113	BRCLR 23,X 80 113	Branch to self until OC1F = 1
0117	LDAA D,X	Increment OC1D to next count
0119	ADDA #8	" " " " "
011B	STAA D,X	" " " " "
011D	BRA 10A	Do another count

In blocks 5 and 6 (program addresses $010C–$010F), the CPU updates the TOC1 register by adding the $T value to it. Thus, the OC1F flag will set after $T free-running counts have elapsed.

Refer to block 7 and program addresses $0110–$0112. The CPU clears the OC1F flag in TFLG1 before going into a polling loop. The CPU uses the BCLR command because OC1F is a read/modify/write bit.

In block 8 (program addresses $0113–$0116), the CPU executes a polling loop from which it escapes when the OC1F flag sets. The OC1F bit sets because of a successful OC1 compare. This successful compare causes the OC1D bits to drive their counterpart output compare pins. Because the user resets the M68HC11 before running the counter program, the OC1D register initially contains $00, i.e., all OC1D bits equal zero. Therefore, the initial modulus thirty-two counter output equals %00000. After escaping the polling loop, the CPU must load the OC1D register with the next modulus thirty-two count.

In blocks 9, 10, and 11 (program addresses $0117–$011E), the CPU updates its OC1D register for the next successful OC1 compare. OC1D bits 7 through 3 form the modulus thirty-two output bits. The CPU cannot simply increment the OC1D register in order to increment the counter outputs. It is untenable to merely increment the OC1D register because the counter output bits constitute the most significant bits of this register. However, the

FIGURE 8-23

OC1D Updates	OC1D7	OC1D6	OC1D5	OC1D4	OC1D3				Counter Output
$00	0	0	0	0	0	X	X	X	%00000
+08									
$08	0	0	0	0	1	X	X	X	%00001
+08									
$10	0	0	0	1	0	X	X	X	%00010
+08									
$18	0	0	0	1	1	X	X	X	%00011
+08									
$20	0	0	1	0	0	X	X	X	%00100
+08									
$28	0	0	1	0	1	X	X	X	%00101
⋮									⋮
$F8	1	1	1	1	1	X	X	X	%11111

CPU can properly increment the counter output bits by adding $08 to the OC1D register after each escape from the polling loop. To understand why, observe the states of the counter output bits each time the CPU adds $08 to the OC1D register. Figure 8–23 shows these state changes. Thus, the CPU reads the current state of OC1D (block 9), adds $08 (block 10), and updates the OC1D register (block 11) to increment the modulus thirty-two counter outputs.

EXERCISE 8.5 Modulus Thirty-Two Counter

REQUIRED EQUIPMENT: Motorola M68HC11EVBU (EVB) and terminal
Dual-trace oscilloscope and probes

GENERAL INSTRUCTIONS: In this laboratory exercise, use the Output Compare 1 (OC1) function to control all five output compare pins (OC1–OC5). Load, demonstrate, and analyze software that turns OC1 into a modulus thirty-two counter, which is capable of counting at a variety of count rates.

PART I: Provide for Oscilloscope Connections

Make provisions to conveniently and rapidly connect the oscilloscope to M68HC11 pins PA7 (27) through PA3 (31). By using both oscilloscope channels, you can observe two counter outputs at a time.[19] Connect oscilloscope channel 1 to the most significant of the two outputs being observed. Trigger the oscilloscope from channel 1.

PART II: Software Development

Refer to Figure 8–22 and its accompanying program list. If you use an EVB, locate your software at address $C000; modify the program's branching instructions accordingly.

---

[19]If a multichannel oscilloscope or logic analyzer is available, use it to observe as many counter outputs as possible simultaneously.

PART III: Debug and Demonstration

Load and debug your software. When sure that it runs properly, demonstrate its operation to the instructor. Show oscilloscope waveforms as requested by the instructor.

_____ *Instructor Check*

PART IV: Laboratory Analysis

Answer the questions below by neatly entering your answers in the appropriate blanks.

1. If addresses $0000–$0001 ($T) contain $0064, what is the counter's rate?

   Count rate = _____ K Counts/second

2. At the count rate established in question 1, what is the duration of each count?

   Count duration = _____ μs

3. Initialize $T ($0000–$0001) to $0064. Run your software. View counter outputs PA7–PA3 with the oscilloscope and complete the counter timing diagram. This diagram already furnishes the PA3 waveform. Accurately and neatly draw the waveforms for PA4, PA5, PA6, and PA7. Show your completed diagram to the instructor.

Modulus &32 Counter Timing Diagram

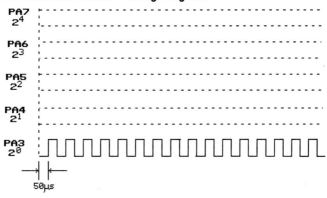

_____ *Instructor Check*

4. Use a systematic, empirical procedure to determine the maximum and minimum count rates at which the modulus thirty-two counter reliably and consistently operates. Also enter the corresponding $T values for these minimum and maximum count rates. Assure that high and low count rates are not isolated. That is, all $T values between T_{MAX} and T_{MIN} should give reliable operation.

   Count rate$_{MAX}$ = _____ K counts/second; T_{MIN} = $_____
   Count rate$_{MIN}$ = _____ counts/second; T_{MAX} = $_____

5. Analyze your counter software and explain why your observed count rate$_{MAX}$ is the highest achievable.

PART V: Laboratory Software Listing

MODULUS &32 COUNTER—EVB

$ADDRESS	CONTENTS	CHARACTER
C000	LDX #1000	Initialize X register
C003	LDD #88F8	Initialize PACTL and OC1M registers
C006	STAA 26,X	" " " " "
C008	STAB C,X	" " " " "
C00A	LDD 0	Load $T

C00C	ADDD 16,X	Add contents of TOC1
C00E	STD 16,X	Store sum to TOC1
C010	BCLR 23,X 7F	Clear OC1F bit in TFLG1
C013	BRCLR 23,X 80 C013	Branch to self until OC1F = 1
C017	LDAA D,X	Increment OC1D to next count
C019	ADDA #8	" " " "
C01B	STAA D,X	" " " "
C01D	BRA C00A	Do another loop

MODULUS &32 COUNTER—EVBU

(See program list accompanying Figure 8–22.)

8.7 JOINT OC1/OCX CONTROL OF AN OUTPUT COMPARE PIN

OC1 can assert control over output compare pins even when individual OC pin functions (OC2–OC5) also control them. In this way, OC1 and individual OC pin mechanisms can jointly control OC pins. A programmer uses joint control to vary the pulse width of a square wave output. Variable pulse width waveforms can be used in such applications as DC voltage regulation or DC motor speed control. This section presents a program that uses joint OC1/OC5 control. This software causes the M68HC11 to output a variable pulse width square wave from its OC5 (PA3) pin. In Exercise 8.6, the reader applies this software to DC motor speed control.

Figures 8–24 through 8–27, along with their accompanying program lists, illustrate joint OC1/OC5 control over the PA3 output compare pin. This software causes the M68HC11 to output a 1 KHz square wave with user-selectable duty cycles. The M68HC11 displays an "INPUT %" message at its terminal. This message prompts the user to enter a two-digit decimal duty cycle value. The INPUT% subroutine generates this prompt and captures the user's key-press data. The microcontroller then converts the user's entry into its equivalent hexadecimal number of free-running counts. The &DYTO$DY subroutine accomplishes this conversion. The software then initializes for joint OC1/OC5 control over the OC5 pin. After this initialization, the microcontroller uses OC5 status polling to generate output rising edges and OC1 interrupt control to generate falling edges.

Refer to Figure 8–24 and its accompanying program list. After initializing its stack pointer (block 1), the CPU calls the INPUT% subroutine (block 2). INPUT% prompts the user to enter two decimal digits representing desired output duty cycle. The user responds to this prompt with two key-presses. INPUT% captures this key-press data. The CPU then calls the &DYTO$DY subroutine (block 3). At 1 KHz, 1% of duty cycle equals 10μs. Ten μs represents twenty free-running counts—at 500ns each. Thus, at 1KHz, each 1% of duty cycle equals $14 free-running counts. The &DYTO$DY subroutine generates a 2-byte number labeled $DUTY. This hex number represents requested duty cycle, expressed as its number of free-running counts. The &DYTO$DY subroutine stores $DUTY to RAM locations $0000–$0001.

Refer to blocks 4 and 5 of Figure 8–24 (program addresses $0109–$0110). In block 4, the CPU initializes its OC1 interrupt pseudovector to $0140, the OC1 Interrupt Service routine's address. In block 5, the CPU points its X register at the control register block.

Refer to blocks 6 and 7, and program addresses $0111–$011C. In these flowchart blocks, the CPU initializes the TCTL1, TMSK1, OC1M, and OC1D registers for proper OC1 and OC5 operation. The CPU initializes these registers as follows:

> TCTL1 = $03; successful OC5 compares cause the PA3/OC5 pin to set, generating positive-going edges.
> TMSK1 = $80, enabling OC1 interrupts.
> OC1M = $08, asserting OC1 control over the PA3/OC5 pin.
> OC1D = $00; successful OC1 compares cause the PA3/OC5 pin to reset, generating negative-going edges.

FIGURE 8-24

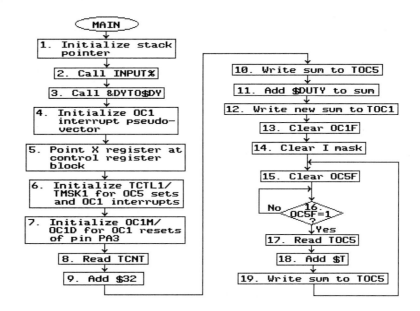

1 KHz Square Wave with Variable Duty Cycle—Memory Allocations

$ADDRESS	CONTENTS
0000–0001	$DUTY
0002	BCD 10's digit
0003	BCD unit's digit
0004	<CR> ASCII
0005	"I" ASCII
0006	"N" ASCII
0007	"P" ASCII
0008	"U" ASCII
0009	"T" ASCII
000A	<SP> ASCII
000B	"%" ASCII
000C	<SP> ASCII
000D	<EOT> ASCII
0100	MAIN program
0140	OC1 Interrupt Service routine
0150	INPUT% subroutine
0170	&DYTO$DY subroutine
01FF	Initial stack pointer

1 KHz Square Wave with Variable Duty Cycle—MAIN Program

$ADDRESS	SOURCE CODE	COMMENTS
0100	LDS #1FF	Initialize stack pointer
0103	JSR 150	Call INPUT% subroutine
0106	JSR 170	Call &DYTO$DY subroutine
0109	LDD #140	Initialize OC1 interrupt pseudovector
010C	STD E0	" " " " " "
010E	LDX #1000	Initialize X register
0111	LDD #380	Initialize TCTL1/TMSK1 for OC5 sets and OC1 interrupts
0114	STAA 20,X	" " " " " " "
0116	STAB 22,X	" " " " " " "
0118	LDD #800	Initialize OC1M/OC1D for OC1 resets of pin PA3
011B	STD C,X	" " " " " " "
011D	LDD E,X	Read TCNT
011F	ADDD #32	Add $32
0122	STD 1E,X	Write sum to TOC5
0124	ADDD 0	Add $DUTY to sum
0126	STD 16,X	Write new sum to TOC1
0128	BCLR 23,X 7F	Clear OC1F
012B	CLI	Clear I mask
012C	BCLR 23,X F7	Clear OC5F
012F	BRCLR 23,X 8 12F	Branch to self until OC5F = 1
0133	LDD 1E,X	Read TOC5
0135	ADDD #7D0	Add $T
0138	STD 1E,X	Write sum to TOC5
013A	BRA 12C	Make another leading edge

Refer to Figure 8–24, blocks 8 through 12, and program addresses $011D–$0127. In these blocks, the CPU initializes TOC5 and TOC1. The CPU initializes these registers with values that will generate initial rising and falling edges. The CPU initializes the TOC5 register by reading the current free-running count from the TCNT register (block 8) and adding $32 (block 9). In block 10, the CPU stores this sum to the TOC5 register. By storing this sum, the CPU delays the first OC5 compare until program execution has reached block 16. Thus, the M68HC11 will generate the first edge of the output square wave when it reaches block 16 in the program. Given the contents of TCTL1, this initial edge will be positive-going.

In block 11, the CPU adds $DUTY to TOC5's contents. Recall that $DUTY represents the pulse width of the square wave—in free-running counts of 500ns each. The CPU writes the sum to TOC1 (block 12). Thus, the OC1 mechanism generates an initial falling edge, $DUTY's worth of free-running counts after the initial rising edge.

Refer to Figure 8–24, blocks 13, 14, and 15, and program addresses $0128–$012E. In these blocks, the CPU clears OC1F, the I mask, and OC5F so that it can generate initial output edges. The CPU clears the OC1F and OC5F flags separately because it must clear OC5F only, as a part of a recurring block 15 to block 19 loop.

Refer to Figure 8–24, blocks 15 through 19, and program addresses $012C–$013B. In this loop, the M68HC11 generates the output waveform's rising edges. After clearing the OC5F bit (block 15), the CPU enters a BRCLR polling loop (block 16), where it reads and re-reads the OC5F flag until it sets. When OC5F sets, it indicates a successful compare between the free-running count and TOC5. This successful compare generates a positive-going edge at the OC5/PA3 pin. Upon escape from the BRCLR loop, the CPU must update its TOC5 register for the next rising edge. The CPU updates TOC5 by reading its contents (block 17), adding $7D0 (block 18), and writing the sum back to TOC5 (block 19). The hex augend $7D0 represents the number of free-running counts—at 500ns each—in the period ($T) of the 1 KHz square wave:

$$T \text{ of } 1\text{KHz} = 1/1\text{KHz} = 1\text{ms}$$
$$1\text{ms}/500\text{ns} = 2000 = \$7D0$$

After updating TOC5, the CPU branches back to block 15 where it clears OC5F and generates another positive-going edge.

While the CPU is in its block 16 BRCLR polling loop, a successful OC1 compare can occur, generating a negative-going output edge and an OC1 interrupt. Since TOC1's contents relate to TOC5's (TOC1 contents = TOC5 + $DUTY), an OC1 compare and interrupt can occur only when the CPU is in the block 16 BRCLR polling loop. If an OC1 interrupt occurred while the CPU was updating TOC5 (blocks 17–19), then TOC5's contents would be indefinite, i.e., could be the old or updated values. Therefore, the updated TOC1 value might also be indefinite, and the output waveform distorted. It takes the CPU thirty-one ($1F) E clock cycles to execute flowchart blocks 17, 18, 19, 15, and 16. Therefore, the minimum $DUTY value should be $31—a duty cycle of 1.55%. To assure consistent, reliable operation, the user should make a duty cycle keyboard entry no smaller than 02%. By entering a duty cycle value no smaller than 2%, the user assures that the CPU is in the BRCLR polling loop when a successful OC1 compare occurs.

Figure 8–25 and its accompanying program list illustrate the OC1 Interrupt Service routine. In this routine, the CPU updates its TOC1 register and clears the OC1F flag in TFLG1. The CPU must clear OC1F as a part of the interrupt service. If the CPU failed to clear the OC1F bit, another OC1 interrupt would occur as soon as the CPU returned from the interrupt (RTI). A continuous string of ill-timed interrupts would disrupt waveform generation. In block 1, the CPU reads the current TOC5 value and adds the $DUTY contents to it (block 2). This sum represents the updated TOC1 value. The CPU returns this updated value to TOC1 (block 3) and clears the OC1F flag (block 4) before returning from the interrupt. Thus, the CPU updates TOC1 so that the next successful OC1 compare will occur $DUTY's worth of free-running counts after the next successful OC5 compare.

Figure 8–26 and its accompanying program list illustrate the INPUT% subroutine. This subroutine prompts the user to enter desired decimal duty cycle value from the keyboard. INPUT% then captures the decimal duty cycle value entered by the user.

Refer to Figure 8–26, blocks 1 and 2, and program addresses $0150–$0155. Also refer to the memory allocations listing that precedes Figure 8–24. In block 1, the CPU points its X register at the "<CR>INPUT<SP>%<SP><EOT>" ASCII data in RAM. The CPU then calls the .OUTSTR0 utility subroutine (block 2). This subroutine displays the message "INPUT %" at the terminal.

FIGURE 8-25

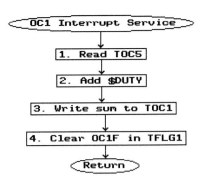

OC1 Interrupt Service

$ADDRESS	SOURCE CODE	COMMENTS
0140	LDD 1E,X	Read TOC5
0142	ADDD 0	Add $DUTY
0144	STD 16,X	Write sum to TOC1
0146	BCLR 23,X 7F	Clear OC1F
0149	RTI	Return

FIGURE 8-26

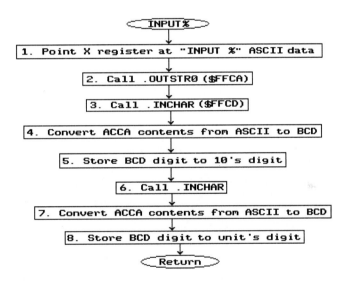

INPUT% Subroutine

$ADDRESS	SOURCE CODE	COMMENTS
0150	LDX #4	Point X register at "INPUT %" ASCII data
0153	JSR FFCA	Call .OUTSTR0
0156	JSR FFCD	Call .INCHAR
0159	ANDA #F	Convert ASCII 10's digit to BCD
015B	STAA 2	Store BCD 10's digit
015D	JSR FFCD	Call .INCHAR
0160	ANDA #F	Convert ASCII unit's digit to BCD
0162	STAA 3	Store BCD unit's digit
0164	RTS	Return

Refer to Figure 8–26, blocks 3, 4, and 5, and program addresses $0156–$015C. After displaying the "INPUT %" message, the CPU calls the .INCHAR utility subroutine (block 3). This subroutine causes the CPU to loop until the user presses a key. When the user presses a key, the CPU reads the resultant ASCII code from the terminal to its A accumulator. In block 4 the CPU converts this ASCII character to BCD by ANDing it with mask $0F. The user enters the % duty cycle ten's digit with the first key-press. Thus, after converting the ASCII character to BCD, the CPU writes it (block 5) to RAM location $0002.

Refer to Figure 8–26, blocks 6, 7, and 8, and program addresses $015D–$0164. In these blocks, the CPU repeats the .INCHAR and ASCII-to-BCD conversion, this time for the second user key-press (blocks 6 and 7). The second key-press is the percent duty cycle unit's digit. Therefore, the CPU writes the BCD unit's digit (block 7) to its assigned RAM location: $0003. The CPU then returns from the INPUT% subroutine to the MAIN program. The MAIN program then calls the &DYTO$DY subroutine.

Figure 8–27 and its accompanying program list show the &DYTO$DY subroutine. This subroutine converts the 10's and unit's BCD digits to a hexadecimal number: $DUTY. $DUTY equals the number of free-running counts in the duty cycle at 1 KHz. Recall that the BCD digits (stored at $0002 and $0003) represent duty cycle as a percentage of the 1 KHz square wave's period. At 1 KHz, 1% of the period equals twenty ($14) free-running counts—where each count equals 500ns. The &DYTO$DY subroutine executes three tasks to derive $DUTY:

1. The CPU must convert the BCD 10's and unit's digits to an equivalent hexadecimal number. This number is the hexadecimal equivalent of requested percent duty cycle.
2. The CPU must multiply the equivalent hexadecimal number by $14—the number of free-running counts in each 1% of duty cycle. This product equals $DUTY, the hex number of free-running counts in the duty cycle (i.e., pulse width) at 1 KHz.
3. The CPU must store the product to allocated $DUTY locations in RAM.

Refer to Figure 8–27, blocks 1 through 4, and program addresses $0170–$0176. In these blocks, the CPU accomplishes the first of the three tasks. It converts percent duty cycle from decimal to hexadecimal. In blocks 1 and 2, the CPU prepares its accumulators for conversion of the BCD 10's digit. The CPU makes this conversion by multiplying the BCD 10's digit by $0A. The product equals the hexadecimal equivalent of the BCD 10's digit

FIGURE 8–27

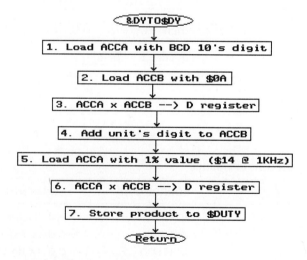

&DYTO$DY Subroutine

$ADDRESS	SOURCE CODE	COMMENTS
0170	LDAA 2	Load ACCA with BCD 10's digit
0172	LDAB #A	Load ACCB with $0A
0174	MUL	ACCA × ACCB → D register
0175	ADDB 3	Add unit's BCD digit to ACCB contents
0177	LDAA #14	Load ACCA with 1% value
0179	MUL	ACCA × ACCB → D register
017A	STD 0	Store product to $DUTY
017C	RTS	Return

value. Thus, the CPU loads its A accumulator with the BCD 10's digit (block 1) and its B accumulator with $0A (block 2). The CPU then executes a MUL command, which multiplies the contents of the two accumulators (block 3). MUL execution returns a 16-bit product to the D register. Since the largest possible BCD 10's digit value is 9, the largest possible product is $005A. Thus, ACCA contains nonsignificant leading zeros. ACCB contains the actual product. In block 4, the CPU completes the conversion by adding the BCD unit's digit to ACCB. The CPU can simply add the BCD unit's digit because hexadecimal units are equal to BCD units. After block 4 execution, ACCB contains the hexadecimal equivalent of percent duty cycle.

Refer to Figure 8–27, blocks 5 and 6, and program addresses $0177–$0179. In these blocks, the CPU accomplishes the second task: multiply the hex equivalent of percent duty cycle by the number of free-running counts in 1% of duty cycle. ACCB contains the hex equivalent of percent duty cycle. In block 5, the CPU loads ACCA with the 1% duty cycle value ($14). In block 6, the CPU executes a MUL command. This command multiplies ACCA by ACCB, and returns the product to the D register. This product represents the hexadecimal number of free-running counts in the pulse width, given the percent duty cycle entered by the user. Thus, this product equals $DUTY.

Refer to Figure 8–27, block 7 and program addresses $017A–$017C. The CPU accomplishes its third task in block 7. In this block, the CPU stores the contents of its D register to $DUTY's designated RAM locations, $0000–$0001. The CPU has completed a &DYTO$DY conversion and returns to the MAIN program.

The program shown in Figures 8–24 through 8–27 uses joint OC1/OC5 control to generate a 1 KHz square wave, with user-selectable duty cycles. In Exercise 8.6, apply this software to DC motor speed control.

EXERCISE 8.6 DC Motor Speed Control

REQUIRED MATERIALS:
- 1 Transistor, 2N3053, with "top hat" heat sink
- 2 Transistors, 2N3904
- 3 Diodes, 1N4001
- 1 Motor, DC, Radio Shack 273-223
- 1 Switch, push-button, push-to-close
- 2 Resistors, 270 Ω, 1/4W
- 1 Resistor, 4.7 KΩ
- 1 Resistor, 10 KΩ
- 1 Resistor, 100 KΩ
- 1 Transformer, filament, 6.3v @ 0.3a
- 1 Fuse, in-line, 0.6a
- 1 Capacitor 4700 µF, 6.3V

SUGGESTED MATERIALS: Use these materials to provide a secure motor mounting:
- 1 Wooden mounting base, 1" × 4", 5"–12" long
- 1 Piece white styrofoam (used in packing materials), nonconductive, approximately 1" × 2.5" × 5/8"
- 2 Rubber bands, 1/4" by 3.5" long

REQUIRED EQUIPMENT: Motorola M68HC11EVBU (EVB) and terminal
Oscilloscope and probe

GENERAL INSTRUCTIONS: This laboratory exercise applies Section 8.7's software to DC motor speed control. It demonstrates how motor speed can vary with the duty cycle of a 1 KHz square wave. You will also enhance the motor control interface and software so that the user can change motor speed more conveniently. This lab incorporates a DC supply to power the motor. This DC power supply requires a 6.3v filament transformer. The primary side of the transformer plugs into a 120vac receptacle. Be sure to completely insulate all primary-side connections and conductors. Also be sure to install an in-line 0.6a fuse in one of the primary-side wires. Take the time to do the job correctly and make the circuitry safe!

PART I: Construct a Secure Motor Mount

Figure 8–28 illustrates a suggested motor mounting. Carefully cut a small concave recess into the styrofoam motor mount. This recess cradles the motor. Make the styrofoam motor mount approximately 1" × 2.5" × 5/8." This 5/8" thickness allows for installation of an infrared LED in a Chapter 9 exercise. Use two stout rubber bands—at least 1/4" wide—to securely mount the motor and styrofoam to the wooden base. Rubber feet, installed on the bottom of the wooden base, keep the base from rocking or vibrating excessively while the motor runs. You may also wish to reserve space on the wooden base to mount the 6.3v filament transformer. If you mount the transformer on the wooden base, attach it well away from the motor to make room for later installation of a phototransistor for a Chapter 9 exercise. Securely solder 22 gauge × 10" wires to the DC motor's connection lugs. You will connect these wires to points A and B as shown in Figure 8–29.

Show your completed motor mounting to the instructor for approval.

_____ *Instructor Check*

PART II: Connect the Motor Control Circuitry

The Figure 8–28 schematic includes a simple power supply to provide DC power to the motor and speed control interface. In Figure 8–29, a dotted line surrounds this power supply. Construct the power supply. Safely and securely mount the filament transformer to a solid mounting base, such as the DC motor's wooden base. Assure that you completely insulate the transformer's primary leads and connections. Be absolutely certain that you connect the filter capacitor's positive (+) lead to the cathodes of the rectifier diodes. **Make no compromises with safety in constructing this power supply.** Before powering up this circuitry, have your instructor inspect it.

Construct the motor control circuitry and connect the DC motor per Figure 8–29.[20] DO NOT connect the control circuitry to the EVBU/EVB. You will make this connection in Part IV. Show your control circuitry to the instructor for approval.

_____ *Instructor Check*

Figure 8–28
Courtesy of Richard J. Helmer.

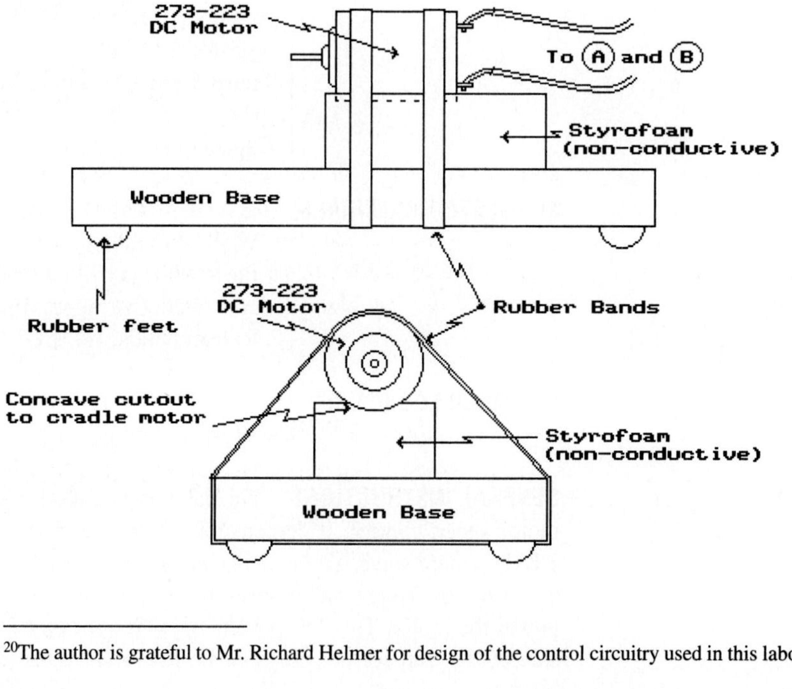

---

[20]The author is grateful to Mr. Richard Helmer for design of the control circuitry used in this laboratory exercise.

Figure 8–29
Courtesy of Richard J. Helmer.

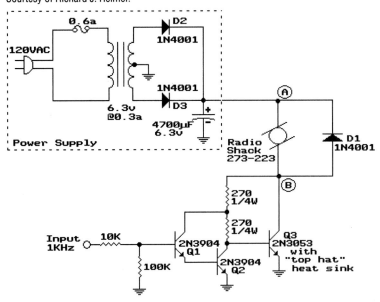

PART III: Load and Debug the Variable Duty Cycle Software

Review Section 8.7 to assure that you understand the strategy of the variable duty cycle software. If you use an EVBU, load your software from the lists accompanying Figures 8–24 through 8–27. If you use an EVB, locate your software routines as shown below:

$ADDRESS	CONTENTS
C000	MAIN program
C040	OC1 Interrupt Service
C050	INPUT% subroutine
C070	&DYTO$DY subroutine

If you use an EVB, adapt the listed software routines for their changed locations.

Connect the oscilloscope to pin 31 of the microcontroller. Run and debug the software. When you are sure that the software runs properly, demonstrate its operation to the instructor. Show your instructor the output waveform on the oscilloscope.

_____ *Instructor Check*

PART IV: Connect the Motor Control Circuitry to the Microcontroller

Connect the input of the motor control circuitry to the PA3/OC5 pin (pin 31) of the M68HC11. Assure that the microcontroller and control circuitry share a common ground. Leave the oscilloscope connected to microcontroller pin 31.

Run and debug the interface. When sure that it operates properly, demonstrate it to the instructor. Explain to the instructor how percent duty cycle correlates to motor speed.

_____ *Instructor Check*

PART V: Laboratory Analysis

Answer the questions below by neatly entering your answers in the appropriate blanks.

1. Explain the relationship between percent duty cycle and motor speed.

2. Refer to Figure 8–29. Describe the states of Q1, Q2, and Q3 when the input square wave is high: Q1 is _____ (on/off); Q2 is _____ (on/off); Q3 is _____ (on/off).

3. Refer to Figure 8–29. Describe the states of Q1, Q2, and Q3 when the input square wave is low: Q1 is _____ (on/off); Q2 is _____ (on/off); Q3 is _____ (on/off).
4. As the percentage of high time increases, the percentage of Q3's off time _____ (increases/decreases), causing V_{ave} supplied to the motor windings to _____ (increase/decrease). This causes motor speed to _____ (increase/decrease).
5. As percent duty cycle increases, the amount of time that Q3 is off _____ (increases/decreases) because the input square wave stays _____ (high/low) a greater percentage of the time.
6. Refer to Figure 8–29. What is D1's purpose?

If you use an M68HC11EVB, answer questions 7, 8, and 9.
7. Describe what happens to motor speed when you reset the EVB.

8. Explain why the motor speed responds as described in question 7.

9. Suggest a way to turn the motor off automatically when the user resets the EVB board.

If you use an M68HC11EVBU, answer questions 10 and 11.
10. Describe what happens to motor speed when you reset the EVBU.

11. Explain why the motor responds as described in question 10.

12. Enhance user control over the interface and motor by adding a pull-up circuit to the M68HC11's $\overline{IRQ}$ pin. Develop an IRQ Interrupt Service routine that accomplishes these tasks:
    - Configure PORTA pin 3 for general-purpose outputs.
    - Write a binary one to pin PA3 to stop the motor.
    - Re-direct program flow to the MAIN program's starting address. This causes the M68HC11 to re-prompt the user for a new, requested percent duty cycle.

MOVE the motor control software to EEPROM. Power down the EVBU/EVB and interface. Connect the IRQ pull-up circuit as shown in Figure 8–30. Develop IRQ interrupt service software that executes the Figure 8–31 flowchart. Enter your software in the blanks provided. Locate your IRQ Interrupt Service routine at address $0180 (EVBU) or $C080 (EVB).

IRQ Interrupt Service Routine

$ADDRESS	SOURCE CODE	COMMENTS
_____	_____	_____
_____	_____	_____
_____	_____	_____
_____	_____	_____
_____	_____	_____
_____	_____	_____

Use the system MM or ASM command to initialize the IRQ interrupt pseudovector at addresses $00EE–$00F0.

Load, debug, and demonstrate the operation of your enhanced interface to the instructor.

_____ *Instructor Check*

FIGURE 8-30

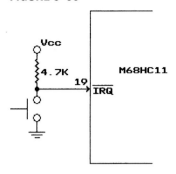

FIGURE 8-31

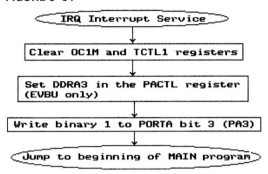

Important Note: You will use your DC motor and control circuitry for later exercises. Leave your DC motor on its mount and motor control circuitry intact.

PART VI: Laboratory Software Listing

MAIN PROGRAM—EVB

$ADDRESS	SOURCE CODE	COMMENTS
C000	LDS #C0FF	Initialize stack pointer
C003	JSR C050	Call INPUT% subroutine
C006	JSR C070	Call &DYTO$DY subroutine
C009	LDD #C040	Initialize OC1 interrupt pseudovector
C00C	STD E0	" " " " " " "
C00E	LDX #1000	Initialize X register
C011	LDD #380	Initialize TCTL1/TMSK1 for OC5 sets and OC1 interrupts
C014	STAA 20,X	" " " " " " " " "
C016	STAB 22,X	" " " " " " " " "
C018	LDD #800	Initialize OC1M/OC1D for OC1 resets of pin PA3
C01B	STD C,X	" " " " " " " " "
C01D	LDD E,X	Read TCNT
C01F	ADDD #32	Add #32
C022	STD 1E,X	Write sum to TOC5
C024	ADDD 0	Add $DUTY to sum
C026	STD 16,X	Write new sum to TOC1
C028	BCLR 23,X 7F	Clear OC1F
C02B	CLI	Clear I mask
C02C	BCLR 23,X F7	Clear OC5F
C02F	BRCLR 23,X 8 C02F	Branch to self until OC5F = 1
C033	LDD 1E,X	Read TOC5
C035	ADDD #7D0	Add $T
C038	STD 1E,X	Write sum to TOC5
C03A	BRA C02C	Make another leading edge

MAIN PROGRAM—EVBU

(See program list accompanying Figure 8–24.)

OC1 INTERRUPT SERVICE ROUTINE—EVB

$ADDRESS	SOURCE CODE	COMMENTS
C040	LDD 1E,X	Read TOC5
C042	ADDD 0	Add $DUTY
C044	STD 16,X	Write sum to TOC1
C046	BCLR 23,X 7F	Clear OC1F
C049	RTI	Return

OC1 INTERRUPT SERVICE ROUTINE—EVBU

(See program list accompanying Figure 8–25.)

INPUT% SUBROUTINE—EVB

$ADDRESS	SOURCE CODE	COMMENTS
C050	LDX #4	Point X register at "INPUT %" ASCII data
C053	JSR FFCA	Call .OUTSTR0
C056	JSR FFCD	Call .INCHAR
C059	ANDA #F	Convert ASCII 10's digit to BCD
C05B	STAA 2	Store BCD 10's digit
C05D	JSR FFCD	Call .INCHAR
C060	ANDA #F	Convert ASCII unit's digit to BCD
C062	STAA 3	Store BCD unit's digit
C064	RTS	Return

INPUT% SUBROUTINE—EVBU

(See program list accompanying Figure 8–26.)

&DYTO$DY SUBROUTINE—EVB

$ADDRESS	SOURCE CODE	COMMENTS
C070	LDAA 2	Load ACCA with BCD 10's digit
C072	LDAB #A	Load ACCB with $0A
C074	MUL	ACCA × ACCB —-> D register
C075	ADDB 3	Add unit's BCD digit to ACCB contents
C077	LDAA #14	Load ACCA with 1% value
C079	MUL	ACCA × ACCB —-> D register
C07A	STD 0	Store product to $DUTY
C07C	RTS	Return

&DYTO$DY SUBROUTINE—EVBU

(See program list accompanying Figure 8–27.)

DISPLAY ASCII DATA—EVB AND EVBU

$ADDRESS	CONTENTS	CHARACTER
0004	0D	\<CR\>
0005	49	"I"
0006	4E	"N"
0007	50	"P"
0008	55	"U"
0009	54	"T"
000A	20	\<SP\>
000B	25	"%"
000C	20	\<SP\>
000D	04	\<EOT\>

IRQ INTERRUPT SERVICE ROUTINE—EVB

$ADDRESS	SOURCE CODE	COMMENTS
C080	CLR C,X	Clear OC1M
C082	CLR 20,X	Clear TCTL1
C084	LDAA #8	Stop motor
C086	STAA 0,X	" "
C088	JMP C000	Jump to MAIN program

IRQ INTERRUPT SERVICE ROUTINE—EVBU

$ADDRESS	SOURCE CODE	COMMENTS
0180	CLR C,X	Clear OC1M
0182	CLR 20,X	Clear TCTL1
0184	LDAA #8	Initialize ACCA to stop motor
0186	STAA 26,X	Set DDRA3 in PACTL
0188	STAA 0,X	Write binary 1 to PA3
018A	JMP 100	Jump to MAIN program

8.8 CHAPTER SUMMARY

An OCx output compare mechanism causes the M68HC11 to take programmed actions when the free-running count matches the contents of its Timer Output Compare Register (TOCx). These actions include the following:

- Set, reset, or toggle an OCx pin.
- Set an OCxF bit in the TFLG1 register.
- Generate a corresponding output compare interrupt.

Upon a match between its TOC contents and free-running count, an output compare pin (PAx/OCx) responds in one of four ways:

- It does nothing.
- It sets.
- It resets.
- It toggles.

Figure 8–1 provides a block diagram of PA6–PA3 pin logic. The pin's logic level comes from one of three possible sources:

1. General-Purpose Output Latch, which contains a write bit from the M68HC11's internal data bus.
2. OC1 Logic, which drives the PAx/OCx pin with the state of its OC1D input.
3. OCx Compare Logic and OCx Output Latch, which cause the PAx/OCx pin to set, reset, or toggle.

OC1 Logic takes precedence over the other two possible sources. The OC1M input asserts to enable OC1 Logic's control over the PAx pin. If OC1M negates, the OCx Compare Logic and OCx Output Latch take precedence. Upon a successful OCx compare, these devices set, reset, or toggle the pin. The OCx Output Latch's Toggle, Clear, and Set inputs determine which pin action takes place. If the OCx Compare Logic's Disable input asserts, the OCx Compare Logic and OCx Output Latch relinquish pin control to the PAx General-Purpose Output Latch.

Refer to Figure 8–2, which details the OCx Compare Logic and OCx Output Latch. OCx Compare Logic strobes transmission gates. These transmission gates send the output of the OCx Output Latch to the PAx pin (Transmission Gate 3), and from the PAx pin to the inputs of the OCx Output Latch (Transmission Gate 23). OCx Compare Logic strobes Transmission Gates 3 and 23 in a complementary fashion. When Transmission Gate 3 strobes on, Transmission Gate 23 strobes off. Thus, when Transmission Gate 3 strobes on, it couples the OCx Output Latch's Q output to the PAx pin. When Transmission Gate 23 is on (i.e., not strobed off), it couples the state of the PAx pin to the inputs of the OCx Output Latch. The OCx Output Latch responds to these inputs, along with the Clear, Set, and Toggle inputs, to assume a new Q output state.

OCx Compare Logic strobes Transmission Gates 3 and 23 only if its Disable input negates (low), and either its OCx Compare or FOCx line asserts (high). PORTA CK synchronizes OCx Compare Logic strobes.

The DDRA3 and I4/O5 bits in the M68HC11E9's PACTL register configure pin PA3 for input captures or output compares. These PACTL bits configure other PA3-associated status and control bits in TCTL1, TCTL2, TFLG1, and TMSK1. DDRA3 and I4/O5 also determine whether the TI4O5 register functions as an input capture (TIC4) or an output compare register (TOC5).

The TCTL1 register allows the user to program PA6–PA3 pin responses to successful compares. Each OC pin is assigned two control bits, OMx and OLx, in the TCTL1 register. These bits control OC2–OC5 pin action as follows:

OMx	OLx	PAx/OCx Pin Action
0	0	No pin action
0	1	Toggle
1	0	Clear
1	1	Set

When the programmer resets both OMx and OLx, their counterpart PAx/OCx pin can be used as a general-purpose output. The M68HC11E9's OC5/PA3 pin can function as either a general-purpose output or input pin, depending upon the state of the PACTL register's DDRA3 bit.

The CPU enables OC2–OC5 interrupts by setting corresponding mask bits in the TMSK1 register. When so enabled, output compare interrupts are generated by successful compares. The I mask provides global interrupt control.

A successful OCx compare sets a corresponding OCxF flag bit in the TFLG1 register. These flags are read/modify/write bits. To clear an OCxF bit, the CPU must write a binary one to it.

A programmer can use output compares to generate a square wave. Square wave programs can use either interrupt control or status polling. The Figure 8–7 program uses interrupt control. This program initializes the TCTL1 register for output toggles, and the TMSK1 register for interrupt control. The program then clears the I mask and enters a loop where it waits for an OC interrupt.

A match between the relevant TOC register and free-running count generates an OC interrupt and an output edge. In the OC interrupt routine, the CPU updates the TOC register for the next edge by adding a $T/2 value to its contents. This augend represents the number of free-running counts in half of the output square wave's period. Thus, the OC output pin toggles two times in each square wave period, generating a 50% duty cycle. After clearing the applicable OCxF bit in TFLG1, the CPU returns to the MAIN program. In the MAIN program, the CPU waits for the next interrupt—which occurs after $T/2's worth of free-running counts elapse.

Figure 8–8 illustrates a program that uses OC status polling to generate a square wave with a 50% duty cycle. In comparison with the Figure 8–7 program (which uses interrupt control), this figure's approach reduces software overhead, increases dynamic efficiency, and thereby increases maximum output frequency from 18.5 KHz to 27.8 KHz.

The interrupt-control strategy illustrated in Figure 8–7 adapts easily to multiple square wave generation. Exercise 8.3 extends the Figure 8–7 approach to four simultaneous square waves, each having a unique frequency. The Exercise 8.3 program employs a single MAIN program and four OC Interrupt Service routines.

Section 8.4 illustrates how the M68HC11 integrates input captures (covered in Chapter 7) with output compares. Section 8.4 presents a program that uses output compares to generate a square wave and input captures to measure its frequency. This software also makes frequency selection and measurement user-friendly.

Refer to Figure 8–10. The OC/IC integration MAIN program generates and measures the output waveform by calling a series of subroutines, and then waiting for output compare interrupts.

The &fINPUT subroutine prompts the user to enter a decimal frequency value from the terminal. As the user enters digits, the CPU captures them, converts them from ASCII to BCD, and stores them in assigned RAM locations.

The CK&BNDS subroutine determines whether the user's frequency entry is within bounds, i.e., from 00032 to 19999. An out-of-bounds entry causes the CPU to re-start the MAIN program.

An inbounds entry causes the CPU to call the &fTO$f subroutine. This subroutine converts the user's decimal frequency request to hexadecimal and stores the results in designated RAM locations.

The CPU then calls the $fTOT/2 subroutine. $fTOT/2 "reciprocates" the hexadecimal frequency value to derive the hex period ($T). This routine then halves $T to find $T/2. $T/2 is expressed in free-running counts.

After deriving $T/2, the CPU calls the OC/IC subroutine. This subroutine generates the initial square wave edges and makes a frequency measurement, using the Input Capture 2 mechanism.

Refer to Figure 8–15. The OC/IC subroutine initializes the TCTL1, TCTL2, and TMSK1 registers for OC2 interrupt control of the output square wave, and IC2 measurement of this waveform—using status polling. As soon as the CPU clears the I mask (Figure 8–15, block 4), the M68HC11 starts generating an output square wave. Each match of the TOC2 register and free-running count toggles the PA6 output pin and generates an OC2 interrupt. The OC2 Interrupt Service routine (Figure 8–17) updates the TOC2 register for the next PA6 pin toggle and OC2 interrupt. As soon as the CPU clears the I mask, it uses status polling to measure the period ($T) of the OC2 output square wave (see Figure 8–15, blocks 5–11).

After measuring $T of the output waveform, the OC/IC subroutine calls a nested subroutine: $TTO&f. This subroutine converts the measured $T value to decimal frequency (&f). Upon return from $TTO&f, the CPU calls another nested subroutine: DISP&f. DISP&f displays measured decimal frequency at the terminal.

Having displayed measured frequency, the CPU returns to the MAIN program where it continues to generate the requested waveform.

Output Compare 1 (OC1) is a special case because it can control all five output compare pins. OC1 asserts this control even when individual OC functions (OC2–OC5) also control their own pins. Thus, OC1 and an individual OCx mechanism can jointly control an OCx pin.

OC1 has its own compare register—Timer Output Compare Register 1 (TOC1). When the free-running count matches the contents of TOC1, pins controlled by OC1 can set or reset, as specified by the OC1M and OC1D registers.

The CPU writes a mask byte to the OC1M register to assert OC1 control over any of the OC1–OC5 pins. By writing binary one to an OC1M bit, the CPU enables OC1 control over the counterpart OC pin. The OC1M register does not necessarily give the OC1 mechanism exclusive control over OC2–OC5 pins. Individual OC2–OC5 mechanisms can share control if enabled by the TCTL1 register. However, the OC1M register does give OC1 exclusive control over pin PA7 if PA7 has been configured as an output by the PACTL register's DDRA7 bit.

Refer to Figure 8–21, which depicts a block diagram of PA7 pin logic. If set, the PACTL register's DDRA7 bit (8) enables the Output Driver. When so enabled, the Output Driver applies the state of the "Cheater" Latch to the PA7 pin. The bit stored by the "Cheater" Latch comes from either the General-Purpose Output Latch or the OC1D register's OC1D7 bit (input 10). OC1 Logic enables Transmission Gate 1 if the OC1M register's OC1M7 bit = 0. OC1 Logic enables Transmission Gate 2 if the OC1M7 bit = 1, and either OC1 Compare or FOC1 asserts. Thus, if the OC1M7 bit enables OC1 control over pin PA7, the state of the OC1D7 bit appears at this pin via Transmission Gate 2. If OC1M7 = 0, the state of the General-Purpose Output Latch appears at pin PA7, via Transmission Gate 1. The PORTA CK (11) strobes Transmission Gate 1 or 2 enables.

The OC1 function uses the TFLG1 and TMSK1 registers in the same manner as the OC2–OC5 mechanisms. TFLG1 register's bit 7 constitutes the OC1F flag. By setting bit 7 (OC1I) of the TMSK1 register, the M68HC11 CPU enables OC1 interrupt control.

Section 8.6 discusses OC1's exclusive control over the five output compare pins, PA7–PA3. This section illustrates exclusive OC1 control with a counter program. This software uses OC pins as modulus thirty-two counter outputs. After initializing PACTL and OC1M for OC1 control over pins PA7–PA3, the program initializes the TOC1 register for the next OC1 compare. The program then clears the OC1F bit in TFLG1 and waits for the next successful OC1 output compare to occur. A successful compare sets OC1F, and drives the OC1–OC5 (PA7–PA3) pins with the contents of OC1D register's bit 7–bit 3. Initially, these bits contain %00000 due to a system reset. The CPU responds to each successful OC1 compare by adding $08 to the OC1D register's contents. Each addition of $08 causes OC1D register's bit 7–bit 3 contents to increment by one. Figure 8–23 illustrates this process. After updating the OC1D register, the program branches back to update TOC1 for the next OC1 output compare. Thus, the OC pins count from %00000 to %11111, then roll over to %00000 and start another counting cycle.

Section 8.7 covers joint OC1/OCx control of an output compare pin. Section 8.7 illustrates this control with software that varies the duty cycle of an output 1 KHz square wave. The software uses OC5 status polling to generate rising edges, and OC1 interrupt control to generate falling edges.

Refer to Figure 8–24. After initializing its stack pointer, the CPU calls the INPUT% subroutine. INPUT% prompts the user to enter a requested duty cycle value. INPUT% captures the user's duty cycle entries and converts them from ASCII to BCD. Upon its return from INPUT%, the CPU calls the &DYTO$DY subroutine. &DYTO$DY converts BCD duty cycle values to a hexadecimal number. This hex value represents the number of free-running counts in the waveforms's pulse width. After returning from &DYTO$DY, the CPU initializes its OC1 interrupt pseudovector and the TCTL1, TMSK1, OC1M, and OC1D registers. The CPU then initializes its TOC5 and TOC1 registers for initial rising and falling output edges. To generate periodic rising edges, the CPU iterates a loop where it uses status polling to detect a successful OC5 compare, then updates its TOC5 register for the next rising edge. While the CPU is in this loop, waiting for the OC5F flag to set, a successful OC1 compare generates an output falling edge and an OC1 interrupt. The OC1 Interrupt Service routine updates TOC1 for the next OC1 interrupt and falling edge. Thus, this software causes the M68HC11 to output a 1 KHz square wave with user-selectable duty cycles.

Chapter 8 demonstrates the impressive adaptability and flexibility of M68HC11 Output Compare functions. This chapter applies OC1–OC5 to the generation of individual and multiple square waves with fixed and variable duty cycles. Chapter 8 also demonstrates OC1–OC5's counting capabilities.

EXERCISE 8.7 Chapter Review

Answer the questions below by neatly entering your responses in the appropriate blanks.

1. An OC mechanism causes the M68HC11 to take programmed actions when the _____ matches the contents of its _____ register.

2. List the "programmed actions" referred to in question 1.

    (1)

    (2)

    (3)

3. List the four possible actions taken by an OC pin upon a successful output compare.

    (1)

    (2)

    (3)

    (4)

4. Refer to Figure 8–1. List the three possible sources of PAx/OCx output pin control.

    (1)

    (2)

    (3)

5. Of the three sources listed in question 4, which has highest priority? _____
   Describe how this source affects the state of the PAx/OCx pin. _____
   Which input (from the left edge in Figure 8–1) enables this source? _____ Which
   source has second priority? _____ Describe how this source affects the state
   of the PAx/OCx pin. _____ Which input (Figure
   8–1, left edge) disables this second-priority source? _____ Describe how the lowest
   priority source affects the state of the PAx/OCx pin.

6. Refer to Figure 8–2. The OCx Compare Logic strobes Transmission Gate 3 _____
   (on/off) and Transmission Gate 23 _____ (on/off). List the input conditions that generate these strobes:

   PORTA CK = binary_____
   OCx Compare or FOCx = binary_____
   Disable = binary_____

7. Refer to Figure 8–2. Describe the purpose of Transmission Gate 3.

8. Refer to Figure 8–2. Describe the purpose of Transmission Gate 23.

9. Refer to Figure 8–2. If the Clear input = binary 1, then Q = binary _____. If the Set
   input = 1, then Q = _____. If the Toggle input = 1, then Q = _____
10. Refer to Figure 8–2. If Disable = 1, then Transmission Gate 3 is _____ (on/off), and
    Transmission Gate 23 is _____ (on/off). If Disable = 1, then Clear = _____, Toggle =
    _____, and Set = _____. The CPU establishes Disable, Clear, Set, and Toggle control
    values in the _____ register.
11. The _____ and _____ bits in the M68HC11E9's PACTL register configure the PA3
    pin as either IC4 or OC5.
12. If the PACTL register's I4/O5 bit = 0, pin PA3 is configured for _____.
13. If the PACTL register's DDRA3 bit = 0, pin PA3 is configured as an _____.
14. After an M68HC11E9 reset, both the DDRA3 and I4/O5 bits are _____ (set/reset). If
    the M68HC11E9's CPU writes to TCTL1 and specifies that successful compares cause
    the PA3/OC5 pin to _____, _____, or _____, the OC5 mechanism overrides the
    _____ bit.
15. To make the PA5 pin set upon successful compares, the M68HC11 CPU should write
    binary _____s to bits _____ and _____ of the _____ register.
16. To enable OC2 and OC5 interrupts, the M68HC11 CPU must write binary _____s to
    bits _____ and _____ in the _____ register.
17. To clear the OC3F bit in TFLG1, a programmer uses the following source code.
    (Specify the offset and mask; assume that the X register contains $1000).

    BCLR _____,X _____

18. To clear the OC5F bit in TFLG1, a programmer uses the following source code.
    (Specify the offset and mask; assume that the X register contains = $1000.)

    BCLR _____,X _____

19. Enter control words that the CPU should write to the listed addresses, in order to configure for output compares as specified.

Specifications	$Address	Control Word
Use OC2, OC4, OC5		
Pin PA3 as input		
Set OC2		
Toggle OC4		

401

Specifications	$Address	Control Word
Reset OC5		
OC2 interrupt control		
OC4 status polling		
OC5 interrupt control		
	1020	$_____
	1022	$_____
	1026	$_____

20. Refer to Figure 8–7 and its accompanying software listing. Explain why this program takes no specific action to initialize the PACTL register for OC5 outputs.

21. Refer to Figure 8–7, block 8. Explain the meaning and application of "$T/2."

22. Compare Figure 8–7 with Figure 8–8. Both of these programs use output compares to generate a square wave. Explain the key difference in their approaches to this task.

    Which of these approaches generates a greater range of frequencies, and how does this approach extend the range?

23. Both Figure 8–7 and Exercise 8.3 use the same strategy to generate output waveforms. However, the Figure 8.7 program generates a maximum frequency of 18.5 KHz, while the Exercise 8.3 approach generates much lower maximum frequencies. Why can the Exercise 8.3 program generate only these lower values?

24. Refer to Figure 8–10. Briefly describe the tasks performed by the routines listed below.

    &fINPUT:

    CK&BNDS:

    &fTO$f:

    $fTOT/2:

    OC/IC:

25. Refer to Figure 8–11. Describe how the CPU converts ASCII characters to BCD digits (block 5).

26. Refer to Figure 8–11. How many times does the CPU execute the block 9 instructions?

27. Refer to Figure 8–12. Explain why the CPU compares the 10,000's BCD digit with 1 (block 2).

28. Refer to Figure 8–12. Explain why the CPU returns from the subroutine if the 1,000's and/or 100's digit are not zero.

29. Refer to Figure 8–13. Explain why the CPU clears the $f locations (block 2).

30. Refer to Figure 8–13. If the CPU uses the MUL instruction to convert the 100's and 10's BCD digits, why does it not use MUL to convert the 10,000's and 1,000's BCD digits?

31. Refer to Figure 8–14. Explain why the author calls the block 1–block 5 process a "reciprocation."

32. Refer to Figure 8–14. Explain why the CPU shifts the $T value right (block 7).

33. Refer to Figure 8–15. Describe the purpose of block 5.

34. Refer to Figure 8–15. Describe the purpose of block 8.

35. Refer to Figure 8–16. To what hex value does the CPU initialize DIFF (block 2)? Why does the CPU initialize DIFF to this value?

36. Refer to Figure 8–16. block 5 execution requires twelve instructions. Explain why this is the case.

37. Refer to Figure 8–17. Describe the purpose of the DISP&f subroutine.

38. Refer to Figure 8–17. Explain why the CPU sets the I mask (block 1).

39. Refer to Figure 8–17. List the two tasks performed by the OC2 Interrupt Service routine.

(1)

(2)

40. Explain how OC1 differs from the OC2–OC5 functions.

41. In the blanks, enter control words that the CPU must write to the PACTL, OC1M, OC1D, and TCTL1 registers to enable the following concurrent pin actions:

    PA7 sets upon successful OC1 compares.
    PA6 sets upon successful OC1 compares.
    PA4 resets upon successful OC1 compares.
    PA6 resets upon successful OC2 compares.
    PA4 sets upon successful OC4 compares.
    PA3 toggles upon successful OC5 compares.

Register $Address	$Control Word
1026	_____
100C	_____
100D	_____
1020	_____

42. Refer to Figure 8–21. Enter zeros or ones in the blanks to indicate the states required to enable Transmission Gate 1 and apply the output of the General-Purpose Output Latch to pin PA7. Enter "N/A" where an input has no bearing on the Transmission Gate 1 or Output Driver enables.

    8 DDRA7 = ____
    11 PORTA CK = ____

6 OC1M7 = ____
12 OC1 Compare or 9 FOC1 = ____

43. Refer to Figure 8–21. Enter zeros or ones in the blanks to indicate the states required to enable Transmission Gate 2 and apply the state of OC1D7 (10) to pin PA7. Enter "N/A" where an input has no bearing on the Transmission Gate 2 or Output Driver enables.

8 DDRA7 = ____
11 PORTA CK = ____
6 OC1M7 = ____
12 OC1 Compare or 9 FOC1 = ____

44. Refer to Figure 8–22. Explain what "$T" stands for (block 4).

45. Refer to Figure 8–22. Explain why the CPU adds $08 to OC1D in blocks 9 through 11.

46. Refer to Figure 8–24. Describe the purpose of the INPUT% subroutine (block 2).

47. Refer to Figure 8–24. Describe the purpose of the &DYTO$DY subroutine (block 3).

48. Refer to Figure 8–24. Explain the purpose of blocks 8, 9, and 10.

49. Refer to Figure 8–24. Describe the tasks accomplished by the block 15–block 19 loop.

50. Refer to Figure 8–25. Describe the tasks accomplished by this interrupt service routine.

51. Refer to Figure 8–26. blocks 3 and 4 capture and convert the ten's digit entered by the _____ from the _____. Blocks 6 and 7 capture and convert the _____ digit.
52. Refer to Figure 8–27. In blocks _____ through _____, the CPU converts % duty cycle from decimal to hexadecimal. In accomplishing this conversion, the CPU first multiplies the BCD _____ digit by $_____, then adds the BCD _____ digit to the product. After completing these tasks, the CPU has the hexadecimal equivalent of percent duty cycle in its _____(A/B) accumulator.
53. Refer to Figure 8–27. Describe the tasks accomplished in blocks 5 and 6.

9 M68HC11 Forced Output Compares, Real-Time Interrupts, and Pulse Accumulator

9.1 GOAL AND OBJECTIVES

Chapter 9 presents three seemingly unrelated M68HC11 features—forced output compares, Real-Time Interrupts, and the Pulse Accumulator. Forced compares allow the CPU to generate output compare pin responses prior to successful output compares. The Real-Time Interrupt feature generates interrupts at one of four periodic rates. The Pulse Accumulator can operate in either of two modes: counting the number of external events over a period of time or measuring the duration of external events. Chapter 9 illustrates how these three features complement each other to enhance M68HC11 timing and control functions. For the most part, this chapter uses DC motor control as the vehicle for illustrating enhanced control. The reader will apply forced output compares to two-speed motor control. This control allows the user to select motor speed with a switch. The reader will apply Real-Time Interrupts and Pulse Accumulator to DC motor speed measurement and regulation. This regulation uses proportional-tachometer control. Thus, Chapter 9 demonstrates how forced compares, Real-Time Interrupts, and the Pulse Accumulator facilitate timing and control in real-world applications.

This chapter's goal is to help the reader apply three M68HC11 features—forced output compares, Real-Time Interrupts, and the Pulse Accumulator.

After studying Chapter 9 and completing its seven exercises, you will be able to do the following:

- Use forced output compares to provide two-speed DC motor speed control.
- Use Real-Time Interrupts to generate periodic time reference pulses.
- Develop source code which configures the Pulse Accumulator for event counting.
- Apply Real-Time Interrupts and the Pulse Accumulator event counting to measure a DC motor's revolutions per minute.
- Apply Real-Time Interrupts and Pulse Accumulator event counting to proportional-tachometer DC motor speed regulation.
- Derive source code which configures the Pulse Accumulator for gated time accumulation.
- Apply gated time accumulation mode to measure a 1 KHz square wave's percent duty cycle.

9.2 M68HC11 FORCED OUTPUT COMPARES

The reader studied M68HC11 output compares in Chapter 8. Output compare mechanisms have one more important feature—forced output compares.

Chapter 8 established that the CPU programs PA7–PA3 pin responses to successful OC1–OC5 compares. A successful compare consists of a match between the free-running count and the contents of a Timer Output Compare (TOCx) register. The CPU programs OCx pin responses by writing control words to the TCTL1, OC1M, and OC1D registers. Using the forced output compare feature, the CPU can force programmed pin responses before a successful compare actually occurs. Applications designers use forced output compares in a wide variety of applications, from camera shutter control to advanced control of building elevators. In its *M68HC11 Reference Manual*, Motorola cites automotive ignition spark advance as an example.[1] In Exercise 9.1, the reader shall apply forced output compares to a two-speed DC motor control.

On-Chip Resources Used by the M68HC11 for Forced Output Compares

Forced output compares extend the CPU's control over its output compare mechanism. To apply forced compares, the CPU uses essentially the same on-chip resources as conventional output compares. In fact, forced compares require only one additional control register. Therefore, following paragraphs remind the reader of the on-chip resources used by conventional output compares and describe how these resources relate to forced compares.

Conventional and forced compares cause output compare pins to toggle, set, or reset.[2] These output compare pins are PORTA pins 3 through 7 (PA3–PA7). Review Chapter 8, Figures 8–1, 8–2, and 8–21, along with the descriptive material that accompanies them. This material describes the structure and operation of PA3–PA7 pin logic. When reviewing this material, pay special attention to the role played by FOCx inputs. An FOCx input generates forced compare responses by an OC pin.

The TCTL1, OC1M, and OC1D registers allow a user to program PA7–PA3 pin responses to conventional and forced output compares. The OC1M and OC1D registers can program any OC pin to set or reset in response to a successful OC1 output compare. A forced OC1 compare will generate these programmed pin responses.

The CPU enables OC1–OC5 interrupts by setting mask bits in the TMSK1 register. When so enabled, successful output compares generate corresponding output compare interrupts. However, forced output compares do not generate output compare interrupts even when interrupts have been enabled in the TMSK1 register.

A successful output compare sets a corresponding OCxF flag in the TFLG1 register. However, forced output compares do not cause corresponding OCxF bits to set.

A successful output compare consist of a match between the free-running count and the contents of a TOCx register. This match causes a corresponding OCx pin response, sets an OCxF flag, and can generate an output compare interrupt. A conventional compare generates these responses even if the CPU has previously forced a set or reset by that pin. Remember, a forced compare generates the programmed pin response, but *does not* generate an OCxF flag or interrupt.

Foregoing paragraphs describe conventional output compare resources, and how they relate to forced compares. How does the CPU actually generate forced output compares?

To generate forced compares, the CPU writes to the Timer Compare Force Register (CFORC, control register block address $100B). Refer to Figure 9–1. Note that each output compare function has an assigned FOCx bit. To force a compare, the CPU simply writes binary one to a corresponding FOCx bit. For example, to force an OC2 compare, the CPU

FIGURE 9–1
Reprinted with the permission of Motorola.

CFORC

B7							B0
FOC1	FOC2	FOC3	FOC4	FOC5			

---

[1]Motorola Inc., *M68HC11 Reference Manual* (1990), p. 10-36.
[2]Recall that output compare pins can also disconnect from their output compare mechanisms, so conventional compares merely set a flag and optionally generate an interrupt.

writes $40 to the CFORC register. To force simultaneous OC3 and OC5 compares, the CPU writes $28 to CFORC. Output compare mechanisms transfer CFORC bit states to the FOCx inputs of OCx pin logic. FOCx = 1 causes its corresponding pin to set, reset, or toggle (see Figures 8–1, 8–2 and 8–21). Response depends upon the states of that pin's Toggle, Set, and Clear inputs. As soon as it forces a pin response, the FOCx bit in the CFORC register resets. Becuase FOCx bit states are transitory, any CFORC read returns only $00. The free-running counter's clock synchronizes OC pin response to a forced compare. This clock can run at the E clock rate (default) or at a rate four, eight, or sixteen times slower than the E clock—depending upon PR1/PR0 bit states in the TMSK2 register. Therefore, an OC pin may respond one, four, eight, or sixteen E clocks after the CPU writes binary one to its corresponding FOCx bit in CFORC.

Thus, by writing binary ones to appropriate FOCx bit(s) in the CFORC register, the CPU can force counterpart OC pin responses before a successful conventional output compare occurs. Forced compares do not set OCxF flags or generate OCx output compare interrupts. Only conventional output compares can do that. OC pin response to a forced compare may delay by as much as sixteen E clocks, if TMSK2 bits PR1 and PR0 are both set.

Two-Speed Motor Control Using Forced Output Compares

In Chapter 8, Exercise 8.6 allows the user to enter duty cycle values from 02% to 98%. Thus, this exercise provides ninety-six different motor speeds. In most applications, such a wide choice is unnecessary. In fact, most applications entail a choice of two or three motor speeds. The practical problem, then, is to design an interface and software that most economically gives the user a choice of two or three speeds. This section presents a simple interface that provides a choice of two motor speeds.

Refer to Figure 9–2. A 1 KHz square wave from the OC5/PA3 pin controls motor speed. This control signal connects to a motor control interface, as constructed for Exercise 8.6. Recall that with this interface, motor speed is inversely related to the 1 KHz control signal's percent duty cycle. In Figure 9–2, the operator uses an SPST switch in a pull-up circuit to select fast or slow motor speed. The operator closes the switch to select the faster motor speed, and opens it to select the slower. A closed switch causes the pull-up circuit to apply a low to PORTA pin PA2. If the CPU reads a low on this pin, it forces output compares that generate the 1 KHz square wave's falling edges. These forced compares reduce the percent duty cycle of the square wave, causing a greater percentage of low time. More low time causes greater motor speed. If the CPU reads a high from pin PA2, it does not force falling edge compares. As a result, the 1 KHz square wave has a larger duty cycle (shorter low time), and the motor runs at a slower speed. Thus, the user can select faster or slower motor speed with the two positions of the SPST switch.

Figure 9–3 and its accompanying program list illustrate the two-speed motor control software. This software uses OC1 and OC5 to generate the 1 KHz square wave's rising and falling edges. To run the DC motor at the slow speed, the software generates a 60% duty cycle. For high speed, the software reduces the output duty cycle to 20%. This software uses status polling to detect successful OC1 and OC5 output compares.

FIGURE 9–2

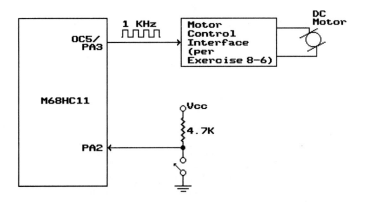

FIGURE 9–3

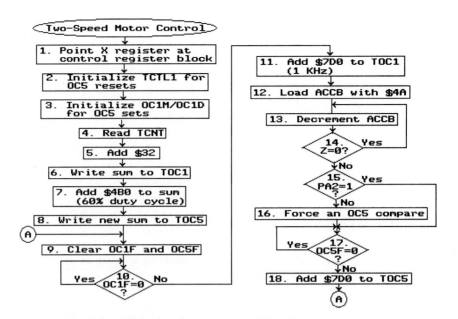

Two-Speed Motor Control Using Forced Output Compares

$ADDRESS	SOURCE CODE	COMMENTS
0100	LDX #1000	Initialize X register
0103	LDAA #2	Initialize TCTL1 for OC5 resets
0105	STAA 20,X	" " " " "
0107	LDD #808	Initialize OC1M/OC1D for OC5 sets
010A	STD C,X	" " " " "
010C	LDD E,X	Read TCNT
010E	ADDD #32	Add $32
0111	STD 16,X	Write sum to TOC1
0113	ADDD #4B0	Add $4B0 to sum (60% duty cycle)
0116	STD 1E,X	Write new sum to TOC5
0118	BCLR 23,X 77	Clear OC1F and OC5F
011B	BRCLR 23,X 80 11B	Branch to self until OC1F = 1
011F	LDD 16,X	Add $7D0 to TOC1 ($T)
0121	ADDD #7D0	" " " "
0124	STD 16,X	" " " "
0126	LDAB #4A	Delay for 20% of $T (less overhead)
0128	DECB	" " " " " "
0129	BNE 128	" " " " " "
012B	BRSET 0,X 4 133	Is 20% duty cycle requested?
012F	LDAA #8	If yes, force an OC5 compare
0131	STAA B,X	" " " " "
0133	BRCLR 23,X 8 133	Branch to self until OC5F = 1
0137	LDD 1E,X	Add $7D0 to TOC5 ($T)
0139	ADDD #7D0	" " " "
013C	STD 1E,X	" " " "
013E	BRA 118	Make more edges

Refer to Figure 9–3, blocks 1 through 3, and program addresses $0100–$010B. In these blocks, the CPU writes $1000 to the X register, $02 to the TCTL1 control register, and $08 to both the OC1M and OC1D registers. Writing $02 to TCTL1 sets that register's OM5 bit and clears its OL5 bit. When OM5 = 1 and OL5 = 0, the OC5 mechanism clears the OC5/PA3 pin upon successful OC5 compares. Writing $08 to OC1M sets that register's

OC1M3 bit. OC1M3 = 1 enables OC1 control over the OC5/PA3 pin. By writing $08 to the OC1D register, the CPU causes the OC5/PA3 pin to set upon successful OC1 compares.

In blocks 4 through 8 (program addresses $010C–$0117), the CPU initializes TOC1 and TOC5 for generation of the 1 KHz square wave's initial edges. This software uses the OC1 mechanism to generate rising edges. Blocks 4, 5, and 6 initialize TOC1 so that the initial OC1 compare occurs while the CPU executes block 10, where it polls the OC1F flag. To initialize TOC1, the CPU reads the current free-running count from the TCNT register (block 4), and adds $32 to this count (block 5). By adding $32 to the current free-running count, the CPU gives itself enough time to execute blocks 7, 8, and 9 before the first successful OC1 compare occurs. In block 6, the CPU writes the sum TCNT + $32 to its TOC1 register.

In blocks 7 and 8, the CPU initializes its TOC5 register for a 60% duty cycle. For this 60% duty cycle, the OC5 mechanism must generate a falling edge $4B0 free-running counts after the initial rising edge:

$$\$4B0 = 1200$$
$$1200 \times 500ns = 600\mu s$$
$$600\mu s = 60\% \text{ of 1ms (the 1 KHz square wave's period)}$$

In block 7, the CPU adds $4B0 to the contents of the TOC1 register. In block 8, the CPU writes this sum to TOC5.

Refer to blocks 9, 10, and 11 (program addresses $0118–$0125). After clearing the OC1F and OC5F flags in the TFLG1 register (block 9), the CPU enters a polling loop (block 10). The CPU escapes this polling loop when the OC1F flag sets, indicating that OC1 has generated a rising edge at pin PA3. In block 11, the CPU updates TOC1 for the next output rising edge. This next rising edge must occur after one-millisecond elapses. Thus, the CPU adds $7D0 to the contents of TOC1:

$$\$7D0 = 2000$$
$$2000 \text{ free-running counts @ } 500ns = 1ms$$

To execute block 11, the CPU reads the TOC1 register, adds $7D0, and writes the sum back to the TOC1 register.

Recall that the interface operator closes a switch in the pull-up circuit to call for the faster of two motor speeds. A closed switch causes the pull-up circuit to apply a low to the microcontroller's PA2 pin. If the CPU reads binary zero at pin PA2, it must generate a 20% rather than a 60% duty cycle. Therefore, the CPU reads pin PA2 and forces an early OC5 compare *if* PA2 = 0. The OC5 mechanism generates falling edges. If PA2 = 0, the CPU must force an OC5 compare 400 free-running counts after a rising edge:

$$400 \times 500ns = 200\mu s = 20\% \text{ of 1ms}$$

In blocks 12–16 (program addresses $0126–$0132), the CPU enters a delay loop (blocks 12, 13, and 14), reads pin PA2 (block 15), and forces an OC5 compare (block 16) if PA2 = 0. The CPU may have to force an OC5 compare after 400 free-running counts have elapsed. The CPU uses 28 of these counts to execute blocks 11, 12, 15, and 16. The CPU expends the remaining 372 counts in a delay loop (blocks 13 and 14). Each delay loop iteration requires five free-running counts. To expend 372 counts, the CPU must execute the loop 74 ($4A) times. Therefore, the CPU loads a loop counter (ACCB) with $4A in block 12. It then executes the delay loop (blocks 13 and 14). Upon escape from the delay loop after seventy-four iterations, the CPU reads pin PA2 (block 15). If this read indicates that PA2 = 0, the CPU forces an OC5 compare at the 20% point (block 16). If PA2 = 1, the CPU branches around the forced compare instructions.

Refer to blocks 17 and 18, and program addresses $0133–$013F. Whether or not it has forced an OC5 compare, the CPU polls the OC5F flag (block 17) until it sets. Remember that forced compares do not set the OC5F flag or generate OC interrupts. A forced OC5 compare merely generates the square wave's falling edge at the 20% rather than the 60% point. When OC5F = 1, the CPU escapes the block 17 polling loop and adds $7D0 to the contents of the TOC5 register (block 18). This addition updates TOC5 for the next possible negative-going output edge, at the 60% point. The next 60% point occurs after one millisecond, i.e., after another $7D0 free-running counts elapse. After executing block 18, the CPU branches back to block 9 to make more 1 KHz square wave edges.

EXERCISE 9.1   Two-Speed DC Motor Control

REQUIRED MATERIALS: Motor control interface, as constructed for Exercise 8.6
DC motor and mounting, as constructed for Exercise 8.6
2 SPST switches
2 Resistors, 4.7 KΩ

REQUIRED EQUIPMENT: Motorola M68HC11EVBU (EVB) and terminal
Oscilloscope and probe

GENERAL INSTRUCTIONS: This laboratory exercise applies software discussed in Section 9.2. The M68HC11 controls motor speed with an interface, as constructed for Exercise 8.6. Likewise, this interface drives a DC motor, as mounted for Exercise 8.6. Review Parts I and II of Exercise 8.6, along with Figures 8–28 and 8–29, to refresh your memory about the motor control interface's construction. Exercise 9.1 also entails a speed-select pull-up circuit that incorporates an SPST switch. Connect this pull-up circuit to the microcontroller's PA2 pin (pin 32).

PART I: Connect the Motor Control Interface and Speed-Select Pull-up Circuit

Refer to Figure 9–2. Connect the interface and pull-up circuit as shown in this figure. Show your connections to the instructor.

_____ *Instructor Check*

PART II: Load and Debug the Two-Speed Motor Control Software.

Review pages 409–411. Assure that you understand the two-speed motor control software. If you use an EVBU, load your software from the list accompanying Figure 9–3. If you use an EVB, locate your software starting at address $C000. If you use an EVB, modify all branching instructions to reflect this program storage. Test and debug the software and interface. When sure that the interface operates properly, demonstrate it to the instructor. Be prepared to show the 1 KHz motor control waveform to the instructor, using the oscilloscope.

_____ *Instructor Check*

PART III: Laboratory Analysis

Answer the questions below by neatly entering your answers in the appropriate blanks.

1. DC motor speed varies _____ (directly/inversely) with low time of the output square wave.
2. Describe the purpose of the pull-up circuit shown in Figure 9–2.

3. Refer to Figure 9–2. To make the motor run at the faster speed, the user should _____ (open/close) the switch.
4. Refer to Figure 9–3. Explain why the CPU must initialize the OC1M and OC1D registers (block 3).

5. Refer to Figure 9–3. Explain why the CPU adds $32 to the TCNT value in blocks 4 and 5.

FIGURE 9–4

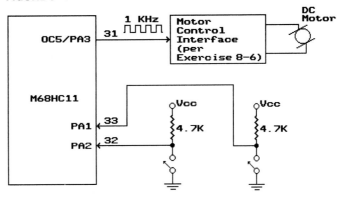

6. Refer to Figure 9–3. Explain why the CPU adds $4B0 to TOC1's contents in block 7.

7. Refer to Figure 9–3. Explain why the CPU loads ACCB with $4A in block 12.

8. Refer to Figure 9–3. List the block(s) not executed if the interface operator requests fast motor speed. _____
9. Refer to Figure 9–3. List the block(s) not executed if the interface operator requests slow motor speed. _____
10. Convert your interface and software for a selection of three motor speeds. Use two pull-up circuits as shown in Figure 9–4. Convert the software for responses as shown in the chart below:

Binary Inputs		M68HC11/Interface Response	
PA2	PA1	Percent Duty Cycle	Motor Speed
0	0	20%	Fast
0	1	20%	Fast
1	0	40%	Medium
1	1	60%	Slow

Use Figure 9–5 as a guide to modifying you software. List this modified software in the blanks below.

$ADDRESS	SOURCE CODE	COMMENTS
_____	_____	_____
_____	_____	_____
_____	_____	_____
_____	_____	_____
_____	_____	_____
_____	_____	_____
_____	_____	_____
_____	_____	_____
_____	_____	_____

Power down your EVB (EVBU) and interface. Add another pull-up circuit to the interface. Load and debug your modified software. When sure that the interface operates properly, demonstrate it to the instructor. Be prepared to show the instructor an oscilloscope display of the 1 KHz motor control waveform.

_____ *Instructor Check*

FIGURE 9–5

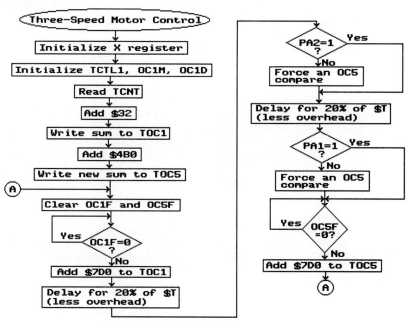

PART IV: Laboratory Software Listing

TWO-SPEED MOTOR CONTROL—EVB

$ADDRESS	SOURCE CODE	COMMENTS
C000	LDX #1000	Initialize X register
C003	LDAA #2	Initialize TCTL1 for OC5 resets
C005	STAA 20,X	" " " " " " " "
C007	LDD #808	Initialize OC1M/OC1D for OC1 sets of pin OC5
C00A	STD C,X	" " " " " " " "
C00C	LDD E,X	Read TCNT
C00E	ADDD #32	Add $32
C011	STD 16,X	Write sum to TOC1
C013	ADDD #4B0	Add $4B0 to sum (60% duty cycle)
C016	STD 1E,X	Write new sum to TOC5
C018	BCLR 23,X 77	Clear OC1F and OC5F
C01B	BRCLR 23,X 80 C01B	Branch to self until OC1F = 1
C01F	LDD 16,X	Add $7D0 to TOC1 ($T)
C021	ADDD #7D0	" " " "
C024	STD 16,X	" " " "
C026	LDAB #4A	Delay for 20% of $T (less overhead)
C028	DECB	" " " " " "
C029	BNE C028	" " " " " "
C02B	BRSET 0,X 4 C033	Is 20% duty cycle requested?
C02F	LDAA #8	If yes, force an OC5 compare
C031	STAA B,X	" " " "
C033	BRCLR 23,X 8 C033	Branch to self until OC5F = 1
C037	LDD 1E,X	Add $7D0 to TOC5 ($T)
C039	ADDD #7D0	" " " "
C03C	STD 1E,X	" " " "
C03E	BRA C018	Make more edges

TWO-SPEED MOTOR CONTROL—EVBU

(See program list accompanying Figure 9–3.)

THREE-SPEED MOTOR CONTROL—EVB

$ADDRESS	SOURCE CODE	COMMENTS
C000	LDX #1000	Initialize X register
C003	LDAA #2	Initialize TCTL1
C005	STAA 20,X	" " "
C007	LDD #808	Initialize OC1M and OC1D
C00A	STD C,X	" " " " "
C00C	LDD E,X	Read TCNT
C00E	ADDD #32	Add $32
C011	STD 16,X	Write sum to TOC1
C013	ADDD #4B0	Add $4B0 to sum
C016	STD 1E,X	Write new sum to TOC5
C018	BCLR 23,X 77	Clear OC1F and OC5F
C01B	BRCLR 23,X 80 C01B	Branch to self until OC1F = 1
C01F	LDD 16,X	Add $7DO to TOC1
C021	ADDD #7D0	" " "
C024	STD 16,X	" " "
C026	LDAB #4A	Delay for 20% of $T
C028	DECB	" " "
C029	BNE C028	" " "
C02B	BRSET 0,X 4 C033	20% duty cycle requested?
C02F	LDAA #8	If yes, force OC5
C031	STAA B,X	" " "
C033	LDAB #4E	Delay for 20% of $T
C035	DECB	" " "
C036	BNE C035	" " "
C038	BRSET 0,X 2 C040	40% duty cycle requested?
C03C	LDAA #8	If yes, force OC5
C03E	STAA B,X	" " "
C040	BRCLR 23,X 8 CO40	Branch to self until OC5F = 1

C044	LDD 1E,X	Add $7D0 to TOC5
C046	ADDD #7D0	" " "
C049	STD 1E,X	" " "
C04B	BRA C018	Make more edges

THREE-SPEED MOTOR CONTROL—EVBU

$ADDRESS	SOURCE CODE	COMMENTS
0100	LDX #1000	Initialize X register
0103	LDAA #2	Initialize TCTL1
0105	STAA 20,X	" " "
0107	LDD #808	Initialize OC1M and OC1D
010A	STD C,X	" " " " "
010C	LDD E,X	Read TCNT
010E	ADDD #32	Add $32
0111	STD 16,X	Write sum to TOC1
0113	ADDD #4B0	Add $4B0 to sum
0116	STD 1E,X	Write new sum to TOC5
0118	BCLR 23,X 77	Clear OC1F and OC5F
011B	BRCLR 23,X 80 11B	Branch to self until OC1F = 1
011F	LDD 16,X	Add $7D0 to TOC1
0121	ADDD #7D0	" " " "
0124	STD 16,X	" " " "
0126	LDAB #4A	Delay for 20% of $T
0128	DECB	" " "
0129	BNE 128	" " "
012B	BRSET 0,X 4 133	20% duty cycle requested?
012F	LDAA #8	If yes, force OC5
0131	STAA B,X	" " "
0133	LDAB #4E	Delay for 20% of $T
0135	DECB	" " "
0136	BNE 135	" " "
0138	BRSET 0,X 2 140	40% duty cycle requested?
013C	LDAA #8	If yes, force OC5
013E	STAA B,X	" " "
0140	BRCLR 23,X 8 140	Branch to self until OC5F = 1
0144	LDD 1E,X	Add $7D0 to TOC5
0146	ADDD #7D0	" " "
0149	STD 1E,X	" " "
014B	BRA 118	Make more edges

9.3 M68HC11 REAL-TIME INTERRUPTS

In many instrumentation and control applications, the M68HC11 must exert precise control over the pace of software execution or interface operation. For example, the M68HC11 may trigger temperature readings at precise intervals. In this case, the microcontroller must output periodic time reference pulses to the temperature reading apparatus. Material on pages 418–420 presents software which generates such time reference pulses. In another example, the M68HC11 counts the number of external events that occur during precisely measured intervals. Pages 429–432 and 441–448 present software that counts DC motor revolutions over such measured time periods. Motorola developed the M68HC11's Real-Time Interrupt (RTI) feature for applications such as these.

The RTI mechanism allows the system designer to schedule interrupts at precisely timed intervals. RTI interrupts have these features:

- The CPU can enable and disable RTI interrupts with the global I mask or with a local TMSK2 register bit.
- RTI interrupts have their own vector address.
- The CPU can promote RTI interrupts to highest I-maskable priority by writing %0111 to HPRIO register bits PSEL3–PSEL0.
- The RTI mechanism can generate interrupts at one of four periodic rates.
- The RTI mechanism has a status bit in the TFLG2 register.

Thus, the Real Time Interrupt mechanism generates an interrupt and sets a status flag at a choice of periodic rates from 4.10ms to 32.77ms.[3]

M68HC11 On-Chip Resources Used for Real-Time Interrupts

Real-time interrupts are simple to implement and program. The M68HC11 uses only three control registers to enable and control RTIs: PACTL, TMSK2, and TFLG2.

Refer to Figure 9–6, which illustrates the PACTL register (control register address $1026). PACTL's RTI Interrupt Rate bits RTR1 and RTR0 control the rate at which the RTI mechanism sets the RTI status flag and generates periodic interrupts. These rates derive from the M68HC11's free-running clock, which runs at the microcontroller's E clock rate. If an 8 MHz crystal connects to the M68HC11's XTAL and EXTAL pins,[4] the free-running clock runs at 2 MHz. Based upon the state of the PACTL register's RTR1/RTR0 bits, the RTI mechanism divides the free-running clock rate by 2^{13}, 2^{14}, 2^{15}, or 2^{16} to establish the periodic rate at which it sets the RTI status flag and generates interrupts:

RTR1	RTR0	Free-Running CK ÷ By	RTI Interrupt Rate
0	0	2^{13}	4.10ms
0	1	2^{14}	8.19ms
1	0	2^{15}	16.38ms
1	1	2^{16}	32.77ms

Thus, PACTL register's RTR1/RTR0 bits allow the CPU to generate RTI status flags and interrupts every 4.1, 8.19, 16.38, or 32.77 milliseconds.

Refer to Figure 9–7, which depicts the TMSK2 register (control register block address $1024). TMSK2's RTII (RTI Interrupt Enable) bit enables RTI interrupts. To enable interrupts, the CPU writes binary one to the RTII bit. By writing zero to RTII, the CPU disables periodic Real-Time Interrupts. The RTII bit's state does not affect the RTI mechanism's generation of RTI status flags. Therefore, the CPU can disable RTI interrupts and poll the RTI status flag to determine when RTI periods have elapsed.

Refer to Figure 9–8, which depicts the TFLG2 register (control register block address $1025). TFLG2's Real-Time (periodic) Interrupt (RTIF) bit sets to indicate that an RTI period has elapsed. Thus, this bit functions as the RTI status bit. If the CPU has enabled interrupts by setting the RTII bit in TMSK2, the RTIF generates an RTI interrupt when it sets. The CPU must reset RTIF as a part of its RTI Interrupt Service routine. If RTIF remains set, the RTI mechanism generates another interrupt as soon as the CPU returns from the interrupt service routine. The CPU would thus be stuck in a continuous interrupt-return-interrupt loop, disrupting program execution.

FIGURE 9–6
Reprinted with the permission of Motorola.

PACTL

B7						RTR1	RTR0

(B0)

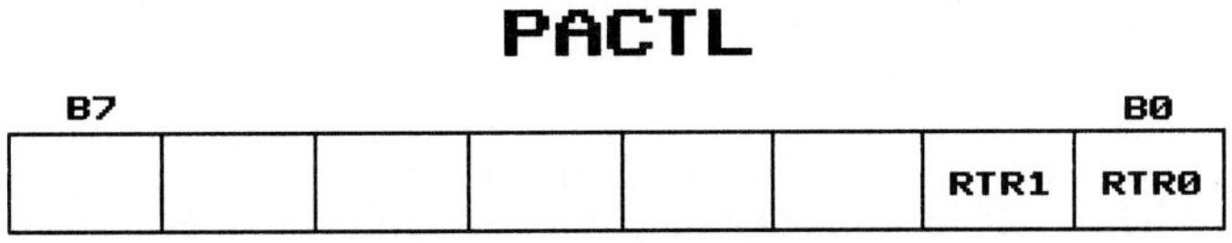

FIGURE 9–7
Reprinted with the permission of Motorola.

TMSK2

B7	RTII						B0

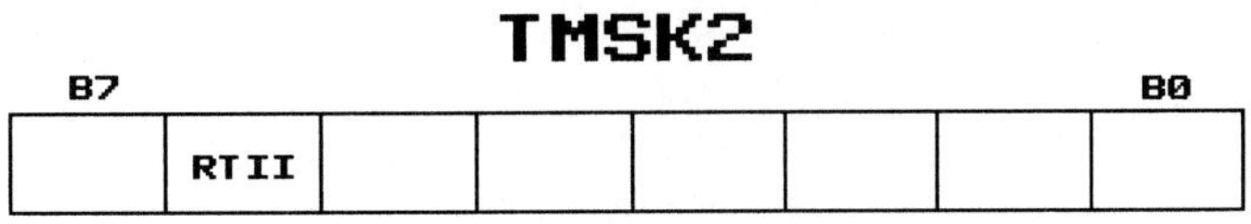

FIGURE 9–8
Reprinted with the permission of Motorola.

TFLG2

B7	RTIF						B0

[3]These rates assume that the M68HC11 connects to an 8 MHz crystal.
[4]Both the EVBU and EVB use an 8 MHz crystal.

The CPU uses 4 bits to enable and control Real-Time Interrupts. PACTL register's RTR1 and RTR0 bits determine the periodic interrupt rate. TMSK2 register's RTII bit enables or disables the RTI interrupts themselves. TFLG2 register's RTIF bit indicates that an RTI period has elapsed.

RTIs Used to Generate Periodic Time Reference Pulses

RTIs facilitate periodic pulse generation. External devices can use these pulses as time references. For example, pulses can trigger periodic collection of environmental data such as temperature, pollutants, or ambient light. This section presents a simple program that generates a one-millisecond positive pulse at a choice of four periodic rates. This program prompts the user to enter a number between zero and three from the terminal keyboard. The CPU converts the user-entered value to hexadecimal and writes it to the PACTL register's RTR1 and RTR0 bits, establishing an RTI rate of 4.10, 8.19, 16.35, or 32.77 milliseconds. Each RTI interrupt causes the CPU to output a one-millisecond pulse. Thus, the user selects the periodic rate of these pulses by entering zero through three from the terminal keyboard. Figures 9–9 and 9–10 with their accompanying program lists illustrate this pulse generation software.

Refer to Figure 9–9, blocks 1 and 2, and program addresses $C000–$C008. Block 1 prompts the user to input "RTI Rate" as a number from zero to three. To execute block 1, the CPU points its X register at ASCII prompt characters "RTI<SP>Rate:<SP><EOT>." The CPU then calls the .OUTSTR utility subroutine, which writes the prompt message to the terminal screen. In block 2, the CPU calls the .INCHAR utility subroutine. .INCHAR captures the user's keyboard entry and reads this ASCII code to ACCA. Thus, in blocks 1 and 2, the CPU prompts the user to enter a requested periodic rate value and then reads the ASCII key-press value to ACCA.

Refer to blocks 3, 4, and 5, and program addresses $C009–$C00E. In block 3, the CPU converts ACCA's key-press value from ASCII to hexadecimal by ANDing the ASCII value with $0F. The CPU then determines whether this user-entered value is inbounds—between zero and three. If the key-press value is out-of-bounds, program flow returns to block 1 and re-prompts the user for another selection. To make an inbounds determination, the CPU compares the converted key-press value with three, the largest permissible entry (block 4). The CPU then tests the C and Z flags to determine if the entry is higher than three (block 5). If the C or Z flag has set, the key-press value is lower than or equal to three and therefore inbounds. If both the C and Z flags have reset, the entry must be larger than three and out-of-bounds. Thus, if C+Z = 0, the CPU branches back to block 1 to re-prompt the user. If C+Z = 1, the CPU proceeds to block 6.

Refer to blocks 6 and 7, and program addresses $C00F–$C013. If the key-press value is inbounds, the CPU executes blocks 6 and 7. In block 6, the CPU points its X register at the control register block. In block 7, the CPU initializes PACTL register's RTR1 and RTR0 bits for the user-requested periodic interrupt rate. To execute block 7, the CPU merely copies ACCA's contents to the PACTL register. The CPU can copy ACCA because this accumulator's least significant bits contain the user-requested value (%00 through %11). RTR1 and RTR0 are the least significant PACTL bits. Therefore, the CPU can copy these bits directly. For example, assume that the user enters "1" from the terminal keyboard in response to the prompt. .INCHAR returns ASCII value $31 to ACCA. When the CPU executes block 3, it converts $31 to $01, i.e., ACCA bits 1 and 0 = %01. In block 7, the CPU copies ACCA's contents to PACTL, making RTR1/RTR0 = %01. Thus, by responding to the "RTI Rate:" prompt with "1," the user selects the 8.19ms periodic rate. Having initialized the PACTL register, the CPU can prepare to generate periodic interrupts (RTIs).

Refer to blocks 8 through 11 (program addresses $C014–$C01F). The software's RTI Interrupt Service routine generates the one-millisecond pulse. In block 8, the CPU initializes the RTI interrupt pseudovector. In block 9, the CPU sets the RTII control bit in TMSK2 to enable RTI interrupts. The CPU clears the global I mask in block 10. Finally, the CPU branches to self while waiting for RTI interrupts (block 11).

The RTI Interrupt Service routine (Figure 9–10) uses the M68HC11's Output Compare 2 mechanism to time falling edges of the output pulses. This routine must also clear the RTIF flag in the TFLG2 register.

FIGURE 9–9

FIGURE 9–10

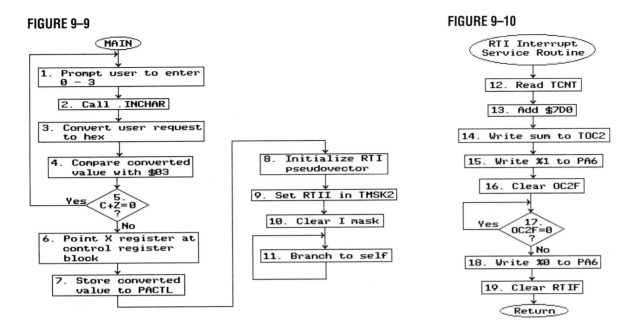

Periodic Time-Reference Pulses

$ADDRESS	MAIN SOURCE CODE	COMMENTS
C000	LDX #0	Point X register at "RTI Rate:" data
C003	JSR FFC7	Call .OUTSTR
C006	JSR FFCD	Call .INCHAR
C009	ANDA #F	Convert input character to hex
C00B	CMPA #3	Compare converted value with $3
C00D	BHI C000	If higher, re-prompt user
C00F	LDX #1000	If equal or lower, point X register at control register block
C012	STAA 26,X	Transfer input character to PACTL
C014	LDD #C030	Initialize RTI pseudovector
C017	STD EC	" " "
C019	LDAA #40	Initialize TMSK2
C01B	STAA 24,X	" " "
C01D	CLI	Clear I mask
C01E	BRA C01E	Branch to self

$ADDRESS	RTI Interrupt Service SOURCE CODE	COMMENTS
C030	LDD E,X	Read TCNT
C032	ADDD #7D0	Initialize TOC2 for 1ms pulse width
C035	STD 18,X	Initialize TOC2 for 1ms pulse width
C037	LDAA #40	Generate rising edge of pulse
C039	STAA 0,X	" " " " "
C03B	BCLR 23,X BF	Clear OC2F
C03E	BRCLR 23,X 40 C03E	Branch to self until OC2F = 1
C042	CLR 0,X	Generate falling edge of pulse
C044	BCLR 25,X BF	Clear RTIF
C047	RTI	Return

$ADDRESS	User Prompt ASCII Data CONTENTS	ASCII CHARACTER
0000	52	"R"
0001	54	"T"
0002	49	"I"
0003	20	<SP>
0004	52	"R"
0005	61	"a"
0006	74	"t"
0007	65	"e"
0008	3A	":"
0009	20	<SP>
000A	04	<EOT>

Refer to Figure 9–10, blocks 12 through 14, and program addresses $CO30–$CO36. In block 12, the CPU reads the current free-running count from the TCNT register. In block 13, the CPU adds the number of free-running counts in one millisecond ($7D0) to the TCNT read value. This sum represents the free-running count at which the time reference pulse should end. In block 14, the CPU writes this sum to its TOC2 register. Thus, in blocks 12–14, the

CPU initializes its TOC2 register with the free-running count at which the time reference pulse should end.

Refer to blocks 15 and 16 (program addresses $C037–$C03D). In block 15, the CPU generates the time reference pulse's rising edge by writing binary one to PORTA pin PA6. Note that a system reset clears the TCTL1 register, making OM2 and OL2 binary zeros. Thus, this reset disconnects pin PA6 from the OC2 mechanism. The CPU generates a rising edge from pin PA6 by writing binary one to it, and generates a falling edge by writing zero. The RTI Interrupt Service routine uses the OC2 mechanism only as a timing device to indicate when it should generate an output falling edge. In block 16, the CPU clears the OC2F flag, so it can use status polling to determine when the free-running count matches the contents of the TOC2 register.

Refer to blocks 17 through 19, and program addresses $C03E–$C047. After clearing its OC2F flag, the CPU enters a polling loop (block 17) in which it branches to self until OC2F sets. OC2F = 1 indicates that one millisecond has elapsed; it is time to generate the time reference pulse's falling edge. Thus, after escaping the polling loop, the CPU generates a falling edge by writing binary zero to PORTA pin PA6 (block 18). In block 19, the CPU clears the RTIF bit in the TFLG2 register before returning from the interrupt routine. Remember that RTIF must be cleared, or the RTI mechanism will generate a spurious interrupt as soon as the CPU returns from the interrupt service routine.

Foregoing paragraphs have shown how a simple program can use RTIs to generate periodic time reference pulses at a 4.1, 8.19, 16.38, or 32.77 millisecond rate. Later sections and laboratory exercises illustrate how the RTI mechanism can establish time references for external event counting.

EXERCISE 9.2   Periodic Time Reference Pulses

REQUIRED EQUIPMENT: Motorola M68HC11EVBU (EVB) and terminal
Oscilloscope and probe

GENERAL INSTRUCTIONS: This laboratory exercise applies the time reference pulse generation software presented in the previous section. This software develops a one-millisecond time reference pulse at a choice of four repetition rates. The interface operator selects a particular repetition rate by entering a number between zero and three from the terminal keyboard. The M68HC11 outputs time reference pulses from its PA6 pin (pin 28).

PART I: Load and Debug the Time Reference Pulse Software

Review the previous section to assure that you fully understand the time reference pulse software. If you use an EVB, load your software from the list accompanying Figures 9–9 and 9–10. If you use an EVBU, locate your MAIN program at address $0100 and the RTI Interrupt Service routine at address $0130. If you use an EVBU, assure that branching instruction destination addresses and the RTI pseudovector reflect these reassigned addresses. Test and debug your software. When sure that the software runs properly, demonstrate it to the instructor. Be prepared to show the instructor an oscilloscope display of the PA6 output wave form.

_____ *Instructor Check*

PART II: Laboratory Analysis

Answer the questions below by entering your responses in the appropriate blanks.

1. List the tasks accomplished by the time reference pulse software's MAIN program.

2. List the tasks accomplished by the RTI Interrupt Service routine.

3. Refer to Figure 9–9. Explain why the CPU calls the .INCHAR subroutine (block 2).

4. Refer to Figure 9–9. Explain why the CPU compares the converted key-press value with $03 (block 4).

5. Refer to Figure 9–9. Explain why the CPU can copy the converted key-press value to the PACTL register (block 7) without modification.

6. Refer to Figure 9–9. Explain why the CPU branches to self continuously in block 11.

7. Refer to Figure 9–10. Describe the task accomplished by the CPU when it executes blocks 12, 13, and 14.

8. Refer to Figure 9–10. Explain why the CPU executes block 15.

9. Refer to Figure 9–10. Explain why the CPU executes block 18.

10. Refer to Figure 9–10. Explain block 17's purpose.

11. Refer to Figure 9–10. Describe the consequences of omitting block 19 from the software.

12. Refer to Figure 9–10 and its accompanying program list. Describe program modifications necessary to generate a 500µs timing pulse width.

13. Describe program modifications necessary to generate a two-millisecond timing pulse width.

421

Make the program modifications described in questions 12 and 13. Demonstrate 500µs and two-millisecond pulse width generation to the instructor. Be prepared to show the instructor an oscilloscope display of these modified pulse widths.

_____ Instructor Check, 500µs PW

_____ Instructor Check, 2ms PW

PART III: Laboratory Software Listing

MAIN PROGRAM—EVB

(See program list accompanying Figure 9–9.)

MAIN PROGRAM—EVBU

$ADDRESS	SOURCE CODE	COMMENTS
0100	LDX #0	Point X register at "RTI Rate" data
0103	JSR FFC7	Call .OUTSTR
0106	JSR FFCD	Call .INCHAR
0109	ANDA #F	Convert input character from ASCII to hex
010B	CMPA #3	Compare converted value with $3
010D	BHI 100	If higher, re-prompt user
010F	LDX #1000	If equal or lower, point X register at control register block
0112	STAA 26,X	Transfer input character to PACTL
0114	LDD #130	Initialize RTI pseudovector
0117	STD EC	" " " "
0119	LDAA #40	Initialize TMSK2
011B	STAA 24,X	" " "
011D	CLI	Clear I mask
011E	BRA 11E	Branch to self

RTI INTERRUPT SERVICE ROUTINE—EVB

(See program list accompanying Figure 9–10.)

RTI INTERRUPT SERVICE ROUTINE—EVBU

$ADDRESS	SOURCE CODE	COMMENTS
0130	LDD E,X	Read TCNT
0132	ADDD #7D0	Initialize TOC2 for 1ms pulse width
0135	STD 18,X	" " " " " "
0137	LDAA #40	Generate rising edge of pulse
0139	STAA 0,X	" " " " " "
013B	BCLR 23,X BF	Clear OC2F
013E	BRCLR 23,X 40 13E	Branch to self until OC2F = 1
0142	CLR 0,X	Generate falling edge of pulse
0144	BCLR 25,X BF	Clear RTIF
0147	RTI	Return

USER PROMPT ASCII DATA—EVB AND EVBU

(See listing accompanying Figure 9–9.)

9.4 INTRODUCTION TO THE M68HC11'S PULSE ACCUMULATOR

The Pulse Accumulator allows the M68HC11 to count either the frequency or duration of external events. If an environmental phenomenon can generate a voltage change from low to high or vice versa, it qualifies as an "external event." Given the wide variety of transducers available, virtually any environmental phenomenon can qualify as an external event. In Exercises 9.4 and 9.5, the reader will use an infrared pair (LED and phototransistor) to translate physical motion into voltage transitions.

An interface designer applies environmentally caused voltage transitions—edges—to the M68HC11's Pulse Accumulator Input (PAI) pin. The Pulse Accumulator can measure the frequency of these edges or the duration of an event that causes them:

- The Pulse Accumulator can count the number of edges and record this count in a special 8-bit register called the Pulse Accumulator Count Register (PACNT). Software then determines frequency of occurrence by relating event count to a specified time period. This mode of operation is called event-counting mode.
- The Pulse Accumulator can determine the time between rising and falling edges to a resolution of 32μs[5]. The Pulse Accumulator records this time in its PACNT register. This mode of operation is called gated time accumulation mode.

Section 9.5 explains event-counting mode and applies this mode to both a DC motor tachometer and proportional-tachometer motor speed regulation. Section 9.6 explains gated time accumulation mode and shows how this mode applies to duty cycle measurement. The reader should find these sections informative and practical.

9.5 M68HC11 PULSE ACCUMULATOR—EVENT-COUNTING MODE

When the Pulse Accumulator operates in event-counting mode, it counts the number of specified edges occurring at the PAI pin. The Pulse Accumulator records the number of edges in the PACNT register. PACNT has an 8-bit capacity, so it can count up to 255 events. However, the Pulse Accumulator mechanism provides a PACNT overflow flag *and* an optional accompanying interrupt. The M68HC11 can therefore increment PACNT extension byte(s) to facilitate event counts of virtually any magnitude. The Pulse Accumulator mechanism also sets a status flag—and can generate an interrupt—each time an external event causes PACNT to increment. Each of these two interrupts—PACNT input edge and PACNT overflow—has a unique vector address. The reader will find event-counting mode easy to understand and simple to operate.

On-Chip Resources Used by the Pulse Accumulator in Event-Counting Mode

When operating in event-counting mode, the Pulse Accumulator uses pin PA7 as its external event detection (PAI) pin. It also uses the resources of four registers: the PACNT register keeps count of external events; the PACTL register controls four important aspects of Pulse Accumulator operation; TFLG2 contains two status bits concerning the Pulse Accumulator, and TMSK2 enables associated interrupts. Following paragraphs explain the relationship of these resources to the Pulse Accumulator as it operates in event-counting mode.

- Pulse Accumulator Input Pin (PAI, PORTA, pin PA7). PA7 is a three-function pin. The M68HC11 uses it for these purposes: General purpose I/O, Output Compare 1 (OC1), and Pulse Accumulator Inputs (PAI). Figure 9–11 illustrates PA7's three purposes.[6] This figure shows that PA7 pin logic consists of three main blocks: Pulse Accumulator/General-Purpose Input Latch, General-Purpose Output Latch, and OC1 Logic. The Pulse Accumulator/General-Purpose Input Latch is of immediate interest. An external event causes an edge at pin PA7(PAI). The Pulse Accumulator/General-Purpose Input Latch[7] captures these edges and sends them to the Pulse Accumulator. When operating in event-counting mode, the Pulse Accumulator increments its PACNT register for each specified edge.

Refer to Figure 9–12, which illustrates how the Pulse Accumulator works and how it connects to PA7 pin logic.[8] The present discussion, and later discussions on gated time

---

[5]This resolution assumes a 2 MHz E clock rate. E CK = 2MHz for both EVBU and EVB.
[6]Motorola Inc., *M68HC11 Reference Manual*, p. 7-10.
[7]The Pulse Accumulator/General-Purpose Input Latch consists of an S-R latch identical to that illustrated and discussed in Chapter 7.
[8]Motorola Inc., *M68HC11 Reference Manual*, p. 11-3.

FIGURE 9–11
Reprinted with the permission of Motorola.

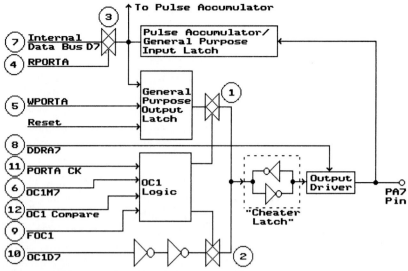

FIGURE 9–12
Reprinted with the permission of Motorola.

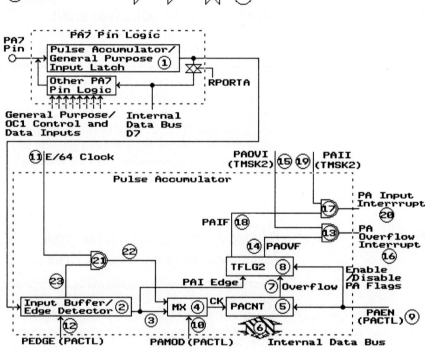

accumulation mode, refer to this figure. If the CPU configures the Pulse Accumulator for event counting, the Pulse Accumulator/General-Purpose Input Latch (1) captures pin PA7 edges. The Pulse Accumulator/General-Purpose Input Latch sends these edges to the Pulse Accumulator's Input Buffer/Edge Detector (2). The Input Buffer/Edge Detector converts specified edges to clock pulses (3) and sends them to the Multiplexer (4). If the CPU has configured the Pulse Accumulator for event-counting mode, the Multiplexer couples clock pulses (3) to the PACNT counter/register (5), which increments its contents. Thus, the Pulse Accumulator translates specified PA7 edges into clock pulses. These clock pulses cause the PACNT register to increment its contents.

- Pulse Accumulator Count Register (PACNT, control register block address $1027). This 8-bit register forms the heart of the M68HC11's Pulse Accumulator. When the CPU configures the Pulse Accumulator for event counting, PACNT increments each time a specified edge occurs at pin PA7.

Refer to Figure 9–12. Note that the CPU can read or write (6) to PACNT at any time. Thus, the CPU can initialize PACNT to any 8-bit value or read its current contents. When its contents roll over from $FF to $00, PACNT sends an overflow signal (7) to the TFLG2 register (8). This overflow signal sets a status flag in TFLG2 and can generate an interrupt—if the CPU has enabled overflow interrupts. The PACNT register can increment its

FIGURE 9–13
Reprinted with the permission of Motorola.

PACTL

B7							B0
DDRA7	PAEN	PAMOD	PEDGE				

contents only if enabled by PAEN = 1 (9). Thus, the 8-bit PACNT register is readable and writable, increments its contents with each specified PA7 input edge, and notifies the TFLG2 register each time its count rolls over.

- Pulse Accumulator Control Register (PACTL, control register block address $1026). In chapters 7 through 9, the reader learned the functions of PACTL's DDRA7, DDRA3, I4/O5, and RTR1/RTR0 bits as these bits relate to input captures, output compares, and Real-Time Interrupts. But Motorola calls this register the Pulse Accumulator Control Register. Indeed, PACTL controls Pulse Accumulator operations. Refer to Figure 9–13. PACTL's DDRA7, PAEN, PAMOD, and PEDGE bits control the answers to four questions:

1. Will PACNT increment when the CPU or OC1 mechanism generates output edges at pin PA7?
2. Will the Pulse Accumulator respond to pin PA7 edges or not?
3. Will the Pulse Accumulator operate in event-counting mode or gated time accumulation mode?
4. Does the Pulse Accumulator respond to rising edges at pin PA7, or does it respond to falling edges?

PACTL bit DDRA7 controls the answer to question 1. Refer to Figure 9–11. A DDRA7 = 1 input from PACTL enables the Output Driver. If DDRA7 = 1, general-purpose or OC1 output edges can be captured by the Pulse Accumulator/General-Purpose Input Latch, sent to the Pulse Accumulator, and counted by the PACNT register. The programmer must reset the PACTL DDRA7 bit to assure that the Pulse Accumulator counts only external events.

PACTL bit PAEN (Pulse Accumulator System Enable) answers question 2. Refer to Figure 9–12. Note that the PAEN bit (9) enables two processes: First, PAEN = 1 causes the PACNT to increment its count in response to clocks from the Multiplexer (4). Second, PAEN = 1 enables the TFLG2 register to set its Pulse Accumulator-oriented status flags when conditions dictate. Of course, the M68HC11 can disable PACNT counting and TFLG2 status bit setting by clearing the PAEN bit in PACTL.

PACTL's PAMOD (Pulse Accumulator Mode) bit answers question 3. By clearing the PAMOD bit, the CPU puts the Pulse Accumulator into event-counting mode. PAMOD = 1 puts the Pulse Accumulator into gated time accumulation mode. Refer to Figure 9–12. The PAMOD input (10) controls whether the Multiplexer (4) will route clocks from Input Buffer/Edge Detector (2), or from E/64 clock 11 (for gated time accumulation). Thus by clearing the PAMOD bit, the CPU causes the Input Buffer/Edge Detector to clock the PACNT register's increments.

Finally, the PEDGE (Pulse Accumulator Edge Control) bit in PACTL answers question 4. Refer again to Figure 9–12. PEDGE bit 12 controls the Edge Detector (2). PEDGE = 0 causes the Edge Detector to output clock pulses (3) when it detects negative-going edges at its input. PEDGE = 1 causes the Edge Detector to output clocks when it detects positive-going edges.

In summary, PACTL bits 7–4 configure pin PA7 as an input or output (DDRA7), enable the Pulse Accumulator (PAEN), determine the Pulse Accumulator's operating mode (PAMOD), and specify the inputs edges to be counted (PEDGE).

- Miscellaneous Timer Interrupt Flag Register 2 (TFLG2, control register block address $1025). Refer to Figure 9–14. TFLG2 register bits 4 and 5 constitute the Pulse Accumulator's status bits. Figure 9–12 shows that event count roll-over causes the PACNT register (5) to send an overflow signal (7) to the TFLG2 register (8). This overflow signal sets the PAOVF (Pulse Accumulator Overflow Flag) bit in the TFLG2 register. Note also that edge detect clock 3 also connects to the TFLG2 register. This edge detect clock sets the PAIF (Pulse Accumulator Input Edge Flag) bit in TFLG2. Thus, the edge detect clock causes PACNT to increment and sets the PAIF bit in TFLG2. PAOVF and PAIF can also generate interrupts if enabled by the TMSK2 register.

FIGURE 9–14
Reprinted with the permission of Motorola.

TFLG2

B7							B0
		PAOVF	PAIF				

FIGURE 9–15
Reprinted with the permission of Motorola.

TMSK2

B7							B0
		PAOVI	PAII				

- Miscellaneous Timer Interrupt Mask Register 2 (TMSK2, control register block address $1024). The Pulse Accumulator can generate two kinds of interrupts: Pulse Accumulator Input Interrupts and Pulse Accumulator Overflow Interrupts. Refer to Figure 9–15. TMSK2 register's PAOVI (Pulse Accumulator Overflow Interrupt Enable) bit enables interrupts that the Pulse Accumulator generates when its PACNT register contents roll over. Refer to Figure 9–12. PACNT signals the TFLG2 register (7) when the event count rolls over from $FF to $00. This signal sets the PAOVF flag in the TFLG2 register (8). The Pulse Accumulator ANDs (13) the PAOVF state (14) with TMSK2's PAOVI bit (15). If PAOVI = 1 AND PAOVF = 1, the Pulse Accumulator generates a PA Overflow Interrupt (16).

    Refer again to Figure 9–15. TMSK2 register's PAII (Pulse Accumulator Input Interrupt Enable) bit enables interrupts which the Pulse Accumulator generates when it detects a specified edge at pin PA7. Refer to Figure 9–12. When the Edge Detector (2) detects an edge as specified by the PEDGE bit (12), it outputs a clock pulse (3). This clock pulse sets the PAIF bit in TFLG2 (8). The Pulse Accumulator ANDs (17) the state of the PAIF bit (18) with TMSK2's PAII bit (19). If both PAIF and PAII are binary one, the AND gate generates a PA Input Interrupt (20).

    This section describes M68HC11 on-chip resources used by the Pulse Accumulator when it operates in event-counting mode. An external event (or internally generated event when DDRA7 = 1) causes a specified edge at pin PA7. The Pulse Accumulator's Input Buffer/Edge Detector detects this edge. The Edge Detector responds by generating a clock pulse, which causes the PACNT to increment. The Edge Detector's output clock pulse also sets the PAIF bit in TFLG2. PAIF = 1 can generate a PA Input Interrupt if PAII (in TMSK2) = 1. If its count rolls over, the PACNT register causes the PAOVF bit to set in the TFLG2 register. PAOVF = 1 can also generate a PA Overflow Interrupt, if PAOVI = 1 in TMSK2.

EXERCISE 9.3   Initialize the M68HC11 for Event-Counting Mode

AIM: Derive BUFFALO one-line assembly source code that configures the M68HC11's Pulse Accumulator for event-counting mode.

INTRODUCTORY INFORMATION: To complete this exercise, develop source code that configures the M68HC11's Pulse Accumulator for specified event-counting operations. This source code initializes the M68HC11's Pulse Accumulator-related control and data registers, along with interrupt pseudovectors.

INSTRUCTIONS: Use the specifications given and refer to pages 423–426 while developing appropriate BUFFALO one-line assembly source code. By executing these instructions, the CPU initializes appropriate interrupt pseudovectors and configures the PACNT, PACTL, and TMSK2 registers for event-counting mode. Specifications may also call for source code that clears the global I mask to enable specified interrupts. Enter your source code in the blanks accompanying each set of specifications. In developing source code, assume that the X register contains $1000. Use indexed mode for all writes to control or data registers.

Use direct mode for pseudovector writes. The Pulse Accumulator Overflow Interrupt pseudovector resides at address $00CD. The Pulse Accumulator Input Edge Interrupt pseudovector resides at address $00CA.

EXAMPLE

Specifications
Event-counting mode
Capture external events
Enable Pulse Accumulator
Count rising edges
Enable Overflow interrupts
Interrupt service routine at address $0130
Clear PACNT
Clear I mask

$ADDRESS	SOURCE CODE	COMMENTS
0100	LDD #130	Initialize overflow interrupt pseudovector
0103	STD CE	" " " " " "
0105	LDAA #20	Initialize TMSK2
0107	STAA 24,X	" "
0109	LDD #5000	Initialize PACTL/PACNT
010C	STD 26,X	" " "
010E	CLI	Clear I mask

Specifications
Event-counting mode
Capture external events
Enable Pulse Accumulator
Count falling edges
Enable Input Edge interrupts
Interrupt service routine at address $C040
Clear PACNT

$ADDRESS	SOURCE CODE	COMMENTS
C000		
___	___	___
___	___	___
___	___	___
___	___	___
___	___	___
___	___	___
___	___	___
___	___	___

Specifications
Event-counting mode
Capture internally generated edges
Enable Pulse Accumulator
Count rising edges
Enable both interrupts
Input Edge Interrupt Service routine at address $0140
Overflow Interrupt Service routine at address $0180
Initialize PACNT to $FF
Clear I mask

$ADDRESS	SOURCE CODE	COMMENTS
010A		

Specifications
Event-counting mode
Capture external events
Enable Pulse Accumulator
Count falling edges
Enable Overflow interrupts
Interrupt service routine at address $B700
Initialize PACNT to $64
Clear I mask

$ADDRESS	SOURCE CODE	COMMENTS
0140		

Specifications
Event-counting mode
Capture external events
Enable Pulse Accumulator
Count rising edges
Clear PACNT
Clear I mask

$ADDRESS	SOURCE CODE	COMMENTS
C030		

DC Motor Speed Control and Tachometer

Given an understanding of PA event-counting mode, the reader can put it to practical use. This section presents software and hardware that combine elements from Chapters 8 and 9 output compares and Real-Time Interrupts (RTIs), along with event-counting mode. By synthesizing these elements, the reader can achieve the following objectives:

- Develop hardware that allows the M68HC11 to count external events—revolutions of a DC motor. This hardware uses a simple optical sensor, consisting of an encoder, infrared LED, and phototransistor.
- Develop software that uses output compares to control DC motor speed. For this purpose, the user shall adapt software from Chapter 8, Section 8.7.
- Develop software that relates event counts to time and calculates motor speed in revolutions per minute (RPM). This software uses RTIs and PA event-counting mode.

Regarding the hardware, this application uses two interfaces. The first controls motor speed via an output compare pin. The second applies an optical sensor signal to the Pulse Accumulator Input (PAI) pin.

Use motor control circuitry as constructed for Exercise 8.6. Figure 9–16 provides a schematic of this circuitry. This section's discussion assumes that the motor control circuit input connects to the M68HC11's OC5/PA3 pin.

Essentially, the DC motor shall be mounted as shown in Figure 8–29. However, the mount incorporates an optical encoder, optoelectronics, and associated circuitry as shown in Figure 9–17. This figure shows that the optical encoder is a simple "propeller-like" piece of cardboard, press-fitted to the motor's output shaft.

The optical encoder interrupts infrared (IR) light transmission from LED to phototransistor. The encoder interrupts IR two times for each revolution of the DC motor.

Figure 9–18 shows the optical sensor circuitry.[9] IR light reception causes the phototransistor to conduct. Phototransistor conduction causes an input low to the Schmitt trigger NAND gate. Note that this NAND gate is connected as an inverter. Thus, the transducer circuitry outputs a high when the phototransistor conducts. When the optical encoder interrupts IR transmission, the phototransistor cuts off, causing an input high to the NAND gate, and an output low at the sensor circuit's output. The Schmitt trigger NAND gate serves to generate clean, noise-free output pulses. When the DC motor runs, the sensor circuitry outputs a square wave—with two pulses for each motor revolution.

Our second task is to develop motor speed control software. This software uses OC1 and OC5 to generate a 1 KHz motor speed control square wave. By storing a control word

FIGURE 9–16
Courtesy of Richard J. Helmer.

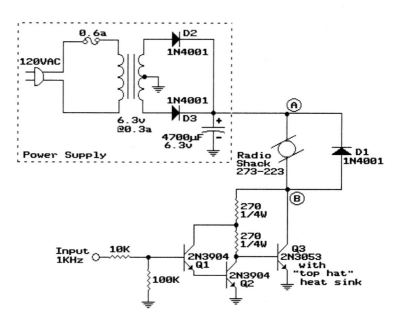

---

[9]Thanks to Mr. Richard J. Helmer for design of the optical sensor circuit and motor control interface.

FIGURE 9–17
Courtesy of Richard J. Helmer.

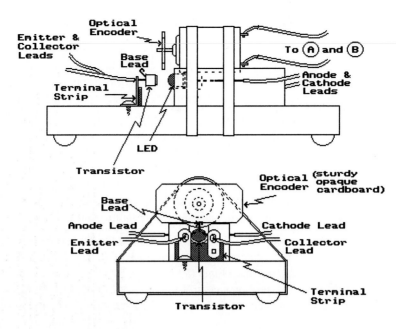

FIGURE 9–18
Courtesy of Richard J. Helmer.

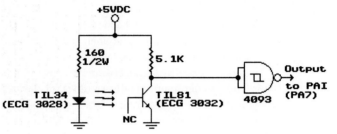

($DUTY) in memory, the user can control the square wave's percent duty cycle. The larger the percent duty cycle, the lower the DC motor's RPM. The M68HC11 outputs the motor speed control square wave from its OC5/PA3 pin. This PA3 output connects to the input of the motor speed control circuitry (Figure 9–16).

Figure 9–19 provides a simplified flowchart of the motor speed control software. The reader should find this flowchart's strategy familiar. It uses OC1F/OC5F status polling to generate an output 1 KHz square wave from pin PA3. Upon its escape from the OC1F polling loop (block 10), the CPU adds $7D0 to TOC1's contents. This action updates TOC1 for the next OC1 compare and output rising edge. Likewise, the CPU adds $7D0 to TOC5 after escaping the block 12 polling loop. Of course, $7D0 equals the number of free-running counts in the 1 KHz square wave's period.

The CPU establishes output pulse width by offsetting TOC5's contents from TOC1's contents (blocks 4–8). Once the CPU initializes TOC1 (blocks 4, 5, and 6), it adds $DUTY to the TOC1 value and writes this sum to TOC5 (blocks 7 and 8). Thus, the CPU establishes a pulse width equal to $DUTY value × 500ns.

To initialize $DUTY, use the BUFFALO MM command. $DUTY resides at addresses $0000–$0001.

To complete task number three, develop software that relates event count to time and derives motor RPM. This software uses Real-Time Interrupts to provide the equation's time component. If the CPU sets PACTL's RTR1, and RTR0 bits, the RTI mechanism generates an interrupt every 32.77 milliseconds. The Pulse Accumulator's PACNT register counts transducer circuitry output edges over a number of RTI periods. Eighteen RTI periods permit a usable sampling of transducer edges, even at slow motor speeds:

$$18 \times 32.77\text{ms} = .59\text{s}$$

Having counted the number of transducer edges in .59 seconds, the CPU extrapolates this count to the number of edges in one minute. One minute consists of 102 periods:

$$60\text{s}/.59\text{s} = 102$$

FIGURE 9-19

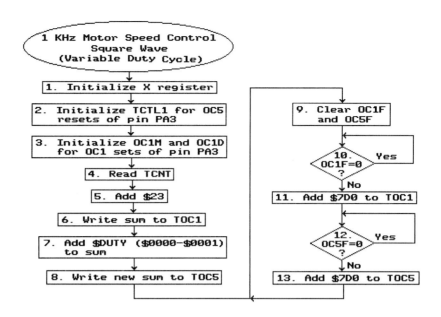

Thus, the CPU can calculate the number of transducer edges in one minute by multiplying the PACNT (after .59 seconds) by 102:

$$\text{\# Transducer edges in 60s} = \text{PACNT} \times 102$$

The transducer circuitry outputs two edges for each motor revolution. Therefore, the product of PACNT × 102 must be halved to derive RPM:

$$\text{RPM} = (\text{PACNT} \times 102)/2$$

Simplify this equation to:

$$\text{RPM} = \text{PACNT} \times 51$$

For example, assume that the Pulse Accumulator counts 115 edges over the period of eighteen RTIs:

$$\text{RPM} = (115 \times 102)/2$$

or

$$= 115 \times 51$$
$$= 5865$$

PACNT counts in hexadecimal. Therefore, express all quantities in hex. For example:

$$\text{RPM} = (\$73 \times \$66)/2$$
$$= \$73 \times \$33$$
$$= \$16E9$$

Figure 9–20 illustrates software that executes task number three. Refer to the MAIN program in Figure 9–20. Since the Pulse Accumulator counts transducer edges over $12 RTI periods, it establishes a memory location as an RTI counter and initializes it to $12 (block 1). After initializing the RTI Interrupt pseudovector (at addresses $00EB–$00ED) in block 2, the CPU initializes the PACTL register for event-counting operations (block 3). In block 4, the CPU clears the PACNT register so it can begin counting transducer edges. To enable RTI interrupts, the CPU sets the RTII bit in TMSK2 (block 5) and clears the global I mask (block 6). The CPU then enters a branch-to-self loop where it waits for recurring RTI interrupts.

Refer to the RTI Interrupt Service routine in Figure 9–20. Recall that in block 1, the CPU initialized the RTI count (RTICNT) to $12. Each time the CPU vectors to the RTI Interrupt Service, it decrements the RTICNT (block 7). The CPU decrements RTICNT so that when this count reaches zero, eighteen RTI periods have elapsed, and the CPU can calculate $RPM. After each RTI decrement, the CPU checks its Z flag (block 8). If Z = 0, the

FIGURE 9-20

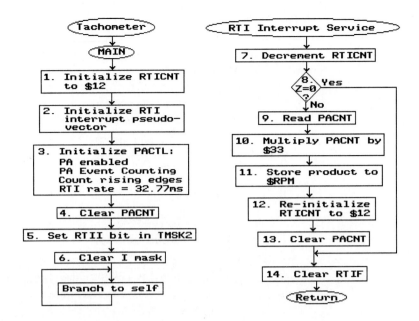

CPU branches around the RPM calculation, clears the RTIF flag (block 14), and returns from interrupt. If Z = 1, then eighteen RTI periods have elapsed, and the CPU proceeds to calculate motor RPM.

To calculate RPM, the CPU reads the number of transducer edges (obtained over .59 seconds) from PACNT (block 9). The CPU reads PACNT to ACCA. In block 10, the CPU loads ACCB with $33, then executes the MUL command to multiply PACNT (ACCA) by $33 (ACCB). The product equals $RPM. The CPU stores this value to a designated memory location (block 11).

The tachometer software causes the M68HC11 to measure RPM continuously. Therefore, in block 12, the CPU re-initializes RTICNT to $12 for the next RPM measurement. In block 13, the CPU clears PACNT for the next measurement.

Whether or not it has calculated RPM, the CPU must clear the RTIF flag (block 13) before returning from the interrupt.

Foregoing paragraphs have shown how the reader can accomplish three objectives:

- Develop an interface that allows the M68HC11 to count external events—DC motor revolutions.
- Develop software that uses output compares to control DC motor speed.
- Develop software that relates event counts to time and calculates $RPM.

In Exercise 9.4, integrate this software and hardware into a unified whole in order to control DC motor speed and measure motor $RPM continuously.

EXERCISE 9.4 DC Motor Speed Control and Tachometer

REQUIRED MATERIALS: Motor control interface, as constructed for Exercise 8.6
DC motor and mounting, as constructed for Exercise 8.6
1 Phototransistor, TIL81 (ECG 3032)
1 LED (IR), TIL34 (ECG 3028)
1 IC-4093B, Schmitt-trigger, quad, two-input NAND
1 Terminal strip, per Figure 9–17
1 Resistor 160 Ω, 1/2W
1 Resistor 5.1 KΩ
1 Optical encoder, per Figure 9–17, constructed of sturdy, opaque cardboard

REQUIRED EQUIPMENT: Motorola M68HC11EVBU (EVB) and terminal
Dual-trace oscilloscope with two probes

GENERAL INSTRUCTIONS: Install an infrared diode and transistor (IR pair) in your DC motor mounting. Press-fit a simple optical encoder onto the DC motor's output shaft. Connect the IR pair to the sensor circuitry. After connecting the motor control interface to the M68HC11's PA3 pin, write a short motor speed control routine as shown in Figure 9–19. Use the motor speed control routine to run the motor at various speeds. At each speed, use the oscilloscope to observe and analyze the transducer circuitry's output waveform. Develop software that synthesizes the Figure 9–19 and 9–20 routines. Analyze and evaluate the operation of your synthesized software. To execute these tasks, follow the detailed instructions given below.

PART I: Install the IR Pair

Dismantle the DC motor mounting. Carefully hollow out a space in the styrofoam motor mount to accommodate the infrared LED, as shown in Figure 9–17. Force the LED's anode and cathode leads through the styrofoam so that they emerge on opposite sides of the motor mount. This procedure physically stabilizes the LED and permits access to its leads. Carefully cut an optical encoder out of thin but durable cardboard. Make a *small* hole in the exact center of the encoder. Assure that this hole is centered and that the blades of the encoder are symmetrical. Careful encoder construction allows vibration-free motor operation. This motor can run at high RPMs! Press fit the encoder onto the motor output shaft as shown in Figure 9–17. Next, solder the phototransistor's emitter and collector leads to the terminal strip as shown in Figure 9–17. With needle-nose pliers, curl the transistor's base lead up and out of the way where it will not contact any wire or object. Attach the terminal strip with phototransistor to the motor mounting board as shown in Figure 9–17. Reassemble the motor mounting. Show your IR pair/optical encoder installation to the instructor.

_____ *Instructor Check*

PART II: Connect the Sensor and Motor Control Circuitry

Connect the IR pair and Schmitt trigger NAND gate as shown in Figure 9–18. Do not connect the NAND gate output to PAI. Instead, connect this output to the oscilloscope. Power up the transducer circuitry. Manually turn the motor shaft so that the optical encoder causes the NAND gate to toggle. Show the instructor this toggling on the oscilloscope display.

_____ *Instructor Check*

Power down the sensor circuitry. Connect the NAND gate's output to the M68HC11's PAI pin. Connect your motor control interface (Figure 9–16) to the microcontroller's PA3 pin. Show your completed connections to the instructor.

_____ *Instructor Check*

PART III: Develop a Motor Speed Control Routine

Refer to Figure 9–19. Develop motor speed control software per this figure. Neatly list your software in the blanks below. Locate this routine at address $0100 (EVBU) or $C000 (EVB).

1 KHz Motor Speed Control

$ADDRESS	SOURCE CODE	COMMENTS
_____	_____	_____
_____	_____	_____
_____	_____	_____
_____	_____	_____

Load, debug, and demonstrate your motor speed control routine to the instructor using a $DUTY value of $03E8. Be prepared to show the instructor the resultant 50% duty cycle waveform on the oscilloscope.

_____ *Instructor Check*

PART IV: Analysis of Motor Speed Control and Sensor Circuit Outputs

Refer to Chart 9–1. Analyze how the speed control waveform's duty cycle relates to motor RPM. Make this comparison by entering calculated values in column 2, measured values from the oscilloscope in column 3, and derived RPM values in column 4. You shall complete columns 5 and 6 in Part VI of this laboratory exercise.

First, complete column 2 of Chart 9–1. Calculate $DUTY for each percent duty cycle listed in column 1. $DUTY equals the number of free-running counts in the speed control waveform's pulse width. For example, calculate $DUTY for a 90% duty cycle. Recall that 100% of the 1 KHz control waveform's period equals 2000 free-running counts. For a 90% duty cycle, the pulse width equals 2000 × .9 free-running counts:

$$PW = 2000 \times .9$$
$$PW = 1800 \ (\$708) \text{ free-running counts}$$

Chart 9–1

1 Duty Cycle	2 $DUTY	3 Oscilloscope Sensor &T (ms)	4 Oscilloscope Calculated &RPM	$RPM	5 Tachometer $RPM	&RPM	6 Percent Error
80	___	___	___	___	___	___	___
70	___	___	___	___	___	___	___
60	___	___	___	___	___	___	___
50	___	___	___	___	___	___	___
40	___	___	___	___	___	___	___
30	___	___	___	___	___	___	___
20	___	___	___	___	___	___	___

Therefore,

$$\$DUTY = \$0708$$

Second, use each $DUTY value to control motor speed. Connect oscilloscope channel two to the sensor circuit's output. For each $DUTY value, use the oscilloscope to carefully measure the sensor output's decimal period (&T) in milliseconds. For each measurement allow the motor to run for at least thirty seconds to allow its speed to stabilize as much as possible. Record your measurements in column 3 of Chart 9–1. Use your recorded &T values to calculate decimal RPM. Enter these calculated &RPM values in column 4 (under the &RPM heading). Recall that the transducer circuit produces two pulses for each motor revolution. Therefore, calculate motor RPM by doubling &T, then dividing this &T × 2 value into sixty (the number of seconds in a minute):

$$\&RPM = 60/(\&T \times 2)$$

Round your column 4 &RPM entries to integers. Convert &RPM entries to hexadecimal and enter under the $RPM heading in column 4.

PART V: Develop Combined Motor Speed Control/Tachometer Software

Review Figures 9–19 and 9–20, along with pages 429–432. Synthesize Figures 9–19 and 9–20 into unified software that controls motor speed and measures motor RPM. Use the Figure 9–21, 9–22, and 9–23 flowcharts (and accompanying memory allocation chart) as guides to software development.

List your software in the blanks below.

MAIN Program

$ADDRESS	SOURCE CODE	COMMENTS

INIT Subroutine

$ADDRESS	SOURCE CODE	COMMENTS

FIGURE 9-21

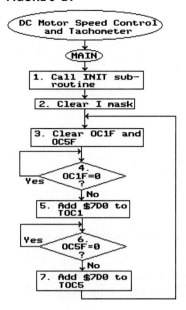

Memory Allocations	
$ADDRESS	**CONTENTS**
0000-0001	$DUTY
0002	RTICNT
0003-0004	$RPM
0100(C000)	MAIN program
0130(C030)	INIT subroutine
0160(C060)	RTI Interrupt Service routine

FIGURE 9-22

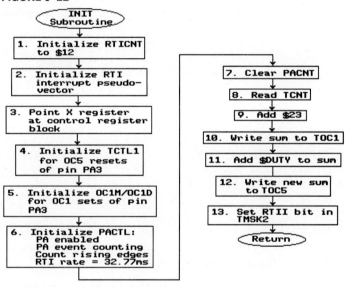

FIGURE 9-23

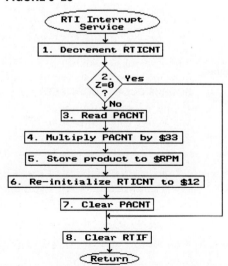

RTI Interrupt Service Routine

$ADDRESS	SOURCE CODE	COMMENTS
_____	_____	_____
_____	_____	_____
_____	_____	_____
_____	_____	_____
_____	_____	_____
_____	_____	_____
_____	_____	_____
_____	_____	_____
_____	_____	_____
_____	_____	_____
_____	_____	_____
_____	_____	_____
_____	_____	_____
_____	_____	_____

Load and debug your software. Demonstrate its operation to the instructor. Show the instructor oscilloscope displays of the 1 KHz speed control waveform (channel 1) and sensor circuit output (channel 2). Also demonstrate the microcontroller's $RPM measurements at addresses $0003–$0004. Base your demonstration on a $DUTY value of $03E8.

_____ *Instructor Check*

PART VI: Analysis of the M68HC11's Tachometer Measurements

Refer to Chart 9–1. Complete column 5 by entering the M68HC11 tachometer's $RPM measurement for each of the listed $DUTY values. For each measurement allow the motor to run for at least thirty seconds to allow its speed to stabilize as much as possible. Calculate decimal percent deviation (Percent Error) for each tachometer $RPM measurement. Enter your results in column 6. Calculate Percent Error as follows:

Step 1. Convert $RPM (tachometer) to decimal (&RPM). Enter this converted value in Chart 9–1 under the &RPM (tachometer) heading.
Step 2. Determine which is less—&RPM (calculated) or &RPM (tachometer).
Smaller value = numerator
Larger value = denominator
Step 3. Divide numerator by denominator.
Step 4. Subtract the quotient from 1.
Step 5. Multiply the difference by 100.
Step 6. Give the product a positive sign if &RPM(tachometer) is greater. Give the product a negative sign if &RPM (calculated) is greater. Enter the signed result in column 6. Round all column 6 entries to three significant figures.

Analyze your Chart 9–1 results by answering the questions below. Enter your answers in the blanks provided.

1. Describe procedural or data gathering difficulties encountered that reduced the reliability and validity of the Chart 9–1 results.

2. In the context of the transducer circuitry and tachometer software, two kinds of error can occur:
   - Linear Error. Linear error causes the tachometer to be inaccurate by a constant percentage over the range of motor speeds.
   - Offset Error. Offset error causes the tachometer to be inaccurate by a constant number of revolutions per minute over the range of motor speeds.

   Calculate the difference between &RPM (tachometer) and &RPM (calculated) for each $DUTY value in Chart 9–1. Enter these differences in the blanks beside corresponding $DUTY values below.

$DUTY	&RPM Difference
0640	_____
0578	_____
04B0	_____
03E8	_____
0320	_____
0258	_____
0190	_____

   Analyze these &RPM difference values. Do they indicate the presence of offset error? _____ Yes _____ No. Explain your answer.

   Analyze your Chart 9–1, column 6 results. Do they indicate the presence of linear error? _____ Yes _____ No. Explain your answer.

3. Refer to Figures 9–21, 9–22, and 9–23. What is the maximum RPM value that the tachometer can measure? $\&RPM_{MAX}$ = _____ (**Hint:** The tachometer's upward range limit relates to maximum PACNT contents.) In the space below, show how you calculated $\&RPM_{MAX}$.

The following paragraphs will introduce software that uses the tachometer to regulate DC motor speed and to hold this speed constant even when motor load changes. This software permits the user to request a particular &RPM rather than $DUTY. The software then varies $DUTY to maintain a constant motor speed. In Exercise 9.5, you will have the opportunity to again evaluate offset error versus linear error, along with the reliability and validity of your results.

PART VII: Laboratory Software Listing

1 KHZ MOTOR SPEED CONTROL—EVB

$ADDRESS	SOURCE CODE	COMMENTS
C000	LDX #1000	Initialize X register
C003	LDAA #2	Initialize TCTL1
C005	STAA 20,X	" " "
C007	LDD #808	Initialize OC1M/OC1D
C00A	STD C,X	" " " "
C00C	LDD E,X	Read TCNT
C00E	ADDD #23	Add $23
C011	STD 16,X	Write sum to TOC1
C013	ADDD 0	Add $DUTY to sum

$ADDRESS	SOURCE CODE	COMMENTS
C015	STD 1E,X	Write new sum to TOC5
C017	BCLR 23,X 77	Clear OC1F and OC5F
C01A	BRCLR 23,X 80 C01A	Branch to self until OC1F = 1
C01E	LDD 16,X	Read TOC1
C020	ADDD #7D0	Add $7D0
C023	STD 16,X	Write sum to TOC1
C025	BRCLR 23,X 8 C025	Branch to self until OC5F = 1
C029	LDD 1E,X	Read TOC5
C02B	ADDD #7D0	Add $7D0
C02E	STD 1E,X	Write sum to TOC5
C030	BRA C017	Make more edges

1 KHZ MOTOR SPEED CONTROL—EVBU

$ADDRESS	SOURCE CODE	COMMENTS
0100	LDX #1000	Initialize X register
0103	LDAA #2	Initialize TCTL1
0105	STAA 20,X	" " "
0107	LDD #808	Initialize OC1M/OC1D
010A	STD C,X	" " "
010C	LDD E,X	Read TCNT
010E	ADDD #23	Add $23
0111	STD 16,X	Write sum to TOC1
0113	ADDD 0	Add $DUTY to sum
0115	STD 1E,X	Write new sum to TOC5
0117	BCLR 23,X 77	Clear OC1F and OC5F
011A	BRCLR 23,X 80 11A	Branch to self until OC1F=1
011E	LDD 16,X	Read TOC1
0120	ADDD #7D0	Add $7D0
0123	STD 16,X	Write sum to TOC1
0125	BRCLR 23,X 8 125	Branch to self until OC5F=1
0129	LDD 1E,X	Read TOC5
012B	ADDD #7D0	Add $7D0
012E	STD 1E,X	Write sum to TOC5
0130	BRA 117	Make more edges

DC MOTOR SPEED CONTROL AND TACHOMETER—MAIN—EVB

$ADDRESS	SOURCE CODE	COMMENTS
C000	JSR C030	Call INIT subroutine
C003	CLI	Clear I mask
C004	BCLR 23,X 77	Clear OC1F and OC5F
C007	BRCLR 23,X 80 C007	Branch to self until OC1F = 1
C00B	LDD 16,X	Add $7D0 to TOC1
C00D	ADDD #7D0	" " "
C010	STD 16,X	" " "
C012	BRCLR 23,X 8 C012	Branch to self until OC5F = 1
C016	LDD 1E,X	Add $7D0 to TOC5
C018	ADDD #7D0	" " "
C01B	STD 1E,X	" " "
C01D	BRA C004	Make more edges

DC MOTOR SPEED CONTROL AND TACHOMETER—MAIN—EVBU

$ADDRESS	CONTENTS	COMMENTS
0100	JSR 130	Call INIT subroutine
0103	CLI	Clear I mask
0104	BCLR 23,X 77	Clear OC1F and OC5F
0107	BRCLR 23,X 80 107	Branch to self until OC1F = 1
010B	LDD 16,X	Add $7D0 to TOC1
010D	ADDD #7D0	" " "
0110	STD 16,X	" " "
0112	BRCLR 23,X 8 112	Branch to self until OC5F = 1
0116	LDD 1E,X	Add $7D0 to TOC5
0118	ADDD #7D0	" " "
011B	STD 1E,X	" " "
011D	BRA 104	Make more edges

DC MOTOR SPEED CONTROL AND TACHOMETER—INIT SUBROUTINE—EVB

$ADDRESS	SOURCE CODE	COMMENTS
C030	LDAA #12	Initialize RTICNT
C032	STAA 2	" " "
C034	LDD #C060	Initialize RTI Interrupt pseudovector
C037	STD EC	" " " " " "
C039	LDX #1000	Initialize X register
C03C	LDAA #8	Initialize TCTL1
C03E	STAA 20,X	" " "
C040	LDD #808	Initialize OC1M/OC1D
C043	STD C,X	" " "
C045	LDD #5300	Initialize PACTL/PACNT
C048	STD 26,X	" " "
C04A	LDD E,X	Read TCNT
C04C	ADDD #23	Add $23
C04F	STD 16,X	Write sum to TOC1
C051	ADDD 0	Add $DUTY to sum
C053	STD 1E,X	Write new sum to TOC5
C055	LDAA #40	Set RTII bit
C057	STAA 24,X	" "
C059	RTS	Return

DC MOTOR SPEED CONTROL AND TACHOMETER—INIT SUBROUTINE—EVBU

$ADDRESS	SOURCE CODE	COMMENTS
0130	LDAA #12	Initialize RTICNT
0132	STAA 2	" " "
0134	LDD #160	Initialize RTI Interrupt pseudovector
0137	STD EC	" " " " " "
0139	LDX #1000	Initialize X register
013C	LDAA #2	Initialize TCTL1
013E	STAA 20,X	" " "
0140	LDD #808	Initialize OC1M/OC1D
0143	STD C,X	" " "
0145	LDD #5300	Initialize PACTL/PACNT
0148	STD 26,X	" " "
014A	LDD E,X	Read TCNT
014C	ADDD #23	Add $23
014F	STD 16,X	Write sum to TOC1
0151	ADDD 0	Add $DUTY to sum
0153	STD 1E,X	Write new sum to TOC5
0155	LDAA #40	Set RTII bit
0157	STAA 24,X	" "
0159	RTS	Return

DC MOTOR SPEED CONTROL AND TACHOMETER—RTI INTERRUPT SERVICE—EVB

$ADDRESS	SOURCE CODE	COMMENTS
C060	DEC 2	Decrement RTICNT
C063	BNE C072	If RTICNT is not 0, clear RTIF and return
C065	LDAA 27,X	If RTICNT = 0, read PACNT
C067	LDAB #33	Multiply PACNT by $33
C069	MUL	" " " "
C06A	STD 3	Store product to $RPM
C06C	LDAA #12	Initialize RTICNT
C06E	STAA 2	" " "
C070	CLR 27,X	Clear PACNT
C072	BCLR 25,X BF	Clear RTIF
C075	RTI	Return

DC MOTOR SPEED CONTROL AND TACHOMETER—RTI INTERRUPT SERVICE—EVBU

$ADDRESS	SOURCE CODE	COMMENTS
0160	DEC 2	Decrement RTICNT
0163	BNE 172	If RTICNT is not 0, clear RTIF and return
0165	LDAA 27,X	If RTICNT = 0, read PACNT
0167	LDAB #33	Multiply PACNT by $33

0169	MUL	" " " "
016A	STD 3	Store product to $RPM
016C	LDAA #12	Initialize RTICNT
016E	STAA 2	" " "
0170	CLR 27,X	Clear PACNT
0172	BCLR 25,X BF	Clear RTIF
0175	RTI	Return

DC Motor Speed Regulation Using Tachometer Control

Chapters 8 and 9 have presented a number of DC motor speed control exercises. Instrumentation/control designers classify these exercises as open-loop, continuous control applications. These applications are continuous because the microcontroller outputs a motor speed control square wave continuously. Review Figure 9–21. This figure shows the motor control software operating in a continuous loop. By resetting the microcontroller and halting motor control program execution, the user relinquishes control over the motor. If the square wave stops, control stops.[10]

As the reader discovered in Exercise 9.4, software exerted poor control over motor speed fluctuations. Exercise 9.4 software's speed control is poor because it exerts open-loop control. Refer to Figure 9–24, part A. Desiring a specific motor speed, the user calculates the necessary setting (1) to generate this speed. This setting is $DUTY. To generate the desired speed, the user initializes $DUTY and starts program execution by the M68HC11 controller (2). The controller generates a continuous 1 KHz motor speed control signal (3). Since $DUTY remains constant, the control signal's duty cycle remains constant, causing the process (motor speed, 4) to approach desired RPM. However, environmental disturbances (5) cause changes in motor speed. These disturbances can include load changes, incidence of vibration, and power fluctuations. Thus, fluctuations in the controlled variable (motor RPM, 6) occur and go uncorrected by the controller. The M68HC11 fails to regulate motor speed fluctuations because the control process is open-loop. Even though the controller measures the controlled variable (tachometer), the software has no provision for feeding back these measurements to the control routine, so that it can take corrective action to regulate motor speed. To effectively counteract disturbances and maintain constant motor speed, the software must close the control loop.

FIGURE 9–24

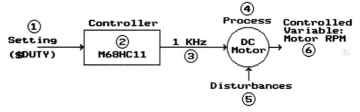

A. Open-Loop Control

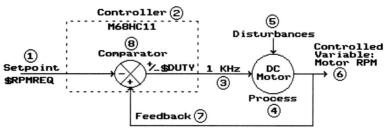

B. Closed-Loop Control

---

[10]EVB users have become particularly aware that a reset concedes control over the DC motor. When the user resets the M68HC11A1, the motor runs ungoverned at its top speed.

Refer to Figure 9–24, part B, which shows how the reader shall close the control loop in Exercise 9.5. The user selects desired $RPM and initializes the control software with this value (the setpoint, 1). The comparator (8) software routine compares current motor RPM (feedback, 7) with the setpoint, and adjusts $DUTY to increase or decrease motor speed as necessary to make the controlled variable (6) match the setpoint. Adjustments in $DUTY cause changes in the duty cycle of the motor speed control signal (3). These duty cycle changes compensate for the disturbances (5) by returning the controlled variable (6) to the setpoint (1). The motor control software feeds back any change in the controlled variable to the comparator routine, which responds by readjusting $DUTY. Thus, the control loop becomes closed and continuous, exerting improved control over motor speed. Following paragraphs show how the reader can introduce closed-loop, continuous DC motor speed control.

Figure 9–25 provides a simplified flowchart of the closed-loop, continuous, DC motor speed control software. The user loads a specified RAM location ($RPMREQ) with a requested hexadecimal RPM value. The user then starts program execution. In block 1, the CPU reads $RPMREQ and converts it to an equivalent PACNT comparison value, labeled $PACNTRQ. $PACNTRQ equals the PACNT register's contents after $12 RTIs *if* the motor runs at *exactly* the requested speed ($RPMREQ). The Comparator routine will compare actual PACNT contents with the $PACNTRQ value and increase or decrease $DUTY accordingly. The tachometer calculates $RPM by multiplying PACNT contents (after $12 RTIs or .59 seconds) by $33. Therefore, the CPU can divide $RPMREQ by $33 to derive $PACNTRQ:

If

$$\$RPM = PACNT \times \$33$$

then

$$\$PACNTRQ = \$RPMREQ/\$33$$

Of course, the CPU stores $PACNTRQ in an assigned RAM location for later use by the Comparator routine.

In block 2, the CPU initializes listed locations and registers for motor speed control and tachometer operation. In block 3, the M68HC11 generates the 1 KHz motor speed control waveform continuously. During each waveform generation loop, the CPU offsets the TOC5 register from TOC1 by $DUTY's worth of free-running counts. Thus, if the comparator changes $DUTY, then the 1 KHz output waveform's pulse width changes accordingly to adjust motor speed.

FIGURE 9–25

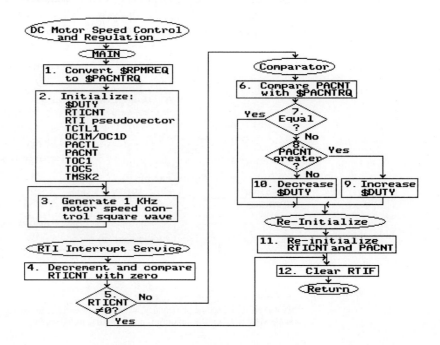

Refer to Figure 9–25, blocks 4–12, which constitute the RTI Interrupt Service routine. The PACNT register counts transducer edges over $12 RTIs. In block 4 the CPU checks the RTICNT. If fewer than $12 RTIs have occurred, the CPU clears the RTIF bit and returns from interrupt. If $12 RTIs have occurred, the CPU executes the Comparator routine, blocks 6–10. In block 6, the CPU compares $PACNTRQ with the actual PACNT contents. PACNT forms the feedback link from the controlled variable ($RPM) to the comparator (see Figure 9–24, part B). If PACNT = $PACNTRQ (block 7), the motor runs at exactly the requested $RPM ($RPMREQ). Therefore, the CPU makes no change in $DUTY and branches from block 7 to block 11. If PACNT is greater than $PACNTRQ (block 8), the motor runs too fast. Therefore, the CPU branches to block 9. In block 9, the CPU increases $DUTY, which slows the motor down. If $PACNTRQ is greater than PACNT, the motor is running too slow. Therefore, the CPU decreases $DUTY (block 10), which speeds the motor up. The Comparator routine increases or decreases $DUTY by an amount equal to twice the difference between PACNT and $PACNTRQ. Thus, $DUTY changes are proportional to motor speed error.

After executing the Comparator routine, the CPU re-initializes RTICNT and clears PACNT (block 11) so that the Pulse Accumulator can count transducer edges over another .59-second period. Whether it has executed the Comparator routine or not, the CPU clears the RTIF flag (block 12) before returning from the RTI interrupt.

Following paragraphs refer to Figures 9–26 through 9–29 and their accompanying software lists. These paragraphs provide a detailed look at the motor speed control and regulation software.

Refer to Figure 9–26 and its accompanying program list. The MAIN program executes three tasks: (1) convert $RPMREQ to $PACNTRQ; (2) call the INIT subroutine; and (3) generate the 1 KHz motor speed control square wave. Figure 9–26, blocks 1 through 3, accomplish the first of these tasks.

Before starting program execution, the interface operator requests a hexadecimal RPM value ($RPMREQ). The operator uses the BUFFALO MM command to store $RPMREQ to RAM locations $0002–$0003 (see Memory Allocations in Figure 9–26). In blocks 1 through 3 and program addresses $0100–$0108, the CPU converts $RPMREQ to the PACNT comparison value: $PACNTRQ. In block 1, the CPU reads $RPMREQ from addresses $0002–$0003. In block 2, the CPU uses the IDIV command to divide $RPMREQ by $33. The X register now contains the $PACNTRQ value. The CPU uses the XGDX command (program address $0106) to transfer the quotient from its X register to its D register. Recall that the IDIV command returns the quotient to the X register and the remainder to the D register. The largest acceptable $RPMREQ value is $32CD (13005). The IDIV command divides $RPMREQ by $0033. Therefore, the largest possible quotient equals $32CD/$0033 = $00FF. Since PACNT contains 8 bits, its comparison value—$PACNTRQ—must also consist of 8 bits. After IDIV execution, this 8-bit comparison value resides in the low byte of the X register. The CPU cannot transfer this low byte directly to the $PACNTRQ storage location. Therefore, the CPU uses the XGDX command to transfer the 16-bit quotient to the D register. This operation leaves the $PACNTRQ value in ACCB. In block 3 (program address $0107), the CPU stores ACCB's contents to the $PACNTRQ storage location.

Refer to Figure 9–26, blocks 4 and 5, and program addresses $0109–$010C. In block 4, the CPU calls the INIT subroutine. This subroutine initializes the RTI interrupt pseudo-vector, $DUTY, RTICNT, and all necessary control registers for motor control, tachometer, and comparator operations. In block 5, the CPU resets the global I mask to permit RTI interrupts.

In Figure 9–26, blocks 6 through 12, the CPU generates the 1 KHz motor speed control square wave (see program addresses $010D–$0126). After clearing OC1F and OC5F (block 6), the CPU branches to self until the OC1F flag sets (block 7). When OC1F sets, the Output Compare 1 mechanism generates a rising edge of the 1 KHz motor control square wave. In block 8, the CPU updates its TOC1 register for the next rising edge, which occurs after $7D0 free-running counts elapse. In block 9, the CPU waits for a successful OC5 compare. This compare generates a square wave falling edge and sets the OC5F flag. OC5F = 1 causes the CPU to escape the block 9 polling loop and to update its TOC5 register for the next falling edge. The CPU updates TOC5 by executing blocks 10, 11, and 12. In block 10, the CPU

FIGURE 9–26

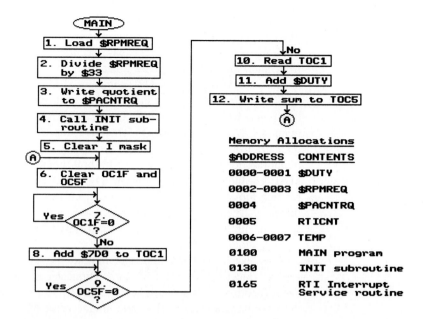

MAIN Program

$ADDRESS	SOURCE CODE	COMMENTS
0100	LDD 2	Load $RPMREQ
0102	LDX #33	Divide by $33
0105	IDIV	" " "
0106	XGDX	" " "
0107	STAB 4	Store quotient to $PACNTRQ
0109	JSR 130	Call INIT
010C	CLI	Clear I mask
010D	BCLR 23,X 77	Clear OC1F and OC5F
0110	BRCLR 23,X 80 110	Branch to self until OC1F = 1
0114	LDD 16,X	Add $7D0 to TOC1
0116	ADDD #7D0	" " " "
0119	STD 16,X	" " " "
011B	BRCLR 23,X 8 11B	Branch to self until OC5F = 1
011F	LDD 16,X	Read TOC1
0121	ADDD 0	Add $DUTY
0123	STD 1E,X	Write sum to TOC5
0125	BRA 10D	Make more edges

reads TOC1, which was updated in block 8. Since the next falling edge must occur $DUTY's worth of counts after a rising edge, the CPU adds the TOC1 and $DUTY values (block 11) and writes the sum to TOC5 (block 12). After executing block 12, the CPU branches back to block 6 to generate more motor control waveform edges. In Figure 9–26, block 4, the CPU calls the INIT subroutine. The next few paragraphs describe this INIT subroutine.

Refer to Figure 9–27 and its accompanying program list, which illustrate the INIT subroutine. In block 1 (program addresses $0130–$0134), the CPU initializes $DUTY to $03E8. This $DUTY value generates a 50% duty cycle. Thus, the MAIN program generates an initial pulse width of $03E8 free-running counts (500μs), causing the motor to run at about half speed. After $12 RTIs, the CPU will make its first $DUTY adjustment, causing motor speed to change toward the requested RPM value.

In block 2 (program addresses $0135–$0138), the CPU initializes RTICNT to $12. Thus, the tachometer measures motor speed every $12 RTIs, (i.e., every .59 seconds), and the comparator adjusts $DUTY accordingly.

FIGURE 9-27

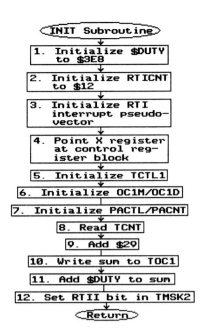

INIT Subroutine

$ADDRESS	SOURCE CODE	COMMENTS
0130	LDD #3E8	Initialize $DUTY
0133	STD 0	" " "
0135	LDAA #12	Initialize RTICNT
0137	STAA 5	" " "
0139	LDD #165	Initialize RTI interrupt pseudovector
013C	STD EC	" " " " " "
013E	LDX #1000	Initialize X register
0141	LDAA #2	Initialize TCTL1
0143	STAA 20,X	" " "
0145	LDD #808	Initialize OC1M/OC1D
0148	STD C,X	" " " "
014A	LDD #5300	Initialize PACTL/PACNT
014D	STD 26,X	" " " "
014F	LDD E,X	Read TCNT
0151	ADDD #29	Add $29
0154	STD 16,X	Write sum to TOC1
0156	ADDD 0	Add $DUTY to sum
0158	STD 1E,X	Write new sum to TOC5
015A	LDAA #40	Set RTII in TMSK2
015C	STAA 24,X	" " "
015E	RTS	Return

After initializing the RTI interrupt pseudovector (block 3) and X register (block 4), the CPU initializes its TCTL1 control register (block 5, program addresses $0141–$0144). By writing $02 to TCTL1, the CPU sets the OM5 bit and resets OL5, programming the Output Compare 5 mechanism to generate falling edges. In block 6 (program addresses $0145–$0149), the CPU writes $08 to both the OC1M and OC1D registers. These writes configure the OC1 mechanism to generate rising edges at pin PA3/OC5. Thus, in blocks 5 and 6, the CPU programs OC1 and OC5 to generate square wave edges.

In block 7, the CPU initializes its PACTL and PACNT registers by writing $53 to PACTL and $00 to PACNT (see program addresses $014A–$014E). By writing $53 to

445

PACTL the CPU configures the Pulse Accumulator and RTI mechanisms for their tachometer functions:

    PAEN = 1   enables the Pulse Accumulator
    PAMOD = 0   selects event-counting mode
    PEDGE = 1   causes the Pulse Accumulator to count input rising edges
    RTR1 and RTR0 are both set, causing the RTI mechanism to generate an interrupt every 32.77 milliseconds.

Of course, the CPU writes $00 to PACNT so that it can feed back an accurate $RPM/$33 count to the Comparator routine.

In Figure 9–27, blocks 8 through 12, the CPU loads TOC1 and TOC5 with initial rising and falling edge compare values. In blocks 8, 9, and 10 (program addresses $014F–$0155), the CPU reads the current free-running count from its TCNT register (block 8), adds $29 (block 9), and writes the sum to TOC1 (block 10). Thus, the CPU assures that the first successful OC1 compare occurs while the CPU executes the OC1 polling loop (block 7, Figure 9–26). In Figure 9–27, blocks 11 and 12 (program addresses $0156–$0159), the CPU initializes TOC5 with the sum of the TOC1 contents plus $03E8 ($DUTY). Thus, the CPU offsets the TOC1 (rising edge) and TOC5 (falling edge) contents by $03E8 free-running counts. Remember that the Comparator routine will begin adjusting this duty cycle .59 seconds after the motor starts running. The comparator will adjust $DUTY every .59 seconds thereafter.

Before returning from the INIT subroutine, the CPU sets the RTII bit in TMSK2 (block 13). This action enables RTI interrupts.

Figures 9–28 and 9–29 (and their accompanying software lists) illustrate the RTI Interrupt Service routine. This routine has four parts:

1. RTICNT Decrement and Test (Figure 9–28).
2. Tachometer Feedback and Comparator (Figure 9–29).
3. RTICNT/PACNT re-initialize (Figure 9–28).
4. Clear RTIF bit (Figure 9–28).

Refer to Figure 9–28, blocks 1, 2, and 17 (program addresses $0165–$0169). These blocks constitute part 1 of the interrupt routine. In block 1, the CPU decrements RTICNT, which was initialized to $12. After $12 RTI interrupts and block 1 decrements, RTICNT equals zero, which sets the Z flag. In block 2, the CPU executes a BNE instruction, which branches program flow to block 17 if RTICNT has not yet decremented to zero. In such a case, the CPU executes block 17, clearing the RTIF bit, and returns from interrupt. If Z = 1, however, the CPU does not branch but instead executes the interrupt routine's second part—the Tachometer Feedback and Comparator routine.

Figure 9–29 illustrates this Tachometer Feedback and Comparator routine. Refer to Figure 9–29, block 3 (program address $016A). In this block, the CPU clears the most significant byte of its D register, to prepare it for a block 10 or block 13 mathematical operation. In block 4 (program address $016B–$016C), the CPU feeds back tachometer information to the Comparator routine by reading PACNT. After $12 RTIs, the PACNT register contains a tachometer value to be compared with $PACNTRQ. Recall that $PACNTRQ contains the "ideal" PACNT value—the value that the PACNT register should hold after $12 RTIs, if the motor runs at exactly the requested RPM.

The CPU compares the setpoint ($PACNTRQ) and feedback (PACNT) values in blocks 5, 6, and 7 (program addresses $016D–$0172). The CPU makes the comparison by subtracting $PACNTRQ from PACNT in block 5. In blocks 6 and 7, the CPU directs program flow based upon comparison results. In block 6, the CPU tests the Z flag to determine whether $PACNTRQ = PACNT. If Z = 1, indicating equality, the motor runs at the requested RPM and requires no speed adjustments. Therefore, the CPU branches to terminal B of the flowchart, taking no corrective action. If Z = 0, indicating inequality, the CPU tests the C flag (block 7). If C = 0, PACNT > $PACNTRQ, and program flow branches from block 7 to block 12. If C = 1, PACNT < $PACNTRQ, the CPU does not branch, but executes block 8. Following paragraphs review this process in more detail.

In Figure 9–29, block 5, the CPU subtracted $PACNTRQ from PACNT and returned the difference to ACCB. If C = 1 (block 7), then $PACNTRQ is greater than PACNT. ACCB contains the difference as a negative number. In block 8 (program address $0173), the CPU two's

FIGURE 9-28

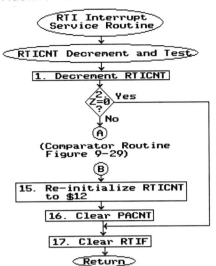

FIGURE 9-29

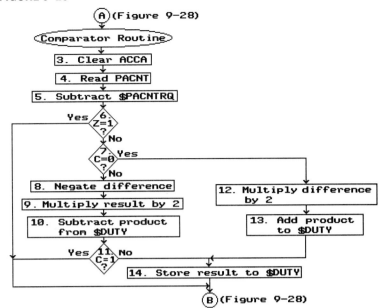

RTI Interrupt Service Routine

$ADDRESS	SOURCE CODE	COMMENTS
	RTICNT Decrement and Test	
0165	DEC 5	Decrement RTICNT
0168	BNE 18A	If not zero, clear RTIF and return
	Comparator Routine	
016A	CLRA	Clear D register MS byte
016B	LDAB 27,X	Read PACNT
016D	SUBB 4	Subtract $PACNTRQ
016F	BEQ 184	If equal, do not adjust $DUTY
0171	BCC 17F	If PACNT>$PACNTRQ, increase $DUTY
0173	NEGB	If PACNT<$PACNTRQ, make difference positive
0174	ASLB	Multiply result by 2
0175	STD 6	Store product to TEMP
0177	LDD 0	Load $DUTY
0179	SUBD 6	Subtract product (TEMP)
017B	BCS 184	If $DUTY rolls over, do not store
017D	BRA 182	Branch to store difference
017F	ASLB	Multiply difference by two
0180	ADDD 0	Add $DUTY
0182	STD 0	Store result to $DUTY
0184	LDAA #12	Re-initialize RTICNT
0186	STAA 5	" " "
0188	CLR 27,X	Clear PACNT
018A	BCLR 25,X BF	Clear RTIF
018D	RTI	Return

complements this difference to make it positive. In block 9 (program address $0174), the CPU doubles this positive value. If $PACNTRQ was greater than PACNT, the motor runs too slowly, and the CPU must compensate by decreasing $DUTY. In block 10 (program addresses $0175–$017C), the CPU subtracts the block 9 result from $DUTY. The block 9 result is an 8-bit value in ACCB. $DUTY is a 16-bit value residing at locations $0000–$0001.

Recall that the CPU cleared ACCA in block 3. Therefore, consider the block 9 result as a 16-bit value—$00XX—residing in the D register. To complete the block 10 subtraction, the CPU stores the subtrahend (block 9 result) to TEMP addresses $0006–$0007. The CPU then reads $DUTY from $0000–$0001 and subtracts the TEMP value.

At this point we must consider the possibility that the motor cannot reach requested RPM, no matter how small the duty cycle. If excessive loading keeps the motor from reaching requested RPM, continuous block 10 subtractions will roll $DUTY over from a very small value to a very large one. In such a case, the motor would respond by either stopping or slowing drastically. $DUTY roll-over occurs because the block 10 subtrahend exceeds its minuend. Thus, roll-over sets the C flag. In block 11 (program addresses $017B–$017C), the CPU branches to terminal B if the C flag has set. If it branches, the CPU avoids storing an invalid $DUTY value (block 14) and prevents the motor from inadvertently slowing or stopping. If block 10 execution resets the C flag, the CPU branches to block 14 where it returns the new $DUTY value from the D register to addresses $0000–$0001.

If PACNT > $PACNTRQ, motor speed is too great, and the CPU must execute blocks 12 and 13 (program addresses $017F–$0181). The block 5 difference is positive. Therefore, the CPU doubles this difference in block 12. In block 13, the CPU adds the current $DUTY value to the block 12 result.

In block 14 (program addresses $0182–$0183), the CPU writes the adjusted $DUTY value to $0000–$0001. The MAIN program uses this new $DUTY value to generate its 1 KHz motor control square wave. After another .59 seconds ($12 RTIs) elapse, the RTI Interrupt Service routine will again adjust $DUTY.

Refer back to Figure 9–28, blocks 15 and 16 (program addresses $0184–$0189). If the CPU has adjusted $DUTY, it must prepare for another round of tachometer operations. Therefore, the CPU re-initializes RTICNT to $12, and PACNT to $00. Before returning from the RTI Interrupt Service routine, the CPU must clear the RTIF bit in TFLG2 (block 17) to avoid a spurious RTI interrupt request.

Foregoing paragraphs have provided a detailed look at the motor speed control and regulation software. This software assumes the same hardware configuration used in Exercise 9.4. Exercise 9.5 gives the reader an opportunity to try out this software and evaluate its performance.

EXERCISE 9.5 DC Motor Speed Regulation Using Tachometer Control

REQUIRED MATERIALS: Motor control interface, as used in Exercise 9.4 DC motor, mounting and sensor circuitry, as used in Exercise 9.4

REQUIRED EQUIPMENT: Motorola M68HC11EVBU (EVB) and terminal
Dual-trace oscilloscope and two probes

GENERAL INSTRUCTIONS: This exercise uses the same interface and connections as Exercise 9.4. The M68HC11 outputs a 1 KHz motor speed control square wave from PORTA pin 3. Apply the sensor circuitry's output waveform to the Pulse Accumulator Input (PAI) pin (PORTA pin 7). In Part I, connect all hardware. In Part II, develop motor speed control and regulation software. In Part III, evaluate interface operation by comparing requested versus measured RPM values. Part III also provides an opportunity to observe the speed regulation process on an oscilloscope.

PART I: Connect the Speed Control and Regulation Interface

Refer to Figures 9–16, 9–17, and 9–18. Assure that all connections shown in these figures are intact and correct. Connect the motor control interface (Figure 9–16) to the microcontroller's PA3 pin. Connect the Figure 9–18 circuit's output to the M68HC11's PAI pin. Show your completed connections to the instructor.

_____ *Instructor Check*

PART II: Develop Motor Speed Control and Regulation Software

Review pages 441–448 to assure that you understand the motor speed control software's strategy. Refer to the software lists accompanying Figures 9–26 through 9–29. If you use an EVBU, load the software as listed. If you use an EVB, allocate memory as follows:

EVB Memory Allocations

$ADDRESS	CONTENTS
0000–0001	$DUTY
0002–0003	$RPMREQ
0004	$PACNTRQ
0005	RTICNT
0006–0007	TEMP
C000	MAIN program
C030	INIT subroutine
C065	RTI Interrupt Service routine

If you use an EVB, assure that the RTI pseudovector, all jumps, and all branches reflect the memory allocations shown above. Connect oscilloscope channel 1 to display the 1 KHz motor speed control square wave. Connect channel 2 to show the sensor circuit's output. Power up the EVBU (EVB) and interface. Load and debug your software. When confident that the software and interface operate properly, demonstrate them to the instructor. Show oscilloscope channel 1 and channel 2 waveforms to the instructor. For this demonstration, use a requested RPM ($RPMREQ) of $1770.

_____ *Instructor Check*

PART III: Laboratory Analysis

Refer to Figure 9–29, blocks 9 and 12 (program addresses $0174 and $017F). In these blocks, the CPU doubles the difference between PACNT and $PACNTRQ. The CPU then adds or subtracts this value from $DUTY. The new $DUTY value causes motor speed to approach the $RPM request value. Given single bytes of program space at addresses $0174 and $017F, we have three options:

> Option 1. Leave the instructions as is, i.e., use the ASLB instruction to *double* the $PACNTRQ/PACNT difference.
> Option 2. Replace the instructions with no-ops (NOPs). The CPU will adjust $DUTY by the value of the $PACNTRQ/PACNT difference.
> Option 3. Replace the instructions with LSRB instructions. The CPU will adjust $DUTY by half the difference between $PACNTRQ and PACNT.

Each of these options affects the regulator's response rate and settling time. The greater the change in $DUTY, the faster the motor will reach or return to the set point speed ($RPMREQ). Thus, the programmer is inclined to maximize response rate. However, response rate relates to other response characteristics—both desirable and undesirable:

> **Overshoot and oscillation.** If we increase rate of change in motor speed, rate change in angular momentum also increases. This exercise's only moving parts are the motor armature and optical encoder. Because of this small mass and the fact that the CPU can change $DUTY only once every .59 seconds, we may expect a damped response. However, if the motor drives an end effector with more mass (e.g., a robot arm or even a cooling fan), moment of inertia assumes a larger role. Moment of inertia becomes a factor in the rate at which motor speed can change in response to control signals—without generating excessive overshoot and oscillation about the setpoint value.
> **Response Gap.** Define Response Gap as the smallest change in the controlled variable that generates a response by the motor speed regulator. Assume that the difference between PACNT and $PACNTRQ equals $01 (%00000001). If the CPU executes a Logic Shift Right (LSR) on this value, it becomes %00000000. Obviously, $DUTY does not change when this value is added (Figure 9–29,

block 13) or subtracted (block 10). Therefore, the PACNT/$PACNTRQ difference must equal at least $02 before the regulator takes effective action to regulate speed. In this instance, the Response Gap = $02 × $33 RPM. Thus, if LSR instructions are used, the smallest change in motor speed that generates a response by the regulation software is 102 revolutions per minute:

$$\$02 \times \$33 = \$66$$
$$\$66 = 102 \text{ RPM}$$

Option 2 (NOPs) and Option 1 (Arithmetic Shifts Left) reduce the Response Gap, since they cause a regulator response to an $01 difference between $PACNTRQ and PACNT.

1. In the chart below, predict option 1, option 2, and option 3 performance on the basis of three characteristics:

    Response Rate. How fast does the motor reach or return to the setpoint RPM ($RPMREQ)?

    Overshoot/Oscillation. How likely are overshoot and consequent oscillation about the setpoint?

    Response Gap. What is the smallest change in $RPM that generates an effective response by the regulator?

    For response rate and overshoot/oscillation, rank the three options for best performance (1), intermediate performance (2), and worst performance (3). Enter rating numbers in the appropriate blanks. Enter calculated hexadecimal Response Gap—in revolutions per minute—into the right-hand column blanks.

    Predicted Performance Characteristics

Option	Response Rate*	Overshoot/ Oscillation*	Response Gap ($RPM)
1	___	___	___
2	___	___	___
3	___	___	___

    * Rate as Best = 1, Intermediate = 2, Worst = 3.

2. Empirically evaluate the actual effects of the three options on regulator performance. Use three $RPMREQ values for your evaluation: $03E8 RPM, $1770 RPM, and $2EE0 RPM. For option 1 evaluation, leave the software as developed for Part II. For option 2 evaluation, change the ASLB instructions (Figure 9–29, blocks 9 and 12) to NOP instructions. For option 3 evaluation, change these instructions to LSRBs. Complete this comparative evaluation by entering your observations and measured data in the chart below. Calculate Percent RPM Error as follows:

    Step 1. Determine which is lesser—&RPM (oscilloscope) or &RPMREQ.
    Smaller value = numerator
    Larger value = denominator
    Step 2. Divide numerator by denominator.
    Step 3. Subtract quotient from 1.
    Step 4. Multiply difference by 100.
    Step 5. Give product a negative sign if &RPMREQ is larger. Give product a positive sign if &RPM (oscilloscope) is larger. Enter your signed result in the Percent RPM Error column. Round entries to three significant figures.

    Observed Performance Characteristics

    $RPMREQ = $03E8
    &RPMREQ = 1000

Option	Response Rate*	Overshoot/ Oscillation*	Measured RPM (Oscilloscope) &RPM	Percent RPM Error
1	___	___	___	___
2	___	___	___	___
3	___	___	___	___

$RPMREQ = $1770
&RPMREQ = 6000

Option	Response Rate*	Overshoot/ Oscillation*	Measured RPM (Oscilloscope) &RPM	Percent RPM Error
1	_____	_____	_____	_____
2	_____	_____	_____	_____
3	_____	_____	_____	_____

$RPMREQ = $2EE0
&RPMREQ = 12000

Option	Response Rate*	Overshoot/ Oscillation*	Measured RPM (Oscilloscope) &RPM	Percent RPM Error
1	_____	_____	_____	_____
2	_____	_____	_____	_____
3	_____	_____	_____	_____

* Rate as Best = 1, Intermediate = 2, Worst = 3.

3. In the blanks, explain which of the three options gives the best overall performance. Explain your reasons for choosing this option.

4. Change the Figure 9–29, blocks 9 and 12 program instructions to the option that gives best performance. Refer to Chart 9–2. Complete for each $RPMREQ value.

5. Analyze your results. Recall that two kinds of speed regulation error can occur:
   - Linear Error. Linear error causes inaccuracy by a constant percentage over motor speed range.

Chart 9–2

1. $RPMREQ	2. Measured RPM (Oscilloscope) &RPM $RPM	3. Measured % Duty Cycle (Decimal, Oscilloscope)	4. Percent RPM Error
03E8	_____	_____	_____
07D0	_____	_____	_____
0BB8	_____	_____	_____
0FA0	_____	_____	_____
1388	_____	_____	_____
1770	_____	_____	_____
1B58	_____	_____	_____
1F40	_____	_____	_____
2328	_____	_____	_____
2710	_____	_____	_____
2AF8	_____	_____	_____
2EE0	_____	_____	_____

- Offset Error. Offset error causes inaccuracy by a constant number of revolutions per minute over motor speed range.

Calculate the difference between $RPM (oscilloscope) and $RPMREQ for each $RPMREQ value in Chart 9–2. Enter these differences in the blanks below.

$RPMREQ	$RPM Difference
03E8	_____
07D0	_____
0BB8	_____
0FA0	_____
1388	_____
1770	_____
1B58	_____
1F40	_____
2328	_____
2710	_____
2AF8	_____
2EE0	_____

Analyze these values. Do they indicate the presence of offset error? _____ Yes _____ No. Explain your answer.

Do the $RPM Difference values indicate the presence of linear error? _____ Yes _____ No. Explain your answer.

PART IV: Laboratory Software Listing

MAIN PROGRAM—EVB

$ADDRESS	SOURCE CODE	COMMENTS
C000	LDD 2	Load $RPMREQ
C002	LDX #33	Divide by $33
C005	IDIV	" "
C006	XGDX	" "
C007	STAB 4	Store quotient to $PACNTRQ
C009	JSR C030	Call INIT
C00C	CLI	Clear I mask
C00D	BCLR 23,X 77	Clear OC1F and OC5F
C010	BRCLR 23,X 80 C010	Branch to self until OC1F = 1
C014	LDD 16,X	Add $7D0 to TOC1
C016	ADDD #7D0	" " "
C019	STD 16,X	" " "
C01B	BRCLR 23,X 8 C01B	Branch to self until OC5F = 1
C01F	LDD 16,X	Read TOC1
C021	ADDD 0	Add $DUTY
C023	STD 1E,X	Write sum to TOC5
C025	BRA C00D	Make more edges

MAIN PROGRAM—EVBU

(See program list accompanying Figure 9–26)

INIT SUBROUTINE—EVB

$ADDRESS	SOURCE CODE	COMMENTS
C030	LDD #3E8	Initialize $DUTY
C033	STD 0	" " "

$ADDRESS	SOURCE CODE	COMMENTS
C035	LDAA #12	Initialize RTICNT
C037	STAA 5	" " "
C039	LDD #C065	Initialize RTI pseudovector
C03C	STD EC	" " " " "
C03E	LDX #1000	Initialize X register
C041	LDAA #2	Initialize TCTL1
C043	STAA 20,X	" " "
C045	LDD #808	Initialize OC1M/OC1D
C048	STD C,X	" " "
C04A	LDD #5300	Initialize PACTL/PACNT
C04D	STD 26,X	" " "
C04F	LDD E,X	Read TCNT
C051	ADDD #29	Add $29
C054	STD 16,X	Write sum to TOC1
C056	ADDD 0	Add $DUTY to sum
C058	STD 1E,X	Write new sum to TOC5
C05A	LDAA #40	Set RTII in TMSK2
C05C	STAA 24,X	" " "
C05E	RTS	Return

INIT SUBROUTINE—EVBU

(See program list accompanying Figure 9–27.)

RTI INTERRUPT SERVICE ROUTINE—EVB

$ADDRESS	SOURCE CODE	COMMENTS
C065	DEC 5	Decrement RTICNT
C068	BNE C08A	If RTICNT not zero, clear RTIF and return
C06A	CLRA	Clear D register MS byte
C06B	LDAB 27,X	Read PACNT
C06D	SUBB 4	Subtract $PACNTRQ
C06F	BEQ C084	If PACNT = $PACNTRQ, do not adjust $DUTY
C071	BCC C07F	If PACNT>$PACNTRQ, increase $DUTY
C073	NEGB	If PACNT<$PACNTRQ, make difference positive
C074	ASLB	Multiply result by 2
C075	STD 6	Store product to TEMP
C077	LDD 0	Load $DUTY
C079	SUBD 6	Subtract product (TEMP)
C07B	BCS C084	If $DUTY rolls over, do not store
C07D	BRA C082	Branch to store difference
C07F	ASLB	Multiply difference by 2
C080	ADDD 0	Add $DUTY
C082	STD 0	Store result to $DUTY
C084	LDAA #12	Re-initialize RTICNT
C086	STAA 5	" " "
C088	CLR 27,X	Clear PACNT
C08A	BCLR 25,X BF	Clear RTIF
C08D	RTI	Return

RTI INTERRUPT SERVICE ROUTINE—EVBU

(See program list accompanying Figures 9–29.)

9.6 M68HC11 PULSE ACCUMULATOR—GATED TIME ACCUMULATION MODE

Section 9.5 covered the Pulse Accumulator's event-counting mode. In this mode, the Pulse Accumulator counts the number of edges appearing at the PAI pin and keeps the count in the PACNT register. Software then relates this event count to a specific time period.

This section describes gated time accumulation mode. In this mode, the Pulse Accumulator records the time between specified input edges appearing at the PAI pin. The Pulse Accumulator measures this time as a count that increments at a thirty-two microsecond rate.[11] The Pulse Accumulator keeps the count in its PACNT register. Exercise 9.7 will give the reader an opportunity to apply gated time accumulation to a practical problem—measurement of percent duty cycle.

On-Chip Resources Used by the Pulse Accumulator in Gated Time Accumulation Mode

For gated time accumulation, the M68HC11 uses the same resources as event counting. Following paragraphs list these resources and describe how the M68HC11 uses them.

- Pulse Accumulator Input Pin (PAI, PORTA, pin PA7). Refer to Figure 9–12. If the M68HC11 configures its Pulse Accumulator for gated time accumulation, the Pulse Accumulator/General-Purpose Input Latch (1) will latch a specified level appearing at the PAI pin. As long as the Pulse Accumulator/General-Purpose Input Latch retains this specified level, gated time accumulation continues. When Input Buffer/Edge Detector 2 detects this specified level at the Pulse Accumulator/General-Purpose Input Latch, it outputs a binary one (23) to AND gate 21. E/64 Clock (11) constitutes the second input to AND gate 21. Having been enabled by the binary one (23), AND gate 21 couples the E/64 Clock (22) to the Multiplexer (4). The Multiplexer couples the E/64 Clock to the PACNT counter/register (5). Thus, for as long as the PAI pin is held at a specified active level, PACNT increments at an E/64 rate, giving the PACNT a resolution of thirty-two microseconds and a range of 8.19 milliseconds—assuming a 2 MHz E clock rate.
- Pulse Accumulator Count Register (PACNT, control register block address $1027). Refer to Figure 9–12. As long as the Pulse Accumulator/General-Purpose Input Latch continues to detect a specified active level at the PAI pin, PACNT register 5 increments at a thirty-two microsecond rate. Note that the M68HC11 can read or write (6) to PACNT at any time. Thus, the CPU can initialize PACNT to any 8-bit value or read the PACNT's current contents. In most gated time accumulation applications, the CPU initializes PACNT to $00 and reads PACNT when an external device drives the PAI pin from its specified active to inactive levels. When its contents roll over from $FF to $00, PACNT sends an overflow signal (7) to the TFLG2 register (8). This overflow signal sets a status flag in TFLG2 and can generate an interrupt. The M68HC11 can use these overflow flags/interrupts to extend PACNT's range by one or more bytes. PACNT can increment its contents only if enabled by PAEN = 1 (9). Thus, the PACNT register is readable and writable, increments its contents at a thirty-two microsecond rate, and notifies TFLG2 each time its count rolls over.
- Pulse Accumulator Control Register (PACTL, control register block address $1026). Refer to Figure 9–13. PACTL bits 4 through 7 control several aspects of the Pulse Accumulator operation in gated time accumulation mode.

    PACTL's DDRA7 bit configures PORTA pin PA7 (PAI) as either an input or an output. If DDRA7 = 1, the Pulse Accumulator/General-Purpose Input Latch can capture general-purpose or OC1 output levels, causing time accumulation by the PACNT register. To time the duration of external events, the programmer must reset DDRA7.

    PACTL's PAEN bit enables the Pulse Accumulator itself. Refer to Figure 9–12. PAEN = 1 (9) causes the PACNT to increment at the E/64 (thirty-two microsecond) rate. PAEN = 1 also enables the TFLG2 register to set its Pulse Accumulator-oriented status flags when conditions dictate. PAEN = 0 disables PACNT counting and TFLG2 status bit setting.

    PACTL's PEDGE bit determines whether a high or low level at the PAI pin initiates time accumulation by the PACNT register. Refer to Figure 9–12. The PEDGE bit (12) controls the Input Buffer/Edge Detector (2). If PEDGE = 0, high levels at the PAI pin generate a binary one (23) to enable AND gate 21. If PEDGE = 1, low levels at this pin enable AND gate 21.

    At the end of the event being timed, an external device drives the PAI pin from its active to inactive level. By driving the PAI pin from its active to inactive level, the external device generates the trailing edge of the pulse being measured. This trailing edge generates a clock pulse (3) which is sent to the Multiplexer (4) and to the TFLG2 register (8).

---

[11]This count rate assumes a 2 MHz E clock frequency.

Recall that PAMOD = 1 causes the Multiplexer to couple only E/64 clocks (22). Therefore clock pulse (3) is not coupled to PACNT. However, clock (3) does cause the TFLG2 register to set its edge detect status bit, and possibly generate an interrupt.

Thus, the PEDGE bit answers two questions:

1. Which PAI pin level causes E/64 counting by the PACNT register?
2. Which input edge—positive-going or negative-going—causes the TFLG2 register to set its input edge detect flag, thus indicating the end of the event?

PEDGE = 0 answers these questions as follows:

1. A high level at the PAI pin causes PACNT time accumulation.
2. A falling edge ends counting and causes the TFLG2 register to set its input edge detect flag.

PEDGE = 1 answers the questions as follows:

1. A low level at the PAI pin causes PACNT time accumulation.
2. A rising edge ends counting and causes the TFLG2 register to set its input edge detect flag.

- Miscellaneous Timer Interrupt Flag Register 2 (TFLG2, control register block address $1025). Refer to Figure 9–14. TFLG2 register's PAOVF and PAIF bits provide status information to the CPU. Refer again to Figure 9–12. When its count rolls over from $FF to $00, the PACNT register (5) sends an overflow indication (7) to the TFLG2 register (8). This signal sets the PAOVF bit in TFLG2. When an external device drives the PAI pin from its active to inactive levels, Input Buffer/Edge Detector 2 sends an edge detect clock (3) to TFLG2. This edge detect clock sets the PAIF bit in TFLG2. The PAOVF and PAIF bits can also generate interrupts if these interrupts are enabled by the TMSK2 register.
- Miscellaneous Timer Interrupt Mask Register 2 (TMSK2, control register block address $1024). Refer to Figure 9–15. TMSK2 register's PAII and PAOVI bits enable interrupts associated with the PAIF and PAOVF flags. The PAOVI bit enables interrupts that the Pulse Accumulator generates when its PACNT register contents roll over. Refer to Figure 9–12. PACNT signals TFLG2 (7) when the PACNT rolls over from $FF to $00. This signal sets the PAOVF flag in TFLG2 (8). The Pulse Accumulator ANDs (13) the PAOVF state (14) with TMSK2's PAOVI bit (15). If PAOVI = 1 AND PAOVF = 1, the Pulse Accumulator generates a PA Overflow Interrupt (16).

Refer to Figure 9–15. TMSK2's PAII bit enables an interrupt that the Pulse Accumulator generates when it detects the trailing edge of a pulse being measured. Refer to Figure 9–12. Edge Detector 2 responds to this specified edge by sending clock pulse 3 to TFLG2 (8). This clock pulse sets the PAIF bit in TFLG2. The Pulse Accumulator ANDs (17) the state of the PAIF bit (18) with TMSK2's PAII bit (19). If both PAIF AND PAII are set, the AND gate generates a PA Input Interrupt (20).

This section explained on-chip resources used by the M68HC11 Pulse Accumulator when it operates in gated time accumulation mode. An external device (or internally generated write when DDRA7 = 1) drives the PAI pin to an active level. The Pulse Accumulator's Input Buffer/Edge Detector detects this active level and routes E/64 clocks to the PACNT register. Until the external device drives the PAI pin from its active to inactive level, PACNT increments at the E/64 (32µs) rate. When the Input Buffer/Edge Detector detects the active to inactive edge, it stops the E/64 clock and outputs a clock pulse that sets the PAIF bit in TFLG2. PAIF = 1 also generates a PA Input Interrupt if TMSK2's PAII bit = 1. If the PACNT rolls over, the PAOVF bit sets in the TFLG2 register. PAOVF = 1 generates a PA Overflow Interrupt if PAOVI = 1 in TMSK2.

EXERCISE 9.6 — Initialize the M68HC11 for Gated Time Accumulation Mode

AIM: Derive BUFFALO one-line assembly source code that configures the Pulse Accumulator for gated time accumulation mode.

INTRODUCTORY INFORMATION: This exercise teaches you how to develop source code that configures the Pulse Accumulator for gated time accumulation. This source code initializes Pulse Accumulator-related control and data registers along with interrupt pseudovectors.

INSTRUCTIONS: Refer to pages 454–455 and the specifications provided while developing appropriate BUFFALO one-line assembly source code. By executing these instructions, the CPU initializes appropriate interrupt pseudovectors and initializes the PACNT, PACTL, and TMSK2 registers for gated time accumulation mode. Specifications may also call for source code that clears the global I mask to enable specified interrupts. Enter your source code in the blanks accompanying each set of specifications. In developing source code, assume that the X register contains $1000. Use indexed mode for all writes to control or data registers. Use direct mode for pseudovector writes. The Pulse Accumulator Overflow Interrupt pseudovector resides at address $00CD. The Pulse Accumulator Input Edge Interrupt pseudovector resides at address $00CA.

EXAMPLE

Specifications
Gated time accumulation mode
Measure an external event
Enable Pulse Accumulator
Active input level is low
Enable Input Edge interrupts
Interrupt service routine at address $0140
Clear PACNT
Clear I mask

$ADDRESS	SOURCE CODE	COMMENTS
0100	LDD #140	Initialize input edge interrupt pseudovector
0103	STD CB	" " " " " "
0105	LDAA #10	Initialize TMSK2
0107	STAA 24,X	" "
0109	LDD #7000	Initialize PACTL/PACNT
010C	STD 26,X	" " "
010E	CLI	Clear I mask

Specifications
Gated time accumulation mode
Measure an external event
Enable Pulse Accumulator
Trailing edge is negative-going
Enable Overflow interrupts
Interrupt service routine at address $C050
Clear PACNT

$ADDRESS	SOURCE CODE	COMMENTS
C000	LDD #C050	Initialize overflow interrupt pseudovector
C003	STD CD	" " " " "
C005	LDAA #20	Initialize TMSK2
C007	STAA 24,X	" "
C009	LDD #6000	Initialize PACTL/PACNT
C00C	STD 26,X	" " "

Specifications
Gated time accumulation mode
Measure an internally generated event
Enable Pulse Accumulator
Active level is high
Enable both interrupts
Input Edge Interrupt Service routine at address $0125
Overflow Interrupt Service routine at address $0157
Initialize PACNT to $CA
Clear I mask

$ADDRESS	SOURCE CODE	COMMENTS
010A		

Specifications
Gated time accumulation mode
Measure an external event
Enable Pulse Accumulator
Trailing edge is positive-going
Enable overflow interrupts
Interrupt service routine at address $B710
Initialize PACNT to $C8
Clear I mask

$ADDRESS	SOURCE CODE	COMMENTS
0140		

Specifications
Gated time accumulation mode
Measure an external event
Enable Pulse Accumulator
Active level is high
Clear PACNT
Clear I mask

$ADDRESS	SOURCE CODE	COMMENTS
C030		

Use Gated Time Accumulation to Measure Percent Duty Cycle

Gated time accumulation mode is ideally suited to pulse width-related measurements. Therefore, a programmer can easily and efficiently apply this mode to duty cycle measurements. This section presents software that executes two tasks:

1. Generate a 1 KHz square wave with programmable duty cycle.
2. Continuously measure and record the square wave's duty cycle. Figures 9–30 and 9–31, along with their accompanying program lists, illustrate the square wave generation/duty cycle measurement software.

Figure 9–30 flowcharts the software's MAIN program. MAIN initializes an interrupt pseudovector and control registers for both square wave generation and measurement. After its initialization phase, MAIN enters a continuous square wave generation loop.

Figure 9–31 flowcharts the Pulse Accumulator Input Edge Interrupt Service routine. Upon each falling edge of the square wave, the CPU vectors to this routine where it measures duty cycle. After measuring duty cycle, the CPU clears its PACNT register and PAIF flag before returning to the 1 KHz square wave generation loop in the MAIN program. Following paragraphs describe Figures 9–30 and 9–31 in more detail.

Refer to Figure 9–30 and its accompanying software list. In block 1 (program addresses $0100–$0104), the CPU initializes its Pulse Accumulator Input Edge Detect Interrupt pseudovector with the interrupt service routine's address. The CPU will vector to this interrupt service routine (Figure 9–31) each time it generates a square wave falling edge.

After initializing its X register (block 2), the CPU writes $02 to the TCTL1 register (block 3, program addresses $0108–$010B). The M68HC11 will output the 1 KHz square wave from its Output Compare 5 pin (PORTA pin 3). The CPU writes a binary one to TCTL1's OM5 bit and a zero to the OL5 bit. This write causes OC5 to generate a falling edge upon each successful OC5 compare.

In block 4 (program addresses $010C–$0110), the CPU configures its Output Compare mechanism to output rising edges from pin OC5 upon successful OC1 compares. To configure OC1, the CPU writes binary ones to the OC1M3 and OC1D3 bits of its OC1M and OC1D registers.

The CPU enables Pulse Accumulator Input Edge Detect Interrupts by writing binary one to TMSK2 register's PAII bit (block 5, program addresses $0111–$0114). The CPU then configures its Pulse Accumulator for duty cycle measurements (block 6, program addresses $0115–$0119). By writing $60 to its PACTL register, the CPU sets the PAEN and PAMOD bits and clears the DDRA7 and PEDGE bits. PAEN = 1 enables the Pulse Accumulator. PAMOD = 1 configures the Pulse Accumulator for gated time accumulation. DDRA7 = 0 configures the PAI pin as an input. The user shall connect the output square wave from PORTA pin 3 to the PAI pin (PORTA pin 7). PEDGE = 0 makes the PAI pin active high and causes the PAIF flag to set upon falling edges. Since the CPU set the PAII bit in TMSK2 (block 5), PAIF = 1 will generate an input edge interrupt. The CPU writes $00 to its PACNT register to initialize it for time accumulation.

The reader should find Figure 9–30, blocks 7 through 17 quite familiar. In these blocks, the CPU initializes its TOC1 and TOC5 registers, then generates a 1 KHz square wave

FIGURE 9-30

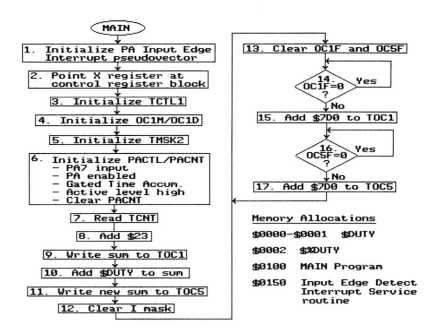

MAIN Program

$ADDRESS	SOURCE CODE	COMMENTS
0100	LDD #150	Initialize input edge pseudovector
0103	STD CB	" " " " "
0105	LDX #1000	Initialize X register
0108	LDAA #2	Initialize TCTL1
010A	STAA 20,X	" " "
010C	LDD #808	Initialize OC1M and OC1D
010F	STD C,X	" " " " "
0111	LDAA #10	Initialize TMSK2
0113	STAA 24,X	" " "
0115	LDD #6000	Initialize PACTL and PACNT
0118	STAA 26,X	" " " " "
011A	LDD E,X	Read TCNT
011C	ADDD #23	Add $23
011F	STD 16,X	Write sum to TOC1
0121	ADDD 0	Add $DUTY to sum
0123	STD 1E,X	Write new sum to TOC5
0125	CLI	Clear I mask
0126	BCLR 23,X 77	Clear OC1F and OC5F
0129	BRCLR 23,X 80 129	Branch to self until OC1F = 1
012D	LDD 16,X	Add $7D0 to TOC1
012F	ADDD #7D0	" " "
0132	STD 16,X	" " "
0134	BRCLR 23,X 8 134	Branch to self until OC5F = 1.
0138	LDD 1E,X	Add $7D0 to TOC5
013A	ADDD #7D0	" " "
013D	STD 1E,X	" " "
013F	BRA 126	Make more edges

continuously. In blocks 7, 8, and 9, the CPU initializes its TOC1 register so that an initial OC1 compare and output rising edge occur while the CPU executes block 14. In blocks 10 and 11, the CPU initializes TOC5 so that a successful OC5 compare occurs $DUTY's worth of free-running counts after the OC1 compare. After clearing the global I mask

FIGURE 9–31

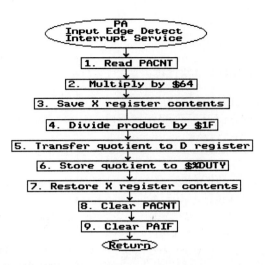

Edge Detect Interrupt Service Routine

$ADDRESS	SOURCE CODE	COMMENTS
0150	LDAA 27,X	Read PACNT
0152	LDAB #64	Multiply PACNT by $64
0154	MUL	" " " "
0155	PSHX	Save X register contents
0156	LDX #1F	Divide product by $1F
0159	IDIV	" " " "
015A	XGDX	Transfer quotient to D register
015B	STAB 2	Store quotient to $%DUTY
015D	PULX	Restore X register contents
015E	CLR 27,X	Clear PACNT
0160	BCLR 25,X EF	Clear PAIF
0163	RTI	Return

(block 12), the CPU enters a continuous loop where it waits for successful output compares and updates its TOC1 and TOC5 registers (blocks 13 through 17). Remember that successful OC5 compares generate output falling edges. These falling edges also generate Pulse Accumulator Input Edge Detect Interrupts.

Refer to Figure 9–31 and its accompanying software list. The output square wave (from PORTA pin 3) connects to the Pulse Accumulator's PAI pin (PORTA pin 7). Output falling edges trigger PA Input Edge Interrupts. The CPU vectors to the Figure 9–31 routine. In this routine, the CPU measures the hexadecimal percent duty cycle ($%DUTY) of the square wave. A rising edge (OC1 compare) toggles the PAI pin to its active level. This active level causes the PACNT register to increment at a thirty-two microsecond rate. A falling edge causes PACNT to stop and hold. Upon vectoring to the interrupt routine, the CPU reads PACNT (block 1) and uses this read value to compute $%DUTY (blocks 2 through 5).

The square wave's period (T) equals one millisecond. To calculate decimal percent duty cycle, divide the pulse width by the period and multiply by 100:

$$\text{percent duty cycle} = PW/T \times 100$$
$$= PW/1ms \times 100$$

When the CPU reads the PACNT, this read value equals the pulse width in thirty-two microsecond units. To arrive at a valid duty cycle figure, the waveform's period must also be expressed in thirty-two microsecond units. A one-millisecond period consists of thirty-one such units. Therefore,

$$\text{percent duty cycle} = PACNT/31 \times 100$$

In hexadecimal,

$$\$\%DUTY = \$PACNT/\$1F \times \$64$$

To avoid floating point complications, modify this equation to

$$\$\%DUTY = (\$PACNT \times \$64)/\$1F$$

In block 1 (program addresses $0150–$0151), the CPU reads PACNT. To make a $%DUTY calculation, the CPU multiplies the PACNT read value by $64 (block 2, program addresses $0152–$0154). To complete its $%DUTY calculation, the CPU must use the IDIV instruction. IDIV entails use of the X register. Since the X register normally points at the control register block, the CPU saves its contents on the stack (block 3, program address $0155). In block 4 (program addresses $0156–$0159), the CPU divides the block 2 product by $1F—the number of thirty-two microsecond increments in the 1 KHz square wave's period. After execution, the quotient resides in the X register. Therefore, in block 5 (program address $015A), the CPU exchanges the D register's contents with the X register's contents. Since percent duty cycle cannot exceed 100 ($0064), the $%DUTY value now resides in ACCB. The CPU writes this $%DUTY value from ACCB to its assigned RAM location (block 6, program addresses $015B–$015C).

After completing a duty cycle measurement, the CPU pulls its saved X register contents from the stack (block 7, program address $015D). Before returning from interrupt, the CPU completes two tasks. First, it clears PACNT's contents for the next pulse width measurement (block 8, program addresses $015E–$015F). Second, the CPU clears the PAIF bit in its TFLG2 register to prevent a spurious interrupt upon its return (block 9, program addresses $0160–$0162).

Exercise 9.7 gives the reader an opportunity to load, execute, and evaluate this software.

EXERCISE 9.7 Measure Percent Duty Cycle

REQUIRED EQUIPMENT: Motorola M68HC11 EVBU (EVB) and terminal
Oscilloscope and probe

GENERAL INSTRUCTIONS: Make an OC5 to PAI connection as shown in Figure 9–32. Load, demonstrate, and analyze software presented in the section above. This software generates a 1 KHz square wave and measures its hexadecimal percent duty cycle. Specify a particular duty cycle value by initializing RAM locations $0000–$0001 ($DUTY) with a number between $0021 and $07AF. $DUTY represents the number of free-running counts in the square wave's pulse width. A $DUTY range of $0021–$07AF generates duty cycle values from two to ninety-eight percent. The software continuously measures hexadecimal percent duty cycle ($%DUTY) and stores it to RAM location $0002. To access this measured $%DUTY value, reset the EVBU (EVB), then use the BUFFALO MD command to display the contents of location $0002.

PART I: Develop Square Wave Generation/ Duty Cycle Measurement Software

Review the previous section to assure you understand this exercise's software strategy. Refer to the software lists accompanying Figures 9–30 and 9–31. If you use an EVBU, load the software as listed. If you use an EVB, allocate memory as follows:

EVB Memory Allocations

$ADDRESS	CONTENTS
0000–0001	$DUTY
0002	$%DUTY
C000	MAIN program
C050	Input Edge Detect Interrupt Service routine

FIGURE 9-32

If you use an EVB, assure that the input edge pseudovector and all branches reflect the memory allocations shown above. Connect the oscilloscope to display the output square wave. Power up the EVBU (EVB). Load and debug your software. When confident that the software runs properly, demonstrate it to the instructor. Use a $DUTY value of $0640 for your demonstration.

_____ *Instructor Check*

PART II: Laboratory Analysis

Evaluate the M68HC11's ability to measure duty cycle accurately over the range of square wave output frequencies. Refer to Chart 9-3. Use the following procedure to complete this chart:

1. Convert the column 1 &DUTY value to hexadecimal and enter in Column 2 ($DUTY). Load this $DUTY value into RAM addresses $0000–$0001.
2. Start program execution. Use an oscilloscope to measure the output square wave's percent duty cycle. Round this measured decimal value to integers and enter in column 5.
3. Reset the M68HC11. Use the BUFFALO MD command to display the contents of RAM location $0002 ($%DUTY). Record this hexadecimal value in column 3.
4. Convert the column 3 hexadecimal value to decimal (&%DUTY). Enter this decimal percent duty cycle value in Column 4.

Chart 9-3

1 &DUTY	2 $DUTY	3 $%DUTY (M68HC11)	4 &%DUTY (M68HC11)	5 &%DUTY (Scope)	6 % Meas. Error
0100	___	___	___	___	___
0200	___	___	___	___	___
0300	___	___	___	___	___
0400	___	___	___	___	___
0500	___	___	___	___	___
0600	___	___	___	___	___
0700	___	___	___	___	___
0800	___	___	___	___	___
0900	___	___	___	___	___
1000	___	___	___	___	___
1100	___	___	___	___	___
1200	___	___	___	___	___
1300	___	___	___	___	___
1400	___	___	___	___	___
1500	___	___	___	___	___
1600	___	___	___	___	___
1700	___	___	___	___	___
1800	___	___	___	___	___

5. Calculate the Column 6 value as follows:

   Step 1. Determine which is lesser—column 5 or column 4: smaller value = numerator; larger value = denominator.

   Step 2. Divide numerator by denominator.

   Step 3. Subtract the quotient from 1.

   Step 4. Multiply the difference by 100.

   Step 5. Give the product a positive sign if column 4 is larger. Give the product a negative sign is column 5 is larger. Round the product to three significant figures and enter in column 6.

Analyze your Chart 9–3 results. Do they indicate the presence of offset error?[12] _____Yes _____ No. Explain your answer.

Analyze your Chart 9–3 results. Do they indicate the presence of linear error?[13] _____Yes _____ No. Explain your answer.

PART III: Laboratory Software Listing

MAIN PROGRAM—EVB

$ADDRESS	SOURCE CODE	COMMENTS
C000	LDD #C050	Initialize input edge pseudovector
C003	STD CB	" " " " " "
C005	LDX #1000	Initialize X register
C008	LDAA #2	Initialize TCTL1
C00A	STAA 20,X	" " "
C00C	LDD #808	Initialize OC1M and OC1D
C00F	STD C,X	" " " " "
C011	LDAA #10	Initialize TMSK2
C013	STAA 24,X	" " "
C015	LDD #6000	Initialize PACTL and PACNT
C018	STAA 26,X	" " " " "
C01A	LDD E,X	Read TCNT
C01C	ADDD #23	Add $23
C01F	STD 16,X	Write sum to TOC1
C021	ADDD 0	Add $DUTY to sum
C023	STD 1E,X	Write new sum to TOC5
C025	CLI	Clear I mask
C026	BCLR 23,X 77	Clear OC1F and OC5F
C029	BRCLR 23,X 80 C029	Branch to self until OC1F = 1
C02D	LDD 16,X	Add $7D0 to TOC1
C02F	ADDD #7D0	" " "
C032	STD 16,X	" " "
C034	BRCLR 23,X 8 C034	Branch to self until OC5F = 1
C038	LDD 1E,X	Add $7D0 to TOC5
C03A	ADDD #7D0	" " "
C03D	STD 1E,X	" " "
C03F	BRA C026	Make more edges

MAIN PROGRAM—EVBU

(See program list accompanying Figure 9–30.)

---

[12]Offset error causes the M68HC11's duty cycle measurements to be inaccurate by a constant amount over the output range.

[13]Linear error causes duty cycle measurements to be inaccurate by a constant percentage over the output range.

PA INPUT EDGE DETECT INTERRUPT SERVICE—EVB

$ADDRESS	SOURCE CODE	COMMENTS
C050	LDAA 27,X	Read PACNT
C052	LDAB #64	Multiply PACNT by $64
C054	MUL	" " " "
C055	PSHX	Save X register contents
C056	LDX #1F	Divide product by $1F
C059	IDIV	" " " "
C05A	XGDX	Transfer quotient to D register
C05B	STAB 2	Store quotient to $%DUTY
C05D	PULX	Restore X register contents
C05E	CLR 27,X	Clear PACNT
C060	BCLR 25,X EF	Clear PAIF
C063	RTI	Return

PA INPUT EDGE DETECT INTERRUPT SERVICE—EVBU

(See program list accompanying Figure 9–31.)

9.7 CHAPTER SUMMARY

This chapter presents three M68HC11 features—forced output compares, Real-Time Interrupts, and the Pulse Accumulator.

Using forced output compares, the CPU generates programmed output compare pin responses—before successful compares actually occur. To apply forced compares, the CPU uses essentially the same on-chip resources as conventional output compares. Forced output compares require only one additional register—the Timer Compare Force Register (CFORC). Figure 9–1 shows that CFORC has a counterpart FOCx bit for each output compare function. To force a particular compare, the CPU simply writes a binary one to its corresponding FOCx bit in CFORC. The free-running counter's clock synchronizes OC pin response to a forced compare. Depending upon the state of the PR1/PR0 bits in TMSK2, the OC pin may respond one, four, eight, or sixteen E clocks after the CPU writes a binary one to an FOCx bit in CFORC. As soon as a set FOCx bit forces an OC pin response, it resets. Thus, any CPU read of CFORC returns only $00. A forced compare does not cause a counterpart OCxF flag to set in the TFLG1 register. Neither can a forced output compare generate an interrupt. However, a subsequent conventional output compare sets its corresponding OCxF flag and can generate an interrupt.

Forced output compares provide a simple way to program for two- or three-speed motor control. This chapter presents an interface and software that allow the user to select from two motor speeds. The operator closes a switch to select the faster speed. This closed switch causes a pull-up circuit to drive pin PA2 low. This low causes the CPU to force OC5 output compares. OC5 compares generate falling edges of a 1 KHz motor speed control square wave. By forcing OC5 compares, the CPU reduces percent duty cycle of the square wave, causing a faster motor speed. An open switch causes the pull-up circuit to apply a high to pin PA2. This high causes the M68HC11 to generate conventional OC5 output compares. Conventional OC5 compares generate a greater percent duty cycle and a slower motor speed.

The Real-Time Interrupt mechanism allows a programmer to schedule RTI interrupts at precisely timed periodic intervals. RTI interrupts have these features:

- The CPU can mask RTI interrupts with the global I mask or with the RTII bit in the TMSK2 register.
- RTI interrupts have a unique vector address.
- The CPU can promote RTIs to highest I-maskable priority—by writing %0111 to HPRIO register bits PSEL3–PSEL0.

- The RTI mechanism can generate interrupts at a selection of four periodic rates. The CPU selects an interrupt rate by writing to the RTR1/RTR0 bits in the PACTL register. These bits allow the RTI mechanism to generate a status flag and interrupt every 4.1, 8.19, 16.38, or 32.77 milliseconds.
- TFLG2 register bit 6 constitutes the RTI status flag (RTIF). This bit sets to indicate that an RTI period has elapsed. If enabled by the RTII bit in TMSK2, the RTIF also generates an interrupt when it sets. The CPU must reset the RTIF as a part of its RTI Interrupt Service routine. Otherwise, the RTIF = 1 state will generate a spurious interrupt as soon as the CPU returns from the current interrupt service.

RTIs facilitate periodic pulse generation by the M68HC11. Pages 418–420 present a simple program that generates a one-millisecond positive pulse at a choice of four periodic rates. This program prompts the user to enter a number from 0 to 3 at the terminal keyboard. The CPU converts a user-entered value to hexadecimal and writes it to PACTL register's RTR1/RTR0 bits, establishing an RTI rate of 4.10, 8.19, 16.35, or 32.77 milliseconds. Each RTI causes the CPU to generate a one-millisecond pulse.

The Pulse Accumulator allows the M68HC11 to count either frequency or duration of external events:

- The M68HC11 can count the number of specified transitions at the Pulse Accumulator Input (PAI) pin and record this count in the Pulse Accumulator Count Register (PACNT).
- The M68HC11 can measure the time between rising and falling edges at the PAI pin—to a resolution of thirty-two microseconds—and record this time in the PACNT register.

When the Pulse Accumulator operates in event-counting mode, it counts the number of specified edges occurring at the PAI pin. The Pulse Accumulator records this number in the PACNT register. PACNT has an 8-bit capacity, which limits event count to 255. However, the Pulse Accumulator mechanism provides a PACNT overflow flag (PAOVF) in the TFLG2 register. If enabled by the PAOVI bit in TMSK2, the PAOVF flag can also generate an interrupt when it sets. The M68HC11 can therefore increment PACNT extension byte(s) to facilitate event counts of virtually any magnitude.

The Pulse Accumulator sets the PAIF status flag in TFLG2 each time it detects a specified edge at the PAI pin. If enabled by the PAII bit in TMSK2, the PAIF flag can also generate an interrupt when it sets.

While operating in event-counting mode, the Pulse Accumulator uses four registers. The PACNT register keeps count of external events. PACTL register bits enable the Pulse Accumulator, specify the edges to be counted, and select the Pulse Accumulator operating mode. PACTL register bit 7 (DDRA7) configures the PAI pin as either an input or output. TFLG2 contains the PAOVF and PAIF status flags. TMSK2 contains PAOVI and PAII interrupt enable bits.

Material on pages 429–432 presents software and hardware that combine output compares, RTIs, and event-counting mode to achieve these objectives:

- Control the speed of a DC motor.
- Measure the DC motor's speed, in revolutions per minute (tachometer)

This application uses two interfaces. The first controls motor speed via an output compare pin. The second applies optical sensor outputs to the M68HC11's PAI pin. When the DC motor runs, the sensor circuitry generates two pulses for each motor revolution.

Motor speed control software uses OC1 and OC5 status polling to generate a 1 KHz motor speed control square wave. The CPU establishes output pulse width by offsetting TOC5's contents from TOC1's contents.

This application's tachometer software relates event count to time and derives motor RPM. It uses RTIs to provide the equation's time component. The CPU sets PACTL's RTR1 and RTR0 bits. Therefore, the RTI mechanism generates an interrupt every 32.77 milliseconds. The Pulse Accumulator counts sensor output rising edges over $12 of these RTI periods (.59 seconds). The Pulse Accumulator records this count in its PACNT register.

One full minute consists of $66 such .59 second periods. The sensor circuitry generates two edges per motor revolution. Thus, the tachometer software can use this equation:

$$\$RPM = (PACNT \times \$66)/\$2$$
$$= PACNT \times \$33$$

Figures 9–21, 9–22, and 9–23 illustrate unified software that controls motor speed and measures motor RPM.

Chapter 8 and 9 motor speed control applications—through Exercise 9.4—exert open-loop control over motor speed. Open-loop control makes no attempt to control motor speed variations caused by disturbances such as motor loading, power fluctuations, or motor vibration. The Pulse Accumulator facilitates closed-loop motor speed regulation. Closed-loop control feeds back motor speed measurements to a comparator, which compensates for environmental disturbances (see Figure 9–24).

Pages 441–448 present software that closes the motor speed control loop. The M68HC11 exerts closed-loop motor speed regulation by comparing actual motor speed—as measured by the Pulse Accumulator—with the user-requested setpoint. Based upon this comparison, the M68HC11 adjusts the duty cycle of its output 1 KHz motor speed control square wave. Adjusted duty cycle compensates for disturbances.

If the CPU configures the Pulse Accumulator for gated time accumulation mode, it records time between specified input edges appearing at the PAI pin. The Pulse Accumulator measures this time as a count that increments at a thirty-two microsecond rate.[14] A transition from inactive to active levels at the PAI pin triggers counting by the PACNT register. A transition from active to inactive levels causes counting to halt.

The PACTL register controls four aspects of gated time accumulation mode. PACTL's DDRA7 bit configures the PAI pin as either an input or output. PACTL's PAEN bit enables counting by the PACNT register, and enables the TFLG2 register to set its PAIF and PAOVF flags. PACTL's PEDGE bit determines whether a high or low level at the PAI pin causes counting to begin, and whether a negative- or positive-going edge causes counting to halt.

Two TFLG2 register bits indicate Pulse Accumulator status when it operates in gated time accumulation mode. The PAOVF bit sets when PACNT rolls over from $FF to $00. The PAIF bit sets when the PAI pin is driven from its active to inactive levels. TMSK2 register bits enable the PAIF and PAOVF flags to generate corresponding interrupts. When the TMSK2 register's PAII bit is set, a PAIF = 1 state generates a PA Input Edge Interrupt. When TMSK2's PAOVI bit is set, a PAOVF = 1 state generates a PA Accumulator Overflow Interrupt.

Software presented on pages 458–461 executes two tasks. First, the software generates a 1 KHz square wave with programmable duty cycle. Second, it continuously measures and records this duty cycle. Figures 9–30 and 9–31 illustrate this software.

Figure 9–30 flowcharts the software's MAIN program. This program initializes the PA Input Edge Interrupt pseudovector and necessary control registers for both square wave generation and measurement. After this initialization phase, the MAIN program enters a continuous square wave generation loop.

Figure 9–31 flowcharts the Pulse Accumulator Input Edge Detect Interrupt Service routine. Each square wave falling edge generates this interrupt. The interrupt service routine measures duty cycle. The square wave's period (T) equals $1F counts of thirty-two microseconds each (32µs = PACNT resolution). To calculate hex percent duty cycle, the CPU effectively divides the pulse width (PACNT contents) by $1F (period) and multiplies the quotient by $64:

$$\$\%DUTY = (\$PACNT/\$1F) \times \$64$$

To avoid floating point complications, the CPU applies this modified equation:

$$\$\%DUTY = (\$PACNT \times \$64)/\$1F$$

After measuring duty cycle, the CPU clears its PACNT register and PAIF flag before returning from the interrupt to the square wave generation loop in the MAIN program.

---

[14]The count rate assumes a 2 MHz E clock frequency.

EXERCISE 9.8   Chapter Review

Answer the questions below by neatly entering your responses in the appropriate blanks.

1. List the on-chip resources used by the M68HC11 for forced output compares.

    (1)

    (2)

    (3)

    (4)

    (5)

    (6)

    (7)

    (8)

2. Of the above listed resources, which is unique to forced output compares?

3. Refer to Figure 8–21 in the previous chapter. Which input allows the CPU to generate a forced PA7 pin response prior to a conventional OC1 compare? _____

4. List the two output compare responses that a forced compare cannot generate.

    (1)

    (2)

5. A successful output compare sets a corresponding OCxF flag and can generate an interrupt, even if the CPU previously _____ a set or reset by the OCx pin.

6. To force simultaneous OC2 and OC4 compares, the CPU writes $_____ to the CFORC register.

7. Any CFORC read by the CPU returns only $_____. Explain why a CFORC read always returns this value.

8. OC pin response to a forced compare is synchronized to _____.
Depending upon the state of the _____ bits in TMSK2, an OC pin may respond _____, _____, or _____ E clocks after the CPU writes binary one to an _____ bit in the _____ register.

9. Refer to Figure 9–2. Describe the purpose of the pull-up circuit connected to pin PA2. What effects do a closed and open switch have on the output from pin OC5/PA3 and the speed of the DC motor?

10. Refer to Figure 9–3, blocks 7 and 8. Explain why the CPU initializes TOC5 with the sum of the TOC1 contents plus $4B0.

11. Refer to Figure 9–3, block 12. Explain why the CPU initializes ACCB to $4A.

12. Refer to Figure 9–3. Which block is not executed if the user selects slow motor speed? _____

13. The _____ mechanism allows for the generation of interrupts at precisely timed intervals.

14. List the two ways of masking RTI interrupts.

    (1)

    (2)

15. Describe how the programmer can promote RTIs to highest I-maskable priority.

16. To program the RTI mechanism for 16.38ms periodic interrupts, the CPU writes $_____ to address $_____.

17. To indicate that an RTI period has elapsed, bit _____ of the _____ register sets. If bit _____ of the _____ register is set, this RTI status bit also generates an interrupt.

18. Refer to Figure 9–9. This program prompts the user to enter a number from 0 through 3 at the terminal. Explain why these particular numbers were selected.

19. Refer to Figure 9–9. Explain why this program branches back from block 5 to block 1, if C+Z = 0.

20. Refer to Figure 9–10. List the tasks accomplished by this RTI service routine.

    (1)

    (2)

    (3)

    (4)

21. Refer to Figure 9–10 and its accompanying program list. Describe program modifications necessary to generate a three-millisecond pulse width.

22. The Pulse Accumulator allows the M68HC11 to count the _____ or _____ of external events. The Pulse Accumulator uses _____ mode to measure the frequency of events and _____ mode to measure their duration.

23. When operating in event-counting mode, the Pulse Accumulator detects and counts specified edges occurring at its _____ pin. The Pulse Accumulator keeps this edge count in the _____ register. Each specified edge causes the _____ flag to set in the _____ register. This flag also generates an interrupt if the _____ bit is set in the _____ register.

24. The CPU establishes Pulse Accumulator operating parameters by setting and clearing control bits in the _____ register.

25. To count external events, enable the Pulse Accumulator, and count rising edges, the CPU must write $_____ to address $_____.

26. Refer to Figure 9–19. Does this software use interrupt control or status polling? _____ In which flowchart blocks does the software establish output pulse width? _____ In which block is the output rising edge generated? _____ In which block is the falling edge generated? _____ In which block does the CPU update the proper TOCx register for the next rising edge? _____ In which block does the CPU update the proper TOCx register for the next falling edge? _____

27. Refer to Figure 9–20. This software uses the equation:

$$\$RPM = \$PACNT \times \$33$$

Explain why this equation uses $33 as its multiplier.

28. Refer to Figure 9–21. List the tasks accomplished by this software.

    (1)

    (2)

    (3)

29. Refer to Figure 9–22, block 1. Explain why the CPU initializes RTICNT to $12.

30. Refer to Figure 9–23. List the tasks accomplished by this routine.

    (1)

    (2)

    (3)

    (4)

31. Describe the major limitation of motor speed control software presented in Chapters 8 and 9 through Exercise 9.4.

32. Explain how the software presented on pages 441–448 overcomes the limitation described in question 31.

33. Refer to Figure 9–25, block 1. Explain how the CPU calculates $PACNTRQ.

34. Refer to Figure 9–25, block 6. Explain why the CPU compares PACNT with $PACNTRQ.

35. Refer to Figure 9–25, block 7. Explain why the CPU branches to block 11 if PACNT = $PACNTRQ.

36. Refer to Figure 9–25, block 10. Explain why the CPU executes this block.

37. Refer to Figure 9–25, block 9. Why does the CPU execute this block?

38. Refer to Figure 9–26. List the three tasks accomplished by this software.

    (1)

    (2)

    (3)

39. Refer to Figure 9–26, block 2. What is the largest acceptable quotient resulting from IDIV execution? $_____. After IDIV execution, the quotient resides in the _____ (LS/MS) byte of the _____ register. The CPU executes the _____ command to transfer this quotient to the _____ (LS/MS) byte of the _____ register. In block 3, the CPU writes this quotient from _____ to the _____ location.

40. Refer to Figure 9–26, blocks 6 through 12. Describe the task accomplished by the CPU when it executes these blocks.

41. Refer to Figure 9–27. Summarize the tasks accomplished by this subroutine.

42. Refer to Figures 9–28 and 9–29. List the tasks accomplished by the RTI Interrupt Service routine.

    (1)

(2)

(3)

(4)

Which of these tasks does the CPU execute if RTICNT does not equal zero? _____ Which tasks does the CPU execute if RTICNT = 0? _____

43. Refer to Figure 9–29. List all blocks executed by the CPU in this figure if PACNT = $PACNTRQ. _____ List all blocks executed by the CPU if PACNT > $PACNTRQ. _____ List all blocks executed by the CPU if PACNT < $PACNTRQ. _____

44. Refer to Figure 9–29, block 8. Explain why the CPU negates the difference generated in block 5.

45. Describe the difference between event-counting mode and gated time accumulation mode.

46. In gated time accumulation mode, PACNT's range equals _____ ms and its resolution equals _____ s. PACNT increments its contents only if enabled by _____. PACNT notifies the _____ register each time its count _____.

47. Explain the functions of PACTL register's DDRA7, PAEN, and PEDGE bits when its PAMOD bit = 1.

    DDRA7:

    PAEN:

    PEDGE:

48. List M68HC11 control/status bits and their states (set or reset) as required to generate a PA Overflow Interrupt.

    (1)

    (2)

    (3)

    (4)

49. List M68HC11 control/status bits and their states (set or reset) as required to generate a PA Accumulator Input Edge Interrupt.

    (1)

    (2)

(3)

(4)

50. To measure the duration of an external event, enable the Pulse Accumulator, and establish PAI pin's active level as low, the CPU must write $_____$ to address $_____$.

51. List the tasks accomplished by the Figure 9–30 software.

   (1)

   (2)

   (3)

52. To calculate percent duty cycle, divide the _____ by the _____ and multiply by decimal _____. Figure 9–31's units of measurement are _____μs counts. In the blank, enter the equation used by the CPU to calculate $%DUTY.

   In this equation, $PACNT represents the output waveform's _____. The $1F value represents the output waveform's _____. The value $64 value represents _____.

10 M68HC11 Analog-to-Digital Conversions and Fuzzy Inference

10.1 GOALS AND OBJECTIVES

This chapter presents the M68HC11's analog-to-digital converter and fuzzy logic. Of all microcontroller features presented in this book, the A/D converter assumes prime importance. Motorola conceived the M68HC11 as a device that reads data about its environment and generates an appropriate response. In most applications, the M68HC11 cannot read data directly from its environment. A system designer must use sensors or transducers to translate analog environmental changes into electrical signals. Such translation usually requires electrical signal conditioning. That is, the system designer must include interface circuitry which matches transducer or sensor outputs to microcontroller inputs. The most prevalent signal conditioning device is the analog-to-digital converter (ADC), because very few sensors or transducers can generate binary data. By integrating A/D conversion into the microcontroller chip itself, Motorola makes the system designer's job much easier. Moreover, integrated A/D conversion makes applications less expensive, more efficient, more reliable, and more compact. The reader shall find the M68HC11's A/D converter to be ingenious, flexible, and easy to program.

This chapter's first goal is to help the reader understand how to connect, configure, and program the M68HC11's analog-to-digital converter. By carefully studying Section 10.2 and completing Exercise 10.1, the reader should achieve these objectives:

- Connect transducers and develop supporting software that cause the M68HC11 to make:
  - Four single-channel A/D conversions and stop.
  - Four multiple-channel A/D conversions and stop.
  - Continuous single-channel A/D conversions.
  - Continuous four-channel A/D conversions.
- Measure the M68HC11 A/D converter's resolution and compare these measured values with Motorola specifications.
- Use the M68HC11's A/D converter to analyze the performance of photoresistor and thermistor circuits.

This chapter's second part applies A/D conversion to a fuzzy logic control application. Fuzzy inference provides a new and potent tool in process control. Fuzzy's perspective differs from conventional Boolean approaches to control software. Boolean analysis assumes every input value to be a member of a category or classification on an "all or none" basis. The question "Is the glass half-full *or* half-empty?" typifies the Boolean approach. From the Boolean perspective, the glass cannot be *both* "half-full" *and* "half-empty." But, fuzzy perspective can recognize that the glass is simultaneously "half-full" *and* "half-empty." Fuzzy inference analyzes environmental conditions in qualitative and human terms. For example, fuzzy logic can recognize that a temperature of 75°F is partly "hot" and partly "warm" simultaneously. This contrasts with the standard Boolean approach, which must classify 75°F as 100% "hot" or 100% "warm." Remarkably, this more human approach generates simpler, less expensive, and more sophisticated control in many applications.

The reader shall discover that fuzzy inference can be a thoroughly digital approach. Thus, digital fuzzy inference requires A/D conversion. The reader shall also discover that fuzzy cannot completely replace conventional control programming. However, fuzzy can replace complex, time consuming, and memory-intensive mathematical algorithms that constrain many digital control applications.

Fuzzy logic is different, and its concepts can seem a bit intimidating. This chapter uses a simple control problem—varying motor speed with changes in environmental temperature and light—to achieve a second goal, which is to help the reader understand how to implement fuzzy control. By carefully studying Sections 10.3 and 10.4, and completing Exercise 9.2, the reader should achieve these objectives:

> Connect an interface and develop software that uses fuzzy inference to vary DC motor speed in response to changes in environmental temperature and light.
> Change DC motor speed controller performance characteristics by modifying fuzzy rules and input membership functions.

Use this chapter to gain vital insight into the practical elements of A-to-D conversion and fuzzy inference.

10.2 THE M68HC11 A/D CONVERTER

The M68HC11's analog-to-digital converter (ADC) scales analog inputs to 8-bit binary equivalents. The ADC scales these analog inputs with respect to two reference voltages: V_{REFH} and V_{REFL}. When possible, the system designer should connect V_{REFH} to the M68HC11's V_{DD} supply voltage and V_{REFL} to ground. If V_{REFH} = +5VDC and V_{REFL} = 0V, the M68HC11's ADC can scale input analog voltages ranging from 0V to +5V. How accurately can it translate these analog values? Given its 8-bit resolution, the ADC scales input analog values into 256 discrete 8-bit binary equivalents. A real-world analog input can vary to an infinite number of levels between 0V and 5V. Since the ADC's output can vary to only 256 levels ($00–$FF), it has a limited ability to recognize changes in the analog input. The reader can calculate the smallest recognizable input change (A/D step size) as follows:

$$\Delta V = V_{REFH} 2^{-n}$$

where

$$\Delta V = \text{A/D step size}$$
$$n = \text{number of bits}$$

Therefore, given the user's connections as described above:

$$\Delta V = V_{REFH} 2^{-n}$$
$$= (5V)2^{-8}$$
$$= 19.531 mV$$

The smallest input analog voltage change that the ADC can recognize is 19.5mV. Following sections describe the M68HC11's ADC in detail. This description begins with a discussion of its operating modes.

A/D Converter Operating Modes

The ADC incorporates eight analog input pins, labeled AN0 through AN7. The M68HC11 offers reasonable flexibility in the way the user can configure these pins for A/D scaling. The microcontroller can scale analog inputs appearing on only one of these analog input pins, or on four of them. A/D control register bits allow the user to select any of the eight analog input pins as a single input. If the user wishes to scale four inputs at a time, then either group AN0–AN3 or group AN4–AN7 must be used.

The M68HC11's ADC makes four conversions at a time, regardless of its input configuration. However, the user can program the ADC to make four conversions and stop, or to make continuous conversions in groups of four. If the ADC makes four conversions and stops, it sets a status flag after the fourth conversion. If the ADC makes continuous conversions, it sets the status flag after each group of four conversions.

Given these options, the user can operate the M68HC11's ADC in any of four operating modes:

1. Scale analog inputs appearing at a single analog input (ANx) pin. Make four conversions, set the status flag, and stop.
2. Scale analog inputs appearing at either pin group AN0–AN3, or group AN4–AN7. Make four conversions, one per pin, set the status flag, and stop.
3. Continuously scale analog inputs appearing at a single ANx pin. After each group of four conversions, set the status flag.
4. Scale analog inputs at either group AN0–AN3 or AN4–AN7. Scale continuously in a round-robin fashion. Set the status flag after each round of four conversions.

Keeping these four operating modes in mind, examine on-chip resources used by the M68HC11 in making analog-to-digital conversions.

On-Chip Resources Used by the M68HC11 in Making A/D Conversions

To convert analog-to-digital, the M68HC11 uses ten pins, two control/status registers, and four result registers. This limited array of resources makes A/D configuration and operation simple.

The ADC requires two scaling reference voltages, V_{REFH} and V_{REFL}. The M68HC11 uses two input pins to accommodate these reference levels. Both M68HC11A1 (EVB) and M68HC11E9 (EVBU) use pin 52 for V_{REFH} and pin 51 for V_{REFL}.[1]

Usually, system designers condition transducer or sensor output voltages to a 0V–5V range. Given this range, the system designer can connect the ADC's V_{REF} pins to the same V_{DD} and V_{SS} sources used by the microcontroller itself (+5VDC and 0V, respectively). These reference voltages make the ADC relatively sensitive (step size = 19.5mV as shown above). Power supply noise may cause undesirable variations in V_{REFH}, interfering with accurate A/D scaling. In such a case, Motorola recommends a low-pass filter network connected between the V_{REFH} and V_{REFL} pins, as shown in Figure 10–1.[2]

Of course, a designer can also specify a separate low-noise power supply for the V_{REFH} connection. Motorola specifies that V_{REFH} should not exceed V_{DD} plus 0.1V, and that V_{REFL} should not be less than V_{SS} minus 0.1V. The minimum acceptable potential difference between V_{REFH} and V_{REFL} is 3V.[3]

Thus, the M68HC11's ADC scales analog inputs with respect to its V_{REFH} and V_{REFL} pin voltages. Commonly, the user connects these reference voltage pins to the M68HC11's V_{DD} and V_{SS} supply voltages. For accurate scaling, the user should establish at least a 3V difference between V_{REFH} and V_{REFL}.

FIGURE 10–1
Reprinted with the permission of Motorola.

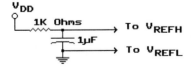

---

[1]The EVBU connects pins 52 and 51 to board V_{DD} and ground, and therefore requires no external reference connections. However, the EVB user must make external reference voltage connections.
[2]Motorola Inc., *M68HC11 Reference Manual* (1990), p. 2-18. The EVBU incorporates this filter; the EVB does not.
[3]Motorola Inc., *MC68HC11E9, HCMOS Single-Chip Microcontroller, Advance Information (MC68HC11E9/D)* (Phoenix, AZ: Motorola Inc, 1988), p. 11-13.

FIGURE 10–2
Reprinted with the permission of Motorola.

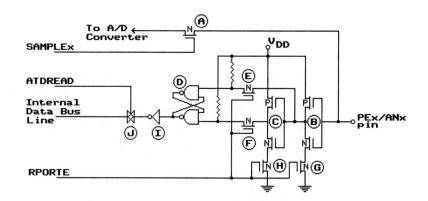

The ADC takes its analog inputs from eight PORTE pins (PE0/AN0–PE7/AN7). These pins can function as either general-purpose input pins or ADC analog input pins. Figure 10–2 diagrams typical PORTE pin logic.[4]

Refer to Figure 10–2. When the microcontroller uses a PORTE pin for A/D purposes, it enables N-channel coupling transistor A. The microcontroller enables this transistor by asserting the SAMPLEx control input for the first twelve clock cycles of a conversion. The ADC requires twelve cycles to properly sample the analog input voltage. After the sampling period, the microcontroller negates SAMPLEx, turning off coupling transistor A. The remaining circuitry shown in Figure 10–2 accommodates general-purpose digital inputs.

P- and N-channel pairs B and C function as inverters. These inverter pairs send their outputs to S-R latch D. Transistor pair B inverts the binary input from the PEx/ANx pin, and sends this inverted value two ways: (1) to inverter C, and (2) via coupling transistor E to latch D. Inverter C re-inverts the binary input and sends it to the latch via coupling transistor F. Inverter pairs B and C operate only if enabled by transistors G and H. Coupling transistors E and F, along with enabling transistors G and H, allow latch D to capture the binary level at pin PEx/ANx. These transistors enable latching when RPORTE assertion turns them on. The microcontroller asserts RPORTE for one fourth of an E clock cycle, when it reads PORTE.

Transistors E, F, G, and H isolate latch D from intermediate analog levels at the PEx/ANx pin. Without such isolation, intermediate analog pin levels can turn on all transistors in inverters B and C, causing large current flows between V_{DD} and ground through these turned-on transistors. Such large current flows disturb the analog signal appearing at the PEx/ANx pin. Even with transistors E, F, G, and H in place, the M68HC11 CPU could inadvertently read PORTE during the twelve-cycle A/D sampling period. Such a read would cause large current flows through inverters B and C. However, RPORTE limits this occurrence to a fourth of an E clock cycle. Obviously, programmers must avoid PORTE reads during A/D sampling periods. Such inappropriate reads would disturb analog levels applied to the PE/AN pins and adversely affect A/D scaling accuracy.

Latch D applies its output to an internal data bus line via inverter gate I and transmission gate J. Gate I inverts the latch output to match the binary level applied to the PEx/ANx pin. The microcontroller asserts ATDREAD when it reads PORTE. When asserted, ATDREAD enables transmission gate J to couple the inverted latch output to the data bus line.

To summarize, PORTE pins PE0/AN0–PE7/AN7 couple analog values to the ADC. To sample an analog value, the M68HC11 asserts SAMPLEx for twelve clock cycles. SAMPLEx turns on a coupling transistor that passes an analog pin level to the ADC. The microcontroller can also use a PORTE pin for general-purpose binary inputs. PORTE pin logic includes inverters, latch, and transmission gate to accommodate general-purpose binary inputs. Special enabling and coupling transistors isolate the binary input circuitry from intermediate analog levels at the PEx/ANx pin. However, this isolation is not effective during RPORTE assertion. The microcontroller restricts RPORTE duration to one fourth of an E clock cycle. Minimal RPORTE duration reduces the possibility of large current flows and consequent distortion due to inadvertent PORTE reads during A/D sampling.

---

[4]Motorola Inc., *M68HC11 Reference Manual*, p. 7-37.

The System Configuration Options Register (OPTION, control register block address $1039) contains 2 bits associated with the A/D converter. Figure 10–3 illustrates these bits, ADPU (A/D Power Up) and CSEL (Clock Select). Following paragraphs explain their function.

The ADC uses a Charge-Redistribution Successive-Approximation A/D Converter. This A/D converter incorporates twelve capacitances. See Figure 10–4, which shows how the ADC configures these capacitances.[5] Motorola expresses capacitance values in ratio-based units rather than in Farads. Capacitances are assigned values of $1/2$, 1, 1.1, 2, 4, 8, and 16. For example, capacitance 16C has sixteen times the capacitance of 1C.

The ADC uses analog switches to connect the capacitances variously to V_{REFH}, V_{REFL}, and the analog input value (ANx). The ADC executes a conversion in three phases. In the first phase, called the sample phase, the ADC connects its capacitive array as shown in Figure 10–4. In this phase, the ADC connects capacitor bottom plates to the analog voltage to be scaled (ANx), and capacitor top plates to V_{REFH}. The capacitors charge to the voltage difference between V_{REFH} and V_{ANx}.[6]

Refer to Figure 10–5, which depicts the ADC's second conversion phase—the hold phase. In this phase, the ADC disconnects both V_{REFH} and analog input voltage (ANx) from the capacitor array. The capacitors conserve their sample phase charges.

Figure 10–6 depicts the ADC's third phase—the conversion phase. In this phase, the analog switches take each capacitance in turn, and switch its bottom plate from V_{REFL} to V_{REFH}. If this connection causes the comparator to output a high, a corresponding bit sets in the Successive Approximation Register (SAR). Also, if the comparator outputs a high, the switched capacitance's bottom plate remains connected to V_{REFH}. If the comparator outputs a low, the SAR's corresponding bit resets, and the capacitance's bottom plate switches back to V_{REFL}. After evaluation of all capacitances, the SAR contains the binary equivalent of the analog input voltage. The ADC transfers the SAR contents to a result register.

Accurate A/D conversion depends upon capacitive charge conservation throughout the three conversion phases. Thus, the ADC requires efficient analog switch operation. To assure

FIGURE 10–3
Reprinted with the permission of Motorola.

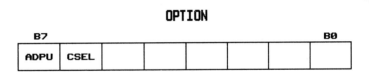

FIGURE 10–4
Reprinted with the permission of Motorola.

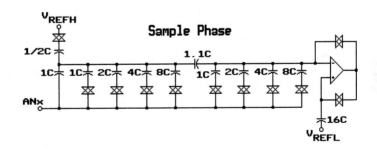

FIGURE 10–5
Reprinted with the permission of Motorola.

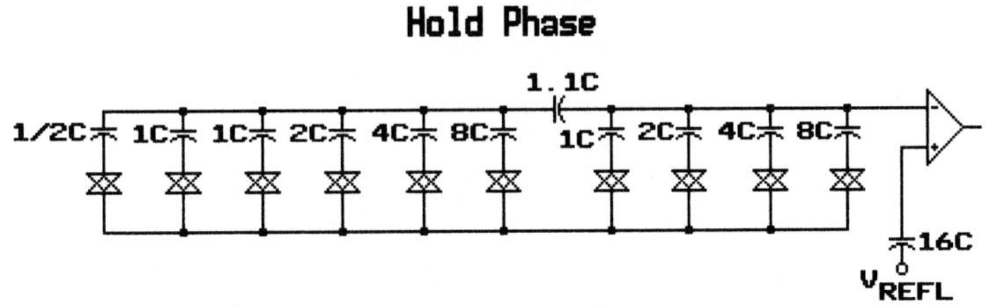

[5]Motorola Inc., *M68HC11 Reference Manual*, p. 12-11.
[6]The $1/2$ C capacitor shifts the ADC's transfer characteristic left by $1/2$ LSB bit value (i.e., $1/2$ step size). If we assume that V_{REFH} = +5V and V_{REFL} = 0V, the first ADC output transition point (from $00 to $01) comes at 9.76mV. Succeeding transition points come at 19.5mV intervals. The series 1.1C capacitor effectively divides the values of the left-hand capacitors by 16. See Motorola Inc., *M68HC11 Reference Manual*, pp. 12-10 and 12-11.

FIGURE 10–6
Reprinted with the permission of Motorola.

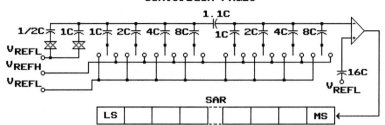

that these switches operate rapidly and efficiently, the M68HC11 drives their gates with seven volts. To develop seven volts, the M68HC11 uses a charge pump. An M68HC11 reset disables the charge pump. The CPU must write binary one to the OPTION register's ADPU (A/D Power Up) bit to start the charge pump. The CPU should write binary one to the ADPU bit as its first task in an A/D conversion routine. By powering up the charge pump first, the CPU gives it time to build a seven-volt output before the first A/D conversion.

The M68HC11 generally uses the E clock to time ADC analog switching and SAR latching. This clock must operate at a high enough rate to keep the charges in the capacitive array from dissipating during the scaling process. Some M68HC11 applications use a relatively low crystal frequency, resulting in E clock rates below 750 KHz. In such cases, a slow A/D switching rate allows capacitive charges to partly dissipate and causes consequent scaling inaccuracies. To deal with this problem, the ADC incorporates an independent RC oscillator as an optional 2 MHz (approximate) clock source. The user should select this independent clock where the E clock rate is less than 750 KHz. However, the independent RC-based clock is asynchronous to general E clock-based microcontroller operations. If the M68HC11 uses the E clock for A/D operations, it can latch A/D comparator outputs during low-noise periods, enhancing accuracy. The M68HC11 cannot do this with the asynchronous RC-based clock. Therefore, a user should select the E clock when possible.

The OPTION register's CSEL (Clock Select) bit allows the user to select either the asynchronous RC-based clock or the E clock as a timing source for A/D operations. By resetting the CSEL bit, the user selects the E clock. By setting CSEL, the user selects the internal RC-based clock.

The CPU's first task in an A/D routine is to write control bits to the OPTION register's ADPU and CSEL bits. ADPU = 1 causes a charge pump to power up. This charge pump generates A/D switching control voltages. The CSEL bit selects from two A/D clocking sources—the E clock or an internal asynchronous RC-based 2 MHz (approximate) clock.

The M68HC11's ADC makes four conversions at a time—from either four analog input pins or one. The ADC automatically stores each digitized result in a separate A/D Result Register (ADR). These four 8-bit registers (ADR1–ADR4) reside at addresses $1031–$1034. ADR registers are read only. The ADC sets a status flag after writing digitized results to the four ADR registers.

The user must select either four-channel or single-channel A/D conversions. The ADC must make four conversions and stop, or make continuous conversions. Further, the user must establish which of the eight ANx inputs will be scaled by the converter. The CPU sets these parameters by writing a control word to the A/D Control/Status Register (ADCTL), control register block address $1030. ADCTL also contains the status bit that sets after four conversions. Figure 10–7 shows the ADCTL register bits.

The ADC's status flag is ADCTL bit 7, the Conversions Complete Flag (CCF). This flag sets after the fourth A/D conversion when all ADR registers contain valid results. The CPU clears the CCF by writing to the ADCTL register. Although the CCF is a read-only bit, an ADCTL write automatically clears it. If the ADC makes continuous conversions, these conversions continue even when the CCF = 1.

FIGURE 10–7
Reprinted with the permission of Motorola.

ADCTL

B7							B0
CCF		SCAN	MULT	CD	CC	CB	CA

The CPU selects either four conversions or continuous conversions by writing a binary one or zero to ADCTL, bit 5, the SCAN (Continuous Scan Control) bit. By writing zero to the SCAN bit, the CPU configures the ADC to make four conversions and stop, putting each result in a separate ADR register. By writing a one to the SCAN bit, the CPU configures the ADC for continuous conversions.

The ADC can scale one analog input channel or four. If the ADC scales a single analog input, it converts this input either four times or continuously—four conversions at a time. Either way, the ADC transfers conversion results to the ADR registers. If the ADC scales four analog inputs, it converts them either one time each, or continuously in round-robin fashion. Either way, the ADC stores the results of four conversions in the ADR registers. The CPU configures the ADC for single-channel or four-channel conversions by writing to the ADCTL register's Multiple Channel/Single Channel Control (MULT) bit. By resetting the MULT bit, the CPU configures the ADC for single-channel conversions. If the CPU sets the MULT bit, the ADC will convert a group of four channels. If the CPU configures for single-channel conversions, it can select any of the eight ANx pins as the single-channel input. If the CPU configures for four-channel scaling, it can scale inputs from pins AN0–AN3 or from pins AN4–AN7.

The CPU uses the ADCTL register's four least significant bits to select either the single-channel ANx pin to be converted or the group of four pins (AN0–AN3 or AN4–AN7).

If the CPU configures for single-channel operation (MULT = 0), it selects the ANx pin by writing ones and zeros to ADCTL's CD, CC, CB, and CA bits:

CD	CC	CB	CA	Selected ANx Pin
0	0	0	0	AN0
0	0	0	1	AN1
0	0	1	0	AN2
0	0	1	1	AN3
0	1	0	0	AN4
0	1	0	1	AN5
0	1	1	0	AN6
0	1	1	1	AN7

If the CPU configures for four-channel operation (MULT = 1), it uses only the ADCTL CD and CC bits to select a group of four ANX pins for scaling:

CD	CC	ANx Pins Selected
0	0	AN0–AN3
0	1	AN4–AN7

If CD and CC = %00, the ADC stores conversion results as follows:

AN0 to ADR1
AN1 to ADR2
AN2 to ADR3
AN3 to ADR4

If CD and CC = %01, results are stored as follows:

AN4 to ADR1
AN5 to ADR2
AN6 to ADR3
AN7 to ADR4

Finally, the reader must understand how the M68HC11 causes A/D conversions to begin. By writing a binary one to the OPTION register's ADPU bit, the CPU causes the 7V charge pump to begin operation. This write does not cause A/D scaling to begin. Rather, the CPU causes A/D scaling to begin by writing to the ADCTL register. This write automatically clears the CCF bit and causes scaling to begin.

If this ADCTL write sets the SCAN bit (continuous conversions), the ADC continues to scale inputs until the microcontroller resets. If the ADCTL's SCAN bit is reset, the ADC makes one group of four conversions, then stops.

FIGURE 10–8
Reprinted with the permission of Motorola.

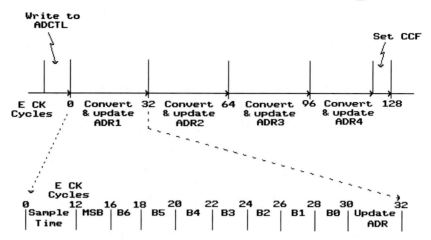

Remember, the ADC sets the ADCTL's CCF flag after each group of four conversions. The CPU resets the CCF by writing a control word to ADCTL. If this write makes SCAN = 0, it generates a new set of four conversions. If SCAN = 1, the ADC continues scaling even if the CPU does not clear the CCF.

A/D Conversion Timing

Figure 10–8 illustrates ADC timing for a cycle of four conversions.[7] Figure 10–8 shows this timing with respect to the M68HC11's E clock. An ADCTL write causes scaling to begin. After this write, each conversion requires 32 E clock cycles. The sequence of four conversions requires 128 E clock cycles. In the final two E clock cycles of each conversion, the ADC updates a result register. Note that the ADC updates result registers in the order ADR1 through ADR4. Note also that along with its ADR4 update, the ADC sets the CCF to indicate that it has completed four conversions.

If the ADC uses the internal RC-based clock in lieu of the E clock (CSEL = 1), four conversions require more than 128 RC clock cycles. Because of the asynchronous RC-based clock, the A/D converter consumes several clock cycles after each conversion in synchronizing its ADR register update with the E clock.

Exercise 10.1 provides the reader with an opportunity to connect, program, operate, and analyze the M68HC11's analog-to-digital converter.

EXERCISE 10.1 Analog-to-Digital Conversions

REQUIRED MATERIALS:
1 Photoresistor, cadmium sulfide, 1MΩ dark
1 Thermistor, 50KΩ @ 25°C
1 Potentiometer, trimmer, 1KΩ
1 Potentiometer, trimmer, 20KΩ
4 Potentiometers, trimmer, 1MΩ
1 Fixed resistor assortment, including three each of the following values: 1K, 1.8K, 2.2K, 3.3K, 4.7K, 10K, 15K, 18K, 22K, 33K, 50K.
1 Capacitor, 1µF (EVB only)
1 Spray Can, circuit chiller ("Freez-It Antistat" or equivalent)

---

[7] Motorola Inc., *M68HC11 Reference Manual*, pp. 12-13 and 12-14.

FIGURE 10-9

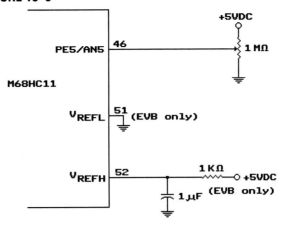

REQUIRED EQUIPMENT: Motorola M68HC11EVBU or EVB, and terminal
Dual-trace oscilloscope and two probes
Digital voltmeter or DMM
Flashlight
Heat gun (hot air blower for heat shrink or hair drying)

GENERAL INSTRUCTIONS: In this exercise, the reader shall connect, program, and operate the M68HC11's ADC in all four of its operating modes. The reader shall also connect, interface, and calibrate light and temperature sensor circuits. You shall use these sensor circuits again in Exercise 10.2. Carefully follow the instructions given in each part of the exercise below.

PART I: Single-Channel A/D Conversions

Power down the EVBU or EVB, and terminal. Make the connections shown in Figure 10–9. EVB users should note the 1KΩ resistor/1μF capacitor network connected between the 5VDC source and V_{REFH}. This network filters power supply noise to increase A/D scaling accuracy. EVBU users should make no connections to pins 51 and 52.

Develop supporting software for the M68HC11's A/D converter. This software causes the M68HC11 to do the following:

1. Make A/D conversions of pin AN5 inputs only.
2. Make four conversions and stop.
3. Display the contents of ADR1–ADR4 at the terminal display, using the BUFFALO .OUT2BS, and .OUTCRL utility subroutines. .OUT2BS converts 2 memory bytes to ASCII characters, beginning with the address pointed at by the X register, then writes these ASCII characters to the terminal screen. After completing these tasks, .OUT2BS leaves the X register pointing at the address following the bytes just encoded. .OUTCRL outputs a carriage return and line feed to the terminal.

Use the Figure 10–10 flowchart as a guide to developing your A/D software. List your program in the blanks provided below. Allocate a RAM memory block for program storage.

Single-Channel A/D
Four Conversions and Stop

$ADDRESS	SOURCE CODE	COMMENTS
_____	_____	_____
_____	_____	_____
_____	_____	_____
_____	_____	_____

FIGURE 10–10

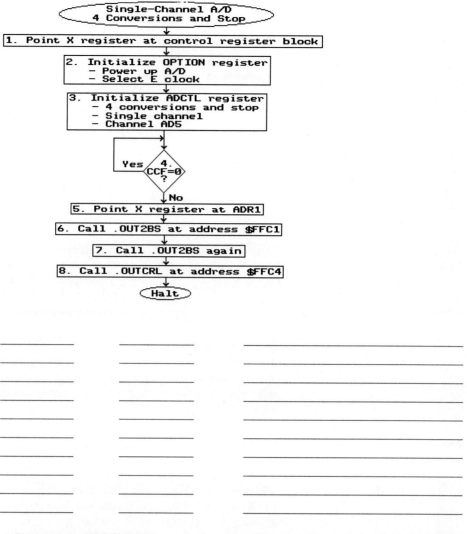

Power up the EVBU (EVB), terminal, and interface circuitry. Load and debug your A/D software and interface.

Use the voltmeter to measure the actual V_{REFH} voltage at pin 52. Enter this measured value in the blank.

V_{REFH} (measured) = _____ V

Carefully adjust the 1MΩ potentiometer for analog input voltages close as possible to the values in the chart below. Use the voltmeter to measure each input voltage as you adjust it. For each input value, follow this procedure:

1. Enter the analog input voltage (voltmeter) in column 1 (V_{IN}, Measured). Round your entry to three significant figures.
2. Run the A/D scaling program. In column 2 ($DATA, Terminal Display), enter the average of the four ADR values displayed on the terminal screen.
3. Translate your column 2 entry into a decimal voltage value (&V). Use the following procedure to make this conversion:

$$\text{Step Size} = V_{REFH}/256$$
$$\&V = \text{Step size} \times \&DATA$$
(&DATA = decimal equivalent of $DATA in column 2).

Enter the decimal voltage value in column 3 (&V, A/D). Round your entry to three significant figures.

4. Compare the column 1 and column 3 values. Are they within Motorola's specified accuracy of ±1 LSB (1 LSB = Step Size)? Put an "X" in the appropriate blank under column 4.

Input Voltage	1. V_{IN} Measured	2. $DATA Terminal Display	3. &V A/D	4. Scaling Accurate to ± 1 LSB? Yes No
0V	_____	_____	_____	___ ___
1V	_____	_____	_____	___ ___
2V	_____	_____	_____	___ ___
3V	_____	_____	_____	___ ___
4V	_____	_____	_____	___ ___
=V_{REFH}	_____	_____	_____	___ ___

Note: The reader must understand that this procedure provides an *invalid* evaluation of ADC accuracy. This procedure uses a digital voltmeter or DMM as the evaluation standard. However, the voltmeter or DMM may well have wider tolerances than the M68HC11. A valid ADC evaluation requires a certified input voltage standard with considerably greater resolution than the average voltmeter or DMM. The present procedure does give the reader an opportunity to exercise the ADC, see tangible results, and analyze these results.

Show your chart results and demonstrate the operation of your interface to the instructor. For this demonstration, adjust the analog input voltage to 3.00V.

_____ *Instructor Check*

Modify the A/D software, so that it causes the M68HC11 to (1) make continuous A/D conversions of analog inputs to the AN5 pin, and (2) periodically display the contents of ADR1–ADR4.

Use the Figure 10–11 flowcharts as guides to program development. List your MAIN program and DELAY subroutine in the blanks provided below. Allocate a block of RAM for software storage. Leave your interface connections as shown in Figure 10–9.

Single-Channel A/D
Continuous Conversions

$ADDRESS	SOURCE CODE	COMMENTS
_____	_____	_____
_____	_____	_____
_____	_____	_____
_____	_____	_____
_____	_____	_____
_____	_____	_____
_____	_____	_____
_____	_____	_____
_____	_____	_____
_____	_____	_____
_____	_____	_____
_____	_____	_____
_____	_____	_____

DELAY Subroutine

$ADDRESS	SOURCE CODE	COMMENTS
_____	_____	_____
_____	_____	_____
_____	_____	_____
_____	_____	_____

FIGURE 10–11

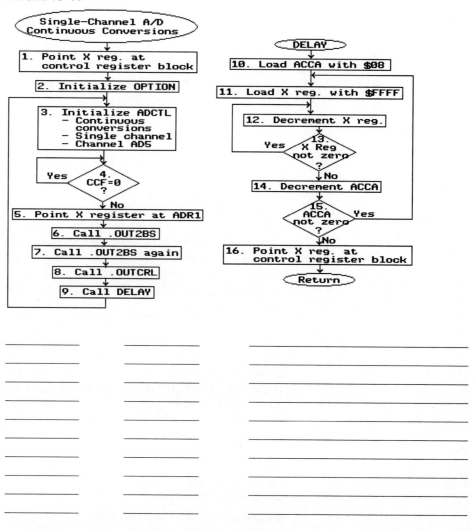

Carefully adjust the 1MΩ potentiometer for input voltages as close as possible to the values specified in the chart below. Use the voltmeter to measure each input voltage as you adjust it. For each input value, follow this procedure:

1. Enter the analog input voltage in column 1. Round your entry to three significant figures.
2. Reset the EVBU (EVB) to stop display scrolling. Enter the average of four sets of displayed values in column 2.
3. Translate your column 2 entry into a decimal voltage value.

$$\&V = \text{Step Size} \times \&DATA$$
(&DATA = decimal equivalent of $DATA in column 2)

Enter this decimal voltage in column 3. Round your entry to three significant figures.

4. Compare the column 1 and column 3 values. Are they within Motorola's specified accuracy of ± 1 LSB? Put an "X" in the appropriate column 4 blank.

Input Voltage	1. V_{IN} Measured	2. $DATA Terminal Display	3. &V A/D	4. Scaling Accurate to ± 1 LSB? Yes / No
0.5V	____	____	____	____ ____
1.5V	____	____	____	____ ____
2.5V	____	____	____	____ ____
3.5V	____	____	____	____ ____
4.5V	____	____	____	____ ____

Show your chart results and demonstrate the operation of your interface to the instructor. For this demonstration, adjust the analog input voltage to 3.50V.

_____ *Instructor Check*

PART II: Four-Channel A/D Conversions

Power down the EVBU (EVB), terminal, and interface. Modify the interface as shown in Figure 10–12 on page 486.

Develop supporting software for the M68HC11's A/D converter. This software causes the M68HC11 to:

1. Make A/D conversions on pins AN4–AN7.
2. Make four conversions and stop.
3. Display the contents of ADR1–ADR4 on the terminal display using the BUFFALO .OUT2BS and .OUTCRL utility subroutines.

Use the Figure 10–10 flowchart as a guide to developing your program. Only two parts of this flowchart must be changed:

1. Change the program's name to Four-Channel A/D, Four Conversions and Stop.
2. Change block 3 to specify the following: convert a four-channel group and select channels AD4–AD7.

Recall that the CPU configures for four-channel operation by setting the MULT bit in ADCTL. When MULT = 1, ADCTL's CD and CC bits select the group of four ANx pins for scaling.

Given these changes, what hexadecimal value must the M68HC11's CPU write to the ADCTL register in order to execute flowchart block 3?

ADCTL = $_____

List your program in the blanks provided below. Allocate a block of RAM for program storage.

Four-Channel A/D
Four Conversions and Stop

$ADDRESS	SOURCE CODE	COMMENTS
_____	_____	_____
_____	_____	_____
_____	_____	_____
_____	_____	_____
_____	_____	_____
_____	_____	_____
_____	_____	_____
_____	_____	_____
_____	_____	_____
_____	_____	_____
_____	_____	_____
_____	_____	_____
_____	_____	_____
_____	_____	_____

Power up the EVBU (EVB), terminal, and interface circuitry. Load and debug your software and interface. Use the voltmeter to measure actual V_{REFH} at pin 52. Enter your measurement in the blank.

V_{REFH} (measured) = _____V

FIGURE 10-12

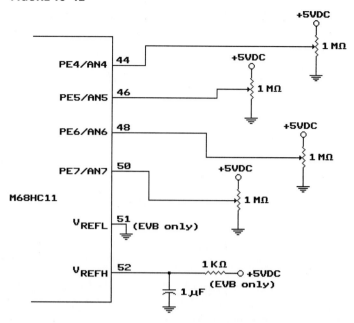

Carefully adjust the four 1MΩ potentiometers for input voltages as close as possible to the values specified in the chart below. Use the voltmeter to measure these voltages as you adjust them. Complete the chart below by carefully following this procedure:

1. Record the four actual input voltages in column 1 of the chart. Round your entries to three significant figures.
2. Run the A/D scaling program four times. In column 2, record the average terminal display $DATA values for each of the four analog inputs.
3. Translate your column 2 entries into decimal voltage values. Use the following procedure to translate column 2 entries.

$$\text{Step Size} = V_{REFH}/256$$
$$\&V = \text{Step size} \times \&DATA$$
(&DATA = decimal equivalent of $DATA from column 2.)

Enter these decimal voltage values in column 3. Round your entries to three significant figures.
4. Compare the column 1 and column 3 values. Are they within Motorola's specified accuracy of ± 1 LSB? Enter "Xs" in the appropriate column 4 blanks.

A/D Channel	Input Voltage	1. $DATA V_{IN} Measured	2. Terminal Display	3. &V A/D	4. Scaling Accurate to ± 1 LSB? Yes No
4	1V	___	___	___	___ ___
5	2V	___	___	___	___ ___
6	3V	___	___	___	___ ___
7	4V	___	___	___	___ ___

Show your chart results and demonstrate the operation of your interface to the instructor. For this demonstration, leave the 1MΩ potentiometers as you had them for the chart analysis.

_____ *Instructor Check*

Modify your A/D software so that it causes the M68HC11 to do the following:

1. Make continuous conversions of voltages appearing at pins AN4–AN7.
2. Periodically display the contents of ADR1–ADR4. Use the Figure 10–11 flowcharts as guides to program development. Only two components of the main program flowchart

must be changed. First, change the program's name to Four-Channel A/D, Continuous Conversions. Second, change block 3 to specify the following: convert four-channel group and select channels AD4–AD7.

Given these changes, what hexadecimal value must the M68HC11 CPU write to the ADCTL register in order to execute flowchart block 3?

ADCTL = $_____

List your A/D program in the blanks provided below. Allocate a block RAM for program storage.

Four-Channel A/D
Continuous Conversions

$ADDRESS	SOURCE CODE	COMMENTS
_____	_____	_____
_____	_____	_____
_____	_____	_____
_____	_____	_____
_____	_____	_____
_____	_____	_____
_____	_____	_____
_____	_____	_____
_____	_____	_____
_____	_____	_____
_____	_____	_____
_____	_____	_____
_____	_____	_____

DELAY Subroutine

$ADDRESS	SOURCE CODE	COMMENTS
_____	_____	_____
_____	_____	_____
_____	_____	_____
_____	_____	_____
_____	_____	_____
_____	_____	_____
_____	_____	_____
_____	_____	_____
_____	_____	_____
_____	_____	_____
_____	_____	_____
_____	_____	_____
_____	_____	_____
_____	_____	_____
_____	_____	_____

Carefully adjust the 1MΩ potentiometers for input voltages as close as possible to the values specified in the chart below. Use the voltmeter to measure these voltages as you adjust them. Complete the chart below by carefully following this procedure:

1. Record the four input voltages in column 1. Round your entry to three significant figures.
2. Reset the EVBU (EVB) to stop display scrolling. In column 2, record the average of four $DATA values from the terminal screen for each of the four A/D channels.
3. Translate your column 2 entries into decimal voltage values. Enter these values in column 3. Round your entries to three significant figures.

4. Compare column 1 and column 3 values. Are they within Motorola's specified accuracy of ± 1 LSB? Put "Xs" in the appropriate column 4 blanks.

A/D Channel	Input Voltage	1. V_{IN} Measured	2. $DATA Terminal Display	3. &V A/D	4. Scaling Accurate to ± 1 LSB? Yes No
4	1.5V	___	___	___	___ ___
5	2.5V	___	___	___	___ ___
6	3.5V	___	___	___	___ ___
7	4.5V	___	___	___	___ ___

Show your chart results and demonstrate the operation of your interface to the instructor. For this demonstration, leave the 1MΩ pots adjusted as you had them for the chart analysis.

_____ *Instructor Check*

PART III: Temperature and Light Sensors

In this part of the laboratory exercise, develop two sensor circuits for upcoming Exercise 10.2. The first of these sensors uses a thermistor rated 50KΩ @ 25°C. The second sensor uses a cadmium sulfide photoresistor rated 1MΩ, dark. Your objective is to determine series fixed and variable resistances. These series resistances should allow calibration of the sensor circuit outputs and also permit widest possible output voltage response. You should have at hand an assortment of fixed resistors varying from 1KΩ to 50KΩ. Substitute and combine these resistances until you achieve the best possible sensor circuit response. Execute the following steps to develop your light and temperature sensor circuits.

Step 1. Refer to Figure 10–13. Power down the EVBU (EVB) and terminal. Connect the circuitry as shown in this figure. Physically separate the photoresistor and thermistor. You will use a hot air gun and circuit chiller to vary the thermistor's temperature. The photoresistor is also sensitive to heat; therefore, mount the photoresistor so you will not inadvertently blow hot air or circuit chiller on it. The series 2.2KΩ (photoresistor) and 35KΩ (thermistor) resistances may not be the best choices for your particular sensors and ambient environmental conditions. However, these resistances probably provide a useful starting point in your search for optimal resistance values. Connect an oscilloscope to M68HC11 pin 45 (Channel 1) and pin 47 (Channel 2).

Step 2. Refer to Figure 10–11. Modify the flowcharts as follows:

- Change the program's name to Four-Channel A/D, Continuous Conversions.
- Change Block 3 to specify the following: convert a four channel group and select channels AD0–AD3.

FIGURE 10–13

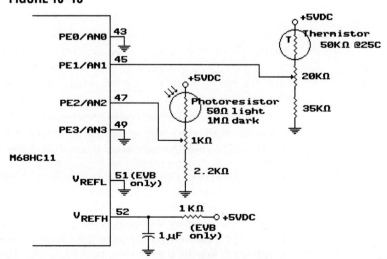

Given these changes, what hexadecimal value must the M68HC11 CPU write to the ADCTL register to execute flowchart block 3?

ADCTL = $_____

List your A/D program in the blanks provided below. Allocate a block of RAM for your program storage.

Four-Channel A/D Continuous Conversions

$ADDRESS	SOURCE CODE	COMMENTS

DELAY Subroutine

$ADDRESS	SOURCE CODE	COMMENTS

Step 3. Power up the EVBU (EVB), terminal, and interface circuitry. Assure that the A/D software runs properly. Establish ambient temperature and light conditions (room lighting and heat/AC thermostat setting) the same as when you will complete upcoming Exercise 10.2. Run your A/D software. Observe the terminal display and oscilloscope. Given your established ambient conditions, you should be able to adjust the 1KΩ and 20KΩ potentiometers so that each sensor circuit outputs 2.5V. Further, at a 2.5V output, the potentiometer shafts should be close to their halfway points between fully clockwise and fully counterclockwise. If you cannot adjust your sensor circuits properly, MOVE your A/D software to EEPROM and power down the EVBU(EVB) and interface. Substitute sensor circuit voltage divider fixed resistance values until you can properly adjust the sensor circuit outputs for ambient conditions. **Be Safe! Power down the EVBU (EVB), terminal, and all interface circuitry whenever you make a fixed resistance change.** After a power down, it only takes a moment to MOVE your program back to RAM from EEPROM. Alternatively, run the A/D software from EEPROM itself.

Step 4. Demonstrate the operation of your sensor circuits and A/D software to the instructor.

_____ *Instructor Check*

Now test the sensor circuits' voltage response by gathering the following materials:

Hot air blower (hair dryer or heat shrink)
Spray can of circuit chiller
Flashlight

Empirically determine the voltage output range of the temperature sensor circuit. Carefully follow the procedural steps listed below.

Step 1. Under ambient temperature conditions, adjust the 20KΩ potentiometer for a 2.5V output from the temperature sensor circuit. Note the hex output (ambient conditions) on the terminal screen and record in the appropriate blank of the chart below.

Step 2. Spray a liberal dose of circuit chiller on the thermistor. Note the lowest resultant hex value on the terminal screen. Record the hex value in the appropriate blank of the chart below.

Step 3. Allow the thermistor's temperature to return to the ambient level as indicated by the terminal screen.

Step 4. Use the hot air gun to heat the thermistor. Use care to avoid heat damage to any nearby components. Note the highest resultant hex value on the terminal screen. Record this hex value in the appropriate blank of the chart below.

Temperature Sensor Circuit

Condition	$OUTPUT (Terminal Screen)
Cold	$_____
Ambient	$_____
Hot	$_____

Empirically determine the voltage output range of the light sensor circuit. Carefully follow the procedural steps listed below.

Step 1. Under ambient light conditions, adjust the 1K Ohm potentiometer for a 2.5V output from the light sensor circuit. Note the hex output (ambient) on the terminal screen, and record in the appropriate blank of the chart below.

Step 2. Shroud the photoresistor to shut out as much light as possible. Note the lowest resultant hex value on the terminal screen. Record this hex value in the appropriate blank of the chart below. Remove the shroud. Allow the light sensor circuit output to return to the ambient level.

Step 3. Hold the flashlight close to the upper surface of the photoresistor. Aim the flashlight's beam directly at this upper surface. Note the highest resultant hex value on the terminal screen. Record this hex value in the appropriate blank of the chart below.

Light Sensor Circuit

Condition	$OUTPUT (Terminal Screen)
Dark	$_____
Ambient	$_____
Bright	$_____

Leave your sensor circuits and connections intact for use in Exercise 10.2. Also, make notes of your recorded light and temperature sensor chart values for use in Exercise 10.2.

PART IV: Laboratory Software Listing

SINGLE-CHANNEL A/D, FOUR CONVERSIONS AND STOP—EVB

$ADDRESS	SOURCE CODE	COMMENTS
C000	LDX #1000	Initialize X register
C003	LDAA #80	Initialize OPTION register
C005	STAA 39,X	" " " "
C007	LDAA #5	Write to ADCTL register
C009	STAA 30,X	" " " "
C00B	BRCLR 30,X 80 C00B	Branch to self until CCF=1
C00F	LDX #1031	Point X register at ADR1 register
C012	JSR FFC1	Call .OUT2BS
C015	JSR FFC1	Call .OUT2BS
C018	JSR FFC4	Call .OUTCRL
C01B	WAI	Halt

SINGLE-CHANNEL A/D, FOUR CONVERSIONS AND STOP—EVBU

$ADDRESS	SOURCE CODE	COMMENTS
0100	LDX #1000	Initialize X register
0103	LDAA #80	Initialize OPTION register
0105	STAA 39,X	" " " "
0107	LDAA #5	Write to ADCTL register
0109	STAA 30,X	" " " "
010B	BRCLR 30,X 80 10B	Branch to self until CCF=1
010F	LDX #1031	Point X register at ADR1 register
0112	JSR FFC1	Call .OUT2BS
0115	JSR FFC1	Call .OUT2BS
0118	JSR FFC4	Call .OUTCRL
011B	WAI	Halt

SINGLE-CHANNEL A/D, CONTINUOUS CONVERSIONS—EVB

$ADDRESS	SOURCE CODE	COMMENTS
C000	LDX #1000	Initialize X register
C003	LDAA #80	Initialize OPTION register
C005	STAA 39,X	" " " "
C007	LDAA #2	Write to ADCTL register
C009	STAA 30,X	" " " "
C00B	BRCLR 30,X 80 C00B	Branch to self until CCF=1
C00F	LDX #1031	Point X register at ADR1 register
C012	JSR FFC1	Call .OUT2BS
C015	JSR FFC1	Call .OUT2BS
C018	JSR FFC4	Call .OUTCRL
C01B	JSR C025	Call DELAY
C01E	BRA C007	Repeat

DELAY Subroutine

C025	LDAA #8	Initialize loop counter
C027	LDX #FFFF	Load X reg. with $FFFF
C02A	DEX	Decrement X register
C02B	BNE C02A	If X register not zero, decrement X register again
C02D	DECA	Decrement loop counter
C02E	BNE C027	If loop counter not zero, do another round
C030	LDX #1000	If loop counter zero, re-initialize X register
C033	RTS	Return

SINGLE-CHANNEL A/D, CONTINUOUS CONVERSIONS—EVBU

$ADDRESS	SOURCE CODE	COMMENTS
0100	LDX #1000	Initialize X register
0103	LDAA #80	Initialize OPTION register
0105	STAA 39,X	" " " "

0107	LDAA #2	Write to ADCTL register
0109	STAA 30,X	" " " "
010B	BRCLR 30,X 80 10B	Branch to self until CCF=1
010F	LDX #1031	Point X register at ADR1 register
0112	JSR FFC1	Call .OUT2BS
0115	JSR FFC1	Call .OUT2BS
0118	JSR FFC4	Call .OUTCRL
011B	JSR 125	Call DELAY
011E	BRA 107	Repeat

DELAY Subroutine

0125	LDAA #8	Initialize loop counter
0127	LDX #FFFF	Load X register with $FFFF
012A	DEX	Decrement X register
012B	BNE 12A	If X register not zero, decrement X register again
012D	DECA	Decrement loop counter
012E	BNE 127	If loop counter not zero, do another round
0130	LDX #1000	If loop counter zero, re-initialize X register
0133	RTS	Return

FOUR-CHANNEL A/D, FOUR CONVERSIONS AND STOP—EVB

$ADDRESS	SOURCE CODE	COMMENTS
C000	LDX #1000	Initialize X register
C003	LDAA #80	Initialize OPTION register
C005	STAA 39,X	" " " "
C007	LDAA #14	Write to ADCTL register
C009	STAA 30,X	" " " "
C00B	BRCLR 30,X 80 C00B	Branch to self until CCF=1
C00F	LDX #1031	Point X register at ADR1 register
C012	JSR FFC1	Call .OUT2BS
C015	JSR FFC1	Call .OUT2BS
C018	JSR FFC4	Call .OUTCRL
C01B	WAI	Halt

FOUR-CHANNEL A/D, FOUR CONVERSIONS AND STOP—EVBU

$ADDRESS	SOURCE CODE	COMMENTS
0100	LDX #1000	Initialize X register
0103	LDAA #80	Initialize OPTION register
0105	STAA 39,X	" " " "
0107	LDAA #14	Write to ADCTL register
0109	STAA 30,X	" " " "
010B	BRCLR 30,X 80 10B	Branch to self until CCF=1
010F	LDX #1031	Point X register at ADR1 register
0112	JSR FFC1	Call .OUT2BS
0115	JSR FFC1	Call .OUT2BS
0118	JSR FFC4	Call .OUTCRL
011B	WAI	Halt

FOUR-CHANNEL A/D, CONTINUOUS CONVERSIONS—EVB

$ADDRESS	SOURCE CODE	COMMENTS
C000	LDX #1000	Initialize X register
C003	LDAA #80	Initialize OPTION register
C005	STAA 39,X	" " " "
C007	LDAA #34	Write to ADCTL register
C009	STAA 30,X	" " " "
C00B	BRCLR 30,X 80 C00B	Branch to self until CCF=1
C00F	LDX #1031	Point X register at ADR1 register
C012	JSR FFC1	Call .OUT2BS
C015	JSR FFC1	Call .OUT2BS
C018	JSR FFC4	Call .OUTCRL
C01B	JSR C025	Call DELAY
C01E	BRA C007	Repeat

DELAY Subroutine

$ADDRESS	SOURCE CODE	COMMENTS
C025	LDAA #8	Initialize loop counter
C027	LDX #FFFF	Load X register with $FFFF
C02A	DEX	Decrement X register
C02B	BNE C02A	If X register not zero, decrement X register again
C02D	DECA	Decrement loop counter
CO2E	BNE C027	If loop counter not zero, do another round
C030	LDX #1000	If loop counter zero, re-initialize X register
C033	RTS	Return

FOUR-CHANNEL A/D, CONTINUOUS CONVERSIONS—EVBU

$ADDRESS	SOURCE CODE	COMMENTS
0100	LDX #1000	Initialize X register
0103	LDAA #80	Initialize OPTION register
0105	STAA 39,X	" " " "
0107	LDAA #34	Write to ADCTL register
0109	STAA 30,X	" " " "
010B	BRCLR 30,X 80 10B	Branch to self until CCF=1
010F	LDX #1031	Point X register at ADR1 register
0112	JSR FFC1	Call .OUT2BS
0115	JSR FFC1	Call .OUT2BS
0118	JSR FFC4	Call .OUTCRL
011B	JSR 125	Call DELAY
011E	BRA 107	Repeat

DELAY Subroutine

$ADDRESS	SOURCE CODE	COMMENTS
0125	LDAA #8	Initialize loop counter
0127	LDX #FFFF	Load X register with $FFFF
012A	DEX	Decrement X register
012B	BNE 12A	If X register not zero, decrement X register again
012D	DECA	Decrement loop counter
012E	BNE 127	If loop counter not zero, do another round
0130	LDX #1000	If loop counter zero, re-initialize X register
0133	RTS	Return

FOUR-CHANNEL A/D, CONTINUOUS CONVERSIONS, TEMPERATURE AND LIGHT SENSORS—EVB

$ADDRESS	SOURCE CODE	COMMENTS
C000	LDX #1000	Initialize X register
C003	LDAA #80	Initialize OPTION register
C005	STAA 39,X	" " " "
C007	LDAA #30	Write to ADCTL register
C009	STAA 30,X	" " " "
C00B	BRCLR 30,X 80 C00B	Branch to self until CCF=1
C00F	LDX #1031	Point X register at ADR1 register
C012	JSR FFC1	Call .OUT2BS
C015	JSR FFC1	Call .OUT2BS
C018	JSR FFC4	Call .OUTCRL
C01B	JSR C025	Call DELAY
C01E	BRA C007	Repeat

DELAY Subroutine

$ADDRESS	SOURCE CODE	COMMENTS
C025	LDAA #8	Initialize loop counter
C027	LDX #FFFF	Load X register with $FFFF
C02A	DEX	Decrement X register
C02B	BNE C02A	If X register not zero, decrement X register again
C02D	DECA	Decrement loop counter
CO2E	BNE C027	If loop counter not zero, do another round
C030	LDX #1000	If loop counter zero, re-initialize X register
C033	RTS	Return

FOUR-CHANNEL A/D, CONTINUOUS CONVERSIONS, TEMPERATURE AND LIGHT SENSORS—EVBU

$ADDRESS	SOURCE CODE	COMMENTS
0100	LDX #1000	Initialize X register
0103	LDAA #80	Initialize OPTION register
0105	STAA 39,X	" " " "
0107	LDAA #30	Write to ADCTL register
0109	STAA 30,X	" " " "
010B	BRCLR 30,X 80 10B	Branch to self until CCF=1
010F	LDX #1031	Point X register at ADR1 register
0112	JSR FFC1	Call .OUT2BS
0115	JSR FFC1	Call .OUT2BS
0118	JSR FFC4	Call .OUTCRL
011B	JSR 125	Call DELAY
011E	BRA 107	Repeat
		DELAY Subroutine
0125	LDAA #8	Initialize loop counter
0127	LDX #FFFF	Load X register with $FFFF
012A	DEX	Decrement X register
012B	BNE 12A	If X register not zero, decrement X register again
012D	DECA	Decrement loop counter
012E	BNE 127	If loop counter not zero, do another round
0130	LDX #1000	If loop counter zero, re-initialize X register
0133	RTS	Return

10.3 USING THE M68HC11 FOR FUZZY INFERENCE

The following sections apply A/D conversion and fuzzy logic to a simple motor control problem. What is fuzzy logic? How does it differ from conventional control technology?

Fuzzy logic applies software and hardware to a control problem. In recent years, designers have developed special fuzzy logic IC chips. These ICs use fuzzy logic to control appliances and consumer goods, such as vacuum cleaners, washing machines, and video cameras. Engineers have also developed experimental ICs that combine fuzzy logic with neural networks. These chips can emulate human intuition and learning in a primitive way. However, most fuzzy logic applications use general-purpose digital computers rather than special ICs to implement fuzzy control.

A digital computer, such as the M68HC11, runs fuzzy inference software. Fuzzy inference is the algorithm that causes hardware to regulate and control a process. Here are some process examples:

- Control an automatic transmission
- Operate a one-button microwave oven
- Fly a helicopter
- Give investment advice
- Operate commuter trains
- Operate elevators
- Control an industrial process, such as cement kiln operations.

How does fuzzy logic differ from more traditional and conventional algorithms? First, consider how conventional algorithms attack control problems. Most processes involve multiple variables that are related in complex ways. As an example, assume that an engineer wishes to control the temperature of a fluid in a vat. Multiple variables affect fluid temperature:

- Inflow rate
- Outflow rate

- Ambient temperature
- Temperature of inflow fluid
- In-vat turbulence (fluid circulation)
- Rate of coolant flow through vat cooling coils

Conventional algorithms usually regulate one of these variables to keep the process under control. Here, the engineer probably chooses to regulate coolant flow.

By selecting a single controlled variable, the engineer allows for a relatively simple control algorithm: Measure fluid temperature. Subtract measured temperature from setpoint temperature to generate an error value. Open or close the coolant inlet valve accordingly.

However, conventional control software must calculate the integral and derivative of the changing error signal to effectively control fluid temperature, decrease response time, and avoid system oscillation. Only recently have affordable computers become fast enough to make such real-time calculations.

To control multiple variables, a conventional control algorithm becomes much more complex. For example, the engineer may wish to control coolant flow, fluid outflow, fluid inflow, and vat turbulence. Because of complex secondary and nonlinear relationships between multiple variables, the engineer often finds it difficult or impossible to develop an algorithm that adequately models the process. For example, many industrial kiln operations defeat mathematical modeling. Accordingly, these operations require highly experienced human operators who intuitively operate multiple open-loop controls. Such human intuition categorizes variables in relativistic and "fuzzy" ways. For example, "Kind of warm and kind of hot," or "Somewhere between medium and fast." Many conventional algorithms resort to categorization of input variables. However, these algorithms classify variables into clearly defined "crisp" categories. For example, a fluid inflow rate of 8.7 gpm might be a 100% member of the category "8–9 gpm." Other clearly defined "crisp" categories might include 7–8 gpm, 8–9 gpm, or 9–10 gpm.

Such strictly defined categories fail to model the intuitive "fuzzy" classifications used by human operators. Crisp classifications often generate abrupt control transitions that may be unworkable in sensitive or subtle control applications, such as kiln operations or piloting a helicopter.

Fuzzy inference models the way a human operator analyzes a situation and decides how to control it. It can successfully model the kiln operator, helicopter pilot, or investment adviser. To this extent, it is a form of artificial intelligence.

Fuzzy inference makes relativistic or "fuzzy" assessments of multiple input variables. For example, fuzzy can categorize inputs as "kind of warm and kind of hot." Also, a fuzzy algorithm generates relativistic and subtle output controls: "Run the motor at somewhere between medium and high speeds."

Fuzzy inference subjects input variables to a three-stage process. This process generates multiple control outputs. The three stages are (1) fuzzification, (2) rule evaluation, and (3) de-fuzzification.

Fuzzification involves simple calculations. These calculations apply graphical analysis to a crisp input value (X axis) versus its membership in categories (Y axis). Graphical representation of these membership categories—or membership functions—forms overlapping trapezoids or triangles. Thus, fuzzification provides a relativistic categorization of each crisp input. For example, fuzzification can determine the extent to which ambient temperature is "cold," "cool," "moderate," "warm," or "hot." Since these membership functions overlap, an ambient temperature of 25°C might be partly "moderate" and partly "warm." Fuzzification categorizes on a scale, e.g., from $00 (no membership) to $FF (full membership). Thus, fuzzification might grade an ambient temperature of 25°C as follows:

Membership Function	Grade (degree) of Membership
cold	$00
cool	$00
moderate	$6E
warm	$91
hot	$00

Fuzzification, then, applies simple analysis to all relevant control inputs and grades them on the extent to which they belong to overlapping—"fuzzy"—membership functions (see the following section). Fuzzification grades membership on a scale from $00 (no membership) to $FF (full membership). Each grade constitutes a fuzzy input. For example, crisp ambient temperature can be fuzzified into five fuzzy inputs (or grades) as shown in the table above. Fuzzification converts all crisp inputs (e.g., inflow rate, inflow temperature, etc.) into such multiple fuzzy inputs.

Next, fuzzy inference subjects fuzzy inputs to rule evaluation. This phase applies common sense and experience-based operational rules. For example, the system designer can interview and observe the system operator, then develop rules based upon this observation. The designer states these rules in an IF/THEN format:

> If ambient temperature is hot,
> and if fluid inflow is small,
> and if fluid inflow is hot,
> and if fluid in-vat temperature is normal,
> then make coolant flow medium-high,
> and then make fluid turbulence moderate.

Depending upon process complexity and control criticality, the designer develops a minimum of six to a maximum of sixty of these rules.

Rule evaluation determines how control outputs must be adjusted to control the process. The example rule given above assumes that fuzzy logic generates two control outputs—coolant flow and turbulence (fluid circulation in the vat). Rule evaluation develops multiple fuzzy outputs for each of these controls. For example,

> The degrees to which coolant flow should be "low," "medium-low," "medium," "medium-high," "high."
> The degrees to which vat turbulence should be "low," "medium," "high."

Thus, fuzzy inference evaluates all if/then rules in the light of all fuzzy inputs. Rule evaluation generates fuzzy outputs, the relative weights of all control adjustment categories.

The final step de-fuzzifies the fuzzy outputs into crisp control output values. These crisp outputs can then drive the control system's end effectors. De-fuzzification accomplishes this task by making center of gravity (COG) calculations. For example, de-fuzzification finds the COG of fuzzy coolant-flow outputs: low, medium-low, medium, medium-high, and high. Likewise de-fuzzification finds a COG for fuzzy turbulence outputs: low, medium, and high. These two COGs constitute crisp control outputs that drive the end effectors.

COG calculation uses graphical analysis of overlapping output membership function triangles or trapezoids. COG calculations require considerable computing power and program complexity. For most applications, the designer can substitute weighted averages for COGs. These calculations use graphical singletons instead of trapezoids or triangles, and they find the weighted average of the singletons. Fuzzy outputs constitute the weights used in these calculations. Weighted averaging of singletons produces outputs very close to COG values without the complicated computations. By using weighted averages instead of COGs, designers can apply relatively simple computers like the M68HC11 to complex control problems with remarkably effective and sophisticated results.

In Exercise 10.2, the reader shall apply fuzzy inference to a simple problem—vary DC motor speed with changes in temperature and light. Following sections explain fuzzy inference using this application as an example. Figure 10–14 provides an overview of the fuzzy motor controller.

Refer to Figure 10–14. A dotted line box surrounds the M68HC11, its resources and control software. The M68HC11's ADC (A) changes 0–5V inputs from the temperature and light sensors into "crisp" digitized data. "Crisp" means that these data have not been processed by the fuzzy inference software. Sensor data vary from 0V to 5V. The temperature sensor outputs 0V under very cold conditions, and 5V under very hot conditions. The

FIGURE 10–14

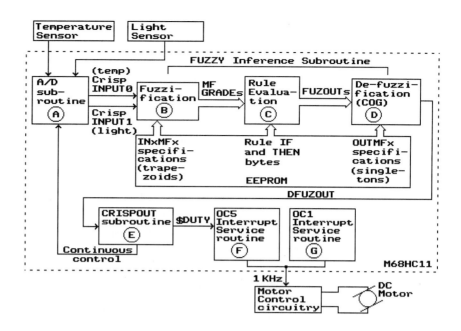

light sensor outputs 0V under dark conditions and 5V under bright light conditions. The ADC translates these sensor values as follows:

ADC Input from Sensor	ADC Crisp Output (INPUT0 or INPUT1)
0V	$00
.	.
.	.
.	.
2.5V	$7F
.	.
.	.
5V	$FF

The A/D subroutine passes crisp digitized sensor data to the Fuzzy Inference subroutine. The Fuzzy Inference subroutine labels these sensor data as INPUT0 (temperature datum) and INPUT1 (light datum). Fuzzy inference subjects INPUT0 and INPUT1 to a three-phase analysis.

1. Fuzzify (B) the INPUTs into membership function GRADEs. These GRADEs represent the degrees to which an INPUT0 datum is "cold," "warm," and "hot," and an INPUT1 value is "dim," "normal," and "bright."
2. Evaluate membership function GRADEs on the basis of nine IF/THEN rules (C) and derive three fuzzy outputs (FUZOUTs). These FUZOUTs represent the degrees to which the motor should run "slow," "medium," and "fast."
3. De-fuzzify (D) the FUZOUTs with reference to output membership function specifications (singletons) and derive a single crisp motor control output (DFUZOUT). DFUZOUT is the weighted average of the three singleton values. The three FUZOUTs are the weights. Following paragraphs discuss fuzzy inference in more detail.

Refer again to Figure 10–14. The system designer stores fuzzy control parameters in the M68HC11's EEPROM. Fuzzy inference software applies these parameters to the crisp data. First, the software fuzzifies (B) INPUT0 and INPUT1. Fuzzification of a crisp temperature input (INPUT0) evaluates degrees (GRADEs) to which it belongs to three membership functions (MFs): cold temperature, warm temperature, and hot temperature. Fuzzification expresses GRADEs of "coldness", "warmness," and "hotness" as values between $00 and $FF. Likewise, the software fuzzifies INPUT1 into "dim," "normal," and "bright" GRADEs between $00 and $FF. The system designer furnishes fuzzification criteria in the form of input Membership Function parameters. Thus, the fuzzification routine (B) calculates six

MF GRADEs—for cold, warm, and hot temperatures; and dim, normal, and bright light. Next, fuzzy inference subjects these six fuzzified MF GRADEs to rule evaluation (C).

A system designer forms fuzzy rules in a simple IF/THEN format. For example:

> If temperature is warm, and If light is dim, then motor speed is slow.

Note that this rule is anything but "fuzzy." To the contrary, it is exact, as are all fuzzy rules in the set. But fuzzy inference arbitrates between the fuzzy rules and evaluates them in a relativistic way. Accordingly, fuzzy rule evaluation actually applies the example rule in this way:

> To the extent that temperature is warm, and to the extent that light is dim, then motor speed should tend to be slow.

The designer encodes these precise fuzzy rules and stores them in EEPROM. In our Figure 10–14 system, fuzzy inference uses nine rules. The Rule Evaluation routine (C) applies these rules to the six MF GRADEs and derives three fuzzy outputs (FUZOUTs). FUZOUTs express the degrees to which the motor should run "slow" (FUZOUT0), "medium" (FUZOUT1), and "fast" (FUZOUT2). Rule evaluation generates these three FUZOUTs as values between $00 and $FF, then passes them to the de-fuzzification routine (D).

De-fuzzification applies the FUZOUTs to output MF specifications (singletons) and derives a weighted average (center of gravity—COG) of the singletons. The FUZOUTs constitute the weights in this calculation. Fuzzy inference passes this weighted average—labeled DFUZOUT—to the CRISPOUT subroutine (E). CRISPOUT converts the DFUZOUT value to $DUTY. $DUTY is the pulse width of a motor speed control square wave in 500ns units.

CRISPOUT passes the $DUTY value to the OC5 Interrupt Service routine (F). This interrupt service routine generates a motor control square wave's falling edges. The OC1 Interrupt Service routine (G) generates the square wave's rising edges. These two interrupt service routines therefore generate a 1 KHz motor speed control square wave. Motor speed varies inversely with $DUTY. The Fuzzy Inference subroutine varies $DUTY in accordance with prevailing temperature and light conditions. The fuzzy controller operates continuously. As soon as the CRISPOUT subroutine passes $DUTY to the OC5 Interrupt Service routine, program flow branches back to the A/D subroutine for another control cycle. The M68HC11's output compare mechanism generates OC1 and OC5 interrupts during the entire fuzzy control cycle. Keeping the overall fuzzy control system (Figure 10–14) in mind, the reader can proceed to an examination of the fuzzy inference process itself.[8]

Fuzzy Inference—Fuzzification

In our example application, fuzzification evaluates the degrees (GRADEs) to which a crisp temperature input is "cold," "warm," and "hot." Likewise, fuzzification evaluates a crisp light input into "dim," "normal," and "bright" GRADEs. To fuzzify a crisp input, the software applies input membership function parameters. A designer develops membership functions in a graphical format. Refer to Figure 10–15, which graphs temperature input membership functions. The graph's X axis represents INPUT0, a crisp hexadecimal temperature input from the M68HC11's ADC. The X axis scales a crisp INPUT0 value from $00 to $FF. Of course, X axis values relate directly to temperature sensor outputs of 0V to 5V. The graph's Y axis scales the degrees (GRADEs) to which an INPUT0 value is a member of the "cold," "warm," and "hot" membership functions (MFs). Each of these three MFs has a trapezoidal shape. Trapezoidal shapes work well in fuzzy applications. INPUT0 is "warm" to some degree if its X axis value lies between $40 and $C0. INPUT0 is "warm" to a maximum degree if its X axis value lies between $60 and $A0 (values which correspond to the top of the "warm" trapezoid). Thus, if INPUT0 lies in the range $60 to $A0 (X axis), it is "warm" to a GRADE of $FF (Y axis). If INPUT0 lies in the range $40 to $60, or $A0 to

---

[8]The fuzzy inference software presented in this chapter is based upon Motorola Inc., Fuzzy Logic Freeware, *M68HC11 Fuzzy Inference Engine*, 1990.

FIGURE 10-15

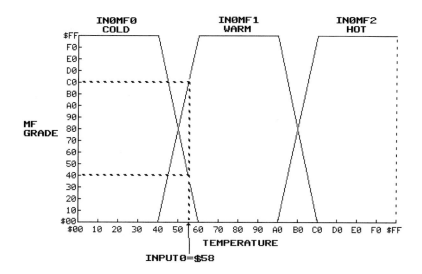

$C0, it is "warm" to some lesser GRADE. For example, if INPUT0 = $58, it is "warm" to a GRADE of $C0. The "warm" MF is assigned a label IN0MF1 (Input 0, Membership Function 1). Therefore, given INPUT0 = $58, IN0MF1 = $C0.

Refer to the trapezoidal shape for input membership function IN0MF0 (cold). INPUT0 is "cold" to a maximum GRADE ($FF) if it lies in the X axis range $00 to $40. INPUT0 is "cold" to a lesser GRADE if its value lies between $40 and $60. Note that our example INPUT0 value of $58 is "cold" to a GRADE of $3F. Therefore if INPUT0 = $58, then its IN0MF0 GRADE is $3F. Thus, our example INPUT0 value of $58 (X axis) is "cold" to a GRADE of $3F (Y axis) and "warm" to a GRADE of $C0 (Y axis).

The trapezoid for IN0MF2 (hot) ranges over INPUT0 X axis values $A0 through $FF. In the range $A0 to $C0, INPUT0 is partly "hot." If its value lies between $C0 and $FF, INPUT0 is "hot" to a maximum GRADE, i.e., IN0MF2 = $FF. Our example INPUT = $58 does not intersect the IN0MF2 (hot) trapezoid at all. If INPUT0 = $58, then IN0MF2 = $00.

QUESTION: If INPUT0 = $D0, what are the GRADEs for IN0MF0, IN0MF1, and IN0MF2?

$$IN0MF0 = \$00$$
$$IN0MF1 = \$00$$
$$IN0MF2 = \$FF$$

QUESTION: If INPUT0 = $A8, what are the resultant membership GRADEs?

$$IN0MF0 = \$00$$
$$IN0MF1 = \$BF$$
$$IN0MF2 = \$3F$$

This discussion illustrates an important difference between fuzzy inference and conventional algorithms. Note that fuzzy inference recognizes "shades of gray" between membership functions. In our example, INPUT0 = $58 is partly "cold" and partly "warm." Conventional algorithms permit no such "shades of gray." To achieve comparable control, a conventional algorithm divides temperature into a larger number of discrete categories. Each category includes only a small range of INPUT0 values, (e.g., $00–$08 or $09–$10). Any INPUT0 value falls into only one of these discrete categories. Increasing INPUT0 values change abruptly from one category to the next, making smooth output control difficult. Triangular or trapezoidal fuzzy membership functions gradually phase in and out, and an INPUT0 value can be a partial member of two categories. Therefore, only three MF categories can generate smooth output control.

Thus, fuzzification translates a crisp INPUT0 value into membership GRADEs labeled IN0MF0, IN0MF1, and IN0MF2. To make these translations, the Fuzzy Inference subroutine must have access to the trapezoid specifications. The designer specifies each trapezoid as 4 bytes of data: Point 1, Slope 1, Point 2, and Slope 2. Figure 10–16 illustrates how points and slopes are established for the IN0MF1 (warm) trapezoid.

FIGURE 10–16

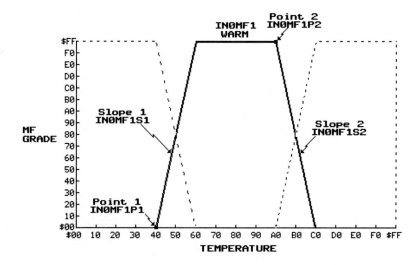

Refer to Figure 10–16. Point 1 (IN0MF1P1) = $40 (X axis), and Point 2 (IN0MF1P2) = $A0 (X axis). Find the value of any *nonvertical* slope as the ratio of change on the Y axis to causal change on the X axis. For example, calculate IN0MF1S1. For a change of $10 in INPUT0 (X axis), GRADE of membership (Y axis) changes by $80. To find the slope, divide the Y axis change by the corresponding X axis change:

$$IN0MF1S1 = \$80/\$10$$
$$= \$08$$

Since the IN0MF1 trapezoid is symmetrical, Slope 2 (IN0MF1S2) is also $08. Thus, the specifications for the IN0MF1 (warm) trapezoid are

$$IN0MF1P1 = \$40$$
$$IN0MF1S1 = \$08$$
$$IN0MF1P2 = \$A0$$
$$IN0MF1S2 = \$08$$

The system designer stores MF point and slope specifications in EEPROM, so that the fuzzification routine can access them. Of course, specifications must also be stored for IN0MF0 and IN0MF2. Figure 10–17 shows specifications for the left trapezoid, IN0MF0. Arbitrarily assign a value of $FF to any vertical slope.[9] For example, IN0MF0S1 = $FF.

For rightmost trapezoid IN0MF2, the specifications would be

$$IN0MF2P1 = \$A0$$
$$IN0MF2S1 = \$08$$
$$IN0MF2P2 = \$FF$$
$$IN0MF2S2 = \$FF$$

Again, assign vertical slope IN0MF2S2 a value of $FF. The system designer stores all of these MF specifications to nonvolatile memory.

The designer must replicate the specification process for INPUT1, light. Figure 10–18 shows a membership function graph for light. Note that it applies triangular membership functions to light. Triangular functions have no particular advantage in our example application. The author uses triangles to illustrate their use. Figure 10–18 shows how the designer derives point and slope specifications from the graph. Use the center triangle as an example. INPUT1 (light), Membership Function 1 (normal), Point 1 (IN1MF1P1) located where the triangle originates at the X axis, closest to the graph's origin. Point 2 (IN1MF1P2) is located at the triangle's peak. Calculate slopes in the same manner as for trapezoids:

$$IN1MF1S1 = Y \text{ axis change}/X \text{ axis change}$$
$$= \$FF/\$50 = \$03$$

---

[9]We know that a vertical slope is actually infinite. However, an arbitrary value of $FF (Y axis maximum) works well for fuzzification.

FIGURE 10–17

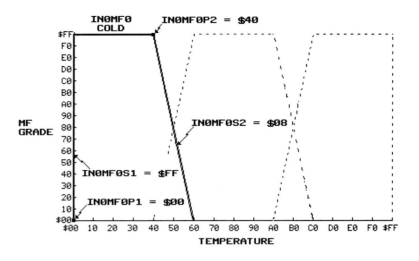

FIGURE 10–18

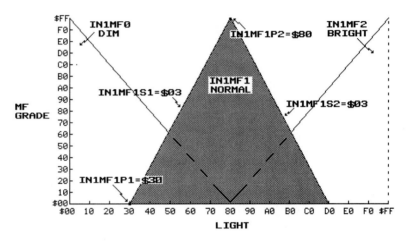

Since Slope 2 is equal to Slope 1:

$$IN1MF1S2 = \$03$$

Thus, IN1MF1's specifications are

$$IN1MF1P1 = \$30$$
$$IN1MF1S1 = \$03$$
$$IN1MF1P2 = \$80$$
$$IN1MF1S2 = \$03$$

Assign an arbitrary value of $FF to vertical slopes. Therefore, the left-hand (dim) triangle's Slope 1 equals $FF. The right-most (bright) triangle's Slope 2 also equals $FF. Calculate the Slope 2 specification for IN1MF0 (dim):

$$IN1MF0S2 = \$FF/\$80 = \$02$$

IN1MF2 (bright) uses the same slope for its IN1MF2S1:

$$IN1MF2S1 = \$FF/\$80 = \$02$$

IN1MF0P1 is located at $00 on the X axis. Therefore,

$$IN1MF0P1 = \$00$$

IN1MF0P2 is also located at $00:

$$IN1MF0P2 = \$00$$

IN1MF2P1 is located at $80 on the X axis:

$$IN1MF2P1 = \$80$$

IN1MF2P2 is located at $FF on the X axis:

IN1MF2P2 = $FF

The list below shows all membership function (MF) specifications, as stored in the M68HC11's EEPROM. Refer to this allocated memory block as the Input Membership Function Block (INMFBLK).

INMFBLK

$ADDRESS	$CONTENTS	LABEL	COMMENT
			TEMPERATURE
B600	00	IN0MF0P1	Cold MF, Point 1
B601	FF	IN0MF0S1	" ", Slope 1
B602	40	IN0MF0P2	" ", Point 2
B603	08	IN0MF0S2	" ", Slope 2
B604	40	IN0MF1P1	Warm MF, Point 1
B605	08	IN0MF1S1	" ", Slope 1
B606	A0	IN0MF1P2	" ", Point 2
B607	08	IN0MF1S2	" ", Slope 2
B608	A0	IN0MF2P1	Hot MF, Point 1
B609	08	IN0MF2S1	" ", Slope 1
B60A	FF	IN0MF2P2	" ", Point 2
B60B	FF	IN0MF2S2	" ", Slope 2
			LIGHT
B60C	00	IN1MF0P1	Dim MF, Point 1
B60D	FF	IN1MF0S1	" ", Slope 1
B60E	00	IN1MF0P2	" ", Point 2
B60F	02	IN1MF0S2	" ", Slope 2
B610	30	IN1MF1P1	Normal MF, Point 1
B611	03	IN1MF1S1	" ", Slope 1
B612	80	IN1MF1P2	" ", Point 2
B613	03	IN1MF1S2	" ", Slope 2
B614	80	IN1MF2P1	Bright MF, Point 1
B615	02	IN1MF2S1	" ", Slope 1
B616	FF	IN1MF2P2	" ", Point 2
B617	FF	IN1MF2S2	" ", Slope 2

The M68HC11 uses MF specifications to calculate GRADEs of membership for each crisp input (INPUT0 and INPUT1). The microcontroller evaluates a crisp input value three times, once for each of its relevant membership functions. To calculate a particular GRADE of membership, the M68HC11 must determine where the crisp input value (X axis) lies with respect to a membership function's trapezoid or triangle. A crisp value can fall into one of three areas, Segment 0, Segment 1, or Segment 2 (SEG0, SEG1, SEG2). If the crisp input (X axis) falls to the left of an MF's Point 1, it falls in SEG0. If the crisp input's X axis value falls between Point 1 and Point 2, it falls in SEG1. If the crisp input falls to the right of Point 2, it is in SEG2. To make these determinations, the CPU compares a crisp input value with an MF's Point 1 and Point 2 specifications.

Figure 10–19 illustrates these segments for the trapezoids on an INPUT0 function graph. This graph's arrows indicate the segment areas for right trapezoid (R), center trapezoid (C), and left trapezoid (L). Note from Figure 10–19 that a crisp input cannot fall into SEG0 for the left trapezoid, or SEG2 of the right trapezoid, since these two segments lie outside the graph's X axis range.

If a crisp input falls in an MF's SEG0, its GRADE of membership in that MF is $00. If this input intersects the trapezoid in SEG1, the microcontroller calculates GRADE of membership using this equation:

$$\text{MF GRADE (SEG1)} = (\text{INPUTx} - \text{INxMFxP1}) \times \text{INxMFxS1}$$

FIGURE 10–19

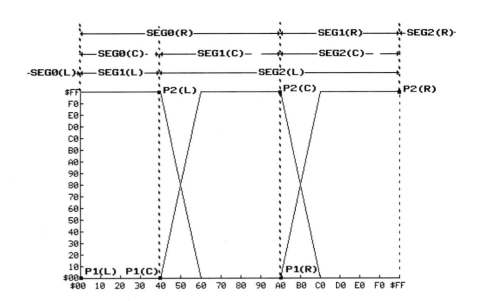

Refer to Figure 10–15, which shows crisp INPUT0 = $58, falling in SEG1 of IN0MF1 (warm). In this example:

$$\begin{aligned}\text{MF GRADE (SEG1)} &= (\text{INPUT0} - \text{IN0MF1P1}) \times \text{IN0MF1S1} \\ &= (\$58 - \$40) \times \$08 \\ &= \$C0\end{aligned}$$

What is INPUT0's MF GRADE in membership function IN0MF0 (cold)? INPUT0 = $58 falls in SEG2 of IN0MF0's trapezoid. If the intersection falls in SEG2, the M68HC11 calculates GRADE using this equation:

$$\text{MF GRADE (SEG2)} = \$FF - [(\text{INPUTx} - \text{INxMFxP2}) \times \text{INxMFxS2}]$$

Thus, the microcontroller calculates INPUT0's GRADE of membership in IN0MF0 follows:

$$\begin{aligned}\text{MF GRADE (SEG2)} &= \$FF - [(\text{INPUT0} - \text{IN0MF0P2}) \times \text{IN0MF0S2}] \\ &= \$FF - [(\$58 - \$40) \times \$08] \\ &= \$FF - (\$18 \times \$08) \\ &= \$FF - \$C0 \\ &= \$3F\end{aligned}$$

What is INPUT0's GRADE of membership in IN0MF2 (hot)? It is $00, because INPUT0 = $58 falls in SEG0 of IN0MF2's trapezoid.

Given the above, fuzzification software must accomplish two tasks for each crisp input versus each relevant membership function. First, the software must determine whether INPUTx falls in SEG0, SEG1, or SEG2, with respect to a trapezoid or triangle. Second, the software must calculate INPUTx's GRADE of membership for that trapezoid or triangle. The calculation procedure depends upon whether INPUTx falls in SEG0, SEG1, or SEG2.

Since our application uses two crisp inputs and uses three MFs per input, the M68HC11 must execute these tasks six times. Refer to Figure 10–20, which flowcharts the first task—determining whether INPUTx falls in SEG0, SEG1, or SEG2 of a particular trapezoid (FINDSEG).

Follow Figure 10–20 execution when the M68HC11 CPU performs the FINDSEG task on INPUT0 = $58 and IN0MF0 (cold). After loading ACCA with $58 (block 1), the CPU compares this INPUT0 value with IN0MF0P1 ($00). Since INPUT0>IN0MF0P1, block 3 execution branches program flow to branch to block 4. In block 4, the CPU compares INPUT0 ($58) with IN0MF0P2 ($40). Since INPUT0>IN0MF0P2, program flow branches from block 5 to the SEG2 operation.

How does the FINDSEG routine flow for INPUT0 = $58 versus IN0MF1? The block 2 comparison of $58 and IN0MF1P1 ($40) causes the CPU to branch from block 3 to block 4.

FIGURE 10-20

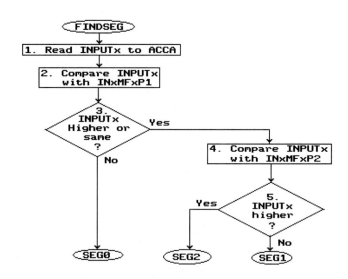

In block 4, the CPU compares $58 with IN0MF1P2 ($A0). Since INPUT0<IN0MF1P2, the CPU takes the "no" path from block 5 and executes the SEG1 routine.

In the case of IN0MF2, the block 2 comparison between $58 and IN0MF2P1 ($A0) causes the CPU to take the "no" path from block 3 and execute the SEG0 routine.

Having determined the segment into which INPUTx falls, the M68HC11 CPU fuzzifies this input by making a membership GRADE calculation. If INPUTx falls in SEG0, its membership GRADE = $00. No calculation is necessary. If INPUTx intersects in SEG1, the CPU calculates GRADE of membership as shown in Figure 10-21. Recall that the CPU uses this equation to calculate a SEG1 GRADE:

$$\text{MF GRADE (SEG1)} = (\text{INPUTx} - \text{INxMFxP1}) \times \text{INxMFxS1}$$

For example, if INPUT0 = $30, and the CPU calculates GRADE for IN0MF0, the CPU executes the Figure 10-21 SEG1 routine as follows.

In block 1, the CPU loads its B accumulator with IN0MF0S1 ($FF). Recall that we arbitrarily assigned this vertical slope a value of $FF. In block 2, the CPU compares IN0MF0S1 with $FF to determine whether it is vertical. If Slope 1 = $FF, INPUT0 must intersect the IN0MF0 trapezoid along the top. INPUT0's GRADE of membership is therefore $FF. In block 2, the CPU branches to block 7 because IN0MF0S1 = $FF. In block 7, the CPU assigns a membership GRADE value of $FF. In block 8, the CPU stores this GRADE to an allocated RAM location labeled IN0MF0GR (INPUT0, Membership Function 0, GRADE).

Consider another example, where INPUT0 = $98. If INPUT0 = $98, and the CPU calculates GRADE for IN0MF1 (the middle trapezoid), the CPU executes the Figure 10-21 SEG1 routine as follows.

In block 1, the CPU loads ACCB with IN0MF1S1 ($08). Since Slope 1 does not equal $FF, program flow proceeds to block 3, where the CPU subtracts IN0MF1P1 ($40) from INPUT0 ($98), generating a difference of $58. In block 4, the CPU multiplies this difference ($58) by IN0MF1S1 ($08), generating a product of $2C0. In block 5, the CPU determines whether this product is less than $100. In this example, the block 4 product is greater than $100. If the product equals or exceeds $100, INPUTx intersects the trapezoid along the top. The CPU must limit the GRADE to $FF. Because the product exceeds $100, the CPU takes the "no" path to block 7 where it assigns a GRADE of $FF. In block 8, the CPU stores this GRADE to the IN0MF1GR RAM location.

Consider a third example. If INPUT0 = $58 and the M68HC11 calculates GRADE for IN0MF1, then its CPU executes blocks 1, 2, 3, 4, 5, and 6. Because its block 4 product ($C0) is lower than $100, the CPU takes the "yes" branch from block 6 *around* block 7. The GRADE is therefore $C0. In block 8, the CPU stores $C0 to the IN0MF1GR location.

If INPUTx intersects a trapezoid in SEG2, the CPU calculates GRADE as shown in Figure 10-22. Recall that the CPU uses this equation to calculate a SEG2 GRADE:

$$\text{MF GRADE (SEG2)} = \$FF - [(\text{INPUTx} - \text{INxMFxP2}) \times \text{INxMFxS2}]$$

FIGURE 10–21

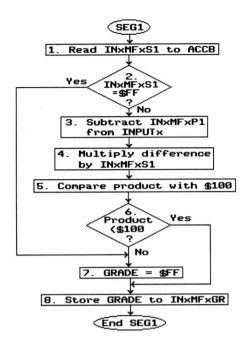

FIGURE 10–22

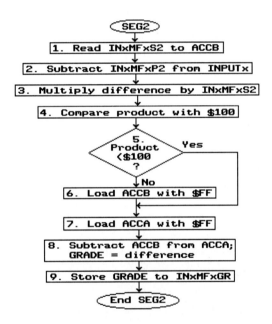

For example, if INPUT0 = $58 and the CPU calculates GRADE for IN0MF0, it executes the Figure 10–22 SEG2 routine as follows.

In block 1, the CPU loads its B accumulator with IN0MF0S2 ($08). In block 2, the CPU uses the A accumulator to subtract IN0MF0P2 ($40) from INPUT0 ($58), generating a difference of $18 in ACCA. In block 3, the CPU multiplies the contents of the two accumulators, returning a product of $08 × $18 = $00C0 to the D register. Note that the least significant hex digits of the product ($C0) reside in ACCB. In blocks 4 and 5, the CPU tests the product to see whether it is lower than $100. In this example, the CPU takes the "yes" branch around block 6 because $C0 < $100. In blocks 7 and 8, the CPU subtracts the product $C0 from $FF to derive GRADE = $3F. In block 9, the CPU stores $3F to the IN0MF0GR location.

Consider a second example. If INPUT0 = $88, the CPU executes all Figure 10–22 blocks and assigns a GRADE = $00 for IN0MF0. In block 2, the CPU subtracts $40 from $88, generating a difference of $48. In block 3, the CPU generates a product of $48 × $08 = $0240. Since this product exceeds $100, the CPU takes the "no" path from block 5 and executes block 6. After the CPU executes blocks 6 and 7, both accumulators contain $FF. The block 8 subtraction leaves a GRADE of $00. This GRADE is correct because INPUT0 = $88 does not intersect the IN0MF0 trapezoid at all.

By finding six GRADE values, the CPU completes fuzzification. The microcontroller must now evaluate these fuzzy inputs (or GRADES) in the light of the IF/THEN rules established by the system designer. This next phase is called Rule Evaluation.

Fuzzy Inference—Rule Evaluation

The system designer formulates if/then rule bytes that describe system operation. The designer codes each rule byte into if fields and a then field. To do rule evaluation, the M68HC11 accesses the relevant fields and uses them to derive fuzzy outputs. Fuzzy rules are easily established and encoded. In the case of our example system, we shall base rules on this general requirement:

The higher the temperature and brighter the light, the faster the motor should run.

Our application uses two crisp inputs, temperature and light, and classifies these inputs into membership functions (MFs) as follows:

Temperature **Light**
Hot (2) Bright (2)
Warm (1) Normal (1)
Cold (0) Dim (0)

Note that a numerical label designates each MF. These labels become fields in the rule bytes. Classify motor response as follows:

Motor Speed
Fast (2)
Medium (1)
Slow (0)

Formulate the above input and motor response categories into rule statements with this syntax:

If _____, and If _____, Then _____.

If statements describe input membership functions, and then statements describe motor response (output) classifications. For example,

If temperature is hot, **and** If light is bright, Then motor speed is fast.

Since the system uses only two inputs and one output, the designer can draw a matrix that implements the general requirement and provides a guide to rule construction. Refer to Figure 10–23, which shows one possible matrix. The labels within the matrix squares refer to motor speed response. If it turns out that system response fails to meet requirements or expectations, we can easily change the rules of operation. For example, we could change the matrix so that the only combination causing a motor fast response is light bright and temperature hot. In Exercise 10.2, the reader will have a chance to change the rules and observe the consequences of these changes. Having constructed the matrix, establish nine rules, one for each square in the matrix:

Rule 1. If temperature is hot (2), and If light is bright (2), Then motor speed is fast (2).[10]

Rule 2. If temperature is hot (2), and If light is normal (1), Then motor speed is fast (2).

Rule 3. If temperature is hot (2), and If light is dim (0), Then motor speed is medium (1).

Rule 4. If temperature is warm (1), and If light is bright (2), Then motor speed is fast (2).

Rule 5. If temperature is warm (1), and If light is normal (1), Then motor speed is medium (1).

---

[10]Recall that we established numerical labels 2, 1, and 0 for temperature and light categories.

FIGURE 10–23

```
                                 Light(1)
                    Bright(2) Normal(1)  Dim(0)
                   +---------+---------+---------+
          Hot(2)   |Fast(2)  |Fast(2)  |Medium(1)|
                   +---------+---------+---------+
Temperature(0)
          Warm(1)  |Fast(2)  |Medium(1)|Slow(0)  |
                   +---------+---------+---------+
          Cold(0)  |Medium(1)|Slow(0)  |Slow(0)  |
                   +---------+---------+---------+
```

Rule 6. If temperature is warm (1), and If light is dim (0), Then motor speed is slow (0).

Rule 7. If temperature is cold (0), and If light is bright (2), Then motor speed is medium (1).

Rule 8. If temperature is cold (0), and If light is normal (1), Then motor speed is slow (0).

Rule 9. If temperature is cold (0), and If light is dim (0), Then motor speed is slow (0).

Having formulated, stated, and labeled the rules, code and store them in memory. Assign a memory byte to each rule. Code each rule byte as shown in Figure 10–24.

Using the scheme presented in Figure 10–24, encode Rule 1 as follows:

`$2A = %00101010`

This rule byte encodes temperature hot (02), light bright (02), and motor fast (02). For Exercise 10.2, encode and store rule bytes in EEPROM as shown below. Note that the final rule byte is $FF, an End-of-Rules marker.

			RULBLK
ADDRESS	**$CONTENTS**	**LABEL**	**COMMENT**
B61B	2A	RUL1	Rule 1
B61C	26	RUL2	Rule 2
B61D	21	RUL3	Rule 3
B61E	1A	RUL4	Rule 4
B61F	15	RUL5	Rule 5
B620	10	RUL6	Rule 6
B621	09	RUL7	Rule 7
B622	04	RUL8	Rule 8
B623	00	RUL9	Rule 9
B624	FF	EOR	End-of-Rules marker

Recall once more that fuzzification calculates a GRADE of membership for each INPUT (temperature and light) versus each relevant membership function. Fuzzy Rule Evaluation applies these GRADEs to the nine rules and derives three byte-length fuzzy

FIGURE 10–24

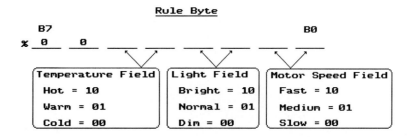

output values—for motor speed slow, motor speed medium, and motor speed fast. Label fuzzy outputs as follows:

FUZOUT0 = motor speed slow
FUZOUT1 = motor speed medium
FUZOUT2 = motor speed fast

Fuzzy rule evaluation software uses a minimum/maximum strategy to derive these three FUZOUT values. The M68HC11 CPU first evaluates a given rule byte's IF fields, and stores the minimum corresponding MF GRADE to a RAM address labeled IFMIN. For example, assume that the CPU evaluates "IF temperature is warm (01)" and "IF light is bright (02)." The corresponding GRADEs for these IF statements are IN0MF1GR (warm) and IN1MF2GR (bright). The CPU stores the lowest (minimum) of these two GRADEs to IFMIN. If IFMIN turns out to be $00, the CPU goes on to the next rule without evaluating the current rule's THEN field. If IFMIN is other than $00, the CPU evaluates the THEN field of the current rule.

To make a THEN field evaluation, the CPU compares the rule's IFMIN value with the applicable corresponding FUZOUT value. For example, assume that the THEN field being evaluated is "THEN motor speed is fast." The corresponding FUZOUT for this statement is FUZOUT2 (fast). If IFMIN is higher than the current corresponding FUZOUT value, the CPU stores IFMIN to FUZOUT. Thus, the CPU stores an IFMIN value to the corresponding FUZOUT, *if* it exceeds IFMIN values from previous rules which have identical THEN statements. After evaluating all nine rules, the CPU has stored maximum applicable IFMIN values to FUZOUT0 (from motor speed slow rules), FUZOUT1 (from motor speed medium rules), and FUZOUT2 (from motor speed fast rules).

Figure 10–25 flowcharts fuzzy rule evaluation. In block 1, the M68HC11 CPU initializes all three FUZOUT RAM locations to $00. The CPU then reads the first rule byte (block 2) and determines whether this byte is End-of-Rules marker $FF (block 3). Since RUL1 is not $FF, the CPU takes the "no" path from block 3 to block 4. In block 4, the CPU uses the rule byte's IF fields to access their corresponding MF GRADEs. In block 5, the CPU compares these two GRADEs and stores the lowest to IFMIN. If IFMIN is zero, the CPU takes the "yes" branch from block 6 back to block 2, where it reads the next rule byte. If IFMIN is not zero, the CPU takes the "no" path from block 6 to block 7. In block 7, the CPU uses the current rule byte's THEN field to access the corresponding FUZOUT. In block 8, the CPU compares IFMIN with the rule's corresponding FUZOUT value. If IFMIN is higher than FUZOUT, the CPU takes the "no" path from block 8 to block 9. In block 9, the CPU replaces the current FUZOUT value with the larger IFMIN value. If the

FIGURE 10–25

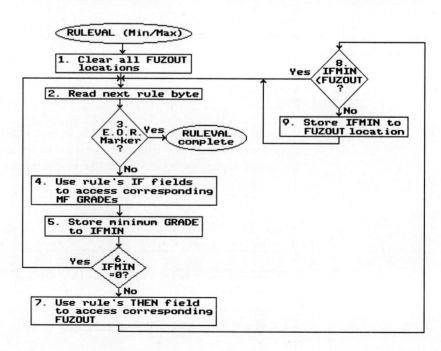

block 8 IFMIN/FUZOUT comparison shows FUZOUT to be higher, the CPU takes the "yes" branch from block 8, back to block 2, skipping block 9. After executing block 9, the CPU branches back to block 2. At block 2, the CPU reads the next rule byte and determines whether or not it is the End-of-Rules marker. If not, the CPU proceeds to block 4 and evaluates the rule. When the CPU reads the End-of-Rules marker, it has evaluated al nine rule bytes. Therefore, the CPU takes the "yes" branch from block 3.

Thus, the CPU applies a minimum/maximum (MIN/MAX) approach to rule evaluation. The CPU first evaluates a given rule's IF fields, and stores the minimum applicable membership function GRADE to the IFMIN in RAM.[11] If IFMIN turns out to be $00, the CPU goes on to the next rule without evaluating the current rule's THEN field. If IFMIN is other than $00, the CPU evaluates the current rule byte's THEN field. Each time the CPU evaluates a THEN field, it may store the IFMIN value to an applicable fuzzy output (FUZOUTx).[12] The CPU stores this IFMIN value if it exceeds corresponding IFMIN values from previous rules having identical THEN statements. Thus, the CPU stores maximum applicable IFMIN values to FUZOUT0 (from motor speed slow rules), FUZOUT1 (from motor speed medium rules), and FUZOUT2 (from motor speed fast rules). After completing this MIN/MAX rule evaluation, the CPU goes on to de-fuzzify the FUZOUT values.

Fuzzy Inference—De-fuzzification

To de-fuzzify, the CPU applies output membership function specifications to the three FUZOUT values and makes a center of gravity (COG) calculation. COG is the de-fuzzified output (DFUZOUT).

For our example system, specify output MF functions as motor speed slow, motor speed medium, and motor speed fast—corresponding to FUZOUT0, FUZOUT1, and FUZOUT2. A designer can specify these output functions with trapezoids or triangles in the same manner as input MFs (see Figure 10–26). However, such trapezoidal or triangular output MFs make the COG calculation quite complex. Designers can often specify output MFs as singletons: selected specific values corresponding to each FUZOUT. A singleton is a specific X axis value, of zero width, which is FUZOUTx units high. See Figure 10–27, which shows singleton values at X axis locations $30 (OUTMF0), $80 (OUTMF1), and $D0 (OUTMF2). The height of each singleton is the Y axis value of its corresponding FUZOUT. Figure 10–27 uses the following FUZOUT values as examples:

$$FUZOUT0 = \$3F$$
$$FUZOUT1 = \$7F$$
$$FUZOUT2 = \$7F$$

FIGURE 10–26

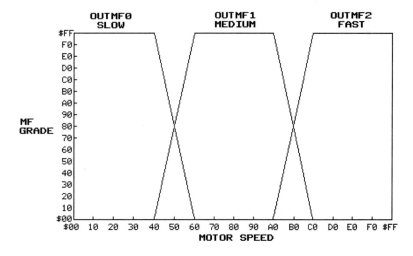

---

[11]Given an If byte "if temperature is hot," the applicable MF GRADE is IN0MF2 (hot).
[12]Given a then byte "then motor speed is medium," the CPU would store the IFMIN value to the applicable fuzzy output—FUZOUT1 (motor speed is medium).

FIGURE 10-27

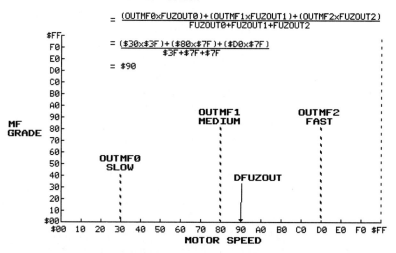

The M68HC11 calculates DUFUZOUT (COG) as the singletons' weighted average. FUZOUT values are the weights. Figure 10-27 shows this COG calculation for our example FUZOUT0, FUZOUT1, and FUZOUT2 values. Figures 10-28 and 10-29 flowchart the software for calculating DFUZOUT (COG) from FUZOUT0, FUZOUT1, and FUZOUT2.

Refer to Figure 10-28, which illustrates how the M68HC11 derives the numerator and denominator of the weighted average calculation. Applicable memory allocations are as follows:

$ADDRESS	RAM	EEPROM	$CONTENTS	LABEL
0002	X		$3F	FUZOUT0 (motor speed slow)
0003	X		$7F	FUZOUT1 (motor speed medium)
0004	X		$7F	FUZOUT2 (motor speed fast)
0005	X		---	DFUZOUT
0007	X		---	FOUTSUMH
0008	X		---	FOUTSUML
0009	X		---	SUMPRODH
000A	X		---	SUMPRODL
000B	X		---	LOOPCOUN
B618		X	$30	OUTMF0 (singleton)
B619		X	$80	OUTMF1 (singleton)
B61A		X	$D0	OUTMF2 (singleton)

Locations $0002–$0004 contain our example FUZOUT values. Locations $B618–$B61A contain the singleton values as shown in Figure 10-27. Locations $0007–$0008 will

FIGURE 10-28

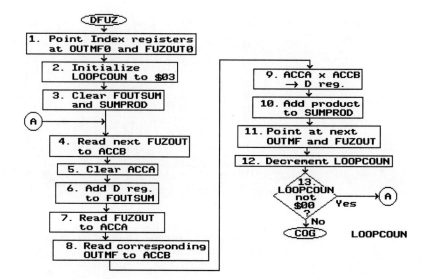

contain the denominator value of FUZOUT0 + FUZOUT1 + FUZOUT2. Locations $0009–$000A will contain the sum-of-products value (OUTMF0 × FUZOUT0) + (OUTMF1 × FUZOUT1) + (OUTMF2 × FUZOUT2).

Refer to Figure 10–28, blocks 1, 2, and 3. Here, the M68HC11 initializes index registers and the LOOPCOUN, FOUTSUM, and SUMPROD RAM locations. In block 1, the CPU points its Y register at the first singleton (OUTMF0) location, and its X register at FUZOUT0. In block 2, the CPU initializes the loop counter—labeled LOOPCOUN. The CPU initializes LOOPCOUN with the number of singletons/FUZOUTs to be processed: $03. In block 3, the CPU clears result locations $0007–$000A. After completing these initializations, the CPU enters a loop where it derives the numerator and denominator for the COG (DFUZOUT) calculation.

In Figure 10–28 (blocks 4, 5, and 6), the CPU accumulates the denominator value (FOUTSUM) in memory locations $0007 and $0008. Recall that this denominator is the sum of the three FUZOUT values. In block 4, the CPU reads the next FUZOUT value to ACCB. In block 5, the CPU clears ACCA. By clearing ACCA, the CPU makes the FUZOUT a 16-bit value, $00XX, contained in the D register. In block 6, the CPU adds FOUTSUM to this 16-bit FUZOUT value, then stores the sum back to RAM addresses $0007–$0008.

In Figure 10–28 (blocks 7–10), the CPU accumulates the numerator value (SUMPROD) in memory locations $0009–$000A. Recall that this numerator is a sum of products: (OUTMF0 × FUZOUT0) + (OUTMF1 × FUZOUT1) + (OUTMF2 × FUZOUT2). In blocks 7 and 8, the CPU reads the current FUZOUT value to ACCA, and its corresponding singleton value (OUTMFx) to ACCB. In block 9, the CPU multiplies these two values. In block 10, the CPU adds the resulting product to the SUMPROD value it is accumulating in RAM locations $0009–$000A.

To properly accumulate the numerator and denominator values, the CPU must make three loops. In blocks 11, 12, and 13, the CPU manages these loop passes and operand accesses. In block 11, the CPU increments its X and Y registers to point at the next OUTMF and FUZOUT locations. In block 12, it decrements LOOPCOUN. If LOOPCOUN = $00 after this decrement, the CPU has made three loop passes and can move on to the COG calculation. If LOOPCOUN is not $00, the CPU takes the "yes" branch from block 13 to block 4 for another pass through the loop. After three loop passes, the CPU has calculated the numerator and denominator of the COG equation. Using our example values of FUZOUT0 = $3F, FUZOUT1 = $7F, and FUZOUT2 = $7F, the value of the numerator and denominator would be the following:

$$\text{Numerator (SUMPROD)} = \$B280$$
$$\text{Denominator (FOUTSUM)} = \$013D$$

Figure 10–29 shows the division of the numerator by the denominator to derive the DFUZOUT (COG) value. Refer to blocks 1, 2, and 3. In these blocks, the CPU integer divides (IDIV) SUMPROD (D register) by FOUTSUM (X register). The IDIV instruction returns the quotient to the X register. As the weighted average of the singletons, the quotient will never exceed the value of the OUTMF2 singleton ($D0 in our example). Therefore, by exchanging the D and X registers (block 4), the CPU transfers the significant digits of the

FIGURE 10–29

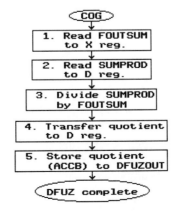

quotient to ACCB. In block 5, the CPU writes the quotient from ACCB to the de-fuzzified output's storage location—labeled DFUZOUT. Given our example values,

$$\text{SUMPROD (numerator)} = \$B280$$
$$\text{FOUTSUM (denominator)} = \$013D$$

the COG routine calculates DFUZOUT as follows:

$$\text{DFUZOUT} = \$B280/\$013D$$
$$= \$90$$

Material on pages 498–512 describes how the Fuzzy Inference subroutine takes two crisp inputs from the A/D converter, and subjects them to fuzzification, rule evaluation, and de-fuzzification:

1. Fuzzify the crisp inputs into six membership function grades (fuzzy inputs).
2. Evaluate these fuzzy inputs on the basis of nine IF/THEN rules, deriving three fuzzy outputs.
3. De-fuzzify the fuzzy outputs with reference to output MF specifications (singletons) and derive a crisp output.

The Fuzzy Inference subroutine can then pass the crisp output (DFUZOUT) to the CRISP-OUT subroutine, which converts DFUZOUT into $DUTY.

10.4 MOTOR CONTROL USING THE M68HC11 AND FUZZY INFERENCE

This section presents complete hardware and software necessary to implement our fuzzy logic application. Recall that its general objective is to vary DC motor speed in response to changes in environmental temperature and light. Refer again to Figure 10–14, which illustrates the following requirements:

1. Interface hardware includes sensor circuits, DC motor, and motor control circuitry. The reader developed sensor circuits in Part 3 of Exercise 10.1. The reader connected motor control circuitry in Chapter 8's laboratory exercises. The M68HC11 will control motor speed with a 1 KHz variable duty cycle square wave, as it did in Chapter 8's exercises. Thus, the motor control circuitry requires no modifications. The reader will connect motor control circuitry to the M68HC11's OC5 pin.
2. The MAIN program uses a modular strategy to manage software execution. MAIN calls four subroutines: INIT, A/D, FUZZY, and CRISPOUT. MAIN's length is only 18 bytes.
3. The INIT subroutine initializes interrupt pseudovectors, TCTL1, OC1M, OC1D, TMSK1, $DUTY, TOC1, and TOC5. The M68HC11 uses these resources to generate the 1 KHz motor speed control square wave.
4. The A/D subroutine causes the M68HC11's A/D converter (ADC) to periodically scale four analog inputs and stop. Our application requires only two of these analog inputs—temperature and light. Therefore, the A/D subroutine reads only the ADR registers corresponding to these inputs. Ground the two other analog input pins.
5. The FUZZY subroutine analyzes the two crisp inputs (temperature and light) passed to it by the A/D subroutine. The FUZZY subroutine also uses a modular approach. It calls three nested subroutines:
   - The FUZZIFY nested subroutine fuzzifies the crisp temperature and light inputs. To accomplish this, FUZZIFY nests further subroutines: FINDSEG, SEG1 and SEG2.
   - The RULEVAL nested subroutine subjects fuzzified inputs to rule evaluation and derives fuzzy outputs.
   - The DFUZ nested subroutine de-fuzzifies the fuzzy outputs and generates a crisp output value—DFUZOUT.
6. The CRISPOUT subroutine converts de-fuzzified output value DFUZOUT to $DUTY. Recall that $DUTY is the number of free-running counts in the 1 KHz motor speed control square wave's pulse width.
7. The OC1 and OC5 Interrupt Service routines generate a 1 KHz motor speed control square wave. The OC1 Interrupt Service routine generates the square wave's rising edges, and the OC5 Interrupt Service routine generates falling edges.

Following sections deal individually with these seven requirements.

FIGURE 10-30

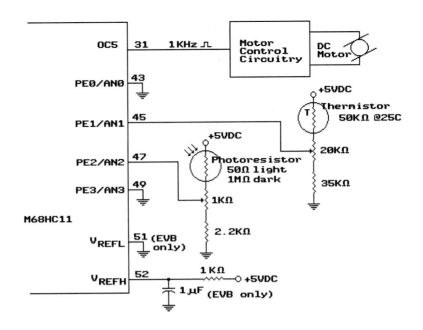

Interface and Connections

Figure 10–30 diagrams fuzzy motor control connections. In Exercise 10.2 below, make these connections using sensor circuits developed for Exercise 10.1 (the ADC exercise above). Motor control circuitry is the same as shown in Chapter 9, Figure 9–16.

MAIN Program and RAM Memory Allocations

Fuzzy motor control software uses 20 bytes of RAM for data storage, as shown in the box entitled RAM Memory Allocations. Store all program routines in EEPROM.

RAM Memory Allocations—EVB and EVBU

$ADDRESS	LABEL	COMMENT
0000	INPUT0	Temperature (crisp)
0001	INPUT1	Light (crisp)
0002	FUZOUT0	Fuzzy output, motor speed slow
0003	FUZOUT1	Fuzzy output, motor speed medium
0004	FUZOUT2	Fuzzy output, motor speed fast
0005	DFUZOUT	De-fuzzified output (crisp)
0006	IFMIN	Minimum IF GRADE
0007	FOUTSUMH	Sum of FUZOUTs MS
0008	FOUTSUML	Sum of FUZOUTs LS
0009	SUMPRODH	Sum of FUZOUTs x OUTMFs MS
000A	SUMPRODL	Sum of FUZOUTs x OUTMFs LS
000B	LOOPCOUN	DFUZ loop counter
000C	$DUTYH	$DUTY MS
000D	$DUTYL	$DUTY LS
000E	IN0MF0GR	Cold temperature GRADE
000F	IN0MF1GR	Warm temperature GRADE
0010	IN0MF2GR	Hot temperature GRADE
0011	IN1MF0GR	Dim light GRADE
0012	IN1MF1GR	Normal light GRADE
0013	IN1MF2GR	Bright light GRADE

FIGURE 10–31

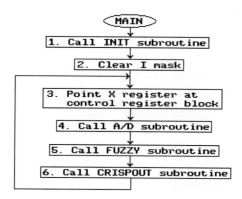

MAIN Program—EVB and EVBU

$ADDRESS	SOURCE CODE	COMMENTS
B62A	JSR B640	Call INIT
B62D	CLI	Clear global I mask
B62E	LDX #1000	Point X register at control register block
B631	JSR B690	Call A/D
B634	JSR B6AA	Call FUZZY
B637	JSR B7A0	Call CRISPOUT
B63A	BRA B62E	Repeat

Figure 10–31 and its accompanying program list illustrate the fuzzy motor controller's MAIN program.

Refer to Figure 10–31. In block 1 (program address $B62A), the M68HC11 CPU calls the INIT subroutine. INIT initializes output compare interrupt pseudovectors and the TCTL1, OC1M, OC1D, TMSK1, TOC1, and TOC5 control registers. Thus, INIT prepares the M68HC11's output compare mechanism to generate the 1 KHz motor speed control square wave. Upon its return from INIT, the CPU clears the global I mask (block 2), which allows square wave generation to begin. The output compare mechanism uses interrupt control to generate the leading and trailing edges of the square wave's pulses. These interrupts can occur at any time the CPU is executing MAIN program blocks 3 through 6, or the subroutines called in these blocks.

After initializing the X register (Figure 10–31, block 3), the CPU calls the A/D subroutine (block 4, MAIN program address $B631). The A/D subroutine scales analog light and temperature inputs from the sensors, and stores the digitized results in RAM locations $0000 and $0001. Thus, these two RAM locations hold the crisp inputs to be fuzzified.

In Figure 10–31, block 5 (program address $B634), the CPU calls the FUZZY subroutine. This subroutine fuzzifies the two crisp inputs into six fuzzy inputs, or GRADEs. The FUZZY subroutine then subjects the GRADEs to rule evaluation and derives three fuzzy outputs (FUZOUTs). The CPU stores these FUZOUTs to RAM addresses $0002–$0004. The FUZZY subroutine then de-fuzzifies the FUZOUTs, deriving a crisp output (DFUZOUT). The CPU stores DFUZOUT to RAM address $0005.

In Figure 10–31, block 6 (program address $B637), the CPU calls the CRISPOUT subroutine. CRISPOUT calculates $DUTY and stores this value to RAM locations $000C–$000D.

To summarize, the MAIN program uses a modular approach by calling four subroutines. Output compare interrupt routines control 1 KHz square wave generation. These interrupt routines may occur any time after the CPU executes MAIN program, block 2. After executing blocks 1 and 2, the CPU enters a continuous loop where it calls and re-calls subroutines that convert crisp values and apply fuzzy inference.

INIT Subroutine

INIT prepares the M68HC11's output compare mechanism for motor speed control square wave generation. The output compare mechanism uses interrupt control to generate

FIGURE 10-32

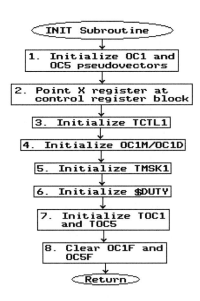

INIT SUBROUTINE—EVB AND EVBU

$ADDRESS	SOURCE CODE	COMMENTS
B640	LDD #B670	Initialize OC1 pseudovector
B643	STD E0	" " " " " "
B645	LDD #B680	Initialize OC5 pseudovector
B648	STD D4	" " " " " "
B64A	LDX #1000	Point X register at control register block
B64D	LDAA #2	Initialize TCTL1
B64F	STAA 20,X	" " "
B651	LDD #808	Initialize OC1M/OC1D
B654	STD C,X	" " " "
B656	LDAA #88	Initialize TMSK1
B658	STAA 22,X	" " "
B65A	LDD #3E8	Initialize $DUTY
B65D	STD C	" " "
B65F	LDD E,X	Initialize TOC1
B661	ADDD #30	" " "
B664	STD 16,X	" " "
B666	ADDD C	Initialize TOC5
B668	STD 1E,X	" " "
B66A	BCLR 23,X 77	Clear OC1F and OC5F
B66D	RTS	Return

this square wave. Figure 10–32 and its accompanying program list illustrate the INIT subroutine.

Refer to block 1 of Figure 10–32 (program addresses $B640–$B649). The M68HC11 uses OC1 interrupts to generate the output square wave's rising edges. Therefore, the microcontroller initializes the Timer Output Compare 1 Interrupt pseudovector address in RAM locations $00E0–$00E1. Since the microcontroller uses OC5 interrupts to generate falling edges, it initializes the OC5 interrupt pseudovector address in RAM locations $00D4–$00D5.

After initializing its X register (block 2), the M68HC11 must initialize TCTL1 (block 3, program address $B64D). The TCTL1 register controls the action an output compare pin takes upon a successful compare. TCTL1 controls this action for pins OC2–OC5. The M68HC11 uses the OC5 mechanism to generate falling edges. Therefore, the CPU writes %10 to TCTL1's OM5 and OL5 bits.

Refer to Figure 10–32, block 4 (program address $B651). The M68HC11 uses the OC1 mechanism to generate rising edges at the OC5 pin. Since the OC1 mechanism must generate rising edges, the CPU writes binary ones to the OC1M5 bit in the OC1M register and the OC1D5 bit in the OC1D register. OC1M and OC1D reside at contiguous addresses in the M68HC11's control register block. Therefore, the CPU loads its D register with two 8-bit control words, one for OC1M and one for OC1D. The CPU then stores the D register to the OC1M location. This write stores ACCA's control byte to OC1M, and ACCB's control byte to OC1D.

Refer to Figure 10–32, block 5 (program address $B656). In this block, the CPU enables OC1 and OC5 interrupts by writing binary ones to the OC1I and OC5I (I4O5I) bits in the TMSK1 register.

In Figure 10–32, block 6 (program address $B65A), the CPU initializes its $DUTY value to $03E8. Recall that $DUTY is the number of free-running counts in the output square wave's pulse width. A $DUTY value of $03E8 runs the motor at about half speed. $DUTY remains at this value ($03E8) only until the FUZZY and CRISPOUT subroutines make their initial executions. Thereafter, FUZZY and CRISPOUT control the magnitude of $DUTY.

In block 7 (program addresses $B65F–$B669), the M68HC11 CPU initializes the Timer Output Compare registers for the first output square wave pulse. The CPU initializes the TOC1 register to TCNT plus $0030, so that the square wave's initial rising edge will occur after the CPU returns from the INIT subroutine. The CPU then adds $DUTY to the TOC1 value and stores the sum to TOC5. Thus, the output compare mechanism will generate an output falling edge $DUTY's worth of free-running counts after a rising edge.

Refer to block 8 (program address $B66A). Here, the CPU clears the OC1F and OC5F (I4O5F) flags in the TFLG1 register. When they set, OC1F and OC5F generate corresponding output compare interrupts. Upon return from the INIT subroutine, the CPU will clear the global I mask, enabling these output compare interrupts to commence. The CPU will then enter a continuous loop where it reads light and temperature data and adjusts $DUTY accordingly.

Motor Control Waveform Generation

While the M68HC11 executes the A/D, FUZZY, and CRISPOUT subroutines, it is interrupted periodically. Each successful OC1 or OC5 compare generates an output edge and an interrupt. The interrupt service routine updates the relevant TOCx register and clears the associated OCxF flag in TFLG1. Figure 10–33 and its accompanying software lists illustrate the output compare interrupt routines.

The M68HC11's OC1 mechanism generates the square wave's rising edges. Each successful comparison between the free-running count and the TOC1 register generates a rising edge and an OC1 interrupt. Refer to the OC1 Interrupt Service in Figure 10–33. This interrupt service routine updates the TOC1 register for the next rising edge, to occur after $07D0 free-running counts elapse. In block 1 of this routine (program address $B670), the M68HC11 re-initializes its X register. The OC1 mechanism may generate this interrupt from any point in the MAIN program loop, or A/D, FUZZY, and CRISPOUT subroutines. The FUZZY subroutine changes the X register's contents. The X register may or may not be pointing at the control register block when the CPU vectors to the OC1 Interrupt Service routine. Therefore, in flowchart block 1, the CPU re-points its X register at the control register block.

In block 2 of the OC1 Interrupt Service routine (program addresses $B673–$B679), the CPU reads the current TOC1 contents to its D register, adds $07D0 to this value, and writes the sum back to TOC1. The value $07D0 equals the number of free-running counts in the period of the 1 KHz motor speed control square wave. Therefore, by adding $07D0, the CPU updates TOC1 for the next output rising edge.

Before returning from the OC1 Interrupt Service routine, the CPU clears the OC1F flag in the TFLG1 register (block 3, program address $B67A). Recall that the CPU must clear this flag to avoid a spurious interrupt request immediately upon return from the current interrupt service routine.

FIGURE 10-33

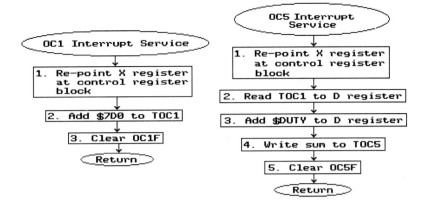

OC1 INTERRUPT SERVICE ROUTINE—EVB AND EVBU

$ADDRESS	SOURCE CODE	COMMENTS
B670	LDX #1000	Re-point X register at control register block
B673	LDD 16,X	Add $7D0 to TOC1
B675	ADDD #7D0	" " "
B678	STD 16,X	" " "
B67A	BCLR 23,X 7F	Clear OC1F
B67D	RTI	Return

OC5 INTERRUPT SERVICE ROUTINE—EVB AND EVBU

$ADDRESS	CONTENTS	COMMENTS
B680	LDX #1000	Re-point X register at control register block
B683	LDD 16,X	Read TOC1
B685	ADDD C	Add $DUTY
B687	STD 1E,X	Write sum to TOC5
B689	BCLR 23,X F7	Clear OC5F
B68C	RTI	Return

The M68HC11's OC5 mechanism generates the motor control square wave's falling edges. Each successful comparison between the free-running count and the TOC5 register generates a falling edge and OC5 interrupt. Refer to the OC5 Interrupt Service flowchart in Figure 10–33. In block 1, the CPU re-points its X register at the control register block. The CPU must re-point the X register because its actual contents are unknown. In block 2 (program address $B683), the CPU reads TOC1's contents to its D register. The most recent OC1 Interrupt Service routine updated TOC1 for the next output rising edge. The next falling edge should occur $DUTY's worth of free-running counts later. Therefore, the CPU adds $DUTY to the TOC1 read value in its D register (block 3, program address $B685). In block 4 (program address $B687), the CPU writes this sum to TOC5. Thus, the CPU updates TOC5 for a falling edge to occur $DUTY's worth of free-running counts after the next rising edge. In block 5 (program address $B689), the CPU prevents a spurious OC5 interrupt by clearing the OC5F bit in TFLG1 before returning from the interrupt.

A/D Subroutine

Refer to Figure 10–30. Note that the temperature sensor applies analog temperature values to the M68HC11's AN1 pin, and the light sensor applies analog light values to AN2. Note also that the AN0 and AN3 pins are grounded. The A/D subroutine converts four analog channels—AN0 through AN3—one time and stops. The CPU then reads the digitized temperature value from ADR2 and the digitized light value from ADR3. The CPU stores these values to the INPUT0 and INPUT1 locations in RAM. The CPU does not read the ADR1

FIGURE 10-34

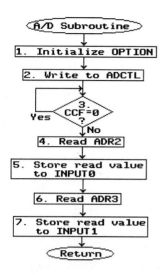

A/D SUBROUTINE—EVB AND EVBU

$ADDRESS	CONTENTS	COMMENTS
B690	LDAA #80	Initialize OPTION
B692	STAA 39,X	" " "
B694	LDAA #10	Write to ADCTL
B696	STAA 30,X	" " "
B698	BRCLR 30,X 80 B698	Branch to self until CCF=1
B69C	LDAA 32,X	Read ADR2
B69E	STAA 0	Store read value to INPUT0
B6A0	LDAA 33,X	Read ADR3
B6A2	STAA 1	Store read value to INPUT1
B6A4	RTS	Return

and ADR4 registers because their corresponding AN pins are grounded and their data irrelevant. Of course, INPUT0 and INPUT1 become the crisp inputs to the FUZZY subroutine.

Refer to Figure 10-34 and its accompanying program list. In block 1 (program address $B690), the M68HC11 CPU initializes its OPTION register by writing a binary one to the ADPU bit and a zero to CSEL. In block 2 (program address $B694), the CPU writes $10 to its ADCTL register. This write value clears the SCAN bit, sets the MULT bit, and selects channels AD0–AD3. SCAN = 0 causes the A/D converter to scale four inputs and stop. MULT = 1 causes the A/D converter to convert a four-channel group. CD = 0 and CC = 0 select channels AD0–AD3. Of course, this write to ADCTL resets the CCF flag and causes conversions to begin.

In block 3 (program address $B698), the CPU loops until the CCF flag sets in ADCTL, indicating that four conversions are complete. The CPU then reads the contents of the ADR2 register (block 4, program address $B69C), and stores this read value to the INPUT0 location in RAM (block 5, program address $B69E). In block 6 (program address $B6A0), the CPU reads ADR3. In block 7 (program address $B6A2), the CPU stores this read value to INPUT1 in RAM. The CPU then returns to the MAIN program from the A/D subroutine. Having obtained and scaled crisp input values from the sensors, the microcontroller must now apply fuzzy inference.

FUZZY Subroutine

The FUZZY subroutine uses a modular approach similar to the MAIN program's. FUZZY calls three nested subroutines: FUZZIFY, RULEVAL, and DFUZ. Thus, the FUZZY subroutine proper is very simple; it makes three subroutine calls, then returns to the MAIN program. See the box below, which contains the FUZZY subroutine source code listing.

FUZZY SUBROUTINE—EVB AND EVBU

$ADDRESS	CONTENTS	COMMENTS
B6AA	JSR B6BA	Call FUZZIFY
B6AD	JSR B720	Call RULEVAL
B6B0	JSR B76A	Call DFUZ
B6B3	RTS	Return

Of course, FUZZY's three nested subroutines execute fuzzy inference's three phases: fuzzification, rule evaluation and de-fuzzification. These nested subroutines execute the flow charts presented on pages 498–512. Now we shall elaborate on these flowcharts and relate them to actual program listings.

The FUZZY subroutine first calls the FUZZIFY nested subroutine. Recall that fuzzification takes two crisp inputs—temperature and light—and converts them to six fuzzy inputs. Each fuzzy input grades the degree to which a crisp input belongs to a relevant membership function. Fuzzification grades a crisp input three times, once for each relevant input function's trapezoid or triangle. To derive a fuzzy input GRADE, fuzzification software must execute two tasks for each membership function (MF):

1. Compare MF Point 1 and Point 2 parameters with the crisp input value. Determine the segment into which the crisp input falls: Segment 0, Segment 1, or Segment 2.
2. Make an appropriate GRADE calculation, based upon whether the crisp input falls in SEG0, SEG1, or SEG2.

The FUZZIFY nested subroutine manages crisp input accesses, MF parameter accesses, and GRADE storage. FUZZIFY calls further nested subroutines, which execute the two fuzzification tasks:

1. The FINDSEG nested subroutine determines the segment in which a crisp input falls.
2. Based upon its segment determination, FINDSEG calls the SEG1 nested subroutine or the SEG2 nested subroutine. These routines make appropriate GRADE calculations. In the case of SEG0, the FINDSEG routine itself assigns a GRADE of $00.

Figure 10–35 provides a coat hanger diagram illustrating the modular relationship between the FUZZY subroutine and its various nested subroutines.

Refer to Figure 10–36, which flowcharts the FUZZIFY nested subroutine. Refer also to its accompanying FUZZIFY source code list. Flowchart blocks 1 and 2 initialize the X and Y registers. The LDY instruction (at program address $B6BA) points the Y register at the first fuzzy input GRADE location in RAM, labeled IN0MF0GR. Fuzzification software will store fuzzy inputs to these GRADE locations as it derives them. The LDX instruction

FIGURE 10–35

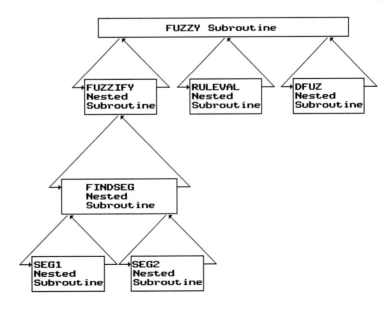

FIGURE 10–36

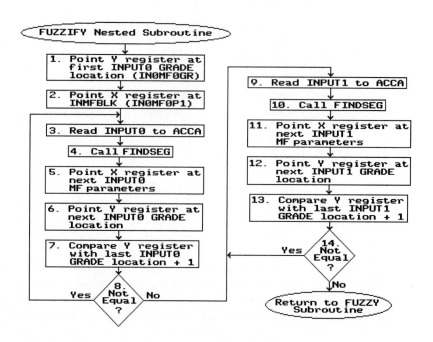

FUZZIFY NESTED SUBROUTINE—EVB AND EVBU

$ADDRESS	CONTENTS	COMMENTS
B6BA	LDY #E	Point Y register at IN0MF0GR
B6BE	LDX #B600	Point X register at IN0MF0P1
B6C1	LDAA 0	Read INPUT0
B6C3	JSR B6E2	Call FINDSEG
B6C6	LDAB #4	Point X register at next IN0MFxP1
B6C8	ABX	" " " " " "
B6C9	INY	Point Y register at next IN0MFxGR
B6CB	CPY #11	Does Y register point at IN1MF0GR?
B6CF	BNE B6C1	If not, calculate another GRADE
B6D1	LDAA 1	If Y register points at IN1MF0GR, read INPUT1
B6D3	JSR B6E2	Call FINDSEG
B6D6	LDAB #4	Point X register at next IN1MFxP1
B6D8	ABX	" " " " " "
B6D9	INY	Point Y register at next IN1MFxGR
B6DB	CPY #14	Does Y register point at IN1MF2GR + 1?
B6DF	BNE B6D1	If not, calculate another GRADE
B6E1	RTS	If all GRADEs calculated, return to FUZZY

(at program address $B6BE) points the X register at the first set of membership function parameters. Of course, the fuzzification software will apply these parameter sets to the crisp temperature and light inputs to derive six GRADEs.

Note that the Figure 10–36 flow chart consists of two loops. The first loop includes blocks 3 through 8. The second loop includes blocks 9 through 14. The first loop fuzzifies INPUT0, the crisp temperature input. The second loop fuzzifies INPUT1—light.

In block 3, the CPU loads ACCA with the crisp temperature value: INPUT0. Block 4 calls the FINDSEG nested subroutine. FINDSEG applies the trapezoidal or triangular function parameters pointed at by the X register. FINDSEG uses these parameters to find the MF segment in which INPUT0 lies. If INPUT0 lies in SEG0, FINDSEG stores $00 to the current GRADE location—pointed at by the Y register. If INPUT0 lies in SEG1 or SEG2, FINDSEG calls the appropriate nested subroutine, SEG1 or SEG2. This nested subroutine calculates and stores a GRADE to the location pointed at by the Y register.

In blocks 5, 6, and 7, the CPU advances the X and Y registers and determines whether it has applied all three relevant MF parameter sets to INPUT0. In block 5, the CPU advances the X register so that it points at the Point 1 parameter of the next MF set. A set consists of four parameters (Point 1, Slope 1, Point 2, Slope 2), each residing at a consecutive address. Therefore, the CPU must add $04 to the X register's contents. The CPU accomplishes this by loading ACCB with $04, then adding this value to the X register's contents. See the LDAB and ABX instructions at program addresses $B6C6–$B6C8.

In block 6, the CPU increments its Y register so that it points at the next fuzzy input GRADE location. The CPU then determines whether it has filled all three INPUT0 GRADE locations. If the CPU has calculated all three INPUT0 GRADEs, the block 6 increment points the Y register at address $0011—the first INPUT1 GRADE location. Thus, to execute block 7, the CPU compares the incremented Y register value with $0011. See the CPY instruction at program address $B6CB. If the CPU has more INPUT0 GRADEs to calculate, it takes the "yes" branch from block 8 back to block 3. If the CPU has calculated all three INPUT0 GRADEs, it takes the "no" path to block 9 in the second FUZZIFY loop.

Blocks 9 through 14 of Figure 10–36 manage the three INPUT1 GRADE calculations. In block 9, the CPU reads the INPUT1 value from its RAM location. The CPU then calls the FINDSEG nested subroutine (block 10). FINDSEG determines the segment in which INPUT1 lies. Based upon its determination, FINDSEG stores a SEG0 GRADE of $00 or calls either the SEG1 or SEG2 nested subroutine. SEG1 or SEG2 makes the appropriate calculation and stores the resultant GRADE. The CPU then returns to FUZZIFY, block 11.

In block 11, the CPU adds $04 to the X register's contents so that it points at the next MF parameters set. In block 12, the CPU increments its Y register to point at the next INPUT1 GRADE location. In block 13, the CPU compares its Y register contents with the last INPUT1 GRADE location, plus one ($0014). If the CPU has more INPUT1 GRADEs to calculate, it takes the "yes" branch from block 14 back to block 9. If it has calculated and stored all INPUT1 GRADEs, the CPU returns to the FUZZY subroutine.

In blocks 4 and 10 (Figure 10–36), the FUZZIFY nested subroutine calls FINDSEG, another nested subroutine. FINDSEG determines the MF trapezoid or triangle segment in which the current crisp input lies. FINDSEG then takes appropriate action to calculate a fuzzy input GRADE. Figure 10–37 and its accompanying program list illustrate the FINDSEG nested subroutine.

Previous material covered the principles of fuzzification, and Figure 10–20 illustrated the FINDSEG program flow. Note the similarity between Figure 10–20 and Figure 10–37. Previous exposure to the FINDSEG strategy should make it easier for the reader to understand.

Refer to Figure 10–37, block 1. Before calling FINDSEG, the FUZZIFY routine loads ACCA with the crisp INPUTx value. Also, FUZZIFY points the X register at the appropriate Point 1 parameter. Thus, the CPU executes Figure 10–37, block 1 by comparing its ACCA contents with the Point 1 value using X-indexed addressing. See the CMPA instruction at program address $B6E2.

If INPUTx (ACCA) is less than the Point 1 value, then INPUTx lies in Segment 0. The CPU takes the "no" path from block 2 to block 3. Recall that the FUZZIFY routine pointed the Y register at the current GRADE location (INxMFxGR). Thus, the CPU uses Y-indexed addressing to clear this GRADE to $00. See the CLR instruction at program address $B6E6.

If INPUTx is the same or higher than the Point 1 value, the CPU must make a further comparison. This comparison determines whether INPUTx lies in Segment 1 or Segment 2. Therefore, if INPUTx equals or exceeds Point 1, the CPU takes the "yes" branch from block 2 to block 4. In block 4, the CPU compares the INPUTx value with the current Point 2 parameter. The CMPA instruction at program address $B6EA executes block 4. If INPUTx is equal to or lower than Point 2, it lies in Segment 1. Therefore, the CPU takes the "no" path from block 5 to block 6. In block 6, the CPU calls the SEG1 nested subroutine. SEG1 calculates and stores the fuzzy input GRADE, then returns to FINDSEG. If INPUTx exceeds the Point 2 value, it lies in Segment 2. The CPU takes the "yes" branch from block 5 to block 7. In block 7, the CPU calls the SEG2 nested subroutine. SEG2 calculates a GRADE, stores it, and returns to FINDSEG. After the fuzzy input GRADE has been calculated and stored, FINDSEG returns program flow to the FUZZIFY nested subroutine. Note the RTS instructions at program addresses $B6E9, $B6F1, and $B6F5.

FIGURE 10–37

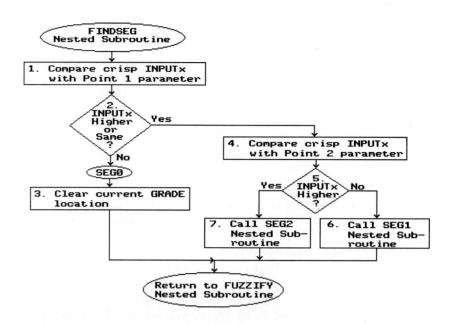

FINDSEG NESTED SUBROUTINE—EVB AND EVBU

$ADDRESS	CONTENTS	COMMENTS
B6E2	CMPA 0,X	Compare INPUTx with INxMFxP1
B6E4	BHS B6EA	If INPUTx is higher or same, it lies in SEG1 or SEG2
B6E6	CLR 0,Y	If INPUTx is lower, assign SEG0 GRADE $00
B6E9	RTS	Return to FUZZIFY
B6EA	CMPA 2,X	Compare INPUTx with INxMFxP2
B6EC	BHI B6F2	If INPUTx is higher, branch to SEG2 call
B6EE	JSR B6F6	If INPUTx is not higher, call SEG1
B6F1	RTS	Return to FUZZIFY
B6F2	JSR B70B	Call SEG2
B6F5	RTS	Return to FUZZIFY

Thus, the FINDSEG nested subroutine may call the SEG1 or SEG2 nested subroutine. Figure 10–38 and its accompanying source code listing illustrate the SEG1 routine. Compare Figure 10–38 with Figure 10–21 and note their similarity. Again, previous exposure should facilitate understanding.

To fully understand Figure 10–38, remember the following:

1. The CPU uses this equation to calculate a Segment 1 fuzzy input GRADE:

$$\text{MF GRADE (SEG1)} = (\text{INPUTx} - \text{INxMFxP1}) \times \text{INxMFxS1}$$

2. The FUZZIFY routine pointed the X register at the current INxMFxP1 (Point 1) parameter in EEPROM. The Slope 1 parameter resides in the memory byte following Point 1.
3. FUZZIFY pointed the Y register at the current fuzzy input GRADE location (INxMFxGR).
4. FUZZIFY loaded ACCA with the current INPUTx value.

In Figure 10–38, block 1, the CPU reads the current Slope 1 parameter to ACCB. The CPU then determines whether Slope 1 is vertical. Recall that we arbitrarily assigned a vale of $FF to all vertical slopes. If the crisp input lies in SEG1, and Slope 1 is vertical, then this input must intersect the MF trapezoid along the top. Therefore, the fuzzy input GRADE must = $FF. In block 2 (program address $B6F8), the CPU compares the Slope 1 parameter with $FF. If Slope 1 = $FF, the CPU takes the "yes" branch to block 7, where it assigns a GRADE of $FF. If Slope 1 is lower than $FF, the CPU takes the "no" path from block 2 to block 3. In block 3, the CPU subtracts the current Point 1 value (pointed at by the X register) from INPUTx (in ACCA). See the SUBA instruction at program address $B6FC.

FIGURE 10-38

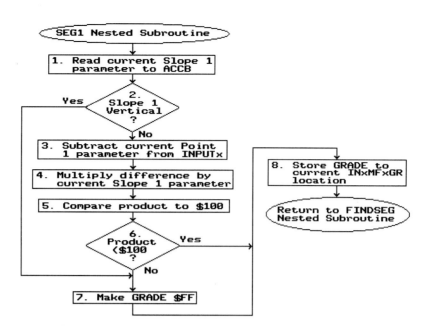

SEG1 NESTED SUBROUTINE—EVB AND EVBU

$ADDRESS	CONTENTS	COMMENTS
B6F6	LDAB 1,X	Load ACCB with Slope 1 parameter
B6F8	CMPB #FF	Is Slope 1 vertical?
B6FA	BEQ B705	If yes, GRADE = $FF
B6FC	SUBA 0,X	If no, subtract Point 1 value from INPUTx
B6FE	MUL	Multiply difference by Slope 1 value
B6FF	CPD #100	Is product lower than $100?
B703	BLO B707	If yes, ACCB holds valid GRADE
B705	LDAB #FF	Make GRADE = $FF
B707	STAB 0,Y	Store GRADE to INxMFxGR location
B70A	RTS	Return to FINDSEG

Recall that the block 1 instruction loaded ACCB with the Slope 1 parameter (see the LDAB instruction at program address $B6F6). Thus, the CPU can use the MUL instruction to execute block 4. This instruction multiplies the Slope 1 value (ACCB) by the difference generated in block 3 (ACCA), and returns the product to the D register. If the block 4 product equals or exceeds $FF, then the GRADE = $FF. A product less than $FF constitutes the actual GRADE value.

In block 5, the CPU compares the product (D register) with $100 to determine whether it should limit this product to $FF. If the product is less than $100 (i.e., the D register contains $00FF or less), the CPU takes the "yes" branch to block 8. If the product is $100 or more, the CPU takes the "no" path to block 7. To execute block 7, the CPU loads ACCB with a fuzzy input GRADE of $FF. See the LDAB instruction at program address $B705.

When the CPU enters block 8, ACCB holds a valid fuzzy input GRADE. If the block 4 product (D register) was $00FF or less, ACCB contains the 8-bit GRADE ($FF or less). If the CPU executed block 7, it loaded ACCB with a GRADE of $FF. Therefore, the CPU executes block 8 by storing the contents of ACCB to the INxMFxGR location pointed at by the Y register. See the STAB instruction at program address $B707. The RTS instruction (program address $B70A) returns program flow to the FINDSEG nested subroutine.

Figure 10-39 and its accompanying source code listing illustrate the SEG2 routine. Again, the reader should note the similarity between Figure 10-39 and Figure 10-22.

To facilitate understanding, remember the following:

1. The CPU uses this equation to calculate a Segment 2 fuzzy input GRADE:

$$\text{MF GRADE (SEG2)} = \$FF - [(\text{INPUTx} - \text{INxMFxP2}) \times \text{INxMFxS2}]$$

FIGURE 10–39

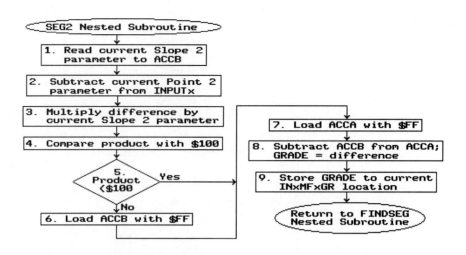

SEG2 NESTED SUBROUTINE—EVB AND EVBU

$ADDRESS	CONTENTS	COMMENTS
B70B	LDAB 3,X	Load ACCB with Slope 2 parameter
B70D	SUBA 2,X	Subtract Point 2 value from INPUTx
B70F	MUL	Multiply difference by Slope 2 value
B710	CPD #100	Is product lower than $100?
B714	BLO B718	If yes, GRADE = $FF – product
B716	LDAB #FF	If no, load ACCB with $FF (GRADE = $00)
B718	LDAA #FF	Load ACCA with $FF
B71A	SBA	GRADE = $FF – ACCB
B71B	STAA 0,Y	Store GRADE to current fuzzy input GRADE location
B71E	RTS	Return to FINDSEG

2. The FUZZIFY routine pointed the X register at the current Point 1 parameter in EEPROM.
3. FUZZIFY pointed the Y register at the current INxMFxGR location.
4. FUZZIFY loaded ACCA with the current crisp INPUTx value.

Refer to Figure 10–39. The CPU executes block 1 by reading the INxMFxS2 parameter to ACCB. Note the LDAB instruction at program address $B70B. This instruction uses an offset of $03 because the Slope 2 parameter lies three addresses from Point 1. Recall that the X register presently contains the Point 1 address. In block 2, the CPU subtracts the Point 2 parameter from the INPUTx value. Note the SUBA instruction at program address $B70D. ACCA contains INPUTx, and the SUBA instruction accesses Point 2 at the X register's contents plus $02. Of course, SUBA execution completes the part of the SEG2 calculation in parentheses.

To complete the part of the calculation in the brackets, the CPU executes block 3. The CPU multiplies the difference (parentheses result) in ACCA by the Point 2 parameter in ACCB. The MUL instruction (program address $B70F) returns the product to the D register.

In block 4, the CPU makes a comparison. If the block 3 result equals or exceeds $100, INPUTx does not intersect the MF trapezoid or triangle under consideration. In fact, INPUTx lies to the right of this MF. Therefore, the fuzzy input GRADE must equal $00. If the block 3 product is $FF or less, the CPU takes the "yes" branch from block 5 to block 7. If the product equals $100 or more, the CPU takes the "no" path from block 5 to block 6.

In block 6, the CPU assures that the fuzzy input GRADE will equal $00. The CPU does this by changing ACCB's contents to $FF. See the LDAB instruction at program address $B716. Remember that the CPU executes block 6 only if the block 3 product (the part of the equation in brackets) equals or exceeds $100.

In Blocks 7 and 8, the CPU completes the calculation by subtracting the equation's brackets result from $FF. In block 7, the CPU loads ACCA with $FF, the minuend. At this point ACCB contains one of two values. If the CPU executed block 6, ACCB contains $FF. If the CPU branched around block 6, ACCB contains the brackets product—a value between $00 and $FF. The CPU executes block 8 by subtracting ACCB's contents from ACCA's contents and returning the difference to ACCA. See the SBA instruction at program address $B71A. Of course, ACCA now holds the SEG2 GRADE.

In block 9, the CPU stores the fuzzy input GRADE to its proper RAM location. The Y register points at this INxMFxGR location. After storing the GRADE, the CPU returns to the FINDSEG nested subroutine. See the RTS instruction at program address $B71E.

Refer to Figure 10–37, the FINDSEG routine. Note that FINDSEG returns program flow to the FUZZIFY nested subroutine (Figure 10–36). After all six GRADEs have been calculated, FUZZIFY returns program flow to the FUZZY subroutine.

The FUZZY subroutine now calls the RULEVAL nested subroutine, which executes the next phase of fuzzy inference.[13] Recall that RULEVAL uses a MIN/MAX strategy to evaluate the six fuzzy input GRADEs in the light of nine rules established by the system designer. Review pages 506–509, which describe fuzzy rule evaluation. In particular, study Figure 10–25, which provides an overview of the rule evaluation process. To help you understand the RULEVAL software below, keep the following points in mind:

1. Nine rule bytes reside at addresses $B61B–$B624 in EEPROM.
2. Each rule byte contains two IF fields, one for temperature and one for light. Two binary bits constitute each IF field.
3. Each rule byte contains one THEN field, which describes motor speed (slow, medium, or fast). A THEN field consists of 2 binary bits.
4. Rule evaluation uses the IF fields to access corresponding fuzzy input GRADEs. Rule evaluation then selects the lowest or minimum (IFMIN) of the two GRADEs.
5. Rule evaluation uses the THEN field to access a corresponding fuzzy output (FUZOUT) value. Rule evaluation compares the IFMIN value with this FUZOUT. If IFMIN is higher than FUZOUT, rule evaluation replaces the current FUZOUT value with IFMIN. This substitution constitutes the MAX part of the MIN/MAX strategy.
6. The software evaluates all nine IF/THEN rules and derives three fuzzy outputs: FUZOUT0 (motor speed slow), FUZOUT1 (motor speed medium), and FUZOUT2 (motor speed fast). FUZOUT0 is the highest IFMIN from all of the "THEN motor speed is slow" rules. FUZOUT1 is the highest IFMIN from the "motor speed is medium" rules. FUZOUT2 is the highest IFMIN from the "motor speed fast" rules.

Figures 10–40 and 10–41, along with their accompanying software lists, illustrate the RULEVAL nested subroutine. Following paragraphs explain this software. Figure 10–40 shows how the M68HC11 accomplishes these three tasks:

1. Clear the fuzzy output (FUZOUT) storage locations in RAM.
2. Read a rule byte and determine whether it is the End-of-Rules marker.
3. Use the rule byte's IF fields to access a corresponding input temperature GRADE (IN0MFxGR) and light GRADE (IN1MFxGR). Load the IN0MFxGR value into ACCA and the IN1MFxGR value into ACCB.

In block 1, the CPU clears the three FUZOUTs. To do this, the CPU clears its D register, then writes the D register contents to FUZOUT0 and FUZOUT1. The CPU then writes the contents of ACCA ($00) to the FUZOUT2 location. See the LDD, STD, and STAA instructions at program addresses $B720–$B726.

In blocks 2 and 3, the CPU initializes its X and Y registers for GRADE and rule byte accesses. In block 2, the CPU points its Y register at the first IF/THEN rule byte address, *minus one*. The RULBLK begins at EEPROM address $B61B so the CPU loads its Y register with $B61A. See the LDY instruction at program address $B727. The RULEVAL routine is a loop, which the CPU executes nine times, once for each rule. At the beginning of each loop, the CPU increments its Y register, then uses it to access the next rule byte. The CPU will increment the Y register before accessing the first rule byte (RUL1). Therefore, the Y register must point at RUL1's address minus one.

---

[13] The Figure 10–35 coat hanger diagram summarizes the relationship of the FUZZY subroutine and its nested subroutines.

FIGURE 10–40

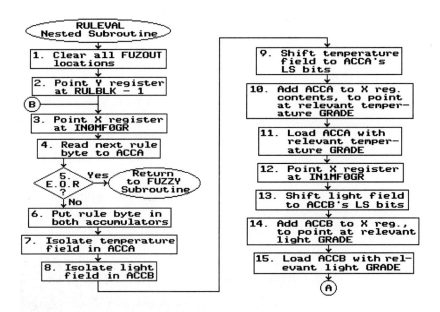

RULEVAL NESTED SUBROUTINE—EVB AND EVBU

$ADDRESS	CONTENTS	COMMENTS
B720	LDD #0	Clear all FUZOUTs
B723	STD 2	" " "
B725	STAA 4	" " "
B727	LDY #B61A	Point Y register at RULBLK–1
B72B	LDX #E	Point X register at IN0MF0GR
B72E	INY	Read next rule byte to ACCA
B730	LDAA 0,Y	" " " "
B733	CMPA #FF	End-of-Rules marker?
B735	BEQ B767	If yes, return to FUZZY subroutine
B737	TAB	If no, put rule byte in both accumulators
B738	ANDA #30	Isolate temperature field in ACCA
B73A	ANDB #C	Isolate light field in ACCB
B73C	LSRA	Shift temperature field to LS bits in ACCA
B73D	LSRA	" " " " " " "
B73E	LSRA	" " " " " " "
B73F	LSRA	" " " " " " "
B740	PSHB	Point X reg. at relevant temperature GRADE
B741	TAB	" " " " " " "
B742	ABX	" " " " " " "
B743	PULB	" " " " " " "
B744	LDAA 0,X	Read temperature GRADE to ACCA
B746	LDX #11	Point X register at IN1MF0GR
B749	LSRB	Shift light field to LS bits in ACCB
B74A	LSRB	" " " " "
B74B	ABX	Point Y reg. at relevant light GRADE
B74C	LDAB 0,X	Read light GRADE to ACCB

In block 3, the CPU points its X register at the first temperature GRADE—IN0MF0GR. See the LDX instruction at program address $B72B.

In block 4, the CPU increments its Y register to point at the next rule byte, then reads this byte to ACCA. See the INY and LDAA instructions at program addresses $B72E–$B732. The CPU must now check to see whether the rule byte is $FF, the End-of-Rules marker. If the rule byte = $FF, the CPU has evaluated all nine IF/THEN rules. Thus, the

CPU takes the "yes" branch from block 5 to an RTS instruction. The RTS returns program flow to the FUZZY subroutine. If the rule byte is other than $FF, the CPU must evaluate the rule and take the "no" path from block 5 to block 6. See the CMPA and BEQ instructions at program addresses $B733–$B736.

In block 6, the CPU transfers the rule byte to ACCB. Both ACCA and ACCB now contain the rule byte. The TAB instruction at program address $B737 causes the transfer.

In blocks 7 and 8, the CPU isolates the rule byte's temperature field in ACCA and its light field in ACCB. To isolate the temperature field (Block 7), the CPU ANDs the rule byte in ACCA with mask $30:

```
Mask = %0 0 1 1 0 0 0 0
```

This mask zeros out all of the bits in the rule byte, except the temperature field (bits 4 and 5). Thus, the temperature field remains intact in ACCA bits 4 and 5. See the ANDA instruction at program address $B738.

To isolate the light field (block 8), the CPU ANDs the rule byte in ACCB with mask $0C. This operation retains the light field in ACCB bits 2 and 3. See the ANDB instruction at program address $B73A.

In Blocks 9, 10, and 11, the CPU uses the temperature field to access the appropriate fuzzy input temperature GRADE. Here, an example may help the reader understand how the CPU does this. Assume, for example, that the temperature field = %01, IF temperature is warm. Because the temperature field = %01, the CPU must access the contents of the warm MF GRADE location, IN0MF1GR. IN0MF1GR resides at RAM address $000F. Block 3 execution pointed the X register at address $000E (IN0MF0GR). If the CPU adds the temperature field value %01 to the X register contents, the X register will then point at $000F, the target address. To accomplish this addition, the CPU must shift the temperature field from bits 4 and 5 in ACCA to its LS bit locations:

```
From %_ _ 0 1 _ _ _ _
to   %_ _ _ _ _ _ 0 1
```

Block 7 execution made all other ACCA bits zero. Therefore, after the temperature field shift, ACCA contains

```
%0 0 0 0 0 0 0 1 or $01
```

The CPU can now add this value ($01) to the X register contents ($000E) to point the X register at the "IF temperature warm" GRADE at address $000F.

Thus, blocks 9 and 10 shift the temperature field into ACCA's LS bits, then add ACCA's contents to the X register to point at the appropriate input MF GRADE. In block 11, the CPU can read the GRADE using X-indexed addressing mode.

Unfortunately, the M68HC11 has no instruction that adds the contents of ACCA to the X register. However, the ABX instruction adds the contents of ACCB to the X register. Therefore, to execute Blocks 9, 10, and 11, the CPU must take these actions:

1. Use four LSRA instructions to shift the temperature field into ACCA's LS bits. See program addresses $B73C–$B73F.
2. Push ACCB's contents onto the stack. This action saves the isolated light field (from block 8). See program address $B740.
3. Transfer ACCA's contents to ACCB. ACCB now contains the isolated, shifted temperature field. See the TAB instruction at program address $B741.
4. Use the ABX instruction to add the contents of ACCB to the X register. See program address $B742. The X register now points at the appropriate IN0MFxGR location in RAM.
5. Pull the isolated light field value from the stack, back to ACCB. See program address $B743.
6. Read the appropriate temperature IN0MFxGR value from RAM to ACCA. See the LDAA instruction at program address $B744.

In flowchart blocks 12–15, the CPU uses the rule byte's light field to access the appropriate corresponding light GRADE. This process uses the same strategy as for temperature, but it is simpler.

In block 12, the CPU loads the X register with $0011, which points at the IN1MF0GR location. See program address $B746.

In block 13, the CPU uses two LSRB instructions to shift the light field into ACCB's LS bits. See program addresses $B749–$B74A.

In block 14, the CPU executes the ABX instruction, which adds the light field value to the X register's contents. See program address $B74B. The X register now points at the appropriate IN1MFxGR location in RAM.

In block 15, the CPU reads the appropriate light IN1MFxGR value from RAM to ACCB.

The CPU must now select the lowest (IFMIN) of the two GRADE values it holds in its accumulators. If this IFMIN GRADE is other than $00, the CPU may then store it to an appropriate FUZOUT location. Remember, however, that the CPU stores this IFMIN GRADE to a FUZOUT *only* if it is larger than previously stored values. Figure 10–41 and its accompanying source code list illustrate this MIN/MAX process.

Refer to Figure 10–41, block 16. The CPU uses the CBA instruction (program address $B74E) to compare the temperature (ACCA) and light (ACCB) GRADEs. The CPU takes the "yes" branch from block 17 to block 19 if the temperature GRADE is less than the light GRADE. If the light GRADE is less, the CPU takes the "no" path from block 17 to block 18. See the BLO instruction at program address $B74F.

FIGURE 10–41

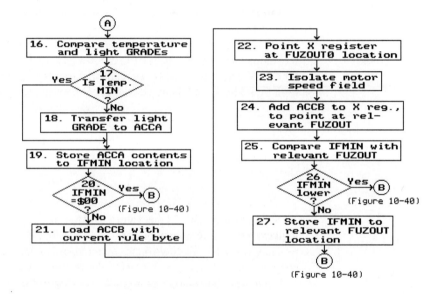

RULEVAL NESTED SUBROUTINE—EVB AND EVBU

$ADDRESS	CONTENTS	COMMENTS
B74E	CBA	Compare temperature and light GRADEs
B74F	BLO B752	If temperature GRADE is MIN, leave it in ACCA
B751	TBA	If light GRADE is MIN, transfer it to ACCA
B752	STAA 6	Store MIN GRADE to IFMIN location
B754	BEQ B72B	If IFMIN = $00, evaluate next rule
B756	LDAB 0,Y	Load ACCB with current rule byte
B759	LDX #2	Point X register at FUZOUT0
B75C	ANDB #3	Isolate motor speed field
B75E	ABX	Point X register at relevant FUZOUT
B75F	CMPA 0,X	Compare IFMIN with FUZOUT
B761	BLO B72B	If IFMIN lower, evaluate next rule
B763	STAA 0,X	If IFMIN higher, store to FUZOUTx location
B765	BRA B72B	Evaluate next rule
B767	RTS	Return to FUZZY subroutine

The CPU executes block 18 if the light GRADE (ACCB) is lower. In this block, the CPU transfers the light GRADE from ACCB to ACCA. See the TBA instruction at program address $B751. Thus, if the temperature GRADE is less, it remains in ACCA. If the light GRADE is less, the CPU transfers it to ACCA.

In block 19, the CPU stores the minimum GRADE (IFMIN) from ACCA to the allocated IFMIN location in RAM. See the STAA instruction at program address $B752. If this IFMIN value = $00, the CPU takes the "yes" branch from block 30 back to block 3 (Figure 10–40). If the CPU takes this "yes" branch, it evaluates the next rule byte. If IFMIN is not $00, the CPU takes the "no" path from block 20 to block 21. See the BEQ instruction at program address $B754.

In blocks 21 through 24, the CPU uses the motor speed field of the current rule byte to access a corresponding FUZOUT. This process resembles the GRADE accesses illustrated in Figure 10–40. In block 21, the CPU reads the current rule byte to ACCB. In block 22, the CPU points its X register at the FUZOUT0 location in RAM. See the LDX instruction at program address $B759.

In block 23, the CPU uses the ANDB instruction, with mask $03, to isolate the rule byte's motor speed field. See this instruction at program address $B75C. ANDB execution leaves the motor speed field in ACCB's least significant bits.

In block 24, the CPU uses the ABX instruction to add ACCB's contents to the X register. See the ABX instruction at program address $B75E. The X register now points at the appropriate FUZOUT location in RAM.

The CPU now executes block 25. Here, the CPU uses X-indexed addressing to compare the FUZOUT value (pointed at by the X register) with the IFMIN value in ACCA. See the CMPA instruction at program address $B75F. If IFMIN is less than FUZOUT, the CPU takes the "yes" branch from block 26 back to block 3 (Figure 10–40). At block 3, the CPU evaluates the next rule byte. However, if IFMIN exceeds FUZOUT, the CPU takes the "no" path from block 26 to block 27. Remember that if IFMIN exceeds the current FUZOUT value, IFMIN becomes the new FUZOUT value. Thus, in block 27, the CPU stores the IFMIN value from ACCA to the current FUZOUT location. Recall that the X register points at this location (per blocks 21–24). See the STAA instruction at program address $B763. After executing block 27, the CPU branches back to block 3 (Figure 10–40), where it evaluates the next rule byte.

After evaluating all rule bytes, the CPU takes the "yes" branch from block 5 (Figure 10–40) back to the FUZZY subroutine. At this point, the M68HC11 has fuzzified the crisp temperature and light inputs and has subjected these fuzzy inputs to rule evaluation. Rule evaluation generated three fuzzy outputs—FUZOUT0 (motor speed slow), FUZOUT1 (motor speed medium), and FUZOUT 2 (motor speed fast). The FUZZY subroutine now calls a nested subroutine that executes de-fuzzification. The DFUZ nested subroutine calculates a weighted average of three output membership singletons. FUZOUTs constitute the weights.

Material found on pages 509–512 provides a comprehensive description of de-fuzzification. Recall from this section that the de-fuzzification software accesses the following RAM and EEPROM locations:[14]

$ADDRESS	LABEL	
0002	FUZOUT0	
0003	FUZOUT1	
0004	FUZOUT2	
0005	DFUZOUT	
0007	FOUTSUMH	
0008	FOUTSUML	
0009	SUMPRODH	
000A	SUMPRODL	
000B	LOOPCOUN	
B618	OUTMF0	(singleton)
B619	OUTMF1	(singleton)
B61A	OUTMF2	(singleton)

---

[14]Page 513 includes a complete listing of RAM locations accessed by all fuzzy control software routines.

FIGURE 10-42

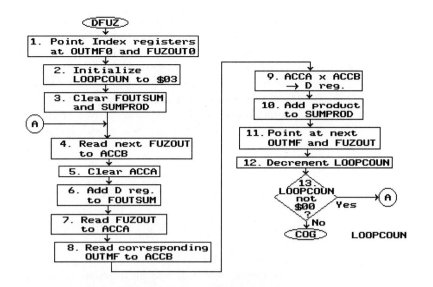

DFUZ NESTED SUBROUTINE—EVB AND EVBU

$ADDRESS	CONTENTS	COMMENTS
B76A	LDY #B618	Point Y register at OUTMF0
B76E	LDX #2	Point X register at FUZOUT0
B771	LDAB #3	Initialize LOOPCOUN
B773	STAB B	" " " "
B775	LDD #0	Clear FOUTSUM and SUMPROD
B778	STD 7	" " " " " "
B77A	STD 9	" " " " " "
B77C	LDAB 0,X	Read next FUZOUT to ACCB
B77E	CLRA	Clear MS byte of D register
B77F	ADDD 7	Add D register contents to FOUTSUM
B781	STD 7	" " " " " " "
B783	LDAA 0,X	Read current FUZOUT to ACCA
B785	LDAB 0,Y	Read current OUTMF (singleton) to ACCB
B788	MUL	FUZOUT X OUTMF —-> D register
B789	ADDD 9	Add product to SUMPROD
B78B	STD 9	" " " " "
B78D	INY	Point Y register at next OUTMF (singleton)
B78F	INX	Point X register at next FUZOUT
B790	DEC B	Decrement LOOPCOUN
B793	BNE B77C	If LOOPCOUN ≠ $00, do another program loop

Figures 10–42 and 10–43, along with their accompanying source code lists, illustrate the DFUZ nested subroutine. This nested subroutine derives a single crisp output, labeled DFUZOUT.

The software illustrated in Figure 10–42 develops the numerator and denominator of the weighted average computation. The numerator is labeled SUMPROD (sum-of-products) and has two assigned RAM locations. The M68HC11 calculates SUMPROD as follows:

$$(OUTMF0 \times FUZOUT0) + (OUTMF1 \times FUZOUT1) + (OUTMF2 \times FUZOUT2)$$

OUTMF0, OUTMF1, and OUTMF2 are output singleton values stored in EEPROM.

The denominator is labeled FOUTSUM (fuzzy outputs sum) and also has two assigned RAM locations. The M68HC11 calculates FOUTSUM as follows:

$$FUZOUT0 + FUZOUT1 + FUZOUT2$$

To calculate numerator and denominator, the CPU must access corresponding singleton and FUZOUT values. In Figure 10–42, block 1 (program addresses $B76A–$B770), the CPU points its X and Y registers at the first of these values. In block 2 (program address $B771), the CPU initializes a program loop counter (LOOPCOUN) to equal the number of FUZOUT/singleton pairs: $03. The CPU will accumulate the numerator and denominator values in the FOUTSUM and SUMPROD locations. Therefore, in block 3, the CPU clears these locations. Having initialized its index registers, FOUTSUM and SUMPROD, the CPU enters a program loop where it calculates the weighted average's denominator and numerator.

Refer to Figure 10–42, blocks 4, 5, and 6 (program addresses $B77C–$B782). In these blocks, the CPU adds a FUZOUT value to FOUTSUM. FOUTSUM constitutes 2 bytes. The individual FUZOUTs are single-byte values. In blocks 4 and 5, the CPU converts an 8-bit FUZOUT value into a 16-bit value residing in the D register. To accomplish this, the CPU reads the current FUZOUT to ACCB (block 4) and clears ACCA (block 5). The CPU adds FOUTSUM to this D register value, then writes the sum to the FOUTSUM RAM locations (block 6).

Refer to Figure 10–42, blocks 7–10, program addresses $B783–$B78C. In these blocks, the CPU calculates a product (OUTMFx × FUZOUTx) and adds this product to SUMPROD. In block 7, the CPU reads the current FUZOUT to its A accumulator. In block 8, the CPU reads the corresponding OUTMF (singleton value) to ACCB. In block 9, the CPU multiplies these two values to derive a product. In block 10, the CPU adds SUMPROD to this 16-bit product and writes the sum to the two SUMPROD RAM locations.

Refer to Figure 10–42, blocks 11 through 13 (program addresses $B78D–$B794). In these blocks, the CPU prepares for the next program loop and checks to see whether it has completed the FOUTSUM and SUMPROD calculations. In block 11, the CPU increments its Y and X registers to point at the next corresponding singleton and FUZOUT values. In block 12, the CPU decrements its loop counter (LOOPCOUN). If this block 12 decrement makes LOOPCOUN equal zero, the CPU has completed its FOUTSUM and SUMPROD calculations; it has executed the Figure 10–42 loop three times. If LOOPCOUN = $00, the CPU proceeds from block 13 to the COG calculation in Figure 10–43. If LOOPCOUN has not decremented to zero yet, the CPU takes the "yes" branch from block 13 to block 4 in order to begin another program loop. To make the COG (weighted average) calculation, the M68HC11 divides the SUMPROD value by the FOUTSUM value.

Refer to Figure 10–43 and its accompanying source code list. The CPU divides SUMPROD by FOUTSUM to complete its COG calculation. The quotient is the weighted average of the three singleton values. To make this calculation, the CPU reads FOUTSUM to its X register (block 14, program address $B795), and reads SUMPROD to its D register (block 15, program address $B797). In block 16 (program address $B799), the CPU executes the IDIV instruction. IDIV divides the D register value by the X register value and then returns the quotient to the X register.

After IDIV execution, the X register's low byte hold the quotient's significant digits. In block 17 (program address $B79A), the CPU transfers the X register contents to its D register. After this transfer, ACCB holds the quotient's significant digits. Therefore, in block 18 (program address $B79B), the CPU writes this quotient from ACCB to the DFUZOUT location in RAM.

Having calculated DFUZOUT—a weighted average (COG)—the CPU returns to the FUZZY subroutine. In turn, the FUZZY subroutine returns program flow to the MAIN program. To complete its pass through the MAIN program's control loop, the CPU must execute only one more subroutine, CRISPOUT.

CRISPOUT Subroutine

CRISPOUT converts DFUZOUT—the crisp output from the FUZZY subroutine—into $DUTY. $DUTY is the number of free-running counts in the motor speed control square wave's positive pulse width. Motor speed varies inversely with $DUTY: as $DUTY increases, motor speed decreases. On the other hand, fuzzy inference calls for faster motor speed by increasing DFUZOUT. Thus, as DFUZOUT increases, $DUTY must decrease.

FIGURE 10–43

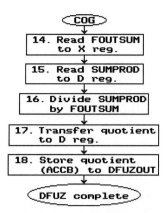

COG ROUTINE—EVB AND EVBU

EVB $ADDRESS	EVBU CONTENTS	COMMENTS
B795	LDX 7	Read FOUTSUM to X register
B797	LDD 9	Read SUMPROD to D register
B799	IDIV	Divide SUMPROD by FOUTSUM
B79A	XGDX	Transfer quotient to D register
B79B	STAB 5	Store quotient to DFUZOUT
B79D	RTS	Return from subroutine

The CRISPOUT subroutine decreases $DUTY as DFUZOUT increases. CRISPOUT accomplishes this task by executing a complement of DFUZOUT, then multiplying this one's complement value by eight. Two examples serve to illustrate the CRISPOUT process.

In the first example, assume that the FUZZY subroutine calls for a relatively slow motor speed by making DFUZOUT = $38. CRISPOUT one's complements this DFUZOUT value to $C7. CRISPOUT then multiplies the one's complement value by $08, deriving a $DUTY of $0638 (1592). $DUTY is the number of free-running counts in the speed control square wave's positive pulse width. Therefore, the pulse width equals 1592 × 500ns, or 796μs. The 1 KHz output square wave's period is 1ms. In this example, then, the square wave's pulse width equals (796μs/1ms) × 100, or 79.6% of the period. This wide positive pulse width causes the motor to run at a proportionally slow speed.

In the second example, assume that the FUZZY subroutine calls for a faster motor speed: DFUZOUT = $C8. CRISPOUT one's complements this value to $37 and multiplies by $08. Thus, $DUTY = $01B8 (440). The positive pulse width now equals [(440 × 500ns)/1ms] × 100, or 22% of the period. This narrow positive pulse width causes the motor run to at a proportionally high speed.

Figure 10–44 and its accompanying source code list illustrate the CRISPOUT subroutine. Refer blocks 1 and 2, and program addresses $B7A0–$B7A2. In these blocks, the M68HC11 CPU reads DFUZOUT to ACCA and one's complements this value. In block 3 (program address $B7A3), the CPU loads its ACCB with the multiplier $08. The CPU then multiplies the contents of ACCA and ACCB, and returns the product to the D register (block 4, program address $B7A5). In block 6 (program address $B7A6), the CPU writes the product from its D register to the $DUTY locations in RAM. The CPU then returns to the MAIN program.

Section 10.4 has provided a detailed analysis of a fuzzy application, which the reader shall implement in Exercise 10.2. This fuzzy application's objective is to vary DC motor speed, in response to changes in environmental temperature and light. This objective entails the following requirements:

1. Interface hardware includes sensor circuits, DC motor, and motor control circuitry.
2. The MAIN program uses a modular strategy to manage overall software execution.
3. The INIT subroutine initializes the M68HC11's output compare mechanism for motor speed control square wave generation.

FIGURE 10-44

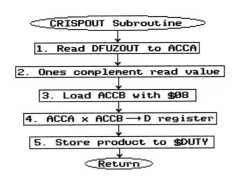

CRISPOUT SUBROUTINE—EVB AND EVBU

$ADDRESS	CONTENTS	COMMENTS
B7A0	LDAA 5	Read DFUZOUT
B7A2	COMA	One's complement read value
B7A3	LDAB #8	Load ACCB with multiplier
B7A5	MUL	ACCA x ACCB → D register
B7A6	STD C	Store product to $DUTY
B7A8	RTS	Return

4. The A/D subroutine periodically scales the temperature and light inputs from the sensor circuits.
5. The FUZZY subroutine uses a modular programming strategy to analyze crisp temperature and light data passed to it by the A/D subroutine. The FUZZY subroutine calls nested subroutines that fuzzify these crisp inputs, subject them to fuzzy rule evaluation, and de-fuzzify the results, which generates a crisp output value: DFUZOUT.
6. The CRISPOUT subroutine converts DFUZOUT to $DUTY.
7. OC1 and OC5 Interrupt Service routines use the $DUTY value to develop a 1 KHz motor speed control square wave.

The reader can now proceed to Exercise 10.2 and implement these requirements. The author believes that the reader's implementation will produce impressive results.

EXERCISE 10.2 Motor Control Using the M68HC11 and Fuzzy Inference

REQUIRED MATERIALS: Temperature sensor circuit, per Part III of Exercise 10.1
Light sensor circuit, per Part III of Exercise 10.1
Motor control circuitry, per Exercise 8.6
DC motor and mount, per Exercise 8.6
1 Spray can, circuit chiller ("Freez-It Antistat" or equivalent)
1 Capacitor, 1µF (EVB only)
1 Resistor, 1K Ohms (EVB only)

REQUIRED EQUIPMENT: Motorola M68HC11EVBU or EVB, and terminal
Oscilloscope and probe
Digital voltmeter or DMM
Flashlight
Heat gun (hot air blower for heat shrink or hair drying)

GENERAL INSTRUCTIONS: This laboratory exercise gives the reader an opportunity to connect, program, operate, and modify a practical fuzzy logic application. Section 10.4 provides a complete software listing and explanation, along with circuit connection instructions. Use Section 10.4 for reference as you connect the interface and load the software. In Part IV of this lab exercise, alter the fuzzy controller's response characteristics by modifying input and output MF parameters. In Part V, modify the controller's nine fuzzy rules to completely reverse its response.

PART I: Connect the Fuzzy Controller Hardware

Power down all controller components. Connect these components as shown in Figure 10–30. After completing your connections and before powering up, have the instructor check your hardware.

_____ *Instructor Check*

PART II: Load the Fuzzy Control Software

Source code lists in Section 10.4 constitute complete fuzzy control software. Enter and store this software in the M68HC11's EEPROM, per the Section 10.4 lists. If you use an EVBU, key the program directly into EEPROM using the BUFFALO one-line assembler. If you use an EVB, key each routine into RAM. After keying a routine into RAM, use the BUFFALO MOVE command to copy the routine to its assigned address block in EEPROM.

After storing all software routines, key fuzzy parameters, rule bytes, and singletons into the M68HC11's EEPROM. Pages 502, 507, and 510 provide listings of these rules, parameters, and singletons. Use the BUFFALO MEMORY MODIFY (MM) command to key these data directly into EEPROM. Use the BUFFALO one-line assembler and MEMORY DISPLAY (MD) command to double check your programs and supporting data.

PART III: DeBug and Demonstrate Fuzzy Controller Operation

Power up the controller. Use the digital voltmeter to adjust each sensor circuit output to 2.5V, under ambient conditions. Connect an oscilloscope to display the 1 KHz motor speed control square wave. Run and debug the controller program and hardware. Use the heat gun, flashlight, circuit chiller, and a light shroud to fully exercise the limits of sensor outputs, and the controller's corresponding response. When sure that hardware and software operate properly, demonstrate the controller to the instructor. Use the heat gun, flashlight, shroud, and circuit chiller in your demonstration.

_____ *Instructor Check*

PART IV: Modify Fuzzy Membership Function Parameters

Refer back to Part III of Exercise 10.1, Analog-to-Digital Conversions, where you recorded the response ranges of the light and temperature sensor circuits. Copy these values from Exercise 10.1 to the blanks below. Use these values for reference as you analyze fuzzy controller performance.

Temperature Sensor		**Light Sensor**	
Cold	$_____	Dark	$_____
Ambient	$_____	Ambient	$_____
Hot	$_____	Light	$_____

Change the fuzzy controller's response to light changes. Modify INPUT1 membership functions per Figure 10–45. Derive input MF parameters from this figure and record them in the $CONTENTS blanks below.

LIGHT

$ADDRESS	$CONTENTS	LABEL
B60C	_____	IN1MF0P1
B60D	_____	IN1MF0S1

FIGURE 10-45

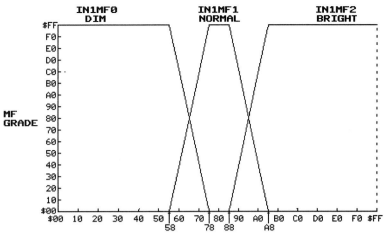

$ADDRESS	$CONTENTS	LABEL
B60E	_____	IN1MF0P2
B60F	_____	IN1MF0S2
B610	_____	IN1MF1P1
B611	_____	IN1MF1S1
B612	_____	IN1MF1P2
B613	_____	IN1MF1S2
B614	_____	IN1MF2P1
B615	_____	IN1MF2S1
B616	_____	IN1MF2P2
B617	_____	IN1MF2S2

Change INPUT1 MF parameters in EEPROM to reflect Figure 10–45. Test the fuzzy controller. Observe how these revised parameters change controller response. Record your observations in the space below.

Again, change the fuzzy controller's response to light changes. Modify INPUT1 MF functions per Figure 10–46. Derive MF parameters from this figure and record them in the $CONTENTS blanks below.

LIGHT

$ADDRESS	$CONTENTS	LABEL
B60C	_____	IN1MF0P1
B60D	_____	IN1MF0S1
B60E	_____	IN1MF0P2
B60F	_____	IN1MF0S2
B610	_____	IN1MF1P1
B611	_____	IN1MF1S1
B612	_____	IN1MF1P2
B613	_____	IN1MF1S2
B614	_____	IN1MF2P1
B615	_____	IN1MF2S1
B616	_____	IN1MF2P2
B617	_____	IN1MF2S2

FIGURE 10-46

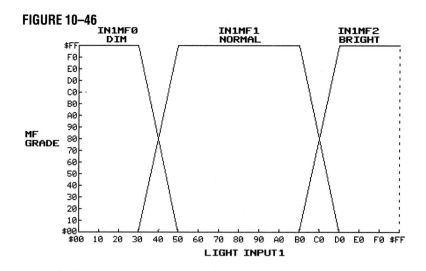

Change MF parameters in EEPROM to reflect Figure 10–46. Test the fuzzy controller. Observe how these revised parameters change controller response. Record your observations in the space below.

Once again, change the fuzzy controller's response. Refer to Figure 10–47. Note that the trapezoids have more gradual slopes. Derive INPUT1 MF parameters from Figure 10–47 and record them in the $CONTENTS blanks below.

LIGHT

$ADDRESS	$CONTENTS	LABEL
B60C	_____	IN1MF0P1
B60D	_____	IN1MF0S1
B60E	_____	IN1MF0P2
B60F	_____	IN1MF0S2
B610	_____	IN1MF1P1
B611	_____	IN1MF1S1
B612	_____	IN1MF1P2
B613	_____	IN1MF1S2
B614	_____	IN1MF2P1
B615	_____	IN1MF2S1
B616	_____	IN1MF2P2
B617	_____	IN1MF2S2

Change INPUT1 MF parameters in EEPROM to reflect Figure 10–47. Test the fuzzy controller. Observe how these revised parameters change controller response. Record your observations in the space below.

These modified parameters should also be suitable for temperature—INPUT0—as indicated by the Figure 10–47 labels in parentheses. Refer to this figure, derive INPUT0 MF parameters, and record them in the $CONTENTS blanks below.

FIGURE 10–47

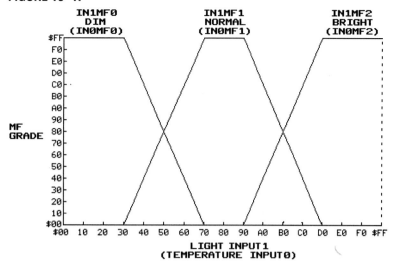

TEMPERATURE

$ADDRESS	$CONTENTS	LABEL
B600	_____	IN0MF0P1
B601	_____	IN0MF0S1
B602	_____	IN0MF0P2
B603	_____	IN0MF0S2
B604	_____	IN0MF1P1
B605	_____	IN0MF1S1
B606	_____	IN0MF1P2
B607	_____	IN0MF1S2
B608	_____	IN0MF2P1
B609	_____	IN0MF2S1
B60A	_____	IN0MF2P2
B60B	_____	IN0MF2S2

Change the INPUT0 MF parameters in EEPROM to reflect Figure 10–47. Test the fuzzy controller to assure that it responds smoothly to both temperature and light changes. When sure that the controller operates properly, demonstrate it to the instructor. Use the heat gun, flashlight, shroud, and circuit chiller in your demonstration.

_____ *Instructor Check*

A system designer can also tailor the DC motor's RPM band to application requirements. The designer can alter RPM band by simply changing the values of the output MF singletons. To illustrate this flexibility, change OUTMFx values in EEPROM as follows:

$ADDRESS	$CONTENTS	LABEL
B618	90	OUTMF0
B619	B0	OUTMF1
B61A	D0	OUTMF2

Test the fuzzy controller and record your observations in the space below.

Return singleton parameters to these values:

$ADDRESS	$CONTENTS	LABEL
B618	30	OUTMF0
B619	80	OUTMF1
B61A	D0	OUTMF2

The author trusts that you have been impressed with fuzzy inference's flexibility. The designer can easily tailor fuzzy response to application requirements by modifying input and output MF parameters.

PART V: Reverse Controller Response by Modifying Fuzzy Rules

Refer to Figure 10–23 and the rules derived from this matrix. Recall that we based these rules upon a general application requirement:

The higher the temperature and brighter the light, the faster the motor should run.

Change this requirement as follows:

The higher the temperature and brighter the light, the slower the motor should run.

Based on this revised requirement, we can make a new rule matrix. Figure 10–48 illustrates this revised matrix. A review of Figure 10–48 shows that Rule 3, Rule 5, and Rule 7 can remain unaltered. However, we must change THEN fields (consequents) in the remaining rule bytes. Where a consequent is "fast," it must be changed to "slow." Where a consequent is "slow," it must be changed to "fast." In the blanks below, enter rule byte data that implement the Figure 10–48 matrix.

RULBLK

$ADDRESS	$CONTENTS	LABEL	COMMENT
B61B	_____	RUL1	Rule 1
B61C	_____	RUL2	Rule 2
B61D	_____	RUL3	Rule 3
B61E	_____	RUL4	Rule 4
B61F	_____	RUL5	Rule 5
B620	_____	RUL6	Rule 6
B621	_____	RUL7	Rule 7
B622	_____	RUL8	Rule 8
B623	_____	RUL9	Rule 9
B624	_____	EOR	End-of-Rules Marker

FIGURE 10–48

		Light		
		Bright(2)	Normal(1)	Dim(0)
Temperature	Hot(2)	Slow(0)	Slow(0)	Medium(1)
	Warm(1)	Slow(0)	Medium(1)	Fast(2)
	Cold(0)	Medium(1)	Fast(2)	Fast(2)

Use this amended RULBLK listing to change the necessary rule bytes in EEPROM. Test the fuzzy controller to assure that its performance meets the new requirement:

The higher the temperature and brighter the light, the slower to motor should run.

When sure that the controller operates in accordance with this new requirement, demonstrate its operation to the instructor. Use the heat gun, flashlight, shroud, and circuit chiller in your demonstration.

_____ *Instructor Check*

This part of Exercise 10.2 demonstrates how easily the user can change the fuzzy controller's operating rules. By making simple rule byte changes, the user can drastically alter controller operation.

Note: Leave fuzzy controller hardware and software intact. A Chapter 11 laboratory exercise uses the fuzzy controller.

10.5 CHAPTER SUMMARY

The M68HC11's Analog-to-Digital Converter (ADC) uses charge redistribution to scale analog inputs into 8-bit binary equivalents. The ADC scales analog inputs with respect to V_{REFH} and V_{REFL} pin voltages. Calculate A/D step size, using this equation:

$$\Delta V = V_{REFH} 2^{-n}$$

where,

$$V_{REFL} = 0V$$
$$\Delta V = \text{A/D Step Size}$$
$$n = \text{number of input bits}$$

The A/D converter operates in four modes:

1. Scale analog voltages appearing at a single ANx pin. Make four conversions, set the status flag, and stop.
2. Scale analog voltages appearing at pins AN0–AN3 or pins AN4–AN7. Make four conversions, one per pin, set the status flag, and stop.
3. Continuously scale analog voltages appearing at a single ANx pin. Set the status flag after each group of four conversions.
4. Scale analog voltages appearing at pins AN0–AN3 or AN4–AN7. Scale continuously in a round-robin fashion. Set the status flag after each round of four conversions.

Refer to Figure 10–2. PORTE pins couple analog inputs to the ADC. To couple an analog value to the ADC, the M68HC11 asserts SAMPLEx for twelve clock cycles. SAMPLEx turns on a coupling transistor. This transistor passes the analog pin level to the ADC. The microcontroller can also use a PORTE pin for general-purpose binary inputs. PORTE pin logic includes inverters, latch, and transmission gate to accommodate general-purpose inputs. Special enabling and coupling transistors isolate the general-purpose input circuitry from intermediate analog levels which may appear at the PEx/ANx pin. However, this isolation is not effective during RPORTE assertion. The microcontroller restricts RPORTE assertion to a one-fourth E clock cycle duration. Minimal RPORTE duration reduces distortion and large current flows resulting from inadvertent PORTE reads during A/D sampling times.

In any A/D routine, the CPU must first write control bits to the OPTION register's ADPU (A/D Power Up) and CSEL (Clock Select) control bits. The ADPU bit powers up a charge pump. This charge pump generates A/D switching control voltages. CSEL selects from two A/D clocking sources—either E clock or an internal, asynchronous, RC-based 2 MHz (approximate) clock.

Whatever its operating mode, the M68HC11's ADC makes four conversions at a time and stores the digitized results in A/D Result Registers ADR1–ADR4.

By writing to its ADCTL register, the M68HC11 CPU causes A/D scaling to begin. This write also configures the ADC. ADCTL's SCAN bit configures the ADC for either four conversions and stop (SCAN = 0) or continuous conversions (SCAN = 1). The CPU configures the ADC for single- or four-channel conversions by writing to ADCTL's MULT bit. MULT = 0 causes single-channel conversions. MULT = 1 causes four-channel conversions. ADCTL's 4 LS bits select AN pins for conversion. If MULT = 0, these bits select one of the eight AN pins as the single analog input. The hex digit formed by the ADCTL low nibble corresponds to the ANx pin selected. If MULT = 1, ADCTL's CD and CC bits select a group of four AN pins to be scaled. If CD/CC = %00, pins AN0–AN3 are selected. If CD/CC = %01, pins AN4–AN7 are selected. In all configurations, the ADC writes a group of four digital results to the ADR1–ADR4 registers, in order.

ADCTL also contains the ADC's status bit—CCF (Conversions Complete Flag). The CCF sets after each group of four conversions. The CPU resets the CCF by writing to ADCTL. If SCAN = 1 (continuous conversions), the ADC continuous to scale analog inputs, even if CCF = 1.

Thus, to make analog-to-digital conversions, the M68HC11 uses the following resources:

Ten pins: V_{REFH}, V_{REFL}, and AN0–AN7
Two control/status registers: OPTION and ADCTL
Four result registers: ADR1–ADR4

If CSEL=0, the M68HC11's ADC consumes 128 E clock cycles to make four conversions, update ADR1–ADR4, and set the CCF flag. If CSEL = 1, the ADC consumes more than 128 RC clock cycles because of synchronization delays.

Sections 10.3 and 10.4 present fuzzy inference. For many applications, fuzzy inference can replace complex conventional algorithms or look-up tables with simple control rules stated in exact and easy-to-understand language. Further, fuzzy inference recognizes the relativity of "real-world" quantities. For example, a temperature of 86°F may be partly "warm" and partly "hot." By grading and analyzing input quantities in this relativistic way, fuzzy inference avoids a problem inherent in conventional algorithms. Conventional algorithms classify inputs into mutually exclusive categories, such as "cold," "warm," or "hot." If a designer bases a conventional control algorithm on too few discreet categories (such as "cold," "warm," and "hot"), the controller generates abrupt, large changes in output response. To smooth output response, the designer must double, triple, or even quadruple the number of input categories. This complicates the control algorithm. In contrast, fuzzy inference can generate smooth, flexible control, based upon a minimal number of input classifications.

Fuzzy inference subjects input quantities to the following analysis:

1. Fuzzify input quantities into membership function GRADEs. GRADEs constitute fuzzy inputs. They represent the degrees to which crisp (un-fuzzified) inputs can be categorized. For example, an input temperature value could be assigned GRADEs of "coldness," "warmness," or "hotness."
2. Evaluate these GRADEs on the basis of IF/THEN rules, deriving fuzzy outputs (FUZOUTs). FUZOUTs represent the degrees to which control output(s) should respond to the crisp inputs. For example, the FUZOUTs may express the degree to which a motor should run "slow," "medium," or "fast." In this example, a FUZOUT represents each of the three motor speed categories.

The system designer forms fuzzy rules in a simple IF/THEN format. For example:

If temperature is warm, and If light is dim, Then motor speed is slow.

Fuzzy inference arbitrates between the fuzzy rules and evaluates them in a relativistic way. Thus, fuzzy rule evaluation actually applies a rule in this way:

To the extent that temperature is warm, and to the extent that light is dim, Then motor speed tends to be slow.

3. De-fuzzify the FUZOUTs with reference to output membership function (MF) specifications (singletons). For example, output MFs might be motor speed "slow," "medium," and "fast." De-fuzzification derives a weighted average of the singletons, using the FUZOUTs as the weights. This weighted average constitutes a fuzzy controller's crisp output.

Chapter 10 applies fuzzy inference to a motor controller. This controller varies motor speed according to changes in environmental temperature (INPUT0) and light (INPUT1). Figure 10–14 summarizes this application.

Material on pages 498–506 discusses the first step in fuzzy inference—fuzzification. In the example application, fuzzification evaluates the degrees (GRADEs) to which the crisp inputs fall into overlapping membership functions (MFs): "cold," "warm," and "hot" temperatures, and "dim," "normal," and "bright" light. To calculate these six GRADEs, the fuzzification routine applies parameters for each of the six MFs. The designer derives these parameters as point and slope specifications from graphical presentations of the MFs.

Refer to Figures 10–15 through 10–18. In a fuzzification graph, the MFs form trapezoids or triangles. The designer specifies each MF in 4 bytes: Point 1, Slope 1, Point 2, Slope 2. Point 1 represents the X axis origin of the trapezoid or triangle. Slope 1 expresses the slope of the trapezoid or triangle's left side. Point 2 defines the trapezoid's upper-right corner as an X axis value. In the case of a triangle, Point 2 defines the X axis value at its apex. Slope 2 expresses the slope of the trapezoid or triangle's right side. Arbitrarily assign a value of $FF to vertical slopes. Express any nonvertical slope as the ratio of Y axis change to causal change on the X axis. Express all slopes as positive values. The designer stores these point and slope parameters in a block of nonvolatile memory.

In the example application, the M68HC11 CPU must derive six fuzzy inputs. That is, the CPU must make six GRADE calculations. The crisp temperature input (INPUT0) must be assigned GRADEs for "cold," "warm," and "hot." The CPU must assign "dim," "normal," and "bright" GRADEs for the crisp light input (INPUT1). To make one of these GRADE calculations, the CPU must execute two tasks:

1. The CPU must determine where the crisp input (INPUTx) falls with respect to a trapezoid or triangle—SEG0, SEG1, or SEG2. Figure 10–19 illustrates these segments.
2. The CPU calculates a GRADE (derives a fuzzy input). GRADE calculation procedure depends upon whether INPUTx falls in SEG0, SEG1, or SEG2. If INPUTx falls in SEG0, the CPU simply assigns a GRADE of $00, since the input's X axis value does not intersect the trapezoid under consideration. If INPUTx falls in SEG1, the CPU uses this equation to calculate GRADE of membership:

$$\text{MF GRADE (SEG1)} = (\text{INPUTx} - \text{Point 1}) \times \text{Slope 1}$$

If INPUTx falls in SEG2, the CPU uses the following equation to calculate GRADE:

$$\text{MF GRADE (SEG2)} = \$FF - [(\text{INPUTx} - \text{Point 2}) \times \text{Slope 2}]$$

The chapter then discusses the second step in fuzzy inference—rule evaluation. The system designer formats and encodes IF/THEN rules that describe system response. To do rule evaluation, the M68HC11 accesses relevant rule bytes, applies them, and derives fuzzy outputs. For this chapter's example application, the designer assigns a rule byte to each of nine fuzzy rules. Code rule bytes as shown in Figure 10–24.

Fuzzy rule evaluation uses a minimum/maximum strategy to derive three fuzzy outputs (FUZOUTs). The CPU first uses a rule byte's two IF fields to access corresponding fuzzy inputs (GRADEs). The CPU stores the minimum of these two GRADEs to a RAM address labeled IFMIN. For example, assume that fuzzy inference evaluates "If temperature is warm" and "If light is bright." The corresponding GRADEs for these If statements are IN0MF1 (warm) and IN1MF2 (bright). The CPU stores the lowest (minimum) of these two GRADEs to IFMIN. If IFMIN turns out to be $00, the CPU goes on to the next rule without evaluating the current rule's THEN byte. If IFMIN is other than $00, the CPU evaluates the THEN byte of the current rule.

To make a THEN byte evaluation, the CPU uses the rule byte's THEN field to access a corresponding fuzzy output (FUZOUT). Then, the CPU compares IFMIN with the FUZOUT. If IFMIN is higher than FUZOUT, the CPU replaces the current FUZOUT value with IFMIN. Thus, the CPU stores an IFMIN value to a corresponding FUZOUT, if it exceeds IFMIN values from previous rules having identical then statements. After evaluating all nine rules, the CPU has stored maximum applicable IFMIN values to FUZOUT0 (motor speed slow rules), FUZOUT1 (motor speed medium rules), and FUZOUT2 (motor speed fast rules). Refer to Figure 10–25 for a summary of fuzzy rule evaluation.

Next, the chapter discusses the final phase of fuzzy inference—de-fuzzification. To de-fuzzify, the CPU applies output membership specifications (singletons) to the FUZOUT values, and makes a weighted average calculation. This weighted average constitutes the de-fuzzified crisp output (DFUZOUT). In our example application, the designer specifies output MF singletons as motor speed slow, motor speed medium, and motor speed fast—corresponding to FUZOUT0, FUZOUT1, and FUZOUT2. A singleton originates at a specific X axis value on an output graph. This graph scales motor speed on its X axis, and FUZOUT values on its Y axis. On the graph, a singleton is as tall as its corresponding FUZOUT (Y axis) value. Figure 10–27 shows example singleton values at X axis locations $30 (OUTMF0), $80 (OUTMF1), and $D0 (OUTMF2).

The CPU calculates DFUZOUT as the weighted average of the singletons. FUZOUTs form the weights. Figure 10–27 shows this calculation for example FUZOUT0, FUZOUT1, and FUZOUT2 values. Figures 10–28 and 10–29 flowchart DFUZOUT calculation software.

Section 10.4 presents the complete hardware and software necessary to implement the example fuzzy motor control application. This application's objective is to vary DC motor speed in response to changes in environmental temperature and light.

Refer to Figure 10–14. This figure illustrates the following fuzzy controller elements.

Element 1. Interface hardware includes sensor circuits, DC motor, and motor control circuitry. The M68HC11 controls motor speed with a 1 KHz variable duty cycle square wave. The reader connects the motor control circuitry's input to the M68HC11's OC5 pin.

Element 2. The MAIN program uses a modular strategy to manage software execution. MAIN calls four subroutines: INIT, A/D, FUZZY, and CRISPOUT.

Element 3. The INIT subroutine initializes two interrupt pseudovectors, TCTL1, OC1M, OC1D, TMSK1, $DUTY, TOC1, and TOC5. The M68HC11 uses these resources to generate the 1 KHz motor speed control square wave.

Element 4. The A/D subroutine causes the M68HC11's ADC to scale four analog inputs and stop. Our application requires only two of these analog inputs—for temperature and light. Therefore, the A/D subroutine reads only the ADR registers that correspond to these inputs. Ground the two other analog input pins.

Element 5. The FUZZY subroutine analyzes the two crisp inputs (temperature and light) passed to it by the A/D subroutine. The FUZZY subroutine also uses a modular approach. It calls three nested subroutines:

1. The FUZZIFY nested subroutine fuzzifies the crisp temperature and light inputs. To accomplish this, FUZZIFY nests further subroutines: FINDSEG, SEG1, and SEG2.
2. The RULEVAL nested subroutine subjects fuzzified inputs to rule evaluation and derives fuzzy outputs.
3. The DFUZ nested subroutine de-fuzzifies the fuzzy outputs and generates a crisp output value—DFUZOUT.

Element 6. The CRISPOUT subroutine converts DFUZOUT to $DUTY. $DUTY is the number of free-running counts in the 1 KHz motor speed control square wave's positive pulse width.

Element 7. The OC1 and OC5 Interrupt Service routines generate a 1 KHz motor speed control square wave. The OC1 Interrupt Service routine generates the square wave's rising edges, and the OC5 Interrupt Service generates falling edges.

Section 10.4 provides a detailed description of these seven fuzzy motor control elements. Exercise 10.2 gives the reader an opportunity to connect, operate, and analyze the example fuzzy motor controller.

EXERCISE 10.3 Chapter Review

Answer the questions below by neatly entering your responses in the appropriate blanks.

1. If $V_{REFH} = 4.8V$, and $V_{REFL} = 0V$, the M68HC11 A/D converter's step size = _____ V.
2. How many analog input pins does the M68HC11 employ? _____ How are these pins labeled? _____.
3. Describe the A/D converter's four operating modes.

   (1)

   (2)

   (3)

   (4)

4. Commonly, the user connects the A/D converter's V_{REFH} pin to _____, and the V_{REFL} pin to _____.
5. For accurate scaling, the user should establish at least a _____ V difference between V_{REFH} and V_{REFL}.
6. List the two purposes of PORTE pins PE0–PE7.

   (1)

   (2)

7. Refer to Figure 10–2. Describe the purposes of the following components.

   Transistor A.

   Device D.

   Devices B and C.

   Transistors E, F, G, and H.

   Gates I and J.

8. Refer to Figure 10–2. Transistor A turns on for _____ clock cycles, when the microcontroller asserts the _____ line. The A/D converter requires these clock cycles to _____ appearing at the _____ pin.
9. Programmers must avoid PORTE reads during _____. Such reads will disturb _____ appearing at the _____ and adversely affect _____.
10. The M68HC11's A/D converter consists of _____ (how many?) capacitances. The A/D converter uses _____ to connect these capacitances variously to _____, _____, and the _____. To assure the switches work efficiently, the M68HC11 drives their gates with _____ V from a _____. To generate this switching voltage, the CPU must write a binary _____ to the bit of the _____ register.

543

11. Explain the function of the OPTION register's CSEL bit.

12. Under what circumstance must the CPU write a binary one to the CSEL bit?

13. What is the disadvantage of the internal RC-based A/D clock?

14. The A/D converter stores its binary results in the _____ registers.
15. The user wishes to configure the A/D converter as follows:

    Make continuous conversions
    Convert pin AN5
    Use the internal RC-based clock

    To do this configuration and cause conversion to begin, the M68HC11 writes $_____ to its OPTION register and $_____ to its ADCTL register.
16. The user wishes to configure the A/D converter as follows:

    Make one conversion per pin and stop
    Convert pins AN4–AN7
    Use the E clock

    To do this configuration and cause conversions to begin, the M68HC11 writes $_____ to OPTION and $_____ to ADCTL. After the CCF flag sets, the CPU could read the digitized pin AN6 value from the _____ register.
17. If the OPTION register's CSEL bit is reset, how many E clock cycles does the A/D converter consume to make a single conversion and update an ADR register? _____. How many E clock cycles does the A/D converter use to make four conversions, update the ADR registers, and set the CCF flag? _____. Why does the A/D converter require more than this number of RC-based clock cycles, when CSEL = 1?

18. For many applications, _____ can replace conventional control algorithms or look-up tables.
19. This approach (question 18) grades and analyzes input quantities in a _____ way.
20. In the blanks, briefly describe the three phases of fuzzy inference.

    Phase 1

    Phase 2

    Phase 3

21. Refer to Figure 10–15. Use this graph to derive the MF GRADEs for INPUT0 = $A8.

    IN0MF0 = $_____
    IN0MF1 = $_____
    IN0MF2 = $_____

FIGURE 10-49

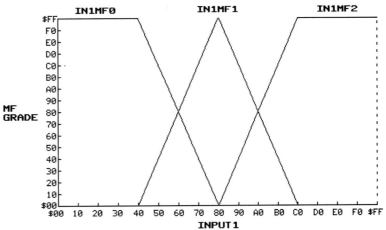

22. Refer to Figure 10-49. Specify IN1MF1 point and slope parameters from the graph.

    IN1MF1P1 = $_____
    IN1MF1S1 = $_____
    IN1MF1P2 = $_____
    IN1MF1S2 = $_____

23. Refer to Figure 10-49. INPUT1 = $72 falls into SEG_____ of IN1MF0, SEG_____ of IN1MF1, and SEG_____ of IN1MF2.

24. Refer to Figure 10-49. Calculate the IN1MF1 GRADE where INPUT1 = $72.

    GRADE = $_____

25. Refer to Figure 10-49. Calculate the IN1MF0 GRADE where INPUT1 = $72.

    GRADE = $_____

26. Refer to Figure 10-49. Calculate the IN1MF2 GRADE where INPUT1 = $72.

    GRADE = $_____

27. Refer to the rule encoding scheme, as illustrated in Figure 10-24. Use this scheme to encode the following rule into a rule byte.

    If temperature is warm, and If light is bright, Then motor speed is slow.

    Rule byte = $_____

28. Refer to Figure 10-25 and assume rule byte values as established on pages 506-509. The CPU is about to evaluate Rule 5. Current conditions are as follows:

    IFMIN = $7F
    FUZOUT0 = $00
    FUZOUT1 = $00
    FUZOUT2 = $7F
    IN0MF0 = $3F
    IN0MF1 = $C0
    IN0MF2 = $00
    IN1MF0 = $00
    IN1MF1 = $7F
    IN1MF2 = $7F

    In the blanks, enter the FUZOUT values *after* Rule 5 evaluation.

    FUZOUT0 = $_____
    FUZOUT1 = $_____
    FUZOUT2 = $_____

29. A fuzzy control application requires five fuzzy outputs—for motor speeds stopped, slow, medium, fast, and maximum. Output singleton values are:

    Motor speed stopped = $00
    Motor speed slow = $40
    Motor speed medium = $80
    Motor speed fast = $C0
    Motor speed maximum = $FF

    Corresponding FUZOUT values are:
    FUZOUT0 (stopped) = $48
    FUZOUT1 (slow) = $88
    FUZOUT2 (medium) = $A0
    FUZOUT3 (fast) = $7F
    FUZOUT4 (max) = $00

    Use these data to calculate DFUZOUT (COG).

    DFUZOUT = $_____

30. Refer to Figure 10–30. Describe the purpose of the RC combination connected to pin 52.

31. Refer to Figure 10–30. Explain the purpose of the 1K Ohm and 20K Ohm potentiometers.

32. Refer to Figure 10–31. Explain why MAIN program blocks 3 through 6 form a continuous loop.

33. Refer to Figure 10–31. Output compare interrupts control _____ generation. These interrupt routines may occur any time after the CPU executes block _____.

34. Refer to Figure 10–32. OC1 interrupts control output square wave _____-edge timing. OC5 (OC4) interrupts control _____-edge timing.

35. Refer to Figure 10–32. Explain why the CPU initializes $DUTY in block 6.

36. Refer to Figure 10–33. Explain why the CPU re-points the X register as its first action in both flowcharts.

37. Refer to Figure 10–33. Describe the general purpose of these interrupt service routines.

38. Refer to Figure 10–33. Explain why the CPU clears the OCxF bit before returning from the interrupt.

39. Refer to Figure 10–34. In which flowchart block does A/D scaling begin? _____.
40. Refer to Figure 10–34. Into which of its four operating modes does the CPU configure the A/D converter?

41. Refer to Figure 10–34. Explain why the CPU does not read and store the contents of ADR1 and ADR4.

42. Refer to Figure 10–34. Describe the general purpose of the A/D subroutine.

43. Refer to Figure 10–36. Summarize the tasks accomplished by the FUZZIFY nested subroutine.

44. Refer to Figure 10–36. The FUZZIFY nested subroutine consists of two loops. The first loop (blocks 3–8) manages the fuzzification of _____. The second loop (blocks 9–14) manages the fuzzification of _____.

45. Refer to Figure 10–37, block 3. Explain why the CPU clears the current GRADE location.

46. Refer to Figure 10–37, block 4. Explain why the CPU compares INPUTx with Point 2.

47. Refer to Figure 10–38. Summarize the two tasks accomplished by the SEG1 nested subroutine.

    Task 1 (blocks 1, 2, 7)

    Task 2 (blocks 3–7)

48. Refer to Figure 10–39. Summarize the two tasks accomplished in blocks 1–6 of the SEG2 nested subroutine.

    Task 1 (blocks 1–3)

    Task 2 (blocks 4–6)

49. Refer to Figure 10–39, blocks 7 and 8. Summarize the task accomplished by the CPU when it executes these blocks.

50. Refer to Figure 10–40. Explain why the CPU clears the FUZOUT RAM locations (block 1).

51. Refer to Figure 10–40, block 5. Describe the task accomplished by the CPU when it executes this block.

52. Refer to Figure 10–40. Summarize the two tasks accomplished by the M68HC11 CPU when it executes blocks 6–15.

    Task 1

    Task 2

53. Refer to Figure 10–41, blocks 16–19. Summarize the task accomplished by the CPU when it executes these blocks.

54. Refer to Figure 10–41, block 20. Summarize the task accomplished by the CPU when it executes this block.

55. Refer to Figure 10–41, blocks 21–24. Describe the task accomplished by the CPU in these blocks.

56. Refer to Figure 10–41, blocks 25–27. Describe the task accomplished by the CPU in these blocks.

57. Refer to Figure 10–42, blocks 1–3. Summarize the three tasks accomplished by the CPU in these blocks.

    Task 1

    Task 2

    Task 3

58. Refer to Figure 10–42, blocks 4–6. Summarize the task accomplished by the CPU when it executes these blocks.

59. Refer to Figure 10–42, blocks 7–10. Summarize the task accomplished by the CPU when it executes these blocks.

60. Refer to Figure 10–42, blocks 11–13. Summarize the two tasks accomplished by the CPU in these blocks.

    Task 1

Task 2

61. Refer to Figure 10–43. Describe the operation performed by the CPU in the COG routine.

62. Refer to Figure 10–44. Explain why the CPU one's complements DFUZOUT in block 2.

11 M68HC11 Expanded Multiplexed Mode

11.1 GOAL AND OBJECTIVES

The EVBU's M68HC11E9 microcontroller incorporates an impressive amount of on-chip memory:

> 512 bytes of RAM, 310 of these bytes available to the EVBU user.
> 512 bytes of EEPROM, all available to the EVBU user.
> 12K bytes of ROM. The EVBU's BUFFALO monitor (operating system) resides in this memory block.

However, some applications require more memory than Motorola can build into a multipurpose chip like the M68HC11. For example, one can imagine a control application requiring thousands of look-up table bytes. Large-scale fuzzy applications may also require more than the available on-chip memory. Expanded multiplexed mode allows the user to expand directly addressable memory by as much as 16K bytes.

To expand memory, the user interfaces external memory chips with the M68HC11's address, data, and control busses. Under ordinary conditions (single-chip mode), the M68HC11 uses these busses internally and does not make them accessible to the user.[1] However, when the M68HC11 operates in expanded multiplexed mode, it connects address, data, and control busses to nineteen of its pins, making them accessible to external devices.

To make address and data busses available, the M68HC11 connects them to PORTB and PORTC pins. Thus, in expanded multiplexed mode, PORTB and PORTC pins become bus pins and cannot be used for parallel I/O operations as described in Chapters 3 and 4. PORTB pins constitute the 8 MS bits of the address bus. The M68HC11 multiplexes 8 LS address bits and 8 data bus bits to PORTC pins.

In expanded multiplexed mode, the M68HC11 connects control signals to its STRA and STRB pins. Recall that the M68HC11 uses STRA and STRB in PORTB/PORTC handshake modes. Since expanded multiplexed mode converts PORTB/PORTC pins to address/data bus pins, the STRA and STRB pins become available for control bus use.

In this chapter, the reader shall learn how to interface memory with the M68HC11 using expanded multiplexed mode. The reader will also learn about a popular SRAM memory chip, the MCM6264C.

---

[1] All discussions and hands-on exercises up to this point have assumed single-chip mode.

Thus, Chapter 11's goal is to help the reader understand how to connect and operate the M68HC11 in expanded multiplexed mode. By carefully studying this chapter and completing its exercises, the reader should achieve these objectives:

Given a hex address, determine which MCM6264 row and columns are selected.
Convert the EVBU's M68HC11E9 to expanded multiplexed mode, and construct an 8K-byte memory expansion.[2]
Move Fuzzy Motor Controller software to expanded memory and demonstrate its operation.

Chapter 11 further demonstrates the M68HC11's impressive flexibility and utility.

11.2 M68HC11 RESOURCES USED FOR EXPANDED MULTIPLEXED MODE

To enter and operate in expanded multiplexed mode, the M68HC11 uses one control register and twenty-one pins. Four control register bits cause the M68HC11 to enter expanded multiplexed mode. At the time of a chip reset, two external pins set/reset these control bits. The remaining nineteen pins serve as control, data, and address bus pins. Following sections describe M68HC11 resources used to enter and operate in expanded multiplexed mode.

MODA/MODB Pins and HPRIO Register

Four bits in the HPRIO (Highest Priority Interrupt and Miscellaneous) Register put the M68HC11 into one of four hardware modes. "Hardware" mode refers to any of four internal hardware configurations. Each configuration gives the M68HC11 unique operating characteristics. Refer to Figure 11–1. The four "hardware mode" control bits constitute the HPRIO register's high nibble.

The four HPRIO high nibble combinations—and resulting hardware modes—are as follows:

HPRIO High Nibble Bit Values	Hardware Mode
%0000	Single Chip
%0010	Expanded Multiplexed
%1101	Special Bootstrap
%0111	Special Test

One can appreciate the significance of these HPRIO bits by reviewing the four hardware modes.

Single-Chip Mode. Up to this point, the reader has operated the M68HC11 in single-chip mode. In this mode, the M68HC11's software resides in on-chip memory. The M68HC11's address and data busses are also internal and unavailable to external devices.

Expanded Multiplexed Mode. In expanded mode, the M68HC11 connects its address bus, data bus, read/write line, and an address strobe to external pins. Expanded mode allows the user to connect external, extended memory for applications requiring large amounts of software and data storage.

FIGURE 11–1
Reprinted with the permission of Motorola.

HPRIO

B7							B0
RBOOT	SMOD	MDA	IRV				

---

[2]Unfortunately, EVB users cannot achieve this objective. The M68HC11A1 on board the EVB operates full-time in expanded multiplexed mode. However, the EVB employs a special chip, called the MC68HC24 Port Replacement Unit, to make the microcontroller appear to operate in single-chip mode. This EVB board configuration makes it impossible to access the data and address busses.

Special Bootstrap Mode. Bootstrap mode changes the way the M68HC11 reacts to a chip reset. Instead of vectoring to a reset routine, the M68HC11 automatically uploads a 256-byte user program via its SCI and executes this program. System designers use bootstrap mode for diagnostic and initialization purposes. Motorola designates bootstrap as a special version of single-chip mode. However, the M68HC11 CPU can connect its internal busses to external pins as a part of the uploaded bootstrap program.[3]

Special Test Mode. Motorola developed test mode to facilitate production testing of M68HC11 chips. Motorola advises that the user approach test mode with caution. Special test mode overrides a number of on-chip protective mechanisms. Consider test mode as a special version of expanded multiplexed mode.[4]

Refer to Figure 11–1. A unique combination of HPRIO high-nibble bits selects each of the hardware modes just discussed. A discussion of these HPRIO bits is now in order.

The RBOOT (Read Bootstrap ROM) bit enables a special routine in on-board ROM. This ROM routine causes the CPU to import a special bootstrap program via the SCI and then execute it. RBOOT = 1 causes the M68HC11 to power up this special section of ROM, and make it a part of the internal memory map. However, RBOOT does not cause the CPU to actually execute this ROM routine. RBOOT = 0 disables bootstrap ROM.

When set, the SMOD (Special Mode) bit either causes the CPU to execute the special bootstrap ROM routine upon reset or puts the microcontroller into special test mode—depending upon the states of the other HPRIO high-nibble bits. If SMOD = 0, the M68HC11 takes neither action, but instead operates in either single-chip or expanded multiplexed mode—again depending upon the states of other HPRIO bits.

The MDA (Mode A Select) bit determines whether the M68HC11 connects its internal busses to external pins. If MDA = 0, internal busses are not connected to external pins. MDA = 1 connects the control, address, and data busses to external pins.

The IRV (Internal Read Visibility) bit causes internal read data to appear on external data bus pins. If IRV = 1 and MDA = 1, internal read data appear externally. If IRV = 0 or MDA = 0, internal read data do not appear externally. IRV can equal one *only* when SMOD = 1.

As a general rule, RBOOT, SMOD, MDA, and IRV are writable during M68HC11 program execution. However, the CPU can write to these bits only under the following conditions:

RBOOT: CPU can write only when SMOD = 1.
SMOD: CPU can write a binary zero only.
MDA: CPU can write only when SMOD = 1.
IRV: CPU can write only when SMOD = 1.

Thus, the CPU can write to RBOOT, MDA, and IRV only when SMOD = 1. And the CPU can only write a zero to SMOD. These conditions have a clear implication. None of these writes are possible unless the SMOD bit was set by an external pin at the time of a chip reset. At the time of a reset, external pins set/reset all four HPRIO MS control bits. These external pins are the MODA and MODB pins.

Table 11–1 shows the relationship between MODB/MODA pin levels at the time of a chip reset and the resultant HPRIO contents.[5] The M68HC11 captures MODB and MODA

TABLE 11–1 **MODA/MODB Pin Levels and HPR1O Bits**

M68HC11 Pins		M68HC11	Latched HPRIO Bits			
MODB (2)	MODA (3)	Hardware Mode	RBOOT	SMOD	MDA	RV
0	0	Special Bootstrap	1	1	0	1
0	1	Special Test	0	1	1	1
1	0	Single-Chip	0	0	0	0
1	1	Expanded multiplexed	0	0	1	0

Reprinted with the permission of Motorola

---

[3]Motorola Inc., *M68HC11 Reference Manual* (1990), p. 3-15.
[4]Motorola Inc., *M68HC11 Reference Manual*, p. 3-14.
[5]Motorola Inc., *M68HC11 Reference Manual*, p. 3-2.

FIGURE 11-2
Reprinted with the permission of Motorola.

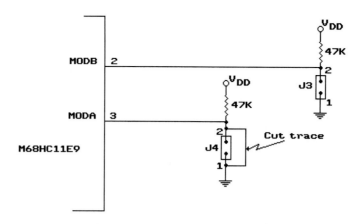

pin levels prior to a rising edge at its $\overline{\text{RESET}}$ pin. The $\overline{\text{RESET}}$ pin's rising edge causes the HPRIO to latch specified RBOOT, SMOD, MDA, and IRV bit levels, as controlled by MODA and MODB.

Thus, the MODB and MODA pins establish M68HC11's hardware mode at the time of a system reset. Up to this point, the reader has operated the EVBU's M68HC11E9 in single-chip mode. The EVBU connects MODB and MODA pins as shown in Figure 11-2. When the EVBU operates in single-chip mode, J3 does not have a jumper, and its pull-up circuit holds the MODB pin high. In single-chip mode, J4 either has a jumper installed or the trace around J4 is continuous. If J4 has a jumper and the trace has been cut, the user can put the EVBU into expanded multiplexed mode by removing the jumper and resetting the EVBU. If J4 has no jumper and the trace has not been cut, the user must do the following:

1. Power down the EVBU.
2. Cut the trace which by-passes J4.
3. Install a J4 jumper to operate in single-chip mode. Remove the jumper to operate in expanded multiplexed mode.

With its trace cut and no jumper, J4's pull-up circuit holds the MODA pin high. Highs at MODB and MODA put the M68HC11 into expanded multiplexed mode upon an EVBU reset. To return the EVBU to single-chip mode, simply install a jumper on the J4 header pins. With the jumper in place, MODB = 1 and MODA = 0, which puts the M68HC11 into single-chip mode.

Most M68HC11 pins have dual or multi-purposes. After a chip reset, the HPRIO register controls hardware mode, and the MODA and MODB pins can serve other purposes. After a reset, MODB can function as a standby power pin for the M68HC11's internal RAM. The user can apply back-up battery power to this pin. When the M68HC11's V_{DD} powers down or fails, this back-up power allows on-chip RAM to retain stored program and data.

As the M68HC11 executes program instructions, MODA functions as an output pin. MODA outputs a low during the first E clock cycle of each instruction and returns high for the remaining cycles. Thus, the MODA pin marks the beginning of each instruction. These markers are useful during system development and debugging.

PORTB Pins (A8–A15)

In expanded multiplexed mode, PORTB pins become address bus output pins A8–A15. During a bus cycle, the M68HC11 drives these pins with the high address byte for a period equal to one E clock cycle. PORTB pins convert to address pins when the HPRIO register's MDA bit sets. Refer to Figure 11-3, which illustrates PORTB pin logic.[6]

When the HPRIO register's MDA bit sets, the M68HC11 connects its internal busses to external pins. MDA = 1 enables AND gate A and disables AND gate B. Being disabled, AND gate B cannot transfer bits from the internal data bus (D0–D7) to D flip-flop C.

---

[6]Motorola Inc., *M68HC11 Reference Manual*, p. 7-12.

FIGURE 11–3
Reprinted with the permission of Motorola.

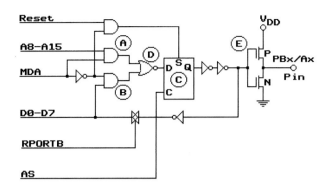

However, AND gate A is enabled and transfers an address bit from an internal address bus line (A8–A15) to D flip-flop C, via NOR gate D. When the address strobe (AS) line is high, flip-flop C is transparent, passing an address bit to totem-pole driver E. The address strobe's falling edge causes flip-flop C to latch the address bit until AS goes high again—starting the next bus cycle.

Thus, when HPRIO register's MDA bit sets, PORTB pins convert to address bus output pins A8–A15. PORTB pins become address pins as follows:

Single-Chip Mode PORTB Pin	Expanded Multiplexed Mode Address Pin
PB0	A8
PB1	A9
PB2	A10
PB3	A11
PB4	A12
PB5	A13
PB6	A14
PB7	A15

PORTC pins function as A0–A7 output pins, but they also serve as data bus I/O pins.

PORTC Pins (AD0–AD7)

Recall from Chapter 3 that PORTC pins are bidirectional. This bidirectionality allows PORTC pins to do double-duty, as both address bus and data bus pins. During the first half of a bus cycle, PORTC pins function as address pins (output only). During the second half of a bus cycle, these pins become data bus pins (I/O). Thus, PORTC pins time-multiplex address and data bits. Figure 11–4 illustrates PORTC pin logic used in expanded multiplexed mode.[7] To drive the PCx/ADx pin with an address or data output bit, the M68HC11 must enable totem-pole output driver A. The M68HC11 enables output driver A by asserting the

FIGURE 11–4
Reprinted with the permission of Motorola.

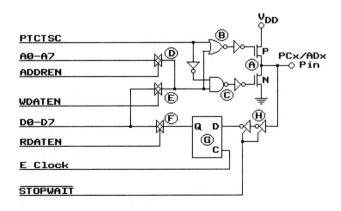

---
[7]Motorola Inc., *M68HC11 Reference Manual*, p. 7-16.

PTCSTC line (making this line low). PTCSTC = 0 enables NOR gate B and NAND gate C. With output driver A and gates B and C enabled, the M68HC11 can drive pin PCx/ADx with either an address bit (A0–A7) or data bit (D0–D7), depending upon which transmission gate (D or E) is enabled. To output an address bit, the M68HC11 asserts the ADDREN (address enable) line. To output a data bit, the microcontroller asserts write data enable (WDATEN).

If its CPU reads from expanded memory, the M68HC11 negates PTCSTC, disabling output driver A. When the E clock goes high, flip-flop G latches a read data bit. The M68HC11 then asserts the RDATEN (read data enable) line, enabling transmission gate F. Gate F drives the M68HC11's internal data bus line with the read data bit. The M68HC11 drives the $\overline{\text{STOPWAIT}}$ line low only when its CPU executes either the WAI or STOP instruction. Driving $\overline{\text{STOPWAIT}}$ low disables input buffers H.

Thus, the M68HC11 asserts ADDREN during the first half of the bus cycle so that an output address bit appears at the PCx/ADx pin. The M68HC11 asserts WDATEN or RDATEN during the last half of a bus cycle. The M68HC11 asserts WDATEN when it writes a data bit to the PCx/ADx pin. The microcontroller asserts RDATEN if it reads a bit from external memory. If the M68HC11 drives the PCx/ADx pin with an output bit—either address or data—it asserts PTCSTC (making it low). If the M68HC11 reads the PCx/ADx pin during the last half of the bus cycle, it negates PTCSTC by driving it high. Remember that in expanded multiplexed mode, PORTC pins are time multiplexed. The M68HC11 drives these pins with address data during the first half of the bus cycle, and makes them I/O pins during the last half of the bus cycle. PORTC pins become address and data pins as follows:

Single-Chip Mode	Expanded Multiplexed Mode	
PORTC Pin	Address Pin	Data Pin
PC0	A0	D0
PC1	A1	D1
PC2	A2	D2
PC3	A3	D3
PC4	A4	D4
PC5	A5	D5
PC6	A6	D6
PC7	A7	D7

The M68HC11's E clock, R/$\overline{\text{W}}$, and AS pins synchronize and control data exchanges with external memory.

Control Bus Pins (E, R/$\overline{\text{W}}$, AS)

Three pins—E, R/$\overline{\text{W}}$, and AS—control timing and duration of data flow between the M68HC11 and external memory.

The E clock pin (pin 5) has a single purpose. It makes the M68HC11's internal E clock available externally for synchronization purposes. The reader shall use the E clock to synchronize chip-level decoding logic.

The read/write-not pin (R/$\overline{\text{W}}$, pin 6) is a dual purpose pin. In single-chip mode, this pin serves as STRB (see Chapter 4). In expanded multiplexed mode, the R/$\overline{\text{W}}$ pin informs external memory whether the bus cycle is a read or a write. To indicate an M68HC11 read, the R/$\overline{\text{W}}$ goes high. To indicate a write, the R/$\overline{\text{W}}$ line goes low.

The address strobe pin (AS, pin 4) is also a dual-purpose pin. In single-chip mode, this pin serves as STRA (see Chapter 4). In expanded multiplexed mode, AS goes high to indicate that the M68HC11 is driving PORTB and PORTC pins with valid address information. Refer to Figure 11–3. The falling edge of the internal AS signal (the same signal that appears on the AS pin) causes the D flip-flop to latch an address bit from the internal address bus. Recall that this flip-flop is transparent while AS is high. Refer to Figure 11–4. The ADDREN signal is the AS signal. When AS/ADDREN = 1, an address bit appears on the PCx/ADx pin. Thus, AS = 1 indicates that valid address information is available on all M68HC11 address pins.

11.3 EXPANDED MULTIPLEXED MODE TIMING

Figure 11–5 shows PORTB and PORTC expanded multiplexed mode timing.[8] Following paragraphs explain the timing signals shown in this figure. The reader should keep Figures 11–3, 11–4, and 11–5 at hand while studying the following paragraphs.

Refer to Figure 11–5, signal A (PORTB, A8-A15), and Figure 11–3. To begin an external bus cycle, the M68HC11 drives its internal address bus with a new address and asserts the address strobe (AS = 1). PORTB flip-flops (Figure 11–3, C) are transparent while AS = 1. Therefore, the MS byte of a new address appears on PORTB pins when AS asserts. When AS negates, its falling edge causes PORTB flip-flops (C) to latch the high address byte. Thus, PORTB pin logic continues to drive pins A8–A15 with this address information until the start of the next bus cycle. Signal A shows that PORTB address information remains valid for the entire bus cycle.

Refer to Figure 11–4 and Figure 11–5, signal B (PORTC, AD0–AD7, all writes). Also refer to the ADDREN, PTCSTC, and WDATEN signals, just above signal B in Figure 11–5. Signal B shows how the M68HC11 multiplexes address and data information to PORTC pins during a write bus cycle. Recall that ADDREN and AS are identical signals. ADDREN assertion enables transmission gates D (Figure 11–4), and low-order address bits appear on PORTC pins. These address bits can appear on PORTC pins because PTCSTC = 0, enabling pin drivers A (Figure 11–4). Note that PTCSTC goes high for a short period when the E clock goes high (point W, Figure 11–5). PTCSTC = 1 disables PORTC pin drivers, and valid address information disappears from PORTC pins. Note also that when PTCSTC goes low (Figure 11–5, point X), WDATEN asserts (point Y). At this same time, the M68HC11 drives its internal data bus with write data. WDATEN = 1 enables transmission gates E (Figure 11–4). Therefore, PORTC output drivers (A) drive PORTC pins with the write data. ADDREN assertion (Figure 11–5, point Z) starts a new bus cycle.

Refer to Figure 11–4 and Figure 11–5, signal C (PORTC, AD0-AD7, external reads). Also refer to the ADDREN signal, and the PTCSTC and RDATEN signals just above signal C in Figure 11–5. Signal C shows multiplexing of address and data information during an external read cycle. Coincident with ADDREN/AS assertion, the M68HC11 makes PTCSTC low (point S, Figure 11–5) and drives its internal address bus with a new address. Since PTCSTC = 0 and ADDREN = 1, PORTC pin logic drives the PORTC pins with this address information. At point T (Figure 11–5), PTCSTC goes high, disabling the PORTC pin drivers. Consequently, address data disappears from the PORTC pins. On the falling edge of the M68HC11's internal PH2 clock, RDATEN asserts (point U, Figure 11–5).

FIGURE 11–5
Reprinted with the permission of Motorola.

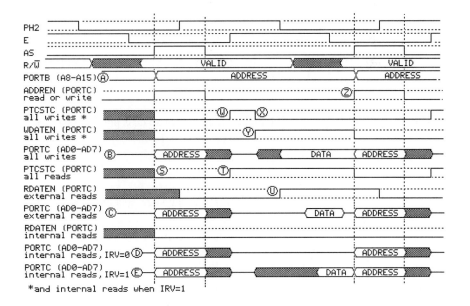

[8]Motorola Inc., *M68HC11 Reference Manual*, p. 7-18.

RDATEN = 1 enables transmission gate F (Figure 11–4). Transmission gate F couples memory read data from flip-flop G to the M68HC11's internal data bus. ADDREN assertion (Figure 11–5, point Z) starts a new bus cycle.

Refer to Figure 11–4 and Figure 11–5, signal D (PORTC, AD0–AD7, internal read, IRV = 0). Also refer to the RDATEN signal just above signal D. Signal D shows how the M68HC11 drives PORTC pins with address information only, during internal reads (IRV = 0). Recall that the HPRIO register's IRV (Internal Read Visibility) bit determines whether or not internal read data appears on PORTC pins. When IRV = 0, read data does not appear on PORTC pins. Note that when IRV = 0, RDATEN remains negated for the entire bus cycle. Therefore, transmission gate F is disabled (Figure 11–4), and the M68HC11 avoids a conflict between internal read data on the internal data bus and PORTC pin levels. However, when ADDREN/AS asserts, PTCSTC also asserts (point S, Figure 11–5). Coincidentally, the M68HC11 drives its internal address bus with a new address. The low byte of this new address appears on PORTC pins because both ADDREN and PTCSTC are asserted. PTCSTC negation (Figure 11–5, point T) disables the PORTC output drivers, and the address information disappears from PORTC pins.

Refer to Figure 11–4 and Figure 11–5, signal E (PORTC, AD0–AD7, internal read, IRV = 1). Refer also to the ADDREN, PTCSTC, and WDATEN for all writes. Note that the labels for PTCSTC and WDATEN (all writes) have asterisks. The asterisk note at the bottom of Figure 11–5 explains that these two signals take effect during internal read bus cycles when IRV = 1. Thus, signal E shows how the M68HC11 multiplexes PORTC pins during internal reads when IRV = 1. When IRV = 1, internal read data also drives PORTC pins. Motorola makes this feature available to the user for debugging and analysis.[9] When the M68HC11 applies a new address to the address bus, it also asserts PTCSTC and ADDREN/AS. Therefore, the address low byte appears on PORTC pins. At point W (Figure 11–5), PTCSTC negates, and address information disappears from PORTC pins. At points X and Y (Figure 11–5), PTCSTC and WDATEN assert (note that ADDREN is negated at this point). Also at this point, the M68HC11 drives the internal data bus with internal read data. Since PTCSTC and WDATEN are asserted, this internal read data appears on PORTC pins.

Figure 11–5 and the above discussion show that in expanded multiplexed mode, valid address data appears on PORTB pins during a complete external bus cycle. However, the M68HC11 time multiplexes address, then data information to PORTC pins. ADDREN, PTCSTC, WDATEN, and RDATEN signals control the PORTC multiplexing process. When ADDREN and PTCSTC assert, address information appears on PORTC pins. When WDATEN and PTCSTC assert, write data appears on PORTC pins. When PTCSTC negates and RDATEN asserts, the M68HC11 can read data applied to PORTC pins by expanded memory. When the HPRIO register's IRV bit is reset, PORTC pin logic applies internal read address bits to PORTC pins. When IRV = 1, PORTC pin logic applies internal read address bits and data bits to PORTC pins.

11.4 74HC373/74LS373 OCTAL LATCH

The M68HC11 multiplexes address bits and then data bits to PORTC pins. A designer must use an external latch to hold address bits after they have disappeared from PORTC pins. An ideal device for this purpose is the 74HC373 (HCMOS) or 74LS373 (LSTTL) octal latch. A 74HC/LS373 uses the layout shown in Figure 11–6. As this figure shows, the 74HC/LS373 employs eight D latches, one for each low-order address bit. The user can connect the M68HC11's AS pin to the latch enable (G) pin of the 74HC/LS373. With this connection, the octal latch captures low-order address bits when AS asserts. The 74HC/LS373's D latches hold address bits A0–A7 and supply them to an external memory device for the remainder of the bus cycle. The user can ground the 74HC/LS373's output enable pin (OE) so that address information is always available to the peripheral memory device.

---

[9]In this mode of operation, the R/$\overline{W}$ pin is high while PORTC pins are driven with internal data. A partially decoded external device might be enabled by the address appearing at PORTB/PORTC pins. R/$\overline{W}$ = 1 would signal this external device to apply read data to the PORTC pins. A data conflict at the PORTC pins would result. Obviously, a user should exercise great caution in setting the IRV bit.

FIGURE 11-6

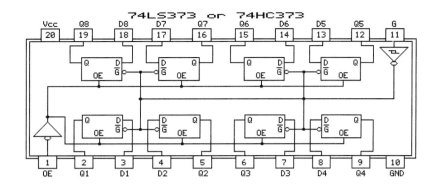

11.5 MCM6264C STATIC RAM

The MCM6264C is a CMOS fast static random access memory (SRAM) chip. Figure 11-7 shows MCM6264C pin assignments.[10] Motorola makes this chip available with a selection of five different access times: 12ns, 15ns, 20ns, 25ns, 35ns.

The MCM6264CP35 (plastic DIP package, 35ns access time) is quite adequate for use with the M68HC11EVBU. This chip uses an 8K × 1 byte memory matrix. Therefore, one MCM6264C chip allows the user to expand the M68HC11's memory by 8K bytes. All of this memory chip's inputs and outputs are compatible with the M68HC11. The MCM6264C requires only +5VDC (±10%) power. Overall, this CMOS chip draws 110mA–150mA of current. Thus, the MCM6264C is an excellent choice for use as an EVBU memory expansion chip.[11]

MCM6264C Memory Matrix

Figure 11-8 illustrates the MCM6264's memory matrix. A user can connect the M68HC11's external address pins to the MCM6264C as shown in the following chart:

M68HC11		MCM6264C	
PORTx Pin Label	**Pin Number**	**Pin Label**	**Pin Number**
PC0/A0*	9	A0	10
PC1/A1*	10	A1	9
PC2/A2*	11	A2	8
PC3/A3*	12	A3	7
PC4/A4*	13	A4	6
PC5/A5*	14	A5	5
PC6/A6*	15	A6	4
PC7/A7*	16	A7	3
PB0/A8	42	A8	25
PB1/A9	41	A9	24
PB2/A10	40	A10	21
PB3/A11	39	A11	23
PB4/A12	38	A12	2

*Connect these pins to the MCM6264C via the 74HC/LS373 octal latch.
Use M68HC11 pins A13–A15 for chip-level decoding.

FIGURE 11-7
Reprinted with the permission of Motorola.

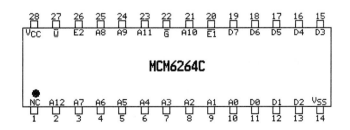

---

[10]Motorola Inc., *Memory Data* (1991), p. 5-120.
[11]Motorola uses MCM6264s as memory expansion chips on its M68HC11EVB.

FIGURE 11–8

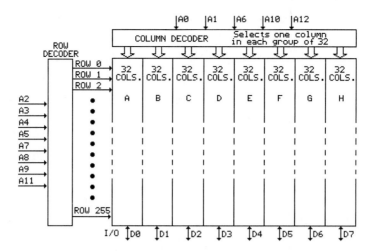

FIGURE 11–9

				A12	A11	A10	A9	A8	A7	A6	A5	A4	A3	A2	A1	A0
1	0	1	0	1	0	1	1	1	1	0	0	1	1	0	1	
$A				$B				$C				$D				

FIGURE 11–10

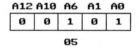

A11	A9	A8	A7	A5	A4	A3	A2
1	1	1	1	0	0	1	1
$F				$3			

FIGURE 11–11

A12	A10	A6	A1	A0
0	0	1	0	1

05

Refer to Figure 11–8. Address bits A11 and A2–A9 select a memory matrix row. Address bits A0, A1, A6, A10, and A12 select eight columns (one from each group of thirty-two).

Using an address decoding chart as shown in Figure 11–9, one can determine exactly which group of eight MCM6264C memory cells are addressed. Assume, for example, that the M68HC11 drives its external address pins with $ABCD, as shown in Figure 11–9. Which group of eight memory cells are selected?

First, which of the memory matrix's 256 rows does address $ABCD select? Figure 11–10 shows bit values applied to the row decoder (Figure 11–8). Figure 11–10 shows that row 243 ($F3) is selected. Note that bit values in Figure 11–10 correspond to Figure 11–9.

Second, determine which of the thirty-two columns in each group is selected. Address bits A0, A1, A6, A10, and A12 are applied to the column decoder. The Figure 11–11 chart shows that column five is selected. Given address $ABCD, the eight MCM6264C memory cells selected are at the intersections of row 243 and column 5 of each column group (5A, 5B, 5C, etc.). Exercise 11.1 gives the reader an opportunity to analyze hexadecimal addresses.

EXERCISE 11.1   MCM6264C Cell Select

AIM: Given a hexadecimal address, determine which memory matrix row and columns are selected.

INTRODUCTORY INFORMATION: By completing this exercise, you will gain familiarity with the MCM6264C's memory matrix.

INSTRUCTIONS: Analyze the hexadecimal addresses listed below. For each address, calculate which matrix row is selected. Then determine which column—within each group of thirty-two—is selected. Enter your decimal answers in the appropriate blanks.

$ADDRESS	ROW (of 256)	COLUMN (of 32)	
1203	64	19	EXAMPLE
77A9	_____	_____	
BA96	_____	_____	
22FA	_____	_____	
345F	_____	_____	
98D7	_____	_____	
4C65	_____	_____	
5D4E	_____	_____	
CB78	_____	_____	
D5CC	_____	_____	

MCM6264C I/O Control

Figure 11–12 illustrates the MCM6264C SRAM's data I/O control.[12] This control consists of I/O logic and tri-state buffers. Two tri-state buffers control data flow for each of the data lines D0–D7. If the MCM6264C reads out a data byte to PORTC pins, I/O logic enables eight output buffers, one for each data line (the bottom buffer in each pair, Figure 11–12). If the M68HC11's CPU writes a data byte to the MCM6264C, the I/O control section enables eight input buffers (the top buffer in each pair). Thus, the MCM6264C drives PORTC pins with read data via eight output buffers, and the MCM6264C stores write data, which it receives via eight input buffers. The MCM6264C's row and column decoders select eight memory cells that hold the read or write data.

In Figure 11–12, the top AND gate enables input (write) buffers. The bottom AND gate enables output (read) buffers. The $\overline{W}$ input connects to both the top and bottom AND gates. The user connects this $\overline{W}$ input to the M68HC11's R/$\overline{W}$ pin. The M68HC11 requests a data read by driving its R/$\overline{W}$ line high. This high at the MCM6264C's $\overline{W}$ input enables the bottom AND gate and disables the top AND gate. If the M68HC11 requests a data write, it drives the R/$\overline{W}$ pin low. This low at the MCM6264C's $\overline{W}$ input enables the top AND gate and disables the bottom.

FIGURE 11–12
Reprinted with the permission of Motorola.

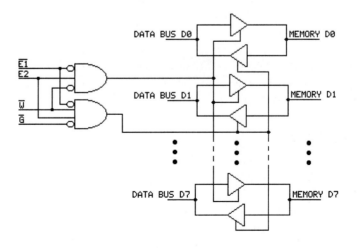

[12]Motorola Inc., *Memory Data* (1991), p. 5-120.

Of course, the memory's $\overline{E1}$, E2, and $\overline{G}$ inputs must be low, high, and low respectively before either the top or bottom AND gate can enable its tri-state buffers. Connect these three MCM6264C inputs to chip-level decoding logic. Connect the M68HC11's A13, A15, and the E clock pins to decoding logic. This decoding logic serves to drive the $\overline{E1}$, E2, and $\overline{G}$ inputs to the proper levels when A13, A14, and A15 indicate that the M68HC11 is addressing the MCM6264C chip. Section 11.6 present a fully decoded 8K-byte memory expansion using the MCM6264C.

11.6 A FULLY DECODED 8K-BYTE MEMORY EXPANSION

Figure 11–13 shows all logic necessary to fully decode an MCM6264C. With this decoding scheme, the MCM6264C resides at addresses $6000–$7FFF. This scheme requires a minimum of chip-level decoding logic, i.e., a single 74LS10 chip. To drive the MCM6264C's E2 input high and $\overline{G}$ input low, the M68HC11's E clock, A13, and A14 pins must all be high. Refer to Figure 11–5. This figure's timing diagram shows that the M68HC11 exchanges data with external memory when the E clock is high. Thus, the E clock serves to synchronize the actual exchange of read or write data. To drive the MCM6264C's $\overline{E1}$ input low, the M68HC11 must drive A15 with logic zero. Address bits A0–A12 select cells on board the MCM6264C. The 74HC373 latches multiplexed A0–A7 address bits from PORTC pins. The M68HC11's address strobe (AS) enables latching by the 74HC373.

Figure 11–13's decoding chart shows why the MCM6264C resides at addresses $6000–$7FFF. Address bits A0–A12 select eight memory cells and can therefore assume any combination of ones and zeros. To enable the MCM6264C's I/O logic, A15 must be low; A14 and A13 must be high. Thus, the lowest possible MCM6264C address is $6000 (all of the "X" bits are zeros), and the highest is $7FFF (all of the "X" bits are ones). $6000–$7FFF amounts to 8K of unique addresses.

Figure 11–14 shows the complete 8K-byte memory expansion interface. Following sections provide a step-by-step description of a read and a write to the MCM6264C. The reader may find Figures 11–5 and 11–14 useful references while studying these descriptions.

Memory Write Cycle

To complete a memory write cycle, the M68HC11 addresses the MCM6264C, then stores a data byte to this memory chip. The following sequence of events occurs during a write cycle:

1. The M68HC11's E clock goes low to indicate the start of the bus cycle.
2. The M68HC11 drives its R/$\overline{W}$ line low to indicate a write operation.
3. The M68HC11 asserts the address strobe (AS) and drives PORTB/PORTC address pins A0–A15 with the memory write address.

FIGURE 11–13

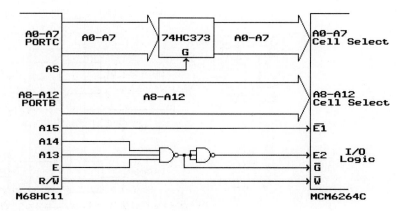

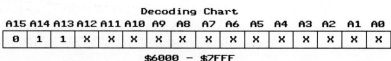

FIGURE 11-14

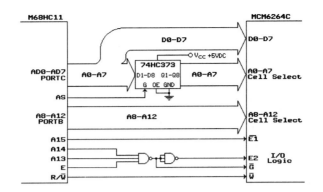

4. The 74HC373 latches address bits A0–A7 and drives the MCM6264C's A0–A7 pins with this information.
5. The M68HC11's E clock goes high, enabling the chip-level decoder and MCM6264C I/O logic.
6. The M68HC11 drives its PORTC pins with D0–D7 data.
7. The MCM6264C stores D0–D7 write data.
8. The M68HC11's E clock goes low, disabling the chip-level decoder and MCM6264C I/O logic. This completes the bus cycle and indicates the start of the next.

Memory Read Cycle

To complete a memory read cycle, the M68HC11 addresses the MCM6264C, then reads a data byte from this memory chip. The sequence of events in the read cycle are as follows:

1. The M68HC11's E clock goes low to indicate the start of the bus cycle.
2. The M68HC11 drives its R/$\overline{W}$ line high to indicate a read operation.
3. The M68HC11 asserts the address strobe (AS) and drives its PORTB/PORTC address pins with the memory read address.
4. The 74HC373 latches address bits A0–A7 and drives the MCM6264C's A0–A7 pins with this information.
5. The M68HC11's E clock goes high, enabling chip-level decoding and MCM6264C I/O logic.
6. The MCM6264C drives its D0–D7 pins with a byte of read data.
7. M68HC11 PORTC pin logic transfers the read data from PORTC pins to the internal data bus. ACCA or ACCB latches this read data byte.
8. The M68HC11's E clock goes low, disabling chip-level decoding and MCM6264C I/O logic. This completes the bus cycle and starts the next.

EXERCISE 11.2   **8K-Byte Memory Expansion**

REQUIRED MATERIALS: 1 MCM6264C HCMOS 8K × 8 SRAM
  1 74HC373 or 74LS373 Octal 3-State D Latch
  1 74LS10 Triple, 3-Input NAND
  1 Capacitor, .1μF
  Fuzzy motor control circuitry, including temperature and light sensor circuits, motor control circuitry, DC motor and mounting, as connected for Exercise 10.2.
  1 Spray can, circuit chiller ("Freez-It Antistat" or equivalent)

REQUIRED EQUIPMENT: Motorola M68HC11EVBU and terminal
  Flashlight
  Heat gun (hot air blower for heat shrink or hair drying)

GENERAL INSTRUCTIONS: In this laboratory, connect and test an 8K-byte EVBU memory expansion.[13] Next, copy fuzzy motor control software from EEPROM to expanded RAM and modify this software for its new location. Finally, demonstrate the operation of this software and the fuzzy motor control hardware.

PART I: Complete a Memory Expansion Connection Diagram

Research pin assignment diagrams for the M68HC11E9 (Quad), MCM6264C, 74LS10, and 74HC/LS373. Using these pin assignment diagrams as a reference, neatly enter pin numbers for all connections shown in Figure 11–15. Ten pin numbers are entered as an example. After entering pin numbers for all connections, show your completed diagram to the instructor.

_____ *Instructor Check*

PART II: Connect the Memory Expansion and Check Fuzzy Motor control Connections

Power down the EVBU. Review pages 552–554 then assure that the EVBU's J4 trace is cut. This trace lies between the two J4 pins, on the bottom of the EVBU board. Assure also that any J4 jumper has been removed.

Check all fuzzy motor control connections. Assure that they are complete and correct. Power up the EVBU and fuzzy motor control circuitry. Check the fuzzy motor control software residing in EEPROM.[14] Make sure that this software is complete and correct. Run the fuzzy motor control software from its EEPROM location. Assure that the controller operates properly.

With the EVBU powered up, check voltages at the J4 pins. With J4's trace cut and the jumper removed, voltages at the J4 pins should be as follows:

J4 pin nearest the microcontroller chip = V_{DD}.
J4 pin farthest from the microcontroller chip = 0V (GND).

If both pins are zero volts, the J4 trace is still continuous.
Power down the EVBU and fuzzy motor control hardware.
Use Figure 11–15 as a reference and carefully connect the memory expansion.
Power up the EVBU and memory expansion hardware. You should now have access to expanded RAM at addresses $6000–$7FFF. Use the BUFFALO Block Fill command to test this memory:

`BF 6000 7FFF AA`   <enter>

Executing this command, the M68HC11 writes $AA to all expansion memory bytes. Use the MD (Memory Display) command to assure that all memory bytes from $6000 to $7FFF now hold $AA. Show the instructor that the microcontroller has stored $AA to all expansion RAM bytes.

_____ *Instructor Check*

PART III: Move Fuzzy Motor Control Software to Expansion RAM

Use the BUFFALO MOVE command to copy the fuzzy motor control software and parameters from EEPROM to expansion RAM. MOVE this software and parameters to addresses $7600–$77A8. You must modify this software before it will run from its expansion

---

[13]EVB users cannot complete this lab. The M68HC11A1 on board the EVB operates full-time in expanded multiplexed mode. However, the EVB employs a special chip, called the MC68HC24 Port Replacement Unit, to make the microcontroller appear to operate in single-chip mode. This EVB board configuration makes it impossible to access the M68HC11A1's address and data busses.
[14]Exercise 10.2 directed the reader to store fuzzy control software and parameters in EEPROM, at addresses $B600–$B7A8.

FIGURE 11-15

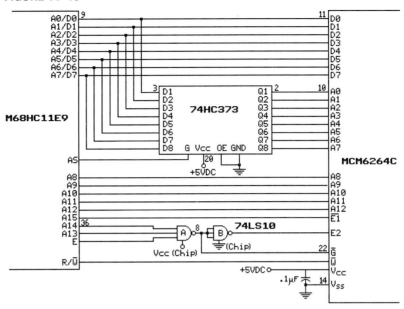

RAM location. Change the following instructions and parameters to reflect the new RAM addresses:

> MAIN program's subroutine calls
> OC1 and OC5 interrupt pseudovectors (INIT subroutine)
> FUZZY subroutine's nested subroutine calls
> FUZZIFY subroutine's nested subroutine calls
> FINDSEG subroutine's nested subroutine calls
> FUZZIFY subroutine's X register loads
> RULEVAL subroutine's Y register load
> DFUZ subroutine's Y register load

Run and debug your revised fuzzy motor control software. When sure that the fuzzy motor controller operates properly, demonstrate its operation to the instructor. Use a flashlight, light shroud, circuit chiller, and heat gun in your demonstration.

_____ *Instructor Check*

11.7 CHAPTER SUMMARY

The M68HC11's expanded multiplexed mode allows a user to expand directly addressable memory by as much as 16K bytes. When the M68HC11 operates in expanded mode, it connects internal address, data, and control bus signals to its external pins. In this mode, PORTB and PORTC pins become bus pins. PORTB pins output eight MS address bits. The M68HC11 time multiplexes eight LS address bits, and then eight data bus bits to PORTC pins. The microcontroller converts its STRA and STRB pins to control signals for external decoding logic and memory.

To enter and operate in expanded multiplexed mode, the M68HC11 uses one control register and twenty-one pins. HPRIO register's four MS bits put the M68HC11 into one of four hardware modes. One of these configurations is expanded multiplexed mode. Under specified conditions, the M68HC11 CPU can write to the four HPRIO control bits during program execution. However, two external pins—MODA and MODB—cause these HPRIO hardware mode control bits to set or reset at the time of a chip reset. Table 11-1 shows the relationship between MODA/MODB pin levels at the time of a chip reset and the resultant HPRIO contents. The $\overline{\text{RESET}}$ pin's rising edge causes the HPRIO to latch specified RBOOT, SMOD MDA, and IRV bit values, as shown in this table.

To put the EVBU's M68HC11E9 into expanded mode, the user must cut the J4 trace as shown in Figure 11–2. When this trace is cut, a pull-up circuit applies a high to the MODA pin. The user can then operate the EVBU in expanded multiplexed mode. To return to single-chip mode, simply install a jumper on the EVBU's J4 header pins. With a jumper installed, the pull-up circuit applies a low to the MODA pin.

In expanded multiplexed mode, PORTB pins convert to address bus output pins. Figure 11–3 illustrates PORTB pin logic. When the HPRIO's MDA bit sets, the M68HC11 connects internal address bus lines to PORTB pins. MDA = 1 enables address outputs from the PBx/Ax pins and disables data outputs from these pins. When the address strobe (AS) goes high, PORTB's D flip-flops are transparent to output address bits. When AS goes low, these flip-flops latch and drive PBx/Ax pins with address information for the duration of the bus cycle.

PORTC pins are bidirectional. This bidirectionality facilitates time multiplexing of address and data information. During the first half of a bus cycle, PORTC pins are address pins (output only). During the second half of a bus cycle, these pins become data bus I/O pins. Figure 11–4 illustrates PORTC pin logic used in expanded multiplexed mode. The M68HC11 asserts ADDREN during the first half of the bus cycle so that address bits appear at the PCx/ADx pins. The M68HC11 asserts WDATEN or RDATEN during the last half of the bus cycle. The M68HC11 asserts WDATEN when it writes data bits to the PCx/ADx pins. The M68HC11 asserts RDATEN if it reads data from external memory. When the M68HC11 drives PCx/ADx pins with output bits—either address or data—it asserts PTCSTC (making it low). PTCSTC = 0 enables the totem-pole output driver. If the M68HC11 reads PCx/ADx pins during the last half of the bus cycle, it negates PTCSTC by driving it high. PTCSTC = 1 disables the output driver.

Three pins—E, R/$\overline{\text{W}}$, and AS—control timing and duration of data flow between the M68HC11 and external memory. The E clock synchronizes chip-level memory decoding logic. The R/$\overline{\text{W}}$ signal informs external memory whether the bus cycle is a read or a write. The AS pin goes high to indicate that the M68HC11 is driving PORTB and PORTC pins with valid address information.

Figure 11–5 shows PORTB and PORTC timing for expanded multiplexed mode. This figure and its accompanying discussion show that valid address data appears on PORTB pins throughout a complete bus cycle. However, the M68HC11 time multiplexes address, followed by data information to PORTC pins. ADDREN, PTCSTC, WDATEN, and RDATEN signals control this multiplexing process. When ADDREN and PTCSTC assert, address information appears on PORTC pins. When WDATEN and PTCSTC assert, write data appears on PORTC pins. When PTCSTC negates and RDATEN asserts, the M68HC11 can read external data applied to PORTC pins. When IRV = 0, PORTC pin logic applies internal read address bits to PORTC pins. When IRV = 1, PORTC pin logic applies internal read address and then data bits to PORTC pins.

A designer must use an external latch to hold address bits after their disappearance from PORTC pins. An excellent device for this purpose is the 74HC373/74LS373 octal latch, as shown in Figure 11–6. This device employs eight D latches. The user connects the 74HC/LS373's latch enable pin (G) to the M68HC11's AS pin. When AS asserts, the 74HC/LS373 latches PORTC address bits. 74HC/LS373's D latches hold these address bits and apply them to the peripheral memory device for the rest of the bus cycle.

The MCM6264CP35 is an 8K × 1 byte fast SRAM, which uses a twenty-eight pin DIP package and has an access time of thirty-five nanoseconds. Refer to Figure 11–8. MCM6264C's address pins A11 and A2–A9 select its memory matrix row. Address pins A0, A1, A6, A10, and A12 select eight memory matrix columns (one column from each group of thirty-two).

Refer to Figure 11–12. MCM6264C's I/O control section consists of I/O logic and tri-state buffers. Two tri-state buffers control data flow for each of the data lines D0–D7. If the MCM6264C reads out a data byte, its I/O logic enables eight output buffers, one for each data line. If the MCM6264C receives a write data byte, its I/O logic enables the eight input buffers. $\overline{\text{E1}}$, E2, and $\overline{\text{G}}$ inputs enable I/O logic. The $\overline{\text{W}}$ input determines whether I/O logic enables the input or the output buffers.

Figure 11–13 shows logic necessary to fully decode the MCM6264C. With this decoding scheme, the MCM6264C resides at address block $6000–$7FFF. This scheme uses two 74LS10 gates for chip-level decoding.

Figure 11–14 shows a complete 8K-byte memory expansion, including data lines. Material on pages 562–563 provide step-by-step descriptions of a read and a write to the MCM6264C.

EXERCISE 11.3 Chapter Review

Answer the questions below by neatly entering your responses in the appropriate blanks.

1. Explain the purpose of expanded multiplexed mode.

2. When the M68HC11 operates in expanded multiplexed mode, it connects internal _____, _____, and _____ busses to pins.

3. In expanded mode, the M68HC11 connects its address bus to PORT _____ and PORT _____ pins. The M68HC11 drives PORT _____ pins with the eight LS address bits and PORT _____ pins with the MS address bits.

4. Describe M68HC11 time multiplexing of address and data information.

5. In expanded mode, the M68HC11 connects control signals to its _____ and _____ pins.

6. Define M68HC11 "hardware" mode.

7. List and describe the M68HC11's hardware modes.

   (1)

   (2)

   (3)

   (4)

8. The M68HC11's hardware mode is determined by _____.

9. List the hardware mode control bits and their functions.

   (1)

567

(2)

(3)

(4)

10. HPRIO's IRV bit can set only when SMOD is _____.
11. If IRV = 1 and MDA = 0, internal read data _____ (does, does not) appear on PORTC pins.
12. The M68HC11's CPU can write to RBOOT, MDA, and IRV only when _____. The CPU can write only binary _____ to SMOD. SMOD can be set only by _____ at the time of a _____. The _____ and _____ pins establish the M68HC11's hardware mode at the time of a system reset.
13. Describe how the user converts the EVBU from single-chip to expanded multiplexed mode.

14. Describe how the user can convert the EVBU from expanded multiplexed mode to single-chip mode.

15. In expanded multiplexed mode, PORTB pins become _____ bus _____ (input/output) pins.
16. Refer to Figure 11–3. MDA = 1 _____ (enables/disables) address outputs from pin PBx/Ax and _____ (enables/disables) data outputs from this pin.
17. Refer to Figure 11–3. When AS = 1, D flip-flop C is _____ to an output address bit. At AS's falling edge, the D flip-flop latches an _____. Thus, when MDA = 1, PORTB pin logic drives each PCx/Ax pin with an address bit for the entire _____.
18. PORTC pins are bi_____, which facilitates time _____ of _____ and _____ information.
19. During the first half of a bus cycle, PORTC pins are _____ pins. During the second half of the bus cycle, these pins become _____ pins.
20. Refer to Figure 11–4. During the first half of the bus cycle, PTCSTC = _____ (1/0), ADDREN = _____ (1/0), WDATEN = _____ (1/0), and RDATEN = _____ (1/0).
21. Refer to Figure 11–4. During the last half of a read bus cycle, PTCSTC = _____ (1/0), ADDREN = _____ (1/0), WDATEN = _____ (1/0), and RDATEN = _____ (1/0).
22. Refer to Figure 11–4. During the last half of a write bus cycle, PTCSTC = _____ (1/0), ADDREN = _____ (1/0), WDATEN = _____ (1/0), and RDATEN = _____ (1/0).
23. Refer to Figure 11–4. Explain the purpose of the E clock input.

24. Refer to Figure 11–4. The M68HC11 drives the $\overline{\text{STOPWAIT}}$ line low when it executes the _____ or _____ instruction. When $\overline{\text{STOPWAIT}}$ = 0, flip-flop G cannot _____.

25. Describe the purposes of the E, R/$\overline{\text{W}}$, and AS pins when the M68HC11 operates in expanded mode.

    E clock:

    R/$\overline{\text{W}}$:

    AS:

26. Refer to Figure 11–5, signal A. PORTB pin logic applies new address bits to PORTB pins when _____.
27. Explain why PORTB pins retain valid address information when AS goes low.

28. Refer to Figure 11–5, signal B. Low-order address bits appear on PORTC pins because ADDREN = _____ (1/0) and PTCSTC = _____ (1/0). Write data bits appear on these pins when PTCSTC goes _____ (high/low) and WDATEN goes _____ (high/low).
29. Refer to Figure 11–5, signal C. The M68HC11 CPU can read data from PORTC pins after PTCSTC goes _____ (high/low) and RDATEN goes _____ (high/low).
30. Refer to Figure 11–5, signal D. PORTC pin logic drives PORTC pins with internal read address bits when ADDREN goes _____ (high/low), PTCSTC goes _____ (high/low), and RDATEN is _____ (negated/asserted). Address information disappears from PORTC pins when _____ negates.
31. Refer to Figure 11–5, signal E. Internal address information appears on PORTC pins when the M68HC11 asserts _____ and _____. Internal read data appears on PORTC pins when PTCSTC (negates/asserts) and WDATEN _____ (negates/asserts). At this point, ADDREN is _____ (negated/asserted).
32. Describe the purpose of the 74HC/LS373 when it is used with the M68HC11 in expanded multiplexed mode.

33. Refer to Figures 11–8 and 11–13. For each pair of row and column numbers given, determine the hex address from which this pair derives. Enter these hex addresses in the appropriate blanks.

    MCM6264C

Row	Column	$Address
239	31	_____
14	14	_____
107	28	_____
3	13	_____
5	17	_____

34. Refer to Figure 11–12. To execute a data write, $\overline{\text{E1}}$ must be _____ (1/0), E2 must be _____ (1/0), $\overline{\text{W}}$ must be _____ (1/0), and $\overline{\text{G}}$ must be _____ (1/0). To execute a data read, $\overline{\text{E1}}$ must be _____ (1/0), E2 must be _____ (1/0), $\overline{\text{W}}$ must be _____ (1/0), and $\overline{\text{G}}$ must be _____ (1/0).
35. Refer to Figure 11–12. To execute a data write, the top AND gate outputs a _____ (high/low), and the bottom AND gate outputs a _____ (high/low). To execute a read, the top AND gate outputs a _____ (high/low), and the bottom AND gate outputs a _____ (high/low).

36. Refer to Figure 11–14. Describe the purpose of the three input NAND gates shown in this figure.

37. Refer to Figure 11–14. Explain why M68HC11 pins AD0–AD7 connect to both the 74HC373 and D0–D7 of the MCM6264C.

38. Number the listed events in the order of their occurrence during a memory write cycle.
    _____ E clock goes high.
    _____ E clock goes low to start bus cycle.
    _____ AS asserts.
    _____ MCM6264C latches write data.
    _____ E clock goes low to complete the bus cycle.
    _____ Data appears on PORTC pins.
    _____ R/$\overline{W}$ goes low.
    _____ 74HC373 latches A0–A7 bits.

39. Number the listed events in the order of their occurrence during a memory read cycle.
    _____ E clock goes low.
    _____ PORTC pin logic transfers data to internal data bus.
    _____ M68HC11 drives PORTB/PORTC pins with memory read address.
    _____ MCM6264C drives its D0–D7 pins with read data.
    _____ 74HC373 latches address bits A0–A7.
    _____ E clock enables chip-level decoding.
    _____ R/$\overline{W}$ goes high.

40. Fuzzy motor control parameters and software consume nearly two, 256-byte pages of memory. How many pages does the 8K-byte expansion memory provide? _____

APPENDIX A:
The M68HC11EVB Board

The M68HC11 Evaluation Board (EVB) incorporates the following features:

- An M68HC11A1 microcontroller operates in expanded multiplexed mode. The M68HC11A1 incorporates 256 bytes of RAM and 512 bytes of EEPROM. The M68HC11A1 has no on-chip ROM or EPROM.
- An MC68HC24 Port Replacement Unit (PRU). When the M68HC11 operates in expanded multiplexed mode, the PRU can make the HC11 appear to operate in single-chip mode. The PRU restores missing PORTB and PORTC parallel I/O functions. Thus, at the EVB board's I/O header, the M68HC11A1 appears to operate in single-chip mode.
- Two UART-based serial I/O ports, RS-232 compatible. One of these ports uses the M68HC11A1's Serial Communications Interface (SCI) to download programs or data to the EVB from a host computer. The other port uses an EVB-mounted UART chip—the M6850 ACIA (Asynchronous Communication Interface Adapter)—to provide communications between the EVB and a user terminal.[1]
- An 8K-byte EPROM contains the BUFFALO Monitor EVB operating system. BUFFALO includes nineteen system commands, a one-line assembler/disassembler, and a number of utility subroutines.
- 8K bytes or 16K bytes of user RAM. Motorola ships each EVB with an 8K-byte SRAM, the MC6264. Each EVB also provides a fully connected socket for a second MC6264 SRAM chip.
- A 60-pin header facilitates connection of a target system to the M68HC11A1/MC68HC24. This header provides convenient access for the hands-on exercises in this book.

Figure A illustrates the M68HC11EVB board's layout.[2] Following paragraphs describe each of the components diagrammed in this figure.

U10 is the M68HC11A1 microcontroller. This device runs in expanded multiplexed mode. The M68HC11A1 drives its PORTC pins with address information during the first half of a bus cycle (E clock low). During the second half of the bus cycle (E clock high), the M68HC11A1 uses PORTC pins as data bus pins. External devices require address information for the entire bus cycle. U2 is an MC74HC373 8-bit latch. When enabled by

---

[1] Motorola provides a freeware KERMIT program. A desk-top computer uses KERMIT to emulate an EVB- or EVBU-compatible terminal. KERMIT configures one of the desk-top computer's serial ports for I/O to the EVB or EVBU board.

[2] Motorola Inc., *M68HC11EVB Evaluation Board User's Manual* (1986), p. 6-5.

FIGURE A
Reprinted with the permission of Motorola.

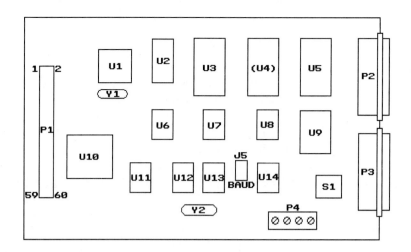

M68HC11A1's address strobe (AS) assertion, this latch captures the PORTC address information. U2 drives the EVB's address bus with this latched information for the bus cycle's duration. Thus, U2 supplies A0–A7 address bits to external devices for a complete bus cycle.

U1 is an MC68HC24 Port Replacement Unit. The PRU replaces microcontroller PORTB and PORTC functions, which are lost when the HC11 operates in expanded multiplexed mode.

U5 is an MC6264 8K-byte user RAM. This RAM's memory cells reside at addresses $C000–$DFFF. (U4) is a socket for an additional MC6264 RAM chip. If installed, this RAM's memory cells reside at addresses $6000–$7FFF.

U3 is an 8K-byte EPROM, which contains the BUFFALO monitor program. The EPROM resides at addresses $E000–$FFFF.

U6 is an MC74HC138 one-of-eight decoder. This device provides chip-select decoding for U3, U4, U5, U9, and U11.

U9 is an MC6850 Asynchronous Communications Interface Adapter (ACIA). The MC6850 provides a serial communications link between the M68HC11A1 and a user terminal.

U8 is an MC1488 RS-232 driver. This driver converts serial output data to RS-232 bus standards. The MC1488 converts serial data from the microcontroller's on-chip SCI and sends this data to a host computer. This MC1488 also converts ACIA serial output data and sends this data to the user's terminal.

U14 is an MC1489 RS-232 receiver. This device converts incoming serial data from RS-232 standards to TTL-compatible serial data, suitable for the SCI and ACIA. Of course, incoming serial data comes from either the user terminal (to the ACIA, then microcontroller) or a host computer (to the M68HC11A1's SCI).

U13 is an MC74HC4040 counter/frequency divider. This chip divides a 2.4576 MHz input signal by 2^8 through 2^{13}, generating output square waves at 300 Hz, 600 Hz, 1200 Hz, 2400 Hz, 4800Hz, or 9600 Hz. These outputs constitute six user-selectable ACIA serial baud rates. A user selects one of these baud rates with a jumper at J5. The jumper connects one of six MC74HC4040 outputs to the ACIA.

U11 is an MC74HC74 dual, D flip-flop. One flip-flop on this chip controls two digital switches. The D flip-flop's Q output controls one switch; the D flip-flop's $\overline{Q}$ output controls the other. Thus, one switch must be open when the other is closed, and vice versa. The microcontroller's PORTD pin 0 (PD0) serves two purposes. This pin can serve as a general-purpose I/O pin, or it can serve as the on-chip SCI's serial input. If the microcontroller's CPU addresses the SCI, the D flip-flop closes one of the digital switches. This switch connects PD0 to the serial receive data line from the host computer. If the CPU addresses general-purpose serial PORTD, the D flip-flop closes the other digital switch. This switch connects PD0 to the user I/O header, i.e., to a target system. The EVB leaves the second MC74HC74 D flip-flop unconnected.

U7 is an MC74HC4066 digital switch. This IC contains four HCMOS switches. The EVB uses these switches for: (1) read/write control line timing, (2) PD0 arbitration (two switches; see the previous paragraph), and (3) external circuit-generated resets.

U12 is an MC74HC14 hex, Schmitt-trigger inverter. The EVB uses five of these inverters. The EVB push-button reset circuit uses one. Two of these inverters, along with 2.4576 MHz crystal Y2, form a Pierce oscillator. This oscillator generates an input to U13—the baud rate frequency divider. The EVB uses two other inverters in the M68HC11's E clock and read/write (R/$\overline{W}$) control lines to make their signals compatible with external chip control inputs.

Y1 is an 8MHz crystal, which forms the frequency determining device of a Pierce oscillator on board the M68HC11A1. This oscillator, along with the M68HC11's clock generator, develop the microcontroller's internal Phase 2 clock, E clock, and address strobe.

P1 is a sixty-pin header which gives the user access to the M68HC11A1 and MC68HC24 PRU. Recall that the PRU emulates PORTB and PORTC single-chip mode operations (general-purpose, parallel I/O). Thus, P1 pins provide access to an M68HC11 microcontroller operating in "single-chip" mode. Many of this book's exercises entail the connection of external peripherals. The reader can use P1 to make these connections.

P2 is a twenty-five-pin connector. The reader should use P2 to connect between the EVB and user terminal.

P3 is also a twenty-five-pin connector. The reader can use P3 to connect between the EVB and a host computer, in order to download programs and data.

Terminal block P4 provides for EVB power connections. The MC1488 (U8) and MC1489 (U9) require −12VDC and +12VDC power. All other EVB components require only +5VDC. Terminal block P4 provides for the connection of −12VCD, +12VDC, and +5VDC power supplies, along with ground.

Finally, S1 is a push-button reset switch. This switch is part of a pull-up circuit. Pressing the button closes the switch and drives the M68HC11A1's $\overline{\text{RESET}}$ pin low, generating a chip reset.

APPENDIX B:
The M68HC11EVBU Board

The M68HC11 Universal Evaluation Board (EVBU) is quite simple as compared to the EVB. In fact, the average EVBU uses only two ICs—the M68HC11E9 microcontroller chip itself and an MC145407 RS-232 driver/receiver. The EVBU incorporates these features:

- An M68HC11E9 microcontroller operates in single-chip mode. The M68HC11E9 incorporates 512 bytes of on-chip RAM, 512 bytes of on-chip EEPROM, and 12K bytes of on-chip ROM. This ROM contains the BUFFALO Monitor operating system. BUFFALO includes nineteen system commands, a one-line assembler/disassembler, and a variety of utility subroutines.
- A UART-based serial I/O port, RS-232 compatible. This port uses the M68HC11E9's Serial Communications Interface (SCI) to communicate with either a user terminal or host computer.[1]
- A sixty-pin header facilitates connection of a target system to the M68HC11E9. This header provides convenient EVBU access for the hands-on exercises in this book.

Figure B illustrates M68HC11EVBU layout.[2] Following paragraphs describe each of the components shown in this figure.

U3 is the M68HC11E9 microcontroller. This device can operate in any of four modes: single-chip, expanded multiplexed, special bootstrap, or special test. Except for Exercise 11.2, all of the hands-on exercises in this book call for operation in single-chip mode. Exercise 11.2 operates the microcontroller in expanded multiplexed mode. For this exercise, the reader shall reconfigure jumper header J4 in order to change the M68HC11E9's operating mode from single-chip to expanded multiplexed.

U4 is an MC145407 RS-232 driver/receiver. This device converts serial output data to RS-232 bus standards. Serial output data emanates from the M68HC11E9's SCI. The MC145407 also converts incoming serial data from a terminal or host computer. The driver/receiver converts this incoming serial data from the RS-232 format to TTL-compatible serial data. The M68HC11E9's SCI receives this data.

U2 is an MC34064 low-voltage detector. This device, along with S1 (the push-button reset switch), are connected to the M68HC11E9's $\overline{\text{RESET}}$ pin. Thus, either S1 or U2 can

---

[1] Motorola provides a freeware KERMIT program. A desktop computer uses KERMIT to emulate an EVB- or EVBU-compatible terminal. KERMIT configures one of the desktop computer's serial ports for I/O to the EVB or EVBU board.
[2] Motorola Inc., *M68HC11EVBU Universal Evaluation Board User's Manual* (1990), p. 6-11.

FIGURE B
Reprinted with the permission of Motorola.

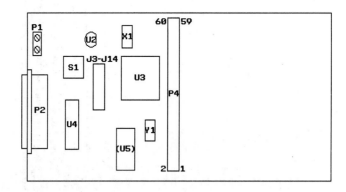

generate a microcontroller reset. S1 generates a reset if the user presses it. U2 generates a reset if it detects that the EVBU board's V_{DD} supply drops below approximately 4.2V.

(U5) is a socket for an MC68HC68T1 real-time clock. This chip keeps track of time and date. It connects to Y1, a 32.768K Hz crystal. The user can configure U5 to operate with battery backup power. To configure for battery power, the user must add diodes to the EVBU board. U5 connects to the M68HC11E9 via the microcontroller's Serial Peripheral Interface (SPI). If U5 has been installed, it will interfere with Chapter 6's hands-on SPI exercises. To complete these exercises, the reader can remove U5 from its socket. Alternately, the reader can cut the J10–J13 shorting traces. The shorting traces are found on the bottom of the EVBU board. These traces short the two pins of each jumper header, J10–J13. To connect U5 to the microcontroller, install jumpers on J10–J13. To use the real-time clock's alarm features, install a jumper on J14. To configure for Chapter 6's exercises, simply remove J10–J14 jumpers.

Motorola has installed a jumper on J7. J7's jumper connects a PORTA pin (PA3/OC5/IC4) to the M68HC11E9's $\overline{\text{XIRQ}}$ interrupt pin. This connection facilitates BUFFALO monitor's TRACE operation, used for debugging and program analysis. Chapters 8, 9, and 10 include hands-on exercises that use the PA3/OC5/IC4 pin. The reader must avoid the TRACE feature while doing these hands-on exercises. To be safe, simply remove the J7 jumper before doing these exercises. Likewise, remove the J7 jumper while doing Chapter 2's exercises on XIRQ interrupts.

X1 is an 8-MHz ceramic resonator, which forms the frequency determining device of a Pierce oscillator on board the M68HC11E9. This oscillator, along with a clock generator, develop the Phase 2 clock, E clock, and address strobe (used in expanded multiplexed mode).

Terminal block P1 provides for EVBU board power connections.

P2 is a twenty-five-pin cable connector. Use P2 to connect the EVBU to a terminal or host computer.

P4 is a sixty-pin header that gives the user access to the M68HC11E9. Many of this book's exercises entail the connection of external devices or memory. The reader can use P4 to make these connections.

APPENDIX C: Connecting the EVB or EVBU to External Circuits

The EVB or EVBU board's sixty-pin header (P1-EVB, P4-EVBU) offers a convenient connection point for external circuits used in this book's exercises. The author suggests a sixty-pin ribbon cable as the means of connection. Refer to Figure C–1. The ribbon cable's length should not exceed six inches. The ribbon cable's shorter length keeps line losses to a minimum. Terminate the ribbon cable with female header connectors as shown in Figure C–1, Part A. Fold the ribbon cable to expose the holes in the outboard connector as shown in Figure C–1, Part B. Mount external components to a protoboard or standard digital trainer. Connect external circuits to the outboard ribbon cable connector using insulated, tinned, single-strand, #20 wires—#20 wire provides a tight, low-resistance fit into the ribbon cable connector and an equally good connection to a protoboard. Always keep these wires as short as possible. Assure that the EVB/EVBU and your external circuit share a common ground.

FIGURE C–1

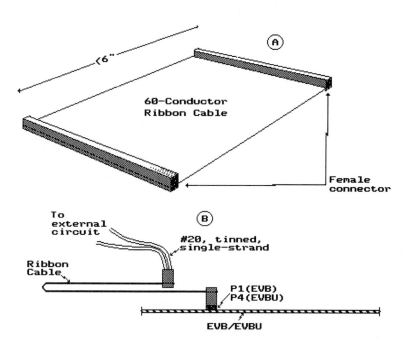

FIGURE C–2

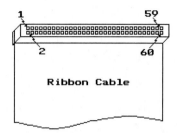

Figure C–2 shows how the holes in the outboard connector relate to header pins. The following list identifies the purpose of each header pin.

Pin 1: EVB or EVBU board ground.

Pin 2: MODE B/$V_{STANDBY}$. The MODE B and MODE A input pins determine microcontroller hardware operating mode. The MODE B pin can also serve as a standby +5VDC input. The standby voltage powers on-chip RAM when regular chip V_{DD} powers down.

Pin 3: MODE A/LIR, EVBU only. Unconnected on the EVB. The MODE A and MODE B pins determine microcontroller hardware operating mode. When the microcontroller fetches and executes program, the MODE A pin becomes an output which drives low during the first E clock cycle of each instruction. "LIR" stands for "load instruction register."

Pin 4: STRA/AS on the EVBU. STRA only on the EVB. STRA (Strobe A) is a parallel I/O handshake input pulse. AS is the address strobe used when the microcontroller operates in expanded multiplexed mode.

Pin 5: E clock output.

Pin 6: STRB or R/$\overline{W}$, on the EVBU. STRB only on the EVB. STRB (Strobe B) is a parallel I/O handshake output pulse. R/$\overline{W}$ is a read/write control output used in expanded multiplexed mode.

Pins 7 and 8: EXTAL/XTAL. A crystal connected across these pins becomes a frequency determining device for an on-chip Pierce oscillator. This oscillator, along with a clock generator, generate the Phase 2 (PH2) and E clock timing signals. Alternately, an external, CMOS-compatible clock can be connected to the EXTAL pin as an E/Phase 2 clock × 4 timing signal. In such a case, a system designer can use the XTAL pin as an output which runs at four times the PH2/E clock frequency. Properly buffered, this signal can drive the EXTAL input of another M68HC11. Both EVB and EVBU use board-mounted crystals or ceramic resonators. Since these circuits are extremely susceptible to noise and loading problems, EVB and EVBU board jumper headers disconnect the EXTAL/XTAL pins from P1/P4.

Pins 9–16: PORT C pins PC0–PC7. In expanded multiplexed mode, these pins become multiplexed address pins A0–A7, then data pins D0–D7. On the EVB, the MC68HC24 PRU makes these pins (at P1) appear to be full-time PC0–PC7 general-purpose parallel I/O pins. The EVB user cannot use these P1 pins as multiplexed address/data pins.

Pin 17: This pin is an input that can generate chip resets. Pin 17, along with the on-board EVB and EVBU reset push-buttons, connect to the microcontroller's $\overline{RESET}$ pin. The $\overline{RESET}$ pin and pin 17 become a low-active output when on-chip circuits generate a microcontroller reset.

Pin 18: $\overline{XIRQ}$ interrupt input pin. On the EVBU, this pin is jumpered to the microcontroller's PA3/OC5/IC4 pin. The EVBU's BUFFALO TRACE feature uses this arrangement.

Pin 19: $\overline{IRQ}$ interrupt input pin.

Pin 20: The user can configure this pin as PORT D, bit 0 (PD0), a general-purpose I/O pin. The user can alternately configure this pin as a serial input pin (RxD) for the Serial Communications Interface (SCI).

Pin 21: The user can configure this pin as general-purpose I/O pin PD1, or as the SCI's serial output pin (TxD).

Pin 22: Depending upon the user's configuration, this pin functions as either general-purpose I/O pin PD2 or the Master In Slave Out (MISO) data pin. The M68HC11's on-board Serial Peripheral Interface (SPI) uses the MISO line.

Pin 23: A user configures this pin as general-purpose I/O pin PD3 or as the SPI's MOSI (Master Out Slave In) data pin.

Pin 24: Pin 24 can operate as general-purpose I/O pin PD4 or as the SPI's System Clock (SCK) pin. The system clock synchronizes SPI serial data exchanges between two M68HC11s or between an M68HC11 and another serial communications device.

Pin 25: This pin can operate as general-purpose I/O pin PD5 or as the SPI's Slave Select ($\overline{SS}$) pin.

Pin 26: V_{DD} (EVB). Not connected (EVBU).

Pin 27: Depending upon user configuration, pin 27 serves three functions. First, this pin can serve as general-purpose I/O pin PA7 (PORTA pin 7). Second, this pin can serve as the Pulse Accumulator's input pin (PAI). The Pulse Accumulator keeps a count of user-specified input edges. Third, this pin can function as the Output Compare 1 (OC1) pin. An output compare mechanism outputs a user-specified edge to an output compare pin upon a successful compare between the M68HC11's free-running counter and the contents of an output compare register.

Pins 28–30: These pins can serve as general-purpose output pins PA6–PA4 or as output compare pins OC2–OC4.

Pin 31: This pin serves two functions on the EVB and three functions on the EVBU. On the EVB's M68HC11A1, pin 31 can serve as general-purpose output pin PA3 or as the Output Compare 5 (OC5) pin. On the EVBU's M68HC11E9, pin 31 can serve the OC5 function. Additionally, pin 31 can serve as an Input Capture 4 (IC4) pin. An input capture mechanism records the M68HC11's free-running count, at the time an external circuit applies a user-specified edge to an input capture pin. On the "E9," pin 31 can also serve as bidirectional, general-purpose I/O pin PA3.

Pins 32–34: The user can configure these pins as general-purpose input pins PA2–PA0 or as Input Capture pins IC1–IC3.

Pins 35–42: On the EVB, these pins serve as general-purpose PORT B output pins PB7–PB0. On the EVBU, these pins operate as PB7–PB0 in single-chip mode and as address pins A15–A8 in expanded multiplexed mode.

Pins 43–50: On both EVB and EVBU, these pins serve two purposes—depending upon their configuration. First, pins 43–50 can serve as the A-to-D converter's analog input pins AN0–AN7. Second, these pins can serve as PORT E general-purpose input pins PE0–PE7. Pins 43–50 are labeled as follows:

```
43 = PE0/AN0
44 = PE4/AN4
45 = PE1/AN1
46 = PE5/AN5
47 = PE2/AN2
48 = PE6/AN6
49 = PE3/AN3
50 = PE7/AN7
```

Pin 51: V_{REFL}. This pin provides a low-level reference voltage for the M68HC11's A-to-D converter. The EVBU grounds this pin on the EVBU board itself. Thus, the EVBU user should leave this pin unconnected. However, the EVB board makes no such on-board ground connection. Therefore, the EVB user must ground this pin when using the M68HC11's A-to-D converter.

Pin 52: V_{REFH}. This pin provides a high-level reference voltage for the A-to-D converter. The EVBU board ties this pin to V_{DD} on the EVBU board itself. Thus, the EVBU user should leave this pin unconnected. The EVB board makes no such connection. The EVB user must tie this pin to the board's V_{DD} supply voltage (+5VDC) when using the M68HC11's A-to-D converter.

Pins 53–56: Unconnected.

Pins 57–58: V_{DD} on the EVBU. Unconnected on the EVB.

Pins 59–60: Ground on the EVBU. Unconnected on the EVB.

APPENDIX D:
AS11 Assembler

D.1 INTRODUCTION

The author chose the BUFFALO one-line assembler/disassembler for use throughout this book's main text. One of this book's reviewers called the BUFFALO assembler "primitive," i.e., without many features common to more elaborate assemblers. The author agrees that, in this sense, the BUFFALO assembler rates as "primitive." However, primitiveness has its benefits. The BUFFALO assembler's simplicity makes it quickly and easily learned. BUFFALO avoids complicated syntax, labels, and assembler directives. The student can therefore spend more time learning about the M68HC11 and less time mastering the assembler.

The BUFFALO assembler entails only a terminal for connection to the EVB or EVBU. Such terminals are widely available and inexpensive. The author's students have often found terminals at give-away prices. Of course the BUFFALO assembler also works with a PC running Motorola's freeware KERMIT program. At the author's school, constant demand for classroom PCs precludes their use for microprocessor/microcontroller courses. At this school, terminals serve a dual function. Students use them for microcontroller classes, and the terminals connect to a file server for use in UNIX classes.

However, more elaborate assemblers such as the AS11 have a number of important advantages. Once a student has learned how to use such an assembler, he or she finds that labels, assembler directives, and error messages make program entry quick and convenient. The AS11 facilitates printing of program lists. Such lists integrate source code, assembled machine code, and valuable analytical features, such as machine cycle counts. Moreover, AS11 proficiency constitutes a marketable skill in and of itself. The AS11 assembler requires a PC, running DOS 2.0 or later versions. This PC must have a serial port available for program downloads to the M68HC11. An AS11 user must have proficiency with a word processor or text processor, basic DOS commands, and communications software such as KERMIT or PROCOMM.

Thus, an instructor who has an adequate number of classroom PCs available may have excellent reasons for basing instruction upon the AS11 or other full-blown assembler. This book's exercises are compatible with the AS11 or other full-featured assemblers. This appendix provides a primer on the AS11, along with illustrations of its use.

D.2 AS11 FUNCTIONS

The AS11 consists of cross-assembler software, which Motorola includes on a floppy disk with each EVB or EVBU. This software translates user-created source code into Motorola S-records. An S-record is an ASCII file suitable for downloading to the M68HC11. An S-record contains format data, executable machine code, memory initialization values, and transmission error check data. At the user's option, the AS11 can also translate source code into a list file, which includes source code, assembled machine code, and three selectable features to facilitate program analysis.

The user employs a PC word processor or text processor to create program source code. After entering this source code, the user converts it to an ASCII file. Most word processors can convert a file to ASCII by executing a simple series of commands. The user then stores this file to the PC storage medium containing AS11 software. The author recommends an .SOU suffix to identify this source code file. For example:

`B:\FILENAME.SOU`

"B:\" identifies the path to a disk containing the AS11, and "FILENAME" is a file designation based upon the program's name, e.g., "DEMO8" or "FUZZCTRL."

To translate a source code file into an S-record, the user invokes the AS11 and identifies the source code file to be translated:

`B:\AS11 FILENAME.SOU`

"B:\" identifies the drive containing both source code file and AS11 software. "AS11" invokes the translation software. Of course, "FILENAME.SOU" identifies the source code file to be translated. In response, the AS11 generates S-record code that contains memory initialization values, executable machine code, format data, and checksum error detection code. Additionally, the AS11 will display a message listing errors it detected in the source file. These errors consist largely of syntax violations. The AS11 appends an .S19 suffix to the S-record file designator:

`B:\FILENAME.S19`

After correcting indicated errors in the source code file and re-executing an S-record translation, the user can download the S-record file to the M68HC11. A serial communications program, such as Motorola's freeware KERMIT program, facilitates an S-record download. KERMIT and DOS commands write the S-record file from the PC's storage medium (for example, the B drive) to the PC's serial port. The user has connected this serial port to the EVBU or EVB via an RS-232 cable. The M68HC11's BUFFALO operating system includes the LOAD T command. This command downloads the S-record file from the PC's serial port to the EVBU or EVB. Thus, the user downloads the S-record from the host PC to the M68HC11 by using a series of KERMIT, DOS, and BUFFALO commands.

At the user's option, the AS11 can also translate a source code file into a list file that contains all source code, resulting machine code, and user-selectable analytical features. If the user wishes to merely display the list file, but not store it, he or she appends a "-l" to the AS11 invocation command:

`B:\AS11 FILENAME.SOU -l`

If the user wants the AS11 to store the list file, he or she uses this command:

`B:\AS11 FILENAME.SOU -l > FILENAME.LST`

Here, "> FILENAME.LST" directs the AS11 to store the list file—designated as FILENAME.LST—to the B drive along with the S-record file.

If the user wishes, the list file can include a machine cycle count for each program instruction. To include cycle counting, the user appends a "c" following the "-l" in the command line:

`B:\AS11 FILENAME.SOU -l c > FILENAME.LST`

At the user's option, the list file can include a symbol table. This symbol table follows the program listing. The table lists all labels used, and the addresses which they reference. To create a symbol table, the user includes an "s" in the command line:

`B:\AS11 FILENAME.SOU -l s > FILENAME.LST`

The list file can also include a cross-reference table. This table lists all labels used and the program line numbers where they occur. To create a cross-reference table, the user includes a "cre" in the command line:

`B:\AS11 FILENAME.SOU -l cre > FILENAME.LST`

Of course, the user can include a "c," "s," "cre," or any combination in the command line:

`B:\AS11 FILENAME.SOU -l c s cre > FILENAME.LST`

Thus, Motorola's AS11 freeware translates user-created ASCII source code files into S-record files and list files. The user can download these S-record files to the M68HC11. Following sections provide detailed advice on creating source code files, translating to S-record files, and using KERMIT and BUFFALO to download S-records.

D.3 CREATING A SOURCE CODE FILE

A source code file consists of source code lines, which make up a program. An AS11 user employs a word or text processor to create a source code file. After creating this file, the user converts it to ASCII. That is, the user assures that the source code file contains only ASCII characters. Most word processors can convert a file to ASCII with a few simple commands.

Source Code Fields

The user enters source code lines one at a time. A source code line includes four fields:

1. Label
2. Operation
3. Operand
4. Comments

The label field identifies constants, operands, addresses, program routines, subroutines, or interrupt routines. Use labels in lieu of direct references to memory addresses or constant values. A label field can contain up to fifteen characters. The label must begin with a letter, a period (.), or an underline (_). Remaining characters in a label may be letters, numbers, periods, dollar signs, or underlines. Avoid spaces (made with the space bar) within a label. Use an underline instead of a space.

The operation field identifies an action to be taken by either the M68HC11 or the AS11 assembler. The operation field can therefore contain either a mnemonic from the M68HC11 instruction set or an assembler directive. An assembler directive does not form a part of the program code. Rather, a directive instructs the AS11 on how to assemble the program code.

The operand field serves four functions. First, the operand field defines an instruction's addressing mode. The operand field employs syntax elements to identify M68HC11 addressing modes. For example, a "#" prefix identifies an immediate-mode operand. Second, the operand field can specify a constant. A numerical entry into the operand field can specify this constant, or the user can identify the constant by entering a previously established label. Third, the operand field can specify an operand's address. A programmer can enter the address itself, or can enter a previously established label identifying the address. Finally, the operand field can identify a program jump or branch destination. Again, the user can enter the address itself or a previously established label.

The comments field allows a programmer to explain a source code line. An assembled S-record will not include comments, but the list file will. Thus, comments are optional. A comment can include any printable characters. Normally, comments follow the operand field in the source code line. However, a comment can follow the operation field, when the instruction or directive requires no operand. For example, an inherent-mode instruction, such as INX, uses no operand. Additionally, the programmer can use a comment to identify program routines or sections. In this case, the programmer enters an asterisk (*) as the first character in the label field. This asterisk informs the AS11 that a comment follows, and that the source code line will include no other fields.

Creating a Source Code Line

To create a source code line, enter the fields in the order discussed above. Start the label field at the left margin and separate fields with a single space (space bar). Do not use spaces within label, operation, or operand fields. The comments field can include spaces. Indicate the end of a source code line with a carriage return ("enter" key).

Numbers and Mathematical Operators

Use care in entering numbers into the operand field. Use the following prefixes to identify decimal, hex, octal, and binary numbers:

> No prefix—decimal (default, e.g., 53927)
> $—hexadecimal (e.g., $D2A7)
> @—octal (e.g., @151247)
> %—binary (e.g., %1101001010100111)

Use the following symbols to identify mathematical operators in the operand field:

> \+ add
> − subtract
> * multiply
> / divide

Assembler Directives

Assembler directives are instructions to the AS11 on how it should assemble an S-record and create a list file. These directives fall into four categories. First, assembler directives may tell the AS11 where to locate program elements in M68HC11 memory. These elements can include program routines, operands, constants, and results. Second, directives may assign labels to numerical values or M68HC11 addresses. Third, directives may establish initial values in M68HC11 memory locations. Fourth, assembler directives control aspects of list file generation. Following paragraphs describe these assembler directives in more detail.

The NAM directive allows a programmer to assign a title to his or her program on the first line of the source code file and list file. Enter the NAM directive into the operation field, then follow with a space and title; for example,

```
NAM DEMO8
```

where "DEMO8" is the program's name.

The ORG (Origin) directive establishes starting addresses for program memory blocks. An ORG directive can establish the starting address of a main program, a subroutine, or an interrupt routine. ORG can establish starting addresses for scratch pad memory blocks, operand storage areas, or results. Enter the ORG mnemonic into the operation field, and the starting address into the operand field. For example,

```
ORG $0100
```

establishes a memory block starting at M68HC11 address $0100. The ORG source code line precedes a section of source code lines relevant to the memory block established. If a programmer enters no ORG directive, the AS11 assumes that the memory block begins at address $0000.

EQU (Equate) directives establish address or operand values, and assign labels to them. An EQU directive does not cause the AS11 to store these values in M68HC11 memory. Enter the label to be assigned into the label field, the EQU mnemonic into the operation field, and the value to be assigned into the operand field. For example,

```
OUTCRL EQU $FFC4
```

equates value $FFC4 to label OUTCRL.

An FCB (Form Constant Byte) directive labels a specified value and assigns it to an M68HC11 memory location. The AS11 selects the memory location. Therefore, the programmer has no direct control over which memory byte will hold the specified value. Of course, a previous ORG directive has established a memory block in which the value will reside. Enter the label to be assigned into the label field, the FCB mnemonic into the operation field, and the value to be assigned into the operand field. For example,

```
AUGEND FCB $9C
```

assigns value $9C to a memory location labeled AUGEND. The AS11 selects the memory location.

If the programmer makes no entry into the operand field, the AS11 stores $00 to the memory location. For example,

```
AUGEND FCB
```

assigns value $00 to a memory location labeled AUGEND.

A programmer can also use an FCB directive to initialize a series of memory locations with 8-bit values. The programmer enters these values into the operand field. Separate operand field values with commas. For example, the following source code line stores $9C, $10, and $8C to successive M68HC11 memory locations. Value $9C resides at a location labeled "AUGENDS." Access value $10 at "AUGENDS+1." Access value $8C at "AUGENDS+2":

```
AUGENDS FCB $9C,$10,$8C
```

If the programmer makes no entry between commas, the AS11 assigns a value of $00. The following would store $9C, $00, and $8C to successive memory locations.

```
AUGENDS FCB $9C,,$8C
```

The following source code line would store $00 to all three memory locations:

```
AUGENDS FCB ,,
```

FDB (Form Double Byte) directives function in the same way as FCBs, except that FDBs store 2 bytes at a time. For example,

```
AUGEND FDB $9C10
```

stores value $9C to a memory location labeled "AUGEND." The AS11 stores $10 to location "AUGEND+1."

Another example:

```
AUGEND FDB
```

stores $00 to a location labeled "AUGEND," and $00 to location "AUGEND+1."

Finally, this source code line stores values $9C, $10, $00, and $00 to successive memory locations. Access $9C at "AUGEND," $10 at "AUGEND+1," $00 at "AUGEND+2," and $00 at "AUGEND+3"

```
AUGEND FDB $9C10,
```

The RMB (Reserve Memory Bytes) directive labels and sets aside a series of M68HC11 memory locations. An RMB does not initialize these memory locations. The AS11 selects memory locations to be reserved. Of course, a previous ORG directive has established a memory block within which reserved bytes will reside. Enter the label to be assigned into the label field, and the RMB mnemonic into the operation field. Enter the number of bytes to be reserved into the operand field. For example,

```
RESULTS RMB 15
```

reserves fifteen successive memory locations. The AS11 labels the first location "RESULTS." Access following locations as "RESULTS+1," "RESULTS+2," etc.

BSZ (Block Storage of Zeros) and ZMB (Zero Memory Bytes) directives cause identical responses from the AS11. BSZ/ZMB operates in a manner similar to the RMB, except that the AS11 stores $00 to all reserved memory bytes. The AS11 selects the series of M68HC11 locations to be reserved and initialized to zero. For example,

```
RESULTS ZMB 20
```

reserves twenty successive memory locations and initializes them to $00. The first location is labeled "RESULTS." Access remaining locations as "RESULTS+1," "RESULTS+2," etc.

The FILL (Fill Memory) directive causes the AS11 to label and reserve a specified number of memory bytes. In addition, the AS11 initializes all of these bytes to a specified value. The AS11 selects the series of memory locations to be reserved. Enter the label to be assigned into the source code line's label field. Enter the FILL mnemonic into the operation field. Enter two values into the operand field. First, enter the 8-bit initialization value, then enter the number of bytes to be reserved and initialized. Separate these two values with a comma. For example,

```
RESULTS FILL $FF,$19
```

reserves a series of twenty-five M68HC11 memory locations and initializes them all to $FF. The first memory location is labeled "RESULTS." Access remaining locations as "RESULTS+1," "RESULTS+2," etc.

The FCC (Form Constant Character String) directive causes the AS11 to label and store an ASCII string into a series of M68HC11 memory locations. Again, the AS11 chooses the memory locations. Enter the label to be assigned into the source code line's label field. Enter the FCC mnemonic into the operation field. Enter the ASCII string into the operand field. Prefix the string with a delimiter. Also add a suffix delimiter at the end of the string. Use identical and printable ASCII characters for the delimiters. The author suggests commas, colons, or semicolons as delimiters. Additionally, assure that the string includes only printable ASCII characters. An example:

```
PROMPT FCC :ENTER_NEXT_OPERAND.:
```

In response to this source code line, the AS11 stores ASCII string "ENTER_NEXT_OPERAND." into a series of M68HC11 memory locations. The first byte in the string is labelled "PROMPT." Access remaining bytes of the string as "PROMPT+1," "PROMPT+2," etc.

The OPT (Assembler Output Options) directive allows a programmer to use source code to create a list file and enable its analytical features. Section D.2, above, describes how the AS11 invocation command line can direct the assembler to create a list file, and include analytical options. This AS11 invocation is a DOS command. By using the OPT directive in a source code line, the programmer can also generate a list file and enable the analytical options. Moreover, the programmer can start or stop list file production and cycle counting at any points in the source file. To enable list file generation, enter the OPT mnemonic into the operation field, and a lower-case "l" into the operand field:

```
OPT l
```

List file generation starts from the point at which this source code is entered. To enable list file generation, cycle counting and a symbol table, use this source code line:

```
OPT l,c,s
```

Note that commas separate the options in the operand field.

Use the following source code line to enable the list file and cross-reference table:

```
OPT l,cre
```

Enter this line into the source file prior to the point at which the AS11 encounters the first label.

To enable list file generation and all options, enter this line:

```
OPT l,c,s,cre
```

Using the OPT directive, a programmer can discontinue list file generation and cycle counting at any point in the program. To discontinue list file generation use the "nol" option:

```
OPT nol
```

List file generation halts from the point of this source code entry. A programmer can resume list file generation at any point by using the "l" option.

Discontinue cycle counting with the "noc" option:

```
OPT noc
```

Cycle counting halts from the point of this source code entry. Use the "c" option to resume cycle counting at any subsequent point in the program.

To discontinue both list file and cycle counting use this entry:

```
OPT nol,noc
```

Finally, the reader should note that OPT source code entries take precedence over list file option entries in the DOS AS11 invocation command.

A PAGE (Top of Page) directive instructs the AS11 to print the next list file line at the top of a new page. This directive does not appear in the list file. Simply enter this mnemonic into the operation field. Make no other entries in the source code line.

A Source Code Example

Refer to Figure D–1 and Figure 1–6 (in Chapter 1). Figure D–1 illustrates an AS11 source code file. This source code will generate object code (machine code) identical to that generated by BUFFALO one-line assembler source code shown in Figure 1–6. Thus, Figure D–1's program will execute the Modified DEMO flowchart shown in Figure 1–6. The author's objective in developing this AS11 source code is to illustrate the use of some assembler directives, some labels, and use of the comments field.

Note that the Figure D–1 source code file contains two sections, separated by two blank lines. The first section establishes a parameter and labels. The second section contains actual program entries.

FIGURE D–1

```
    NAM MODEMO
XREG EQU $0001
AUGEND FCB
SUMS RMB 5


* PROGRAM BEGINS HERE
 ORG $100
START LDX #XREG POINT X REG AT SUMS
 LDAA #$2 LOAD ADDEND
ADDLOOP ADDA AUGEND ADD AUGEND
 STAA 0,X STORE SUM TO NEXT SUMx
INX INCREMENT X REG
 CPX #SUMS+5 COMPARE X REG WITH LAST SUMS ADDRESS + 1
 BNE ADDLOOP IF NOT EQUAL, ADD AGAIN
 WAI
```

The source code file's first section includes no ORG statement. Therefore, the AS11 establishes a memory block at the default address: $0000. The first line of this section names the program "MODEMO," which stands for "Modified DEMO." The second line equates the label "XREG" with value $0001. This equate does not establish a memory storage location. The third line establishes (FCB) a memory location labeled "AUGEND." Since this is the first memory byte established by the AS11, it will reside at address $0000. Note that the author assigned no value to this location.[1] An EVB or EVBU operator will use the BUFFALO MM command to initialize the "AUGEND" location just prior to program execution. The fourth line reserves (RMB) five memory bytes. The AS11 will establish the first of these bytes at address $0001. The author assigned the label "SUMS" to this memory location. The remaining reserved memory bytes reside at "SUMS+1," "SUMS+2," etc. The memory location following the last reserved byte (SUMS+4) would therefore course be "SUMS+5." Thus, the first section of source code has assigned three labels: "XREG," AUGEND," and "SUMS." This section has also established six memory locations at M68HC11 addresses $0000 through $0005.

The source code file's second section includes ten source code lines. In the first line, the author has entered a comment that identifies this second section as the program area. In the second line, an ORG statement establishes a memory block beginning at address $0100.

Note that the author has labeled the third line as "START." This label simply identifies the first actual program instruction (LDX). This instruction specifies immediate addressing mode with a "#" in its operand field. Thus, this instruction uses immediate mode to load the X register with a value labeled "XREG." Recall that "XREG" equates with $0001.

The fourth line again specifies immediate mode with "#" in its operand field. The instruction loads ACCA with constant value $02. This source code line illustrates that labels are often unnecessary.[2] Use labels only where they facilitate source code entry, or where they make the source code more understandable.

Note that the author labeled the fifth source code line as "ADDLOOP." This source code line is the destination of a branch instruction. Thus, this line's label facilitates entry of the branch instruction (BNE) in the ninth line. In the fifth line, the ADDA instruction adds a value, stored in a location labeled "AUGEND," to ACCA's contents. The user will employ the BUFFALO MM command to initialize this memory location just prior to program execution.

The sixth source code line causes the M68HC11's CPU to store the contents of ACCA to a sum location using indexed mode. This line uses no label. The seventh line uses an inherent mode INX instruction. This source code line therefore includes no operand field.

The eighth line again uses a "#" symbol in its operand field to specify immediate mode. Here, the M68HC11's CPU compares the contents of its X register with value "SUMS+5." This value equals the address labeled "SUMS," plus five. Recall that the label "SUMS" refers to a reserved memory byte's address. An RMB directive does not specify the contents of reserved locations.

The ninth line specifies a BNE instruction. This line's operand field specifies the destination address with label "ADDLOOP." The author assigned this label to the ADDA instruction in the fourth line. Thus, the BNE instruction's destination is the ADDA instruction's address.

The tenth source code line, of the second section, specifies the inherent mode WAI instruction. This instruction halts M68HC11 program execution.

D.4 THE LIST FILE AND S-RECORDS

The previous section demonstrated the creation of a source code file—MODEMO.SOU. We now wish to invoke the AS11 assembler and translate this source code into S-records and a list file. Let us assume that we wish to include cycle counts in our list file. Let us

---

[1]The AS11 assigns a default value of $00.
[2]Indeed, the "XREG" label in this program is unnecessary. The author used this label to show how to assign and apply it in conjunction with an EQU directive.

FIGURE D-2

```
0001                                                NAM  MODEMO
0002 0001                              XREG         EQU  $0001
0003 0000 00                           AUGEND       FCB
0004 0001                              SUMS         RMB  5
0005
0006
0007                                   * PROGRAM BEGINS HERE
0008 0100                                           ORG  $100
0009 0100 ce 00 01       [  3 ]        START        LDX  #XREG        POINT X REG AT
                                                                     SUMS
0010 0103 86 02          [  2 ]                     LDAA #$2          LOAD ADDEND
0011 0105 9b 00          [  3 ]        ADDLOOP      ADDA AUGEND       ADD AUGEND
0012 0107 a7 00          [  4 ]                     STAA 0,X          STORE SUM TO
                                                                     NEXT SUMx
0013 0109 08             [  3 ]                     INX               NCREMENT X REG
0014 010a 8c 00 06       [  4 ]                     CPX  #SUMS+5      COMPARE X REG
                                                                     WITH LAST SUMS
                                                                     ADDRESS + 1
0015 010d 26 f6          [  3 ]                     BNE  ADDLOOP      IF NOT EQUAL,
                                                                     ADD AGAIN
0016 010f 3e             [ 14 ]                     WAI               HALT
```

further assume that the AS11 software resides on a disk in the PC's B drive, and that MODEMO.SOU was also stored to this disk.

To invoke the AS11 and create S-records and list file, use this DOS command from the B:\ prompt:

`B:\AS11 MODEMO.SOU -l c > MODEMO.LST`

Figure D-2 shows the resulting list file.[3] This file's left-most column contains source code line numbers. The next column lists M68HC11 memory addresses. The third, fourth, and fifth columns from the left contain assembled M68HC11 machine code, including opcode, address and operand bytes. The sixth column lists machine cycle counts in brackets. Of course, the "c" entry in the AS11 invocation command generated these counts. The seventh column is the source code's label field. The next column contains the operation field, and the ninth column is the operand field. Finally, the right-most column contains the comments field.

Figure D-3 shows file MODEMO.S19, which contains S-records assembled by the AS11. The first line contains an S1 data record as indicated by its two left-most characters. The next two characters ($04) indicate record length, the number of following S-record character pairs containing address information and data, including a pair of checksum characters. The next four characters ($0000) indicate the M68HC11's memory address where this S-record's data bytes will be stored. The next two characters ($00) are the actual data to be stored. The two right-most characters form a checksum. A checksum is the least significant byte of a value. This value is the one's complement of the sum of the values of all record length, address, and data character pairs. In the case of this S-record, the sum equals $04:

$04 + $00 + $00 + $00 = $04

The one's complement of $04 equals $FB.

FIGURE D-3

```
S104000000FB
S1130100CE000186029B00A700088C000626F63E5E
S9030000FC
```

---

[3]The AS11 does not produce a list file with the neatly aligned columns shown in Figure D-2. The author cleaned up the columns with a word processor.

MODEMO.S19's second line also contains an S1 record. This record follows the same pattern as the first line. This record contains nineteen ($13) character pairs, including checksum data. Data will be stored to M68HC11 address $0100. The data includes sixteen pairs of machine code program characters. The checksum equals $5E, the least significant byte of the one's complement of $4A1. $4A1 represents the total of all record length, address, and data character pairs.

The third line contains a termination (S9) record. This record includes $03 character pairs, including two pairs of address characters and a checksum. $0000 indicates the M68HC11 address where the first S-record's data will be stored. An S9 record includes no data bytes. Checksum value $FC represents the one's complement of sum $03 + $00 + $00.

D.5 USE KERMIT TO DOWNLOAD S-RECORDS TO THE M68HC11

Motorola's KERMIT freeware offers effective means for downloading S-records to the M68HC11. This section provides step by step instructions for downloading file MODEMO.S19 to an EVB or EVBU. Let us assume that both MODEMO.S19 and KERMIT are stored on a disk in the B drive. Let us further assume that the PC's COM2 serial port connects to the EVB or EVBU. Follow these steps to complete the download.

Step 1. From the B prompt, use this DOS command to invoke KERMIT:

`B:\>KERMIT`   <enter>

Step 2. From the KERMIT prompt, identify the serial port to be used in the download:

`Kermit-MS>SET PORT 2`   <enter>

Step 3. From the KERMIT prompt, establish the download baud rate:

`Kermit-MS>SET BAUD 9600`   <enter>

Step 4. From the KERMIT prompt, connect KERMIT to the serial port:

`Kermit-MS>CONNECT`   <enter × 2>

Step 5. Reset the EVB or EVBU board, then press the Enter key:

                                                                                                                   <enter>

Step 6. From the BUFFALO prompt, enter this BUFFALO system command:

`>LOAD T`   <enter>

Step 7. Return to KERMIT by holding down the Ctrl key while typing a "]" then "C":

`(Ctrl)]C`

Step 8. At the KERMIT prompt, tell KERMIT to transfer the S-records to the serial port:

`Kermit-MS>PUSH`   <enter>

Step 9. At the B prompt, use a DOS command to transfer the MODEMO.S19 file:[4]

`B:\>TYPE MODEMO.S19 > COM2`   <enter>

Step 10. At the B prompt, use an EXIT command to return to KERMIT:

`B:\>EXIT`   <enter>

---

[4]If KERMIT resides on a different drive than the S-records to be downloaded, specify the path to the S19 file. For example, assume that KERMIT resides on the C drive, and MODEMO.S19 resides on the B drive. Use this DOS TYPE command:

`C:\>TYPE B:\MODEMO.S19 > COM2`

where "B:\" indicates the path to the MODEMO.S19 file.

Step 11. At the KERMIT prompt, reconnect KERMIT to the M68HC11:

`Kermit-MS>CONNECT` <enter>

Step 12. Reset the EVB or EVBU board, and press the Enter key:

<enter>

Step 13. From the BUFFALO system prompt, use BUFFALO system commands to view program storage areas. Assure that the download was successful.

Note: To exit BUFFALO and return to KERMIT use:

(Ctrl)]C

To exit KERMIT and return to DOS, use:

`Kermit-MS>EXIT` <enter>

D.6   LIST FILE FOR A MORE COMPLEX PROGRAM

Modified DEMO program (Figure 1–6) has served to illustrate the AS11 assembly and downloading process. However, a more complex program's list file may also help answer some questions that have occurred to the reader. Refer to Chapter 1, Exercise 1.5. This exercise includes a DEMO8 flowchart and memory allocations (Figure 1–18). Also refer to Part IV, Laboratory Software Listing, of this exercise. Review the list file below. Source code in this list file generates object code identical to that generated by BUFFALO one-line assembly code in the Laboratory Software Listing.

```
0001                              NAM DEMO8
0002 0100                         MAIN     EQU $100
0003 0180                         CLRSTOR  EQU $180
0004 0125                         HGHR$64  EQU $125
0005 0130                         LOSAM$4B EQU $130
0006 013a                         GTREQ$E7 EQU $13A
0007 0145                         LESS$F6  EQU $145
0008 0150                         GRTR$1E  EQU $150
0009 0160                         DISPLAY  EQU $160
0010 ffc4                         OUTCRL   EQU $FFC4
0011 ffbe                         OUT1BS   EQU $FFBE
0012 0000                         TESTOPS  EQU $0
0013 0005                         RESULTS  EQU $5
0014
0015                            * MAIN
0016 0100                         ORG MAIN
0017 0100 bd 01 80    [ 6 ]       JSR CLRSTOR
0018 0103 ce 00 00    [ 3 ]       LDX #TESTOPS
0019 0106 a6 00       [ 4 ]       EVALOOP LDAA 0,X
0020 0108 bd 01 25    [ 6 ]       JSR HGHR$64
0021 010b bd 01 30    [ 6 ]       JSR LOSAM$4B
0022 010e bd 01 3a    [ 6 ]       JSR GTREQ$E7
0023 0111 bd 01 45    [ 6 ]       JSR LESS$F6
0024 0114 bd 01 50    [ 6 ]       JSR GRTR$1E
0025 0117 08          [ 3 ]       INX
0026 0118 8c 00 05    [ 4 ]       CPX #TESTOPS+5
0027 011b 26 e9       [ 3 ]       BNE EVALOOP
0028 011d bd 01 60    [ 6 ]       JSR DISPLAY
0029 0120 3e          [14 ]       WAI
0030
```

```
0031                        * CLRSTOR SUBROUTINE
0032 0180                     ORG CLRSTOR
0033 0180 ce 00 05   [ 3 ]    LDX #RESULTS
0034 0183 6f 00      [ 6 ]    CLRLOOP CLR 0,X
0035 0185 08         [ 3 ]    INX
0036 0186 8c 00 1e   [ 4 ]    CPX #RESULTS+25
0037 0189 26 f8      [ 3 ]    BNE CLRLOOP
0038 018b 39         [ 5 ]    RTS
0039
0040                        * HGHR$64 SUBROUTINE
0041 0125                     ORG HGHR$64
0042 0125 81 64      [ 2 ]    CMPA #$64
0043 0127 23 02      [ 3 ]    BLS RET1
0044 0129 a7 05      [ 4 ]    STAA $5,X
0045 012b 39         [ 5 ]    RET1 RTS
0046
0047                        * LOSAM$4B SUBROUTINE
0048 0130                     ORG LOSAM$4B
0049 0130 81 4b      [ 2 ]    CMPA #$4B
0050 0132 22 02      [ 3 ]    BHI RET2
0051 0134 a7 0a      [ 4 ]    STAA $A,X
0052 0136 39         [ 5 ]    RET2 RTS
0053
0054                        * GTREQ$E7 SUBROUTINE
0055 013a                     ORG GTREQ$E7
0056 013a 81 e7      [ 2 ]    CMPA #$E7
0057 013c 2d 02      [ 3 ]    BLT RET3
0058 013e a7 0f      [ 4 ]    STAA $F,X
0059 0140 39         [ 5 ]    RET3 RTS
0060
0061                        * LESS$F6 SUBROUTINE
0062 0145                     ORG LESS$F6
0063 0145 81 f6      [ 2 ]    CMPA #$F6
0064 0147 2c 02      [ 3 ]    BGE RET4
0065 0149 a7 14      [ 4 ]    STAA $14,X
0066 014b 39         [ 5 ]    RET4 RTS
0067
0068                        * GRTR$1E SUBROUTINE
0069 0150                     ORG GRTR$1E
0070 0150 81 1e      [ 2 ]    CMPA #$1E
0071 0152 2f 02      [ 3 ]    BLE RET5
0072 0154 a7 19      [ 4 ]    STAA $19,X
0073 0156 39         [ 5 ]    RET5 RTS
0074
0075                        * DISPLAY SUBROUTINE
0076 0160                     ORG DISPLAY
0077 0160 ce 00 00   [ 3 ]    LDX #TESTOPS
0078 0163 bd ff c4   [ 6 ]    DISLOOP JSR OUTCRL
0079 0166 c6 05      [ 2 ]    LDAB #$5
0080 0168 bd ff be   [ 6 ]    DISLOOP1 JSR OUT1BS
0081 016b 5a         [ 2 ]    DECB
0082 016c 26 fa      [ 3 ]    BNE DISLOOP1
0083 016e 8c 00 1e   [ 4 ]    CPX #RESULTS+25
0084 0171 26 f0      [ 3 ]    BNE DISLOOP
0085 0173 39         [ 5 ]    RTS
```

APPENDIX E:
M68HC11 Instruction Set[1]

INSTRUCTIONS, ADDRESSING MODES, AND EXECUTION TIMES

Source Forms	Operation	Boolean Expression	Addressing Mode for Operand	Machine Coding (Hexadecimal) Opcode / Operand(s)	Bytes	Cycles	Condition Codes S X H I N Z V C
ABA	Add Accumulators	A + B → A	INH	1B	1	2	— — Δ — Δ Δ Δ Δ
ABX	Add B to X	IX + 00:B → IX	INH	3A	1	3	— — — — — — — —
ABY	Add B to Y	IY + 00:B → IY	INH	18 3A	2	4	— — — — — — — —
ADCA (opr)	Add with Carry to A	A + M + C → A	A IMM A DIR A EXT A IND,X A IND,Y	89 ii 99 dd B9 hh ll A9 ff 18 A9 ff	2 2 3 2 3	2 3 4 4 5	— — Δ — Δ Δ Δ Δ
ADCB (opr)	Add with Carry to B	B + M + C → B	B IMM B DIR B EXT B IND,X B IND,Y	C9 ii D9 dd F9 hh ll E9 ff 18 E9 ff	2 2 3 2 3	2 3 4 4 5	— — Δ — Δ Δ Δ Δ
ADDA (opr)	Add Memory to A	A + M → A	A IMM A DIR A EXT A IND,X A IND,Y	8B ii 9B dd BB hh ll AB ff 18 AB ff	2 2 3 2 3	2 3 4 4 5	— — Δ — Δ Δ Δ Δ

Source Forms	Operation	Boolean Expression	Addressing Mode for Operand	Machine Coding (Hexadecimal) Opcode / Operand(s)	Bytes	Cycles	Condition Codes S X H I N Z V C
ADDB (opr)	Add Memory to B	B + M → B	B IMM B DIR B EXT B IND,X B IND,Y	CB ii DB dd FB hh ll EB ff 18 EB ff	2 2 3 2 3	2 3 4 4 5	— — Δ — Δ Δ Δ Δ
ADDD (opr)	Add 16-Bit to D	D + M:M + 1 → D	IMM DIR EXT IND,X IND,Y	C3 jj kk D3 dd F3 hh ll E3 ff 18 E3 ff	3 2 3 2 3	4 5 6 6 7	— — — — Δ Δ Δ Δ
ANDA (opr)	AND A with Memory	A • M → A	A IMM A DIR A EXT A IND,X A IND,Y	84 ii 94 dd B4 hh ll A4 ff 18 A4 ff	2 2 3 2 3	2 3 4 4 5	— — — — Δ Δ 0 —
ANDB (opr)	AND B with Memory	B • M → B	B IMM B DIR B EXT B IND,X B IND,Y	C4 ii D4 dd F4 hh ll E4 ff 18 E4 ff	2 2 3 2 3	2 3 4 4 5	— — — — Δ Δ 0 —

---

[1] Motorola Inc., *M68HC11 E Series Programming Reference Guide*, pp. 19–35. Reprinted with the permission of Motorola.

Source Forms		Operation	Boolean Expression	Addressing Mode for Operand	Machine Coding (Hexadecimal) Opcode Operand(s)	Bytes	Cycles	Condition Codes S X H I N Z V C
ASL	(opr)	Arithmetic Shift Left		EXT IND,X IND,Y A INH B INH	78 hh ll 68 ff 18 68 ff 48 58	3 2 3 1 1	6 6 7 2 2	— — — — ∆ ∆ ∆ ∆
ASLD		Arithmetic Shift Left Double		INH	05	1	3	— — — — ∆ ∆ ∆ ∆
ASR	(opr)	Arithmetic Shift Right		EXT IND,X IND,Y A INH B INH	77 hh ll 67 ff 18 67 ff 47 57	3 2 3 1 1	6 6 7 2 2	— — — — ∆ ∆ ∆ ∆
BCC	(rel)	Branch if Carry Clear	? C = 0	REL	24 rr	2	3	— — — — — — — —
BCLR	(opr) (msk)	Clear Bit(s)	M • (mm) → M	DIR IND,X IND,Y	15 dd mm 1D ff mm 18 1D ff mm	3 3 4	6 7 8	— — — — ∆ ∆ 0 —
BCS	(rel)	Branch if Carry Set	? C = 1	REL	25 rr	2	3	— — — — — — — —
BEQ	(rel)	Branch if = Zero	? Z = 1	REL	27 rr	2	3	— — — — — — — —
BGE	(rel)	Branch if ≥ Zero	? N ⊕ V = 0	REL	2C rr	2	3	— — — — — — — —
BGT	(rel)	Branch if > Zero	? Z + (N ⊕ V) = 0	REL	2E rr	2	3	— — — — — — — —
BHI	(rel)	Branch if Higher	? C + Z = 0	REL	22 rr	2	3	— — — — — — — —
BHS	(rel)	Branch if Higher or Same	? C = 0	REL	24 rr	2	3	— — — — — — — —

Source Forms		Operation	Boolean Expression	Addressing Mode for Operand	Machine Coding (Hexadecimal) Opcode Operand(s)	Bytes	Cycles	Condition Codes S X H I N Z V C
BITA	(opr)	Bit(s) Test A with Memory	A • M	A IMM A DIR A EXT A IND,X A IND,Y	85 ii 95 dd B5 hh ll A5 ff 18 A5 ff	2 2 3 2 3	2 3 4 4 5	— — — — ∆ ∆ 0 —
BITB	(opr)	Bit(s) Test B with Memory	B • M	B IMM B DIR B EXT B IND,X B IND,Y	C5 ii D5 dd F5 hh ll E5 ff 18 E5 ff	2 2 3 2 3	2 3 4 4 5	— — — — ∆ ∆ 0 —
BLE	(rel)	Branch if ≤ Zero	? Z + (N ⊕ V) = 1	REL	2F rr	2	3	— — — — — — — —
BLO	(rel)	Branch if Lower	? C = 1	REL	25 rr	2	3	— — — — — — — —
BLS	(rel)	Branch if Lower or Same	? C + Z = 1	REL	23 rr	2	3	— — — — — — — —
BLT	(rel)	Branch if < Zero	? N ⊕ V = 1	REL	2D rr	2	3	— — — — — — — —
BMI	(rel)	Branch if Minus	? N = 1	REL	2B rr	2	3	— — — — — — — —
BNE	(rel)	Branch if Not = Zero	? Z = 0	REL	26 rr	2	3	— — — — — — — —
BPL	(rel)	Branch if Plus	? N = 0	REL	2A rr	2	3	— — — — — — — —
BRA	(rel)	Branch Always	? 1 = 1	REL	20 rr	2	3	— — — — — — — —

Source Forms		Operation	Boolean Expression	Addressing Mode for Operand	Machine Coding (Hexadecimal) Opcode Operand(s)	Bytes	Cycles	Condition Codes S X H I N Z V C
BRCLR	(opr) (msk) (rel)	Branch if Bit(s) Clear	? M • mm = 0	DIR IND,X IND,Y	13 dd mm rr 1F ff mm rr 18 1F ff mm rr	4 4 5	6 7 8	— — — — — — — —
BRN	(rel)	Branch Never	? 1 = 0	REL	21 rr	2	3	— — — — — — — —
BRSET	(opr) (msk) (rel)	Branch if Bit(s) Set	? (M) • mm = 0	DIR IND,X IND,Y	12 dd mm rr 1E ff mm rr 18 1E ff mm rr	4 4 5	6 7 8	— — — — — — — —
BSET	(opr) (msk)	Set Bit(s)	M + mm → M	DIR IND,X IND,Y	14 dd mm 1C ff mm 18 1C ff mm	3 3 4	6 7 8	— — — — ∆ ∆ 0 —
BSR	(rel)	Branch to Subroutine	See Special Ops	REL	8D rr	2	6	— — — — — — — —
BVC	(rel)	Branch if Overflow Clear	? V = 0	REL	28 rr	2	3	— — — — — — — —
BVS	(rel)	Branch if Overflow Set	? V = 1	REL	29 rr	2	3	— — — — — — — —
CBA		Compare A to B	A − B	INH	11	1	2	— — — — ∆ ∆ ∆ ∆
CLC		Clear Carry Bit	0 → C	INH	0C	1	2	— — — — — — — 0
CLI		Clear Interrupt Mask	0 → I	INH	0E	1	2	— — — 0 — — — —
CLR	(opr)	Clear Memory Byte	0 → M	EXT IND,X IND,Y	7F hh ll 6F ff 18 6F ff	3 2 3	6 6 7	— — — — 0 1 0 0
CLRA		Clear Accumulator A	0 → A	A INH	4F	1	2	— — — — 0 1 0 0

Source Forms	Operation	Boolean Expression	Addressing Mode for Operand	Machine Coding (Hexadecimal) Opcode Operand(s)	Bytes	Cycles	Condition Codes S X H I N Z V C
CLRB	Clear Accumulator B	$0 \rightarrow B$	B INH	5F	1	2	— — — — 0 1 0 0
CLV	Clear Overflow Flag	$0 \rightarrow V$	INH	0A	1	2	— — — — — — 0 —
CMPA (opr)	Compare A to Memory	$A - M$	A IMM A DIR A EXT A IND,X A IND,Y	81 ii 91 dd B1 hh ll A1 ff 18 A1 ff	2 2 3 2 3	2 3 4 4 5	— — — — Δ Δ Δ Δ
CMPB (opr)	Compare B to Memory	$B - M$	B IMM B DIR B EXT B IND,X B IND,Y	C1 ii D1 dd F1 hh ll E1 ff 18 E1 ff	2 2 3 2 3	2 3 4 4 5	— — — — Δ Δ Δ Δ
COM (opr)	1's Complement Memory Byte	$\$FF - M \rightarrow M$	EXT IND,X IND,Y	73 hh ll 63 ff 18 63 ff	3 2 3	6 6 7	— — — — Δ Δ 0 1
COMA	1's Complement A	$\$FF - A \rightarrow A$	A INH	43	1	2	— — — — Δ Δ 0 1
COMB	1's Complement B	$\$FF - B \rightarrow B$	B INH	53	1	2	— — — — Δ Δ 0 1
CPD (opr)	Compare D to Memory 16-Bit	$D - M:M+1$	IMM DIR EXT IND,X IND,Y	1A 83 jj kk 1A 93 dd 1A B3 hh ll 1A A3 ff CD A3 ff	4 3 4 3 3	5 6 7 7 7	— — — — Δ Δ Δ Δ

Source Forms	Operation	Boolean Expression	Addressing Mode for Operand	Machine Coding (Hexadecimal) Opcode Operand(s)	Bytes	Cycles	Condition Codes S X H I N Z V C
CPX (opr)	Compare X to Memory 16-Bit	$IX - M:M+1$	IMM DIR EXT IND,X IND,Y	8C jj kk 9C dd BC hh ll AC ff CD AC ff	3 2 3 2 3	4 5 6 6 7	— — — — Δ Δ Δ Δ
CPY (opr)	Compare Y to Memory 16-Bit	$IY - M:M+1$	IMM DIR EXT IND,X IND,Y	18 8C jj kk 18 9C dd 18 BC hh ll 1A AC ff 18 AC ff	4 3 4 3 3	5 6 7 7 7	— — — — Δ Δ Δ Δ
DAA	Decimal Adjust A	Adjust Sum to BCD	INH	19	1	2	— — — — Δ Δ Δ Δ
DEC (opr)	Decrement Memory Byte	$M - 1 \rightarrow M$	EXT IND,X IND,Y	7A hh ll 6A ff 18 6A ff	3 2 3	6 6 7	— — — — Δ Δ Δ —
DECA	Decrement Accumulator A	$A - 1 \rightarrow A$	A INH	4A	1	2	— — — — Δ Δ Δ —
DECB	Decrement Accumulator B	$B - 1 \rightarrow B$	B INH	5A	1	2	— — — — Δ Δ Δ —
DES	Decrement Stack Pointer	$SP - 1 \rightarrow SP$	INH	34	1	3	— — — — — — — —
DEX	Decrement Index Register X	$IX - 1 \rightarrow IX$	INH	09	1	3	— — — — — Δ — —
DEY	Decrement Index Register Y	$IY - 1 \rightarrow IY$	INH	18 09	2	4	— — — — — Δ — —

Source Forms	Operation	Boolean Expression	Addressing Mode for Operand	Machine Coding (Hexadecimal) Opcode Operand(s)	Bytes	Cycles	Condition Codes S X H I N Z V C
EORA (opr)	Exclusive OR A with Memory	$A \oplus M \rightarrow A$	A IMM A DIR A EXT A IND,X A IND,Y	88 ii 98 dd B8 hh ll A8 ff 18 A8 ff	2 2 3 2 3	2 3 4 4 5	— — — — Δ Δ 0 —
EORB (opr)	Exclusive OR B with Memory	$B \oplus M \rightarrow B$	B IMM B DIR B EXT B IND,X B IND,Y	C8 ii D8 dd F8 hh ll E8 ff 18 E8 ff	2 2 3 2 3	2 3 4 4 5	— — — — Δ Δ 0 —
FDIV	Fractional Divide 16 by 16	$D/IX \rightarrow IX; r \rightarrow D$	INH	03	1	41	— — — — — Δ Δ Δ
IDIV	Integer Divide 16 by 16	$D/IX \rightarrow IX; r \rightarrow D$	INH	02	1	41	— — — — — Δ 0 Δ
INC (opr)	Increment Memory Byte	$M + 1 \rightarrow M$	EXT IND,X IND,Y	7C hh ll 6C ff 18 6C ff	3 2 3	6 6 7	— — — — Δ Δ Δ —
INCA	Increment Accumulator A	$A + 1 \rightarrow A$	A INH	4C	1	2	— — — — Δ Δ Δ —
INCB	Increment Accumulator B	$B + 1 \rightarrow B$	B INH	5C	1	2	— — — — Δ Δ Δ —

Source Forms		Operation	Boolean Expression	Addressing Mode for Operand	Machine Coding (Hexadecimal) Opcode / Operand(s)		Bytes	Cycles	Condition Codes S X H I N Z V C
INS		Increment Stack Pointer	SP + 1 → SP	INH	31		1	3	— — — — — — — —
INX		Increment Index Register X	IX + 1 → IX	INH	08		1	3	— — — — — Δ — —
INY		Increment Index Register Y	IY + 1 → IY	INH	18 08		2	4	— — — — — Δ — —
JMP	(opr)	Jump	See Special Ops	EXT IND,X IND,Y	7E hh 6E ff 18 6E ff	ll	3 2 3	3 3 4	— — — — — — — —
JSR	(opr)	Jump to Subroutine	See Special Ops	DIR EXT IND,X IND,Y	9D dd BD hh AD ff 18 AD ff	ll	2 3 2 3	5 6 6 7	— — — — — — — —
LDAA	(opr)	Load Accumulator A	M → A	A IMM A DIR A EXT A IND,X A IND,Y	86 ii 96 dd B6 hh A6 ff 18 A6 ff	ll	2 2 3 2 3	2 3 4 4 5	— — — — Δ Δ 0 —
LDAB	(opr)	Load Accumulator B	M → B	B IMM B DIR B EXT B IND,X B IND,Y	C6 ii D6 dd F6 hh E6 ff 18 E6 ff	ll	2 2 3 2 3	2 3 4 4 5	— — — — Δ Δ 0 —

Source Forms		Operation	Boolean Expression	Addressing Mode for Operand	Machine Coding (Hexadecimal) Opcode / Operand(s)		Bytes	Cycles	Condition Codes S X H I N Z V C
LDD	(opr)	Load Double Accumulator D	M → A, M + 1 → B	IMM DIR EXT IND,X IND,Y	CC jj DC dd FC hh EC ff 18 EC ff	kk ll	3 2 3 2 3	3 4 5 5 6	— — — — Δ Δ 0 —
LDS	(opr)	Load Stack Pointer	M:M + 1 → SP	IMM DIR EXT IND,X IND,Y	8E jj 9E dd BE hh AE ff 18 AE ff	kk ll	3 2 3 2 3	3 4 5 5 6	— — — — Δ Δ 0 —
LDX	(opr)	Load Index Register X	M:M + 1 → IX	IMM DIR EXT IND,X IND,Y	CE jj DE dd FE hh EE ff CD EE ff	kk ll	3 2 3 2 3	3 4 5 5 6	— — — — Δ Δ 0 —
LDY	(opr)	Load Index Register Y	M:M + 1 → IY	IMM DIR EXT IND,X IND,Y	18 CE jj 18 DE dd 18 FE hh 1A EE ff 18 EE ff	kk ll	4 3 4 3 3	4 5 6 6 6	— — — — Δ Δ 0 —

Source Forms		Operation	Boolean Expression	Addressing Mode for Operand	Machine Coding (Hexadecimal) Opcode / Operand(s)		Bytes	Cycles	Condition Codes S X H I N Z V C
LSL	(opr)	Logical Shift Left	C ← [b7...b0] ← 0	EXT IND,X IND,Y A INH B INH	78 hh 68 ff 18 68 ff 48 58	ll	3 2 3 1 1	6 6 7 2 2	— — — — Δ Δ Δ Δ
LSLD		Logical Shift Left Double	C ← [b15...b0] ← 0	INH	05		1	3	— — — — Δ Δ Δ Δ
LSR	(opr)	Logical Shift Right	0 → [b7...b0] → C	EXT IND,X IND,Y A INH B INH	74 hh 64 ff 18 64 ff 44 54	ll	3 2 3 1 1	6 6 7 2 2	— — — — 0 Δ Δ Δ
LSRD		Logical Shift Right Double	0 → [b15...b0] → C	INH	04		1	3	— — — — 0 Δ Δ Δ
MUL		Multiply 8 by 8	A × B → D	INH	3D		1	10	— — — — — — — Δ
NEG	(opr)	2's Complement Memory Byte	0 − M → M	EXT IND,X IND,Y	70 hh 60 ff 18 60 ff	ll	3 2 3	6 6 7	— — — — Δ Δ Δ Δ
NEGA		2's Complement A	0 − A → A	A INH	40		1	2	— — — — Δ Δ Δ Δ
NEGB		2's Complement B	0 − B → B	B INH	50		1	2	— — — — Δ Δ Δ Δ
NOP		No Operation	No Operation	INH	01		1	2	— — — — — — — —

Source Forms	Operation	Boolean Expression	Addressing Mode for Operand	Machine Coding (Hexadecimal) Opcode Operand(s)	Bytes	Cycles	Condition Codes S X H I N Z V C
ORAA (opr)	OR Accumulator A (Inclusive)	A + M → A	A IMM A DIR A EXT A IND,X A IND,Y	8A ii 9A dd BA hh ll AA ff 18 AA ff	2 2 3 2 3	2 3 4 4 5	— — — — Δ Δ 0 —
ORAB (opr)	OR Accumulator B (Inclusive)	B + M → B	B IMM B DIR B EXT B IND,X B IND,Y	CA ii DA dd FA hh ll EA ff 18 EA ff	2 2 3 2 3	2 3 4 4 5	— — — — Δ Δ 0 —
PSHA	Push A onto Stack	A ↑ Stk, SP = SP − 1	A INH	36	1	3	— — — — — — — —
PSHB	Push B onto Stack	B ↑ Stk, SP = SP − 1	B INH	37	1	3	— — — — — — — —
PSHX	Push X onto Stack (Lo First)	IX ↑ Stk, SP = SP − 2	INH	3C	1	4	— — — — — — — —
PSHY	Push Y onto Stack (Lo First)	IY ↑ Stk, SP = SP − 2	INH	18 3C	2	5	— — — — — — — —
PULA	Pull A from Stack	SP = SP + 1, A ↓ Stk	A INH	32	1	4	— — — — — — — —
PULB	Pull B from Stack	SP = SP + 1, B ↓ Stk	B INH	33	1	4	— — — — — — — —
PULX	Pull X from Stack (Hi First)	SP = SP + 2, IX ↓ Stk	INH	38	1	5	— — — — — — — —

Source Forms	Operation	Boolean Expression	Addressing Mode for Operand	Machine Coding (Hexadecimal) Opcode Operand(s)	Bytes	Cycles	Condition Codes S X H I N Z V C
PULY	Pull Y from Stack (Hi First)	SP = SP + 2, IY ↓ Stk	INH	18 38	2	6	— — — — — — — —
ROL (opr)	Rotate Left	(diagram: C ← b7…b0 ← C)	EXT IND,X IND,Y A INH B INH	79 hh ll 69 ff 18 69 ff 49 59	3 2 3 1 1	6 6 7 2 2	— — — — Δ Δ Δ Δ
ROR (opr)	Rotate Right	(diagram: b7…b0 → C)	EXT IND,X IND,Y A INH B INH	76 hh ll 66 ff 18 66 ff 46 56	3 2 3 1 1	6 6 7 2 2	— — — — Δ Δ Δ Δ
RTI	Return from Interrupt	See Special Ops	INH	3B	1	12	Δ ↓ Δ Δ Δ Δ Δ Δ
RTS	Return from Subroutine	See Special Ops	INH	39	1	5	— — — — — — — —
SBA	Subtract B from A	A − B → A	INH	10	1	2	— — — — Δ Δ Δ Δ
SBCA (opr)	Subtract with Carry from A	A − M − C → A	A IMM A DIR A EXT A IND,X A IND,Y	82 ii 92 dd B2 hh ll A2 ff 18 A2 ff	2 2 3 2 3	2 3 4 4 5	— — — — Δ Δ Δ Δ
SBCB (opr)	Subtract with Carry from B	B − M − C → B	B IMM B DIR B EXT B IND,X B IND,Y	C2 ii D2 dd F2 hh ll E2 ff 18 E2 ff	2 2 3 2 3	2 3 4 4 5	— — — — Δ Δ Δ Δ

Source Forms	Operation	Boolean Expression	Addressing Mode for Operand	Machine Coding (Hexadecimal) Opcode Operand(s)	Bytes	Cycles	Condition Codes S X H I N Z V C
SEC	Set Carry	1 → C	INH	0D	1	2	— — — — — — — 1
SEI	Set Interrupt Mask	1 → I	INH	0F	1	2	— — — 1 — — — —
SEV	Set Overflow Flag	1 → V	INH	0B	1	2	— — — — — — 1 —
STAA (opr)	Store Accumulator A	A → M	A DIR A EXT A IND,X A IND,Y	97 dd B7 hh ll A7 ff 18 A7 ff	2 3 2 3	3 4 4 5	— — — — Δ Δ 0 —
STAB (opr)	Store Accumulator B	B → M	B DIR B EXT B IND,X B IND,Y	D7 dd F7 hh ll E7 ff 18 E7 ff	2 3 2 3	3 4 4 5	— — — — Δ Δ 0 —
STD (opr)	Store Accumulator D	A → M, B → M + 1	DIR EXT IND,X IND,Y	DD dd FD hh ll ED ff 18 ED ff	2 3 2 3	4 5 5 6	— — — — Δ Δ 0 —
STOP	Stop Internal Clocks		INH	CF	1	2	— — — — — — — —
STS (opr)	Store Stack Pointer	SP → M:M + 1	DIR EXT IND,X IND,Y	9F dd BF hh ll AF ff 18 AF ff	2 3 2 3	4 5 5 6	— — — — Δ Δ 0 —

Source Forms		Operation	Boolean Expression	Addressing Mode for Operand		Machine Coding (Hexadecimal)		Bytes	Cycles	Condition Codes						
						Opcode	Operand(s)			S	X	H	I	N	Z	V C
STX	(opr)	Store Index Register X	IX → M:M + 1		DIR	DF	dd	2	4	—	—	—	—	Δ	Δ	0
					EXT	FF	hh ll	3	5							
					IND,X	EF	ff	2	5							
					IND,Y	CD EF	ff	3	6							
STY	(opr)	Store Index Register Y	IY → M:M + 1		DIR	18 DF	dd	3	5	—	—	—	—	Δ	Δ	0
					EXT	18 FF	hh ll	4	6							
					IND,X	1A EF	ff	3	6							
					IND,Y	18 EF	ff	3	6							
SUBA	(opr)	Subtract Memory from A	A − M → A	A	IMM	80	ii	2	2	—	—	—	—	Δ	Δ	Δ Δ
				A	DIR	90	dd	2	3							
				A	EXT	B0	hh ll	3	4							
				A	IND,X	A0	ff	2	4							
				A	IND,Y	18 A0	ff	3	5							
SUBB	(opr)	Subtract Memory from B	B − M → B	B	IMM	C0	ii	2	2	—	—	—	—	Δ	Δ	Δ Δ
				B	DIR	D0	dd	2	3							
				B	EXT	F0	hh ll	3	4							
				B	IND,X	E0	ff	2	4							
				B	IND,Y	18 E0	ff	3	5							
SUBD	(opr)	Subtract Memory from D	D − M:M + 1 → D		IMM	83	jj kk	3	4	—	—	—	—	Δ	Δ	Δ Δ
					DIR	93	dd	2	5							
					EXT	B3	hh ll	3	6							
					IND,X	A3	ff	2	6							
					IND,Y	18 A3	ff	3	7							
SWI		Software Interrupt	See Special Ops		INH	3F		1	14	—	—	—	1	—	—	— —
TAB		Transfer A to B	A → B		INH	16		1	2	—	—	—	—	Δ	Δ	0 —
TAP		Transfer A to CCR	A → CCR		INH	06		1	2	Δ	↓	Δ	Δ	Δ	Δ	Δ Δ

Source Forms	Operation	Boolean Expression	Addressing Mode for Operand	Machine Coding (Hexadecimal)		Bytes	Cycles	Condition Codes						
				Opcode	Operand(s)			S	X	H	I	N	Z	V C
TBA	Transfer B to A	B → A	INH	17		1	2	—	—	—	—	Δ	Δ	0 —
TEST	TEST (Only in Test Modes)	Address Bus Counts	INH	00		1	*	—	—	—	—	—	—	— —
TPA	Transfer CC Register to A	CCR → A	INH	07		1	2	—	—	—	—	—	—	— —
TST (opr)	Test for Zero or Minus	M − 0	EXT	7D	hh ll	3	6	—	—	—	—	Δ	Δ	0 0
			IND,X	6D	ff	2	6							
			IND,Y	18 6D	ff	3	7							
TSTA		A − 0	A INH	4D		1	2	—	—	—	—	Δ	Δ	0 0
TSTB		B − 0	B INH	5D		1	2	—	—	—	—	Δ	Δ	0 0
TSX	Transfer Stack Pointer to X	SP + 1 → IX	INH	30		1	3	—	—	—	—	—	—	— —
TSY	Transfer Stack Pointer to Y	SP + 1 → IY	INH	18 30		2	4	—	—	—	—	—	—	— —
TXS	Transfer X to Stack Pointer	IX − 1 → SP	INH	35		1	3	—	—	—	—	—	—	— —
TYS	Transfer Y to Stack Pointer	IY − 1 → SP	INH	18 35		2	4	—	—	—	—	—	—	— —
WAI	Wait for Interrupt	Stack Regs and WAIT	INH	3E		1	**	—	—	—	—	—	—	— —
XGDX	Exchange D with X	IX → D, D → IX	INH	8F		1	3	—	—	—	—	—	—	— —
XGDY	Exchange D with Y	IY → D, D → IY	INH	18 8F		2	4	—	—	—	—	—	—	— —

NOTES:

Cycle:
- * = Infinity or until reset occurs
- ** = 12 cycles are used beginning with the opcode fetch. A wait state is entered which remains in effect for an integer number of MCU E-clock cycles (n) until an interrupt is recognized. Finally, two additional cycles are used to fetch the appropriate interrupt vector (total = 14 + n).

Operands:
- dd = 8-bit direct address $0000–$00FF. (High byte assumed to be $00.)
- ff = 8-bit positive offset $00 (0) to $FF (255) added to index.
- hh = High order byte of 16-bit extended address.
- ii = One byte of immediate data.
- jj = High order byte of 16-bit immediate data.
- kk = Low order byte of 16-bit immediate data.
- ll = Low order byte of 16-bit extended address.
- mm = 8-bit mask (set bits to be affected).
- rr = Signed relative offset $80 (−128) to $7F (+ 127). Offset relative to the address following the machine code offset byte.

Condition Codes:
- — = Bit not changed
- 0 = Always cleared (logic 0).
- 1 = Always set (logic 1).
- Δ = Bit cleared or set depending on operation.
- ↓ = Bit may be cleared, cannot become set.

APPENDIX F:
Parts/Equipment Listing

Equipment

    Motorola M68HC11EVBU or M68HC11EVB
    Terminal with keyboard, or personal computer running KERMIT software
    Power supply, +5VDC, 500mA
    Power supply, +12VDC and −12VDC, 500mA (EVB only)
    Power supply, variable, 0–5VDC, 500mA
    Voltmeter, 0–10VDC
    Milliammeter, 0–200mA
    Oscilloscope, dual trace, with probes
    Flashlight
    Heat gun (for heat shrink or hair drying)

Resistors/Potentiometers

1	160Ω, 1/2W
2	270Ω, 1/4W
3	1KΩ
3	1.8KΩ
3	2.2KΩ
3	3.3KΩ
3	4.7KΩ
4	5.1KΩ
3	10KΩ
3	15KΩ
3	18KΩ
3	22KΩ
3	33KΩ
3	50KΩ
1	100KΩ
1	20KΩ potentiometer
1	1KΩ trimpot
1	20KΩ trimpot
4	1MΩ trimpot
1	Photoresistor, cadmium sulfide, 1MΩ dark
1	Thermistor, 50KΩ @ 25°C

Capacitors

- 1   .1µF
- 1   1µF (EVB only)
- 1   4700pF, 6.3V

Transistors/Diodes

- 1   NPN, 2N3053 (ECG128), with "top hat" heat sink
- 2   NPN, 2N3904 (ECG123AP)
- 1   Phototransistor, TIL81 (ECG3032)
- 3   Diode, 1N4001 (ECG116)

LEDs

- 1   TIL34 (ECG3028), infrared
- 1   DL-1414, alphanumeric display

Integrated Circuits

- 1   74LS10 or 74HC10, triple, 3-input NAND
- 1   74LS20 or 74HC20, dual, 4-input NAND
- 1   74LS74A or 74HC74A, dual, D flip-flop
- 1   74LS91N, 8-bit, serial in-serial out, shift register
- 2   74LS241N, octal, 3-state driver
- 1   74LS373 or 74HC373, octal, 3-state D latch
- 1   4093B, Schmitt-trigger, quad, 2-input NAND
- 1   MCM6264CP35, HCMOS, 8K × 8 SRAM

Miscellaneous Devices

- 4   Push-button switch, push-to-close
- 1   DIP switch, 4 switches
- 1   Motor, DC, Radio Shack 273-223
- 1   Transformer, filament, 6.3V @ 0.3a
- 1   Fuse holder, in-line
- 1   Fuse, 0.6a
- 1   Terminal strip (see Figure 9–17)
- 1   Spray can, circuit chiller ("Freeeze-It Antistat" or equivalent)

Parts and Equipment Sources

All Electronics Corp., P.O. Box 567, Van Nuys, CA 91408, 1-800-826-5432

Circuit Specialists Inc., P.O. Box 3047, Scottsdale, AZ 85271-3047, 1-800-528-1417

Digi-Key Corporation, 701 Brooks Ave. South, P.O. Box 677, Thief River Falls, MN 56701-0677, 1-800-344-4539

Herbach and Rademan Company, 18 Canal St., P.O. Box 122, Bristol, PA 19007-0122, 1-800-848-8001 (orders only), 215-788-5583

Jameco, 1355 Shoreway Road, Belmont, CA, 94002-4100, 1-800-831-4242

Newark Electronics; branches in major cities; see local phone directories, 312-784-5100 (administrative offices)

INDEX

1N4001 diode, 391, 600
2N3053 transistor, 391, 600
2N3904 transistor, 391, 600
4093B quad 2-input, Schmitt-trigger NAND, 430, 432, 600
74HC20/74LS20 dual four-input NAND, 181–182, 600
74HC241/74LS241 octal 3-state driver, 202–203, 206, 286, 290–291, 296, 600
74HC373/74LS373 octal latch, 558–559, 562–566, 571, 600
74LS10 triple three-input NAND, 562–565, 600
74LS74A dual D flip-flop, 281, 283–284, 600
74LS91N 8 bit serial in-serial out shift register, 281–286, 295, 600

ABA instruction, 48, 76, 593
ACCA, 4–10, 73
ACCB, 4–10, 73
ADCA instruction, 48–52, 76, 593
ADCB instruction, 48–49, 76, 593
ADCTL register, 478–480, 485, 487, 489, 540
  CCF bit, 478, 480, 540
  CD, CC, CB, CA bits, 479, 485, 540
  MULT bit, 479, 485, 540
  SCAN bit, 479–480, 540
ADDA instruction, 21, 31, 48–50, 76, 593

ADDB instruction, 48, 76, 593
ADDD instruction, 48, 76, 593
Address bus, 4, 9, 73, 551–559, 562–563, 565–566, 572, 578–579
Address latch, 4–5, 9–10, 73
Address strobe (AS), 9–10, 552, 555–558, 562–563, 566, 572–573, 576, 578
Addressing modes, 29–32, 74–75
  Direct addressing, 5, 21, 30–31, 74
  Extended addressing, 5, 30, 74
  Immediate addressing, 21, 30, 74
  Indexed addressing, 5, 22, 31
  Inherent addressing, 22, 29–30, 74
  Relative addressing, 5, 31–32, 75
ADR registers, 478–479, 481–483, 485–486, 540
ALU, 4–10, 73–74
Ammeter, 108, 120–122, 599
ANDA instruction, 42–48, 75, 593
ANDB instruction, 42–44, 75, 593
AS11 assembler, 581–592
  Analytical features, 581–583, 586–587
  Assembler directives, 584–587
  BSZ/ZMB directives, 586
  Checksums, 582, 589–590
  Comments field, 583–584
  Cross-reference table, 583, 587
  Cycle counts, 582–583, 586–587
  DOS commands, 582, 590–591
  EQU directive, 585

  Error messages, 582
  FCB directive, 585
  FCC directive, 586
  FDB directive, 585
  FILL directive, 586
  Label field, 583
  List file example, 588–589
  List files, 581–582, 586–589
  NAM directive, 584
  Numbers and mathematical operators, 584
  Operand field, 583
  OPT directive, 586–587
  ORG directive, 584–585
  PAGE directive, 587
  RMB directive, 585–586
  Source code file, 582, 587–588, 591–592
  S-records, 582, 588–590
  Symbol table, 583, 586–587
  Operation field, 583
ASCII codes, 94, 155–156, 158–159, 184, 223
ASL instruction, 41–42, 75, 594
ASLA instruction, 41–42, 44–48, 75, 594
ASLB instruction, 7, 41–42, 75, 594
ASLD instruction, 41–42, 75, 594
ASR instruction, 42, 75, 594
ASRA instruction, 42, 75, 594
ASRB instruction, 42, 75, 594
Asynchronous communications, 175, 213, 221, 225, 254, 293

A-to-D converter (ADC), 2–3, 73, 90, 129, 473–494, 496–497, 539–540, 579–580
  AN pins, 474–476, 478–479, 481, 485–486, 488, 539–540, 579
  Charge pump, 477–478, 539
  Charge redistribution, 477–478, 539
  Conversion phase, 477–478
  Conversion timing, 480, 540
  Hold phase, 477
  Low-pass filter, 475, 480–481, 486, 488, 513
  Operating modes, 474–475, 539
  RC oscillator, 478, 539
  Sample phase, 477
  Step size, 474, 482–484, 486, 539
  Successive approximation register (SAR), 477

BAUD register, 221, 227–229, 236, 255
  SCP bits, 227–228
  SCR bits, 227–228
BCC instruction, 62, 76, 594
BCD numbers, 9, 53–54, 74
BCLR instruction, 40, 44–48, 75, 229–230, 236–238, 303–305, 308, 338, 349, 594
BCS instruction, 62, 76, 594
BEQ instruction, 62, 76, 594
BGE instruction, 62, 65–73, 76, 594
BGT instruction, 62, 76, 594
BHI instruction, 62, 65–73, 76, 594
BHS instruction, 62, 76, 594
Binary numbers, 6
BITA instruction, 59–60, 76, 242, 594
BITB instruction, 59–60, 76, 594
BLE instruction, 63, 65–73, 76, 594
BLO instruction, 62, 76, 594
BLS instruction, 62, 65–73, 76, 594
BLT instruction, 62, 65–73, 76, 594
BMI instruction, 62, 76, 594
BNE instruction, 31–32, 62, 76, 594
Boolean analysis, 473–474
Boolean expressions, 11
BPL instruction, 62, 76, 229, 594
BRA instruction, 32, 61, 76, 594
BRCLR instruction, 63, 76, 388, 594
BRSET instruction, 63, 76, 594
BSET instruction, 40, 75, 229–230, 236–238, 304, 594
BSR instruction, 35, 61, 63–64, 76, 594
BUFFALO ROM, 2–3, 6, 74, 101, 571, 575, 583, 590–591
BUFFALO ROM system commands, 11–29, 32–33, 74

Assembler/disassembler (ASM), 14–16, 25–26, 32, 74, 581, 587
Block fill (BF), 14, 25, 74, 564
Breakpoint set (BR), 17–18, 27–29, 74
Bulk erase (BULKALL), 88, 95–96
Go (G), 14, 23, 27, 29, 74
Load (LOAD), 582, 590
Memory display (MD), 11–12, 23, 74, 534, 564
Memory modify (MM), 12–13, 23, 74, 351, 382, 430, 534
Move (MOVE), 14, 24–25, 74, 372–373, 394, 534, 564
Proceed/continue (P), 18, 28, 74
Register modify (RM), 16–17, 23–24, 74
Stop at address (STOPAT), 11
Trace (T), 17, 24, 74, 576, 578
Verify (VERF), 11
BUFFALO ROM system prompt, 18–19
BUFFALO ROM utility subroutines
  .INCHAR, 363, 389–390, 418–419
  .OUT1BS, 65, 67, 69, 72, 320–321, 371–372
  .OUT2BS, 115–116, 371–372, 481–482, 484–485
  .OUTCRL, 65, 67, 69, 72, 481–482, 484–485
  .OUTRHL, 94, 113–114, 210–211, 332
  .OUTSTR, 115–116, 203, 287, 289, 332–333, 418–419
  .OUTSTR0, 320–321, 363, 371–372, 388–389
BVC instruction, 62, 594
BVS instruction, 62, 594
Byte, definition, 1

CFORC register, 408–409, 464
  FOCx bits, 408–409, 464
CLC instruction, 59, 594
CLI instruction, 59, 100, 594
CLR instruction, 39, 75, 594
CLRA instruction, 39, 75, 594
CLRB instruction, 39, 75, 595
CLV instruction, 59, 595
CMPA instruction, 60–61, 76, 595
CMPB instruction, 60–61, 76, 595
COM instruction, 42–48, 75, 595
COMA instruction, 42–43, 75, 595
COMB instruction, 42–43, 75, 595
Condition code register (CCR), 4–6, 10–11, 61–63, 73, 76, 99, 102, 104

C flag, 4–9, 11, 48–49, 59–62, 73–76, 305–306
H flag, 4, 9–10, 73–74
I mask, 4, 6, 9–10, 59, 76, 100–106, 108, 129–130, 179, 184, 199, 268, 308, 340, 353, 416, 426–428, 456–458
N flag, 4–5, 8–9, 11, 59–63, 73
S bit, 4, 6, 9, 59, 76, 104, 106
V flag, 4–5, 7–9, 11, 59–63, 73–74, 76
X mask, 4, 6, 9–10, 59, 76, 99, 101, 104–106, 110, 130–131
Z flag, 4–5, 8–9, 11, 59–63, 73–74
Conditional branches, 31–32, 61–63, 65–73, 75–76
CONFIG register, 88, 90, 95–96
  NOCOP bit, 88, 90, 104, 129
Control bus, 4, 10, 551–553, 557, 565
Control, closed-loop, 441–442, 466
Control, open-loop, 441, 466
Control register block, 87, 90, 129, 142
COPRST register, 88–89, 129
CPD instruction, 60–61, 76, 595
CPU, 3–6, 9–10, 73
  Address handling, 4–5, 73
  Context, 35, 85, 97, 101–102, 129–130, 271–272
  Data handling, 4–5, 73–74
  Instruction handling, 4, 9–10, 73–74
CPX instruction, 30, 60–61, 76, 595
CPY instruction, 60–61, 76, 595

DAA instruction, 9, 53–54, 74, 76, 595
Data bus, 4–5, 9, 10, 73–74, 551–553, 555–558, 563, 565–566, 578
DDRC register, 142, 147–149, 158, 171, 176–178, 183, 185, 191–192, 196–200, 202
DDRD register, 235–236, 264, 274–275, 278–280, 282, 287, 289, 295
DEC instruction, 39, 75, 595
DECA instruction, 39, 75, 595
DECB instruction, 39, 75, 595
Decoder/controller/sequencer, 4, 9–10, 74
Decoding logic, 562–565, 567
DES instruction, 39, 75, 595
DEX instruction, 39, 75, 595
DEY instruction, 39, 75, 595
DIP switch, 168–169, 171, 600
DL-1414 alphanumerical display, 155–161, 163–171, 181–184, 208–209, 211, 600
  Addressing, 155

602

Pins, 156
Timing, 156
D register, 4–6, 8–10, 73
Duty cycle, 343, 351, 365, 386–388, 390–391, 393, 398, 400, 407, 409, 411, 430, 458–464

E clock, 86, 88, 96, 227–228, 265, 268, 270–272, 294–295, 302, 310, 338, 345, 381, 409, 417, 464, 554, 556–557, 562–563, 566, 571, 573, 576, 578
EORA instruction, 42–44, 75, 595
EORB instruction, 42–44, 75, 595
Error, linear, 438, 451–452, 463
Error, offset, 438, 452, 463
EVB, 2, 90, 93, 571–573
   P1, 573, 577–580
EVBU, 2–3, 86–87, 93, 575–576
   J10–J13, 576
   J14, 576
   J3, 554
   J4, 554, 564, 566
   J7, 576
   P4, 576, 577–580
Expanded multiplexed mode timing, 557–558
EXTAL pin, 86, 578

FDIV instruction, 52–53, 76, 595
Free-running counter, 4, 90, 129, 301–311, 313, 338, 344–347, 397, 399–400, 408–409, 417, 579
   Extension bytes, 302, 304–307, 338
   Resolution and range, 303, 338
   Specific time-out, 306–311, 339
Free-running counts
   Per millisecond, 411
   Per minute, 319, 322
   Per second, 321–322
Frequency measurement, 343, 360–379, 398
Full-duplex mode, 221–222, 254, 293
Full-handshake I/O, 175, 191–214, 213
   Interlocked, 192–194, 196, 201–202, 213
   Pulsed, 192, 194–196, 201–202, 213
   Three-state variation, 192, 196–198, 208–213
Fuzzy Inference, 2, 473–474, 494–542, 551
   Neural networks, 494
   COG, 496, 498, 509–510, 531–532
   Crisp inputs, 495–499, 502–506, 517–520

Crisp outputs, 496–498, 519, 541
De-fuzzification, 495–498, 509–512, 529–532, 541–542
De-fuzzified output (DFUZOUT), 509–512, 531–533, 542
Fuzzification, 495–506, 512–525, 540–541
Fuzzy inputs, 496–500, 502–509, 522–523, 540–541
Fuzzy outputs, 496–498, 506–511, 525, 528–531, 540–542
Grade of membership, 495–500, 502–509, 519–529, 540–541
IC chips, 494
INMFBLK, 502
Input membership functions, 498–504, 519, 521–524, 534–538, 541
MIN/MAX, 509, 525, 528, 541–542
Output membership functions, 509–511, 530–531, 542
RULBLK, 507, 525–526, 538–539
Rule evaluation, 495–498, 506–509, 525–529, 540–541
Singletons, 496–498, 509–511, 530–531, 541–542
vs. conventional algorithms, 494–495, 499, 540
Fuzzy motor controller, 512–539, 541–542, 552, 564–565

Global interrupt masks, 99
Greater numbers, 62–63
Ground pin, 580

Half-duplex mode, 221–222, 254
Handshakes, 175, 213
Hardware modes, 552–554
   Expanded multiplexed, 4, 551–567, 571
   Single chip, 551–556, 571, 575
   Special bootstrap, 552–553
   Special test, 552–553
Hexadecimal numbers, 6
Higher numbers, 7, 62
HPRIO register, 102–103, 129–130, 335, 416, 464, 552–555, 558, 565–566
   IRV bit, 553, 557–558, 566
   MDA bit, 553–555, 566
   PSEL3–PSEL0 bits, 102–103, 130, 335, 416, 464
   RBOOT bit, 553, 566
   SMOD bit, 553, 566

IDIV instruction, 52–53, 76, 595
Illegal opcode trap interrupts, 85, 90, 98–99, 112–116, 124, 129
INC instruction, 40, 75, 595
INCA instruction 40, 75, 595
INCB instruction, 40, 75, 595
Index registers, 4–5, 9, 10, 73, 75
INIT register, 89–90, 129
   RAM3–RAM0 bits, 89–90
   REG3–REG0 bits, 89, 142
Input capture interrupts, 98, 315, 319, 323, 327, 339, 347–348
   General purposes, 331–338, 340
Input captures, 4, 301, 312–331, 339, 360–379, 579
   Edge specifications, 315, 325, 340
   IC pin noise margin, 313
   IC pins, 312–313, 331–332, 335, 339, 344–348, 579
INS instruction, 11, 21, 30, 40, 75, 596
Instruction set, 593–598
Interrupt control, 176–179, 184–186, 231–233, 268, 355–356
Interrupt priorities, 102–103, 129–130, 335
Interrupt service routines, 85, 97, 100–101, 129
Interrupts, introduction, 85–86, 97, 103
INX instruction, 30, 40, 75, 596
INY instruction, 40, 75, 596
I/O, 1–3
IRQ interrupts, 85, 98–100, 105–109, 120–122, 129–130, 177–179, 184–185, 198–200, 205, 207, 281–186, 288–292, 295–296, 395–397
$\overline{\text{IRQ}}$ pin, 98–100, 105–106, 120, 129–130, 179, 200, 202–203, 394, 578

JMP instruction, 63, 76, 91, 101, 596
JSR instruction, 35, 63–73, 76, 596

KERMIT program, 581–583, 590–591, 599

LDAA instruction, 9, 33, 75, 596
LDAB instruction, 33, 75, 596
LDD instruction, 33, 75, 596
LDS instruction, 33, 75, 596
LDX instruction, 21, 30, 33, 75, 596
LDY instruction, 11, 33, 75, 596
Lesser numbers, 62–63
Local interrupt masks, 99
Lower numbers, 7, 62

603

LSL instruction, 41–42, 75, 596
LSLA instruction, 41–42, 75, 596
LSLB instruction, 41–42, 75, 596
LSLD instruction, 41–42, 75, 596
LSR instruction, 42, 75, 596
LSRA instruction, 42, 44–48, 75, 596
LSRB instruction, 42, 75, 596
LSRD instruction, 42, 75, 596

M6801 microprocessor, 2, 4
M68B50P ACIA, 246, 571–572
M68HC11
   Instructions, 32–73, 593–598
   Introduction, 2
M68HC11A1, 2, 571–573
M68HC11E9, 2, 73, 575–576
Machine code vs. BUFFALO assembly code, 20–23
Machine cycles, 10
MC145407 driver/receiver, 245, 575
MC1488 RS-232 driver, 572–573
MC1489 RS-232 receiver, 572–573
MC34064P undervoltage sensor, 87, 93, 575–576
MC68HC24 PRU, 552, 571–573, 578
MC68HC68T1 real-time clock, 576
MC74HC138 one-of-eight decoder, 572
MC74HC14 hex Schmitt-trigger inverter, 573
MC74HC4040 counter/frequency divider, 572–573
MC74HC4066 digital switch, 572
MC74HC74 dual D flip-flop, 572
MCM6264C SRAM, 551–552, 559–566, 571–572, 600
   I/O control, 561–562, 566
   Memory matrix, 559–561, 566
Memory, 1
   DRAM, 1
   EEPROM, 1–3, 73, 88, 106, 130, 361, 373, 394, 489, 497–498, 500, 502, 507, 534, 551, 564, 571, 575
   EPROM, 1–2, 571–572
   RAM, 1–3, 73, 90, 91, 129, 551, 554, 571–572, 575
   Read cycle, 563, 567
   ROM, 1–3, 73, 90–91, 100–101, 551, 553, 571, 575
   SRAM, 1, 551, 559–565, 571
   Write cycle, 562–563, 567
Microcomputer elements, 1–2
Microcontroller, definition, 1
Microprocessor, 1–2

MODA pin, 552–554, 565–566, 578
MODB pin, 552–554, 565–566, 578
Motor speed control, 379, 386–397, 400, 407, 409–416, 423, 429–441
   Circuitry, 391–394, 412–413, 429, 432–433, 448, 512–513, 532–533, 542, 563–564
Motor speed measurement, 407, 416, 423, 429–441, 466
Motor speed regulation, 441–453, 466
MUL instruction, 50–52, 76, 391, 596
Multiplexing, 141, 555–558, 565–566

NEG instruction, 42–43, 75, 596
NEGA instruction, 42–43, 75, 596
NEGB instruction, 42–43, 75, 596
Nested interrupts, 101
Nested subroutines, 64, 67, 76
NMI interrupts, 110, 131
NOP instruction, 104, 114, 130–131, 596
NRZ, 3, 222–225, 254
Numerical prefixes, 6, 584

OC1D register, 379–380, 382–384, 386–387, 399–400, 408
   OC1Dx bits, 380, 382, 399
OC1M register, 379–380, 382–383, 386–387, 399–400, 408
   OC1Mx bits, 380–382, 399
One's complement, 42–43, 75
Opcode fetch, 10
Opcodes, 1, 9–10, 112–113, 593–598
Optical sensor, 429–430, 432–433, 448, 465
OPTION register, 87–89, 93, 95–96, 105, 130, 477, 539–540
   ADPU bit, 477–478, 539
   CME bit, 87
   CR1/CR0 bits, 87–89, 93, 95–96, 129
   CSEL bit, 477–478, 539–540
   DLY bit, 87, 105
   IRQE bit, 87, 107, 130
ORAA instruction, 42–48, 75, 597
ORAB instruction, 42–44, 75, 597
Oscilloscope, 91, 93, 353–354, 357–358, 372, 375–376, 384–385, 391, 412, 420, 433, 448, 461–462, 481, 533, 599
Output compare interrupts, 98, 315, 344, 347–353, 355–362, 371–372, 382, 387–389, 398–400, 408–409

Output compares, 4, 301, 343–400, 408–409
   Forced, 345, 382, 399, 407–416, 464
   Joint OC1/OCx control, 386–397, 400
   OC1, 344–345, 379–397, 399–400, 408–411, 425, 579
   OC2–OC5, 343–351, 379, 397, 408–411
   OC pins, 344–349, 379–387, 397–400, 407–408, 579
Overshoot and oscillation, 449–451

PACNT register, 423–428, 430–432, 454–455
   Extension bytes, 423, 454–458, 465–466
PACTL register, 315–317, 339, 346–350, 352–353, 380–383, 397–399, 417–418, 423, 425–428, 431, 445–446, 454, 456–458, 465–466
   DDRA3 bit, 314, 339, 347–348, 352–353, 382, 397–398, 425
   DDRA7 bit, 381–382, 399, 425, 426, 454, 465–466
   I4/O5 bit, 314–315, 339, 347–349, 352–353, 382, 397–398, 425
   PAEN bit, 424–425, 446, 454, 466
   PAMOD bit, 424–425, 446, 455
   PEDGE bit, 424–426, 446, 454, 466
   RTR bits, 417–418, 425, 430, 446, 465
Parallel I/O introduction, 141, 170
Parallel vs. serial I/O, 141, 170
Parity, 222–226, 229
Parts/equipment sources, 600
Parts listing, 599–600
Phase 2 clock (PH2), 177, 192–200, 213, 557, 573, 576, 578
Photoresistor, 473, 480, 488–490, 513
Pierce oscillator, 86, 105, 573, 576, 578
PIOC register, 90, 176–181, 183–185, 191–202, 208, 213
   CWOM bit, 201
   EGA bit, 180, 197, 200
   HNDS bit, 180, 197, 199–200
   INVB bit, 180, 200
   OIN bit, 200
   PLS bit, 200
   STAF bit, 176–180, 183–185, 191–200, 203, 208, 213
   STAI bit, 178, 180, 199–200
Polling, 103

PORTA, 3, 142, 344–346, 380–382, 397, 399, 408, 576, 579
PORTB, 3–4, 142–143, 145–146, 156–161, 163–171, 176–179, 182–185, 191–192, 213, 551, 554–559, 562–563, 565–566, 571–573, 579
   Pin logic, 145–146
   Reads, 142–143, 146–147
   Resets, 145–146
   Timing, 145–146
   Writes, 142–143, 146
PORTC, 3–4, 142–145, 147–149, 156–161, 163–171, 176–177, 181–186, 191–198, 200–201, 213, 551, 555–559, 561–563, 565–566, 571–573, 578
   Pin logic, 147–149
   Reads, 142–145, 147–148
   Timing, 147–149
   Writes, 142, 145, 147–148
PORTCL, 149, 176–180, 182–185, 191–203, 213
PORTD, 3, 142, 222, 235–236, 265, 268, 272, 274, 279–280, 294–295, 572, 578–579
   RxD pin, 222, 226–227, 230, 232, 254–255, 578
   TxD pin, 222, 226, 231, 233–236, 254–255, 579
PORTE, 3, 476, 539, 579
Power consumption, 104, 106, 109, 121–123, 130
Prebytes, 9, 112–113
PROCOMM program, 581
Program counter, 4–5, 10, 32, 35, 73, 75–76, 90–91, 100–101
Programmer's model, 10–11
Programs/routines
   0–1000 Decimal Count, 160–168, 171
   1 KHz Motor Speed Control Square Wave (Variable Duty Cycle), 429–430, 436, 438–439, 465
   Check $\overline{SS}$ Pin Status, 273
   COG, 511–512, 531–532
   CONV&HUNDSEC, 325–326, 330–331
   COP Test, 94–97
   CRISPOUT, 498, 512, 514, 531–533, 542
   DC Motor Speed Control, 386–396, 400
   DC Motor Speed Control and Regulation, 442–453, 466

DC Motor Speed Control and Tachometer, 429–441, 465
DEMO, 19–20
DEMO2, 35–39
DEMO3, 44–48
DEMO4, 49–50, 54–57
DEMO5, 50–51, 55, 57
DEMO6, 51–52, 55–58
DEMO7, 52–53, 56, 58
DEMO8, 64–73, 591–592
DFUZ, 510–512, 518, 529–531, 542, 565
DL-1414 Display, 156
Extend Free-Running Counter Bit Capacity and Range, 304–306, 338
Fifteen-Second Time-Out and Display, 308–311
FINDSEG, 503–504, 512, 519–522, 525, 542, 565
Four-Channel A/D, Continuous Conversions, 486–488, 492–493
Four-Channel A/D, Continuous Conversions, Temperature and Light Sensors, 488–489, 493–494
Four-Channel A/D, Four Conversions and Stop, 485, 492
Free-Running Counter Specific Time-Out, 306–311, 339
Full-Handshake I/O Laboratory, 203–208
Full-Handshake/Three-State Variation, 209–213
FUZZIFY, 512, 518–521, 524–525, 542, 565
Fuzzy Inference Subroutine (FUZZY), 497, 512, 514, 518–531, 533, 542, 565
Fuzzy Motor Control, A/D, 512, 514, 517, 532, 542
Fuzzy Motor Control, INIT, 512, 514–516, 532, 542
Fuzzy Motor Control, MAIN, 512–514, 532, 542, 565
Fuzzy Motor Control, OC1/OC5 Interrupt Services, 512, 516–517, 533, 542, 565
Generate Four Simultaneous Square Waves with OC2–OC5, 357–360, 398
ILLOP Demonstration, 113–116, 124, 127, 131
Integrate Input Captures with Output Compares, 360–379, 398
IRQ Demonstration, 107–109, 120–122, 125–126, 130–131

Keypad Read and Display, Interrupt Control, 184–188, 213–214
Keypad Read and Display, Status Polling, 182–184, 186–187, 197–198, 213
Measure Percent Duty Cycle, 458–464, 466
Modified DEMO, 26–27, 32, 587–590
Modulus 32 Counter, 382–386, 399
MSGRX, 241–244
MSGTX, 238–240
Parity Generator, 224
Periodic Time Reference Pulses, 418–422, 465
PORTC Parallel I/O, 144–145
Read PORTB, 142–143
Read PORTC, 143–144
Read PORTC/Write PORTB, 168–170, 171
RULEVAL (Min/Max), 508–509, 512, 518, 525–529, 542, 565
SCI Initialization, 236
SCI "Please Input Data"–"Memory Full," 249–254
SCI "Press 3 Keys," 246–248
SCI Receive, 244–245
SCI Transmit, 240–241
SEG1, 504–505, 512, 519–523, 542
SEG2, 504–506, 512, 519–525, 542
Send an Idle Character Between Messages, 234–235
Single-Channel A/D, Continuous Conversions, 483–485, 491–492
Single-Channel A/D, Four Conversions and Stop, 481–483, 491
SPI Master and Slave Shift Register, 282–286
SPI Master and SPI Slave, 287–293
SPI Multiple-Master System, 275–279
SWI Demonstration, 116–120, 125, 128, 131–132
Tachometer, 430–432, 436, 439–440, 465–466
"TAXI," 158–160, 171
Three-Speed Motor Control, 413–416
Two-Speed Motor Control, 409–415, 464
Use Input Capture Pins for General-Purpose Interrupts, 331–338, 340
Use OC5 to Generate a Square Wave, 351–357, 398

Use TOFs to Generate a Delay, 333, 337
Using Input Captures to Measure Time Between Events, 318–331, 340
Write PORTB, 143
XIRQ Demonstration, 110–112, 122–123, 126–127, 131
Pseudovectors, 91, 100–101, 129
PSHA instruction, 34–39, 75, 597
PSHB instruction, 34–35, 75, 597
PSHX instruction, 34–35, 75, 597
PSHY instruction, 34–35, 75, 597
PULA instruction, 34–35, 75, 597
PULB instruction, 34–39, 75, 597
Pull-up circuit, 107
Pulse accumulator, 4, 73, 407, 422–466
  Event counting mode, 407, 423–453, 465
  Gated time accumulation mode, 407, 423, 425, 453–466
  Input edge interrupts, 98, 423–424, 426, 455, 458, 460–461, 464–466
  Overflow interrupts, 98, 423–424, 426, 454–455, 465–466
  PAI pin, 4, 423–426, 453–455, 465, 579
PULX instruction, 34–35, 597
PULY instruction, 34–35, 597

Read/modify/write bits, 303, 308, 315, 338–339, 349, 382
Real time (RTI) interrupts, 98–99, 129, 301, 407, 416–422, 429–444, 446–449, 453, 464–465
Receive data register buffer (RDR), 222, 227, 229, 232, 255
Relative address, 32, 75
$\overline{\text{RESET}}$ pin, 86–91, 93, 104–106, 128–130, 554, 566, 573, 575, 578
Resets
  Clock monitor, 86–87, 90–91, 128–129
  Computer operating properly (COP), 85–86, 88–96, 104, 128–129
  External circuit, 86–87, 90–91, 93, 101–102, 104–105, 110, 128–130, 572
  Introduction, 85–86
  Power-on (POR), 85–87, 89–93, 101–102, 104–105, 110, 128–130
Reset service routines, 85–87, 89–91, 129, 553

Reset switch, 92, 95
Response gap, 449–450
ROL instruction, 41, 49–50, 75, 597
ROLA instruction, 41, 75, 597
ROLB instruction, 41, 75, 597
ROR instruction, 41, 75, 597
RORA instruction, 41, 75, 597
RORB instruction, 41, 75, 597
RTI instruction, 35, 76, 101–102, 108, 110, 129, 179, 597
RTS instruction, 35, 64–73, 76, 597
R/$\overline{\text{W}}$ line, 9, 10, 552, 556, 561–563, 566, 573, 578

SBA instruction, 50, 76, 597
SBCA instruction, 50, 76, 597
SBCB instruction, 50–51, 76, 597
SCCR1 register, 221, 226, 229–231, 236, 255
  M bit, 229–230, 232, 234, 255
  R8 bit, 229–232, 255
  T8 bit, 226, 229–231, 234–235, 255
  WAKE bit, 230–231, 255
SCCR2 register, 221, 230, 233–236, 255
  ILIE bit, 233, 256
  RE bit, 233–235, 256
  RIE bit, 233, 256
  RWU bit, 235, 256
  SBK bit, 235, 256
  TCIE bit, 233, 256
  TE bit, 233–235, 256
  TIE bit, 233, 256
SCDR register, 226, 229, 231–233, 255
SCI, 3, 104, 221–262, 293, 553, 571–572, 575, 578–579
  Address mark wake up, 230–231, 241–242
  Frames, 223, 225–226, 229, 232, 234, 254–255
  Framing bits, 141, 221–223, 225–227, 229, 254–255, 263
  Idle line wake up, 230
  Initialization, 236–238
  Serial communications errors, 227, 231–233, 255
  Serial system interrupts, 97, 222, 233, 256
  Start bit qualification, 232
  Transmit data register buffer (TDR), 222, 226–227, 229, 231, 255
  Voltage levels, 223
  Wake-up, 222, 229–231, 235, 254
SCSR register, 221, 231–233, 255
  FE bit, 231, 233, 255

IDLE bit, 231–233, 255–256
NF bit, 231, 233, 255
OR bit, 231–232, 255
RDRF bit, 231–233, 255–256
TC bit, 231–233, 255–256
TDRE bit, 231, 233, 255–256
SEC instruction, 59, 597
SEI instruction, 59, 100, 597
Sensors, temperature and light, 488–490, 493–494, 512–513, 532–533, 542, 563
SEV instruction, 59, 597
Sign bit, 8
Simple-strobed I/O, 175–187, 213
Software (SWI) interrupts, 17, 85, 98–99, 116–120, 125, 129
SPCR register, 264–280, 282, 284–285, 287, 289, 294
  CPHA bit, 266–268, 270–273, 284–286, 290–292, 294–296
  CPOL bit, 266–268, 270–273, 284–285, 291–292, 294
  DWOM bit, 268, 279, 294–295
  MSTR bit, 268, 270–274, 278–279, 294–295
  SPE bit, 268, 273, 294–295
  SPIE bit, 268, 270–271, 274, 294–295
  SPR bits, 265–267, 294
SPDR register, 264–267, 272–273, 275–278, 280–283, 286, 288–290, 293–295
  Read data buffer, 264–265, 293
  Shift register, 264–267, 272, 293
SPI, 3, 104, 263–299, 576, 579
  Baud rates, 263
  Clock generator, 265
  Data exchange beginning and end, 263–268, 270–271, 293–295
  Double-buffering, 265
  Master and slaves, 263–268, 270–283, 286–296
  MISO, 263–268, 272–286, 288–290, 293, 295–296, 579
  MOSI, 263–268, 272–275, 277–284, 286, 288–290, 293, 295–296, 579
  Multiple-master system, 274–279
  SCK, 263–268, 270–275, 277–280, 282–284, 286, 290, 293–296, 579
  Serial transfer complete interrupts, 98, 268, 274, 276–279, 295
  Single buffering, 265, 272
  $\overline{\text{SS}}$ pin, 266, 268, 272–280, 282–283, 286, 288, 290, 294–296, 579

SPSR register, 264, 267, 273–280, 282–283, 288–290, 295
   MODF bit, 272–274, 276–277, 279, 295
   SPIF bit, 267–268, 270–274, 276–278, 282–283, 286–290, 294–296
   WCOL bit, 272–273, 278–279, 295
STAA instruction, 21–22, 30–31, 33, 75, 597
STAB instruction, 33, 75, 597
Stack, 5, 34–39, 61, 63–64, 75–76, 101–102, 107, 110, 129, 275
Stack pointer (SP), 4–5, 10, 33–34, 73, 75, 101–102, 110, 275
Status polling, 103, 176–178, 182–184, 186, 197–198, 231–232, 268, 271, 355–356, 398–400
STD instruction, 33, 51–52, 75, 347, 597
STOP instruction, 6, 9–10, 91, 94, 103–106, 110–112, 122–123, 130–131, 302, 597
Storage media, 1
Strobe A (STRA), 176–183, 191–203, 208, 213, 551, 556, 565, 578
   Active level vs. active edge, 196–198, 200, 208
Strobe B (STRB), 176–178, 180–183, 191–203, 208, 213, 551, 556, 565, 578
STS instruction, 33, 75, 597
STX instruction, 33, 75, 598
STY instruction, 33, 75, 598
SUBA instruction, 50, 76, 598
SUBB instruction, 50–51, 76, 598
SUBD instruction, 50, 76, 598
Subroutines, 61, 63–73, 76
Synchronous communications, 175, 263, 293

TAB instruction, 33, 75, 598
TAP instruction, 59, 76, 99, 598
TBA instruction, 33, 75, 598
TCNT register, 302, 338–339, 372, 387
   Double-byte reads, 302, 307–308, 338

TCTL1 register, 347–353, 382, 386–387, 397–398, 400, 408
   OMx/OLx bits, 348–349, 352, 382, 398
TCTL2 register, 315–317, 331–333, 340, 347–348
   EDGxB/EDGxA bits, 315, 331–333, 340, 347, 397
TFLG1 register, 314, 331–333, 339–340, 347–350, 380, 382, 397–399, 408, 464
   I4O5F bit, 315, 339, 347–349, 352
   ICxF bits, 314–315, 331–333, 339–340
   OCxF bits, 349, 382–383, 397–400, 408–409, 420, 464
TFLG2 register, 303–305, 338, 416–418, 423–426, 454–455, 465–466
   PAIF bit, 424–426, 455, 465–466
   PAOVF bit, 424–426, 455, 465–466
   RTIF bit, 417–418, 432, 465
   TOF bit, 303–308, 338–339
Thermistor, 473, 480, 488–490, 513
Three-state (tri-state) buffers, 10, 561–562
TI4O5 register, 314, 339, 346–348, 398
TIC registers, 313–315, 339
TIL34 LED (IR), 430, 432–433, 600
TIL81 phototransistor, 430, 432–433, 600
Time protection, 89, 107, 130, 142, 303
Timer overflow interrupts, 98, 303, 307–308, 311, 319, 322–323, 327, 338–339
TMSK1 register, 315–317, 331–333, 339, 347–352, 380, 382, 386–387, 397–400, 408
   I4O5I bit, 315, 339, 347–349, 352
   ICxI bits, 315, 331–333, 339–340
   OCxI bits, 349, 382, 399
TMSK2 register, 302–303, 316–317, 338, 409, 416–418, 423, 426–428, 455, 456–458, 464–465
   PAII bit, 424, 426, 455, 465–466
   PAOVI bit, 424, 426, 455, 465–466
   PR bits, 302–323, 338, 409, 464

   RTII bit, 417–418, 431, 464–465
   TOI bit, 303, 307, 316, 338
TOC1 register, 379–380, 383, 387, 399–400
   Double-byte store, 347
TOC2–TOC5 registers, 343–344, 346–348, 387, 397–400, 420
   Double-byte stores, 347
TPA instruction, 59, 76, 598
TSTA instruction, 59, 76, 598
TSTB instruction, 59, 76, 598
TSX instruction, 33–34, 75, 114, 598
TSY instruction, 33–34, 75, 114, 116, 598
Two's complement, 7–8, 42–43, 75
TXS instruction, 33–34, 75, 598
TYS instruction, 33–34, 75, 598

UART, 3, 73, 246, 571, 575
Unconditional branches, 31–32, 61, 75–76
Undervoltage detector, 85–86, 92–93

V_{DD}, 92–93, 108, 554, 564, 576, 579–580
Voltmeter, 91–92, 120–122, 481–483, 533, 599
V_{PPBULK}, 106
V_{REFH}, 474–475, 482, 485–486, 488, 539–540, 580
V_{REFL}, 474–475, 481, 486, 488, 539–540, 579

WAI instruction, 22, 30, 35, 103–104, 106, 110–112, 122–123, 130–131, 598

XGDX instruction, 34, 75, 598
XGDY instruction, 34, 75, 598
XIRQ interrupts, 85, 99, 104, 106, 110–112, 129–131, 576
$\overline{\text{XIRQ}}$ pin, 99, 104–106, 110, 122, 129–131, 576, 578
XTAL pin, 86, 578